Ellingson

Radio Systems Engineering

Using a systems framework, this textbook provides a clear and comprehensive introduction to the performance, analysis, and design of radio systems for students and practicing engineers. Presented within a consistent framework, the first part of the book describes the fundamentals of the subject: propagation, noise, antennas, and modulation. The analysis and design of radios including RF circuit design and signal processing is covered in the second half of the book. The former is presented with minimal involvement of Smith charts, enabling students to grasp the fundamentals more readily. Both traditional and software-defined/direct sampling technology are described, with pros and cons of each strategy explained. Numerous examples within the text involve realistic analysis and design activities, and emphasize how practical experiences may differ from theory or taught procedures. End-of-chapter problems are provided, as are a password-protected solutions manual and lecture slides to complete the teaching package for instructors.

Steven W. Ellingson is an Associate Professor at Virginia Tech. He received his PhD in Electrical Engineering from the Ohio State University. He held senior engineering positions at Booz-Allen & Hamilton, Raytheon, and the Ohio State University ElectroScience Laboratory before joining the faculty of Virginia Tech. His research focuses on wireless communications and radio frequency instrumentation, with funding from the National Science Foundation, the National Aeronautics and Space Administration, the Defense Advanced Research Projects Agency, and the commercial communications and aerospace industries. Professor Ellingson serves as a consultant to industry and government on topics pertaining to RF system design and is an avid amateur radio operator (call sign AK4WY).

"*Radio Systems Engineering* offers a comprehensive introduction to the architecture and components of radio systems. It reviews all the fundamentals that students need to understand today's wireless communication systems, including modern modulation schemes, radio wave propagation, and noise impact. It also covers all the blocks of modern radio transmitter and receiver systems, such as antennas, filters, amplifiers, and signal processing. This textbook gives engineering students a complete overview of radio systems and provides practicing wireless engineers with a convenient comprehensive reference."

Patrick Roblin, *Ohio State University*

Radio Systems Engineering

STEVEN W. ELLINGSON

Virginia Polytechnic Institute and State University

Shaftesbury Road, Cambridge CB2 8EA, United Kingdom

One Liberty Plaza, 20th Floor, New York, NY 10006, USA

477 Williamstown Road, Port Melbourne, VIC 3207, Australia

314–321, 3rd Floor, Plot 3, Splendor Forum, Jasola District Centre, New Delhi – 110025, India

103 Penang Road, #05–06/07, Visioncrest Commercial, Singapore 238467

Cambridge University Press is part of Cambridge University Press & Assessment,
a department of the University of Cambridge.

We share the University's mission to contribute to society through the pursuit of
education, learning and research at the highest international levels of excellence.

www.cambridge.org
Information on this title: www.cambridge.org/9781107068285

First published 2016

A catalogue record for this publication is available from the British Library

Library of Congress Cataloging-in-Publication data
Names: Ellingson, Steven W., 1965– author.
Title: Radio systems engineering / Steven W. Ellingson, Virginia Polytechnic
Institute and State University.
Description: Cambridge, United Kingdom ; New York, NY : Cambridge University Press, 2016.
Identifiers: LCCN 2016024246| ISBN 9781107068285 (Hardback)
Subjects: LCSH: Radio. | Radio–Transmitters and transmission–Design and
construction. | Radio–Receivers and reception–Design and construction. |
Radio circuits–Design and construction. | Radio frequency modulation.
Classification: LCC TK6550 .E475 2016 | DDC 621.384–dc23 LC record
available at https://lccn.loc.gov/2016024246

ISBN 978-1-107-06828-5 Hardback

Additional resources for this publication at www.cambridge.org/Ellingson

CONTENTS

ILLUSTRATIONS

TABLES

PREFACE

The topic of this book is the analysis and design of radio systems and their constituent subsytems and devices. Chapters 1–6 cover the basics of radio: antennas, propagation, noise, and modulation. Chapters 8–18 cover radio hardware: circuits, device technology where relevant, and the basics of digital signal processing. The intervening Chapter 7 ("Radio Link Analysis") serves as the bridge between these two parts of the book by demonstrating how system-level characteristics lead to particular hardware characteristics, and vice versa. Appendix B serves as a summary of commonly-encountered radio systems in this context. Appendix A addresses modeling of path loss in terrestrial channels, building on the fundamentals presented in Chapter 3 ("Propagation").

This book presumes the reader is interested primarily in communications applications, however, most of the material is relevant to other applications of radio including navigation, radar, radio astronomy, geophysical remote sensing, and radio science.

The use of the term "systems engineering" in the title of this book bears some explanation. In the parlance of communications engineers, the scope of this book is limited to considerations at the physical layer and below, with little coverage of issues at higher layers of the protocol stack. Further, some readers may judge that only Chapters 1, 7, and Appendix B (or maybe *just* Chapter 7) address systems engineering, in the sense of complete "end-to-end" system analysis and design. Nevertheless, the title "Radio Systems Engineering" seems appropriate because each topic – especially those in Chapters 8–18 – is presented in the context of system-level design, with extensive cross-referencing to associated topics in the first half of the book and to Chapter 7.

Intended uses of this book This book is has been written with three possible uses in mind.

- *Primary textbook in senior- and beginning graduate-level courses in radio systems engineering.* This book can serve as a single textbook for a two–three semester course sequence. Associated courses typically have titles such as "radio communications systems," "wireless engineering," "wireless systems," etc. The recommended syllabus would cover the chapters in present order, from beginning to end. For a two-semester sequence a few chapters or sections of chapters would likely need to be bypassed; in this case Chapters 12 ("Antenna Integration"), 18 ("Digital Implementation of Radio Functions"), and sections from the last half of Chapter 6 ("Digital Modulation") are suggested for omission. Alternatively, this book may serve as a textbook for a traditional one-semester radio communications systems course, using Chapters 1–7; or a traditional one-semester radio circuits course, using Chapters 8–18. In either case the other half of the book is useful as context or supplementary material.

- *Reference for practicing radio systems engineers and engineers who require a reference on topics outside of their primary area.* I imagine that I am not too different from many practising engineers who find themselves proficient in a few areas, but who occasionally need to function in realms which are less familiar. In this case it is useful to have a single reference in which all of the most common topics are summarized in a consistent way, and to a comparable level of detail. To be clear, the intent is not to replace books on anyone's bookshelf – but it is hoped that this book will be one of those that you keep at arm's reach, has the answers to the questions that most often arise, and points in a productive direction for those topics which fall outside the scope of this book.
- *Supplementary reference in "special topics" courses pertaining to radio.* At many institutions, radio and wireless topics are not taught in systems-oriented courses, but rather through a set of courses in specialized topics. These courses have titles such as "antennas," "propagation," "digital communications," "amplifiers," "frequency synthesis," and so on – essentially, the titles of chapters of this book. It is possible that instructors will find this book suitable as the primary textbook for such specialized courses, especially if several such courses are arranged into "tracks" which students are encouraged to follow in sequence. If the instructor finds that this book does not provide sufficient depth of coverage for a particular special-topic course, this book can serve as a supplementary reference, providing the student with some higher-level context and a quick reference for associated topics.

Assumptions about the reader This book assumes the reader has university-level training in electrical engineering (EE), including elementary circuit theory, signals and systems, probability and statistics, basic analog and digital electronics, and electromagnetics. In most four-year EE programs these topics are covered within the first three years. Practicing electrical engineers who have been away from academia for a long time should have no difficulty following this book; also those working in related fields such as physics or engineering science should find this material accessible with little or no additional engineering training.

Except for a few parenthetical references, I have avoided the use of the Smith chart in this book. It is not possible to avoid reference to the real–imaginary plane on which the Smith chart is superimposed. However, I feel the value of knowing how to maneuver the circles, arcs, and rotations of the Smith chart is not really apparent until one is comfortable with the fundamentals of RF circuit analysis and design. No doubt some experienced readers and some instructors would prefer – for good reasons – to see Smith charts introduced early and used throughout. I am sympathetic to that point of view, and would like those persons to know that there is no barrier to augmenting this book with Smith charts; the places to do this (primarily in Chapters 8–10) should be obvious.

Acknowledgements Here's a list of talented and helpful people who have contributed to this book: Undergraduate student Joe Brendler built and tested many amplifiers and oscillators in order to identify a few that could be presented in this book as examples and problems. The survey and characterization of MMIC amplifiers summarized in Figures 11.3 and 11.5 was the work of PhD student R.H. (Hank) Tillman. Paul Swetnam and Ken Ayotte of Dynamic Sensor Systems LLC graciously provided the phase noise data that appears in Figure 16.15. I am grateful to Louis Beex, Harpreet Dhillon, Allen MacKenzie, and Mike Ruohoniemi for their review of various portions of this book. I have subjected many students at my home institution of Virginia Tech to early versions of the chapters within this book, and they have provided many useful suggestions. Undergraduate student Jay Sheth volunteered to review some chapters and to work the problems as an independent study project, which was very helpful. A university is a pretty good place to write a book, and the Bradley Department of Electrical and Computer Engineering at Virginia Tech was an especially good place to write this particular book. I appreciate the support of the Department and its staff while working on this project. It has been a delight to work with Julie Lancashire, Sarah Marsh, Heather Brolly, Claire Eudall, Rachel Cox, Beverley Lawrence, and Caroline Mowatt of Cambridge University Press on this project. Finally, I appreciate the support of my wife, Karen Peters Ellingson, who also served as photographer and is responsible for the uncredited pictures appearing in this book.

1 Introduction

1.1 RADIO: WHAT AND WHY

Radio is the use of unguided propagating electromagnetic fields in the frequency range 3 kHz and 300 GHz to convey information. Propagating electromagnetic fields in this frequency range are more commonly known as *radio waves*. In a radio communication system, a transmitter converts information in the form of analog signals (e.g., voice) or digital signals (i.e., data) to a radio wave, the radio wave propagates to the receiver, and the receiver converts the signal represented by the radio wave back into its original form. Radio systems engineering encompasses the broad array of topics pertaining to the analysis and design of radio communications systems.

Radio is not unique in its ability to transfer information using unguided electromagnetic radiation. As shown in Table 1.1, the electromagnetic spectrum also includes infrared (IR), optical, ultraviolet (UV), X-rays, and γ-rays. In principle, the only fundamental difference between these phenomena is the associated range of wavelength λ, which is related to frequency f as follows:

$$\lambda f = c \tag{1.1}$$

where c is the speed of light; approximately 3.0×10^8 m/s in free space.

In practice, the various forms of electromagnetic radiation are quite different, with their behavior being determined by wavelength relative to the sizes of structures in the environment in which they propagate, and the nature of the media in which propagation occurs. Radio propagates relatively efficiently through air and most building materials, and tends to be scattered – as opposed to being absorbed and dissipated – by structures larger than a wavelength. This is in contrast to IR, optical, and UV, all of which tend to be much more easily dissipated by propagation through air or building materials. X-rays and γ-rays are not as limited by this problem, but are relatively difficult to create and capture, and are relatively dangerous to human health. So, whereas wireless communication is certainly possible using electromagnetic radiation in any of these regimes, radio is especially convenient in that it is simultaneously easy to use, has good propagation characteristics, and is relatively safe.

This is not to say that radio does not also have disadvantages compared to other forms of electromagnetic radiation. Perhaps the most important of these is that bandwidth is relatively limited. This is for two reasons: First, the span of available frequencies is limited. Although 300 GHz may seem like a lot, all of the advantages described above are most pronounced at the low end of the spectrum, and for this reason the vast majority of radio systems operate at frequencies below about 15 GHz. In contrast, the optical range of frequencies is about 2 PHz

Table 1.1. The electromagnetic spectrum. Note that the indicated boundaries are arbitrary but consistent with common usage. "Span" is the ratio of highest to lowest frequency.

Regime	Frequency Range	Wavelength Range	Span
γ-Ray	$>3 \times 10^{19}$ Hz	<0.01 nm	
X-Ray	3×10^{16} Hz–3×10^{19} Hz	10–0.01 nm	10^3
Ultraviolet (UV)	2.5×10^{15}–3×10^{16} Hz	120–10 nm	10^1
Optical	4.3×10^{14}–2.5×10^{15} Hz	700–120 nm	$10^{0.5}$
Infrared (IR)	300 GHz–4.3×10^{14} Hz	1 mm–700 nm	10^3
Radio	3 kHz–300 GHz	100 km–1 mm	10^8

(2 million GHz!) wide, and a *single* optical communications channel commonly has bandwidth at least as wide as 15 GHz. The relatively small amounts of spectrum available at radio frequencies motivate the use of relatively sophisticated modulation schemes in order to increase *spectral efficiency*; that is, the rate of information that can be transferred effectively over a specified bandwidth. These more sophisticated schemes require higher-performance hardware, resulting in the imposition of stringent performance requirements on radios. Comparable measures are not required in higher-frequency portions of the electromagnetic spectrum, due to the relative abundance of spectrum.

1.2 THE RADIO FREQUENCY SPECTRUM

The radio segment of the electromagnetic spectrum covers eight orders of magnitude in frequency (or wavelength), with significant variation in properties and utilization over that span. Therefore it is useful to further subdivide the radio spectrum into bands. One common scheme for defining and naming the bands is the scheme promulgated by the International Telecommunications Union (ITU), shown in Table 1.2. The acronyms "VLF," "LF," "MF" stand for "very low frequency," "low frequency," "medium frequency," and so on. An alternative partitioning of the spectrum is by bands designated by the IEEE (Institute of Electrical and Electronics Engineers) as shown in Table 1.3. In both cases band names are primarily historical, and do not convey any particular technical information. Furthermore, the assignment of band names to frequency ranges is arbitrary, and one occasionally encounters definitions using somewhat different frequency ranges. Nevertheless, the schemes shown in Tables 1.2 and 1.3 are widely used and facilitate concise engineering discussion.

Use of the radio spectrum is determined by both technical and legal considerations. While this book is concerned primarily with technical considerations, it is useful to know something about the legal framework within which radio systems operate. A framework for the use of the radio spectrum has been established through a system of international treaties developed

Table 1.2. **The radio frequency spectrum, with ITU band designations. WLAN: Wireless local area network, LMR: Land mobile radio, RFID: Radio frequency identification.**

Band	Frequencies	Wavelengths	Typical Applications
EHF	30-300 GHz	10–1 mm	WLAN (60 GHz), Data Links
SHF	3–30 GHz	10–1 cm	Terrestrial and Satellite Data Links, Radar
UHF	300–3000 MHz	1–0.1 m	TV Broadcasting, Cellular, WLAN
VHF	30–300 MHz	10–1 m	FM and TV Broadcasting, LMR
HF	3–30 MHz	100–10 m	Global terrestrial communications, CB Radio
MF	300–3000 kHz	1000–100 m	AM Broadcasting
LF	30–300 kHz	10–1 km	Navigation, RFID
VLF	3–30 kHz	100–10 km	Navigation

Table 1.3. **The radio frequency spectrum by IEEE band designations. In addition, the term "P-band" is sometimes used to indicate frequencies around 300 MHz (1 m wavelength).**

Band	Frequencies	Wavelengths
W	75–110 GHz	0.400–0.273 cm
V	40–75 GHz	0.750–0.400 cm
Ka	27–40 GHz	1.111–0.750 cm
K	18–27 GHz	1.667–1.111 cm
Ku	12–18 GHz	2.500–1.667 cm
X	8–12 GHz	3.75–2.50 cm
C	4–8 GHz	7.5–3.75 cm
S	2–4 GHz	15–7.5 cm
L	1–2 GHz	30–15 cm

through the auspices of the ITU. An international framework of regulation is important for at least two reasons. First, some forms of radio communication extend across national boundaries: two prominent examples being HF-band broadcasting and satellite communications. Second, it is in the common interest to standardize the technical characteristics of radio systems in order to allow them to be used internationally; two prominent examples here being personal cellular telecommunications and air traffic control systems.

Within the international regulatory framework, national governments create and enforce additional regulations to further elaborate on permitted uses of the spectrum. In the USA, the federal government's use of spectrum is regulated by the National Telecommunications and Information Administration (NTIA), whereas non-federal (i.e., commercial, amateur, and passive scientific) use of spectrum is regulated by the Federal Communications Commission (FCC). FCC regulations concerning use of the spectrum are codified in Title 47 of the US Code of Federal Regulations (CFR).

Table 1.2 lists a few applications associated with each ITU band and which are common to all nations. The vast majority of radio communications systems operate in the span from 500 kHz (the lower edge of the 500 kHz–1800 kHz AM broadcast band) to about 15 GHz, although a few important applications exist at lower and higher frequencies.

EXAMPLE 1.1

What is the free-space wavelength in the center of the US AM broadcast band? What ITU frequency band does this fall into?

Solution: The US AM broadcast band is 540–1600 kHz, so the center frequency is 1070 kHz. The wavelength at that frequency is $c/$ (1070 kHz) $\cong$ 280 m. This falls in the MF band.

EXAMPLE 1.2

For a particular application, the free-space wavelength is 3 m. What application could this be, and what ITU frequency band does this fall into?

Solution: The frequency at that wavelength is (3 m) $/c \cong$ 100 MHz. In most of the world, this frequency is used for FM broadcasting (in the US, the FM broadcast band is 88–108 MHz). This falls in the VHF band.

1.3 RADIO LINK ARCHITECTURE

A radio link is a system employing radio waves to convey information between locations. At the highest level, radio links can be classified in terms of *directionality*, *topology*, and *multiple access technique*. Some examples are provided in Table 1.4. It is useful to be familiar with these concepts while considering the lower-level details which are the focus of this book.

Directionality refers to the intended direction of information flow. There are three primary types, as illustrated in Figure 1.1: simplex, half-duplex, and full-duplex. A simplex link moves data in one direction only;[1] prime examples being the link between any AM, FM, or TV broadcasting station and a receiving radio. Another example is wireless telemetry, in which a device continuously transmits information about itself to a central location, where the

[1] This is the formal definition. As will be pointed out shortly, "simplex" may also refer to topology as opposed to directionality.

Table 1.4. **Examples of radio links to demonstrate the concepts of directionality, topology, and multiple access. See text for elaboration on the topology of LMR and cellular systems.**

	Directionality	Topology	Multiple Access
Backhaul	full-duplex	point-to-point	(none)
TV Broadcast	simplex	broadcast	(none)
Telemetry	simplex	point-to-point	(depends)
LMR	half-duplex	broadcast	usually PTT ("manual" TDMA)
Cellular (Voice)	full-duplex*	point-to-point	FDMA/TDMA/CDMA

*Typically FDD.

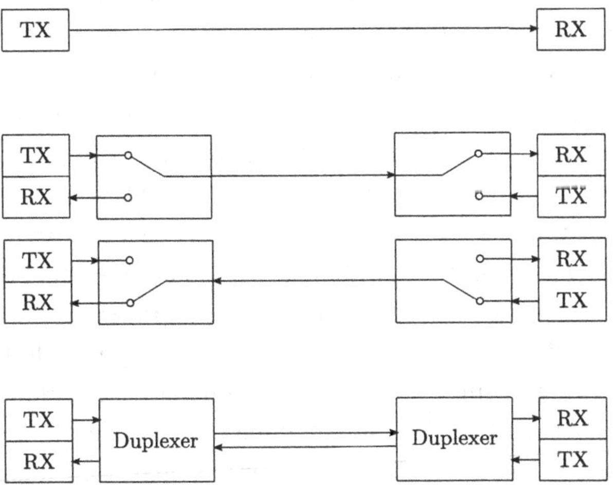

Figure 1.1. Top to bottom: Simplex, half-duplex, and full-duplex radio links. "TX" represents a transmitter, "RX" represents a receiver.

information is interpreted and there is no response. In a half-duplex link, information is sent in both directions, but in only one direction at a time. When the direction is controlled by users, this scheme is referred to as "push to talk" (PTT). An important class of PTT half-duplex systems is land mobile radio (LMR). In LMR, the user at one end sends a message, then the user at the other end of the link responds, and there is no overlap. In contrast, a full-duplex link accommodates simultaneous transmission of information in both directions; a good example being modern cellular telephony.[2]

Directionality impacts the design of radios as follows: Simplex radios are either transmitters or receivers, but not both. Half-duplex radios have both transmitters and receivers, but only one requires access to the radio's antenna at any given time. True full-duplex radios also have both a transmitter and receiver, and both are potentially active at the same time. Simultaneous receive and transmit normally requires *frequency division duplexing* (FDD), which means receive and transmit occur at separate, normally widely-spaced frequencies. The receiver and transmitter in

[2] Not to confuse the issue, but: Cellular telephone systems can also be used to implement (what appear to the users as) PTT-type links, providing an LMR-type service to its customers without the need for a physical LMR system.

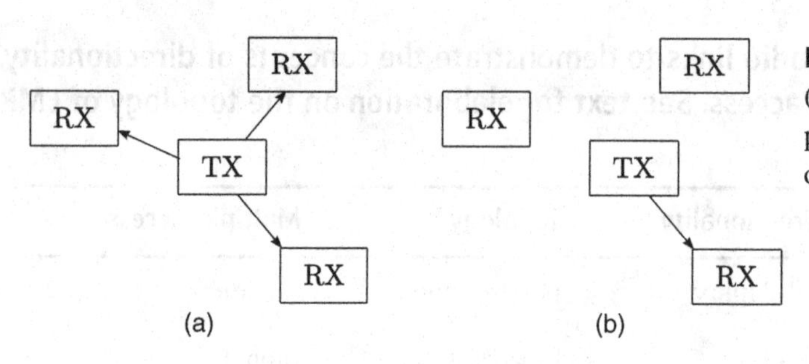

Figure 1.2. Topologies: (a) broadcast, (b) point-to-point. This is a physical perspective; what users perceive depends on the multiple access technique.

a true full-duplex radio share the antenna through a device known as a *duplexer*, which serves to isolate the receiver from the high power generated by the transmitter. High-performance duplexers are difficult to implement, so FDD is sometimes augmented or replaced with *time-division duplexing* (TDD). A TDD system is fundamentally half-duplex; i.e., the transmitter and receiver take turns using the antenna; however, users perceive the link as full duplex. TDD offers some relief from the challenges inherent in simultaneous transmit and receive, but introduces the problem of coordinating the transmit/receive transition between the ends of the link.

Topology refers to the geometry of the radio link, and there are only two primary types: broadcast, and point-to-point. These are shown in Figure 1.2. A broadcast link conveys information from a particular radio to all radios within range, typically over a region which completely surrounds the transmitter. Obviously, AM, FM, and TV broadcasts are examples of links with broadcast topology. LMR, too, uses a broadcast topology: Any user transmits with the expectation of being received by any other user within range. A point-to-point link is a link which nominally conveys information from one radio to a *particular* radio. A good example of a point-to-point link is a "backhaul" link used in telecommunications systems to form high-bandwidth connections between two particular sites, such as a cellular base station and a call switching center (see Section 7.7 for an example). Note that the term "topology" refers to the geometry of the *intended* flow of information: Thus, a transmitter may well transmit radio waves in all directions, but the associated link is considered point-to-point if there is only one intended recipient.

There are cases in which the distinction between broadcast and point-to-point links is not clear-cut. A good example is an LMR system employing *repeaters*, crudely illustrated in Figure 1.3. A repeater is an LMR radio which receives user transmissions and simultaneously retransmits them on a different frequency in order to avoid interference with the original transmission. Repeaters are typically fixed stations which can employ better antennas and higher power; therefore this scheme increases the size of the geographical region over which an LMR system may operate. In contrast to repeater-less LMR systems, an LMR link utilizing a repeater actually consists of two radio links: a point-to-point link from transmitting user to repeater, and a broadcast link from repeater to receiving users. It should be noted that it is common practice to refer to LMR systems that do *not* use a repeater as being "simplex" – a bit confusing, but keep in mind that this is actually a reference to the topology as opposed to the directionality, which is half-duplex in either case.

Cellular telecommunications systems also have "multiple personalities" in this respect: Whereas *user* information flow is always point-to-point, cellular networks require control

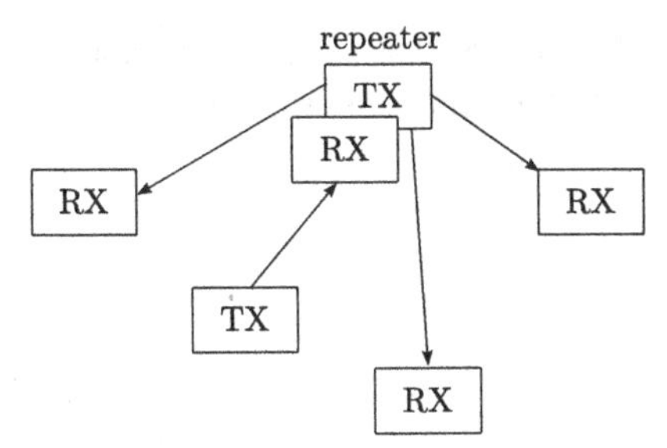

Figure 1.3. A repeater topology: Point-to-point to the repeater, broadcast from the repeater.

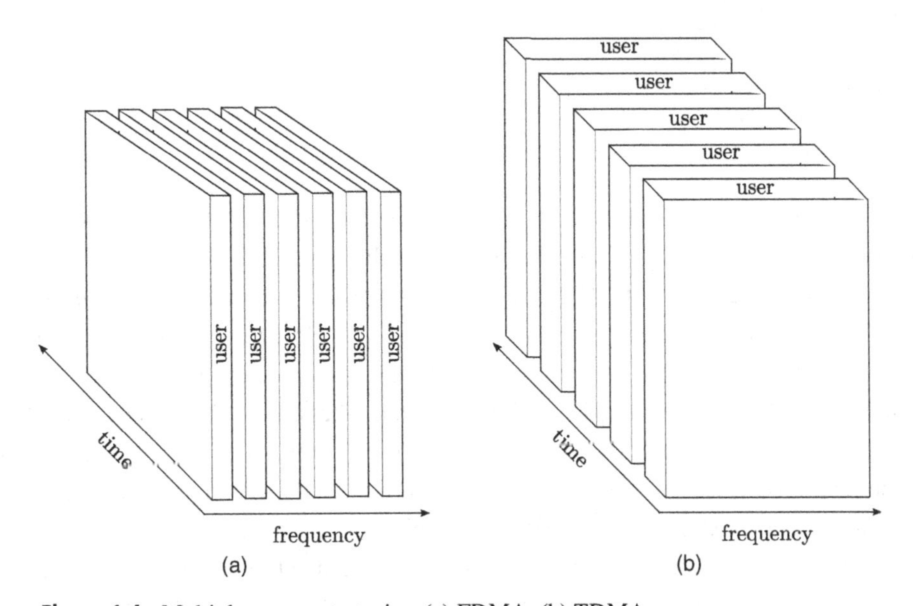

Figure 1.4. Multiple access strategies: (a) FDMA, (b) TDMA.

signals which must be broadcast. For example, cellular phones must broadcast a control signal to make themselves known to base stations within range, and base stations broadcast information to all cell phones within range, including time and service provider identification.

Multiple access refers to the method used to manage access to spectrum when the potential exists for multiple users to require simultaneous access. Not all links require this: For example, AM/FM/TV broadcast systems transmit continuously, so there is no need to manage access; and LMR/PTT systems have no explicit multiple access methodology; users are expected to monitor the channel and to refrain from transmission while it is in use. In contrast, multiple access is a paramount consideration in other systems, including cellular telephony, WLAN, and satellite communications systems. There are three principal categories of multiple access techniques: *frequency-division multiple access* (FDMA), *time-division multiple access* (TDMA), and *code-division multiple access* (CDMA). In FDMA (Figure 1.4(a)), links are assigned to dedicated channels (in this case meaning subdivisions of the available spectrum) so that multiple links may be active simultaneously. In TDMA (Figure 1.4(b)), radios take turns transmitting; i.e., each user gets access to increased spectrum while other users wait. The tradeoff is self-evident: increased spectrum available to each user, but not available with 100% duty cycle, and requiring some increased sophistication to precisely coordinate transmission times.

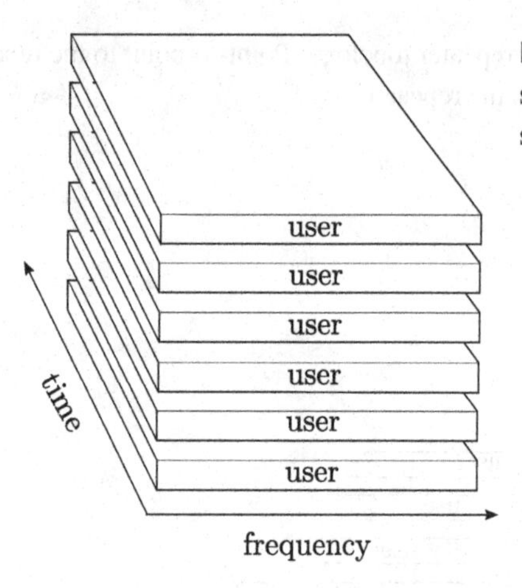

Figure 1.5. CDMA. In DSSS form, users are separated by spreading code; in FHSS form users are separated by hopping sequence.

In CDMA (Figure 1.5), radios transmit simultaneously using the same spectrum, and interference is mitigated using a *spread spectrum* technique. The spread spectrum technique most commonly associated with CDMA is *direct sequence spread spectrum* (DSSS; see Section 6.18), commonly known simply as "spreading." Spreading renders the signal as a noise-like transmission which is uniformly distributed over the spectrum available to all transmitters. Spreading is done in a deterministic manner that allows the signal to be "despread" at the receiving end. The DSSS form of CDMA is commonly used in cellular mobile telecommunications systems. Alternatively, a form of CDMA can be implemented using *frequency hopping spread spectrum* (FHSS), in which the center frequency of a user changes many times per second according to a predetermined sequence that is unique to the transmitter. In either DSSS or FHSS forms, the principal advantage of CDMA is flexibility: In principle any number of users can simultaneously access the entire available spectrum, and the penalty for increasing the number of users is an apparent increase in the in-band noise.

It should be noted that there are additional multiple access techniques beyond the "big three" of FDMA, TDMA, and CDMA. For example: In satellite communications systems it is common to use the orthogonal polarizations of a radio wave to implement two links simultaneously without interference. Some wireless data networks employ *carrier sense multiple access* (CSMA), which is essentially an unscheduled form of TDMA in which radios are allowed to transmit if they are unable to sense an active signal in the channel. Also, in some cellular systems, FDMA is employed to divide the spectrum into channels, and TDMA is to accommodate multiple users within each channel.

While we are on the topic of cellular telecommunications systems: The defining feature of these systems is multiple access in the geographical sense; such systems consist of *cells*, which are contiguous geographical regions, typically on the order of kilometers in dimension, containing a single base station and all mobile users nominally being serviced by that base station. This is illustrated in Figure 1.6. The partitioning of the coverage area into cells allows the same frequencies to be reused in non-adjacent cells, provided there is sufficient distance between cells to maintain acceptably low interference. Thus cellular networks implement what is essentially a spatial form of multiple access. This scheme facilitates reuse of frequencies within a cellular network, which is essential given the limited availability of suitable spectrum.

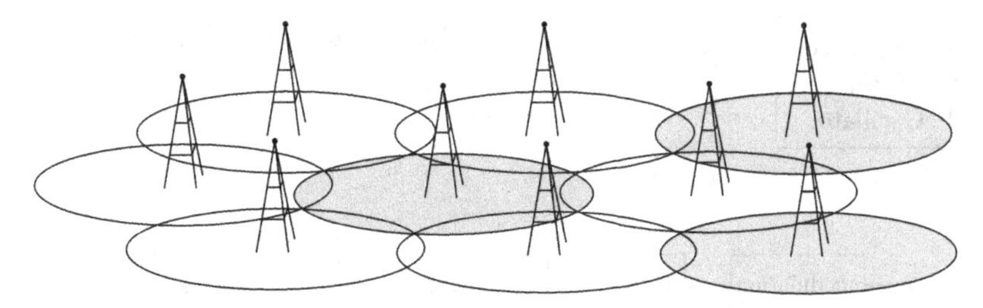

Figure 1.6. A cellular telecommunications system. Each circle represents a coverage cell served by a base station. Sets of cells which do not overlap (e.g., the cells shaded in this figure) are in principle able to use the same frequencies without interference.

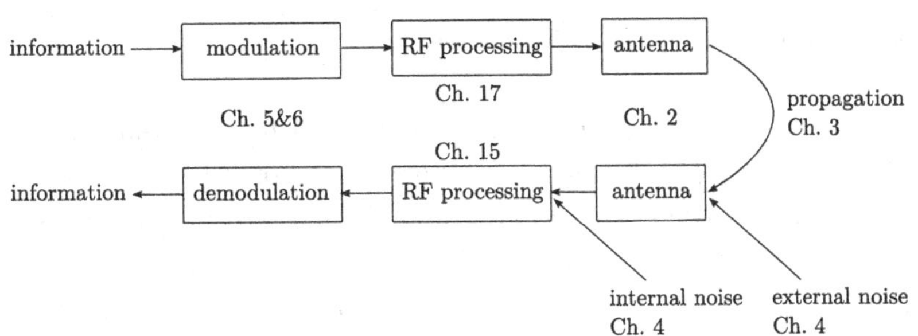

Figure 1.7. A high-level representation of a radio link (chapter numbers are indicated for quick access to discussions of these particular topics).

1.4 ELEMENTS OF A RADIO LINK

Figure 1.7 shows the elements that commonly appear in a radio link. The transmitter accepts the information and generates a representation of that information that is suitable for radio transmission. This process is known as *modulation*, covered in Chapters 5 (modulation of analog information) and 6 (modulation of digital information). Frequently, additional processing of the modulated signal is necessary; for example, to modify the center frequency or to increase power; these details are addressed in Chapter 17 ("Transmitters"). Antennas are used to convert the resulting electrical signals to radio waves and vice versa, as explained in Chapter 2. Between antennas, the radio wave is subject to several forms of loss in addition to a plethora of possible distortions, including multipath interference. We refer to these effects collectively as *propagation*, which is the topic of Chapter 3. The signal processed by the *receiver* is further degraded by environmental noise; i.e., noise originating from outside the receiver; as well as internal noise; i.e., noise generated by the receiver itself. Environmental and internal noise are addressed in Chapter 4. The remaining steps are counterparts of processing in the transmitter: RF processing to modify center frequency and apply gain to overcome propagation losses, and demodulation to convert the recovered signal to recognizable information, typically in its original form. Demodulation itself may involve quite a few steps; glance at Figure 6.1 for a preview.

Before plunging into the details of radio system analysis and design, it is useful to know a little more about typical architectures of transmitters and receivers. Let's start with transmitters. Figure 1.8 shows a simple and very common transmitter architecture. Here, the key component is an *oscillator*, which functions as the modulator. An oscillator is a device which produces sinusoidal output at a specified magnitude and frequency. In this transmitter, the information signal controls the oscillator, resulting in an *amplitude-modulated* (AM) or *amplitude-shift*

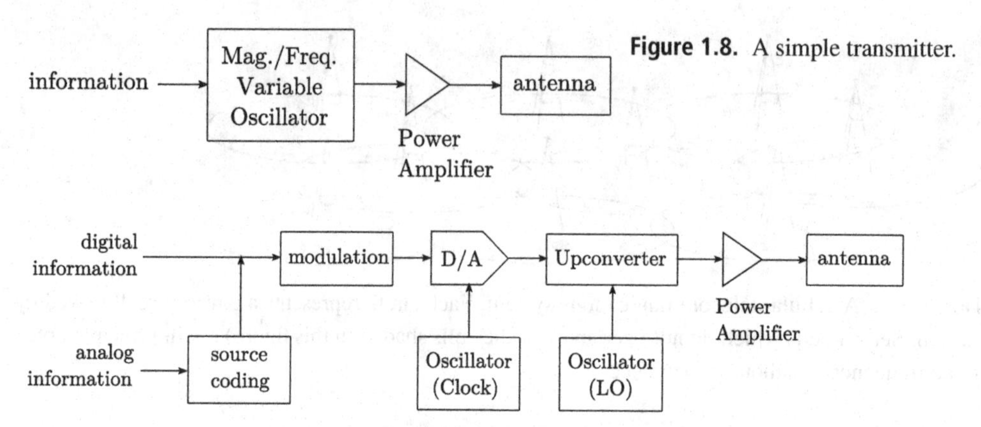

Figure 1.8. A simple transmitter.

Figure 1.9. A transmitter employing digital baseband processing.

keying (ASK) signal when the magnitude is controlled, or a *frequency-modulated* (FM) or *frequency shift keying* (FSK) signal when the frequency is controlled. The RF processing consists of a power amplifier (PA), whose purpose is to amplify the signal so as to increase range. This architecture appears in a variety of applications including some AM and FM broadcast transmitters, wireless keyless entry devices, and certain other devices. When used as shown in Figure 1.8, the variable oscillator is sometimes known as an *exciter*.

Whereas the architecture of Figure 1.8 is well-suited to AM/ASK and FM/FSK modulations, the more sophisticated modulations required for improved spectral efficiency are considerably more difficult to implement in this architecture. This includes modern digital broadcasting services (both audio and television) and the mobile radios employed in modern cellular telecommunications; e.g., your cell phone. For these systems it is typically necessary to perform modulation in the digital domain (i.e., using digital devices), and then to convert the result to analog form. This approach is shown in Figure 1.9. Here, the information is digital; either inherently so (i.e., the information is digital data) or via conversion from an analog signal, such as voice. The conversion from analog form to a digital representation is known as *source coding* (Section 6.2).[3] In either case, modulation converts the data into a discrete-time representation of what we would like to transmit. To get a radio frequency signal in the analog domain, we first convert the digital representation of the modulator output to an analog signal using a digital-to-analog converter, which is commonly referred to as a "DAC" or "D/A" (Section 17.3). It is usually not possible to do the D/A conversion at radio frequencies; in this case the conversion is performed at a lower frequency and then *upconverted* to the RF frequency at which it will be transmitted. The upconverter accomplishes this by combining the signal with one or more *local oscillator* signals, as explained in Chapter 14. In some cases it is possible and desirable to perform the D/A conversion at the intended transmit frequency, in which case the upconverter can be eliminated. See Figure 17.1 and associated text for a more detailed overview of the architectural possibilities.

When digital-domain processing is performed using computers (i.e., in software) or in reprogrammable logic devices (i.e., in firmware), such architectures may be referred to as *software*

[3] In some quarters "source coding" is a synonym for data compression, and does not necessarily imply A/D conversion. In this book we'll use the term more generally to include A/D conversion when the source information is analog.

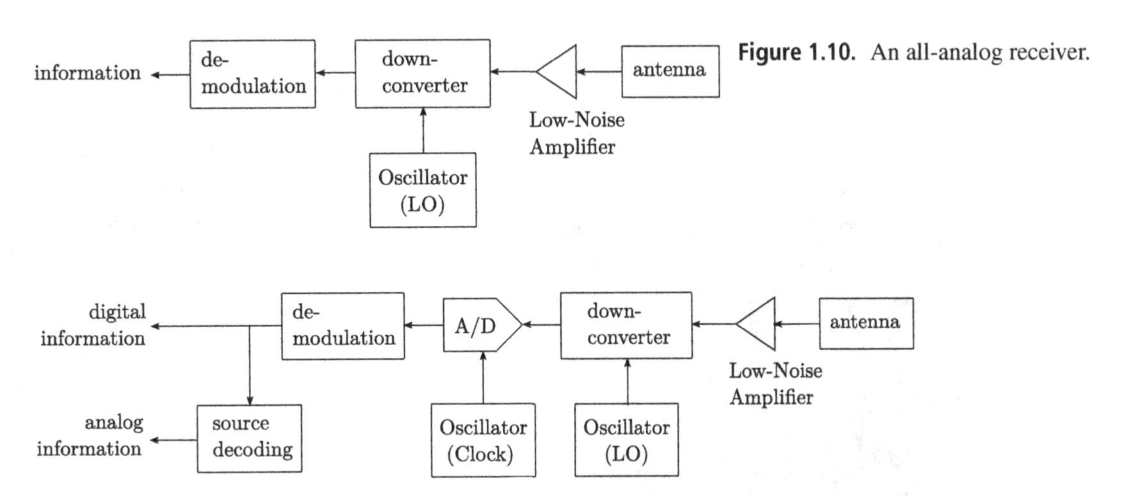

Figure 1.10. An all-analog receiver.

Figure 1.11. A receiver employing digital baseband processing.

radio or *software-defined radio* (SDR) architectures. Generally, however, the term "software radio" is reserved specifically for applications in which the flexibility offered by software or reprogrammable firmware is intended to be exploited to allow changes to the design after the radio is put into use. This incurs some additional challenges, as discussed in Section 18.6.

Receiver architectures analogous to the "simple" and "digital baseband processing" transmit architectures of Figures 1.8 and 1.9 are shown in Figures 1.10 and 1.11. In both architectures we employ a *low-noise amplifier* (LNA), whose job it is to overcome propagation loss while not adding too much additional internal noise (Section 10.5). Then we downconvert – that is, shift the center frequency from the radio band to a lower frequency that is easier to process. Like upconversion, this is done by combining the input signal with a radio frequency signal generated by an oscillator. The downconverter will typically need to apply a lot of additional gain, because the gain of the LNA alone will usually not be enough. The reasons for this will become apparent in Chapters 10, 11, and 15. Also, the downconverter will need to apply some filtering (Chapter 13) to reject interference (later, we will refer to this filtering as *selectivity*). Finally, there is demodulation. Analog demodulators are common, but are typically limited to AM and FM for analog information and ASK and FSK for data. For more sophisticated modulations, we typically follow the approach shown in Figure 1.11: Digitize the signal using an analog-to-digital converter ("ADC" or "A/D"; Section 15.2) and perform the demodulation digitally. In cases where the demodulation is implemented in software or reprogrammable firmware, we may refer to this as a software radio architecture. Also, as in the case of transmitters, it is sometimes possible and desirable to digitize the received signal "directly," without downconversion; for details on this, see Sections 15.6.1–15.6.2.

1.5 MODERN RADIO DESIGN: LEVELS OF INTEGRATION

Separate from the topic of architecture, we would like to consider the issue of *integration*. As recently as the early 1970s, radios were typically composed entirely from discrete components; i.e., separate resistors, capacitors, inductors, transistors, transformers, and so on. Some radios

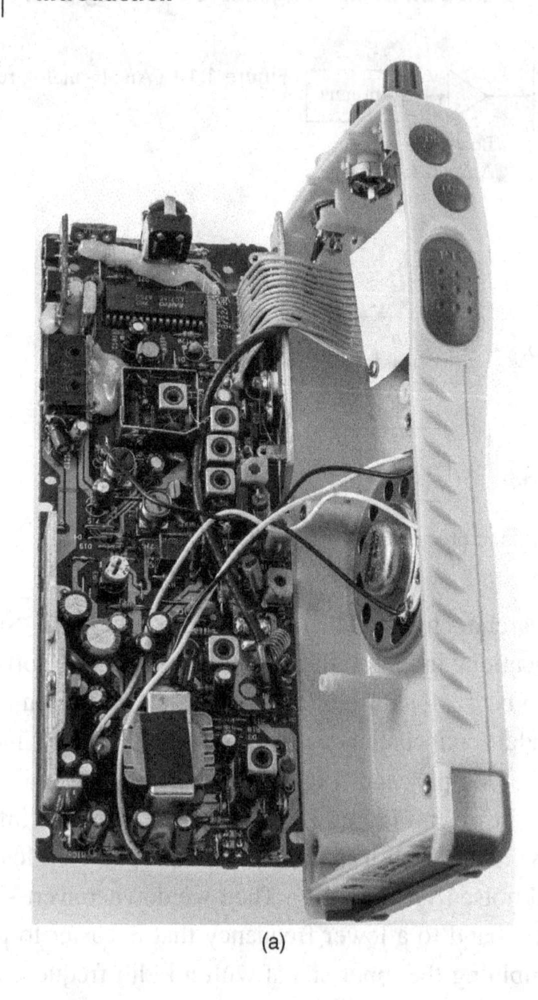

(a)

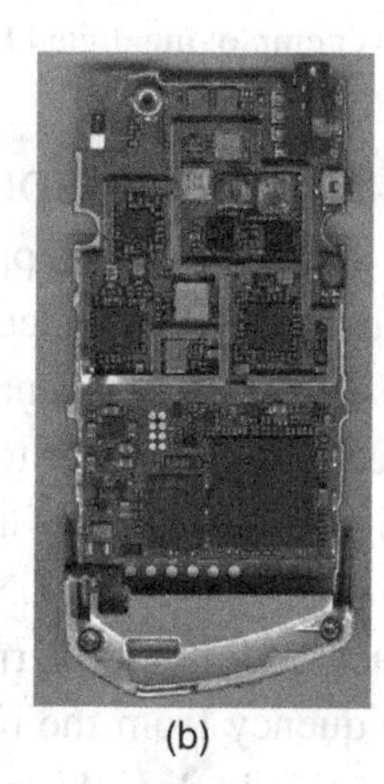

(b)

Figure 1.12. Levels of integration: (a) An all-analog Citizen's Band (CB) handheld radio exhibiting a relatively low level of integration; (b) A modern cellular handheld phone with digital baseband processing, exhibiting a relatively high level of integration. Shown approximately to scale.

continue to be designed and manufactured in this manner; see e.g., Figure 1.12(a). Presently most radios, including nearly all new designs, consist of a just a few integrated circuits ("ICs" or "chips"), with a sprinkling of discrete components required only to perform a few functions that remain difficult to integrate: These include resonators, bypass capacitors, certain bandpass filters, and baluns; see e.g. Figure 1.12(b). These ICs are sometimes referred to as radio frequency ICs (RFICs). The development and proliferation of RFIC-based radio design has evolved for a number of reasons: size reduction for certain, but also cost reduction and improved ability to implement very complex circuits that are both reliable and repeatable.

The engineer's ability to design using discrete components continues to be an important skill, for several reasons. First, RFICs are themselves collections of discrete components. Second, RF functions in some applications continue to be best handled using discrete component design. Prominent examples include LNAs for cellular base station receivers, which frequently benefit from the use of highly-optimized discrete transistors; and filters and impedance matching circuits generally, since these typically require inductive reactance which is difficult and often impractical to implement in RFIC form.

Thus, aspiring radio engineers should understand both the fundamentals of radio engineering, which are best described in terms of discrete circuits, and should also be familiar with the unique issues and challenges associated with RFIC design. In this book, we will approach the material from the perspective of design that will seem initially to pertain to a low level of integration, and will point out special considerations associated with IC implementations in a separate section in chapters where this topic becomes particularly relevant.

1.6 SPECIFICATIONS IN MODERN RADIO DESIGN

Table 1.5 illustrates key specifications – and thus the key challenges – in radio design. The design of receivers can be boiled down to six specifications: Sensitivity, linearity, and selectivity, all of which we would like to maximize; and size/weight, power consumption, and cost, all of which we would like to minimize. Level of integration comes to bear primarily on size/weight and cost requirements, and power consumption considerations are certainly not unique to radio design. Let us therefore focus for the moment on the first three receiver specifications.

Sensitivity refers to the strength of the signal arriving at the receiver that is required to achieve a specified minimum signal-to-noise ratio (SNR) at the output. Improved ("higher") sensitivity implies that the receiver is able to deliver a specified minimum SNR at the output given a *weaker* signal. Within this framework there remain many ways to specify sensitivity, each being more or less appropriate in particular cases; these details are addressed in Section 7.5.

Linearity refers to the maximum strength of the signal of interest, or of signals at nearby frequencies, which can be processed without unacceptable levels of distortion. While "sensitivity" addresses the ability to process weak signals, "linearity" addresses the ability to process strong signals. Linearity and sensitivity are typically contradictory goals, and most receiver designs represent a tradeoff between these goals which is appropriate for the particular application. Like sensitivity, there are many ways to specify linearity; a primer is provided in Section 11.2.

Table 1.5. **Types of specifications for radios.**

Receivers:	Transmitters:
Sensitivity	ERP
Linearity	Linearity
Selectivity	Spectral Purity
Both Receivers and Transmitters:	
Size/Weight	
Power Consumption	
Cost	

Selectivity refers to the ability of a receiver to prevent signals at nearby frequencies from interfering with the processing of the intended signal. A receiver with high selectivity is able to process the signal of interest when its frequency is very close to that of an interfering signal, whereas the output of a receiver with low selectivity will be degraded under similar conditions. Selectivity is related to bandwidth, since is it necessary for a receiver to access a segment of spectrum which is at least as large as the bandwidth of the signal of interest. However, some modern receivers are designed to process multiple signals simultaneously, so selectivity and bandwidth are not necessarily the same thing. Selectivity refers to mitigation of the effects of signals which are separated in frequency from the desired signal, so the width of the passband is not necessarily the issue. Selectivity is further addressed in Section 15.5.

Similar considerations apply for transmitters. In place of sensitivity, the analogous specification for transmitters is **effective radiated power** (ERP) – a measure of how much power the transmitter can deliver in the form of a radio wave, which, combined with the sensitivity of the distant receiver, determines the range of a radio link; see Section 7.3. In place of selectivity, the analogous concern for transmitters is **spectral purity** – that is, the ability of a transceiver to avoid producing undesired signals at unintended frequencies. **Linearity** – the ability to process strong signals without excessive distortion – also applies to transmitters, but in the transmit case the strong signal is the one generated locally, and typically no other signals are relevant.

The overall challenge of radio design can be summarized as follows: There are typically goals or constraints (minimum specifications) for each of the characteristics shown in Table 1.5, and it is typically difficult or impossible to meet all of these requirements simultaneously. In this book, we'll focus primarily on the radio-specific requirements (as opposed to size/weight, power consumption, and cost), and we'll find even among these that there are typically painful tradeoffs required to achieve a realizable design. The reasons for this, and techniques for managing the associated tradeoffs, will be pointed out along the way.

1.7 ORGANIZATION OF THIS BOOK

Above we have reviewed the elements of radio systems engineering at the highest level, and pointed out some sections in this book in which these elements are discussed. Now a brief overview of the organization of this book may be helpful.

The first half of this book provides a background in four of the five primary aspects of radio systems engineering: antennas (Chapter 2), propagation (Chapter 3), noise (Chapter 4), and modulation (Chapters 5 and 6). The fifth aspect, radios (Chapters 8–18), is the focus of most of the second half of the book.

The connection between these two parts is made in Chapter 7 ("Radio Link Analysis"), in which examples of contemporary radio systems are presented in the context of the material presented in the preceding chapters. An outcome of Chapter 7 should be an understanding of the requirements that radios must satisfy in different applications: For example, transmitter power (e.g., why is 100 mW suitable in one application yet 1 KW is required in another), receiver sensitivity (e.g., why is this specified to be very high in some applications and relatively low in others).

The later chapters of this book focus on radio design, with an emphasis on RF circuit analysis and design, as well as architectural approaches. We begin with introductions to two essential tools of RF circuit design: Two-port theory (Chapter 8) and impedance matching (Chapter 9). The remaining chapters address the details of the various elements – both analog and digital – of modern radio analysis and design.

Problems

1.1 List advantages and disadvantages of radio for the transfer of information using unguided electromagnetic radiation, in contrast to other electromagnetic phenomena.

1.2 For each of the "typical applications" identified in Table 1.2, briefly explain the application, identify the specific frequency range used for the application, and characterize the range over which radio links in the application operate.

1.3 For the following frequencies, determine the free-space wavelength, the ITU frequency band, and identify an application that typically uses that frequency: (a) 450 MHz, (b) 2 GHz.

1.4 For the following wavelengths, determine the frequency (assuming free-space), the ITU frequency band, and identify an application that typically uses that wavelength: (a) 33 m, (b) 3 cm.

1.5 For the following applications, indicate the application-specific frequency range, the associated range of wavelengths, the associated ITU band(s), directionality, topology, and multiple access technique(s): (a) U.S. Global Positioning System (GPS), (b) IEEE 802.11b (WLAN).

1.6 What determines whether a particular analog or digital modulation is well-suited to implementation using an analog modulator? Give examples.

1.7 Explain the difference between digital information, digital modulation, digital signal processing, and software radio.

2 Antenna Fundamentals

2.1 INTRODUCTION

An antenna is the interface between a radio and the unguided electromagnetic wave that transfers information between radios. This chapter presents the fundamentals of antenna engineering. This chapter is organized as follows: Section 2.2 explains how antennas create radio waves, culminating in an equivalent circuit model for a transmitting antenna. Section 2.3 addresses the reception of radio waves, culminating in an equivalent circuit model for a receiving antenna. Sections 2.4 and 2.5 explain the principles of antenna pattern and polarization. Section 2.6 addresses the integration of antennas with radios; in particular impedance matching and baluns (this topic continues as the primary focus of Chapter 12). Sections 2.7–2.12 describe classes of antennas most relevant to radio communications applications, including dipoles, monopoles, patches, beam antennas, reflector antennas, and arrays.

2.2 CREATION OF RADIO WAVES

The creation of a radio wave using an antenna is difficult to explain concisely in a completely rigorous manner. However, those aspects of the theory which are most relevant to radio systems engineering are relatively simple, and are presented here. We begin with a rudimentary physics-based explanation, and conclude with a few key concepts that are sufficient for understanding the antennas described later in this chapter.

2.2.1 Physical Origins of Radiation

Here are the two essential concepts: First, a radio wave is a coupling of electric and magnetic fields that transfers power over a distance and which persists in the absence of its source. This is the phenomenon of propagation, which is the topic of Chapter 3. Second, the source of a radio wave is a time-varying current.

To make sense of these concepts, let us briefly review some principles from undergraduate-level electromagnetics. Initially, we shall use real-valued quantities, as opposed to phasors. Consider an electric charge which is concentrated at one point in space; i.e., a point charge. When this charge is stationary, it gives rise to an electric field $\mathcal{E}$ which is static. If the charge is allowed to move with a constant velocity – that is, at a constant speed and with unchanging direction – then we have a steady current $\mathcal{J}$, which gives rise to a static magnetic field $\mathcal{B}$.

Now consider what happens if the electric charge is accelerated. The acceleration might be a change in speed with time, or a change in direction with time, or both. The result is that $\mathcal{J}$ is no longer constant, but rather time-varying. As a result, $\mathcal{B}$ is also time-varying. It is known that a time-varying magnetic field gives rise to an electric field and the governing equation is one of Maxwell's Equations:

$$\nabla \times \mathcal{E} = -\frac{\partial}{\partial t}\mathcal{B} \tag{2.1}$$

where "$\nabla\times$" is the curl operator. This equation indicates that the variation in $\mathcal{E}$ over space is related to the variation in $\mathcal{B}$ over time.

Completely independently, $\mathcal{E}$ should be time-varying because $\mathcal{E}$ decreases with increasing distance from the charge, and the distance to any particular point in space becomes closer or further away as the charge is moving. A time-varying electric field gives rise to a magnetic field which is described by another of Maxwell's Equations:

$$\frac{1}{\mu}\nabla \times \mathcal{B} = \mathcal{J} + \epsilon\frac{\partial}{\partial t}\mathcal{E} \tag{2.2}$$

where μ and ϵ are the permeability and permittivity of the material in which the fields exist. In radio engineering, propagation normally occurs in air, which is often well-modeled as a medium with ϵ and μ equal to the free space values. This equation indicates that the variation in $\mathcal{B}$ over space depends not only on the original charge (now represented as the current $\mathcal{J}$), but also to the variation in $\mathcal{E}$ over time.

Now, consider what happens if the charge stops moving, or simply vanishes. In either case $\mathcal{J} \to 0$, but Equations (2.1) and (2.2) seem to indicate that $\mathcal{E}$ and $\mathcal{B}$ may continue to be non-zero. In fact, these equations indicate that $\mathcal{E}$ and $\mathcal{B}$ are coupled in such a way that the time-varying $\mathcal{B}$ becomes the source of $\mathcal{E}$, time-varying $\mathcal{E}$ becomes the source of $\mathcal{B}$, and that this relationship may continue independently of $\mathcal{J}$. In other words, time-varying current has given rise to a propagating electromagnetic field – a radio wave – that may persist in the absence of the current.

Equations (2.1) and (2.2) are simultaneous partial differential equations that, when combined with Maxwell's equations for the divergence of $\mathcal{E}$ and $\mathcal{B}$, can be reduced to a partial differential equation for $\mathcal{E}$ and $\mathcal{B}$ individually, known as the *wave equation*. The solution of the wave equation for $\mathcal{E}$ is the propagating electric field intensity in response to an impressed current $\mathcal{J}$. We now leap directly to that solution.

2.2.2 Radiation from Linear Antennas; Far-Field Approximations

We now consider how a transmitting antenna gives rise to a propagating electric field. Figure 2.1 shows an example that demonstrates the principle. We begin with a current source attached to a section of parallel-wire transmission line that is open-circuited on the opposite end. Note that the current source cannot have a DC component, since the DC impedance of the open-circuited transmission line is infinite. However, it is possible for the current source to instead provide a steady sinusoidally-varying current, because in this case the transmission line is able to support a "standing wave" which satisfies the boundary condition requiring that the current

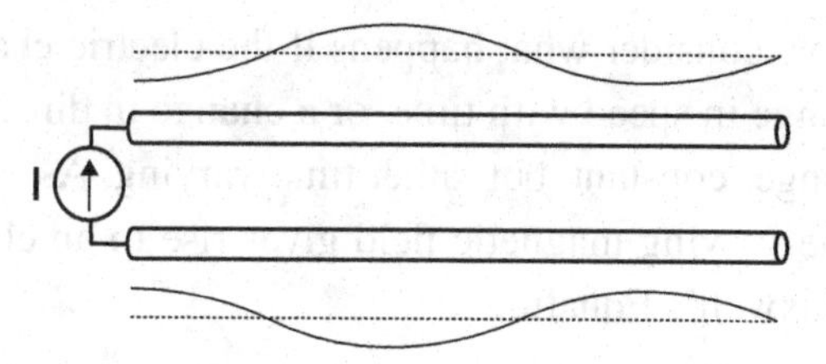

(a) Parallel-conductor transmission line with open circuit termination.

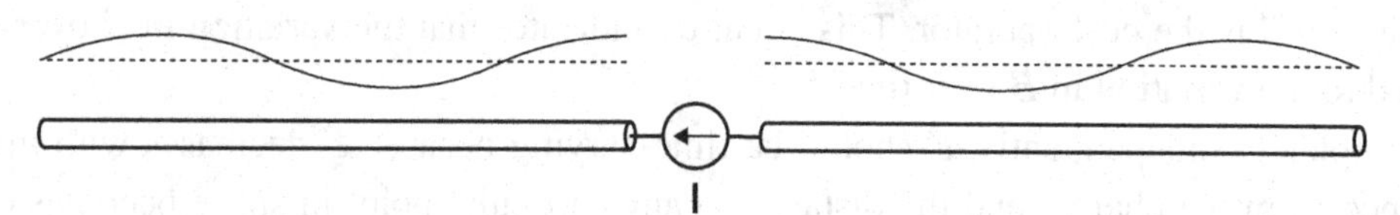

(b) Same, bent into a dipole.

Figure 2.1. Evolution from a transmission line to a dipole. A possible current distribution is shown along side the conductors.

at the end of the transmission line must be zero. Thus, the current at other points along the transmission line can be non-zero, but only if the source is time-varying.

At first glance, it may appear that the structure in Figure 2.1(a) should radiate, but in fact it does not. The reason is that the currents on the two conductors flow in opposite directions, so as long as the spacing between the conductors is much less than a wavelength, the vector sum of the currents at any point on the transmission line is effectively zero. However, if we bend the conductors such that they are collinear as opposed to parallel, as shown in Figure 2.1(b), the vector sum of currents is no longer zero. We have formed a dipole antenna. Since there is now significant time-varying current at each point along the length of the structure, there is potential for efficient creation of a radiating electromagnetic field.

The electromagnetic field radiated by any antenna depends on the distribution of current on the antenna. Continuing under the assumption of sinusoidally-varying excitation, we shall employ phasor quantities from this point forward. The simplest imaginable distribution of current, illustrated in Figure 2.2(a), is the *point current moment*

$$\mathbf{J}(\mathbf{r}) = \hat{\mathbf{z}}\, I\, \Delta l\, \delta(\mathbf{r}) \tag{2.3}$$

where $\mathbf{J}(\mathbf{r})$ is the current moment at $\mathbf{r} = \hat{\mathbf{r}}r$, $I\,\Delta l$ is the scalar current moment, having units of $\mathrm{A}\cdot\mathrm{m}$; and $\delta\,(\mathbf{r})$ is the Dirac delta function. The Dirac delta function is defined by the following properties: $\delta\,(\mathbf{r}) = 0$ for $r > 0$, and $\iiint \delta\,(\mathbf{r})\,dv = 1$ over any volume including $\mathbf{r} = 0$. Thus, the point current moment is non-zero only at the origin ($\mathbf{r} = 0$). I and Δl do indeed refer to a current and a length respectively, but initially this is not relevant because the point current moment is non-zero only at a single point. For $r \gg \lambda$, the electric field intensity due to the point current moment is

$$\mathbf{E}_{\Delta}\,(\mathbf{r}) \cong \hat{\theta}\, j\eta \frac{I\,(\beta\Delta l)}{4\pi}\,(\sin\theta)\,\frac{e^{-j\beta r}}{r} \tag{2.4}$$

where η is the wave impedance (about $120\pi \cong 377\ \Omega$ for free space), and β is the wavenumber $2\pi/\lambda$ (see e.g., [76, Sec. 2.3]).

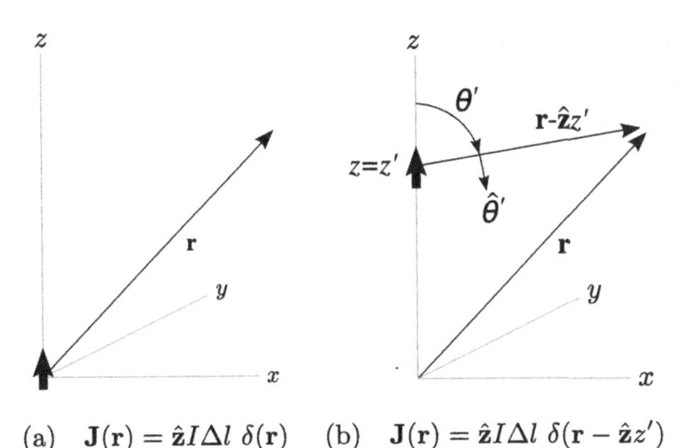

Figure 2.2. The $\hat{\mathbf{z}}$-directed point current moment located (a) at $\mathbf{r} = 0$, (b) at $\mathbf{r} = \hat{\mathbf{z}}z'$.

A broad class of antennas of particular interest in radio systems engineering can be represented as a distribution of current along a line. Consider the displaced point current moment shown in Figure 2.2(b). In this case the electric field intensity can be written as a function of the position $\hat{\mathbf{z}}z'$ of the current moment by modification of Equation (2.4) as follows:

$$\mathbf{E}_{\Delta}\left(\mathbf{r}, z'\right) \cong \hat{\theta}'\, j\eta \frac{I(z')\,(\beta\,\Delta l)}{4\pi}\left(\sin\theta'\right)\frac{e^{-j\beta\left|\mathbf{r}-\hat{\mathbf{z}}z'\right|}}{\left|\mathbf{r}-\hat{\mathbf{z}}z'\right|} \tag{2.5}$$

where θ' and $\hat{\theta}'$ are defined as shown in Figure 2.2(b). Additional point current moments may be distributed along the z-axis, and the resulting fields can be calculated as the sum of the contributions of these individual point current moments.

Equation (2.5) is dramatically simplified if $I(z')$ can be assumed to be non-zero only over some length L around the origin; i.e., if $I(z') = 0$ for $\left|z'\right| > L/2$; certainly this is true for all practical antennas. Then we note that in most radio engineering problems, $L \ll r$: That is, the size of the antenna is much less than the distance between antennas. Under this assumption, the vectors $\mathbf{r}$ and $\mathbf{r} - \mathbf{z}z'$ become approximately parallel, and therefore $\hat{\theta}' \cong \hat{\theta}$ and $\theta' \cong \theta$. Furthermore, the magnitude dependence can be simplified by noting that $z' \ll r$, so

$$\frac{1}{\left|\mathbf{r}-\hat{\mathbf{z}}z'\right|} \cong \frac{1}{r} \tag{2.6}$$

Also, the phase dependence can be simplified by taking advantage of $\theta' \cong \theta$:

$$e^{-jk\left|\mathbf{r}-\hat{\mathbf{z}}z'\right|} \cong e^{-jk(r - z'\cos\theta)} \tag{2.7}$$

Using these approximations, we obtain

$$\mathbf{E}_{\Delta}\left(\mathbf{r}, z'\right) \cong \hat{\theta}\, j\eta \frac{I(z')\,(\beta\,\Delta l)}{4\pi}\left(\sin\theta\right)\left(e^{+j\beta z'\cos\theta}\right)\frac{e^{-j\beta r}}{r} \tag{2.8}$$

The two simplifying assumptions we have made so far are that the distance to the observation point is large compared to a wavelength ($r \gg \lambda$) and that the size of the antenna is small compared to the distance to the observation point ($L \ll r$). These assumptions are

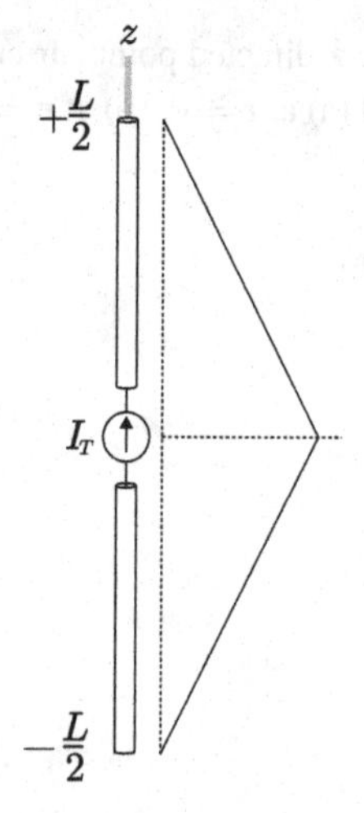

Figure 2.3. Current distribution for the electrically-short dipole (ESD).

collectively referred to as the *far field conditions*, and are typically well-satisfied in practical radio engineering problems.[1]

Let us now define a differential current moment by letting $I(z')\Delta l \rightarrow I(z')dz'$. The associated electric field is

$$d\mathbf{E}\left(\mathbf{r}, z'\right) \cong \hat{\theta}\, j\eta \frac{I(z')\left(\beta dz'\right)}{4\pi} (\sin\theta) \left(e^{+j\beta z' \cos\theta}\right) \frac{e^{-j\beta r}}{r} \tag{2.9}$$

Now the electric field for any current distribution $I(z)$ satisfying the far field conditions can be calculated as

$$\mathbf{E}\left(\mathbf{r}\right) = \int_{-L/2}^{+L/2} d\mathbf{E}\left(\mathbf{r}, z'\right) \tag{2.10}$$

which expands as follows:

$$\mathbf{E}\left(\mathbf{r}\right) \cong \hat{\theta}\, j\frac{\eta}{2} (\sin\theta) \frac{e^{-j\beta r}}{r} \left[\frac{1}{\lambda} \int_{-L/2}^{+L/2} I(z') e^{+j\beta z' \cos\theta} dz'\right] \tag{2.11}$$

Two practical antennas that fit this description and that have high relevance to radio engineering are the electrically-thin, electrically-short dipole (ESD) and the electrically-thin half-wave dipole (HWD). By electrically-thin, we mean the maximum cross-sectional dimension is $\ll L$, and therefore $\ll \lambda$. In this case, the current can be assumed to be uniformly distributed in circumference and varying only with z.

Let us begin with the electrically-thin ESD, shown in Figure 2.3. The ESD is defined as a straight dipole for which $L \ll \lambda$. We know that $I(z')$ must be zero at the ends, and equal to the terminal current I_T at the center. Since $L \ll \lambda$, we also know that the variation in current over the length of the dipole must be very simple. In fact, $I(z')$ should increase monotonically from zero at the ends to I_T at the antenna terminals. To a good approximation, this can be modeled as the "triangular" distribution

[1] A commonly-cited third criterion is $r > 2L^2/\lambda$, which addresses specifically the validity of the "parallel rays" approximation $\theta' \cong \theta$. This is typically not an issue except for antennas which are very large compared to a wavelength, like large reflector antennas.

$$I(z') \cong I_T \left(1 - \frac{2}{L}\left|z'\right|\right) \tag{2.12}$$

Further, since $L \ll \lambda$, the factor $e^{+j\beta z' \cos\theta}$ inside the integral must be $\cong 1$ over the limits of integration. Evaluating Equation (2.11) we then find

$$\mathbf{E}_{ESD}\left(\mathbf{r}\right) \cong \hat{\theta}\, \frac{j}{8\pi}\eta I_T \left(\beta L\right)\left(\sin\theta\right)\frac{e^{-j\beta r}}{r} \tag{2.13}$$

This is the expression we seek: the electric field intensity generated by an electrically-thin ESD in the far field.

Note that our result is in phasor form. If we ever need to know the actual, physical electric field intensity, we may calculate it in the usual way:

$$\mathcal{E}_{ESD}\left(\mathbf{r}\right) = \mathcal{R}e\left\{\mathbf{E}_{ESD}\left(\mathbf{r}\right)e^{j\omega t}\right\} \tag{2.14}$$

where $\omega = 2\pi f$ and f is frequency. We obtain:

$$\mathcal{E}_{ESD}\left(\mathbf{r}\right) \cong \hat{\theta}\, \frac{1}{8\pi}\eta \left|I_T\right|\left(\beta L\right)\left(\sin\theta\right)\frac{\cos\left(\omega t - \beta r + \pi/2 + \psi\right)}{r} \tag{2.15}$$

where ψ is the phase of I_T.

The corresponding solution for the electrically-thin HWD is presented in Section 2.7.

2.2.3 Equivalent Circuit Model for Transmission

For the purposes of radio systems analysis, it is useful to model the transmitting antenna as a passive load attached to the transmitter, as shown in Figure 2.4. In this equivalent circuit model, the antenna is represented as an impedance Z_A equal (by definition) to the ratio of the voltage across the antenna terminals to the current through the antenna terminals. The power dissipated in this load represents the power radiated by the antenna, plus any power dissipated due to losses internal to the antenna. Radiated power is maximized by conjugate matching the transmitter output to Z_A, so we are motivated to quantify Z_A for the antennas we intend to use.

As a starting point, let us return to the open-circuited stub illustrated in Figure 2.1(a). Because there is no dissipation of power within the stub and no transmission of power beyond the stub, the input impedance has no real-valued component and thus is purely imaginary-valued. For dipole antennas (Figure 2.1(b), including ESDs and HWDs), we expect this impedance to be different in two ways. First, we expect the imaginary part of the impedance will be different because the geometry has changed. Let us label the associated reactance X_A. Second, we now expect the impedance to have a non-zero real-valued component, because

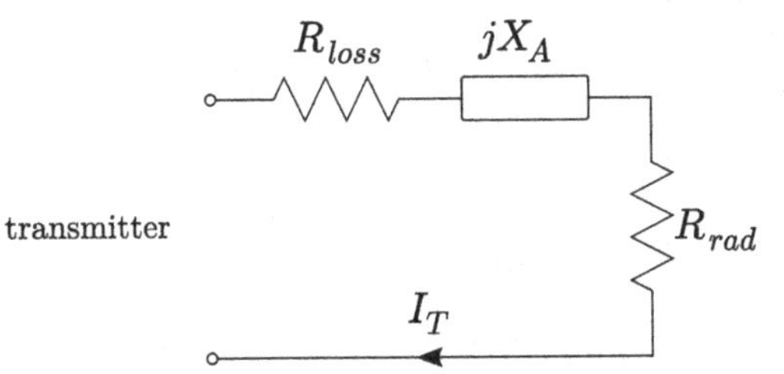

Figure 2.4. Equivalent circuit representing an antenna in transmission.

there is power flow from source to antenna and subsequently into a radio wave. Assuming the antenna structure itself is lossless, this power must be equal to the power carried away by the radio wave. In recognition of this fact, we refer to the real part of this lossless antenna's input impedance as the *radiation resistance* R_{rad}. Thus, the input impedance of any lossless antenna can be written

$$Z_A = R_{rad} + jX_A \tag{2.16}$$

If the antenna is not lossless, then some fraction of the power provided by the source is dissipated by the antenna and is not radiated. In this case, the real part of the antenna's input impedance is $R_{loss} + R_{rad}$, where R_{loss} is the loss resistance. Thus in general we have

$$Z_A = R_{rad} + R_{loss} + jX_A \tag{2.17}$$

as shown in Figure 2.4.

To determine R_{rad} for the ESD, we proceed as follows: Equation (2.13) tells us the radiated electric field as a function of the applied terminal current I_T. At a distance $r \gg L$ from the antenna, the radius of curvature of the expanding phase front is very large, and so the field is well-modeled locally as a plane wave. Using the plane wave approximation, the power density is

$$S_{rad}(\mathbf{r}) \cong \frac{|\mathbf{E}_{ESD}(\mathbf{r})|^2}{2\eta} \cong \frac{1}{128\pi^2}\eta\,|I_T|^2\,(\beta L)^2\,(\sin\theta)^2\,\frac{1}{r^2} \tag{2.18}$$

assuming all quantities are in peak (as opposed to root mean square) units. Note that $S_{rad}(\mathbf{r})$ is the *spatial* power density; i.e., units of W/m^2. The total power radiated by the antenna can be determined by integrating the above expression over a sphere of radius $r \gg L$ surrounding the antenna:

$$P_{rad} = \int_{\theta=0}^{\pi}\int_{\phi=0}^{2\pi} S_{rad}(\mathbf{r})\,r^2\,(\sin\theta)\;d\theta\;d\phi \tag{2.19}$$

Evaluating this integral, we obtain:

$$P_{rad} \cong \frac{1}{48\pi}\eta\,|I_T|^2\,(\beta L)^2 \tag{2.20}$$

Note that there is no dependence on the distance from the antenna (r). This is expected, since the total power passing through any sphere enclosing the antenna must be equal to the power applied to the antenna minus the power dissipated within the antenna, if any. In fact, from the equivalent circuit model of Figure 2.4, we have

$$P_{rad} = \frac{1}{2}\,|I_T|^2\,R_{rad} \tag{2.21}$$

Setting Equations (2.20) and (2.21) equal and solving for R_{rad} we obtain

$$R_{rad} \cong \eta\frac{(\beta L)^2}{24\pi} = \frac{\eta\pi}{6}\left(\frac{L}{\lambda}\right)^2 \tag{2.22}$$

In free space $\eta \cong 120\pi\ \Omega$, so we find

$$R_{rad} \cong 20\pi^2\left(\frac{L}{\lambda}\right)^2 \tag{2.23}$$

This is the radiation resistance of an electrically-thin ESD.

The reactive component of the impedance is more difficult to derive. A reasonable approximation for ESDs made from conductors having circular cross-section of radius $a \ll L$ is known to be [32, Ch. 4]

$$X_A \cong -\frac{120\Omega}{\pi L/\lambda}\left[\ln\left(\frac{L}{2a}\right) - 1\right] \tag{2.24}$$

(Another approximation is described in Section 2.7.5.) Note that $X_A \to -\infty$ with decreasing frequency. This is to be expected, as the antenna is becoming increasingly similar to a zero-length open-circuited transmission line, which has infinite negative reactance.

Finally, we consider the loss resistance R_{loss}. The resistance ΔR_{loss} of a conducting cylinder of length Δl and radius a, assuming constant current over the length of the cylinder, is

$$\Delta R_{loss} = R_S \frac{\Delta l}{2\pi a} \tag{2.25}$$

where R_S is the referred to as the *surface resistance*. For most metals at radio frequencies, the skin depth is $\ll a$. In this case

$$R_S \cong \sqrt{\frac{\pi f \mu}{\sigma}} \tag{2.26}$$

where σ is the conductivity of the material (units of S/m). Using these expressions we find the power dissipation in a segment of length Δl at position $\hat{\mathbf{z}}z'$ along a linear antenna (such as the electrically-thin ESD) is

$$\Delta P_{loss}(z') = \frac{1}{2}\left|I(z')\right|^2 \Delta R_{loss} \tag{2.27}$$

Now R_{loss} can be calculated by integration over the length of antenna. For the electrically-thin ESD, we find

$$R_{loss} \cong R_S \frac{1}{2\pi a}\frac{L}{3} \tag{2.28}$$

EXAMPLE 2.1

Estimate the impedance of a straight dipole having length 1 m and radius 1 mm over the HF band. Assume the dipole is constructed from an aluminum alloy having conductivity 3×10^6 S/m.

Solution: From the problem statement, we have $L = 1$ m, $a = 1$ mm, and $\sigma = 3 \times 10^6$ S/m. The HF band is 3–30 MHz (see Figure 1.2). The largest wavelength in this range is about 100 m. Note $L \ll \lambda$ over this range, and the dipole can be considered to be electrically-short. Also, $L/a = 1000$, so this dipole can also be considered to be electrically-thin. Therefore the equations derived above for the electrically-thin ESD apply. Figure 2.5 shows the three components of the antenna's impedance Z_A, namely R_{rad}, X_A, and R_{loss}, calculated using Equations (2.23), (2.24), and (2.28), respectively.

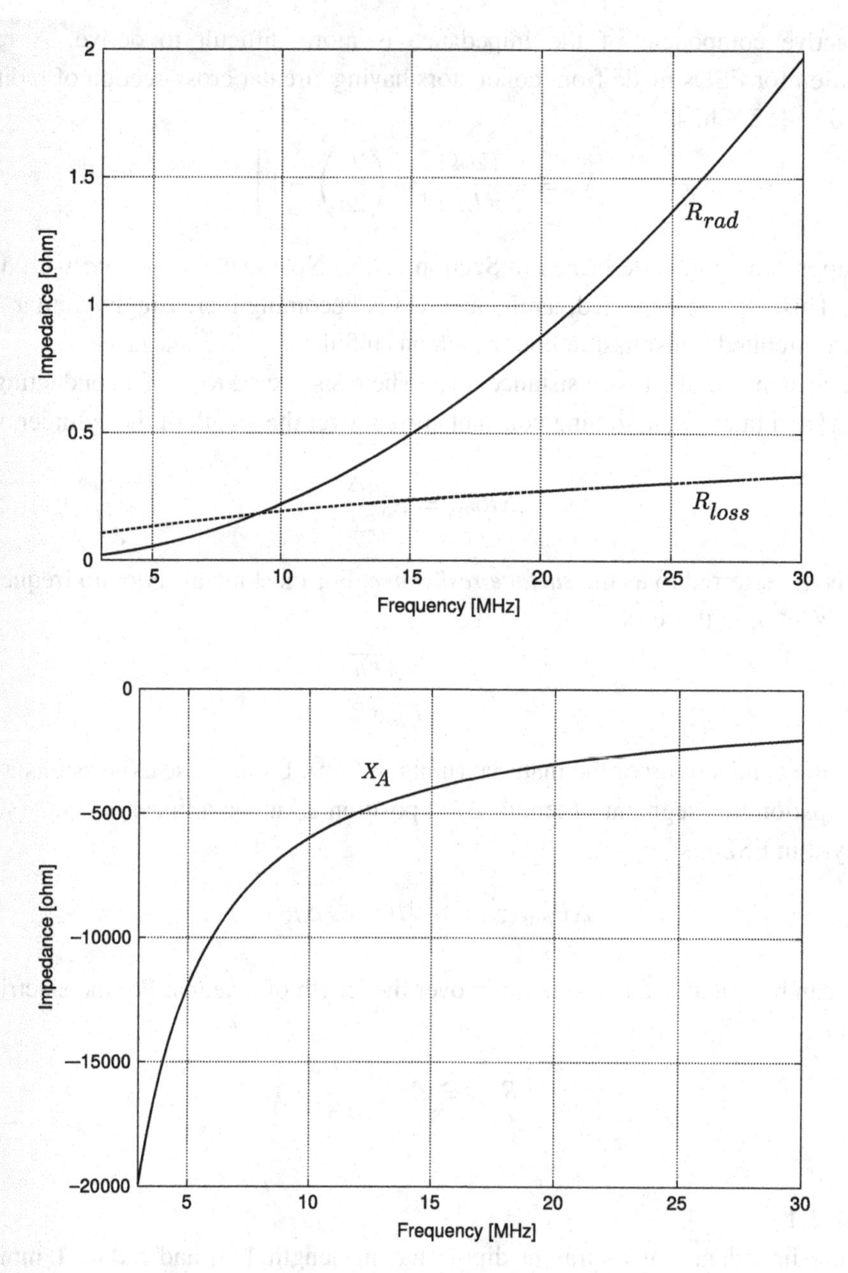

Figure 2.5. Example 2.1: Impedance of an electrically-thin ESD ($L = 1$ m, $a = L/1000$, and $\sigma = 3 \times 10^6$ S/m) in the HF-band.

Note that for an electrically-thin ESD, $|X_A| \gg R_{rad}$. Thus, effective transfer of power from transmitter to radio wave using an ESD is not possible without some form of impedance matching to mitigate the reactance. Also note that R_{loss} becomes comparable to or greater than R_{rad} at the lowest frequencies. Although the power transferred from transmitter to radio wave is $\frac{1}{2}\,|I_T|^2\,R_{rad}$, the total power transferred from transmitter to antenna includes the power dissipated (not radiated) by the antenna, which is $\frac{1}{2}\,|I_T|^2\,R_{loss}$. The *radiation efficiency* ϵ_{rad} is defined to be the ratio of power radiated to power delivered to the antenna. From the above considerations, we see that the radiation efficiency of *any* antenna may be calculated as

$$\epsilon_{rad} = \frac{R_{rad}}{R_{rad} + R_{loss}} \tag{2.29}$$

EXAMPLE 2.2

What is the radiation efficiency of the electrically-thin ESD in Example 2.1?

Solution: Using Equation (2.29) we find ϵ_{rad} ranges from about 16% to about 86%, increasing with increasing frequency. In this case the increase in efficiency with frequency is because R_{rad} increases in proportion to f^2, whereas the increase in R_{loss} is proportional to $\sqrt{f}$.

2.2.4 The Impedance of Other Types of Antennas

To this point we have considered only the electrically-thin ESD. The same procedures for finding R_{rad} and R_{loss} can also be applied to other types of linear antennas, such as the HWD (Section 2.7). Unfortunately, estimating X_A for antennas other than the electrically-thin ESD and HWD is quite difficult except in some special cases. In general, the methods described above to calculate any of the components of the impedance Z_A become intractable for many important categories of antennas. This is because the current distribution for these antennas cannot be accurately estimated using simple mathematical expressions.

So how does one determine Z_A for antennas in general? Typically, a numerical electromagnetic simulation is used to determine the current distribution. A very common simulation approach is the *method of moments*. The method of moments is a numerical solution of Maxwell's Equations in integral frequency-domain form, yielding the currents over the entire structure in response to a voltage applied to the antenna terminals. Given a numerical solution for the current distribution, Z_A can then be computed as the applied voltage divided by the terminal current as determined by the simulation. This method is best suited to narrowband analysis of antennas that are composed of wires or can be modeled as grids of wires.

A popular alternative to the method of moments is the *finite-difference time-domain* (FDTD) method. In this method, the antenna and its surroundings are modeled as a grid of small surface or volume elements, and the associated current distribution is determined using a discrete time-domain version of Maxwell's Equations. This technique has fewer restrictions on the types of antennas that can be analyzed and yields a time-domain (hence broadband) solution, but also entails a greater computational burden.

2.3 RECEPTION OF RADIO WAVES

In the receive case, the incident radio wave gives rise to currents on the antenna, which are delivered to the receiver through the antenna terminals. We now consider the equivalent circuit applicable in the receive case, and use this to define two new quantities: effective length and effective aperture.

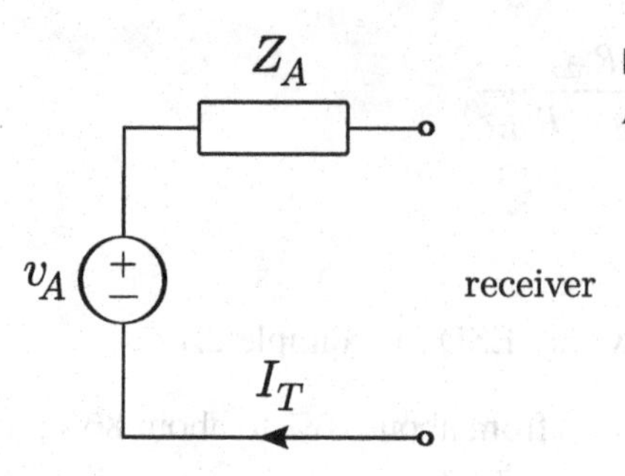

Figure 2.6. Equivalent circuit representing an antenna in reception. As in transmission, $Z_A = R_A + R_{loss} + jX_A$.

2.3.1 Equivalent Circuit Model for Reception; Effective Length

We noted in Section 2.2.3 that a transmit antenna can be described as an electrical load having an impedance Z_A. The analogous description for the receive case is shown in Figure 2.6. The antenna is modeled as a Thévenin equivalent circuit with series impedance Z_A being the same impedance Z_A that appears in the transmission model. The open-circuit voltage v_A is given by

$$v_A = \mathbf{E}^i \cdot \mathbf{l}_e \tag{2.30}$$

where $\mathbf{E}^i$ is the electric field intensity of the radio wave incident on the antenna, $\mathbf{l}_e$ is the *vector effective length* of the antenna, and "·" denotes the dot product. Since v_A has units of volts (V) and $\mathbf{E}^i$ has units of V/m, we see $\mathbf{l}_e$ has units of length. Clearly, $\mathbf{l}_e$ is a property of the antenna. Also, we see that $|v_A|$ is proportional to $|\mathbf{l}_e|$ and depends on how closely the orientation of $\mathbf{l}_e$ is aligned with the polarization of $\mathbf{E}^i$.[2]

To determine $\mathbf{l}_e$ for a particular antenna, one could perform an experiment in which the antenna is illuminated with a known linearly-polarized electric field $\mathbf{E}^i$ while the antenna was open-circuited. In this case, v_A is equal to terminal voltage, which can in principle be directly measured. This procedure can be repeated while rotating the polarization of $\mathbf{E}^i$ in order to determine the direction of $\mathbf{l}_e$, since the correct polarization should maximize the terminal voltage. Note that the magnitude and phase of $\mathbf{l}_e$ should also depend on the direction from which $\mathbf{E}^i$ arrives as well as its polarization, so this process would need to be repeated for all directions of interest. In practice such measurements are quite difficult and tedious, and are rarely done. Even when such measurements are done, the results are usually expressed in terms of directivity and polarization (as explained in Section 2.4) as opposed to vector effective length. Nevertheless the concept of vector effective length is quite useful for a variety of reasons, and so we seek a method by which to calculate it.

A method for calculating vector effective length can be derived from the reciprocity theorem of electromagnetics. This method is suitable for antennas which can be described in terms of linear current distributions, such as the ESD, although the method can be generalized. First, one

[2] The reader is warned that, even though Equation (2.30) is universal, the definition of vector effective length presented in this section is not precisely the same as the definition that appears elsewhere; see e.g. [76, Sec. 4.2]. The difference is in whether one considers the polarization of $\mathbf{l}_e$ to represent the polarization of $\mathbf{E}^i$ (so that the purpose of the dot product in Equation (2.30) is merely to convert a vector to a scalar), or that of the antenna (as it is here). In most applications where the concept finds application – e.g., co-polarized power transfer in the direction of greatest directivity – the distinction is not important.

applies a current source to the antenna terminals and determines the resulting distribution of current on the antenna structure. Let this current source have the value $I_T^{(t)}$ (using the superscript to distinguish it from the receive current) and let the resulting current be $I^{(t)}(z')$. Once this current distribution is known, the vector effective length may be calculated as

$$\mathbf{l}_e = \hat{\mathbf{z}} \frac{1}{I_T^{(t)}} \int_{z=-L/2}^{+L/2} I^{(t)}(z')dz' \tag{2.31}$$

Note that for straight wire antennas, $\mathbf{l}_e$ is oriented precisely as the antenna is oriented.

We can now calculate the vector effective length for the ESD using the current distribution given by Equation (2.12). We find

$$\mathbf{l}_e \cong \hat{\mathbf{z}} \frac{L}{2} \tag{2.32}$$

(The details are left as an exercise; see Problem 2.3.) Note that the effective length turns out to be about one-half of the physical length of the antenna. Generalizing, one finds that the effective length is proportional to the physical length, and is less by an amount having to do with the uniformity of the current distribution.

Let us now consider the transfer of power from an incident radio wave to a receiver. Continuing with the equivalent circuit model, the receiver can be modeled as a passive load having an input impedance Z_L attached to the antenna terminals. Then the voltage v_L across the antenna terminals and current i_L through the receiver are

$$v_L = v_A \frac{Z_L}{Z_A + Z_L} \tag{2.33}$$

$$i_L = \frac{v_A}{Z_A + Z_L} \tag{2.34}$$

Once again assuming phasor quantities in peak (as opposed to root mean square) units, the power delivered to the receiver is

$$P_L = \frac{1}{2} \mathcal{R}e \left\{ v_L i_L^* \right\} = \frac{1}{2} \mathcal{R}e \left\{ \left| \mathbf{E}^i \cdot \mathbf{l}_e \right|^2 \frac{Z_L}{|Z_A + Z_L|^2} \right\} \tag{2.35}$$

Of interest is the maximum power $P_{L,max}$ that can be delivered to the receiver. First, let us assume $R_{loss} \ll R_{rad}$ so that we may neglect any power dissipated by losses internal to the antenna. Next, we note that P_L is maximized if $\mathbf{l}_e$ is perfectly aligned with $\mathbf{E}^i$; in this case $\left| \mathbf{E}^i \cdot \mathbf{l}_e \right| = \left| \mathbf{E}^i \right| l_e$ where $l_e = |\mathbf{l}_e|$, known simply as the effective length. We also know from elementary circuit theory that power transfer is maximized when $Z_L = Z_A^*$; in this case $Z_A + Z_L = 2R_{rad}$. Thus we find that the maximum power that can be transferred to the receiver is

$$P_{L,max} = \frac{\left| \mathbf{E}^i \right|^2 l_e^2}{8R_{rad}} \tag{2.36}$$

Of course if the antenna is not conjugate-matched to the receiver (i.e., $Z_A \neq Z_L^*$), then $P_L < P_{L,max}$. We have already seen in the case of the electrically-short ESD that conjugate matching may not be straightforward or even practical, so for that class of antennas we might reasonably expect $P_L \ll P_{L,max}$ is the best we can do in practice. For other antennas encountered later in this chapter conjugate matching is not out of the question, but is difficult to achieve over a large bandwidth. These issues often make the power loss due to impedance mismatch a major consideration in the design of antennas and their integration with receivers: More on this topic in Chapter 12.

Finally, the concept of *effective height* bears mention. Effective height is effective length for an antenna which is intended to be vertically-polarized (although the term is sometimes used as a synonym for effective length even when the antenna is not necessarily used in this manner). The concept most often finds application in AM broadcasting, which uses predominantly monopole-type antennas attached directly to earth ground.

2.3.2 Effective Aperture

Alternatively, the incident radio wave is described in terms of power density S^i, having units of W/m^2, and which is related to the incident electric field intensity by

$$S^i = \frac{|\mathbf{E}^i|^2}{2\eta} \tag{2.37}$$

In this case it is common to define an *effective aperture* A_e for the antenna, which defined through the expression

$$P_{L,max} = A_e S^i \tag{2.38}$$

so A_e has units of m^2 and quantifies the power delivered to a receiver by a lossless antenna for which $\mathbf{l}_e$ is perfectly aligned with $\mathbf{E}^i$ and for which Z_A is conjugate-matched to Z_L. Combining Equations (2.36), (2.37), and (2.38), we find

$$A_e = \frac{\eta l_e^2}{4R_{rad}} \tag{2.39}$$

To determine A_e for an ESD, we note $l_e \cong L/2$, and use R_{rad} from Equation (2.22), yielding a maximum value (i.e., for broadside orientation) of

$$A_e \cong 0.12\lambda^2 \quad \text{(electrically-thin ESD)} \tag{2.40}$$

Remarkably, the result does not depend on length. In other words, The maximum power that can be delivered by an ESD to a conjugate-matched receiver depends only on frequency, and not at all on the length of the antenna. At first glance this seems to mean ESDs can be made arbitrarily small, which would be marvelous news for developers of small radios. However, this is not quite the whole story, because – once again – the antenna must be conjugate-matched to the receiver in order to achieve this condition.

Equation (2.40) was derived in a manner which is specific to electrically-thin ESDs; however, effective aperture can be determined for any type of antenna. For other kinds of antennas, A_e may or may not depend on physical dimensions and frequency in a simple way. A class of antennas for which A_e *does* increase in a simple way with increasing size is reflectors, addressed in Section 2.10.2.

2.4 PATTERN AND RECIPROCITY

Clearly, an antenna used for transmission delivers more power in some directions than others. The same antenna used for reception captures more power from some directions than others.

These "best directions" are the same. In this section we quantify the performance of antennas as a function of direction using the concepts of *directivity* and *pattern*, and establish the equivalence of these concepts in the transmit and receive cases.

2.4.1 Transmit Case

The variation in power density as a function of direction can be characterized by the directivity $D(\theta, \phi)$, defined as the ratio of the power density (i.e., W/m^2) in the specified direction (θ, ϕ) to the mean power density over all directions. Since the total power radiated by the antenna is not a function of distance, $D(\theta, \phi)$ is not a function of distance.

To demonstrate the concept, let us determine the directivity for the electrically-thin ESD. The total radiated power P_{rad} is given by Equation (2.20). The mean power density $\langle S_{rad} \rangle$ at the some distance r is therefore P_{rad} divided by the area of a sphere of radius r; i.e., $4\pi r^2$. Therefore

$$\langle S_{rad} \rangle \cong \frac{1}{192\pi^2} \eta \, |I_T|^2 \, (\beta L)^2 \, \frac{1}{r^2} \tag{2.41}$$

and, now using Equation (2.18),

$$D(\theta, \phi) = \frac{S_{rad}}{\langle S_{rad} \rangle} \cong 1.5 \sin^2 \theta \tag{2.42}$$

In general, directivity is a two-dimensional function in spherical coordinates which can be difficult to visualize. For this reason it is common to describe directivity in terms of *E-plane* and *H-plane* directivity. The E-plane is the plane in which the electric field vector, determined in the direction of maximum directivity, lies. The H-plane is the plane perpendicular to the E-plane in which the directivity is maximum; this is also a plane in which the magnetic field vector lies. The directivity of the ESD is illustrated in Figure 2.7. For the ESD, the maximum directivity is 1.5 (1.76 dB) as determined in Equation (2.42), and is achieved in all directions in the $z = 0$ plane. Thus, any plane of constant ϕ can be considered to be the E-plane of the ESD. The H-plane is $\theta = \pi/2$, or, equivalently, the x–y plane. Note that for the ESD, $D(\theta = \pi/2, \phi)$ is constant (and equal to maximum directivity); that is, the H-plane directivity of the ESD is constant. The E-plane directivity of the ESD goes to zero for $\theta = 0$ and $\theta = \pi$; these are referred to as *nulls*, and indicate that no power is radiated in these directions.

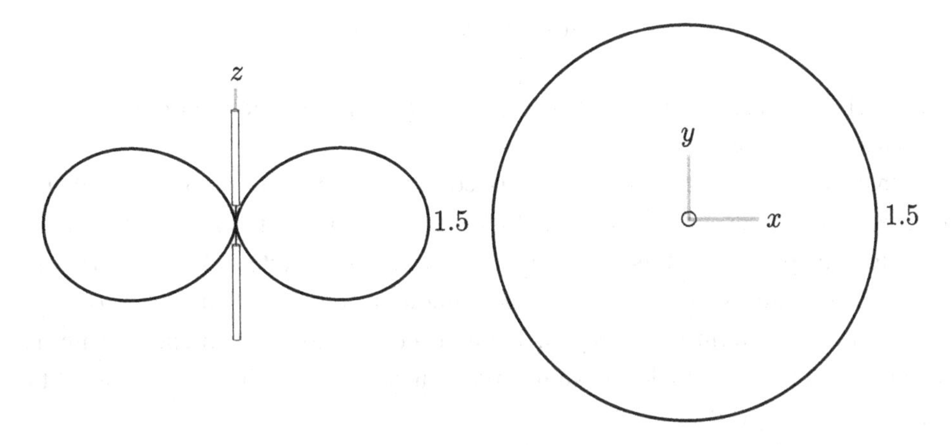

Figure 2.7. Directivity of the ESD in the E-plane (left) and H-plane (right). Linear scaling.

It is frequently of interest to characterize the directivity associated with a particular polarization. For example: The electric field radiated by the ESD is always $\hat{\theta}$-polarized (see Equation (2.13)); thus we might say that we previously calculated "$D_\theta(\theta, \phi)$," and that "$D_\phi(\theta, \phi)$" was identically zero.

If the antenna is intended to be mounted in a fixed position and orientation, then the best mounting orientation is probably that for which the other end of the link is in the direction of maximum directivity. Thus, antennas are frequently specified in terms of maximum directivity (a single number), as opposed to directivity as a function of direction (a pattern). In fact this usage is so common that the term "directivity," when used as a specification, is normally interpreted to mean the directivity in the direction in which it is maximum, and for the polarization anticipated to be used by the antenna at the other end of the link. For example: For the ESD, the directivity would typically be stated as 1.76 dB, since this is the maximum value, corresponding to $\theta = \pi/2$ and $\hat{\theta}$ polarization of the electric field.

Note that the minimum possible value of the maximum directivity of any antenna is 1. This is because the maximum value of the directivity cannot be less than the mean value of the directivity. Furthermore, if the maximum directivity is 1, then the directivity must be a constant value of 1 in all possible directions. Such an antenna is said to be *isotropic*. An isotropic antenna transmits equal power density in all directions. Truly isotropic antennas do not exist in practice; in fact, it is quite difficult to develop a practical antenna with directivity significantly less than that of the ESD; i.e., 1.76 dB. However, the concept is useful as a reference for comparing practical antennas. For example, an antenna having a maximum directivity of D_0 is often said to have a directivity of $10 \log_{10} D_0$ dBi, where the "i" means "relative to isotropic." Again using the example of the ESD, we would say its directivity is 1.76 dBi.

The directivity of antennas is often specified as *gain*. This can be very confusing for the uninitiated. The term "gain" used in this context is not referring to an increase in power; this should be obvious since antennas (as we have been describing them) are passive devices. Instead, "gain" refers to increase in power density. However, there is an additional distinction in that the use of the term "gain" also implies that losses associated with the antenna have been taken into account. Thus, we may formally define the gain $G(\theta, \phi)$ as follows:

$$G(\theta, \phi) = \epsilon_{rad} D(\theta, \phi) \tag{2.43}$$

and, similarly to directivity, it is common to use the term "gain" to refer to the maximum gain for the intended polarization.

One final note on this terminology: It is common to use the term "gain" in lieu of the term "directivity" independently of how losses contributing to radiation efficiency are taken into account. In particular, this practice is employed in Chapters 3 and 7. This confusing state of affairs requires the engineer to be careful to identify what precisely is meant by "gain" whenever the term is employed: It might be directivity, it might be gain as defined in Equation (2.43), or it could be gain with some other combination of losses taken into account.

EXAMPLE 2.3

Characterize the maximum gain of the electrically-thin ESD considered in Example 2.1.

Solution: The maximum directivity of this antenna is 1.76 dBi. ϵ_{rad} for this antenna was determined in Example 2.2 to vary from 16% to 86% for 3–30 MHz. So, the maximum gain of this antenna following the formal definition (Equation (2.43)) is −6.2 dB to +1.1 dB (or "dBi" if you prefer), increasing with increasing frequency.

Characterizations such as in Figure 2.7 (another example is Figure 2.19) are generally referred to as *patterns*. Broadly speaking, there are four categories of patterns. A pattern which is uniform in all directions is said to be isotropic, as explained above. Although no antenna is truly isotropic, many antennas are uniform in one plane; such antennas are said to be *omnidirectional*. For example, the ESD is omnidirectional because its H-plane pattern is uniform. Antennas which are not omnidirectional can be informally categorized as *low gain* or *high gain*. Unfortunately this is another example of inconsistent nomenclature: Almost always the terms "low gain" and "high gain" are referring to directivity as opposed to gain, and thus are not intended to reflect any particular level of radiation efficiency. Antennas are typically considered "low gain" if their maximum directivity is less than about 4 dBi, whereas "high gain" antennas have maximum directivities greater than about 4 dBi, with the threshold of 4 dBi given here being completely arbitrary. High-gain antennas have relatively complex patterns; we shall address this in more detail in Section 2.10.

2.4.2 Receive Case

In Section 2.3.2 we found that the power delivered by a receive antenna to a receiver varies as a function of direction and polarization. We saw in the previous section – for the transmit case – that directivity has a similar functional dependence. For the ESD, for example, the open-circuit voltage v_A on receive is maximum for an electric field which arrives with polarization aligned with $\mathbf{l}_e$, and this polarization and direction of arrival also correspond to the maximum directivity of the ESD on transmit. This suggests a more general relationship between the receive and transmit characteristics. The derivation is beyond the scope of this text; however, a useful result is worth noting:

$$D(\theta,\phi) = \frac{4\pi}{\lambda^2} A_e(\theta,\phi) \tag{2.44}$$

where $A_e(\theta,\phi)$ is defined similarly to A_e except that the assumed direction of incidence is (θ,ϕ) as opposed to the direction which maximizes power transfer. Note that the relationship between D and A_e depends only wavelength; that is, frequency. In this sense, we may consider directivity to be a characteristic of receive antennas, even though we developed the concept from transmit considerations. Consequently, we may refer to "transmit antenna patterns" described in the previous section also as being the "receive antenna patterns".

2.5 POLARIZATION

Polarization refers to the orientation of the electric field vector radiated by an antenna in transmit, or the orientation of the incident electric field vector to which a receive antenna is most sensitive. By reciprocity we expect these orientations to be the same for transmit and receive, and so it suffices to address the transmit case only.

A linearly-polarized antenna generates an electric field whose orientation for any path taken from the antenna is fixed (relative to the antenna) as a function of time. Many antennas – including dipoles – exhibit linear polarization. For the ESD, this can be confirmed using Equation (2.13): The orientation of the radiated electric field of a $\hat{\mathbf{z}}$-oriented dipole is $\hat{\theta}$ and does not vary with time.

The other category of polarization commonly employed in radio systems is *circular polarization.* The orientation of a circularly-polarized electric field rotates in the plane transverse to the direction of propagation. This is perhaps most easily seen by example, as follows: A common method to generate circular polarization is to feed two orthogonally-oriented dipoles with signals that are identical except that a 90° phase shift applied to one of the dipoles. This is shown in Figure 2.8. The 90° phase shift might be implemented either as a quarter-wavelength of additional transmission line, or using an electronic phase shift circuit (see Figure 14.9 for a simple example). When the phase of a sinusoidal signal applied to the phase shifter is zero, the magnitude of the (real-valued, physical) electric field from dipole A is maximum, and the magnitude of electric field from dipole B is zero. Similarly when the phase of a sinusoidal signal applied to the phase shifter is 90°, the magnitude of the (real-valued, physical) electric field from dipole A is zero, and the magnitude of electric field from dipole B is maximum. At all other times, the magnitudes of electric field radiated by each dipole are intermediate between the maximum positive value and the maximum negative value. Thus, the orientation of the total radiated electric field vector rotates as a function of time. Furthermore, the direction of rotation may be reversed simply by swapping the connections between the phase shifter and the dipoles.

The two possible directions of rotation of the electric field vector are identified as *right circular* and *left circular*. These two conditions are orthogonal in precisely the same sense that the linear polarizations of individual orthogonally-oriented dipoles are orthogonal. Thus, an

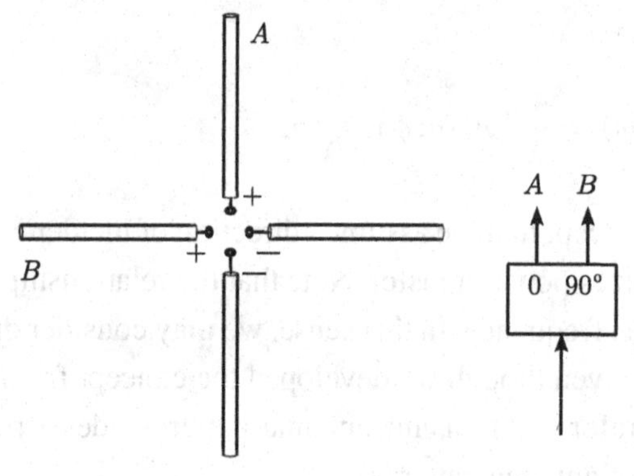

Figure 2.8. Generation of right-circular polarization using orthogonally-oriented dipoles fed 90° out of phase.

antenna system which is intended to optimally receive left-circular polarization must itself be designed for left-circular polarization, and the response of an antenna system with right-circular polarization to the same signal would be zero. This makes it pretty important that everyone involved agrees on the definitions of "right" vs. "left" circular polarization. The definition most commonly employed in radio engineering is the standard promulgated by the IEEE: Point the thumb of your right hand in the direction of propagation; then right-circular polarization is rotation of the electric field vector in the direction in which your fingers curl. Thus, the antenna in Figure 2.8 produces right-circular polarization. If the electric field vector rotates in the opposite direction, you have left-circular polarization.

Circular polarization is useful in applications in which it is impossible to predict the mutual orientation of the transmit and receive antennas. This is commonly the case in communications links with non-geosynchronous satellites, since these satellites may precess in orientation relative to the ground as they orbit the Earth, and the orientation of the antennas used in mobile ground stations might not be predictable. Using circular polarization as opposed to linear polarization makes these uncertainties irrelevant, as long as both ends of the link use the same (i.e., left or right) circular polarization. The penalty for attempting to receive circular polarization using a linearly-polarized antenna (or vice versa) is a factor of one-half in power (i.e., 3 dB).

When the orientation of the electric field vector rotates but also changes in magnitude as function of orientation, we have *elliptical polarization*. Elliptical polarization is typically not desirable, but occurs naturally when the direction of propagation (or direction of incidence) is a direction in which the pattern gains of the constituent linear polarizations are not equal. For example, the polarization of the dual-dipole antenna shown in Figure 2.8 transitions from circular to elliptical as the direction of interest moves from away from the common broadside axis toward the plane in which the dipoles lie. In the plane in which the dipoles lie, the polarization reaches the extreme limit of elliptical polarization, which is linear polarization.

2.6 ANTENNA INTEGRATION

When antennas are connected to radios, two issues immediately arise: impedance matching and current mode compatibility. A sufficiently complete discussion of these issues requires an introduction to the fundamentals of two-port theory and impedance matching presented in Chapters 8 and 9 respectively, as well as an introduction to noise theory which is provided in Chapter 4. With these fundamentals covered we may then productively address the topic of antenna integration, which is presented in Chapter 12. For now, a brief introduction to these issues is sufficient to get us through the next few chapters.

2.6.1 Impedance Matching

Most antennas do not naturally provide a real-valued impedance corresponding to one of the convenient "standard" impedances commonly used in radio engineering, such as $50 + j0\ \Omega$ or $75 + j0\ \Omega$. In Chapters 9 and 12 we consider methods that may be used to match the impedances of antennas and radios, and find that it is often not possible to obtain a close match over the desired bandwidth. Therefore it is important to be able to characterize the impact of the mismatch. This characterization is typically in terms of *voltage standing wave ratio (VSWR)* and sometimes in terms of *transducer power gain*. In the context of transmission, VSWR is a measure of the degree of reflection from the interface between the transmitter and antenna. When there is reflection from the antenna due to impedance mismatch, the incident and reflected fields in the transmission line connecting transmitter to antenna combine to form a standing wave pattern. The ratio of the maximum to minimum voltages in the standing wave pattern is the VSWR. This is most commonly expressed as

$$\text{VSWR} = \frac{1+|\Gamma|}{1-|\Gamma|} \tag{2.45}$$

where Γ is the *voltage reflection coefficient*

$$\Gamma = \frac{Z_A - Z_S}{Z_A + Z_S} \tag{2.46}$$

where Z_S is the source impedance. Reflectionless impedance matching requires $Z_A = Z_S$, which yields $\Gamma = 0$ and VSWR $= 1$. To the extent that Z_A is not equal to Z_S, $|\Gamma| > 0$ and VSWR > 1. Typically – but certainly not always – antennas are designed to have VSWR close to 1 at some nominal (perhaps center) frequency, and are specified to have VSWR < 3 or so over the bandwidth over which they are designed to operate. The concept of VSWR also applies to the receive-mode operation of an antenna, in the sense that increasing VSWR implies that an increasing fraction of the arriving power is scattered by the antenna as opposed to being delivered to the receiver.

Sometimes it is useful to have a more direct assessment of power delivered to the antenna (when transmitting) or to the receiver (when receiving). This is known as *transducer power gain* (TPG). The concept of TPG is addressed in detail in Section 8.4; for now it suffices to define TPG as the power P_L delivered to the load relative to the power P_A "available" from the source. Available power P_A is defined as the power that the source can deliver to a conjugate-matched load (*not* the actual load), which is the maximum power that the source could deliver to a load, period. Let us consider the transmit case. In this case, the transmitter can be modeled as a voltage source in series with a source impedance Z_S, attached in series with the antenna impedance. In this case we find after a short bit of circuit analysis:

$$\text{TPG} \equiv \frac{P_L}{P_A} = \frac{4R_S R_A}{|Z_S + Z_A|^2} \tag{2.47}$$

where R_S is the real part of Z_S. If $X_S = 0$ or $X_A = 0$, then one finds

$$\text{TPG} = 1 - |\Gamma|^2 \text{, if } X_S \text{ or } X_A \text{ (or both) are equal to zero} \tag{2.48}$$

The above equation may be familiar from an undergraduate electromagnetics course.

Finally, we note that optimum VSWR requires reflectionless matching ($Z_A = Z_S$) whereas optimum TPG requires conjugate matching ($Z_A = Z_S^*$). If the associated reactances are negligible, then optimization of VSWR implies optimization of TPG, and vice versa. This is the most commonly encountered condition in radio systems engineering. However, if either of the associated reactances is non-negligible, then optimization of VSWR and optimization of TPG may be contradictory goals. Said differently, optimum TPG may entail unacceptably high VSWR. Resolution of this issue depends on the application, and is addressed in detail in Chapter 12.

2.6.2 Current Mode Matching; Baluns

Figure 2.3 indicates that dipoles are intrinsically *balanced* devices exhibiting a *differential* current mode. That is, current flowing into one terminal is equal to the current flow out of the other terminal. However, many circuits and transmission lines (coaxial cable, in particular) utilize a scheme in which one terminal is grounded (hence, a current sink) and the other terminal is exclusively used to convey the signal. Such devices are said to be *unbalanced* in the sense that they exhibit a *single-ended* current mode. Ideally antennas and radios are not only impedance-matched, but also current mode-matched. So, for example, the proper transmission line for a dipole is a *balanced* transmission line, such as a parallel-wire transmission line. If an unbalanced (intrinsically single-ended) transmission line such as coaxial cable is used instead, then the current mode mismatch results in a fraction of the current traveling on the *outside* of the cable, which in effect makes the cable radiate and receive as part of the antenna system. This is typically not desirable.

Nominally the interface between a coaxial cable and a dipole would be accomplished using a *balun*,[3] which converts a single-ended current mode to a differential mode, and vice versa. Baluns have applications beyond antennas and are discussed in detail in Section 12.7. For a primer on the theory of differential (vs. single-ended) circuits and signals, see Section 8.7.

2.7 DIPOLES

In previous sections we have used the electrically-thin ESD as an example for demonstrating antenna concepts. The broader class of *dipoles* includes many practical antennas which vary in length, thickness, feed location, and exist in a number of useful topological variants.

2.7.1 General Characteristics

To begin, a convenient generalization of the current distribution for $L/a \gg 1$ (reminder: a is the radius) in response to a terminal current I_T is

$$I(z) \approx I_T \frac{\sin\left(\beta\left[L/2 - |z|\right]\right)}{\sin\left(\beta L/2\right)} \tag{2.49}$$

Note that this expression yields the required values $I(0) = I_T$ and $I(\pm L/2) = 0$, regardless of L. Also note that this expression yields the expected nearly-triangular distribution for the ESD ($L \ll \lambda$, Equation (2.12), and Figure 2.3). Although Equation (2.49) is presented here as an "educated guess," the validity of this approximation is easily confirmed using rigorous

[3] This term is often said to be a contraction of the term "balanced-to-unbalanced"; although baluns are typically fully bi-directional.

Table 2.1. Characteristics of straight thin dipoles of the indicated lengths. All radiation resistances should be considered approximate.

L/λ	D	R_{rad} at center	R_{rad} at current max.	E-plane HPBW
$\ll 1$ (ESD)	1.76 dB	Eq. (2.23)	same	90°
0.50 (HWD)	2.15 dB	73 Ω	same	78°
0.75	2.75 dB	400 Ω	186 Ω	64°
1.00 (FWD)	3.82 dB	$\rightarrow \infty$	200 Ω	48°
1.25	5.16 dB	160 Ω	106 Ω	32°

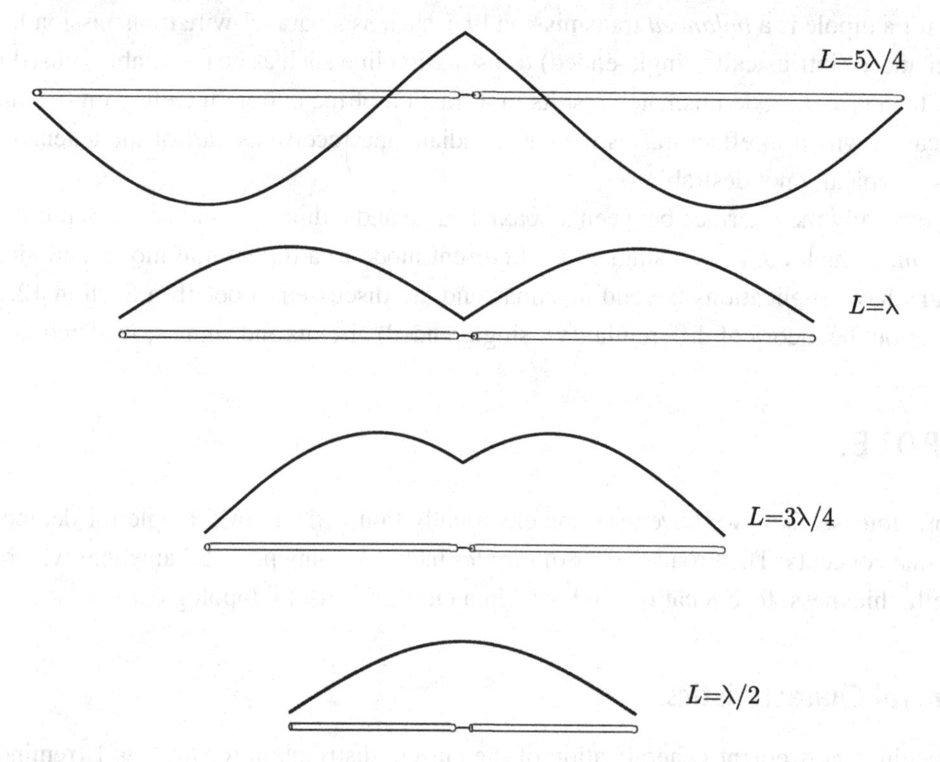

Figure 2.9. Current distributions for straight thin dipoles of the indicated lengths, computed using Equation (2.49).

theoretical and/or numerical methods. Figure 2.9 shows current distributions computed using Equation (2.49) for a variety of dipole lengths.

Electrically-thin straight dipoles of arbitrary length can be analyzed using the methods of Sections 2.2.2–2.2.3. Results for the dipoles shown in Figure 2.9 (as well as for the ESD) are shown in Table 2.1. (For R_{rad}, only the "at center" column is relevant for the moment.) In all cases the patterns are omnidirectional and very similar to those of the ESD (Figure 2.7); the primary difference is a reduction in angular extent of the E-plane pattern. This is quantified in terms of the half-power beamwidth (HPBW) in Table 2.1. HPBW is defined as the angle over which power density is at least one-half the maximum power density.

Note that dipoles longer than $L = 5\lambda/4$ are not shown in Table 2.1, as these suffer from bifurcation of the main lobe and are therefore generally less useful in radio applications.

As for the ESD, the reactance X_A depends on radius a and is relatively difficult to compute; numerical methods such as the moment method or FDTD are typically used instead. Figure 2.10 shows Z_A (both real (R_{rad}) and imaginary (X_A) components) plotted as a function of dipole length L in wavelengths. (Again, only the "terminals at center" results are relevant for the moment.) Note that this plot may alternatively be interpreted as Z_A as function of frequency for a fixed-length dipole.

We now consider a few special cases of thin straight dipoles of particular interest.

2.7.2 The Electrically-Thin Half-Wave Dipole

The electrically-thin half-wave dipole (HWD), has length $L = \lambda/2$. The resulting current is proportional to one-half of a sinusoid, as shown in Figure 2.9 (bottom). This distribution can be described mathematically as

$$I(z) = I_T \cos\left(\pi \frac{z}{L}\right) = I_T \cos\left(2\pi \frac{z}{\lambda}\right) \tag{2.50}$$

(This is, of course, accurate only at the frequency at which $L = \lambda/2$ *exactly*, and is only an approximation for nearby frequencies.) Using this in Equation (2.31), the effective length is found to be λ/π, which is about 64% of the physical length. Note that this is somewhat larger than the value of 50% for the ESD. The directivity of the HWD is found to be

$$D(\theta, \phi) \cong 1.64 \frac{\cos\left(\frac{\pi}{2}\cos\theta\right)}{\sin\theta} \tag{2.51}$$

The maximum directivity is 1.64 = 2.15 dBi, which is slightly larger than that of the ESD.

The maximum effective aperture of the HWD is nearly the same as an ESD (see Equation (2.40) and associated text):

$$A_e \cong 0.13\lambda^2 \quad \text{(electrically-thin HWD)} \tag{2.52}$$

as might have been guessed from the lack of length dependence in the derivation of A_e for the ESD.

The radiation resistance can be determined using the same procedure employed in Section 2.2.3 for the ESD; the result is $R_{rad} \cong 73\ \Omega$. Note that this is much larger than the radiation resistance of the largest dipole which can be considered electrically short.

The reactance X_A of an infinitesimally-thin HWD is about $+42.5\ \Omega$. X_A can be "tuned" to zero by slightly reducing length (as is apparent from examination of Figure 2.10) or by slightly increasing radius. Either modification also slightly reduces R_{rad}, so the impedance of a lossless dipole operating near its half-wavelength resonance is typically about $70 + j0\ \Omega$. The fact that the magnitude of the reactance of the HWD is small compared to the radiation resistance makes impedance matching for maximum power transfer much easier than for the ESD.

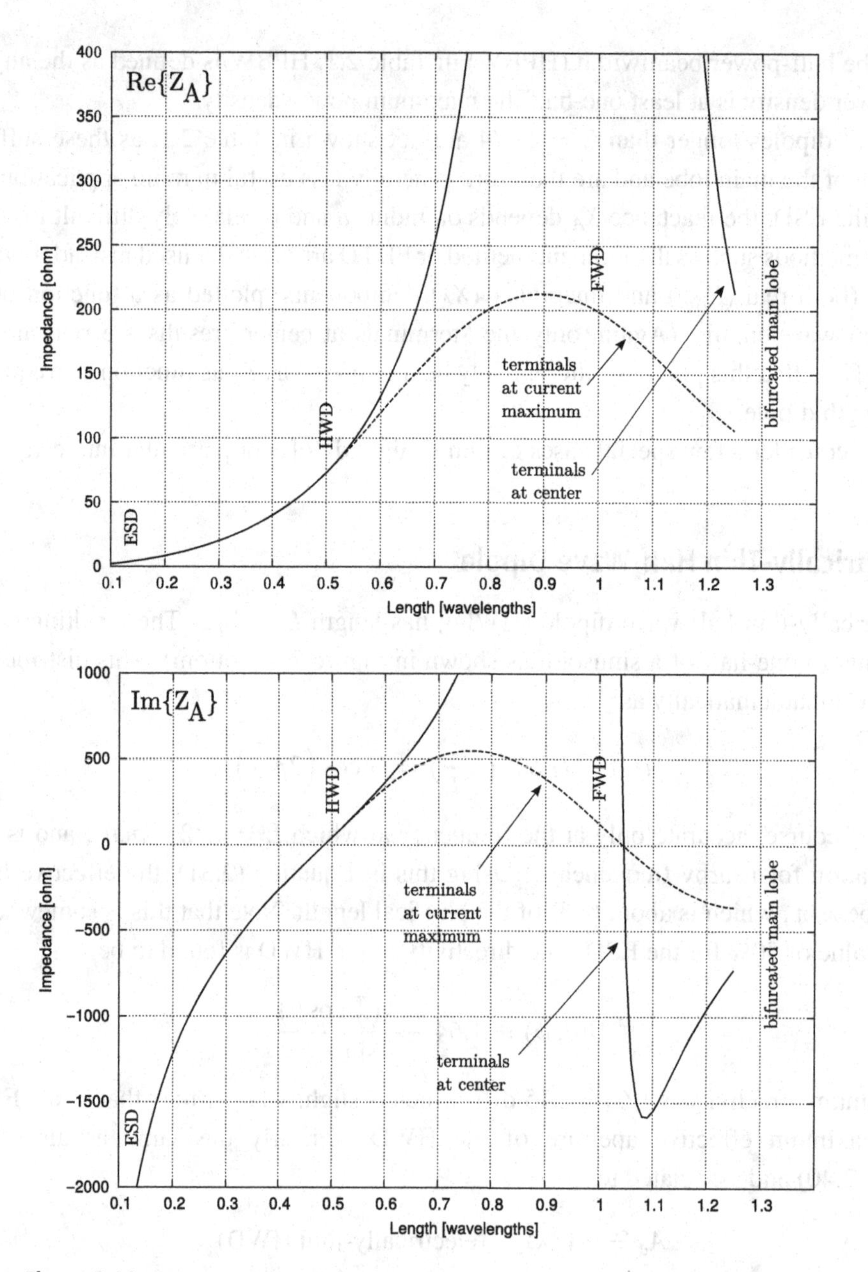

Figure 2.10. Impedance as a function of frequency for a straight dipole with $L/a = 10\,000$ (quite thin). *Top:* real component; *bottom:* imaginary component.

The loss resistance for the HWD is found to be

$$R_{loss} \cong R_S \frac{1}{2\pi a}\frac{\lambda}{4} \tag{2.53}$$

For the metals typically used in the construction of HWDs, $R_{loss} \ll R_{rad}$, and therefore $\epsilon_{rad} \cong 1$ is typically a good approximation for HWDs in free space.

2.7.3 Electrically-Thin Dipoles with $\lambda/2 < L \leq \lambda$; Off-Center-Fed Dipoles

As length increases beyond $\lambda/2$, directivity and radiation resistance continue to increase and HPBW continues to decrease, as indicated in Table 2.1. However, the current distribution

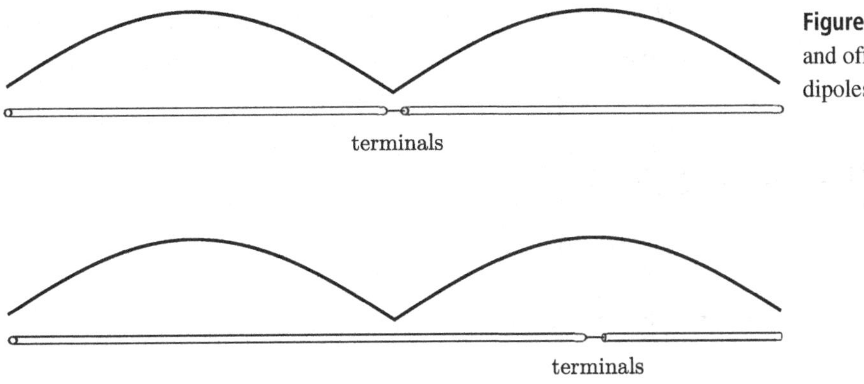

Figure 2.11. Center-fed and offset-fed full-wave dipoles.

changes such that the center of the dipole is no longer a current maximum, as is evident in Figure 2.9. This reaches an extreme condition when $L = \lambda$, for which the current at the center of the dipole goes to zero. This is a full-wave dipole (FWD). Since the terminal current of a FWD is zero, the radiation resistance of a FWD is infinite. At first glance, this would seem to preclude the use of a FWD in any practical application, since there appears to be no possibility of power transfer to or from the antenna. Furthermore, examination of Figure 2.10 seems to indicate that dipoles throughout the range $0.7\lambda < L < 1.2\lambda$ have R_{rad} and X_A so large as to make impedance matching intractable.

A simple modification that makes these dipoles useable is to simply move the terminals from the center of the dipole to the position of a current maximum. This is shown for the FWD in Figure 2.11. With this modification the impedance of a FWD is much easier to accommodate, and there is no change in the current distribution or characteristics other than the impedance. The impedance for terminals located at the current maximum is indicated in Table 2.1 and Figure 2.10. Such dipoles are commonly-referred to as being *off-center-fed*.

In principle dipoles of *any* length can be off-center-fed. In general, moving the terminals to a position of lower current increases the radiation resistance. Thus, shifting the position of the terminals may be used to assist in matching the dipole to its associated electronics.[4]

2.7.4 The Electrically-Thin 5/4-λ Dipole

As L increases beyond $5\lambda/4$, the main lobe of the dipole bifurcates into two lobes and a null emerges in the broadside direction. This pattern has diminished applications in most radio applications. However, $L = 5\lambda/4$ – i.e., just before pattern bifurcation – is of great interest because this dipole achieves the greatest broadside directivity and minimum HPBW of any straight dipole, and is relatively easy to match even when center-fed. This dipole finds applications in base station applications in the UHF range and above, where the length (2.5 times that of a HWD) is of little consequence.

[4] One might consider using this trick to increase the radiation resistance of an ESD. The limitation in practice is that the gap associated with the separation between the terminals becomes significant compared to the length of the dipole, so the results are typically disappointing.

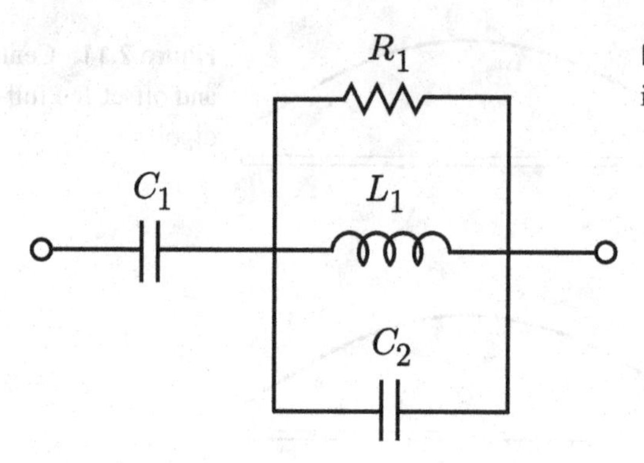

Figure 2.12. An equivalent circuit representing the impedance of a lossless straight dipole, from [77].

2.7.5 Equivalent Circuits and Numerical Methods for Straight Dipoles of Arbitrary Length and Radius

It is often convenient to have an equivalent circuit model for the dipole impedance Z_A, consisting of discrete resistors, capacitors, and inductors. Such a model has the advantage that $Z_A = R_{rad} + jX_A$ can then be computed as a single quantity using simple circuit theory methods, as opposed to having separate procedures for determining R_{rad} and X_A from relatively complex electromagnetic theory. Equivalent circuits also provide a relatively simple means to account for a variety of dipole thicknesses L/a.

A variety of solutions to this problem exist; here we present two. The equivalent circuit proposed by Tang, Tieng & Gunn (1993) [77] is shown in Figure 2.12.[5] The component values are given below in terms of the length L in meters and the thickness parameter $b \equiv L/a$:

$$C_1 = \frac{6.0337L}{\log_{10} b - 0.7245} \text{ pF} \tag{2.54}$$

$$C_2 = L\left(\frac{0.89075}{(\log_{10} b)^{0.8006} - 0.861} - 0.02541\right) \text{ pF} \tag{2.55}$$

$$L_1 = 0.1L\left[(1.4813 \log_{10} b)^{1.012} - 0.6188\right] \mu\text{H} \tag{2.56}$$

$$R_1 = 0.41288\,(\log_{10} b)^2 + 7.40754b^{-0.02389} - 7.27408 \text{ k}\Omega \tag{2.57}$$

This method yields R_{rad} which is accurate to within about 1% and X_A which is accurate to within about 15% for $L < 0.6\lambda$ and $L/a > 300$.

For dipoles longer than about 0.6λ, a different equivalent circuit model is needed. One possibility among many is the model of Hamid & Hamid (1997) [24], which models the impedance of dipoles having $L > \lambda/2$ using circuits of increasing numbers of components with increasing length. The center-fed impedance is determined, which must subsequently be adjusted if the dipole is intended to be off-center-fed as described in Section 2.7.3.

When the dipole is very thick (e.g., L/a less than about 100) or consists of less-good conductors (such as brass), or when accurate pattern or directivity data are also required, it

[5] It is worth noting that this model is single-ended, whereas a dipole is actually differential. In applications where a differential model is necessary, see Section 8.7 for tips on how to modify the equivalent circuit.

is somewhat more common to use a computational electromagnetics technique such as the moment method or FDTD. This approach is typically *required* when the dipole is located close (e.g., within a few wavelengths) to other antennas or structures.

2.7.6 Planar Dipoles; Dipoles on Printed Circuit Boards

Dipoles are routinely implemented on printed circuit boards (PCBs); an example is shown in Figure 2.13. This is obviously convenient when possible, since typically most or all of the electronics in a radio are also implemented on a PCB. However, a PCB dipole has two features not previously considered: First, the proximity of dielectric material. As long as the dielectric is electrically thin (e.g. comparable to a) and there is no ground plane on the opposite side, the effect of the dielectric is typically very small. Second, the dipole is necessarily planar, as opposed to cylindrical, in cross-section. As long as the maximum dimension of the cross-section is very small compared to both L and λ, a planar dipole having width w is approximately equivalent to a cylindrical dipole having radius $a = w/4$ [8]. In all other respects, PCB dipoles behave similarly to cylindrical dipoles in free space.

Of course it is possible to create "free-standing" planar dipoles, independent of PCB implementation, in which case the above rule-of-thumb also applies.

2.7.7 Other Dipole-Type Antennas

Practical dipole-type antennas exist in an astounding number of variants – some barely recognizable as dipoles. Although the scope of this book precludes a detailed discussion, Figure 2.14 shows a few variants that are too popular not to mention.

- A *sleeve dipole* (Figure 2.14(a)) is an HWD formed from coaxial cable: The center conductor forms one arm of the dipole, and the shield is turned back on itself to form the other arm of the dipole. When properly designed, the doubled-back shield section also serves as a quarter-wave balun which properly mitigates the difference between the unbalanced current mode of the coaxial cable and the balanced mode of the HWD.

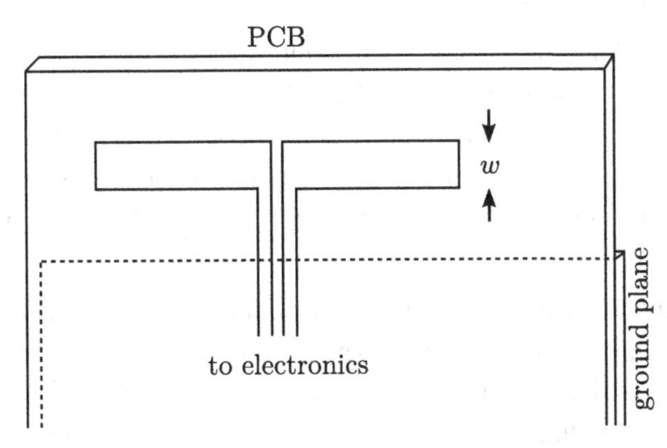

Figure 2.13. A planar dipole implemented on a printed circuit board.

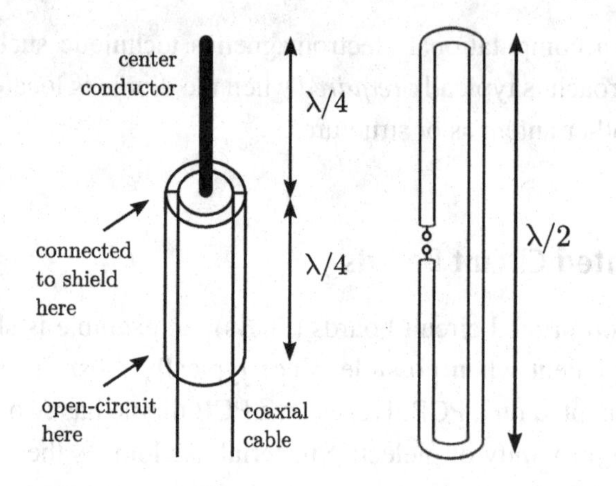

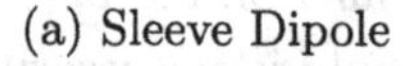

(b) Folded Dipole

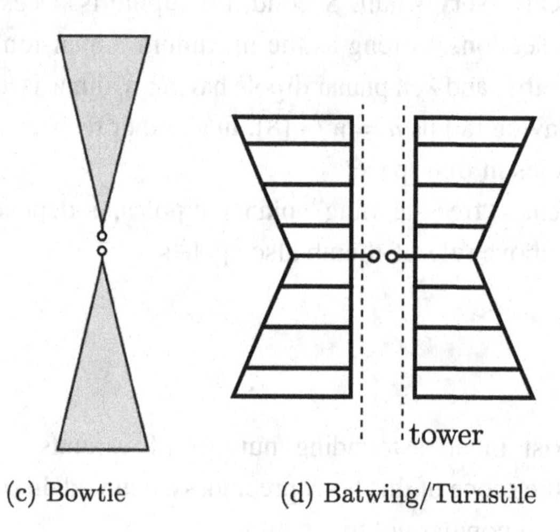

(c) Bowtie (d) Batwing/Turnstile

Figure 2.14. Other commonly-encountered dipole-type antennas.

- A *folded dipole* (Figure 2.14(b)) is formed from two half-wavelength sections: One is a more-or-less traditional HWD, whereas the other is essentially a short-circuited HWD. The sections are connected at the ends. The spacing between sections is $\ll \lambda$, so there is strong interaction between the sections. The result is effective length and pattern similar to a HWD, but with increased bandwidth and Z_A which is about four times greater than that of a HWD; i.e., roughly $292 + j168\ \Omega$. This higher impedance is more convenient for certain types of transmission line; in particular "parallel conductor" line which has a nominal characteristic impedance of 300 Ω. The folded dipole is also convenient for four-element arrays because four folded dipoles fed in parallel have a combined impedance which is close to the impedance of common transmission line types (i.e., 50 Ω and 75 Ω); more on this in Section 2.11.
- The *bowtie* (Figure 2.14(c)) and comparable antennas are examples of "fat dipoles": dipoles with very small "effective" L/a, but not necessarily exhibiting the cylindrical cross-section presumed to this point. The principal advantage of this topology is increased bandwidth, which can be up to about 25% depending on the particular definition of bandwidth employed.

These antennas are typically constructed from wire grid as opposed to solid sheet so as to reduce weight and wind loading.

- The *batwing* antenna (Figure 2.14(d)) is a wire grid fat dipole employed primarily for broadcasting in the VHF and UHF bands. Batwing antennas are typically arranged as *turnstile* arrays consisting of pairs of batwing antennas arranged collinearly along a tower; see Section 2.11.

The pattern and impedance of these antennas is typically determined by experience, measurements, or by numerical methods such as the method of moments or FDTD, especially when response over a range of frequencies is required. Additional details about these and various other useful dipole variants is well-covered in a number of antenna textbooks and handbooks including [9, 76].

2.8 MONOPOLES

In many applications where a dipole might be used, one or both of the following problems emerge: (1) the electrical interface to the radio is single-ended (e.g., one terminal is grounded), (2) mounting of the dipole is awkward due to space constraints or the proximity of a large conducting surface which affects the dipole's performance. Monopoles are typically used in these cases. Let us first describe the monopole, and then discuss why monopoles are often favored when conditions (1) or (2) apply.

2.8.1 General Characteristics

The fundamental concept of the monopole is illustrated in Figure 2.15. Beginning with a dipole, an ideal monopole is obtained by replacing one "arm" with a flat, perfectly-conducting ground plane that extends to infinity in all directions. From electromagnetic image theory, we know that the currents on the remaining arm should be identical to the currents on the corresponding arm of the original dipole, and that the electromagnetic fields radiated by the monopole above the ground plane should be identical to those of the dipole in the same region. Below the ground plane, however, the fields radiated by the ideal monopole must be identically zero.

These facts allow us to quickly determine the directivity and impedance of a monopole, given the characteristics of the corresponding dipole. Recall that directivity is power density divided by mean power density over a sphere enclosing the antenna. Below the ground plane, $D(\theta, \phi) = 0$ since no power can be radiated into this region. Above the ground plane, $D(\theta, \phi)$ is twice the value for the corresponding dipole, since the power density is equal whereas mean power density is reduced by one-half because $D(\theta, \phi) = 0$ below the ground plane. Similarly the total power radiated by a transmitting ideal monopole is one-half that for the corresponding dipole given the same terminal current I_T, so from Equation (2.21) we see R_{rad} for the ideal monopole must be one-half the corresponding value for the dipole. In fact, it is possible to show (although beyond the scope of this text) that Z_A (i.e., both real and imaginary components of the impedance) for the ideal monopole is one-half Z_A for the corresponding dipole.

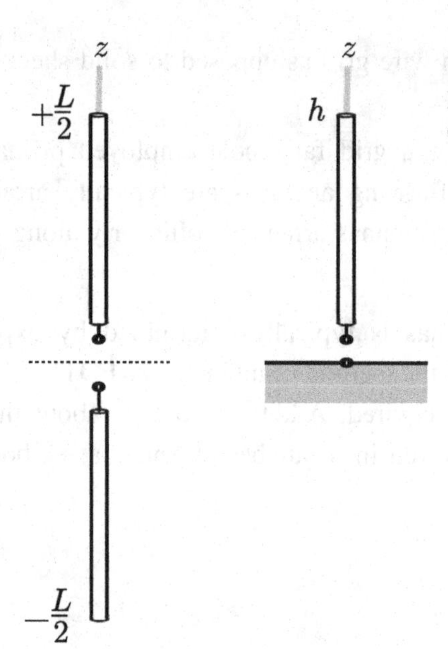

Figure 2.15. Relationship between dipoles and monopoles from the perspective of image theory. In this figure, the midpoint of the dipole and ground plane of the monopole lie in the $z = 0$ plane.

It should be apparent that the electrical interface to the monopole is intrinsically single-ended, where the ground plane can be interpreted as the datum. This is particularly convenient when the connection between antenna and radio uses coaxial cable, which is also intrinsically single-ended.

2.8.2 The Ideal Electrically-Thin Electrically-Short Monopole

Let us refer to the ideal monopole which corresponds to the electrically-thin ESD as an "electrically-short monopole" (ESM). From the above considerations, we find that the maximum directivity and impedance of an ESM are 4.77 dBi and

$$Z_A \cong 40\pi^2 \left(\frac{h}{\lambda}\right)^2 + R_S \frac{1}{2\pi a} \frac{h}{3} - j \frac{30\Omega}{\pi\,(h/\lambda)} \left[\ln\left(\frac{h}{a}\right) - 1\right] \tag{2.58}$$

respectively. The effective aperture and effective length are the same as those for the corresponding dipole.

2.8.3 The Ideal Electrically-Thin Quarter-Wave Monopole

The monopole which corresponds to the HWD is a *quarter-wave monopole* (QWM). In this case we find that the maximum directivity and impedance are 5.16 dBi (twice the HWD value) and

$$Z_A \cong 36.5 + j21\ \Omega \tag{2.59}$$

(half the HWD value). Just as in the case of the HWD, we can zero the reactive impedance of the QWM by slightly reducing length or slightly increasing thickness, yielding $Z_A \approx 35\ \Omega$.

2.8.4 The 5/8-λ Monopole

Around 800 MHz a wavelength is about 37 cm and a QWM is just 9 cm long. Base station and vehicular installations can easily accommodate antennas much larger than this. A larger antenna is beneficial since increasing h increases effective length and therefore directivity. A convenient modification from the ideal QWM in this case is simply to increase h to about $5\lambda/8$. This corresponds to a dipole having $L = 5\lambda/4$; see Section 2.7.4 for a discussion as to why this is the longest dipole that is typically considered acceptable. Although a 5/8-λ monopole is 2.5 times longer than a QWM, this is only about 23 cm at 800 MHz. Referring to Table 2.1, and keeping in mind the principles of dipole–monopole duality, we see that the broadside directivity of a 5/8-λ monopole is expected to be 8.16 dB – very large compared with every other antenna considered so far. Thus monopoles of this type prevail over QWMs at frequencies in the upper UHF range and above.

It is usually not practical to feed a monopole off-center, since this would require terminals above the ground plane. For this reason "half-wave monopoles" (corresponding to FWDs) typically do not appear in practical applications. Similarly it is uncommon to modify a monopole's impedance using the off-center feeding technique.

2.8.5 Practical Monopoles

Monopoles are favored over dipoles when operating in the proximity of large conducting surfaces because a monopole can use the conducting surface as its ground plane. An example of this concept is shown in Figure 2.16(a), where the roof and trunk of a vehicle are being used as ground planes. The same approach is also frequently employed for aircraft antennas. For both automobiles and aircraft it is apparent that the structure which has been co-opted for use as a ground plane is far from flat and infinite in extent. Nevertheless, such installations work well enough as long as the radius of curvature of the ground plane is large compared to a wavelength, and as long as the minimum dimension of the ground plane is at least $\lambda/2$ or so. In fact it is not really even necessary for the ground plane to be a plane: Figure 2.16(b) shows a common variant of the QWM in which the ground "plane" is formed from four quarter-wavelength radials. Of course, significant variations from the impedance and pattern of the ideal QWM are to be expected when the ground plane is implemented in this manner.

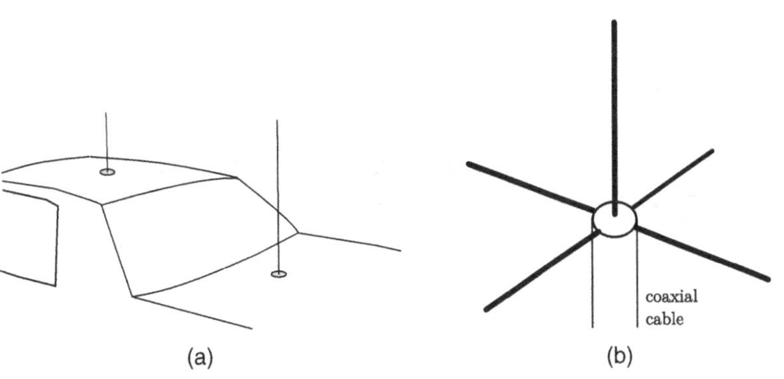

Figure 2.16. Practical monopoles: (a) mounted on the roof and trunk of a sedan, (b) a mobile radio base station QWM using four radials in lieu of a ground screen.

As in the case of the dipole, the number of important variants is large and a summary of these variants is beyond the scope of this book; recommended starting points for additional information include [9] and [83]. However, it is worth keeping in mind that essentially every dipole variant has a corresponding monopole variant.

2.9 PATCH ANTENNAS

A patch antenna consists of a planar metallic radiator that is separated from a ground plane by a slab of dielectric material having thickness $\ll \lambda$. Figure 2.17(a) illustrates the design of a typical patch antenna; in this case, for a handheld GPS receiver. The patch may be fed using a conductor which protrudes through the ground plane to some point on the patch (as in Figure 2.17(a)), or alternatively using a microstrip transmission line connected to an edge of the patch. In either case an electric field is created between the patch and the ground plane, which subsequently radiates from the edges of the patch.

Analysis of patches is difficult compared with dipoles and monopoles, and is beyond the scope of this text. However, a few generally-applicable statements are in order. The patch antenna exhibits greatest directivity in the direction perpendicular to the ground plane. The maximum directivity of traditional patches is usually within a few dB of 8 dBi. This may be confirmed by the following informal analysis: Assume the radiation from the patch is well modeled as the radiation from two parallel dipoles separated by the width of the patch. Then each dipole contributes a directivity of roughly 2 dBi, which is increased by roughly 3 dB by the presence of the ground plane, since power remains constant whereas accessible solid angle is reduced by one-half. Combining these dipoles increases the directivity by about 3 dB, yielding 8 dBi.

The impedance and polarization are determined by the dimensions and shape of the patch, and the point of connection. Square patches fed from near the center of the patch (as in Figure 2.17(a)) are typically resonant (i.e., have reactance which is low relative to the radiation

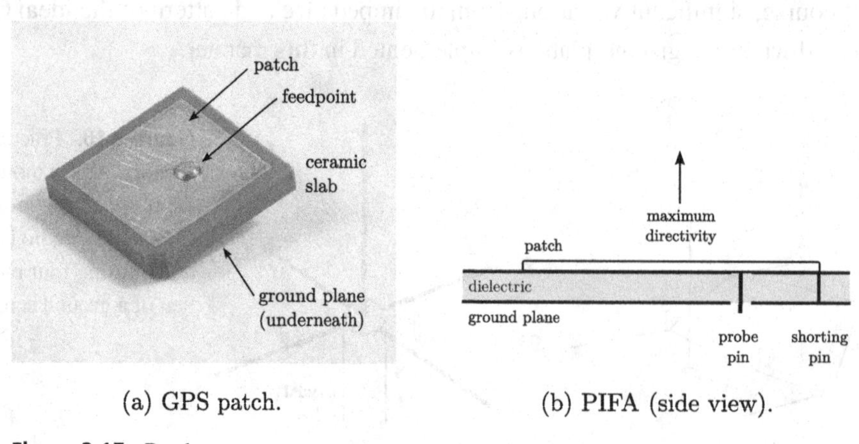

(a) GPS patch. (b) PIFA (side view).

Figure 2.17. Patch antennas.

resistance) when the perimeter of the patch is about $\lambda/\sqrt{\epsilon_r}$, where ϵ_r is the relative permittivity of dielectric spacing material. The point of connection is slightly off-center, resulting in circular polarization. Square patches which are fed from an edge are typically resonant when the electrical length of the perimeter is $2\lambda/\sqrt{\epsilon_r}$, and are linearly-polarized. The impedance may be further modified using shorting connections, which connect the patch to the ground plane at additional locations.

EXAMPLE 2.4

The patch shown in Figure 2.17(a) is 20 mm $\times$ 21 mm in size and uses a ceramic material having relative permittivity $\cong$ 5.4. Confirm that this patch is suitable for GPS.

Solution: The perimeter of the patch is 82 mm and therefore the patch should be resonant when $\lambda/\sqrt{5.4} \cong 82$ mm. This yields $\lambda \cong 19.4$ cm and thus $f \cong 1574.4$ MHz. This is very close to the GPS L1 center frequency of 1575.42 MHz.

Because patch antennas consist of metalization above a ground plane separated by a dielectric, they are a natural choice when implementation on a printed circuit board (PCB) is desired. In particular, it is quite simple to add a patch antenna to a radio circuit which has been implemented on a PCB.

Patches have the additional attractive feature that their size can be reduced by increasing the dielectric constant of the spacing material. For a typical ϵ_r of 4.5 (corresponding to a popular type of fiberglass epoxy material commonly used in PCBs), a square patch can be made to resonate efficiently when its side length is only about $\lambda/10$. This makes patch antennas extremely popular in consumer electronics, including mobile phones and mobile computing devices.

A commonly-encountered patch antenna is the planar inverted-F antenna (PIFA), shown in Figure 2.17(b). The PIFA is formed by shorting one side of the patch such that the patch, feed conductor, and shorting section form an "F" shape as viewed from one side. Many of the antennas used in modern mobile electronics operating at UHF or above utilize some variant of the PIFA.

EXAMPLE 2.5

Figure 2.18 shows some examples of PIFAs used in cellular mobile phones. The convoluted shape of the PIFA patch combined with parasitic (unconnected) elements results in low reactance at multiple frequencies, facilitating multiband operation.

The principal drawbacks of patch antennas, relative to dipoles and monopoles, are radiation efficiency and bandwidth. Patches typically have poor radiation efficiency partially because the dielectric spacing material has some loss, but also because the antenna may interact strongly with nearby lossy materials when embedded in devices such as mobile phones. In mobile phones in particular, radiation efficiency for a resonant PIFA is on the order of 50%, and

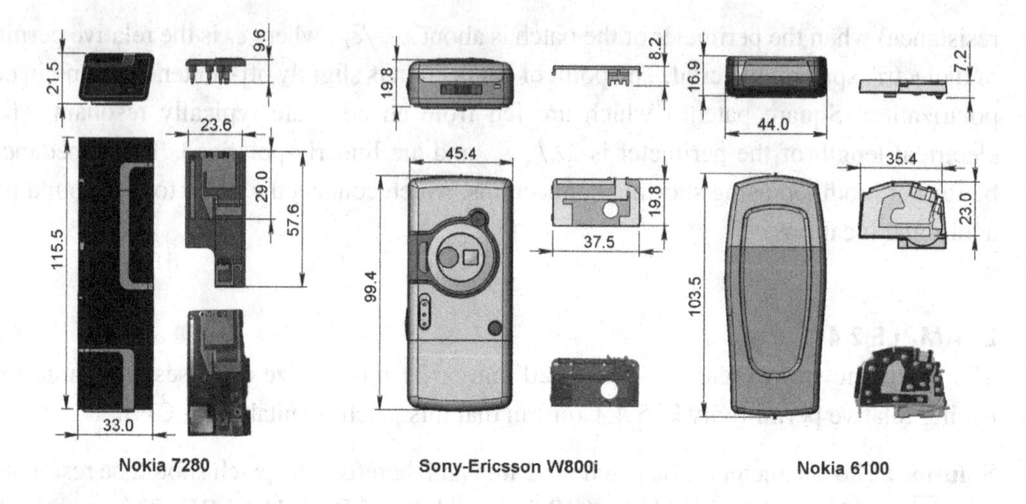

Figure 2.18. Mobile phones for 900/1800/1900 MHz GSM cellular systems using PIFA-type antennas. All dimensions are in mm. ©2012 IEEE. Reprinted, with permission, from [67].

sometimes much lower.[6] The bandwidth issue arises because resonant patches are much smaller than resonant dipoles or monopoles. As we shall see in Chapter 12, the bandwidth that an antenna can accommodate is inversely related to its overall dimensions relative to a wavelength.

2.10 HIGH-GAIN ANTENNAS

Dipoles, monopoles, and patches are low-gain antennas. Sometimes, higher directivity is required or desired. There are at least three broad classes of high-gain antennas: beam antennas (Section 2.10.1), reflector antennas (Section 2.10.2), and certain types of arrays (Section 2.11).

From a transmit perspective, higher directivity is achieved by increasing the power density in a specified direction. Because an antenna is passive, the total radiated power is constant. Therefore increasing directivity in one direction results in reduced directivity in other directions.

The pattern of a typical high-gain antenna is shown in Figure 2.19. High directivity inevitably results in a more complex pattern consisting of a *main lobe*, *sidelobes*, and (sometimes) a *backlobe*. As explained earlier in this chapter, *half-power beamwidth* (HPBW) is the angular distance between points on the main lobe for which the directivity is one-half that of the maximum directivity. Since E-plane and H-plane patterns may be different, E-plane HPBW may be different from H-plane HPBW.

Lobes other than the main lobe are referred to as sidelobes. Typically, the sidelobes adjacent to the main lobe have the greatest directivity. The ratio of the directivity of a sidelobe (typically the largest sidelobe) to that of the main lobe is known as the *sidelobe level*. A backlobe is a sidelobe which is directly opposite the main lobe. Not all high-gain antennas have an identifiable backlobe. When a backlobe exists, the ratio of maximum (i.e., main lobe) directivity to the maximum directivity within the backlobe is known as the *front-to-back ratio*.

[6] As will be pointed out in Chapter 12, some part of this loss may be deliberately introduced as a technique for improving bandwidth.

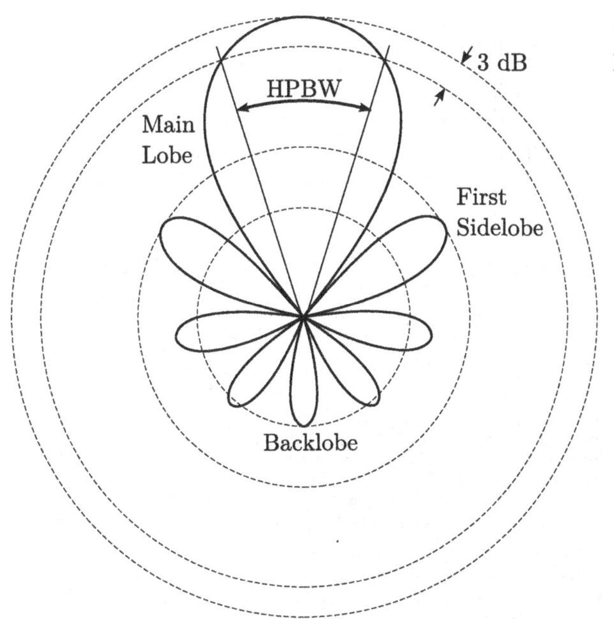

Figure 2.19. Features of the pattern of a typical high-gain antenna.

The principle of reciprocity dictates that the receive directivity and pattern should equal the transmit directivity and pattern (see Section 2.4.2). Thus, high-gain transmit antennas are also high-gain receive antennas.

2.10.1 Beam Antennas; The Yagi

Beam antennas consist of a *driver* antenna whose directivity is increased through electromagnetic ("parasitic") interaction with elements similar to the driver.[7] A beam antenna consisting of a HWD driver and dipole-type parasitic elements is commonly known as a *Yagi-Uda antenna* or simply as a *Yagi*; see Figure 2.20(d) for an example. In a Yagi, the parasitic elements consist of a *reflector* and some number of *directors*, and the direction of maximum directivity is in the direction of the directors. The driver, reflector, and directors are joined using a *beam*, which exists only for mechanical purposes and normally has only a small effect on the electromagnetic characteristics of the antenna. Beam antennas can alternatively be constructed from loops and certain other kinds of low-gain elements.

The operation of the Yagi depends on electromagnetic interactions between the driver, reflector, and directors that cannot be described using simple closed-form equations. However, it is useful to know some basic design concepts. Beginning with a HWD driver, the addition of a reflector which is slightly longer than the driver and which is separated by 0.1λ–0.25λ, as shown in Figure 2.20(b), results in an increase in directivity in the direction opposite the reflector by about 3–5 dB, to about 5–7 dBi. A similar improvement can be achieved by adding a director which is slightly shorter than the driver and with a similar spacing

[7] Here we should note another common example of potentially-confusing nomenclature: The term "beam antenna" is sometimes used to mean *any* high-gain antenna; i.e., "beam" is taken to mean that the main lobe dominates the pattern. By this definition, reflectors and arrays are also "beam antennas." That is not quite what we mean here.

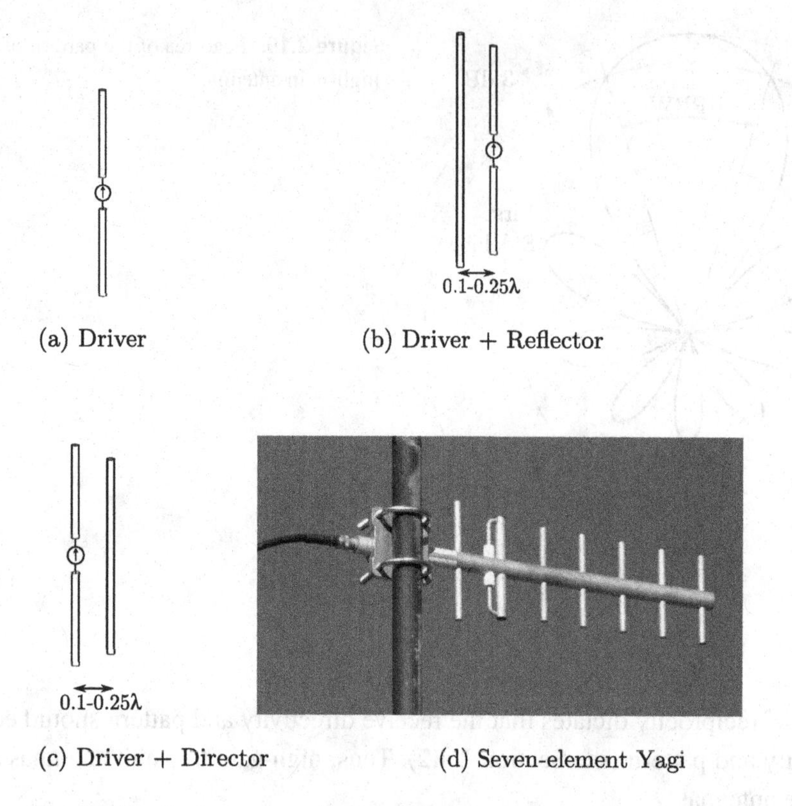

(a) Driver (b) Driver + Reflector

(c) Driver + Director (d) Seven-element Yagi

Figure 2.20. Evolution from dipole to a beam antenna: (a) dipole, (b) adding a reflector, (c) adding a director. A practical example of a beam antenna having multiple directors to further increase gain is shown in (d).

from the driver, as shown in Figure 2.20(c). These effects can be combined; a properly-designed three-element Yagi consisting of a driver, reflector, and director typically has a gain of about 9 dBi. Further increases in gain are achieved using additional equally-spaced directors. Figure 2.20(d) shows an example of a Yagi using five directors, having directivity of about 11.5 dBi. In principle directivity can be made arbitrarily large using increasing numbers of directors; albeit with diminishing returns. In practice, the reflector antenna (Section 2.10.2) is often a more practical and compact solution when directivity much greater than 10 dBi is required.

The impedance, directivity, and pattern of practical beam antennas are quite difficult to determine without resorting to a numerical method (once again, method of moments and FDTD are popular choices) or to empirical solutions (e.g., tables showing results for particular designs). For an introduction to practical analysis and design techniques for beam antennas, reference [76] is recommended.

Beam antennas are extremely popular for applications requiring moderate gain at frequencies in the UHF band and below. These applications include reception of terrestrial broadcast television and wireless data communications between fixed points. Beam antennas are less common above UHF, because (as noted above) greater directivity can typically be achieved using reflector antennas (the topic of the next section) having smaller overall size and a more convenient mechanical design.

A common method for feeding a beam antenna using a coaxial cable is to use a *gamma match*. An example is shown in Figure 2.21. In this scheme the shield is grounded to the center of the driven element, and the inner conductor is connected to the driven element a short distance from the center via a capacitive tuning stub. This allows the driven element and the shield to be directly connected to the metallic boom, yeilding a simple and sturdy interface that obviates the need for a balun.

2.10.2 Reflectors

A paraboloidal reflector can be used to dramatically increase the directivity of a low- to moderate-gain antenna, known in this application as a *feed*. The concept is shown in Figure 2.22. The method of operation is most easily understood in terms of *geometrical optics*, which is applicable if the minimum dimension and surface radii of curvature of the reflector are much greater than a wavelength. According to geometrical optics (also known as *ray optics*) the electromagnetic field scattered from each point on the reflector is approximately equal to the field reflected by an infinite conducting plane tangent to the surface. If the feed is located

Figure 2.21. A gamma match being used to interface coaxial cable to the driven element of a Yagi.

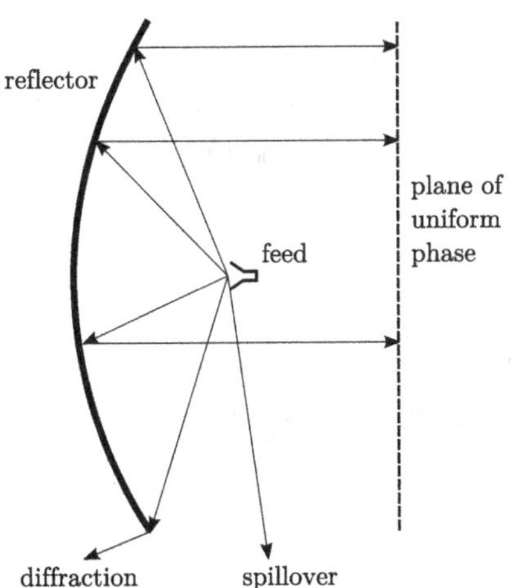

Figure 2.22. Use of a paraboloidal reflector to increase the directivity of a low-gain antenna (the *feed*).

at the focus of the paraboloid defining the reflector surface, then the field reflected from each point on the reflector travels in the same direction; that is, in parallel with the axis of revolution of the paraboloid. Furthermore, it can be shown that the phase of the reflected contributions is equal in any plane perpendicular to the axis of the paraboloid. Thus the total reflected field is a plane wave traveling along the axis.

The directivity of a reflector antenna system is limited by two factors: First, *spillover*; that is, some fraction of the feed pattern does not illuminate the reflector. Second, *diffraction* from the edges of the paraboloid results in some fraction of the power from the feed being scattered in other directions.

A good engineering approximation of the far-field pattern of a reflector antenna can be obtained using the following tenet of antenna theory: The electric field in a planar aperture is related to the far-field pattern by the two-dimensional Fourier transform. In this case, we define a circular planar aperture in front of the dish, and note that according to the geometrical optics approximation, the electric fields within that aperture have uniform phase, and that the electric fields outside this aperture are identically zero. The pattern is therefore the Fourier transform of a two-dimensional function having constant-phase within a circular aperture. The details of this derivation are beyond the scope of this text, but the results can be summarized as follows. For a circular aperture of radius a, the HPBW is found to be

$$\text{HPBW} \cong \frac{\lambda}{2a} \quad \text{(electrically-large circular aperture)} \tag{2.60}$$

and the directivity is found to be

$$D \cong \pi^2 \left(\frac{2a}{\lambda}\right)^2 \quad \text{(electrically-large circular aperture)} \tag{2.61}$$

We can calculate the effective aperture using the above expression in Equation (2.44); we find

$$A_e \cong \pi a^2 \quad \text{(electrically-large circular aperture)} \tag{2.62}$$

Note that πa^2 is the area of a circle of radius a. This area is referred to as the *physical aperture*, A_{phys}; so more concisely we may say that the effective aperture of a reflector is nominally equal to its physical aperture.

In practice, A_e is significantly reduced by factors including spillover, diffraction, and variations in the magnitude and phase of the reflected field over the aperture due to feed antenna limitations. The reduction in A_e relative to the nominal effective aperture A_{phys} is quantified by the *aperture efficiency*, ϵ_a, as follows:

$$A_e = \epsilon_a A_{phys} \tag{2.63}$$

The aperture efficiency of practical reflector antenna systems is typically between 0.5 and 0.8 (50%–80%). Aperture efficiency applies independently of radiation efficiency: that is, ϵ_a is a modification to directivity, whereas both ϵ_a and ϵ_{rad} apply to gain. Thus:

$$D = \frac{4\pi}{\lambda^2} A_e = \epsilon_a \frac{4\pi}{\lambda^2} A_{phys} \tag{2.64}$$

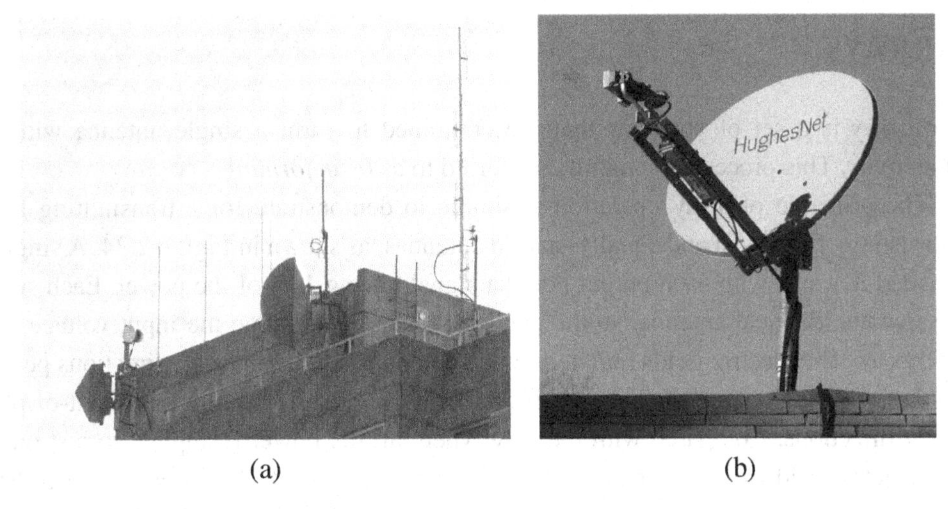

(a) (b)

Figure 2.23. Practical reflector antennas: (a) SHF-band terrestrial microwave data link antennas (feed and reflector surface covered by a radome). (b) A small offset-fed Ku-band antenna used for satellite communications in support of retail operations.

and

$$G = \epsilon_{rad} D = \epsilon_{rad} \epsilon_a \frac{4\pi}{\lambda^2} A_{phys} \tag{2.65}$$

In a reflector antenna, ϵ_{rad} is ordinarily attributed entirely to the feed.

As high-gain antennas, reflectors have several advantages over beam antennas. First, reflector antennas have large bandwidth, since the mechanism of reflection has no bandwidth dependence; at least in the sense that geometrical optics applies. Instead, the bandwidth of a reflector antenna is normally limited by the bandwidth of the feed antenna, and typical feed antennas have larger bandwidth than high-directivity beam antennas. Second, unlike a beam antenna, the directivity of a reflector antenna of a given size increases with increasing frequency. Beginning in the high UHF range, it is usually possible to devise a reflector with greater directivity than a comparably-sized beam antenna. (Of course the opposite is true at lower frequencies.)

Examples of reflector antennas in action are shown in Figure 2.23.

EXAMPLE 2.6

Find the physical aperture, effective aperture, directivity, and gain of a 3 m diameter reflector antenna at 1.5 GHz. Assume that spillover and diffraction effects result in aperture efficiency of 55%, and assume ideal radiation efficiency.

Solution: Here, the aperture is defined by a circle of diameter 3 m, so the physical aperture $A_{phys} = 7.07$ m^2. Thus the aperture efficiency $A_e = \epsilon_a A_{phys} = 3.89$ m^2. At 1.5 GHz, $\lambda = 0.2$ m, so $D = 4\pi A_e/\lambda^2 \cong 1221 \cong 30.9$ dBi. Assuming ideal radiation efficiency, $\epsilon_{rad} = 1$ so in this case $G \cong D \cong 30.9$ dBi.

2.11 ARRAYS

An *array* is a set of antennas that are combined to form a single antenna with increased directivity. This process is sometimes referred to as *beamforming*.

The principle of array operation is simple to demonstrate for a transmitting linear array consisting of identical and equally-spaced antennas, as shown in Figure 2.24. A single signal is divided N ways, with each output being a signal having $1/N$ of the power. Each signal drives a separate, identical antenna, so the applied voltage is less than the input voltage by a factor of $1/\sqrt{N}$. The electric fields radiated by the antennas add in-phase in directions perpendicular to the array (e.g., along the horizon, if this array is vertical), and normally out-of-phase in all other directions. Compared with the case where all the power is applied to just one antenna, the electric field from the array has increased by a factor of $N/\sqrt{N} = \sqrt{N}$ broadside to the array. Consequently the radiated power density in the broadside direction has increased by a factor of $(\sqrt{N})^2 = N$. The total radiated power is the same regardless of whether we apply the available power to one antenna or N antennas; therefore the power density averaged over all directions is the same in both cases. Since the maximum power density has increased by a factor of N, and the average power density is independent of N, the broadside directivity of the array is increased by N relative to the directivity of a single antenna. The factor of N in this case is referred to as the *array factor* (AF).

For AF $= N$, as determined in the above example, the interelement spacing d must be selected appropriately. From our experience with dipoles and reflectors, one might expect large d would be preferable, as this would increase the overall dimension of the array and therefore maximize directivity. On the other hand, it is clear that $d \geq \lambda$ is problematic; for example, $d = \lambda$ implies that the signals from each antenna also "phase up" along the axis of the array, manifesting as additional lobes in the pattern which must degrade directivity. Although the

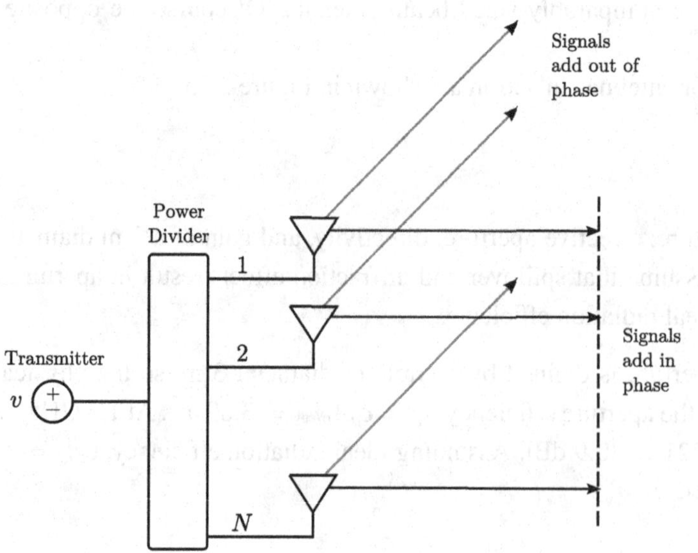

Figure 2.24. Use of multiple antennas in a linear array to increase directivity (transmit case).

derivation is beyond the scope of this book, it can be shown that the optimum spacing is slightly less than λ. For the $N = 4$ array considered here, the optimum spacing – which achieves AF $\cong N$ – is about $\cong 0.85\lambda$.[8]

It can be said that the directivity of an array is equal to the array factor, which depends only on the number and position of the antennas, times the directivity of a single constituent antenna. This is known as the principle of *pattern multiplication*. By reciprocity the above argument also applies for arrays operating in receive mode, just as we have assumed for other antenna systems examined in this chapter.

EXAMPLE 2.7

A common example of the type of array described above is the vertical array shown in Figure 2.23(a) (look toward the right edge of the photo). In this case the array consists of four folded dipoles. Estimate the broadside directivity of this array and characterize its pattern.

Solution: Here, $N = 4$. Array broadside corresponds to the H-plane of the constituent dipoles. The directivity of a folded dipole is about 2 dBi. Therefore the broadside directivity of the array is $\cong 10 \log_{10} 4 + 2$ dBi $\cong 8$ dBi. The array has an omnidirectional pattern with high directivity along the horizon in all directions, and nulls toward the zenith and nadir. This array finds application in certain broadcast and LMR applications.

Another commonly-encountered array is the planar patch array. This array consists of patch antennas arranged vertically on a common ground plane. The directivity and pattern characteristics of such arrays can be anticipated by the property of pattern multiplication. Modern cellular base stations often use vertical stacks of patches to achieve high gain over a specified *sector*, which is an angular region typically covering about 120° in azimuth. Many of the white panels in Figure 2.25 are individual patch arrays having gain in the range 10–20 dBi. Some of these panels are arrays using dipoles as opposed to patches.

A planar array is an array consisting of antennas distributed over a planar surface as opposed to being distributed along a line. This results in a main lobe which is highly directional in both dimensions; similar to that of a beam or reflector antenna. Planar arrays of patches are commonly employed when a high-gain antenna is required but space considerations require the antenna to be conformal to an existing surface. A common application is in mobile very small aperture terminal (VSAT) satellite communications systems installed in vehicles such as buses and aircraft, in which the beam can be "electronically" steered to track the satellite and in which a mechanically-steered beam or reflector antenna would be impractical. Planar patch arrays are also increasingly used as "low profile" wall-mountable high-gain antennas in outdoor point-to-point WLAN links.

[8] Full disclosure: It is possible for AF to be greater – perhaps even much greater – than N for N sufficiently large. The applications in radio engineering are limited, however, since it is rare for the number of elements appearing in a vertical array to be more than four or so due to space and mechanical considerations.

Figure 2.25. The antenna tower of a modern cellular cite, supporting several UHF and SHF-band cellular systems.

2.12 OTHER COMMONLY-ENCOUNTERED ANTENNAS

The previous sections have described a few classes of antennas that see widespread use in radio applications. This section describes a few more classes of antennas that are also commonly-encountered, but for which a more detailed discussion is beyond the scope of this text.

The *horn* antenna, shown in Figure 2.26(a), is a moderately-directive antenna which is frequently employed as a feed in reflector antenna systems. The horn antenna also appears in a variety of applications requiring a compact, efficient antenna with more gain than a dipole or patch, but less gain than a beam or reflector. Horn antennas typically employ a monopole-type antenna located in the "throat" of the horn. The use of a monopole facilitates a simplified interface to coaxial cable, which is desirable in many applications. In transmission, the power radiated from the monopole is directed to an opening having a much larger cross-sectional area than the throat. From this opening, the source of the radio wave can be interpreted to be the electric field distribution which has been created in the opening aperture. As in the case of the

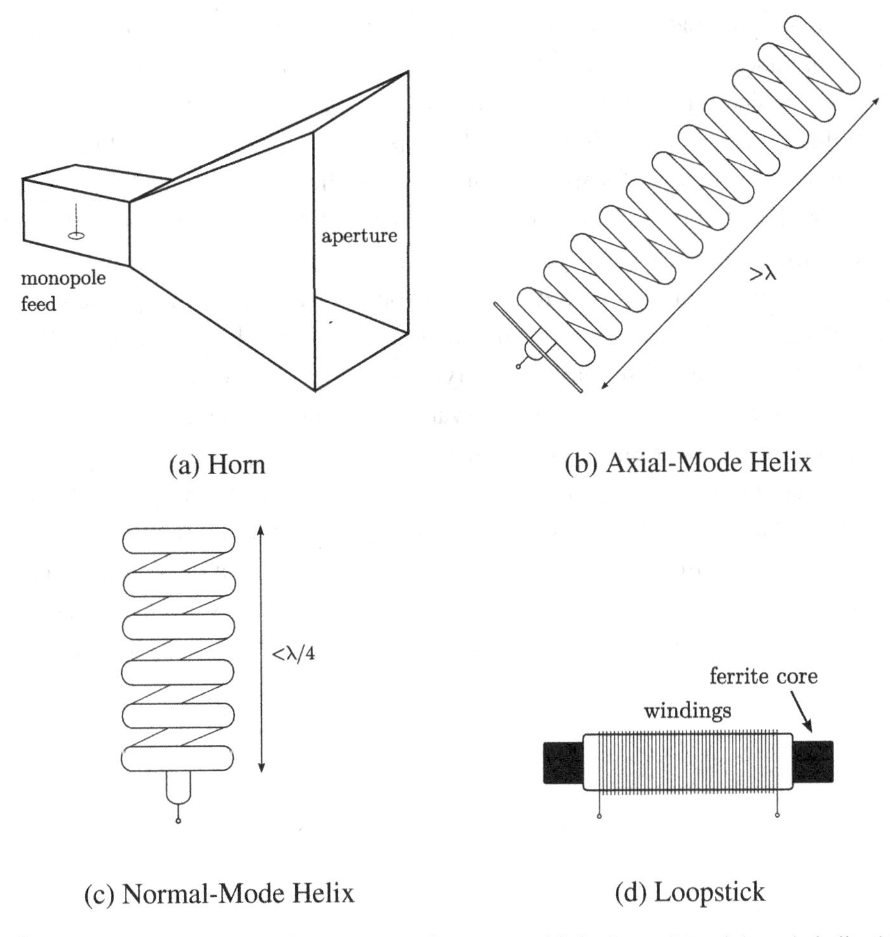

Figure 2.26. Other commonly-encountered antennas: (a) the horn, (b) axial-mode helix, (c) normal-mode helix, and (d) loopstick.

reflector antenna, the nominal directivity is roughly $4\pi A_{phys}/\lambda^2$ where A_{phys} is the physical area of the aperture. Also as in the case of the reflector, the maximum directivity is less due to non-uniformity in the magnitude and phase of the electric field created in the aperture.

The *axial-mode helix* antenna, shown in Figure 2.26(b), is a high-gain antenna that is qualitatively similar to a beam antenna. This antenna is essentially a coil having a circumference of about 1λ and a pitch of about 0.25λ at the desired center frequency of operation. Each turn of the coil plays a role analogous to a director within a beam antenna. An important difference is that this antenna is circularly-polarized, which is a requirement in certain applications – e.g., space communications – where the relative orientation of the antennas on either end of the link is unknown or variable. An advantage of this antenna is relatively large bandwidth (typically $\cong 20\%$) compared to most beam antennas.

The *normal-mode helix* antenna, shown in Figure 2.26(c), is a coil whose circumference and overall length are both less than or much less than $\lambda/4$. This results in a dipole-like pattern which is omnidirectional in the plane perpendicular to the axis of the coil. The normal-mode helix has an advantage over electrically-short monopoles of comparable size in that they tend to have somewhat higher radiation resistance. For this reason they are sometimes used as external antennas in handheld radios. The coil is typically enclosed in a plastic or rubber-type material for protection, which often gives the false impression that the antenna is a short monopole.

We conclude this chapter with the humble *loopstick* antenna, shown in Figure 2.26(d). The loopstick appears in various forms in applications in the MF bands and lower; in particular, in portable AM and shortwave broadcast receivers. At these frequencies, electrically-short monopoles, dipoles, helices, and high-ϵ_r patches are all unacceptably large and have hopelessly narrow bandwidths. To achieve small size and roughly frequency-independent performance, the loopstick uses an alternative approach quite unlike that used by any other antenna discussed previously. This antenna is essentially a coil-type inductor. On receive, the magnetic component of the incident radio wave induces a time-varying current in the coil which is proportional to the inductance. This current is delivered with very low impedance but can be sensed and transformed into a moderate impedance signal using an appropriate amplifier (typically known as a "transimpedance" amplifier). However, very high inductance is required in order to obtain currents of usable magnitude, which requires that the coil have a very large number of turns around a ferrite core. This in turn means high ohmic loss, and thus very low radiation efficiency, which is the primary disadvantage of the loopstick compared to other electrically-small antennas.

Problems

2.1 A monopole having $h = 3.25$ cm, $a = 1$ mm, and made from aluminum (assume $\mu = \mu_0$ and $\sigma = 3 \times 10^6$ S/m) is being considered for use in a 433 MHz radio telemetry transmitter. Calculate the impedance, radiation efficiency, directivity, and gain.

2.2 Repeat the analysis shown in Figure 2.5 for the antenna described in the above problem. Consider the range 300–500 MHz.

2.3 Derive Equation (2.32) (vector effective length of an electrically-short dipole).

2.4 Derive the result that $A_e = 0.12\lambda^2$ for the electrically-short dipole, starting with the expressions for A_e and R_{rad} for this antenna (Equations (2.39) and (2.22), respectively).

2.5 Equation (2.44) ($D = 4\pi A_e/\lambda^2$) is presented without derivation. But it can be checked. Confirm that this equation is consistent with the independently-derived expressions for the directivity and effective aperture of an electrically-short dipole.

2.6 Using the same method employed for the electrically-short dipole, show that the radiation resistance of a half-wave dipole is about 73 Ω.

2.7 A popular alternative to the quarter-wave monopole is the 5/8-λ monopole. As the name implies, $h = 5\lambda/8$. Sketch the current distribution for this antenna in same manner shown in Figure 2.3. What advantages and disadvantages does this antenna have relative to the quarter-wave monopole? Consider size and gain.

2.8 Calculate the directivity of a z-aligned half-wave dipole at $\theta = 45°$. Give your answer in dBi. Repeat for the electrically-short dipole, and both corresponding monopole versions. Compare.

2.9 A particular handheld receiver design requires that the GPS patch from Example 2.4 be reduced in area by 20%. You may modify the size of the patch and the dielectric constant, but not the height of the dielectric slab. What are the required dimensions and dielectric constant for the new design?

2.10 A reflector intended for satellite communications at 4 GHz is circular with a diameter of 3 m. The aperture efficiency is 75% and the feed has a radiation efficiency of 95%. What is the gain of this antenna?

2.11 A certain patch antenna has a gain of 4.2 dBi. Seven of these are arranged into a vertical stack array. What is the directivity of this array?

3 Propagation

3.1 INTRODUCTION

In Chapter 2, a radio wave was defined as an electromagnetic field that transfers power over a distance and persists in the absence of its source. In this chapter our primary concern is what happens to radio waves between antennas, and how the information borne by radio waves is affected. This chapter is organized as follows. In Section 3.2, we begin with the simplest possible case – propagation in free space – and introduce the concept of path loss. In Section 3.3 we review reflection and transmission, including the important problems of reflection from the Earth's surface and scattering by buildings and terrain. In Section 3.4 we derive the path loss associated with ground reflection. In Section 3.5 we consider the effect of scattering in the terrestrial propagation channel, develop models for the channel impulse response, and introduce metrics used to concisely describe these models in terms useful for practical radio systems engineering. Then we get specific about the nature of propagation in 30 MHz–6 GHz (Section 3.6), 6 GHz and above (Section 3.7), and 30 MHz and below (Section 3.8). This chapter concludes with a discussion of a few additional mechanisms for propagation that are sometimes important to consider (Section 3.9).

3.2 PROPAGATION IN FREE SPACE; PATH LOSS

Consider an antenna transmitting a sinusoidal signal in free space; that is, in a region in which there is no ground, terrain, buildings, or other objects capable of interacting with the radio wave. Assuming the antenna is located at $r = 0$ (i.e., the origin of the coordinate system), the radio wave has the following general form in the far field of the transmitter:

$$\mathbf{E}(r,\theta,\phi) = \hat{\mathbf{e}}(\theta,\phi)\, V_0(\theta,\phi)\, \frac{e^{-j\beta r}}{r} \tag{3.1}$$

where $\hat{\mathbf{e}}(\theta,\phi)$ is a unit vector describing the polarization of the electric field and $V_0(\theta,\phi)$ represents the antenna pattern as well as the magnitude and phase of the signal applied to the antenna. Let us temporarily ignore the phase response of the antenna; i.e., let us assume for the moment that $\partial V_0/\partial\theta = \partial V_0/\partial\phi = 0$. In this case Equation (3.1) is a spherical wave; that is, surfaces of constant phase are concentric spheres centered on the antenna at $r = 0$. However, for large r the radius of curvature of the phase front becomes so large that it appears "locally" to be a plane wave; that is, from the perspective of an observer far from the antenna, the phase appears nearly constant over the plane perpendicular to the direction of propagation.[1] For antennas of practical interest, $\partial V_0/\partial\theta$ and $\partial V_0/\partial\phi$ change slowly with both θ and ϕ, so this effect remains after we reintroduce the phase due to the antenna.

A corollary to the derivation presented in Section 2.2.2 is that the far-field electric field intensity is always polarized in the plane perpendicular to the direction of propagation; i.e., $\hat{\mathbf{e}}(\theta,\phi)\cdot\hat{\mathbf{r}} = 0$. Therefore one may write:

$$\hat{\mathbf{e}}(\theta,\phi) = \hat{\theta}e_\theta(\theta,\phi) + \hat{\phi}e_\phi(\theta,\phi) \tag{3.2}$$

where

$$\sqrt{e_\theta^2(\theta,\phi) + e_\phi^2(\theta,\phi)} = 1 \tag{3.3}$$

This fact will become useful when we consider reflection, since it reduces the number of possible orthogonal components of polarization from three to just two, both perpendicular to the direction of propagation. In fact $\hat{\theta}$ and $\hat{\phi}$ are not the only possible choices for these two components, nor are they necessarily the most convenient choices, as we shall see in the next section.

Once the field is expressed in the form of Equation (3.1), it is easy to compute the power density S_{ave} (W/m^2) at any point (r,θ,ϕ):

$$S_{ave}(r,\theta,\phi) = \frac{|\mathbf{E}(r,\theta,\phi)|^2}{2\eta} = \frac{|V_0(\theta,\phi)|^2}{2\eta r^2} \tag{3.4}$$

where we have assumed $\mathbf{E}$ is expressed in peak (as opposed to RMS) units. Note that the power density decreases by r^2 over the same distance. This weakening is typically the dominant mechanism of "loss" in a radio link, and is attributable entirely to the "spreading out" of power with increasing distance from the antenna.

Usually the quantity S_{ave} is not quite what we need. The usual problem is this: Given a certain power P_T (units of W) applied to the transmitting antenna, what is the power P_R (also units of W) captured by the receiving antenna? A further wrinkle is that one typically prefers an answer in which the effect of the patterns of the transmit and receive antennas can be separated from the loss due to spreading and other mechanisms, for more complex channels. The reduction in power density with spreading is *path loss*, which we shall now derive for free space conditions.

Referring to Figure 3.1, imagine that P_T is being delivered to an antenna that subsequently radiates this power isotropically (that is, uniformly in all directions), with $|V_0(\theta,\phi)|$ that is constant with respect to θ and ϕ. We no longer require an explicit global coordinate system, so let us represent the distance between between antennas using a different symbol, R. S_{ave} at a distance R from the transmit antenna is simply P_T divided by the area of a sphere of radius R; i.e., $S_{ave} = P_T/(4\pi R^2)$. If the antenna is not isotropic but instead has gain G_T in the direction of the receiver, then the power density is increased by this factor, so $S_R = G_T P_T/(4\pi R^2)$. Assuming the polarization of the receive antenna is perfectly matched to the polarization of the incident wave, P_R is related to S_R by the effective aperture A_e of the receive antenna. A_e is given by $D_R\lambda^2/4\pi$ (Equation (2.44) of Chapter 2), where D_R is the directivity (i.e., not accounting for loss) of the receive antenna in the direction of the incident signal. Thus, the relationship between P_R and P_T can be written succinctly as follows:

[1] A good analogy is the fact that the curvature of the Earth's surface is not apparent to an observer on the ground; i.e., the Earth appears "locally" to be flat.

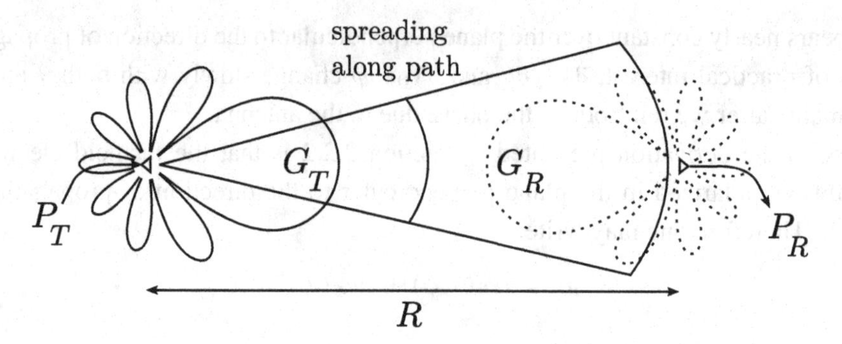

Figure 3.1. Scenario for the development of the Friis transmission equation.

$$P_R = P_T G_T \left(\frac{\lambda}{4\pi R}\right)^2 G_R \tag{3.5}$$

This is known as the *Friis transmission equation for free space propagation*. The third factor is the reciprocal of the free space path loss L_{p0}; i.e.,

$$L_{p0} \equiv \left(\frac{\lambda}{4\pi R}\right)^{-2} \tag{3.6}$$

A suitable definition of free space path loss, inferred by Equation (3.5), is that L_{p0} is the ratio of of transmit power to receive power (i.e., P_T/P_R) when the transmit and receive antennas have unit gain (i.e., $G_T = G_R = 1$) along the path of propagation. When we consider path loss in non-free space conditions later, the definition will remain the same.

The utility of the path loss concept is two-fold: (1) P_R is directly related to P_T without having to determine power density first, and (2) the effect of antenna pattern (i.e., gain as a function of direction) is conveniently factored out as a separate consideration. Note, however, that path loss is *not* a description of propagation alone: It applies specifically to the power captured by an antenna, and is not simply a ratio of the received power density to transmit power density. In fact, the contribution of the receive antenna in Equation (3.6) is represented by a factor of $4\pi/\lambda^2$, and the remaining factor of $4\pi R^2$ is describing spherical spreading.

EXAMPLE 3.1

A satellite at an altitude of 800 km transmits a signal to a ground station directly below it. The satellite transmits 5 W into an antenna having a gain of 6 dBi, and the ground station's antenna has a gain of 10 dBi. The center frequency of the transmission is 1.55 GHz. What is the path loss and the power delivered to the receiver?

Solution: From the problem statement we have $P_T = 5$ W, $G_T = 6$ dBi, $G_R = 10$ dBi, and $R = 800$ km. It seems reasonable to assume free space propagation in this case, and if we also assume antennas which are well-matched in impedance and polarization then Equation (3.5) applies. Since $\lambda = 19.4$ cm at 1.55 GHz, the free space path loss L_{p0} is 154.3 dB. Thus we have $P_R = 7.38 \times 10^{-14}$ W, which is -131.3 dBW or -101.3 dBm.

Obviously, not all situations of practical interest are well-represented by Equation (3.5). A first step in generalizing this result is to account for the possibility of path loss L_p which is different from that encountered in free space; thus:

$$P_R = P_T G_T L_p^{-1} G_R \tag{3.7}$$

Figuring out ways to estimate L_p in scenarios that are not well-modeled as free space is a primary objective of this chapter. We'll take the Friis transmission equation one step further in Chapter 7 (Section 7.2), where we will make further modifications to account for other effects such as antenna impedance mismatch and polarization mismatch.

Before moving on, a caveat about terminology: One should be aware that the generic term "path loss" is often used in lieu of a specific quantity such as free space path loss, *average* (*mean* or *median*) path loss, path *gain* (i.e., the reciprocal of path loss), and so on. In practice, it is usually possible to infer the specific meaning from context, so this is typically not a barrier to understanding.

3.3 REFLECTION AND TRANSMISSION

Aside from free-space propagation, the simplest thing that can happen to a radio wave in free space is that it can encounter a planar boundary with some semi-infinite homogeneous dielectric material as depicted in Figure 3.2. This is a pretty good description of two important cases: (1) a radio wave incident on the ground; and (2) a radio wave at frequencies in the upper VHF band or higher encountering certain kinds of terrain, or the wall, floor, or side of a building. When a radio wave encounters such a boundary, we expect a reflected wave with magnitude and polarization determined by the material properties of the surface. We now work this out using some results from basic electromagnetic theory.

3.3.1 Reflection from a Planar Interface

We established in Section 3.2 that a radio wave in free space, in the far field of the antenna which created it, can be approximated locally as a plane wave. Figure 3.2 depicts the situation when this plane wave encounters a semi-infinite homogeneous and potentially lossy dielectric

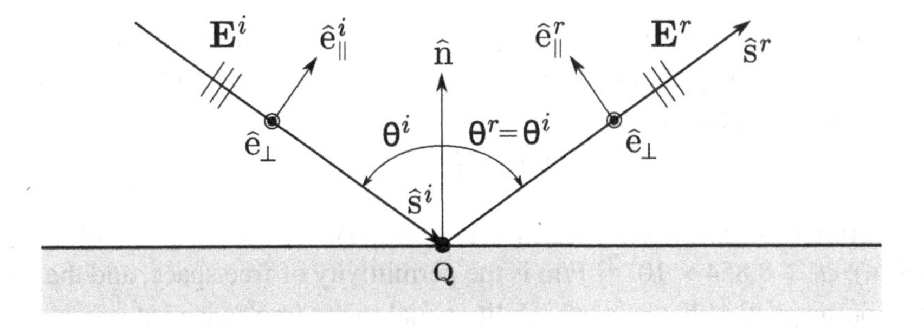

Figure 3.2. Plane wave incident on a planar boundary between free space (top) and a lossy dielectric (bottom). Note $\hat{\mathbf{e}}_\perp$ points toward the viewer.

medium which meets free space at a planar boundary. The incident electric field intensity at a point of incidence **Q** is most conveniently written in the following form:

$$\mathbf{E}^i(\mathbf{Q}) = \hat{\mathbf{e}}_\perp E^i_\perp(\mathbf{Q}) + \hat{\mathbf{e}}^i_\parallel E^i_\parallel(\mathbf{Q}) \tag{3.8}$$

where $\hat{\mathbf{e}}_\perp$ is a unit vector which is perpendicular to the *plane of incidence*, defined as the plane containing the unit normal vector to the planar boundary, $\hat{\mathbf{n}}$, and the direction of incidence, $\hat{\mathbf{s}}^i$, as follows:

$$\hat{\mathbf{e}}_\perp \equiv \frac{\hat{\mathbf{s}}^i \times \hat{\mathbf{n}}}{|\hat{\mathbf{s}}^i \times \hat{\mathbf{n}}|} \tag{3.9}$$

and $\hat{\mathbf{e}}^i_\parallel$ is defined to be orthogonal to both $\hat{\mathbf{e}}_\perp$ and $\hat{\mathbf{s}}^i$, as follows:

$$\hat{\mathbf{e}}^i_\parallel \equiv \hat{\mathbf{e}}_\perp \times \hat{\mathbf{s}}^i \tag{3.10}$$

Note that the "⊥" and "∥" components of the incident field correspond to "horizontal" and "vertical" polarization, respectively, when $\theta^i \approx \pi/2$, as is the case in terrestrial radio links.

In terms of Equation (3.1),

$$\hat{\mathbf{e}}(\theta,\phi) = \hat{\mathbf{e}}_\perp e_\perp(\theta,\phi) + \hat{\mathbf{e}}^i_\parallel e^i_\parallel(\theta,\phi) \tag{3.11}$$

which is essentially Equation (3.2) but now written in terms of a different set of unit vectors. The reason for expressing the incident field in this particular way, as opposed to using $\hat{\theta}$ and $\hat{\phi}$ as in Equation (3.2), is that this particular decomposition of the incident field leads to a relatively simple solution for the reflected field at **Q**:

$$\mathbf{E}^r(\mathbf{Q}) = \hat{\mathbf{e}}_\perp \Gamma_\perp E^i_\perp(\mathbf{Q}) + \hat{\mathbf{e}}^i_\parallel \Gamma_\parallel E^i_\parallel(\mathbf{Q}) \tag{3.12}$$

where

$$\hat{\mathbf{e}}^r_\parallel \equiv \hat{\mathbf{e}}_\perp \times \hat{\mathbf{s}}^r \tag{3.13}$$

$\hat{\mathbf{s}}^r$ is the direction in which the reflected plane wave travels, and $\Gamma_\perp$ and $\Gamma_\parallel$ are the *reflection coefficients*. Symmetry requires that $\hat{\mathbf{s}}^r$ lies in the same plane as $\hat{\mathbf{s}}^i$ and $\hat{\mathbf{n}}$, and *Snell's Law* requires that $\theta^r = \theta^i$; together these determine $\hat{\mathbf{s}}^r$. The reflection coefficients are given by

$$\Gamma_\perp = \frac{\cos\theta^i - \sqrt{\tilde{\epsilon}_r - \sin^2\theta^i}}{\cos\theta^i + \sqrt{\tilde{\epsilon}_r - \sin^2\theta^i}} \tag{3.14}$$

$$\Gamma_\parallel = \frac{\tilde{\epsilon}_r\cos\theta^i - \sqrt{\tilde{\epsilon}_r - \sin^2\theta^i}}{\tilde{\epsilon}_r\cos\theta^i + \sqrt{\tilde{\epsilon}_r - \sin^2\theta^i}} \tag{3.15}$$

In the above expressions, $\tilde{\epsilon}_r$ is the *relative complex permittivity* of the material upon which the plane wave is incident. This quantity represents both the permittivity and conductivity of the material and is defined as follows:

$$\tilde{\epsilon}_r \equiv \epsilon_r - j\frac{\sigma}{\omega\epsilon_0} \tag{3.16}$$

where ϵ_r is the (physical, real-valued) relative permittivity, σ is the conductivity (typically expressed in S/m), $\epsilon_0 \cong 8.854 \times 10^{-23}$ F/m is the permittivity of free space, and the media are assumed to be non-magnetic (that is, permeability equal to the free space value).

The derivation of this solution for the reflected field at **Q** is beyond the scope of this text (a recommended reference is [42, Ch. 6]); however, it is straightforward to check this solution for the special case in which the reflecting media is perfectly-conducting material. For a perfect conductor, $\sigma \rightarrow \infty$. In this case, Equation (3.16) indicates $\tilde{\epsilon}_r \rightarrow -j\infty$, so the reflection coefficients $\Gamma_\perp \rightarrow -1$ and $\Gamma_\parallel \rightarrow +1$ and are independent of θ^i. These are as expected from basic electromagnetic theory, and a quick check of the result in this case is to note that it satisfies the electromagnetic boundary condition that the component of the total electric field $(\mathbf{E}^i(\mathbf{Q}) + \mathbf{E}^r(\mathbf{Q}))$ that is tangent to the surface of the perfect conductor must be zero.

After reflection, the wave continues to propagate as a spherical wave in the direction $\hat{\mathbf{s}}^r$ according to

$$\mathbf{E}^r(d) = \mathbf{E}^r(\mathbf{Q})\, \frac{e^{-j\beta d}}{d} \tag{3.17}$$

where d is the distance from $\mathbf{Q}$ along the ray $\hat{\mathbf{s}}^r$.

Before moving on, note that if either $|\Gamma_\perp|$ or $\left|\Gamma_\parallel\right|$ are significantly less than 1, then some of the incident power is transmitted into the medium as opposed to being reflected. We won't need these, but the relevant expressions for the transmitted field are readily available in basic texts on electromagnetics and propagation (again, [42] is recommended).

3.3.2 Reflection from the Surface of the Earth

Reflection coefficients for reflection from the surface of the Earth depend on frequency, permittivity, conductivity, and angle of incidence. The relative permittivity of the surface of the Earth depends on the specific material composition, which depends on geography; and moisture content, which depends on weather. A typical value for ground (i.e., soil) is $\epsilon_r \sim 15$. The conductivity of the ground is normally in the range 10^{-1} S/m for moist soil, such as farmland, to 10^{-3} S/m for dessicated soil typical of rocky or sandy conditions. Assuming typical values $\epsilon_r = 15$ and $\sigma = 10^{-2}$ S/m in Equation (3.16), we obtain

$$\tilde{\epsilon}_r \sim 15 - j\frac{180}{f_{\text{MHz}}} \tag{3.18}$$

where f_{MHz} is the frequency in MHz.

Let us now consider the effect of frequency for this particular case. When $f_{\text{MHz}} = 12$, $\tilde{\epsilon}_r \sim 15 - j15$; i.e., the real and imaginary components are equal; thus the ground can be characterized as a highly lossy dielectric. When $f_{\text{MHz}} \gg 12$, $\tilde{\epsilon}_r \sim 15$; i.e., the real component dominates over the imaginary component, so the ground behaves as a high-permittivity dielectric with negligible loss. When $f_{\text{MHz}} \ll 12$, $\tilde{\epsilon}_r \sim -j180/f_{\text{MHz}}$ and the ground is not at all like a dielectric, and instead behaves as an imperfect conductor. In terms of defined frequency bands:

- In the upper VHF band and above, the surface of the Earth is well-approximated as a lossless dielectric with $\epsilon_r \sim 15$. The reflection is a plane wave in the same sense that the incident field is locally a plane wave.
- In the MF band and below, the Earth is an imperfect conductor with $\tilde{\epsilon}_r$ that depends on frequency. However, in this regime $\lambda > 100$ m, so antennas are typically located within a small fraction of a wavelength of the ground, and therefore it is not reasonable to assume that the radiated phase front can be interpreted as a plane wave incident on the ground. A more sophisticated analysis is required in this case. It turns out that the dominant mechanism is

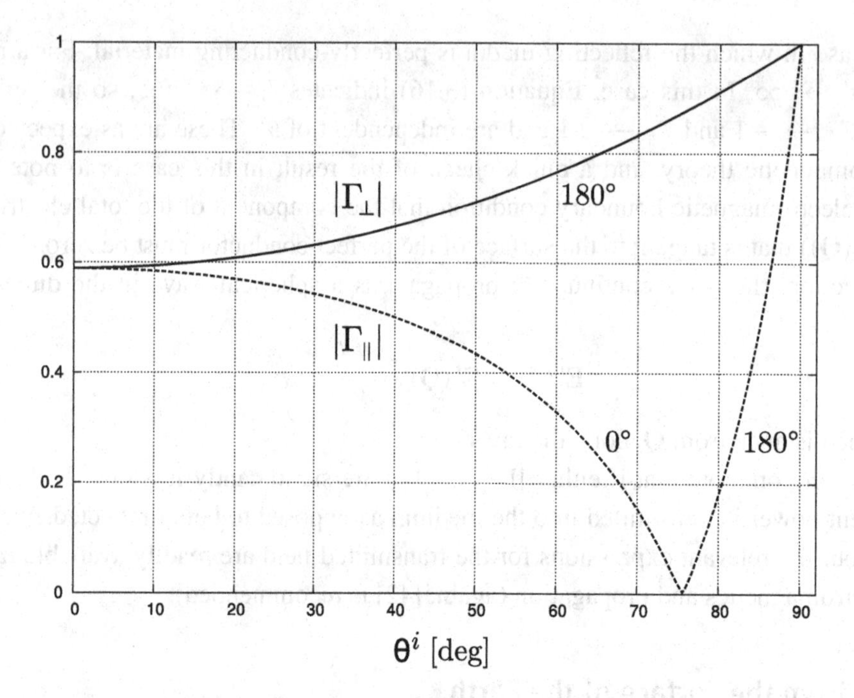

Figure 3.3. Magnitude of reflection coefficients for typical Earth ground ($\tilde{\epsilon}_r = 15$) at frequencies in the upper VHF band and above. Phase is also indicated.

neither "direct" nor reflected, but rather *groundwave* propagation. See Section 3.8 for more about this.

- Behavior in the HF and lower VHF bands is transitional between the above cases.

Whereas the situation is a bit murky for HF and below, we may productively continue by restricting our attention to frequencies in the upper VHF band and above, making the assumption that the ground is well-modeled as a low-loss dielectric having $\tilde{\epsilon}_r \sim 15$. The associated reflection coefficients are shown in Figure 3.3. The key finding here is that, for θ^i close to 90°, both $\Gamma_\perp$ and $\Gamma_\parallel$ are approximately equal to -1. This finding holds true over a wide range of constitutive parameters and frequencies. (Problem 3.4 offers the opportunity to explore this.) This approximation greatly simplifies the analysis of VHF terrestrial propagation involving ground reflection, as we shall see in Section 3.4.

3.3.3 Scattering from Terrain and Structures

The results of Section 3.3.1 can be used to analyze reflection from certain terrain features, such as mountains; as well as artificial structures, such as interior and exterior walls, ceilings, and floors. However, there are some caveats and additional considerations.

First, we must be careful to restrict our attention to frequencies at which the associated wavelength is small relative to the area available for reflection, and the surface must be approximately smooth and flat relative to a wavelength; otherwise the scattering is not well-modeled as plane wave reflection. Mountains often meet the "size" and "smooth and flat" requirements in the lower VHF band, but fail the latter requirement at higher frequencies. Buildings may have dimensions on the order of tens of meters and are smooth to within a centimeter or so, so the analysis of Section 3.3.1 is valid in the UHF and SHF ranges if the surface radius of curvature is large relative to a wavelength.

The material composition of buildings is often complex. Interior and exterior walls are typically constructed from various combinations of concrete,[2] brick, wood, gypsum board, and glass; and sometimes from granite, marble, and other rock-type materials. All of these materials have ϵ_r in the range 1.5–12, with relatively low loss; i.e., with $\text{Im}\{\tilde{\epsilon}_r\} \ll \text{Re}\{\tilde{\epsilon}_r\}$. For the purposes of the analysis we are about to do, it is a reasonable approximation to consider them to be low-loss dielectric materials with negligible variation with frequency. The resulting reflection coefficients behave similarly to those identified in Figure 3.3 (which was calculated for $\tilde{\epsilon}_r = 15$). The principal effects of smaller values of ϵ_r is to reduce the magnitude of the reflection coefficients for values of θ^i which are not close to grazing incidence ($\theta^i \to \pi/2$), and to shift the value of θ^i at which $\Gamma_{\|} = 0$ to lower values of θ^i. Also, it should be noted that because walls are vertically-oriented, the mapping of horizontal and vertical polarizations to $\perp$ and $\|$ polarizations is opposite the mapping for ground reflection: Assuming terrestrial propagation, $\perp$ corresponds to vertical polarization and $\|$ corresponds to horizontal polarization.

From a radio wave propagation perspective, floors and ceilings fall into two broad categories. In residential structures and certain commercial structures, walls and floors are constructed primarily of wood, which has $\epsilon_r \sim 2$. Thus wood floors and ceilings reflect in a similar manner as the ground, but with a much lower ϵ_r. In most commercial structures, the space between levels is constructed from various combinations of steel and steel-reinforced concrete, often with sufficient additional metallic content from HVAC ducts and other utility structures to result in behavior that is closer to that of a perfectly-conducting ground plane.

There are two additional considerations which often foil attempts to model scattering in and around buildings as simple reflection (see Figure 3.4). First, because walls have limited thickness, transmission into the structure may be important. In particular, a radio wave may penetrate into a structure and then be subsequently scattered back out of the structure with significant power density remaining. At frequencies in the lower UHF band and below, absorption losses internal to buildings may result in out-scattered power density that is significant relative to the power density reflected from the exterior. Predicting out-scattered propagation is extraordinarily difficult. The second additional consideration is *diffraction*, which is scattering from edges and corners. In contrast to reflection, diffraction scatters power in all directions. Thus, significant power density may be diffracted into exterior regions which are not accessible to the field reflected from exterior walls. Estimating the diffracted field is possible, but is an advanced topic beyond the scope of this text. (A recommended introduction to this topic and to propagation in and around buildings generally is [5].) It should also be noted that diffraction from terrain – in particular, mountain summits and ridges – is also frequently important, for similar reasons [42].

[2] Actually, structural concrete is typically reinforced with steel rods ("rebar"); however, this is a complication that we may overlook for our immediate purposes.

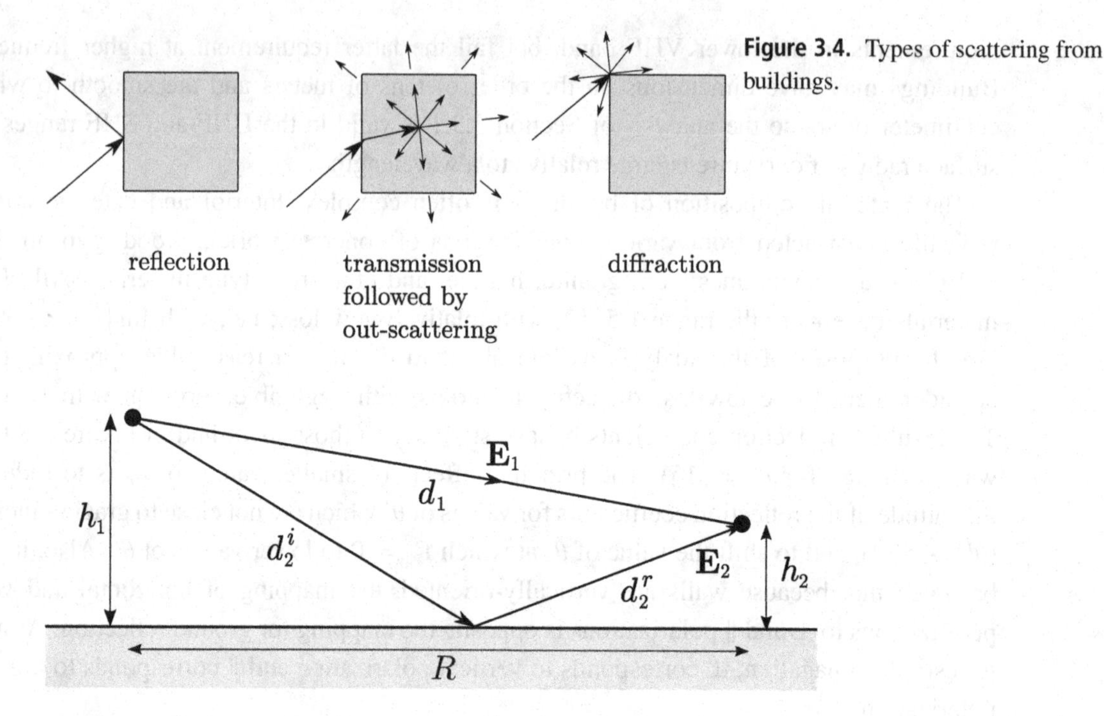

Figure 3.4. Types of scattering from buildings.

Figure 3.5. Geometry for two-ray propagation associated with ground reflection.

3.4 PROPAGATION OVER FLAT EARTH

A broad class of terrestrial propagation scenarios can be modeled as shown in Figure 3.5, in which the total electric field intensity $\mathbf{E}(R)$ arriving at the receiving antenna consists of a "direct path" component $\mathbf{E}_1(d_1)$ experiencing free space propagation over a path length d_1, plus a component $\mathbf{E}_2(d_2)$ that has been singly-reflected from the ground and travels over a total path length $d_2 = d_2^i + d_2^r$. Thus, $\mathbf{E}(R) = \mathbf{E}_1(d_1) + \mathbf{E}_2(d_2)$. This is the simplest and most common example of the phenomenon commonly referred to as *multipath.* Let us now derive expressions for $\mathbf{E}(R)$ and consider the implications for power density and path loss.

3.4.1 A General Expression for the Wave Arriving at the Receiving Antenna

First, note that we may simplify the problem by decomposing $\mathbf{E}(R)$ into its $\perp$- and $\|$-polarized components, resulting in simpler scalar expressions. Assuming R is much greater than both antenna heights h_1 and h_2, θ^i will be very close to $\pi/2$. In this case we know from Section 3.3.2 that the reflection coefficient is approximately -1 for both polarizations. Therefore it suffices to perform the derivation in scalar form for just one polarization, since the results must be the same for both polarizations. Following Equation (3.1), we describe the electric field intensity incident on the receiving antenna via the direct path as follows:

$$E_1(d_1) = V_0 \frac{e^{-j\beta d_1}}{d_1} \tag{3.19}$$

The reflected path consists of two segments: A segment of length d_2^i from transmitting antenna to point of reflection, and segment of length d_2^r from point of reflection to receiving antenna. The field incident on the point of reflection is

$$E_2^i(d_2^i) \cong V_0 \frac{e^{-j\beta d_2^i}}{d_2^i} \tag{3.20}$$

This expression is approximate because the pattern of the transmit antenna will lead to a slight difference in the value of V_0 between Equation (3.19) and (3.20). However, this difference will usually be small for antennas of practical interest, which typically have patterns that vary slowly in elevation; and will certainly be negligible when R is much greater than h_1 and h_2.[3]

The reflected field at the point of reflection is Equation (3.20) multiplied by the reflection coefficient Γ. To account for propagation from the point of reflection to the receiving antenna there must be additional factors that account for phase accrued along the remaining path length as well as the continuing reduction of power density. The appropriate expression is:

$$E_2^r(d_2^r) = E_2^i(d_1^i)\Gamma \frac{e^{-j\beta d_2^r}}{\left(d_2^i + d_2^r\right)/d_2^i} \tag{3.21}$$

The justification for the factor of $\left(d_2^i + d_2^r\right)/d_2^i$ in the denominator may not be obvious. The argument for this factor is as follows: Because the ground is flat and has $\Gamma \cong -1$, image theory applies.[4] Since the field incident on the ground is a spherical wave, the field reflected from the ground must therefore be a spherical wave which appears to emanate from the image of transmitting antenna. The total path length from the image of the transmitting antenna to the actual receiving antenna is $d_2^i + d_2^r$, so the product of the factor of d_2^i in the denominator of Equation (3.20) and the associated factor in Equation (3.21) must be $d_2^i + d_2^r$. Therefore, this factor in Equation (3.21) must be $\left(d_2^i + d_2^r\right)/d_2^i$.

Substituting Equation (3.20) into Equation (3.21) we obtain:

$$E_2^r(d_2^r) \cong V_0\Gamma \frac{e^{-j\beta\left(d_2^i+d_2^r\right)}}{d_2^i + d_2^r} = V_0\Gamma \frac{e^{-j\beta d_2}}{d_2} \tag{3.22}$$

Now the total field incident on the receiving antenna is the sum of Equations (3.19) and (3.22):

$$E = E_1(d_1) + E_2^r(d_2^r) \cong V_0 \left[\frac{e^{-j\beta d_1}}{d_1} + \Gamma \frac{e^{-j\beta d_2}}{d_2}\right] \tag{3.23}$$

EXAMPLE 3.2

In an LMR system operating at 450 MHz, the base station antenna is at a height of $h_1 = 15$ m, and the mobile station antenna is at a height of $h_2 = 2$ m. Plot the magnitude of the electric field intensity at the receiving antenna as a function of R when $V_0 = 10$ V rms. Assume the material properties of the ground are given by Equation (3.16) with $\epsilon_r = 15$ and $\sigma = 10^{-2}$ S/m.

Solution: The solution is shown in Figure 3.6, which uses Equation (3.23) with reflection coefficients calculated using Equations (3.14) and (3.15).

[3] Problem 3.6 gives you the opportunity to test this and other approximations.

[4] See Section 2.8 for a reminder of what image theory entails.

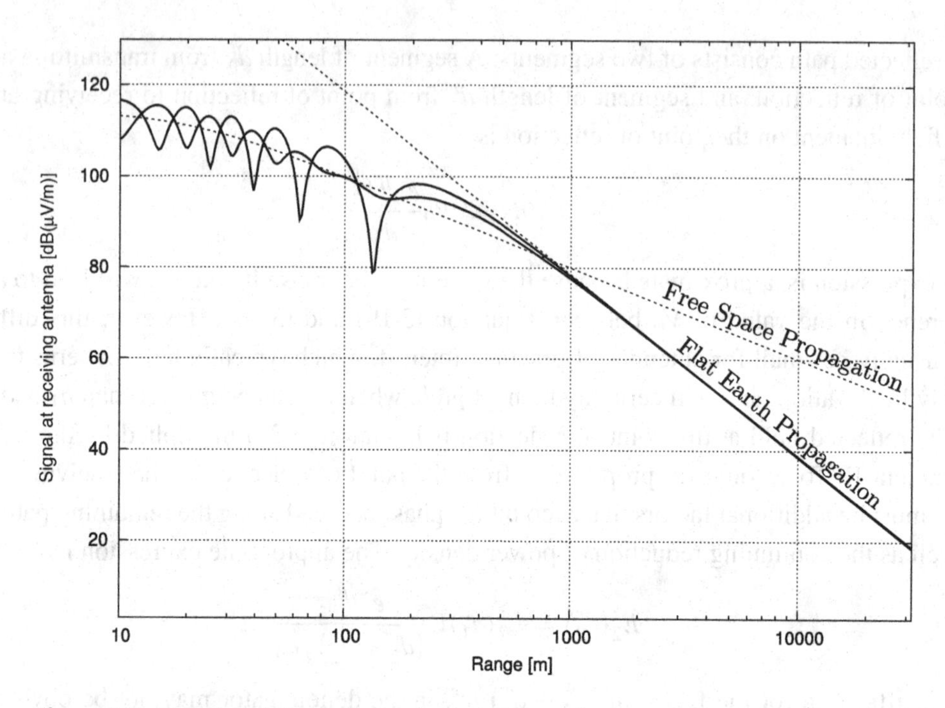

Figure 3.6. Electric field intensity (rms) at a mobile station antenna 2 m above the ground in response to a base station antenna 15 m above the ground, at 450 MHz. The two solid curves correspond to the ⊥ and ∥ polarizations (Solution to Example 3.2). The two dashed curves are solutions computed assuming free space propagation (Equation (3.19)) and approximate flat Earth (two-ray) propagation (Equation (3.36)).

EXAMPLE 3.3

Repeat Example 3.2 with the modification that *both* antennas are at a height of 2 m.

Solution: The solution is shown in Figure 3.7.

Comparing the solutions of the two examples above, we see that a decrease in antenna height leads to a significant increase in path loss, particularly at longer ranges. This is the primary motivation for mounting antennas as high as possible. Close to the transmitter, we observe a large, quasi-periodic variation around the prediction of the free-space model. This variation is constructive and destructive interference as the relative phase of the ground-reflected component varies with increasing R. Beyond a certain distance, we observe that the field strength drops at a rate of 40 dB for each factor-of-10 increase in R; i.e., path loss is proportional to R^4. The value of R at which we transition to R^4 behavior is known as the *breakpoint*. We see in Figures 3.6 and 3.7 that the breakpoint occurs at $\approx$ 800 m and $\approx$ 100 m for $h_1 = 15$ m and 2 m, respectively. Radio links of this type have usable ranges up to tens of kilometers, so the range dependence of path loss beyond the breakpoint is of primary importance. We consider this in more detail in the next section.

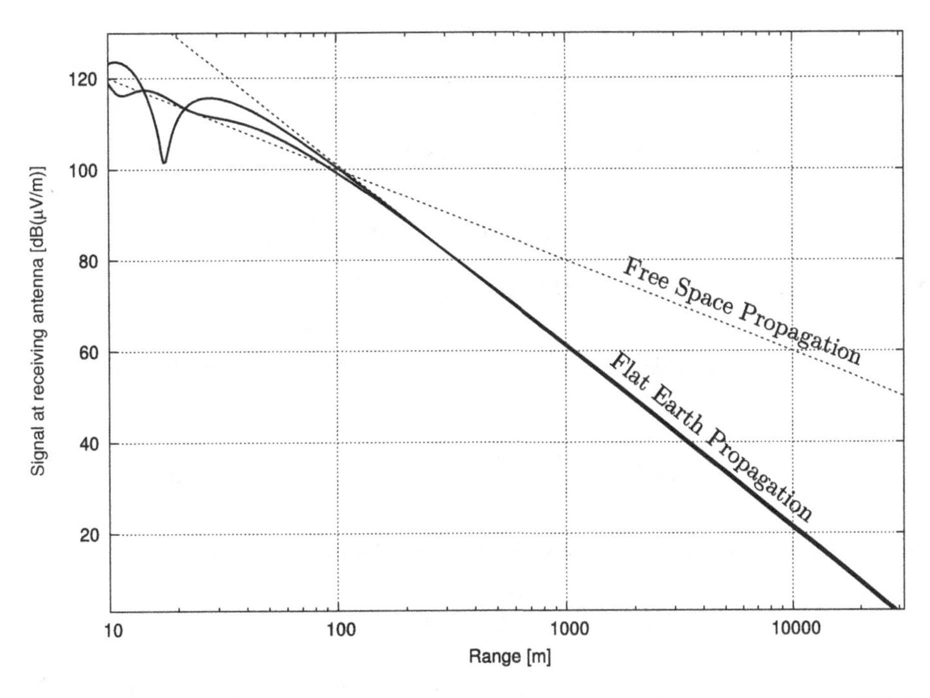

Figure 3.7. Same as Figure 3.6, except now with both antennas 2 m above the ground (Solution to Example 3.3).

3.4.2 Flat Earth Path Loss; Breakpoint Analysis

In this section we derive a simple approximate expression for field strength beyond the breakpoint in flat Earth propagation, which allows us to then derive a simple expression for the breakpoint distance.

To begin, we observe that the factors of d_1 and d_2 in the denominators of the two terms in Equation (3.23) are both approximately equal to R when R is much greater than h_1 and h_2. Thus:

$$E \cong \frac{V_0}{R}\left[e^{-j\beta d_1} + \Gamma e^{-j\beta d_2}\right] \tag{3.24}$$

Note that it is *not* valid to apply the same approximation to the phase factors in the numerators; see Section 2.2.2 for a reminder of why this is so. Next, it will be convenient to extract a factor of $e^{-j\beta d_1}$ from within the square brackets in the above expression:

$$E \cong \frac{V_0}{R}e^{-j\beta d_1}\left[1 + \Gamma e^{-j\beta(d_2-d_1)}\right] \tag{3.25}$$

Now we seek an expression for $d_2 - d_1$, as it appears in the above expression, in terms of R, h_1, and h_2. This is accomplished as follows. First, note that d_1 and d_2 can be expressed exactly as follows:

$$d_1 = \sqrt{(h_1 - h_2)^2 + R^2} = R\sqrt{1 + \frac{(h_1 - h_2)^2}{R^2}} \tag{3.26}$$

$$d_2 = \sqrt{(h_1 + h_2)^2 + R^2} = R\sqrt{1 + \frac{(h_1 + h_2)^2}{R^2}} \tag{3.27}$$

The square root appearing in the above expressions can be linearized using the *binomial approximation*:

$$(1 + x)^n \cong 1 + nx \quad \text{for} \quad x \ll 1 \tag{3.28}$$

Taking $n = 1/2$ and $x = (h_1 \pm h_2)^2 / R^2$, we obtain:

$$d_1 \cong R\left[1 + \frac{1}{2}\frac{(h_1 - h_2)^2}{R^2}\right] \quad \text{for} \quad (h_1 - h_2)^2 \ll R^2 \tag{3.29}$$

$$d_2 \cong R\left[1 + \frac{1}{2}\frac{(h_1 + h_2)^2}{R^2}\right] \quad \text{for} \quad (h_1 + h_2)^2 \ll R^2 \tag{3.30}$$

We now subtract the above expressions. After some algebra, we find:

$$d_2 - d_1 \cong \frac{2h_1h_2}{R} \quad \text{for} \quad R^2 \gg 3h_{max}^2 \tag{3.31}$$

since $(h_1 + h_2)^2 = h_1^2 + h_1h_2 + h_2^2 \leq 3h_{max}^2$, where h_{max} is the greater of h_1 and h_2. Substituting this expression back into Equation (3.25), we obtain

$$E \cong \frac{V_0}{R}e^{-j\beta d_1}\left[1 + \Gamma e^{-j\beta(2h_1h_2/R)}\right] \tag{3.32}$$

The above expression can be simplified further, as follows: We expand $e^{-j\beta(2h_1h_2/R)}$ in a Taylor series around $h_1h_2/R = 0$:

$$e^{-j\beta(2h_1h_2/R)} = 1 - j\beta\left(\frac{2h_1h_2}{R}\right) + \cdots \tag{3.33}$$

where the ellipses represent terms including the factor $(h_1h_2/R)^n$ with $n \geq 2$, and which are therefore negligible for $h_{max}^2 \ll R^2/3$, as we have already assumed. Thus:

$$E \cong \frac{V_0}{R}e^{-j\beta d_1}\left[1 + \Gamma - j\Gamma\beta\left(\frac{2h_1h_2}{R}\right)\right] \tag{3.34}$$

Next we note once again that for $h_{max}^2 \ll R^2/3$, θ^i is very close to $\pi/2$ and therefore $\Gamma \cong -1$ for both the $\perp$ and $\parallel$ polarizations, assuming realistic ground conditions. Thus:

$$E \cong j\beta V_0\left(\frac{2h_1h_2}{R^2}\right)e^{-j\beta d_1} \tag{3.35}$$

Substituting $\beta = 2\pi/\lambda$ and taking the magnitude, we find

$$|E| \cong |V_0|\,\frac{4\pi}{\lambda}\frac{h_1h_2}{R^2} \tag{3.36}$$

This equation is used to compute the curve labeled "Flat Earth Propagation" in Figures 3.6 and 3.7; note that the approximation is quite accurate for sufficiently large R.

Finally, we compute the power density:

$$S_{ave} = \frac{|E|^2}{\eta} \cong \frac{|V_0|^2}{\eta}\frac{16\pi^2}{\lambda^2}\frac{h_1^2h_2^2}{R^4} \tag{3.37}$$

(here, $|E|$ in rms units). Path loss is inversely proportional to power density, so we are now ready to make some conclusions about path loss L_p for flat Earth propagation when $R^2 \gg 3h_{max}^2$. First, note that $L_p \propto R^4$, as opposed to R^2 for free space propagation, as was originally noted in the discussion following Examples 3.2 and 3.3. Next, note that $L_p \propto h_1^{-2}h_2^{-2}$; i.e., path loss increases as the inverse square of antenna height. This is also apparent by comparing Examples 3.2 (Figure 3.6, $h_1 = 15$ m) and 3.3 (Figure 3.7, $h_1 = 2$ m). We clearly see here strong motivation for mounting base station antennas at the greatest possible height.

It is often useful to have an estimate of the breakpoint; that is, the distance R beyond which Equation (3.36) applies, and the above conclusions about path loss are valid. The derivation indicates that these conclusions hold for $R^2 \gg 3h_{max}^2$, but what we would prefer is a particular value that does not require a subjective estimate of what it means to be "much greater than". Such a value can be obtained by solving for the value of R for which the magnitude of the free space electric field intensity (Equation (3.19)) equals the value predicted by Equation (3.36). This gives:

$$|V_0| \frac{1}{d_1} \equiv |V_0| \frac{4\pi}{\lambda} \frac{h_1 h_2}{R_b^2} \tag{3.38}$$

where $R = R_b$ is the breakpoint. Since $d_1 \cong R$ for the purposes of magnitude calculation, we find

$$R_b \cong \frac{4\pi}{\lambda} h_1 h_2 \tag{3.39}$$

EXAMPLE 3.4

Determine the breakpoint in Examples 3.2 and 3.3.

Solution: Equation (3.39) gives $R_b \cong 565$ m and 75 m for Examples 3.2 and 3.3, respectively. These values are roughly 25% less than the value of R_b that might be determined "by eye". This level of error is typical of the uncertainty associated with other aspects of propagation analysis, and so Equation (3.39) turns out to be quite useful as a rule of thumb.

3.5 MULTIPATH AND FADING

In the previous section we noted that terrain and structures may create significant scattering which may be subsequently perceived as multipath by the receiver. This impacts not only path loss, but can also distort the temporal and spectral characteristics of the transmitted signal. These effects are prevalent in terrestrial propagation in all bands from VHF and above. These effects are also prevalent in HF propagation, where the primary culprit turns out to be the ionosphere, as we shall see in Section 3.8.

3.5.1 Discrete Multipath Model for Terrestrial Propagation

For the moment it will be convenient to describe a single path as a time-domain electric field being used to convey a message signal $s(t)$. Either polarization of the signal at a reference point close to the transmit antenna can be described as

$$\mathcal{E}_t(t) = \mathcal{E}_0 s(t) \tag{3.40}$$

If there is only one path and we neglect polarization effects, the wave incident on the receive antenna may be written as

$$\mathcal{E}_r(t) = \left[L_p(R)\right]^{-\frac{1}{2}} \mathcal{E}_t(t-\tau) \tag{3.41}$$

where τ is the delay associated with path length. Next we envision a scenario where there is not just one, but rather $N \geq 1$ viable paths between antennas. At least one of these paths is likely to involve reflection from the ground. The additional paths are the result from scattering from buildings and terrain features. The nth such path can be expressed as $\left[L_{p,n}(R_n)\right]^{-1/2} \mathcal{E}_t(t-\tau_n)$, where the subscript n is used to denote parameters unique to the nth path. The total received signal is simply the sum over the N paths:

$$\mathcal{E}_r(t) = \sum_{n=1}^{N} \left[L_{p,n}(R_n)\right]^{-1/2} \mathcal{E}_t(t-\tau_n) \tag{3.42}$$

To simplify further, we now define *average path loss* $\overline{L_p}(R)$ such that the path loss for the nth path may be expressed exactly as

$$\left[L_{p,n}(R_n)\right]^{-1/2} = \left[\overline{L_p}(R)\right]^{-1/2} a_n \tag{3.43}$$

for each n, where a_n are unitless, path-specific values. (We'll return to what specifically we mean by "average path loss" later.) We may now write Equation (3.42) as follows:

$$\mathcal{E}_r(t) = \mathcal{E}_0 \left[\overline{L_p}(R)\right]^{-1/2} \sum_{n=1}^{N} a_n\, s(t-\tau_n) \tag{3.44}$$

We now wish to separate the signal $s(t)$ from the effect of the propagation channel, represented by a_n and τ_n. This can be done by rewriting the above expression as follows:

$$\mathcal{E}_r(t) = \mathcal{E}_0 \left[\overline{L_p}(R)\right]^{-1/2} s(t) * \left[\sum_{n=1}^{N} a_n\, \delta(t-\tau_n)\right] \tag{3.45}$$

where "$*$" denotes convolution; i.e.

$$s(t) * \delta(t-\tau_n) \equiv \int_{-\infty}^{+\infty} s(u)\, \delta(t-\tau_n-u)\, du \tag{3.46}$$

which is just $s(t-\tau_n)$. The factor in square brackets in Equation (3.45) is known as the *channel impulse response* (CIR). It is convenient to assign the CIR its own symbol:

$$h(t) = \sum_{n=1}^{N} a_n\, \delta(t-\tau_n) \tag{3.47}$$

so now Equation (3.45) may be written compactly as:

$$\mathcal{E}_r(t) = \mathcal{E}_0 \left[\overline{L_p}(R)\right]^{-1/2} s(t) * h(t) \tag{3.48}$$

3.5.2 The Static Channel: Channel Impulse Response

Although we arrived at the notion of CIR through the development of a model, the concept applies generally. In practice, CIR is typically defined so as to exclude mean path loss and the transfer functions of the antennas, as we have done above. By "impulse response", we mean that the CIR is the response when the input is an impulse. Since we defined the channel as consisting of N discrete paths in the previous section, the CIR of Equation (3.47) consists of the sum of N impulses.

The utility of decomposing the propagation channel into separate factors associated with $\overline{L_p}$, which is constant over a relatively large area, and CIR, which captures the magnitude and phase behavior of the channel over small distances and times, is that CIR behaves in a qualitatively similar way independently of R and across a wide variety of scenarios. Furthermore, differences between CIRs can be conveniently expressed using a concise statistical description as opposed to deterministic functions. These features turn out to be quite convenient, as we shall see later.

CIRs are often depicted as *power delay profiles* (PDPs), which are essentially plots of the squared-magnitude of the CIR. Figure 3.8 shows a typical PDP for a UHF-band cellular link, obtained by measurement. (For other examples of measured CIRs, an old but good reference is [12].) To facilitate some numerical analysis, consider the simulated PDPs shown in Figure 3.9. Figure 3.9(a) is the type of PDP that would typically be associated with radio links between base stations and mobile stations in rural and suburban scenarios, where the CIR tends to be dominated by a few scatterers close to the mobile station, and ground reflection. Exponential decay is typical. Often, the number of significant close-in scatters is so large that the PDP appears not as discrete impulses, but rather as a continuous (but "lumpy") exponential decay, as suggested by the dashed line. Figure 3.9(b) is a type of PDP that would typically be associated with environments consisting of tall buildings or hilly terrain. In these environments, the situation depicted in Figure 3.9(a) is repeated several times, with each instance associated with

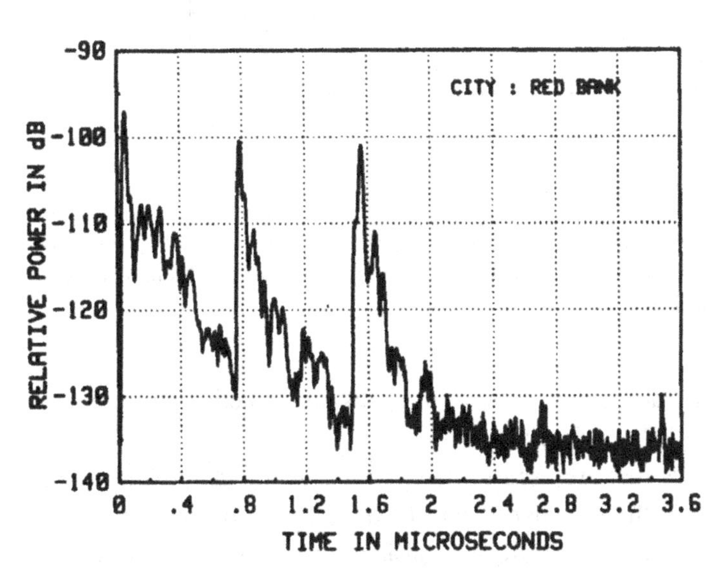

Figure 3.8. A measured power delay profile (PDP) for a 850 MHz link between antennas at heights 9.1 m and 1.8 m, separated by about 1 km in a small US city. ©1988 IEEE. Reprinted, with permission, from [15].

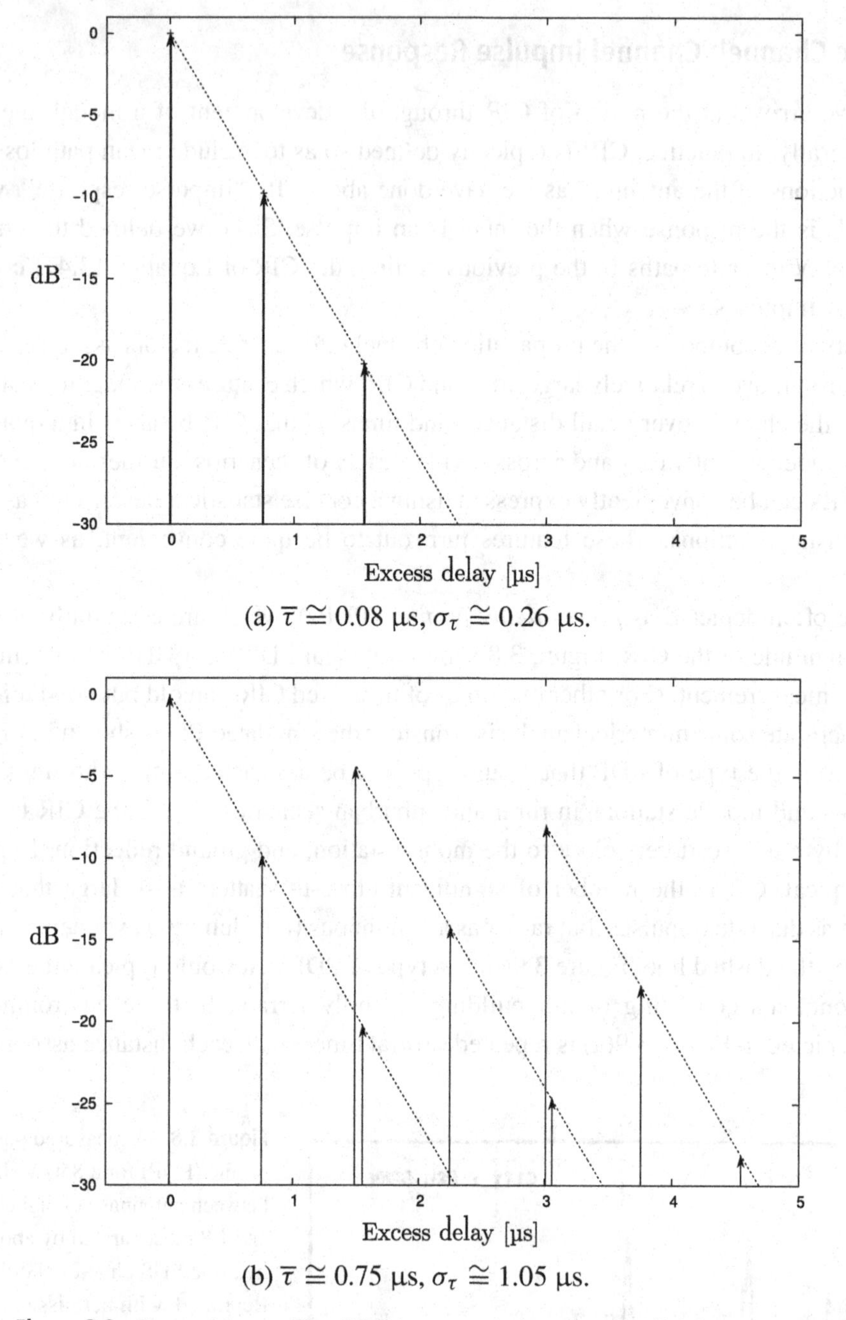

Figure 3.9. Power delay profiles (PDPs) representative of a radio link between a base station and a mobile station in (a) rural and suburban environments and (b) environments consisting of tall buildings or hilly terrain. See Examples 3.5 and 3.6.

a particular building or terrain feature. These instances are sometimes referred to as *fingers*, and are also evident in the PDP of Figure 3.8.

The two primary metrics that are used to describe a CIR are *delay spread* and *coherence bandwidth*. Delay spread is the effective length of the CIR; in practice, the time span over which multipath contributions to the CIR are considered to be significant. Therefore, a CIR consisting of $N = 1$ path has zero delay spread, whereas a CIR that has significant multipath contributions with delays ranging over a duration $\Delta\tau$ is said to have a delay spread of $\Delta\tau$. The difficulty in the latter case is determining what constitutes a "significant" contribution. An arbitrary but commonly-used metric is *root mean square (RMS) delay spread* σ_τ, defined as the standard deviation of the PDP-weighted delay:[5]

$$\sigma_\tau \equiv \sqrt{\frac{\int_0^\infty (t-\overline{\tau})^2 \, |h(t)|^2 \, dt}{\int_0^\infty |h(t)|^2 \, dt}} \tag{3.49}$$

where $\overline{\tau}$ is the mean of CIR-weighted delay:

$$\overline{\tau} \equiv \frac{\int_0^\infty t \, |h(t)|^2 \, dt}{\int_0^\infty |h(t)|^2 \, dt} \tag{3.50}$$

The above definitions are generally valid. If the CIR can be reasonably approximated using the discrete multipath model (Equation (3.47)), the integrals reduce to sums over the paths as follows:

$$\sigma_\tau = \sqrt{\frac{\sum_{n=1}^N (\tau_n - \overline{\tau})^2 \, |a_n|^2}{\sum_{n=1}^N |a_n|^2}} \tag{3.51}$$

and

$$\overline{\tau} = \frac{\sum_{n=1}^N \tau_n \, |a_n|^2}{\sum_{n-1}^N |a_n|^2} \tag{3.52}$$

EXAMPLE 3.5

The CIR depicted in Figure 3.9(a) is given by ($\tau_1 = 0$ μs, $a_1 = 1$), ($\tau_2 = 0.75$ μs, $a_2 = 0.325$), and ($\tau_3 = 1.55$ μs, $a_3 = 0.098$). Calculate the rms delay spread.

Solution: From Equation (3.52), $\overline{\tau} \cong 0.08$ μs. From Equation (3.51), the delay spread $\sigma_\tau \cong 0.26$ μs.

EXAMPLE 3.6

The first finger of the CIR depicted in Figure 3.9(b) is identical to the CIR of Figure 3.9(a) (see Example 3.5). Subsequent fingers are identical to the first finger, but delayed by 1.5 μs and 3.0 μs, and scaled by 0.6 and 0.4, respectively. Calculate the rms delay spread.

Solution: From Equation (3.52) (now with $N = 9$), $\overline{\tau} \cong 0.75$ μs. From Equation (3.51), $\sigma_\tau \cong 1.05$ μs.

These delay spread statistics are typical for mobile radio systems in the VHF band and above. More on this in Section 3.6.

Let us now consider the frequency response $H(\omega)$ of the propagation channel. This is a concern because the propagation channel has the potential to act as a filter that distorts the

[5] Magnitude brackets ("|...|") are included to accommodate the option (not yet discussed) to express the CIR in complex-valued "baseband" form, in which the a_ns may be complex-valued.

signal $s(t)$ which is sent through it. The frequency response of the channel is simply the Fourier transform of the CIR; i.e., $H(\omega) \equiv \mathcal{F}\{h(t)\}$. For the discrete multipath model, we find

$$H(\omega) = \sum_{n=1}^{N} a_n \, e^{-j\omega\tau_n} \tag{3.53}$$

For an $N = 1$ channel, $H(\omega) = a_1$; that is, constant with frequency, and the frequency response is said to be *flat*. In other words, if there is only one path, there can be no distortion of $s(t)$ aside from simple delay. For $N \geq 2$, $H(\omega)$ has the potential to vary with frequency. If, on the other hand, the delay spread is sufficiently small, then the variation with frequency will be small; perhaps even negligible relative to the bandwidth of the signal. Thus, it is useful to have some metric that indicates the bandwidth over which the variation in $H(\omega)$ can be assumed to be negligible, and to have this be related to the delay spread.

The metric we seek is *coherence bandwidth*, B_c. Like all specifications of bandwidth, a formal definition depends on some criterion which is typically arbitrary; e.g., the difference relative to a nominal value that is used to determine frequency limits. In the analysis of CIR, one instead typically settles for an informal estimate in the form

$$B_c \approx \frac{1}{m\sigma_\tau} \tag{3.54}$$

where typical choices for values of m are 5, 10, and 50; all based on empirical experience combined with some preference for how much variation is considered acceptable. The usefulness of the concept of coherence bandwidth is limited to identifying whether channel frequency variation is expected to be significant, or not, over the span of one's signal. A useful determination of the *extent* to which the variation is significant requires a more careful analysis in terms of the specific modulation and CIR characteristics (more on this in Chapters 5 and 6).

EXAMPLE 3.7

Compute $|H(\omega)|$ for the CIR of Example 3.5 and use this result to assess Equation (3.54).

Solution: $|H(\omega)|$ is shown in Figure 3.10(a), evaluated in a 1 MHz span around 871 MHz. This frequency is chosen for demonstration purposes only; there is no expectation of significant differences at higher or lower frequencies. We previously determined $\sigma_\tau \cong 0.26$ μs, so B_c is estimated to be in the range 77.4 kHz to 774 kHz for $m = 50$ and $m = 5$, respectively. The associated variation is $|H(\omega)|$ is seen to be less than 1 dB or so for $m = 50$, and up to a few dB for $m = 5$.

EXAMPLE 3.8

Repeat the previous example for the CIR of Example 3.6.

Solution: $|H(\omega)|$ is shown in Figure 3.10(b), again evaluated in a 1 MHz span around 871 MHz. We previously determined $\sigma_\tau \cong 1.04$ μs, so B_c is estimated to be in the range 19.1 kHz to 191 kHz for $m = 50$ and $m = 5$, respectively. The variation in $|H(\omega)|$ corresponding to a given value of m is found to be similar to that found in the previous example.

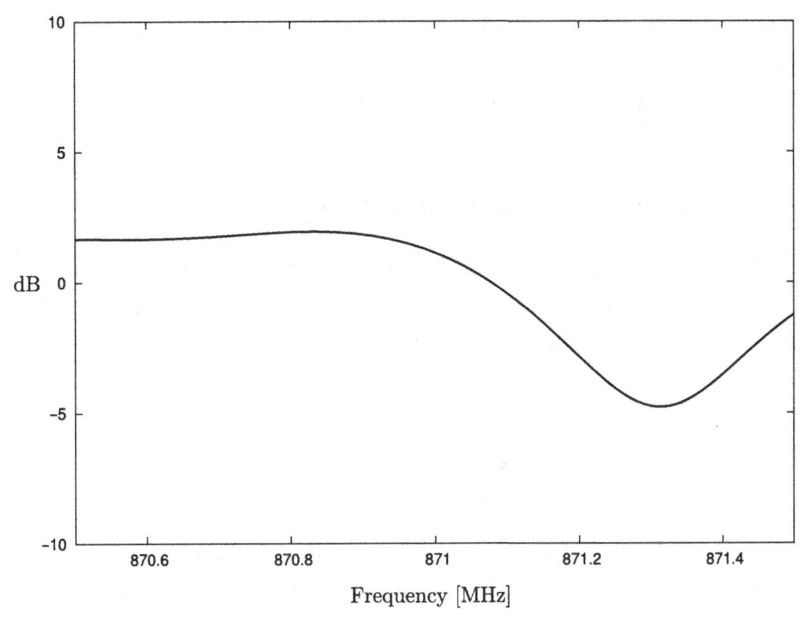

(a) $B_c \cong 77.4$ kHz, 774 kHz for $m = 50, 5$; respectively.

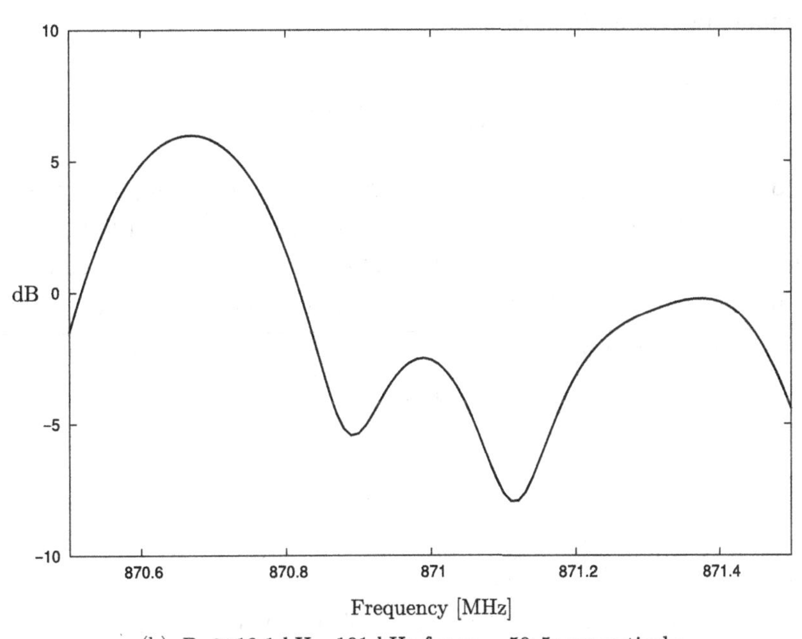

(b) $B_c \cong 19.1$ kHz, 191 kHz for $m = 50, 5$; respectively.

Figure 3.10. Frequency response $|H(\omega)|$ corresponding to the CIRs depicted in Figure 3.9, plotted in dB relative to mean value over the displayed span.

Comparing the above examples, the consequence of increasing delay spread is evident as an increase in the variability of $|H(\omega)|$. This situation is commonly known as *frequency-selective fading*. Note that the interpretation of a propagation channel as being "frequency-selective" depends on the bandwidth of interest B_m:[6] That is, *any* channel for which B_c is significantly

[6] Note that B_m might be either the bandwidth of the signal or the bandwidth of the receiver, but typically it is the former.

less than B_m is "frequency-selective", whereas any channel for which B_c is significantly greater than B_m is "flat".

For an example of actual observed frequency-selective fading, see Figures 6.38 and 6.41.

3.5.3 The Dynamic Channel: Doppler Spread and Fading

To this point we have assumed that neither the transmitter, receiver, nor scattering structures are in motion. Let us now consider the consequences of motion. We begin with a modified version of the discrete multipath model (Equation (3.47)) in which the path parameters – including the number of paths – are now functions of time:

$$h(t) = \sum_{n=1}^{N(t)} a_n(t)\, \delta(t - \tau_n(t)) \tag{3.55}$$

Let us limit our attention to the behavior of the channel on the shortest timescales over which we expect to see significant variation. If there is any motion, then $\tau_n(t)$ is continuously changing, whereas $N(t)$ and $a_n(t)$ might reasonably be assumed to vary much more slowly. Therefore $\tau_n(t)$ should be the initial focus of our attention, and we assume that over some time frame it is reasonable to express Equation (3.55) as

$$h(t) \cong \sum_{n=1}^{N} a_n\, \delta(t - \tau_n(t)) \tag{3.56}$$

Now we need to know something about the form of $\tau_n(t)$. Referring to Figure 3.11, note that a transmitter, receiver, or scatterer moving at a constant speed v over a distance $d = v\Delta t$ causes the path length τ_n to change by $d \cos\phi_n$, where ϕ_n is the angle between the direction of motion and the incident wave. Next, let us assume a linearly-varying delay of the form $\tau_n(t) = \tau_{0,n} + u_n t$, where $\tau_{0,n}$ is τ_n at some earlier reference time common to all n, and

$$u_n \equiv \frac{\partial \tau_n}{\partial t} \cong \frac{d \cos\phi_n / c}{\Delta t} = \frac{v}{c} \cos\phi_n \tag{3.57}$$

Thus we have

$$h(t) \cong \sum_{n=1}^{N} a_n\, \delta\left([1 - u_n]\, t - \tau_{0,n}\right) \tag{3.58}$$

Note that acceleration and other higher-order forms of motion are also possible and could be modeled in a similar way. However, for our purposes the present assumptions will suffice to reveal the behaviors that dominate the behavior of realistic channels.

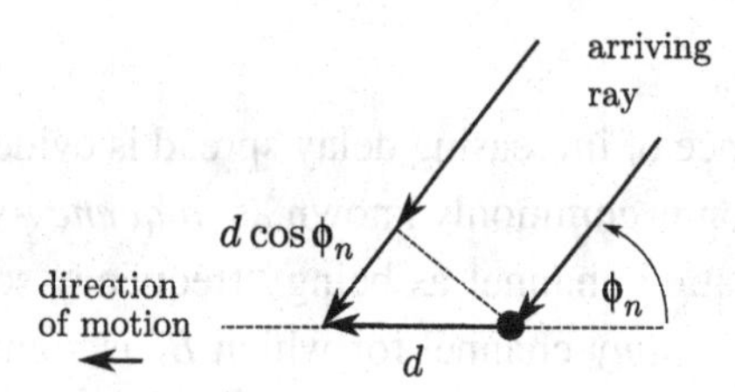

Figure 3.11. Determining the change of path length due to a transmitter, receiver, or scatterer moving at a constant velocity. Note we assume that the elapsed time Δt is vanishingly small such that ϕ_n remains approximately constant.

If the bandwidth of $s(t)$ is small relative to B_c, then the effect of the channel is well-described as a time-domain variation. This is known as *flat fading*, and applies to many signals of practical interest, so it worth considering here. This is most easily demonstrated by representing $s(t)$ as an analytic signal modulating a sinusoidal carrier

$$s(t) = \mathcal{R}e\left\{m(t)e^{+j\omega_c t}\right\} \tag{3.59}$$

where $m(t)$ is the *message signal*, centered at zero frequency and assumed to have bandwidth $B_m \ll \omega_c$, the complex-valued exponential factor is known as the *carrier*, and ω_c is the *carrier frequency*. Now applying the CIR to obtain $y(t) = s(t) * h(t)$:

$$y(t) = \mathcal{R}e\left\{\sum_{n=1}^{N} a_n\, m\left([1-u_n]\,t - \tau_{0,n}\right)\, e^{+j\omega_c([1-u_n]t-\tau_{0,n})}\right\} \tag{3.60}$$

In any scenario of practical interest, $v \ll c$ with considerable overkill. Therefore

$$m\left([1-u_n]\,t - \tau_{0,n}\right) \cong m(t - \tau_{0,n}) \tag{3.61}$$

since the time scaling of the message signal will be undetectable in the limited time frame already established for this analysis. We may invoke the *narrowband assumption* $\tau_{0,n} \ll B_m^{-1}$ to further simplify:

$$m\left(t - \tau_{0,n}\right) \cong m(t)e^{-j\omega\tau_{0,n}} \tag{3.62}$$

i.e., we assume that the phase variation in the spectrum of $m(t)$ associated with $\tau_{0,n}$ (relative to the other $\tau_{0,n}$s) is sufficiently small that it may be approximated as a constant phase shift over the bandwidth. This is already assured because we have assumed that $B_m \ll B_c$. For the carrier we have

$$e^{+j\omega_c([1-u_n]t-\tau_{0,n})} = e^{+j(\omega_c-\omega_{d,n})t}\, e^{-j\omega_c\tau_{0,n}} \tag{3.63}$$

where

$$\omega_{d,n} \equiv \omega_c \frac{v}{c}\cos\phi_n = 2\pi\frac{v}{\lambda}\cos\phi_n \tag{3.64}$$

and where λ is the wavelength at ω_c. The quantity $\omega_{d,n}$ is recognized as a *Doppler shift*. Equation (3.60) may now be written as

$$y(t) \cong \mathcal{R}e\left\{\sum_{n=1}^{N} a_n m(t)\, e^{+j(\omega_c-\omega_{d,n})t}\, e^{-j\omega_c\tau_{0,n}}\right\} \tag{3.65}$$

The result is N copies of $s(t)$, each scaled in magnitude by a_n, rotated in phase by $\omega_c\tau_{0,n}$, and appearing to be at a slightly different carrier frequency $\omega_c - \omega_{d,n}$. The associated spectrum will therefore consist of N copies of the spectrum of $s(t)$, each displaced by $\omega_{d,n}$ in frequency. This effect is known as *Doppler spreading*.

We may now rearrange Equation (3.65) as follows:

$$y(t) \cong \mathcal{R}e\left\{m(t)\,\gamma(t)\,e^{+j\omega_c t}\right\} \tag{3.66}$$

where we have made the definition

$$\gamma(t) \equiv \sum_{n=1}^{N} a_n\, e^{-j\omega_{d,n}t}\, e^{-j\omega_c\tau_{0,n}} \tag{3.67}$$

Note that $\gamma(t)$ describes the CIR in a manner that is completely independent of $m(t)$. Comparing Equations (3.59) and (3.66), we find that the effect of the CIR can be represented as a multiplicative process, as opposed to convolution. Specifically, the CIR transforms $m(t)$ into $m(t)\gamma(t)$, and therefore the variation in the magnitude of $y(t)$ due to the CIR is proportional to $|\gamma|$.

Let us now examine the behavior of $|\gamma(t)|$ using an example.

EXAMPLE 3.9

Simulate $|\gamma(t)|$ for a typical UHF mobile radio propagation channel at 871 MHz ($\lambda \cong 36$ cm) and a typical vehicle speed of $v = 24$ km/h. Use Equation (3.67) with $N = 10$, a_ns equal to 1, and $\omega_{d,n}$s randomly distributed over $\pm 2\pi v/c$.

Solution: See Figure 3.12.

The striking feature of this result is the prevalence of deep nulls – up to tens of dB, occurring roughly every $\lambda/2$ on average. This particular aspect of flat fading is known as *fast fading*, with the term "fast" referring to the fact that the variation over distance is much "faster" than the R^2 to R^4 variation associated with the average path loss $\overline{L_p}$.

It is often useful to know the statistical distribution of $|\gamma(t)|$ as well as something about the rate at which $|\gamma(t)|$ changes. First, let us consider the statistical distribution. The *cumulative distribution function* (CDF) of a random variable X is defined as $P(X \leq q)$ for some q. For present purposes it is convenient to define X as a sample taken from the deterministic quantity

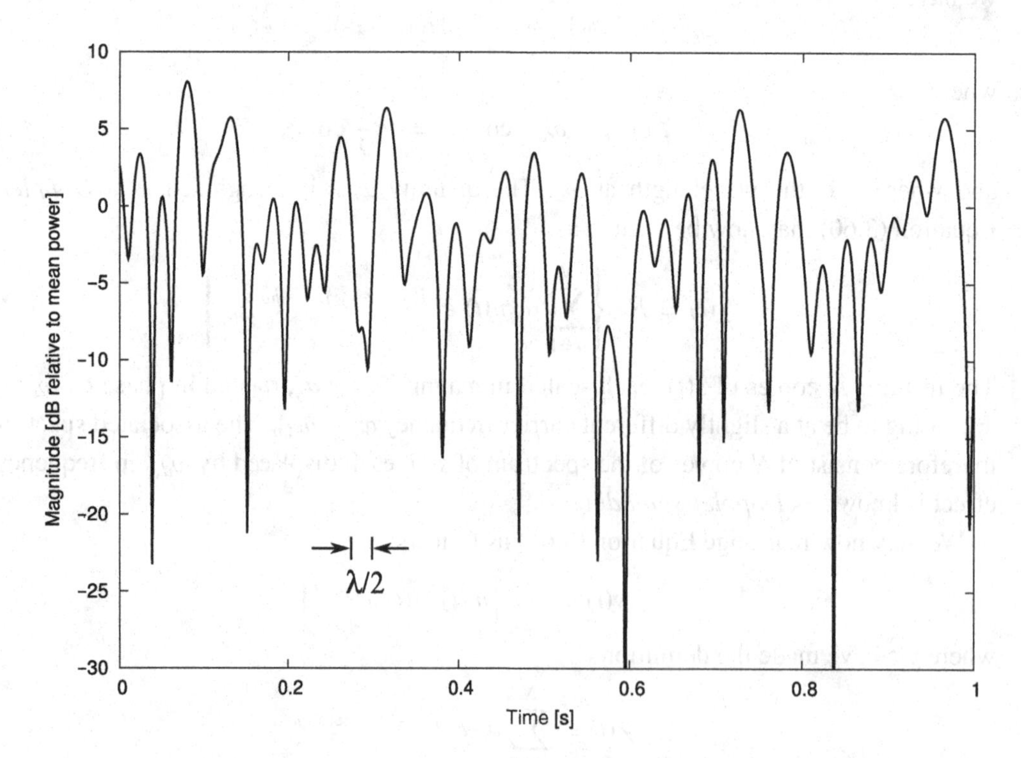

Figure 3.12. A simulation of $|\gamma(t)|$ for $f = 871$ MHz and $v = 24$ km/h (see Example 3.9).

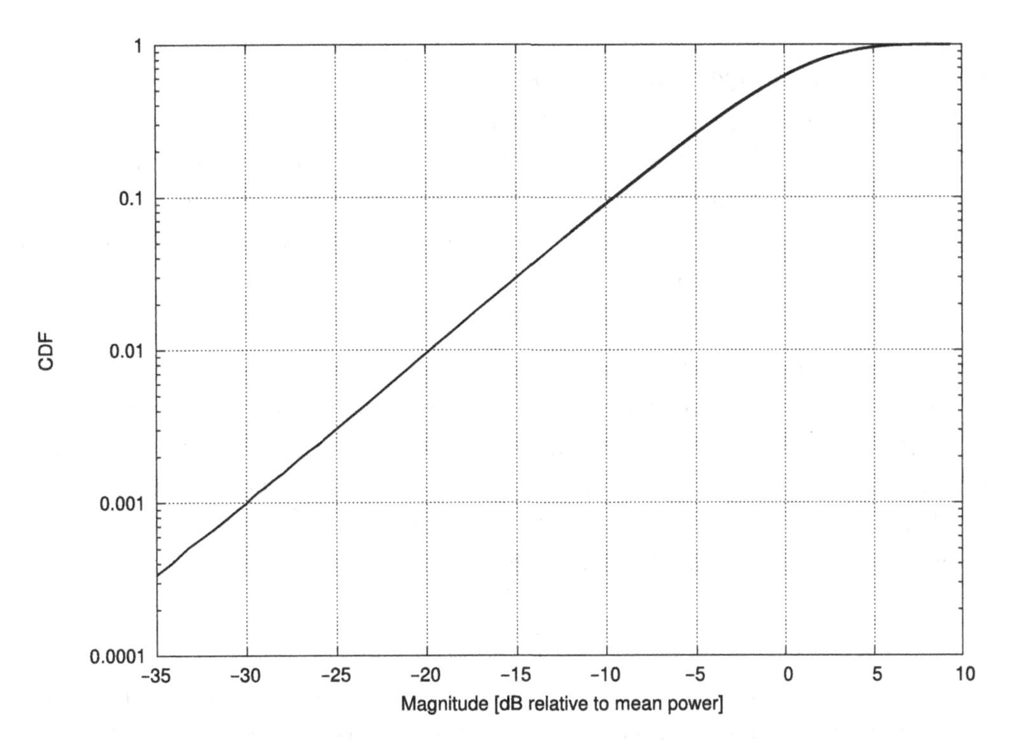

Figure 3.13. CDF of observed values of $|\gamma|/\overline{\gamma}$ from the example described in Figure 3.12.

$|\gamma|/\overline{\gamma}$, where $\overline{\gamma}$ is defined as the mean value of $|\gamma(t)|$. That is, we define the CDF of X as the probability that $|\gamma(t)|$, normalized to have a mean of 1, is less than or equal to the threshold q. Figure 3.13 shows the CDF of X, computed directly from the same data used to generate the results shown in Figure 3.12. Since $\gamma(t)$ has "voltage-like" units, we have elected to plot the horizontal ("q") axis in dB.

Note that the probability that $|\gamma(t)|$ at any moment is $\leq$ 10 dB below its mean value is $\cong$ 10%. Said differently, the fraction of time that the instantaneous power is $\leq$ 10 dB below its mean value is $\cong$ 10%. Similarly, note that the fraction of time that the instantaneous power is $\leq$ 20 dB below its mean value is $\cong$ 1%. This particular statistical distribution is known as the *Rayleigh distribution*, and so fading which is well-modeled by Equation (3.67) with large N is known as *Rayleigh fading*. The CDF of any random variable drawn from a Rayleigh distribution is given by

$$P(X \leq q) = 1 - e^{-q^2} \tag{3.68}$$

as can be confirmed by spot-checking values plotted in Figure 3.13.

EXAMPLE 3.10

Field measurements indicate an average power of -105 dBm is received by a particular radio as it moves about in a region in which the channel is believed to exhibit Rayleigh fast fading. The bandwidth of the receiver is much less than the coherence bandwidth. What fraction of the time can the received power be expected to drop below -123 dBm?

Solution: The bandwidth of the receiver is B_m. From the problem statement, $B_m \ll B_c$, so flat fading may be assumed. The Rayleigh flat fading distribution is given by Equation (3.68). In this case, we wish to know the fraction of time the received power is -123 dBm $-$ (-105 dBm) $=$ 18 dB below the mean. Therefore $q = -18$ dB $\cong 0.126$. Applying Equation (3.68), we find $P(X \leq 0.126) = 0.016$: The received power is expected to be below the threshold 1.6% of the time.

Coherence time T_c is the duration over which variation in the CIR can be considered insignificant. The rate at which the CIR varies overall is roughly the same as the rate at which the CIR varies within a coherence bandwidth, so T_c is roughly the same as the time over which variation in $\gamma(t)$ can be considered to be insignificant. A more precise definition of T_c is problematic, because different applications are more or less sensitive to this time variation. In wireless communications applications, coherence time is typically calculated as

$$T_c \equiv \frac{1}{\mu f_{d,max}} = \frac{\lambda}{\mu v} \tag{3.69}$$

where $f_{d,max} = v/\lambda$ is the maximum Doppler shift and μ depends on the application. Note that this definition corresponds to the time required to traverse a distance equal to λ/μ, whereas the null-to-null distance in Rayleigh fading is roughly $\lambda/2$. Therefore T_c computed using Equation (3.69) corresponds to a duration over which the variation is very small only if $\mu \gg 2$. Most mobile wireless communications systems are robust to large variability in the CIR, so values of μ tend to be in the range $1-6$ in those applications ($\mu = 1$, 2.36, and 5.58 are popular choices; for details, see [61]). In other applications a "tighter" definition with larger μ may be more appropriate.

EXAMPLE 3.11

What is the coherence time in Example 3.9 (Figure 3.12) assuming $\mu = 5$?

Solution: Using Equation (3.69), $T_c \cong 10.3$ ms, which is the time required to traverse 0.4λ.

3.5.4 Spatial Autocorrelation and Diversity

Fading of the type depicted in Figure 3.12 imposes severe limitations on the performance of radio communications systems, and one is motivated to find ways to mitigate its effects. It is instructive to consider the worst case scenario: This is that the channel variation should slow or halt while $|\gamma|$ is at or near a minimum. In this scenario, the receiver may experience a signal-to-noise ratio that appears to be stuck at a level tens of dB below the nominal mean power.

A solution in this case is *spatial diversity*, in which two or more antennas separated by some distance d are employed to create an effective channel with more favorable fading characteristics. To see how this might work, first note that the curve traced out in Figure 3.12

could equally well represent $|\gamma|$ while varying d at some instant in time. The relationship between elapsed time Δt and d is

$$\frac{d}{\lambda} = \frac{v\Delta t}{\lambda} = f_{d,max}\Delta t \tag{3.70}$$

A distance of $d = \lambda/2$ is shown for scale on Figure 3.12. Clearly d for effective space diversity should be at least this large. Further, it makes sense that larger d should be more effective than smaller d, since one expects the correlation between $|\gamma(t)|$ and $|\gamma(t+\Delta t)|$ to decrease with increasing Δt. Thus, the autocorrelation of the function $\gamma(\cdot)$ as a function of d/λ, or equivalently as a function of $f_{d,max}\Delta t$, is of interest. The latter is easier to work out since we already have a suitable time-domain expression; namely Equation (3.67). The autocorrelation of $\gamma(t)$ in the time-domain is:

$$\rho(\Delta t) \equiv \lim_{T\to\infty} \frac{1}{T} \int_{-T/2}^{+T/2} \gamma(t)\gamma^*(t-\Delta t)dt \tag{3.71}$$

After substitution and some math, we find that for large N:

$$\frac{\rho(\Delta t)}{\rho(0)} = J_0\left(\omega_{d,max}\Delta t\right) = J_0\left(2\pi\frac{d}{\lambda}\right) \tag{3.72}$$

where $J_0(\cdot)$ is the zero-order Bessel function of the first kind.

This result is plotted in Figure 3.14. Observe that spatial correlation decreases with increasing separation, but does so in an oscillatory manner within a slowly decaying envelope.

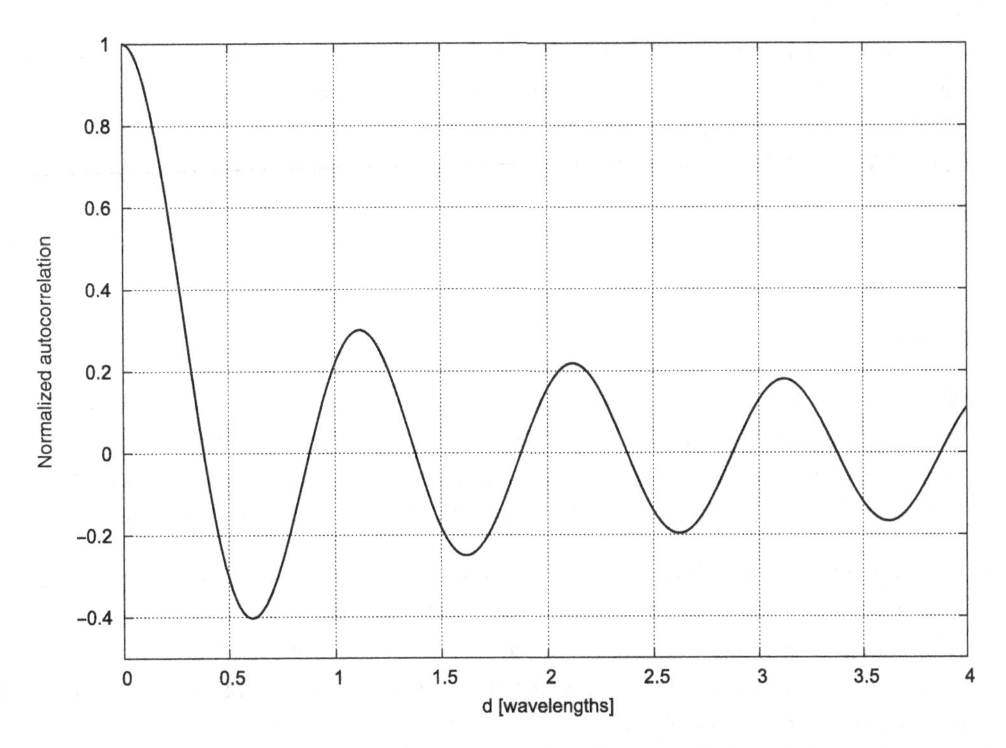

Figure 3.14. Spatial autocorrelation in Rayleigh fading. Note that this is equivalently the temporal autocorrelation using the change of variables indicated by Equation (3.70).

Also note the result is consistent with Figure 3.12, which suggests that antenna separation less than about $\lambda/4$ always results in highly-correlated signals.

Now an important caveat: This result is specific to the Rayleigh fading model on which it is based. When fading is better described using a different model, the spatial autocorrelation – and specifically the positions of the zero-crossings – will be different. Therefore there is no point in attempting to "tune" antenna spacing to correspond to zeros in the spatial autocorrelation. Nevertheless, some form of the decaying oscillation exhibited in Figure 3.14 is generally observed in channels which exhibit flat fading. Therefore the most effective strategy is simply to use the largest possible antenna separation, and separations on the order of 10λ or so are well-justified.

Given two or more antennas with separation sufficient to realize channels with low correlation, there are a number of schemes for combining them so as to mitigate the effect of magnitude fading. Principal among these are *selection combining* and *maximal ratio combining* (MRC). In selection combining, one simply uses the antenna whose output is currently largest. This is demonstrably suboptimal, but has the advantage of being very simple to implement. In MRC, the output is a linear combination of the antennas with combining coefficients that are selected to maximize the signal-to-noise ratio (SNR) of the output. MRC is optimal but is more difficult to implement. For additional discussion of MRC including expected performance, see Section 6.14.3. For additional information on spatial diversity and spatial characteristics of the propagation channel generally, [31] and [41] are recommended.

EXAMPLE 3.12

Spatial diversity is commonly employed on the uplink (mobile station to base station link) of cellular telecommunications systems. Assuming frequencies around 850 MHz, $\lambda \approx 35$ cm. To satisfy the 10λ criterion suggested above in this frequency band requires antennas spaced apart by about 3.5 m. This is close to the limit of what a conventional cell tower can accommodate; see Figure 2.25 for an example.

3.5.5 Summary

In this section we have found that the prevalence of scattering in terrestrial radio links results in propagation channels which are potentially complex and liable to significantly distort radio signals in a variety of ways. Before moving on, let us take a moment to organize what we have found. The propagation channel can be characterized in terms of the following metrics.

- The coherence bandwidth B_c, which is inversely proportional to (thus, essentially the same information as) the delay spread σ_τ. The former indicates the largest message bandwidth (B_m) that can be accommodated without spectral distortion; the latter will prove useful in assessing the extent to which delayed copies of the message signal create interference for earlier copies of the message signal.[7]

[7] In Chapter 6 we will refer to this as *intersymbol interference*.

- The coherence time T_c, which is inversely proportional to (thus, essentially the same information as) the Doppler spread. T_c is useful to know for the analysis and design of modulations which are sensitive to the temporal variation of the channel. The time variation of the channel is due to motion, so T_c can be equivalently be represented as a *coherence distance* equal to $T_c v$, which is on the order of $\lambda/2$.
- The statistical distribution of $|\gamma|$. The two distributions identified so far are the constant distribution associated with free space propagation, and the Rayleigh distribution associated with the discrete multipath model, but known to be a reasonable description of actual channels. Other possibilities will be identified in Section 3.6.3. This distribution influences the analysis and design of modulations, transmitter power control, receiver automatic gain control (AGC), and antenna diversity schemes.
- Now coming full circle: the average path loss $\overline{L_p}$. We can now be more specific about what we mean by "average": We mean the average at one location over some time $\gg T_c$, or, equivalently, averaged over a region of size $\gg$ the coherence distance. The point is that $\overline{L_p}$ represents the variation in path loss over relatively large temporal and spatial scales, whereas $|\gamma|$ represents the variation in path loss over relatively small temporal and spatial scales.

3.6 TERRESTRIAL PROPAGATION BETWEEN 30 MHZ AND 6 GHZ

The free-space wavelengths at 30 MHz and 6 GHz are 10 m and 5 cm, respectively. This is a regime in which the ground, terrain features, buildings, and vehicles are potentially large relative to a wavelength, and therefore may be efficient scatterers of radio waves. In this case, Sections 3.4 and 3.5 provide a cogent description of terrestrial propagation as is. In this section we need simply to elaborate and fill in a few relevant details. The situation is significantly different at lower frequencies ($<$ 30 MHz) and higher frequencies ($>$ 6 GHz), so we shall consider these regimes separately, in Sections 3.8 and 3.7 respectively.

3.6.1 Radio Horizon

With just a few exceptions (see Section 3.9), the useful range of terrestrial signals at frequencies in the VHF band and above are limited by the curvature of the Earth. The limit itself is the *radio horizon*; that is, the range at which the line of sight from an antenna becomes tangent to the sphere representing the surface of the Earth, as shown in Figure 3.15. This range is

$$R \cong (4.12\text{ km})\sqrt{\frac{h}{1\text{ m}}} \tag{3.73}$$

where h is height of the antenna above the average terrain height [41]. Since the range to the radio horizon increases with h, antenna height is important in determining the maximum effective range of a terrestrial radio link in this frequency regime.

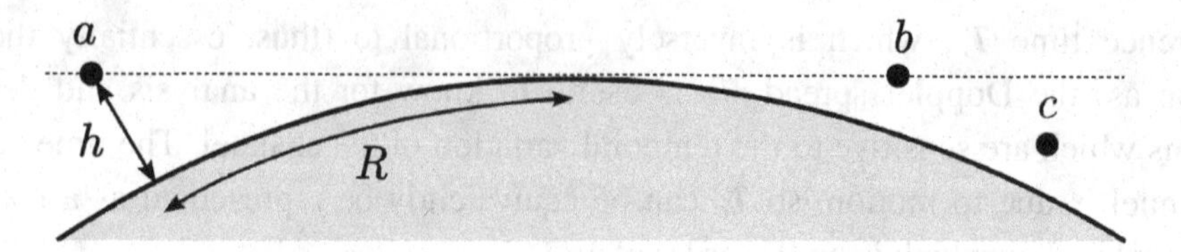

Figure 3.15. Radio horizon for an antenna located at point a. Point b shares a radio horizon with point a, whereas point c is below the radio horizon of point a.

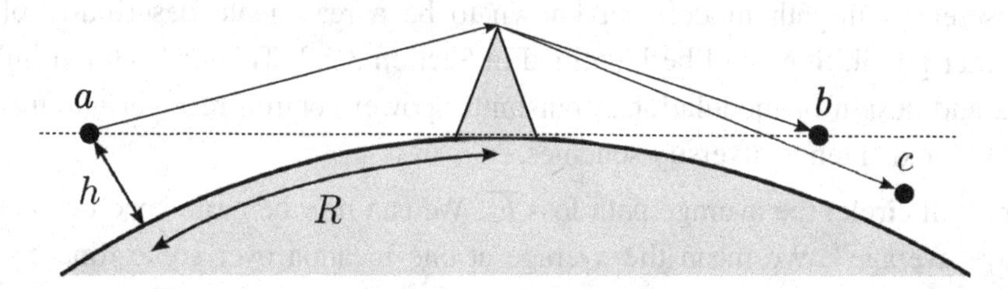

Figure 3.16. Extension of a radio link beyond radio horizons by diffraction from intervening terrain.

EXAMPLE 3.13

A UHF television broadcast antenna is mounted at a height of 100 m. Assuming flat terrain and receive antennas which are very close to the ground, what is the maximum effective range of this antenna?

Solution: From Equation (3.73), the distance to the radio horizon is about 41 km, so this is the maximum effective range for receiving antennas at ground level.

Effective range can be increased by increasing the height of the receiving antenna. A receiving antenna has its own radio horizon, and the link can be closed if the radio horizon of the transmitter overlaps with the radio horizon of the receiver.

Of course the surface of the Earth is not smooth. The presence of terrain or buildings results in maximum range that may be less than – or greater than – the range implied by radio horizons. When terrain or buildings obstruct free space propagation, the route to the receiver must involve some combination of transmission through the obstacle or diffraction around the obstacle, greatly increasing the path loss. On the other hand, propagation beyond the radio horizon is sometimes possible by diffraction from the peak of intervening terrain and structures, as shown in Figure 3.16. Although the diffraction itself is weak, this is effective because there may be reduced scattering along the path to and from the peak.

3.6.2 Delay Spread and Coherence Bandwidth

The PDPs shown in Figure 3.9 are representative of PDPs commonly encountered in this frequency regime. The form depicted in Figure 3.9(a) is typical of rural or suburban environments when at least one of the antennas is mounted well above the height of any scattering objects. The delay spread in this case is found from measurement studies to be $\mathcal{O}(1\ \mu s)$ or less. Using Equation (3.54), B_c is found to be in the range ~10 kHz to ~1 MHz for the $m = 5$ and $m = 50$ definitions of delay spread, respectively. The form depicted in Figure 3.9(b) is typical of scenarios in which paths include efficient scattering from nearby tall buildings or hilly terrain; each finger being associated with one of these paths. The delay spread associated with each finger may be similar to the delay spreads in the previous scenario, but the overall delay spread depends on the separation between the fingers, which may extend to $\mathcal{O}(10\ \mu s)$. The resulting coherence bandwidth in such cases is typically less than 100 kHz.

Similar results emerge in indoor environments, but only at somewhat higher frequencies. This is because reflection from walls and other indoor features becomes efficient for wavelengths of roughly 1 m or less, corresponding to a low-frequency limit on the order of 300 MHz as opposed to 30 MHz. Because the distance between walls and ceilings is on the order of meters, indoor delay spreads are typically on the order of tens to hundreds of nanoseconds, and the associated coherence times are on the order of hundreds of kHz to tens of MHz.

3.6.3 Fading Statistics and Coherence Time

As in Section 3.5.3, let us now restrict our attention to a bandwidth $\leq B_c$ and consider fading and coherence time. We observed earlier that the discrete multipath model gives rise to fading with Rayleigh magnitude statistics and with T_c given by Equation (3.69). This model is found to be so widely applicable that it is common to simply assume this behavior in the analysis and design of modulation schemes and radio links. There are, however, at least three important cases in which a different model may be more appropriate.

- *Two-Ray Fading.* In scenarios in which an unobstructed line of sight path exists and other scattering mechanisms are negligible, the propagation channel may be well-modeled using the discrete multipath model with $N = 2$, corresponding to the direct path plus the ground-reflected path. Two examples where this occurs are in communications between high-directivity antennas both mounted high above ground, and in air-to-ground communications when the ground station is located in a region of low scattering. Since ground-reflected paths have reflection coefficients very close to -1, the received signal exhibits a near-perfect null every half-wavelength, resulting in magnitude statistics somewhat more severe than those of the Rayleigh distribution.
- *Ricean Fading.* Sometimes the CIR will be more or less as anticipated by the discrete multipath model, except that one of the paths will dominate over all others. This is commonly encountered in rural and suburban environments when one or both antennas is much higher than surrounding terrain and buildings. This leads to fading magnitude statistics which are less severe than Rayleigh by an amount which depends on the ratio of the power in the dominant path to the combined power of all other paths. This type of fading is well-described by the "Rice-K" distribution, and so is commonly referred to as Ricean fading.

- *Nakagami Fading.* When the CIR consists of multiple well-resolved fingers (as in Figure 3.9(b)), one might assume that the fading associated with each finger individually has Rayleigh statistics. In this case, the fading statistics for all fingers combined will have magnitude distributed according to the "Nakagami-*m*" distribution. This hypothesis is supported by measurements of actual CIRs, which indicate that Nakagami distributions are a somewhat better fit to the observed magnitude statistics, especially for channels with longer delay spreads. The Nakagami-*m* distribution is essentially a generalization of the Rayleigh and Rice distributions, and the distinction becomes important in certain problems in which the results may be sensitive to the precise form of the fading distribution.

In all cases, T_c is roughly the same and Equation (3.69) is applicable. Since wavelengths in this frequency regime range from 10 m to 5 cm, and assuming speeds ranging from 1 m/s (e.g., walking) to 30 m/s (typical highway speed), we find coherence times ranging from 333 μs to 2 s, respectively. Any practical system covers a much smaller frequency range; for example, a PCS cellular telecommunications system operates at roughly 2 GHz and therefore must accommodate T_c in the range 1–30 ms.

3.6.4 Average Path Loss

We noted in Section 3.4.2 that path loss in the flat Earth model was proportional to R^4 beyond the breakpoint distance $R = R_b$, and at shorter distances was roughly proportional to R^2, albeit with additional fading. The situation is dramatically more complex in terrestrial channels, because terrain and buildings may effectively block and scatter radio waves in the 30 MHz–6 GHz regime. Let us now consider some ways in which average path loss may be estimated.

Breakpoint – Power Law Model. In this model, we simply assume free space propagation ($\overline{L_p}(R) = L_{p0}(R)$) up to the breakpoint distance, and then

$$\overline{L_p}(R) = \overline{L_p}(R_b)\left(\frac{R}{R_b}\right)^n \quad , \quad R > R_b \tag{3.74}$$

for greater distances. In this model, n is known as the *path loss exponent.* This model strikes a useful balance between simplicity and the ability to describe a wide range of commonly-encountered propagation conditions. In particular, $n = 2$ corresponds to free space propagation, and increasing n to 4 provides a "piecewise log-linear" approximation to the flat Earth model. Similarly, it is found that a very wide variety of actual propagation channels are well-characterized by appropriate choices of R_b and n. Appropriate values of n are found to be in the range 2–5. Sometimes, a better fit is obtained using multiple breakpoint distances with the intervening distances each being assigned the appropriate path loss exponent. To be clear, this approach has order-of-magnitude accuracy at best; on the other hand, differences between more sophisticated models and actual propagation measurements are often this large. Thus, this approach is sufficient in many applications.

EXAMPLE 3.14

The link from a base station antenna to a mobile user operating a compact handheld phone operates at 1900 MHz. The mobile is 10 km from the base station. The propagation path is well-characterized by Equation (3.74) with $\overline{L_p}(R_b) = 128.3$ dB, $R_b = 1$ km, and $n = 3.2$. What is the mean path loss at the mobile?

Solution: From Equation (3.74) we have $\overline{L_p}(10 \text{ km}) = 160.3$ dB.

Location-Specific Path Loss Models. Sometimes the power law approach is unable to capture aspects of the environment that have a big impact on path loss. A prominent example can be found in the engineering of broadcast radio and television systems. Such systems use very high antennas, resulting in radio horizon distances on the order of tens to hundreds of kilometers. In this case attenuation and scattering by terrain are important considerations, so geography must be considered. In the location-specific approach, one gathers details about the geometry and material composition of the buildings and terrain at a specific site and along a particular path, and then the problem is solved as an electromagnetic scattering problem. Since practical scenarios are typically too complex to analyze rigorously, various simplifications are made to make the calculation tractable. A popular model using this approach is the *Longley–Rice model* (also known as the *irregular terrain model* (ITM)).[8]

Empirical path loss models. An empirical model consists of a mathematical function which is fit to measurements. There is no expectation that the resulting model will yield accurate estimates of $\overline{L_p}$ on a case-by-case basis; however the model will typically be good enough on a statistical basis to permit informed design of specific radio links and to facilitate engineering analyses of radio links generally. Popular empirical models include the Hata model [22], the COST231 extension to the Hata model [36], and the Lee model [41]. Appendix A provides an introduction to empirical path loss modeling, including a detailed description of the Hata model.

There is one additional aspect of propagation in the 30 MHz–6 GHz regime which is important to consider: *slow fading*. Slow fading is additional variation in average path loss with R which is much "slower" than fast (i.e., $|\gamma|$) fading. This variation is normally attributable to shadowing by terrain and large structures, and therefore is not always observed. In environments in which it is prevalent (and in which location-specific modeling is not desirable or possible), the additional fading can be modeled as following a log-normal distribution [5].

3.7 PROPAGATION ABOVE 6 GHZ

The principal difference between propagation at frequencies in the SHF and higher bands, compared to propagation at lower frequencies, is the primacy of path loss in determining link range. The increased path loss is sufficiently large that many systems require high-directivity

[8] The ITM has somewhat diffuse origins. Suggested references include (in chronological order): [78], [46], [28], and [55].

antennas or beamforming systems, and are otherwise limited to relatively short ranges. The increased path loss is attributable to two distinct issues: Reduction in effective aperture of receive antennas (Section 3.7.1), and media losses including atmospheric attenuation and rain fade (Section 3.7.2–3.7.4). While these phenomena certainly exist at all frequencies, they are normally negligible at frequencies below SHF. And, just to be clear, there is nothing special about 6 GHz in this respect: Others could be justified in arguing that some lower or higher frequency represents a more appropriate threshold for this distinction.

Separate from the path loss issue, propagation considerations in this regime are essentially the same as in the 30 MHz–6 GHz regime. Phenomena affecting propagation in mobile radio channels scale as expected with frequency. On the other hand, the use of highly-directive antennas in point-to-point applications limits the potential for scattering from objects not close to the line of sight. This may result in a propagation which is very similar to the free-space case, except perhaps for the additional path loss associated with absorption.

3.7.1 Increased Path Loss Due to Diminished Effective Aperture

This first issue bearing on the problem of path loss at high frequencies has essentially nothing to do with propagation, but instead concerns the characteristics of antennas and, subsequently, the limitations of power amplifiers. Returning to Equations (3.5) and (3.6), recall that path loss is proportional to $(R/\lambda)^2$, and that the dependence on λ is associated with the effective aperture of the receiving antenna.[9] In other words, we say path loss increases with frequency $(= c/\lambda)$ squared because this is the rate at which the effective aperture of the receiving antenna decreases.

In one sense this is merely a consequence of the way we define "path loss," and one could imagine alternative definitions where the frequency dependence of the receive antenna's effective aperture is included elsewhere. Nevertheless, the associated problem is real. For example, even if one is able to implement an antenna having constant gain over frequency,[10] one finds P_R/P_T (the ratio of received power to transmit power) decreases with frequency squared.

To see how this impacts practical radio systems, consider the following example: You obtain acceptable P_R at 6 MHz using $P_T = 100$ mW. Increasing frequency to 60 MHz while keeping the gains of the transmit and receive antennas constant, P_R/P_T is reduced by a factor of 100, so P_T must be increased to 10 W to maintain the same range. Similarly, increasing frequency to 600 MHz requires increasing P_T to 1 kW – already this is impractical in many applications, and explains why UHF systems commonly have less range than VHF systems in comparable applications. Now increasing to 6 GHz, P_T increases to 100 kW, which in many applications is impractically large. If one wishes to operate a 6 GHz communication link using P_T on the order of 100 mW, then the maximum range is probably going to be decreased by a factor of about 100 relative to what could be achieved at 600 MHz, and decreased by a factor of about 10^6 relative to what could be achieved at 6 MHz, using the same transmit power.

9 If this is a surprise, review the derivation in the paragraph preceding Equation (3.5).

10 Not out of the question: For example, this can be done with a suitably-designed horn antenna over a surprisingly large frequency range.

3.7.2 Increased Path Loss Due to Media Losses; Attenuation Rate

Media loss refers to the tendency of the medium to extract power from the radio wave, thereby eroding power density with increasing distance. These losses are observed in all media other than free space. These losses are even observed in air, although the effects are normally not apparent at frequencies in the UHF band and below. In radio systems, the loss mechanisms of primary concern are *atmospheric absorption* and *rain fade*. These mechanisms are sometimes collectively referred to as *extinction*.

Generally, the effect of media loss is to reduce the electric field intensity of a radio wave traveling over a distance R by an additional factor of $e^{-\alpha R}$, where α is the *attenuation constant*. Consequently the reduction in power density is a factor of $(e^{-\alpha R})^2 = e^{-2\alpha R}$. (Be aware that attenuation constant is sometimes defined such that $e^{-\alpha R}$ is the reduction in power density, as opposed to electric field intensity. This is not wrong; however, it is not consistent with undergraduate electromagnetics textbooks, and so is likely to create confusion.)

Attenuation rate is defined as this quantity expressed as a loss in units of dB per unit length; i.e.,

$$10 \log_{10} e^{-2\alpha R} = -8.69\alpha R \text{ dB} \tag{3.75}$$

so, for example,

$$\text{attenuation rate} = 8.69\alpha \text{ dB/km} \tag{3.76}$$

where α has units of km^{-1}. In other words, attenuation rate is the increase in path loss, per unit distance, beyond the path loss that would be expected in ideal free space.

Note that media loss accrues exponentially, and that the exponent is $2\alpha R$. So, the impact on path loss is small when this exponent is less than 1; however, the impact on path loss becomes important when the exponent increases beyond 1, and very quickly overtakes the impact of spreading (i.e., R^n) loss as the exponent becomes much greater than 1. Therefore a rule of thumb for when media loss becomes important is $R > 1/2\alpha$, or equivalently when $\alpha > 1/2R$.

3.7.3 Atmospheric Absorption

An important form of media loss is atmospheric absorption. Absorption in the Earth's atmosphere is dominated by the effect of two constituent gases: water vapor (H_2O) and oxygen (O_2). At UHF and below the associated attenuation rate is $\ll$ 1 dB/km, decreasing with frequency, and is therefore typically negligible. In the SHF and EHF regime this rate can reach a few dB/km when humidity and temperature are sufficiently high, although quiescent rates are typically at least an order of magnitude lower.

In a span of roughly 10 GHz around 60 GHz, O_2 absorption becomes the dominant loss mechanism, increasing attenuation rate to 10–15 dB/km independently of humidity or temperature. The especially high path loss makes this band unsuitable for outdoor communications in most applications. However, this same attribute makes this band attractive for unlicensed indoor wireless networks, where the extraordinarily high path loss can be exploited to reduce interference between nearby systems.

EXAMPLE 3.15

At a range of 300 m, how much additional path loss should one one expect at 60 GHz due to atmospheric absorption?

Solution: At 60 GHz, O_2 absorption is the dominant media loss mechanism. The associated attenuation rate is between 10 dB/km and 15 dB/km. The increase in path loss at a range of 300 m is therefore between 3 dB and 4.5 dB.

A weaker absorption mechanism is present at about 22 GHz, associated with water vapor. The attenuation rate associated with this mechanism peaks at about 0.2 dB/km, which is not much greater than the attenuation rate at frequencies above and below the peak. Nevertheless, this band would not be a wise choice for long-range terrestrial communication.

There is a popular myth that a similar absorption feature exists at 2.4 GHz, associated with water vapor. The myth appears to be associated with the observation that microwave ovens operate at 2.4 GHz; however, the mechanism at work in a microwave oven is *dielectric heating*, not absorption by water.[11]

The characterization of atmospheric absorption in this section leaves out details which may be important in certain applications. For the whole story, [44] is recommended.

3.7.4 Rain Fade

Rain fade is a temporary increase in the apparent attenuation rate due to scattering and absorption by rain along the path of propagation. Like atmospheric absorption, the effect is normally negligible at UHF and below, and emerges into significance under certain conditions in the SHF and higher bands.

The 10–30 GHz band is commonly used for satellite communications. In this band, rain may increase the apparent attenuation rate by 0.1–10 dB/km, depending on frequency and rainfall rate. This accounts for the degraded performance of direct broadcast satellite (DBS) systems, which operate at $\approx$ 12 GHz, in the presence of heavy rain.

A popular empirical model for characterization of rain fade is given in ITU-R Rec. P.828 [30].

3.8 TERRESTRIAL PROPAGATION BELOW 30 MHZ

In the HF band and below, mechanisms for radio propagation are significantly different from those prevalent in the VHF band and above. An important difference is that buildings and terrain features are not necessarily large relative to a wavelength ($\lambda > 10$ m), and therefore tend not to scatter radio waves as effectively as they do at higher frequencies. Another important difference

[11] In fact, the reason microwave ovens operate at 2.4 GHz is to keep them from interfering with licensed radio systems in other bands – the same reason unlicensed WLAN systems operate in that band.

in this frequency regime – noted in Section 3.3.2 – is that the magnitude of the reflection coefficient of the ground may be significantly less than 1. Furthermore, a new possibility emerges: skywave propagation. Let us now sort through these considerations.

We already noted in Section 3.3.2 that the material properties of the Earth's surface cause groundwave propagation to dominate at frequencies $\ll$ 12 MHz. Thus groundwave propagation is important for applications in the MF band and below, such as broadcast AM radio (0.5–1.8 MHz). In this mode of propagation, the $\parallel$ ($\approx$vertical) component of radio wave is bound to the surface of the Earth and the $\perp$ ($\approx$horizontal) component does not propagate effectively; therefore antennas must be vertically polarized. Because the groundwave is bound to the surface of the Earth, it spreads primarily in azimuth and not in elevation, which tends to reduce path loss associated with spreading. However, the wave is partially embedded in the ground, which is behaving as an imperfect conductor, which tends to increase path loss. A more rigorous analysis of groundwave propagation is relatively difficult and somewhat beyond the scope of this text ([42] is recommended).

Groundwave propagation can play a role in the HF band, but is typically not the dominant mechanism. Most of the power radiated by an HF antenna is able to propagate through free space. HF radio waves which follow paths close to the ground are limited by blocking and scattering in ways similar to radio waves at higher frequencies. However, a far more efficient mode of propagation available in the HF band is *skywave propagation*, illustrated in Figure 3.17. In this case, skywave propagation is refraction from the ionosphere. The ionosphere is a region of free electrons 85–600 km above the surface of the Earth, created by the ionization of the atmosphere by radiation from the Sun. The gradient in the electron density of the ionosphere with altitude tends to efficiently refract HF radio waves back toward the Earth. The extreme height of the ionosphere enables propagation with low path loss between points separated by distances far in excess of the radio horizon. Furthermore, returning skywaves may be efficiently reflected by the surface of the Earth, leading to two or more "skips" and greatly increasing range. Under the right conditions, worldwide communication is possibly using just a few watts of transmit power – much less than is used by radio horizon-limited transmitters operating at higher frequencies.

The efficiency of ionospheric refraction is dependent on a number of factors [14]. Usually ionospheric refraction is effective only for frequencies below about 30 MHz, improving

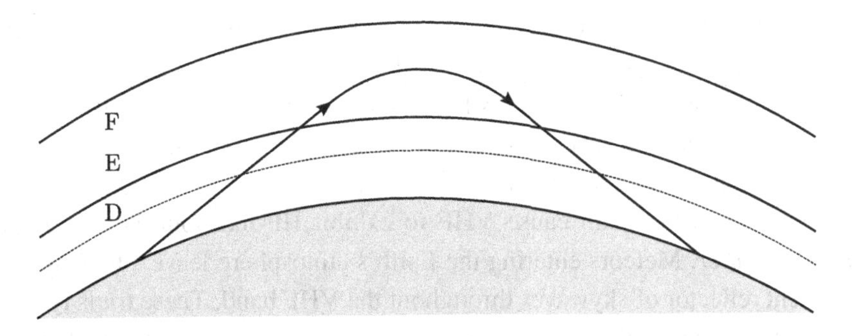

Figure 3.17. Skywave propagation in the HF band. Letters indicated named layers of the ionosphere.

with decreasing frequency. The maximum frequency at which this mechanism is considered effective is known as the *maximum useable frequency* (MUF). The MUF varies with latitude, time of day, season, and phase within the 11-year solar cycle. Over a daily cycle, MUF typically varies from ~10 MHz to ~30 MHz.

The ionospheric skywave channel is normally not quite as simple as Figure 3.17 implies. The electron density in the ionosphere is irregular on many spatial scales, which creates delay spread. Ionospheric electron densities are constantly evolving, so path lengths vary, leading to fading. Thus it is possible and convenient to describe the channel using the same terminology established for terrestrial channels at higher frequencies in Section 3.5. For the ionospheric skywave channel:

- Delay spread typically $\ll$ 1 ms, increasing to $\mathcal{O}$ (10 ms) during times of severe ionospheric disturbance.
- Coherence bandwidth $\sim$ $\mathcal{O}$ (10 kHz), shrinking to $\mathcal{O}$ (100 Hz) during times of severe disturbance.
- When signal bandwidth < coherence bandwidth, fading statistics follow the Rayleigh distribution to a good approximation.
- Coherence time $\sim$ $\mathcal{O}$ (10 min), except around sunset and sunrise, and during other periods of disturbance.

At night (and below the MUF), ionospheric skywave propagation has very low loss, so that path loss is limited primary by spreading, which is not much greater than experienced in free space conditions. During the day, the "D" (lowest) layer of the ionosphere emerges. Unlike the higher (E and F) layers of the ionosphere, the D layer tends to absorb HF radio waves. D-layer absorption may contribute as much as tens of dB to the path loss, but is time-variable and its effect is notoriously difficult to estimate.

3.9 OTHER MECHANISMS FOR RADIO PROPAGATION

There are a few other propagation mechanisms which are important in certain applications.

Occasionally, ionospheric conditions become disturbed in a way that allows skywave propagation to prevail even at VHF frequencies. For example, *sporadic E* ("E_s") is a condition in which a geographically-limited region of the ionosphere is temporarily but dramatically enhanced. When this happens, the low end of the VHF band behaves very similarly to HF below MUF. The MUF for E_s conditions is often as high as 70 MHz, and occasionally much higher. E_s is most common in the northern hemisphere during daylight hours from April through July, and events range in length from a few minutes to a few hours.

Another phenomenon which can cause VHF to exhibit HF-like ionospheric propagation behavior is *meteor scatter*. Meteors entering the Earth's atmosphere leave a trail of ionization which is an efficient reflector of skywaves throughout the VHF band. These trails typically last only for seconds, so the channel created by any one meteor is only briefly available. On the other hand, there is a steady stream of meteors entering the atmosphere, so this propagation mechanism is always available. Meteor scatter is sufficient to allow communications at VHF frequencies over continental distances [71].

Tropospheric scatter is a skywave propagation mechanism in which radio waves in the UHF and lower SHF bands are scattered from the upper leves of the troposphere, roughly 7–20 km above the surface of the Earth depending on latitude. Because the scattering is not very efficient, this method requires high transmit power and antennas with very high directivity (such as reflectors), and range is typically limited to a few hundred kilometers.

Ducting is a somewhat different mechanism of propagation, in which radio waves in the VHF and lower UHF bands become trapped by gradients in the index of refraction between adjacent layers of the atmosphere. This allows long-range propagation following a path of roughly constant height above ground, and with greatly reduced spreading of power in the elevation plane. This mechanism depends on weather and climate, and is only intermittently available. A common symptom of ducting is the ability to receive broadcast FM and television stations at distances many times greater than the radio horizon.

Problems

3.1 Equation (2.11) is the far-field electric field intensity for an antenna consisting of a thin wire oriented along the z-axis. Write this equation in the form of Equation (3.1). What are $\hat{\mathbf{e}}(\theta,\phi)$ and $V_0(\theta,\phi)$?

3.2 A data link is established from an aircraft cruising at 10 km altitude and a satellite directly above it in geosynchronous orbit. The aircraft transmits 10 W into an antenna having a gain of 5 dBi, and the satellite's antenna has a gain of 20 dBi. The center frequency of the transmission is 10 GHz. What is the path loss in dB and the power in dBm delivered to the receiver?

3.3 Derive Equation (3.18).

3.4 Reproduce the result shown in Figure 3.3, but now include the effect of conductivity assuming: (a) $\sigma = 10^{-3}$ S/m at 30 MHz, (b) $\sigma = 10^{-1}$ S/m at 30 MHz, (c) $\sigma = 10^{-3}$ S/m at 300 MHz, and (d) $\sigma = 10^{-1}$ S/m at 300 MHz. Summarize: How good is the assumption that the ground is a lossless dielectric in the VHF band?

3.5 A vertically-polarized 2 GHz radio wave is incident on a 10 m × 10 m brick wall from a cellular base station 1 km distant. Assume the brick has $\epsilon_r = 4$ and $\sigma = 0.4$ S/m. (a) Confirm that it is reasonable to model this scenario using the methodology of Section 3.3.1. (b) What is the minimum and maximum possible loss (in dB) incurred in the resulting reflection, and under what conditions do they occur? Here "loss" means power density of the reflected field relative to power density of the incident field, both at the point of reflection.

3.6 In Section 3.4 it is assumed that $\Gamma_\perp = \Gamma_\parallel = -1$ because $\theta^i \cong \pi/2$, and that variation of the elevation plane pattern of the transmit antenna is negligible over the angular separation between the two paths. How reasonable are these assumptions? Consider a scenario in which the antennas are vertically-polarized half-wave dipoles mounted at heights of 10 m and 1 m. (a) Find the minimum separation between antennas for which $\theta^i \geq 85°$. (b) Find the minimum separation between antennas for which the variation in the antenna pattern is less than 1 dB.

3.7 A 10 km radio link consisting of antennas at heights of 30 m and 3 m just barely achieves the minimum required received power. (a) Determine whether it is reasonable to analyze this scenario using the methods of Section 3.4.2. (b) What must be the height of the currently 30 m antenna in order to achieve the minimum required received power at a range of 25 km? (c) Explain why this result does not depend on frequency.

3.8 Consider a link consisting of antennas, both mounted at heights of 5 m, operating at 915 MHz. What is the breakpoint distance?

3.9 In Example 3.9, N was arbitrarily chosen to be 10. What's the minimum value of N for which a good approximation to a Rayleigh fading distribution can be expected? (a) As a reference, repeat Example 3.9 and reproduce Figures 3.12 and 3.13 for $N = 10$, and overlay the simulated CDF with a plot of Equation (3.68). (b) Repeat for $N = 2$, and comment. (c) Repeat for the minimum value of N that yields reasonable agreement between the simulated and Rayleigh CDFs.

3.10 In a Rayleigh flat fading channel, what fraction of the time can the incident power density be assumed to be (a) below the mean value, (b) less than 15 dB below the mean value, (c) greater than 27 dB below the mean value?

3.11 Land mobile radio (LMR) systems operate at frequencies of roughly 30 MHz to 1 GHz, and LMR radios can be expected to move at speeds from 1 m/s to 30 m/s. Calculate the coherence time for each of the four extreme cases (minimum frequency and minimum speed, minimum frequency and maximum speed, etc.) assuming $\mu = 5$.

3.12 A UHF television antenna is mounted at a height of 150 m. Assume flat terrain. (a) What is the maximum effective range assuming worst-case antenna height? (b) What is the minimum receive antenna height for a maximum effective range of 60 km?

3.13 The calculation in Example 3.14 didn't depend on frequency. Why not?

3.14 Repeat Example 3.14 assuming (a) free space propagation over the entire path, (b) R^4 path loss dependence after the breakpoint.

3.15 Rain significantly impacts the path loss in direct broadcast satellite (DBS) television systems, which operate around 12 GHz. Through how much rain must a DBS signal propagate in order to be attenuated by 3 dB? Indicate a range of values and the associated assumptions.

4 Noise

4.1 INTRODUCTION

As discussed in Chapter 1, it is noise which ultimately limits the sensitivity of a radio system. Noise is present in a variety of forms, both internal and external to the receiver. This chapter addresses the fundamentals of noise, beginning with some basic theory (Sections 4.2 and 4.3), and continuing with concepts pertaining to the characterization of internal noise (Section 4.4), and external noise (Section 4.5).

4.2 THERMAL NOISE

The atoms and electrons comprising any material experience random displacements with an associated kinetic energy that increases with physical temperature. In an electric circuit, the combined displacement of charge-carrying particles – electrons in particular – is observed as a current. The net difference in electrical potential over some distance can be interpreted as a voltage. The voltage and current resulting from random displacement of charged particles due to temperature is known as *thermal noise*. This specific noise mechanism is also known as *Johnson Noise* or *Johnson–Nyquist Noise*.

Because the displacements of charged particles due to temperature are random, the resulting noise voltage and current waveforms can only be described statistically. The instantaneous magnitude of noise waveforms are found to well-modeled as a Gaussian distribution with zero mean.[1] To determine the relevant characteristics of this distribution, let us consider the example of a resistor.

Figure 4.1(a) shows a Thévenin equivalent circuit[2] for a resistor, consisting of an open-circuit voltage v_n in series with an ideal (noiseless) resistance R. v_n represents the thermal noise generated by the resistor. To learn something about $v_n(t)$, we imagine this circuit being terminated by an ideal load resistor also having ideal resistance R as shown in Figure 4.1(b). The voltage across the load resistor in this circuit is $v_L(t) = v_n(t)/2$, whereas the current flowing through this resistor is $i_L(t) = v_n(t)/2R$. The power dissipated in the load resistor is therefore

[1] This follows fairly directly from the Central Limit Theorem of probability theory.

[2] It is also reasonable to use a Norton (current source) equivalent circuit here, and in fact some physical noise mechanisms more closely resemble the Norton model. For the purposes of this book, however, we can get everything we need from the Thévenin model.

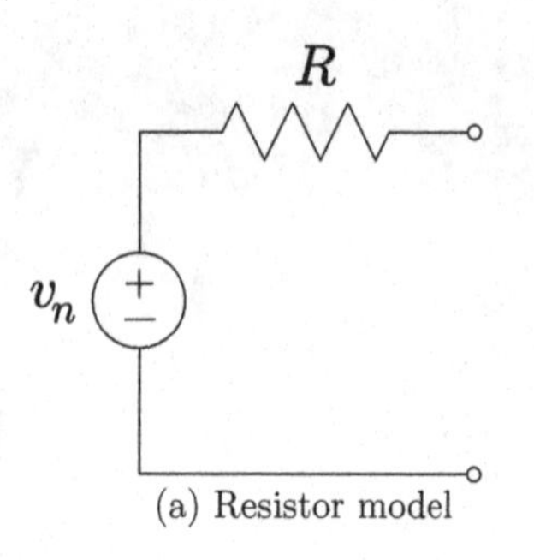

(a) Resistor model

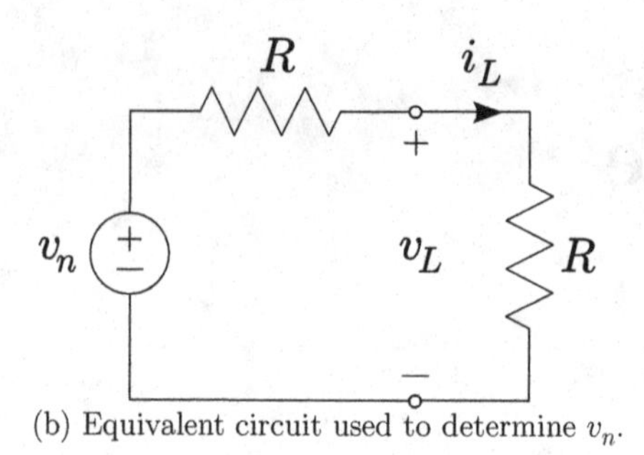

(b) Equivalent circuit used to determine v_n.

Figure 4.1. Thévenin model of a realistic (noisy) resistor.

$$P_L = \lim_{\tau\to\infty} \frac{1}{\tau} \int_t^{t+\tau} v_L(t) i_L(t) dt \tag{4.1}$$

$$= \frac{1}{4R} \left[\lim_{\tau\to\infty} \frac{1}{\tau} \int_t^{t+\tau} v_n^2(t) dt \right] = \frac{\sigma_n^2}{4R} \tag{4.2}$$

where σ_n^2 is (by definition) the variance of the random variable $v_n(t)$.

The power P_L can be determined independently from thermodynamics: If both resistors are at the same physical temperature T_{phys}, then[3]

$$P_L = kT_{phys}B \tag{4.3}$$

where B is bandwidth and k is *Boltzmann's constant*

$$k \cong 1.38 \times 10^{-23} \text{ joules per kelvin (J/K).} \tag{4.4}$$

The joule is unit of energy, whereas the kelvin is unit of temperature; so Boltzmann's constant represents the energy associated with the temperature of the medium, as described in the first paragraph of this section. We seek σ_n, which is not only the square root of σ_n^2, but also the root mean square value of $v_n(t)$. Since Equations (4.2) and (4.3) must be describing the same quantity, we may set them equal and solve for σ_n. We find

$$\sigma_n = \sqrt{4kT_{phys}RB} \tag{4.5}$$

EXAMPLE 4.1

Describe the noise voltage of a 1 kΩ resistor at a physical temperature of 20 °C, assuming a measurement with 1 MHz bandwidth.

Solution: Here, $R = 1000\ \Omega$ and $T_{phys} = 20\ °\text{C} \cong 293$ K. Therefore

$$\sigma_n \cong \left(4.02 \frac{\text{nV}_{rms}}{\sqrt{\text{Hz}}}\right) \sqrt{B} \tag{4.6}$$

If we were to measure the noise voltage in a 1 MHz bandwidth, we would find $\cong 4.02\ \mu\text{V}_{rms}$.

The quantity in parentheses in Equation (4.6) is the *noise voltage spectral density*, which is of interest because it describes the noise voltage purely in terms of physical parameters of the

[3] If we are to be rigorous, it should be noted that this is not exact, but rather is using the Rayleigh–Jeans approximation to Planck's Law. In practice, the associated error is negligible for frequencies up to tens of GHz. Above that, caution about the validity of the approximation is appropriate.

resistor (i.e., R and T_{phys}) and does not depend on quantities which are measurement-related and thus not intrinsic to the resistor (i.e., B).

Finally, we should note that the Thévenin equivalent circuit of Figure 4.1(a) is certainly not the only way to model the thermal noise associated with a passive device. Norton equivalent circuits also have applications, and in some cases it is useful to model noise sources as a combination of series voltage sources and parallel current sources.

4.3 NON-THERMAL NOISE

Thermal noise is certainly not the only type of noise encountered in radio circuits. *Non-thermal noise* is an "umbrella" term for all types of noise which are not thermal in origin. In this section we shall consider two common types of non-thermal noise – namely, *flicker noise* and *shot noise* – and then see how these types of noise can sometimes be represented within the same mathematical framework as thermal noise.

The fundamental origin of flicker noise is the time variability of the voltage–current relationship in electronic devices. Although this relationship is normally assumed to be utterly constant, this is never exactly the case, and so the apparent resistance (being the ratio of voltage to current) is variable. Although all electronic devices suffer from this problem at some level, this effect is particularly apparent in semiconductor devices due primarily to time-varying "clumping" of charge near media junctions. The resulting noise spectrum is "colored," with power spectral density which is highest near DC, and which decreases in proportion to $1/f$ where f is frequency. For this reason, flicker noise is also referred to as "$1/f$" *noise*.

Shot noise is associated with the discrete nature of charge carriers. The description of thermal noise in Section 4.2 assumes continuous distributions of charge. However, the smallest charge encountered is that of a single electron ($\cong 1.6 \times 10^{-19}$ coulomb). Any particular accumulation of charge must consist of an integer number of electrons; thus charge distributions and the associated currents and voltages are never exactly continuous. The discrete nature of noise becomes apparent when very low levels of current are considered. At sufficiently low levels, currents and voltages no longer exhibit Gaussian statistics. Such low levels of current are occasionally encountered in semiconductor electronics.

In the previous section we determined the thermal noise voltage of a resistor. We now consider how we might represent the noise voltage of any one-port device, regardless of the physical origin of the noise. The applicable concept is *equivalent noise temperature*, illustrated in Figure 4.2. The impedance $Z_S = R_S + jX_S$ is the output ("source") impedance of the device, v_S is the open-circuit voltage of the device in the absence of noise, and v_n is the open-circuit noise voltage. Let P_L be the power delivered to a load attached to the device. P_L is maximized when the impedance of the load is conjugate-matched to Z_S; that is, when the load impedance is $Z_S^* = R_S - jX_S$. If we now set $v_S = 0$, P_L is due entirely to v_n. The equivalent noise temperature T_{eq} is now defined as follows:

$$P_L = kT_{eq}B \tag{4.7}$$

Comparing Figures 4.1 and 4.2, we see that T_{eq} can be interpreted as the physical temperature of a resistor having resistance R_S that would deliver the same power P_L to a load impedance

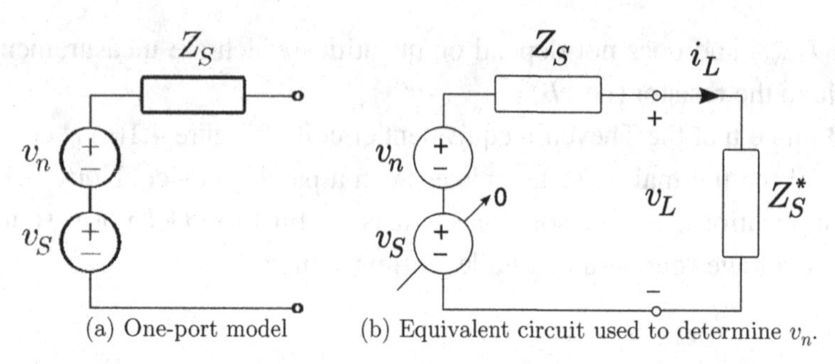

Figure 4.2. Circuit models used to develop the concept of equivalent noise temperature.

equal to R_S. If the device being considered actually *is* a resistor, then $T_{eq} = T_{phys}$. For any other device, the product kT_{eq} is just an alternate way to describe the noise power spectral density P_L/B. In this way, T_{eq} can be used to describe noise which is thermal, non-thermal, or any combination of the two.

EXAMPLE 4.2

A laboratory noise generator is designed to have an output impedance of 50 Ω. The voltage at the output of the noise generator when turned on but unterminated is found to 5 $\text{nV}_{rms}/\sqrt{\text{Hz}}$. What is the equivalent noise temperature of this device?

Solution: Since the noise generator is open-circuited, we have measured $v_n = 5\ \text{nV}_{rms}/\sqrt{\text{Hz}}$. Since $R = 50\ \Omega$, the power spectral density P_L/B delivered to a 50Ω load resistor would be $v_n^2/4R = 1.25 \times 10^{-19}$ W/Hz. Since $P_L/B = kT_{eq}$, $T_{eq} = 9058$ K.[4]

Note that equivalent noise temperature is not necessarily related in any particular way to physical temperature: If the physical temperature of the noise generator in the above example were equal to its equivalent temperature, then the device would be instantly incinerated! Only for thermal noise is $T_{eq} = T_{phys}$.

Before moving on, one comment about the generality of this approach to quantifying non-thermal noise. For this approach to be valid, the noise must be approximately white and approximately ergodic, as is the Gaussian noise associated with thermal noise. By "approximately white" we mean the variation in noise power spectral density should be small within the bandwidth of interest. This does not mean that "colored" noise such as flicker noise cannot be modeled in this way; it simply means that the bandwidth of interest must be small enough that the "$1/f$" trend does not significantly affect the analysis. By "approximately ergodic" we mean that the statistical distribution of v_n should be approximately constant with time. For example, noise which arrives in pulses separated by relatively less-noisy intervals is not appropriately described in terms of an equivalent temperature, except within the duration of the pulse or for periods much longer than the duration of a pulse. For most applications in radio engineering at the system level, ergodic white noise tends to dominate over other ("defective") forms of noise, so the equivalent temperature approach is widely used.

[4] A note to those who might be inclined to try this: This is not the recommended way to determine T_{eq} for RF devices in general, because practical devices might not function properly when open-circuited.

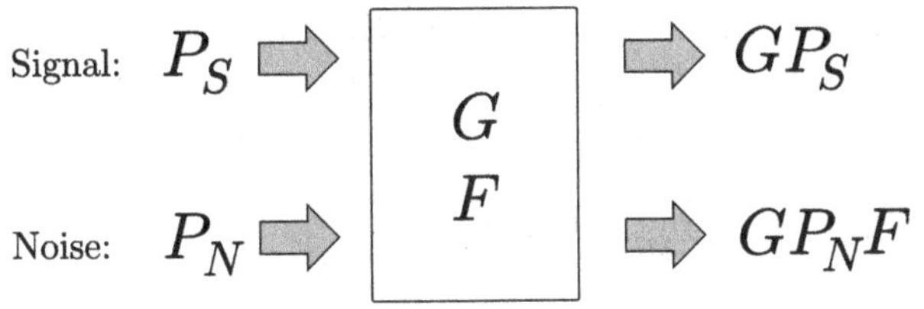

Signal: P_S ⇨ [G / F] ⇨ GP_S

Noise: P_N ⇨ ⇨ $GP_N F$

Figure 4.3. Noise characterization of a two-port device in terms of noise figure.

4.4 NOISE CHARACTERIZATION OF TWO-PORT DEVICES; NOISE FIGURE

In the previous chapter we considered the noise characteristics of one-port devices; that is, devices that have one output and no inputs. We now consider the characterization of noise associated with devices with two ports; that is, an input and and output. We shall demonstrate that the "noisiness" of such devices is aptly characterized in terms of the decrease in the signal-to-noise ratio (SNR) from input to output.

4.4.1 Single Two-Port Devices

Our model is shown in Figure 4.3. In this model, the input is perfectly impedance-matched to a source and the output is perfectly impedance-matched to a load, such that there is a unidirectional flow of power from left to right with no reflections. The gain of the two-port is G, which may be either ≥ 1 (normally indicating an amplifier or other active device) or ≤ 1 (normally indicating a passive device having loss). Under these conditions, input power P_S applied by the source results in output power GP_S delivered to the load. We also have the possibility of noise arriving at the input of the two-port, with associated noise power P_N. If the two-port contributed no noise, the noise power delivered to the load would be GP_N. We choose to model the noise contribution of the two-port as a factor $F \geq 1$, resulting in actual noise power $GP_N F$ delivered to the load. Using these definitions, the SNR delivered to the input of the two-port is

$$\text{SNR}_{in} = \frac{P_S}{P_N} \tag{4.8}$$

the SNR delivered to the load is

$$\text{SNR}_{out} = \frac{GP_S}{GP_N F} = \frac{\text{SNR}_{in}}{F} \tag{4.9}$$

and thus we find

$$F = \frac{\text{SNR}_{in}}{\text{SNR}_{out}} \tag{4.10}$$

The quantity F is formally known as the *noise factor*. Noise factor is customarily expressed in dB, in which case it usually referred to as *noise figure*.[5]

[5] It is admittedly a bit odd that the same quantity should have different names depending on whether it is expressed in linear units or dB; nevertheless, this is the custom. It is also common (perhaps even customary) to refer to the *concept* of F generally as "noise figure," as I do in this textbook.

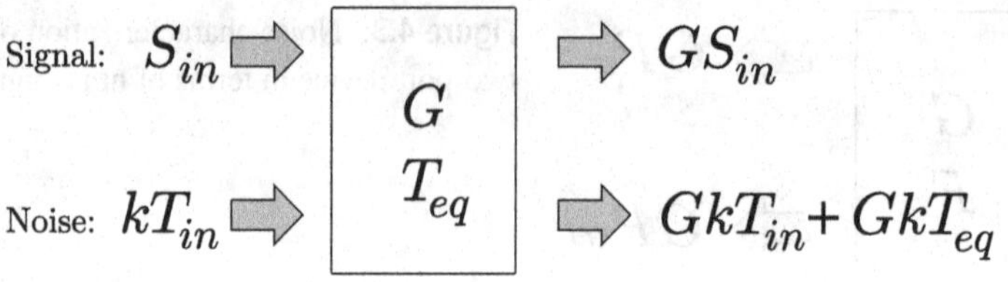

Figure 4.4. Noise characterization of a two-port device, now in terms of noise temperature.

It is important to understand the relationship between noise figure and equivalent noise temperature, as defined in the previous section. Figure 4.4 shows the same two-port device under the same conditions, but now interpreted in terms of noise temperature. Because noise temperature is most closely related to power spectral density (i.e., W/Hz) as opposed to power (W), all power quantities are now represented as power spectral densities. The input signal power P_N is replaced by the power density S_{in}, which becomes GS_{in} at the output. The input input noise power P_N is replaced by the equivalent noise power density kT_{in} which becomes GkT_{in} at the output. The noise contributed by the device is represented in terms of the equivalent noise temperature T_{eq}, which is applied at the input and thus appears as GkT_{eq} at the output. Since we are representing the noise contribution of the device as noise added at the input, the quantity T_{eq} is known as the *input-referred equivalent noise temperature* of the device.

Now we can once again determine the SNRs at the input and output of the device, and subsequently the noise figure. The SNR delivered to the input of the two-port is

$$\text{SNR}_{in} = \frac{S_{in}}{kT_{in}} \tag{4.11}$$

the SNR delivered to the output is

$$\text{SNR}_{out} = \frac{GS_{in}}{G(kT_{in} + T_{eq})} = \frac{S_{in}}{k(T_{in} + T_{eq})} \tag{4.12}$$

and we find

$$F = \frac{\text{SNR}_{in}}{\text{SNR}_{out}} = \frac{T_{in} + T_{eq}}{T_{in}} = 1 + \frac{T_{eq}}{T_{in}} \tag{4.13}$$

Thus we see that F can be described in terms of T_{eq}, but only if T_{in} is specified. This gives rise to a minor quandary. On one hand, the input-referred equivalent noise temperature is a reasonable and complete description of the "noisiness" of a two-port device. On the other hand, noise figure frequently turns out to be more convenient, but is not a complete description of the noisiness of a two-port because it depends on the input noise power. For this reason, it is customary to define noise figure not in terms of the *actual* input noise power density, but rather in terms of a *reference* noise power density. In other words, the noise figure is normally defined instead as follows:

$$F = 1 + \frac{T_{eq}}{T_0} \tag{4.14}$$

where T_0 is a *reference equivalent noise temperature*, or simply *reference temperature*. It should be emphasized that the choice of T_0 is both intrinsic to the definition of noise figure, and utterly arbitrary. Most often T_0 is selected to be 290 K, which is the standard value promulgated by the Institute of Electrical & Electronics Engineers (IEEE). However, T_0 is sometimes alternatively specified to be 293 K and – to be clear – there is no "wrong" value. The potential for ambiguity in T_0 makes it good practice to state the value of T_0 in use whenever noise figure is employed.

The reader should take note that the choice of T_0 is not only arbitrary, but also utterly unrelated to physical temperature. This is often confusing to the uninitiated, perhaps because 290 K $\approx$ 17 °C $\approx$ 62 °F is a value that is not too far from room temperature (albeit a bit on the brisk side). This probably also explains the enthusiasm for selecting $T_0 = 293$ K $\approx$ 20 °C $\approx$ 68 °F, which gives a nice round number in Celsius and which might be considered "room temperature" in a typical laboratory.

EXAMPLE 4.3

An amplifier has a noise temperature of 250 K. What is the amplifier's noise figure?

Solution: Presumably 250 K is the amplifier's input-referred equivalent noise temperature. If we take T_0 to be 290 K, then

$$F = 1 + \frac{T_{eq}}{T_0} = 1 + \frac{250\ K}{290\ K} = 1.86 = 2.7 \text{ dB} \tag{4.15}$$

In other words, the noise figure is 2.7 dB for a reference temperature of 290 K.

An important special case of the above analysis concerns passive devices with two ports, such as transmission lines which have only loss. In this case, the additional noise contributed by the device is Johnson noise. Figure 4.4 applies, although $G < 1$. The noise power spectral density delivered to the load is still $Gk(T_{in} + T_{eq})$; however, in this particular case we also know that the output noise power spectral density must be equal to kT_{phys}, assuming the two-port is in thermodynamic equilibrium with its surroundings. (This is the same argument we employed in Section 4.2 in order to determine the noise voltage of a resistor.) Therefore

$$Gk(T_{in} + T_{eq}) = kT_{phys} \tag{4.16}$$

Solving for T_{eq} we find

$$T_{eq} = \frac{1}{G}T_{phys} - T_{in} \tag{4.17}$$

The noise figure of a passive lossy two-port is obtained by substituting Equation (4.17) into Equation (4.14). One obtains:

$$F = \frac{1}{G}\frac{T_{phys}}{T_{in}} \tag{4.18}$$

If we choose T_{in} to be equal to the physical temperature, then

$$F = \frac{1}{G} \quad (\text{when } T_{in} = T_{phys}) \tag{4.19}$$

Now we see a possible motivation for setting the reference temperature equal to the physical temperature: This makes the noise factor of a lossy passive two-port equal to its loss.

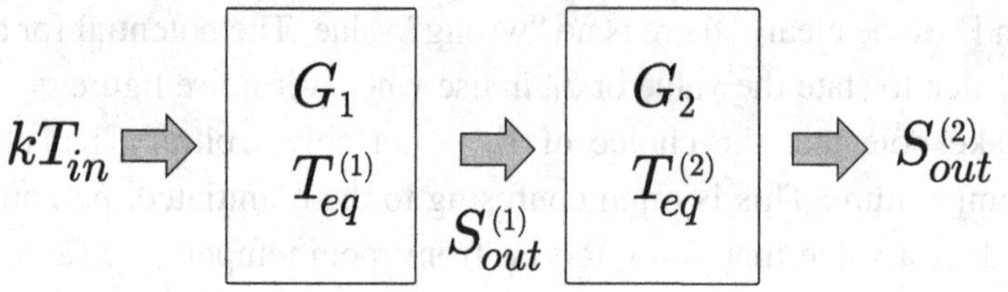

Figure 4.5. A two-stage cascade to be analyzed in terms of noise temperature.

EXAMPLE 4.4

Consider a coaxial cable which has a loss of 2 dB and a physical temperature of 19 °C (66 °F). What is its noise figure if T_{in} is equal to (a) T_0 (the IEEE standard value of 290 K) and (b) the physical temperature?

Solution: A loss of 2 dB means $G = -2$ dB $= 0.631$, and 19 °C = 292.1 K. For $T_{in} = 290$ K, we have $F = 1.60 = 2.03$ dB. For $T_{in} = T_{phys}$, we have $F = 1/G = 2.00$ dB.

The small difference found in the above example demonstrates that Equation (4.19) is a pretty good approximation for a wide range of physical temperatures around what most people would consider to be room temperature. However, electronics tend to get very hot; so in each new application it is a good idea to consider whether T_{phys} is sufficiently close to T_0 for the approximation to apply.

4.4.2 Cascades of Two-Port Devices

Practical radio systems consist of cascades of two-port devices connected in series. Therefore it is useful to have a method to determine the noise characteristics of a cascade from the noise characteristics of its constituent two-ports. A particularly simple derivation of the noise figure of a two-stage cascade begins by determining the noise temperature of the cascade.[6] The scenario is shown in Figure 4.5, which is a simple extension the single-stage problem first considered in Figure 4.4. In the earlier single-stage problem, the input noise spectral density kT_{in} appears at the output as a noise spectral density

$$S_{out} = kT_{in}G + kT_{eq}G = k\left(T_{in} + T_{eq}\right)G \tag{4.20}$$

In the two-stage problem, this is the output from the first stage:

$$S_{out}^{(1)} = kT_{in}G_1 + kT_{eq}^{(1)}G_1 = k\left(T_{in} + T_{eq}^{(1)}\right)G_1 \tag{4.21}$$

where $T_{eq}^{(1)}$ is the equivalent noise temperature of the first stage. This becomes the input to the second stage. Assuming perfect impedance matching the output of the second stage is

$$S_{out}^{(2)} = S_{out}^{(1)}G_2 + kT_{eq}^{(2)}G_2 = k\left(T_{in} + T_{eq}^{(1)} + \frac{T_{eq}^{(2)}}{G_1}\right)G_1G_2 \tag{4.22}$$

[6] The advantage of going this route is that it avoids the hassle of having to consider a "signal" – which is extraneous anyway, since the noise figure does not depend on signal level.

Comparing Equations (4.20) and (4.22), we find that the noise temperature of the cascade is simply

$$T_{eq} = T_{eq}^{(1)} + \frac{T_{eq}^{(2)}}{G_1} \tag{4.23}$$

Already, there are two profound conclusions to be made: (1) The noise temperature of a cascade cannot be less than the noise temperature of the first stage, and (2) the contribution of a subsequent stage can be made arbitrarily small by making the gain of the preceding stage sufficiently large.

Driving on to the goal of an expression for the cascade noise figure, we recall that

$$F = 1 + \frac{T_{eq}}{T_0} \tag{4.24}$$

where T_0 is an arbitrarily determined reference noise temperature. Setting $T_0 = T_{in}$ and substituting Equation (4.23), we obtain

$$F = 1 + \frac{T_{eq}^{(1)} + T_{eq}^{(2)}/G_1}{T_0} = F_1 + \frac{T_{eq}^{(2)}}{T_0 G_1} \tag{4.25}$$

Applying Equation (4.24) to the second term above, we find

$$F = F_1 + \frac{F_2 - 1}{G_1} \tag{4.26}$$

which is the expression we seek. Once again, we find that the noise performance of the cascade cannot be better than the noise performance of the first stage. The noise figure of cascades having more than two stages can be determined by repeated application of the two-stage expression. The resulting expression is:

$$F = F_1 + \frac{F_2 - 1}{G_1} + \frac{F_3 - 1}{G_1 G_2} + \frac{F_4 - 1}{G_1 G_2 G_3} + \cdots \tag{4.27}$$

EXAMPLE 4.5

The Mini-Circuits GALI-74 monolithic microwave integrated circuit (MMIC) amplifier has gain 25.0 dB and noise figure 2.7 dB at the frequency of interest. The Mini-Circuits GVA-81 MMIC amplifier has gain 10.5 dB and noise figure 7.4 dB at the same frequency. Input and output impedances are 50 Ω throughout. When used in cascade, what is the combined noise figure?

Solution: Here, G_1 = 25.0 dB = 316, F_1 = 2.7 dB = 1.86, and F_2 = 7.4 dB = 5.50. From Equation (4.26) we find F = 1.88 = 2.73 dB. Note F is only slightly degraded by the large noise figure of the second stage, thanks to the large gain of the first stage.

EXAMPLE 4.6

In the above problem, a long coaxial cable is inserted between the amplifiers. The loss of the cable is 4.1 dB. Now what is the cascade noise figure?

Solution: The "gain" of the cable is G_2 = −4.1 dB = 0.39, and we assume impedances are matched at both ends. If we assume the physical temperature of the cable is equal to the reference temperature (e.g., 290 K), then the noise figure of the cable is $F_2 = 1/G_2 = 4.1$ dB = 2.57. Now we have what we need to work the problem using either Equation (4.27) or by two iterations of Equation (4.26). Either way, we find $F = 1.90 = 2.8$ dB. Remarkably, the cascade noise figure is only 0.1 dB worse than the noise figure of the first amplifier.

4.5 EXTERNAL NOISE

In this section we consider sources of noise that originate from outside the radio and its antenna. Such sources fall into two categories: anthropogenic and natural. Anthropogenic noise is associated with technology, emanating from machines, electronic devices, power lines, and so on. Natural noise is a feature of the environment independent from anthropogenic contributions, and which includes thermal and non-thermal components from a variety of sources. Let us first consider the concept of antenna temperature, which is a convenient metric for characterization of external noise that can be modeled as thermal.

4.5.1 Antenna Temperature

In Section 4.2 (specifically, in Figure 4.1) we observed that a resistor could be modeled as a noise voltage source in series with an ideal resistance. In Section 2.3.1 (specifically, in Figure 2.6) we observed that an antenna could be modeled as a voltage source v_A representing the received radio wave in series with the antenna impedance Z_A. Combining these ideas, we obtain a model for the noise captured by an antenna, shown in Figure 4.6. The antenna is modeled as a resistance R_{rad} (the radiation resistance, as defined in Chapter 2) whose physical temperature is, in this case, represented by the quantity T_A. We refer to T_A as the *antenna temperature*. If we terminate the antenna into a conjugate-matched load Z_A^* in thermodynamic equilibrium, then the power spectral density delivered to the load is kT_A.

Thus a formal definition is as follows: T_A is the physical temperature of a resistor which delivers a power spectral density to a matched load equal to that delivered by a specified antenna to a conjugate-matched load. A somewhat simpler but equivalent definition is that kT_A is the power spectral density due to external noise that is delivered by an antenna to a conjugate-matched load.

Implicit in the antenna temperature concept is the assumption that external noise is either thermal in origin, or non-thermal but well-modeled as thermal noise. This is not the first time we have made this kind of assumption – recall from Section 4.3 that one is often justified in

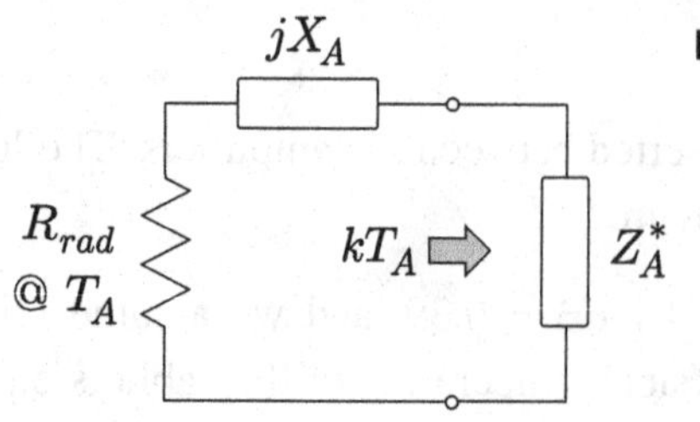

Figure 4.6. Antenna temperature concept.

treating non-thermal internal noise as if it were thermal. However, the same caveats apply, and there are certain situations where the non-thermal origins of external noise become apparent and may have a significant effect.

Before moving on, let us consider one such example with the goal of developing the ability to recognize when the "thermal noise assumption" might not be reasonable. An example of non-thermal radio noise which may or may not be well-modeled as thermal is the noise from electrical power lines. Carefully installed and well-maintained power lines produce negligible radio noise.[7] Over time, however, the devices used to insulate the power line from other conducting materials degrade, resulting in *arcing*, the generation of closely-spaced radio pulses having duration on the order of microseconds. The bandwidth of a radio pulse having a length of 1 μs is on the order of 1 MHz. Therefore, a receiver having a bandwidth much less than 1 MHz "smears" the pulses beyond recognition, making them appear – to a good approximation – to be a continuous and ergodic source of noise. In this case, power line noise is well-modeled as an increase in antenna temperature. On the other hand, a receiver having a bandwidth much greater than 1 MHz does not smear the pulses, in which case the power line source appears as an pulsating interfering signal, and is not well-modeled as an increase in antenna temperature.

Other types of impulsive noise that can also be viewed as either an increase in antenna temperature or as a distinct interfering signal – depending on the bandwidth of the receiver – include internal combustion engines (spark plug discharges) and many forms of digital electronics (clock signals, other periodic or quasi-periodic signals).

Before moving on, it should be noted that antenna temperature can be alternatively represented as a noise figure. The relationship between antenna temperature and antenna noise figure is exactly the same as the relationship between equivalent noise temperature and noise figure for internal electronics, as addressed in Section 4.4. Just as with internal electronics, it is necessary to specify a reference temperature T_0 in order to calculate the antenna noise figure. Just as in the case of electronic devices, the reference temperature used to calculate antenna noise figure is utterly unrelated to the physical temperature. In particular – and as we shall see below – we have no particular expectation that the antenna temperature should be 290 K, even if the physical temperature of the antenna is 290 K.

4.5.2 Natural Sources of Noise

Sources of radio noise that can be important in radio systems are summarized in Table 4.1. These sources can be broadly categorized as either well-modeled as thermal noise, or not. Sources which can be well-modeled as thermal noise can be characterized in terms of a *brightness temperature* T_B. Brightness temperature can be defined as the antenna temperature measured by an antenna with infinite directivity and which is pointed at the source of the emission. To a good approximation, T_B is equal to the antenna temperature measured by a high-gain antenna whose beamwidth is less than the angular extent of the source. In general, the relationship between antenna temperature and brightness temperature is

[7] Neglecting here the fact that power lines are sometimes also used for data communications, as in "broadband over power line" (BPL) systems. BPL systems *do* generate significant levels of radio noise, which we are not considering here.

Table 4.1. **Natural sources of noise that may affect radio systems, arranged by increasing distance from a receiver. The brightness temperature T_B is indicated for sources which can be well-modeled as thermal.**

Source	T_B	Frequencies	Remarks
Terrain	< 300 K	All	Thermal
Lightning		VHF and below	Infrequent
Water vapor	1–300 K	UHF and above	Thermal
Atmospherics		MF and below	Infrequent
Moon	≅ 200 K	All	Thermal; typ. negligible T_A
Quiet sun	10^3–10^6 K	All	Thermal; typ. negligible T_A
Solar flare		All	Infrequent
Jupiter		10–40 MHz	Infrequent
Galaxy	See Eq. (4.29)	All	Synchrotron radiation
CMB	2.7 K	All	

$$T_A = \frac{1}{4\pi} \int_{\theta=0}^{\pi} \int_{\phi=0}^{2\pi} T_B(\theta,\phi)\, D(\theta,\phi) \sin\theta\, d\theta\, d\phi \tag{4.28}$$

where $D(\theta,\phi)$ is the directivity of the antenna and $T_B(\theta,\phi)$ is the brightness temperature in the same direction.

Sources which are well-modeled as thermal noise include terrain, atmospheric gases, the Moon, the "quiet" sun, Galactic "synchrotron" radiation, and the cosmic microwave background (CMB). The noise from terrain is truly thermal, and T_B is different from the physical temperature by an amount that depends on angle of incidence and polarization. Thus, terrain can be a significant source of noise. Noise from atmospheric gases is due primarily to absorption by water vapor (H_2O), and depends on frequency, zenith angle, and environmental factors. Noise from the Moon and quiescent Sun is also thermal, and the latter is of course spectacularly hot; however, both Moon and Sun are so small in angular extent ($\approx 0.5°$ as viewed from the surface of the Earth) compared with the beamwidth of most antennas that they usually contribute negligibly to T_A. The Galactic synchrotron background is the sum radio emission from relativistically-accelerated electrons throughout our Galaxy. Unlike lunar and solar emission, Galactic emission arrives from all directions on the sky. The equivalent brightness temperature of this emission is quite high; for low-gain antennas the associated antenna temperature can be roughly estimated by

$$T_A \approx 800\ \text{K} \cdot \left(\frac{f}{100\ \text{MHz}}\right)^{-2.55},\ \text{for}\ f \geq 5\ \text{MHz} \tag{4.29}$$

with some variation depending on time of day; i.e., which parts of the Galaxy are above the horizon. Finally, the CMB is radiation originating from the sudden transition of the Universe from a uniform plasma to a gaseous state about 300 000 years after the Big Bang. This radiation becomes important relative to the Galactic synchrotron "foreground" at frequencies above ~1 GHz, and is actually the dominant source of external noise for some applications in this frequency regime.

Natural sources which are not well-modeled as thermal noise include lightning, atmospherics, solar flares, and Jupiter "decametric" emission. Each of these sources is bursty with complex time-domain and frequency-domain behaviors, making the distinction been noise and interference ambiguous in each case. These sources are also quite infrequent; for example, the time between significantly-strong solar flares may be days or longer. Of these sources, lightning probably has the greatest significance for radio communications. The noise from a single nearby lightning strike is significant at all frequencies. However, at frequencies in the VHF range and below where radio noise from thunderstorms can travel over great distances, lightning noise can be a nearly continuous source of significant noise.

4.5.3 Anthropogenic Sources of Noise

In radio systems utilizing low-gain antennas, anthropogenic noise can be modeled as an antenna temperature. As we shall see in Chapter 7, this antenna temperature is often relevant in the design of radio systems, and it is useful to have some idea of what to expect. For this purpose the International Telecommunications Union (ITU) provides a widely-used empirically-derived model for anthropogenic noise, which can be stated as follows:[8]

$$T_A = b\left(\frac{f}{1\ \mathrm{MHz}}\right)^a \ \mathrm{K} \tag{4.30}$$

where f is frequency and a and b are given in Table 4.2. The resulting antenna temperature vs. frequency curves are shown in Figure 4.7. In this scheme, T_A is a fit to measurements of the median antenna temperature conducted at locations deemed to be of the indicated types. Worth noting is the significant variation in T_A depending on location type, and the variation of T_A with frequency, which is both dramatic and approximately independent of locale. Some limitations of this scheme should be noted. First, the uncertainties in b are quite large; i.e., an order of magnitude or more. A significant contribution to this uncertainty is the variation of T_A as a function of position within a given locale. Furthermore, this model assumes low-gain omnidirectional antennas at ground level, and specifically does not account for variation in T_A with antenna height. Another consideration is that the models given in Table 4.2 are based on measurements made during the 1970s [74]. Although there is no particular reason to expect a to change, whether b is significantly changing (either increasing or decreasing) over time

[8] This is the noise factor model from ITU-R R.372-10 [29], expressed as an antenna temperature in order to remove the dependence on an arbitrary reference noise temperature.

Table 4.2. Empirically-derived models for antenna temperature associated with anthropogenic noise (see Equation (4.30)); f_1 and f_2 are the lowest and highest frequencies at which each model is considered to be valid.

Environment	b K·MHz^{-a}	a	f_1 MHz	f_2 MHz	T_A at 30 MHz
City	1.39×10^{10}	−2.77	0.3	250	1 130 000 K
Residential	5.16×10^{9}	−2.77	0.3	250	418 000 K
Rural	1.52×10^{9}	−2.77	0.3	250	123 000 K
Quiet Rural	6.64×10^{7}	−2.86	0.3	30	3960 K

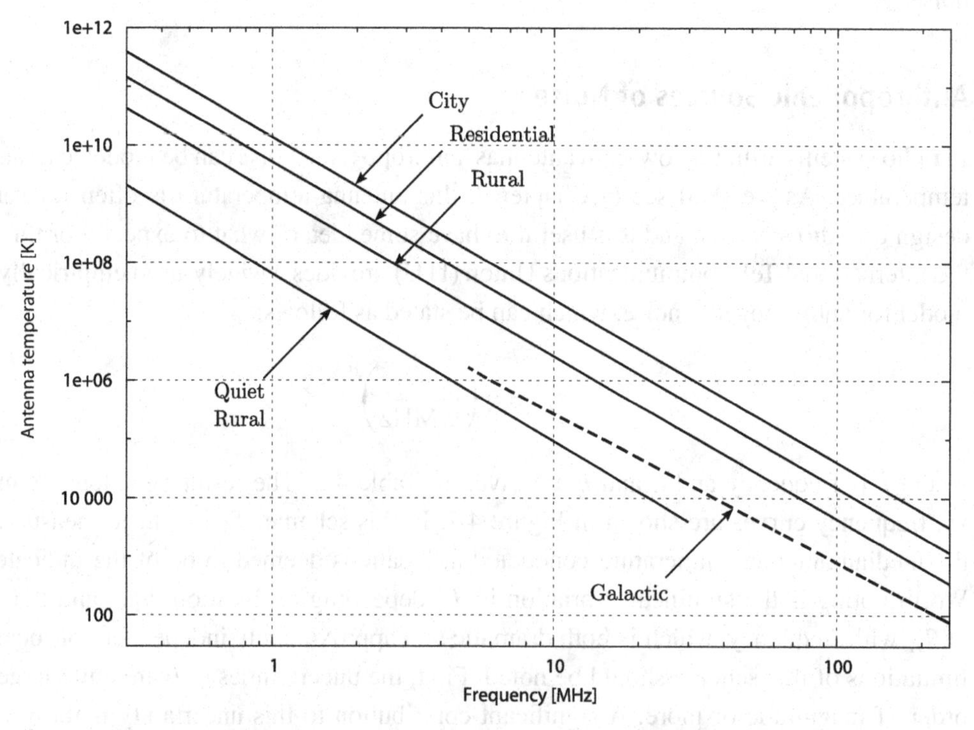

Figure 4.7. Anthropogenic noise modeled as median antenna temperature using the parameters given in Table 4.2. Also shown is the Galactic background antenna temperature ($b = 4.60 \times 10^7$ K·MHz^{-a}, $a = -2.3$, computed from Equation (4.29)), for comparison.

due to changes in technology is a subject of debate, with only very sparse data available as evidence. For all of these reasons, it best to view the models as crude estimates; certainly not an alternative to measurements if accurate values are required, but probably suitable as rough guidelines for the purpose of setting system specifications and requirements.

Problems

4.1 Determine the rms noise voltage spectral density of a 47 kΩ resistor at a physical temperature of 25 °C. Then, calculate the rms noise voltage in a bandwidth of 250 kHz.

4.2 A laboratory noise generator produces an equivalent noise temperature of 30 000 K into a 75 Ω termination. What is the power spectral density delivered to the termination? Characterize the statistics (distribution, mean, rms, variance) of voltage samples measured across the termination.

4.3 What is the input-referred equivalent noise temperature of an amplifier having a gain of 20 dB and noise figure of 7.3 dB? What is the power spectral density of the noise produced by the amplifier, as measured at its output? Assume $T_0 = 290$ K.

4.4 Consider a coaxial cable which has a loss of 13 dB. The physical temperature of the cable and everything attached to it is 25 °C. The input is terminated into a matched load. What is noise power spectral density delivered to a matched output? To the best of your ability to determine, what is the noise figure? What is the noise figure if T_{in} is assumed to be 290 K?

4.5 In Example 4.6, what is the maximum cable loss for which the cascade noise figure is less than 3 dB?

4.6 Repeat (a) Example 4.5 and (b) Example 4.6 using *only* noise temperatures. In each case, calculate the input-referred noise temperature and show that it is consistent with the noise figures computed in the example.

4.7 Consider a low-gain antenna having a directivity of $\cong$ 3 dB over the entire sky, and identically zero elsewhere. What is the expected contribution of the Sun to the antenna temperature? The Sun is about 0.5° across, as viewed from the surface of the Earth. Repeat the calculation for an antenna having a beamwidth of 5°.

4.8 Calculate the expected antenna temperatures for all four categories of anthropogenic noise indicated in Table 4.2, as well as Galactic noise (as a fifth category), at 88 MHz (close to the low end of the US FM broadcast band). For comparison, determine the noise figure of amplifiers that produce the same input-referred noise power spectral density, assuming $T_0 = 290$ K.

5 Analog Modulation

5.1 INTRODUCTION

Modulation may be defined generally as the representation of information in a form suitable for transmission. One might define *analog modulation* as the modulation of analog information, and predominantly voice and other audio signals. However, this is a bit vague since *digital modulation* – the topic of the next chapter – may be used for precisely the same purpose. A possible distinction between analog and digital modulation might be made in the following way: Digital modulation requires the information to be transmitted to be represented in digital form at some point in the modulation process, whereas this is not required in analog modulation. Therefore, analog modulation might be best defined simply as modulation that is not better described as digital modulation.

This chapter provides a brief but reasonably complete introduction to analog modulation as it is used in modern communications systems. First, some preliminaries: Section 5.2 provides a brief overview of the overarching concept of sinusoidal carrier modulation. Then Sections 5.3 and 5.4 introduce the concept of complex baseband representation, which provides useful insight into the analysis of sinusoidal carrier modulation, including the effects of propagation and noise. Most commonly-used analog modulations may be classified as either amplitude modulation, single sideband modulation, or frequency modulation; these are described in Sections 5.5, 5.6, and 5.7, respectively. This chapter concludes with a brief discussion of techniques which are employed to improve the quality of audio signals in analog modulation systems.

5.2 SINUSOIDAL CARRIER MODULATION

Generally speaking, modulation schemes (now referring to both analog and digital modulations) fall predominantly into just two classes: *baseband modulation* and *sinusoidal carrier modulation*. Examples of baseband modulation include *pulse modulation* – used primarily in wired communication systems such as Ethernet – and *impulse modulation* – the basis for most ultrawideband (UWB) wireless communications systems. In impulse modulation, information is represented by the spacing between impulses, and the impulses are extremely narrow so as to make the bandwidth of the signal extraordinarily large. Present-day UWB impulse radio systems are limited to short-range, low-power applications beyond the scope of this book; the reader is referred to [64] for more information.

Table 5.1. **Representation of common analog sinusoidal carrier modulations in the form $A(t)\cos[\omega_c t + \phi(t)]$. $m(t)$ is the message signal. A_c, k_a, k_p, and k_f are constants.**

Modulation	$A(t)$	$\phi(t)$	Classifications
DSB-LC AM	$A_c[1 + k_a m(t)]$	0	Linear
DSB-SC AM	$A_c m(t)$	0	Linear
SSB	AM with suppressed sideband		Linear, Quadrature
VSB	AM with partial sideband		Linear, Quadrature
PM	A_c	$k_p m(t)$	Angle
FM	A_c	$2\pi k_f \int_{t_0}^{t} m(\tau)d\tau$	Angle

Sinusoidal carrier modulation is employed by the vast majority of radio communications systems. The idea is to convey information by manipulating the magnitude and/or phase of a sinusoidal "carrier". All such modulations have the form:

$$s_{RF}(t) = A(t)\cos[\omega_c t + \phi(t)] \tag{5.1}$$

where $\omega_c = 2\pi f_c$ is the *carrier frequency*, $A(t)$ is the magnitude, and $\phi(t)$ is phase in excess of that associated with the $\omega_c t$ term. Table 5.1 shows how various sinusoidal carrier modulations are represented in this form. The vast majority of digital modulations are also representable in this form, as we shall see in Chapter 6.

5.3 COMPLEX BASEBAND REPRESENTATION

Equation (5.1) can be written equivalently as:

$$s_{RF}(t) = \mathcal{R}e\left\{\left[A(t)e^{j\phi(t)}\right]e^{j\omega_c t}\right\} \tag{5.2}$$

as can readily be verified by application of *Euler's Formula*:

$$e^{j\theta} = \cos\theta + j\sin\theta \tag{5.3}$$

The factor inside the square brackets of Equation (5.2) is commonly referred to as the *complex baseband* representation of $s_{RF}(t)$:[1]

$$s(t) \equiv A(t)e^{j\phi(t)} \tag{5.4}$$

Although not strictly necessary, this representation has two characteristics that make it very compelling. First, $s(t)$ describes behavior of the RF signal which is determined exclusively by

[1] A note on terminology: This is also sometimes referred to as *baseband representation*; i.e., without the qualifier "complex". From a systems engineering perspective, this is sometimes ambiguous because the term "baseband" can also refer to the part of a radio that performs modulation or demodulation, regardless of the representation of the signal. Thus, the qualifier "complex" here reminds us that we are referring to a signal representation, and not merely a functional block within a radio.

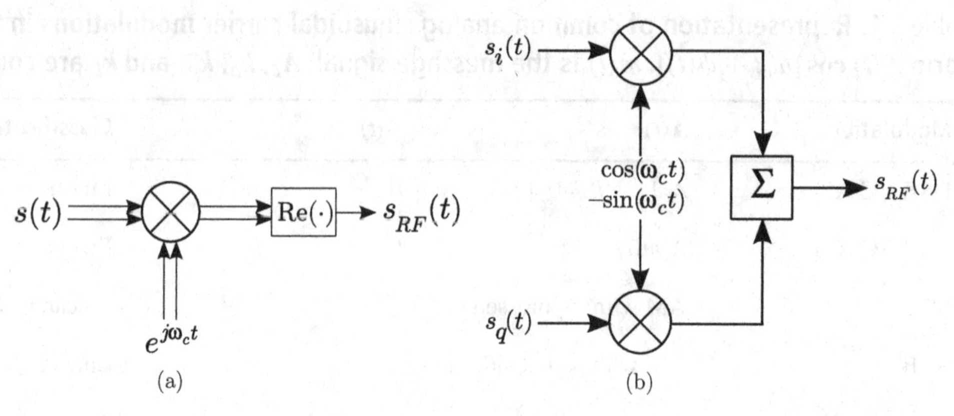

Figure 5.1. Equivalent representations of the conversion of the complex baseband signal $s(t)$ to $s_{RF}(t)$. (a) From Equation (5.2) (double lines are used to indicate the in-phase and quadrature components of complex-valued signals); (b) from Equation (5.6).

the modulation, separate from the contribution of the carrier. This greatly simplifies the analysis of sinusoidal carrier modulations.[2] Second, this representation suggests an implementation approach for transmitters and receivers that is of great practical utility.

Before elaborating on these advantages, let us consider some basic concepts and "how to" for complex baseband representation. First, we recognize that Equation (5.4) is a phasor written in polar form. This expression may alternatively be written in Cartesian form:

$$s(t) = s_i(t) + js_q(t) \tag{5.5}$$

where the real-valued functions $s_i(t)$ and $s_q(t)$ are referred to as the *in-phase* ("I") and *quadrature* ("Q") components of $s(t)$. Substitution of this form of the phasor into Equation (5.2) followed by application of Euler's Formula yields the following expression:

$$s_{RF}(t) = s_i(t)\cos\omega_c t - js_q(t)\sin\omega_c t \tag{5.6}$$

which in turn suggests the implementation shown in Figure 5.1(b).

Recovery of $s(t)$ from $s_{RF}(t)$ is accomplished as shown in Figure 5.2. This can be confirmed by first working out the output of the complex multiplier in Figure 5.2(a):

$$s_{RF}(t) \cdot 2e^{-j\omega_c t} = A(t)\cos\left[\omega_c t + \phi(t)\right] \cdot 2\left(\cos\omega_c t - j\sin\omega_c t\right)$$

Applying basic trigonometric identities one obtains

$$= A(t) \cdot 2\left[\frac{1}{2}\cos\phi(t) + j\frac{1}{2}\sin\phi(t)\right] + \text{terms having frequency } 2\omega_c$$

The terms centered at $2\omega_c$ are eliminated by the lowpass filters. Combining the remaining terms, we obtain

$$A(t)e^{+j\phi(t)} = s(t)$$

as expected.

Now we return to the question of why complex baseband representation is worth the trouble. First, ask yourself what is the easiest way to determine $A(t)$ for a sinusoidal carrier modulation.

[2] It is ironic that something called "complex" is invoked to simplify analysis. Nevertheless, it is true!

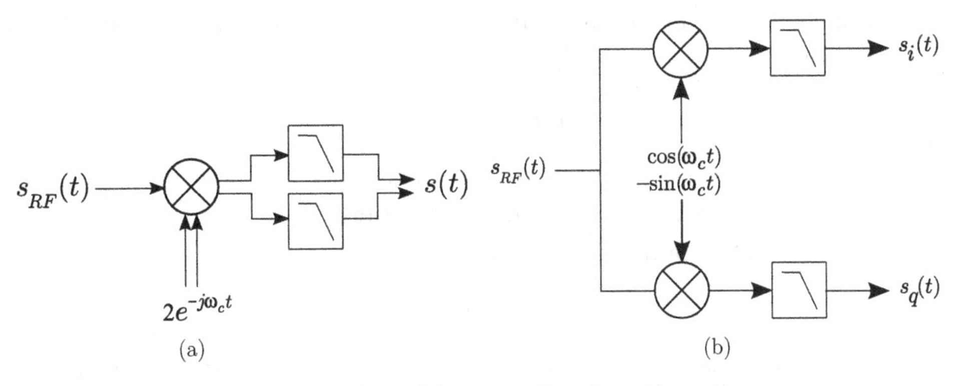

Figure 5.2. Equivalent representations of the conversion of $s_{RF}(t)$ to $s(t)$.

Using $s_{RF}(t)$, you need to observe the signal for long enough to accurately identify a peak, or to collect samples so that you can solve for $A(t)$ from the data. Using $s(t)$, however, $A(t)$ can be determined from just one sample as follows:

$$A(t) = |s(t)| = \sqrt{s_i^2(t) + s_q^2(t)} \tag{5.7}$$

Similarly, determination of $\phi(t)$ from $s_{RF}(t)$ requires collecting a sufficient number of samples with appropriate spacing and duration to unambiguously solve for $\phi(t)$; whereas determination of $\phi(t)$ from $s(t)$ is possible in just one sample, as follows:

$$\phi(t) = \angle s(t) = \arctan\left(s_q(t), s_i(t)\right) \tag{5.8}$$

where "arctan(y, x)" is the four-quadrant inverse tangent function. These and other properties of complex baseband signals not only simplify analysis, but also shed some valuable insight into behaviors of signals and the systems that manipulate them.

The potential advantage of complex baseband representation as an implementation method is suggested by Figures 5.1(b) and 5.2(b). Figure 5.1(b) indicates that RF signals may be first generated as in-phase and quadrature signals centered at DC, which are subsequently "upconverted" to the desired center frequency. Figure 5.2(b) indicates that RF signals may be processed by first generating the in-phase and quadrature components of the associated baseband signal. As an implementation technique, the method suggested by Figures 5.1 and 5.2 is known as *direct conversion*, and we shall have much more to say about this in Chapters 14, 15, and 17.

5.4 COMPLEX BASEBAND REPRESENTATION OF NOISE

In order for complex baseband representation to be truly useful for analysis, it is important to know how noise in complex baseband representation relates to noise in the physical RF signal. Let $n_{RF}(t)$ be the noise at RF, and let us assume it is zero mean. The power associated with $n_{RF}(t)$ is simply the variance of $n_{RF}(t)$:

$$\sigma_{n,RF}^2 \equiv \left\langle n_{RF}^2(t) \right\rangle \equiv \lim_{\tau \to \infty} \frac{1}{\tau} \int_t^{t+\tau} n_{RF}^2(t)dt \tag{5.9}$$

where the angle brackets denote averaging over time as indicated above. Following Figure 5.2, the noise power of the in-phase component of the output of the multiplier is

$$\left\langle [n_{RF}(t) \cdot 2\cos\omega_c t]^2 \right\rangle = 2\sigma^2_{n,RF} \tag{5.10}$$

as can be readily verified by application of the appropriate trigonometric identity followed by integration, which eliminates the sinusoidal term. Similarly, the noise power of the quadrature component of the output of the multiplier is

$$\left\langle [n_{RF}(t) \cdot (-2\sin\omega_c t)]^2 \right\rangle = 2\sigma^2_{n,RF} \tag{5.11}$$

Finally, we note that the in-phase component of this noise is uncorrelated with the quadrature component of this noise; that is,

$$\lim_{\tau\to\infty} \frac{1}{\tau} \int_t^{t+\tau} [n_{RF}(t) \cdot 2\cos\omega_c t]\,[n_{RF}(t) \cdot (-2\sin\omega_c t)]\,dt = 0 \tag{5.12}$$

which is readily verified by application of the appropriate trigonometric identity and evaluating the integral. The total noise at the input to the lowpass filters in Figure 5.2, accounting for both the in-phase and quadrature components, is therefore $2\sigma^2_{n,RF} + 2\sigma^2_{n,RF} = 4\sigma^2_{n,RF}$. This is because the power of the sum of uncorrelated signals is simply the sum of the powers of the signals.

The effect of the lowpass filter is to reduce the noise power by a factor of two, since the lowpass filter removes the terms associated with $2\omega_c$ which is half the total power. Summarizing: (1) The noise powers in the in-phase and quadrature components of the complex baseband signal are equal to each other and $\sigma^2_{n,RF}$; (2) the in-phase and quadrature components of the noise are uncorrelated; and (3) the total noise power in the complex baseband signal is $\sigma^2_{n,RF} + \sigma^2_{n,RF} = 2\sigma^2_{n,RF}$.

5.5 AMPLITUDE MODULATION (AM)

AM is a class of analog modulations in which information is represented by variation in the magnitude of the carrier. Confusingly, "AM" is also the name commonly used to refer to a particular type of amplitude modulation, alternatively referred to *conventional AM*, *double-sideband transmitted carrier (DSB-TC) AM*, or *double-sideband large carrier (DSB-LC) AM*. All three terms mean precisely the same thing.

AM is sometimes referred to as a *linear modulation*, in the sense that the modulated signal is, in some sense, proportional to the message signal. This has consequences for the design of amplifiers used in AM transmission, as we shall see in Chapter 17.

5.5.1 Modulation and Spectrum

For "conventional" (DSB-LC) AM, the RF signal may be written

$$s_{RF}(t) = A_c\,[1 + k_a m(t)]\cos\omega_c t \tag{5.13}$$

where $m(t)$ is the message (e.g., audio) signal and k_a is the *AM modulation index*. The modulation index is typically set so that the maximum value of $|k_a m(t)|$ is somewhere in

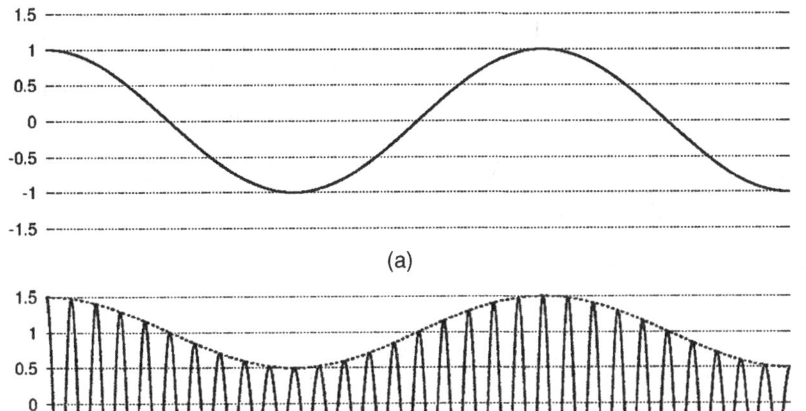

Figure 5.3. DSB-LC AM waveform $s_{RF}(t)$ for $m(t) = \cos(\omega_m t)$ and $k_a = 0.5$: (a) $m(t)$, (b) $s_{RF}(t)$, with the envelope indicated by a dashed line.

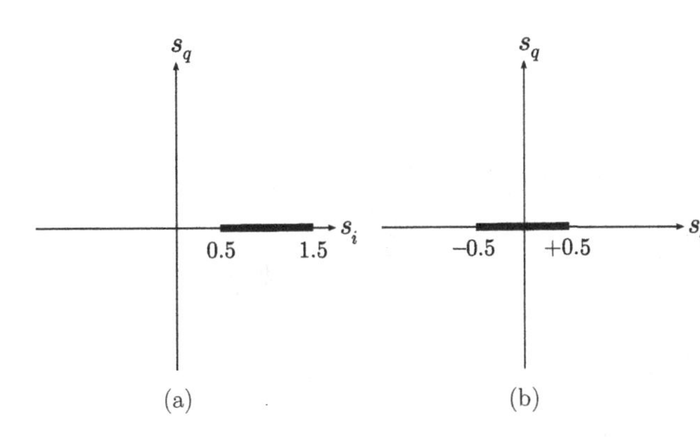

Figure 5.4. Representation of AM signals in complex baseband form in the complex plane: (a) the DSB-LC signal in Figure 5.3, (b) the DSB-SC signal in Figure 5.6.

the range 0.5–1, resulting in the waveform shown in Figure 5.3(b). The complex baseband representation of AM is simply

$$s(t) = A_c\left[1 + k_a m(t)\right] \tag{5.14}$$

which is depicted graphically in Figure 5.4(a).

An important consideration for any modulation is its spectrum. The spectrum is determined by taking the Fourier transform of Equation (5.13):

$$\begin{aligned} S_{RF}(\omega) &= \mathcal{F}\left\{A_c\left[1 + k_a m(t)\right]\cos\omega_c t\right\} \\ &= \mathcal{F}\left\{A_c \cos\omega_c t + A_c k_a m(t)\cos\omega_c t\right\} \\ &= \frac{1}{2}A_c\delta\left(\omega - \omega_c\right) + \frac{1}{2}A_c\delta\left(\omega + \omega_c\right) + \frac{1}{2}A_c k_a M\left(\omega - \omega_c\right) + \frac{1}{2}A_c k_a M\left(\omega + \omega_c\right) \end{aligned} \tag{5.15}$$

where $M(\omega)$ is the spectrum of $m(t)$; i.e., $\mathcal{F}\{m(t)\}$. This is illustrated in Figure 5.5. Since $s_{RF}(t)$ is real-valued, $S_{RF}(-\omega) = S^*_{RF}(+\omega)$ (i.e., is "conjugate symmetric") and therefore it suffices to show only the positive frequencies.

At this point two key observations can be made. First, the bandwidth B of the AM signal is twice the lowpass bandwidth W of $m(t)$, i.e., $B = 2W$. It is in this sense that this form of AM is "double-sideband": The signal is composed of both the positive and (redundant, in this

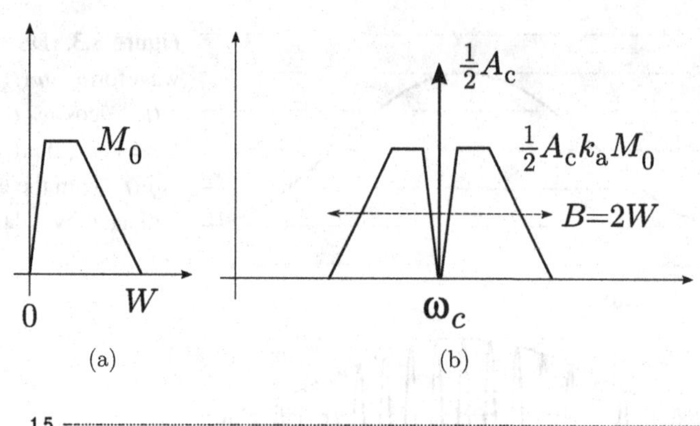

Figure 5.5. DSB-LC spectra. (a) An example message signal spectrum, (b) spectrum of the associated DSB-LC AM signal, $|S_{RF}(\omega)|$.

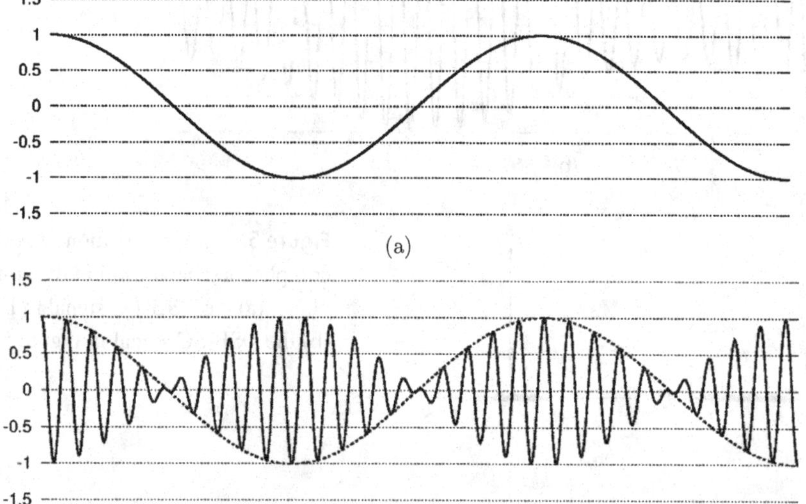

Figure 5.6. DSB-SC AM waveform $s_{RF}(t)$ for $m(t) = \cos(\omega_m t)$: (a) $m(t)$, (b) $s_{RF}(t)$, with the envelope indicated by a dashed line.

case) negative frequencies of $M(\omega)$. The second observation to be made is that a considerable amount of power is present in the form of a tone located at ω_c which is associated with the carrier and unrelated to $m(t)$.

DSB-LC AM is typically used to transmit voice and low-fidelity music, which require $W \approx$ 4 kHz. Therefore DSB-LC AM signals tend to have $B \approx 8$ kHz. DSB-LC is used in AM broadcasting with stations separated by about 10 kHz – somewhat greater than 8 kHz in order to provide sufficient separation to accommodate the poor selectivity of low-cost receivers.

A natural question to ask might be: "Why isn't AM simply

$$s_{RF}(t) = A_c m(t) \cos \omega_c t \tag{5.16}$$

as opposed to having the form shown in Equation (5.13)?" The modulation represented by Equation (5.16) is known as *double-sideband suppressed carrier (DSB-SC) AM*, and has the advantage that 100% of the power of $s_{RF}(t)$ is in $M(\omega)$. A DSB-SC AM waveform is shown in Figure 5.6(b), and the complex baseband representation is shown in Figure 5.4(b). The spectrum is identical to the DSB-LC spectrum shown in Figure 5.5(b) except without the carrier ("$\frac{1}{2}A_c$") term. The short answer to the question is that DSB-LC is easier to demodulate than DSB-SC, which again accommodates simple, low-cost receivers. To understand why DSB-LC is easier to demodulate, it is useful to understand the effect of propagation on modulated signals.

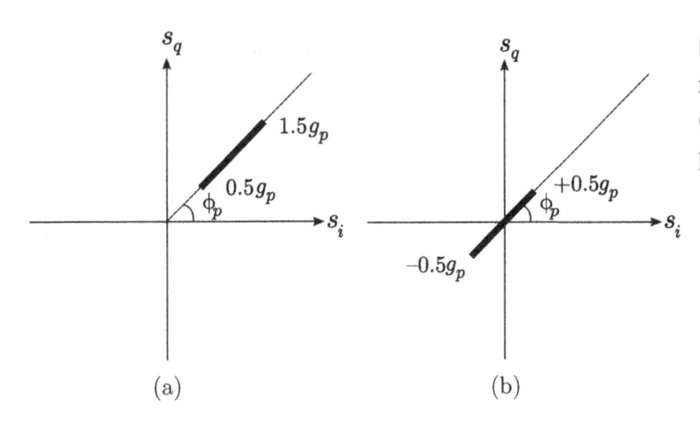

Figure 5.7. Complex baseband representation of received (a) DSB-LC and (b) DSB-SC AM signals, accounting for propagation, for maximum $|k_a m(t)| = 0.5$.

5.5.2 Effect of Propagation

Because AM conveys information as changes in the magnitude of a carrier, it performs poorly in flat fading environments in which the coherence time T_c is comparable to or less than $1/B$. For example: For $B = 8$ kHz, T_c must be $\gg$125 μs; otherwise fast fading might be interpreted as part of the message signal. This must be true even when one or both ends of the link are traveling at high speeds. This has historically limited the use of AM to broadcast applications – in which T_c is very large – and to mobile applications in the MF, HF, and lower VHF bands – where T_c is large because it is proportional to wavelength (see Equation (3.69)). In either scenario the propagation channel is well-modeled as a real-valued positive gain g_p and phase ϕ_p. Thus the received DSB-LC AM signal may be written

$$s_{RF}(t) = g_p A_c \left[1 + k_a m(t)\right] \cos\left(\omega_c t + \phi_p\right) \tag{5.17}$$

or, in complex baseband representation, as

$$s(t) = g_p A_c \left[1 + k_a m(t)\right] e^{j\phi_p} \tag{5.18}$$

with the understanding that g_p and ϕ_p are also functions of time, but varying much more slowly than $m(t)$. The complex baseband representation of this signal is depicted in Figure 5.7(a). The corresponding result for DSB-SC AM is shown in Figure 5.7(b).

5.5.3 Incoherent Demodulation

Demodulation of DSB-LC AM is possible by either *incoherent* or *coherent* methods. We consider incoherent demodulation first, because it is simple to understand, simple to implement, and has continuing applications.

Incoherent demodulation of DSB-LC AM is known specifically as *envelope detection.* Envelope detection applied to a DSB-LC AM signal in complex baseband form is shown in Figure 5.8. The first step is simply to take the absolute value of the signal, which strips off the "$e^{+j\phi_p}$" factor. The resulting signal – the instantaneous magnitude of the complex baseband representation – is known as the *envelope*. Next a DC block is used to remove the carrier term. DC blocking might be as simple as subtracting the mean value of the signal computed over a few cycles of $m(t)$. The signal is lowpass-filtered to exclude noise and spurious signals beyond the desired lowpass bandwidth W. Finally, gain is applied to bring the output to the desired

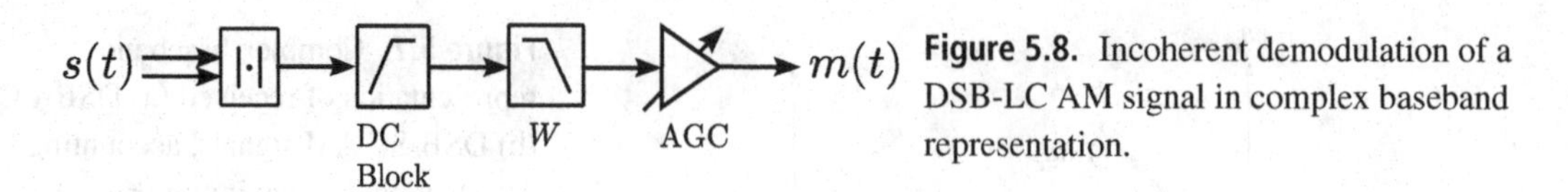

Figure 5.8. Incoherent demodulation of a DSB-LC AM signal in complex baseband representation.

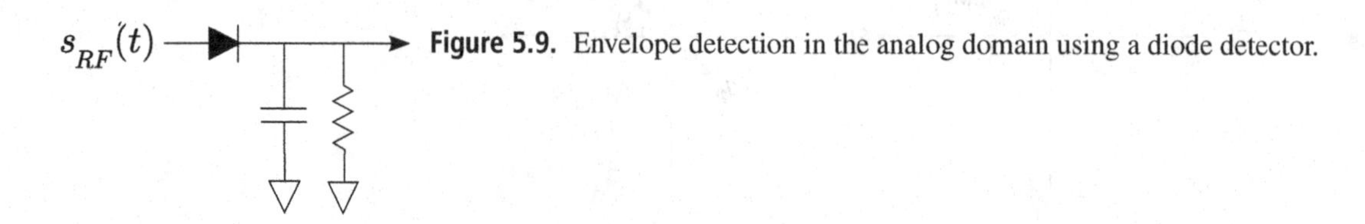

Figure 5.9. Envelope detection in the analog domain using a diode detector.

level while compensating for the variable gain of the propagation channel. This step is a form of *automatic gain control* (AGC); the word "automatic" here conveys the notion that the gain of the amplifier is determined by the power measured at the output of the amplifier. In practice, this function may be implemented as shown, or may be implemented in some earlier stage of the receiver (see Section 15.8), or distributed across both.

It is not necessary to convert the DSC-LC AM signal to complex baseband form to apply envelope detection. If the receiver is to be implemented entirely in the analog domain, it is typically easier to implement the "|·|" block in Figure 5.8 using a *diode detector* as shown in Figure 5.9. In this scheme, a Shottky-type diode is used to rectify the modulated signal – making all values positive – and then an RC-type lowpass filter is used to smooth the result, yielding a reasonable approximation to the envelope. The signal level must be sufficiently large to force the diode into and out of forward-active bias on each cycle of the modulated signal, which requires a voltage on the order of 0.7 V. At the same time, the signal must not be so large as to be distorted by the intrinsically non-linear *I*-*V* characteristic of the diode. Thus, the range of input signals for which the diode detector is effective is limited to within roughly one order of magnitude around 1 V.

Advantages of incoherent demodulation were cited at the beginning of this section. Let us now identify some of the limitations. First, this scheme will not work for *overmodulated* signals; that is, signals for which $|k_a m(t)|$ may be greater than 1. This is because the envelope of overmodulated DSB-LC signals intermittently changes sign, and the sign change is lost when the "|·|" operation is used to detect the envelope. For the same reason, envelope detection cannot be used to demodulate DSB-SC AM, since the envelope changes sign on every zero-crossing of $m(t)$. You can also see this by comparing Figures 5.7(a) and (b): Envelope detection applied to DSB-SC AM will rectify $m(t)$ as well as the associated carrier. Finally, incoherent demodulation is less sensitive than coherent demodulation: That is, for reasonable levels of input SNR, coherent demodulation yields higher output SNR. This is most easily understood as a consequence of the properties of coherent demodulation, which we consider next.

5.5.4 Coherent Demodulation

The optimal scheme for recovery of $m(t)$ from any AM signal (DSB-LC or -SC) is by *coherent demodulation*. Here, we mean "optimal" in the sense described in the last paragraph of the previous section: Coherent demodulation yields the best possible output SNR.

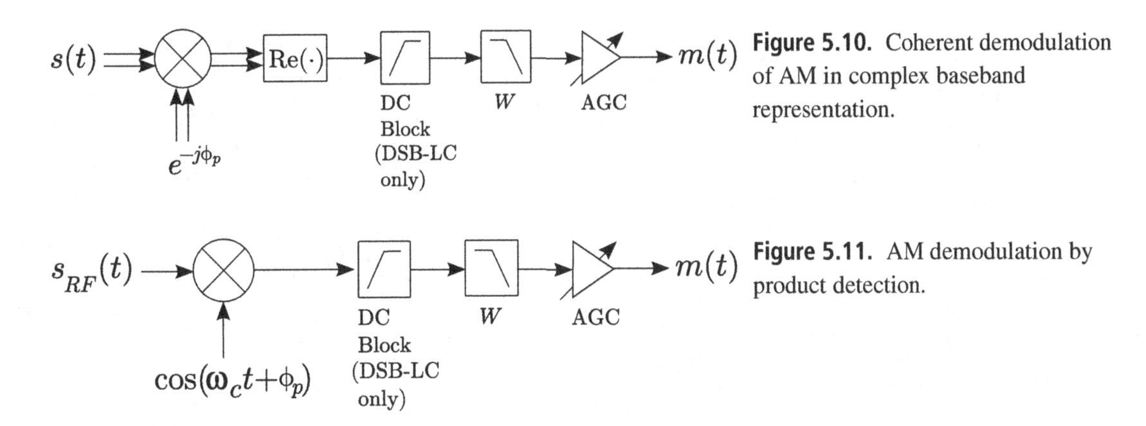

Figure 5.10. Coherent demodulation of AM in complex baseband representation.

Figure 5.11. AM demodulation by product detection.

Coherent demodulation may be implemented on signals in either bandpass or complex baseband form; the implementations are mathematically equivalent. Figure 5.10 shows the implementation in complex baseband representation. We first multiply the input signal by $e^{-j\phi_p}$, which removes the contribution of propagation to the phase of the signal. This is sometimes referred to as *derotation*, since it is essentially a rotation of the signal in the complex (s_i–s_q) plane such that the signal becomes entirely real-valued: The situation shown in Figure 5.7 has been transformed into that shown in Figure 5.4. The matter of how we obtain ϕ_p will be addressed a little later. From this point the process is no different than that employed in envelope detection.

Coherent demodulation of an AM signal in bandpass form is known as *product detection.* An example is shown in Figure 5.11. Here, $s_{RF}(t)$ is multiplied with a tone having frequency ω_c and phase ϕ_p, resulting in two terms: one centered $2\omega_c$, and the desired term centered at $\omega = 0$. The former is removed by lowpass filtering, leaving the latter which is the desired output. Note this is essentially equivalent to the operation shown in Figure 5.2 but excluding the computation of $s_q(t)$, since it is not required.

Finally we consider the matter of how to find ϕ_p. For DSB-LC AM signals in complex baseband representation we see from Figure 5.7(a) that $\phi_p = \arctan\left(s_q, s_i\right)$. Given that $s(t)$ is likely to contain significant noise, it is usually best to do some averaging; i.e.,

$$\hat{\phi}_p \leftarrow \left\langle \arctan\left(x_q(t), x_i(t)\right) \right\rangle \tag{5.19}$$

where $\hat{\phi}_p$ is the desired estimate of ϕ_p, $x(t)$ is the received version of $s(t)$ (i.e., $s(t)$ plus noise), and the integration time should be $\ll T_c$.

Unfortunately this does not work for DSB-SC AM or overmodulated DSB-LC because the phase of the modulation experiences phase shifts of π radians whenever the envelope changes sign. In this case a slight modification to the procedure is required, as shown in Figure 5.12. The squaring operation eliminates the envelope sign reversals that would otherwise confuse the phase estimation. The lowpass filters serve to both smooth the rectified envelope, and to average the noise. Finally, the result of the inverse tangent calculation must be divided by two to account for the phase shift associated with the squaring operation.

The methods described by Equation (5.19) and Figure 5.12 are referred to as *open-loop* estimators. While these methods are simple to implement, they have the disadvantage of being vulnerable to sudden changes in the propagation channel and to the effects of interference.

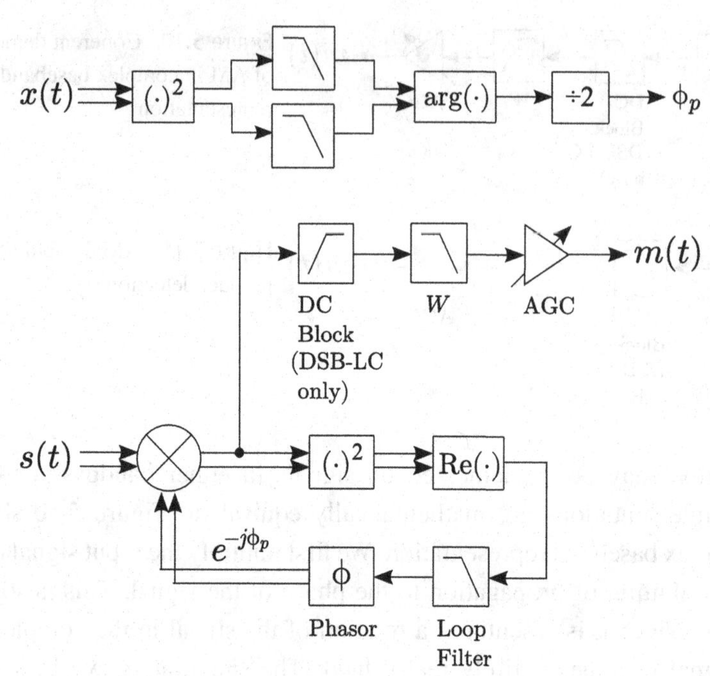

Figure 5.12. Open-loop estimation of carrier phase, suitable for all forms of AM. Here, $x(t) = s(t) \cdot g_p e^{j\phi_p} +$ noise, and "arg(·)" is shorthand for the four-quadrant arctan function.

Figure 5.13. Costas loop method for coherent demodulation of AM, in complex baseband representation.

In this sense, open-loop estimators are not considered to be *robust*. It is often possible to do better in this respect using a feedback approach in which the present estimate of carrier phase is guided by the results of earlier estimates of carrier phase. A popular method that embodies this idea is the *Costas loop demodulator* shown in Figure 5.13. This approach is in some sense a combination of the methods of Figures 5.10 and 5.12, but now the input to the carrier phase estimator is the result of the application of an earlier estimate of ϕ_p to the input signal, and we never explicitly calculate ϕ_p. Instead, we note that the imaginary component of the output of the squaring operation is proportional to $\sin 2(\phi_p - \hat{\phi}_p)$. So, if $|\phi_p - \hat{\phi}_p| < \pi/4$, then this quantity varies monotonically with the carrier phase error and therefore can be used as a phase control signal for the device (the "phasor") which generates the $e^{-j\hat{\phi}_p}$ signal. Specifically, the phasor chooses $\hat{\phi}_p$ so as to minimize future values of the magnitude of this control signal. As long as the error magnitude and rate of change of ϕ_p are not too large, $\hat{\phi}_p$ will converge to ϕ_p and subsequently "track" ϕ_p as it changes.

The Costas loop demodulator is an example of a broad class of devices known as *phase-locked loops* (PLLs). The PLL concept will reappear several times in our discussion of analog and digital modulations, and we'll have more to say about PLLs in Chapter 16. For now, however, it is appropriate to summarize some issues that arise in the design of Costas loop demodulators, and to point out that these issues stem from the underlying PLL architecture. First, we note that if $|\phi_p - \hat{\phi}_p| > \pi/4$, then $\sin 2(\phi_p - \hat{\phi}_p)$ is not going to be a monotonic function of the phase error, and the demodulator will become confused. There are two ways this can happen: At start-up, since the value of ϕ_p is unknown even approximately, and whenever there is a sufficiently large and sustained change in the input signal; for example, due to a sudden change in the propagation channel or due to interference. In either case the demodulator must "acquire" the carrier phase, which is a somewhat different problem than simply tracking the carrier phase. There are a number of ways to implement acquisition, including using an

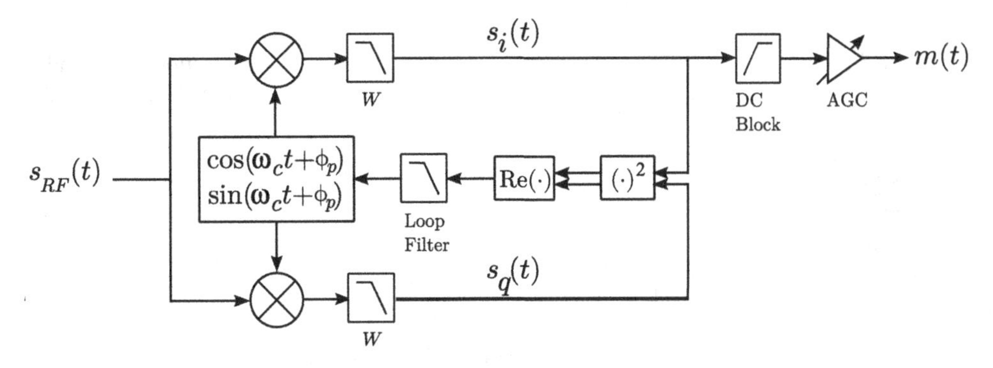

Figure 5.14. Costas loop method for coherent demodulation of AM, operating on the received signal in bandpass form.

open-loop estimate to bootstrap the process. Other matters that must be considered in a PLL-based demodulator have to do with the design of the loop filter and the design of the algorithm that converts the phase control signal into an estimate of ϕ_p. These matters lie somewhat outside the scope of the present chapter, but are considered in a little more detail in Chapter 16.

Finally, we note that the Costas loop demodulator – and coherent demodulation generally – need not necessarily be implemented entirely in the complex baseband domain. Figure 5.14 shows the Costas loop demodulator implemented in "product detector" form, operating directly on the bandpass representation of the received signal. This structure can be interpreted as a conversion to complex baseband representation in which the LO is synthesized by a variable-frequency oscillator, controlled by the baseband phase error estimator. This oscillator is traditionally identified as a *voltage-controlled oscillator* (VCO) or a *numerically-controlled oscillator* (NCO), where the former implies an analog-domain implementation and the latter implies a digital-domain implementation. In either case, the idea is that derotation occurs as part of the conversion to baseband, as opposed to being an operation on a complex baseband signal. Note there is one other distinction in this implementation: The explicit tracking of carrier *frequency* as well as carrier phase. Because frequency is simply the time derivative of phase, control of either phase or frequency implies control of both; however, in practice it is convenient that frequency error contaminating the complex baseband representation is minimized.

5.5.5 Sensitivity of Coherent and Incoherent Demodulation

Now we address the claim made in the previous section that coherent demodulation is more sensitive than incoherent demodulation. This is most easily seen in complex baseband representation. Figure 5.15 shows a single sample of $s(t)$ as it appears at the input of the receiver. Also shown is the associated sample of additive noise $n(t) = n_i(t) + jn_q(t)$, which sums vectorially with $s(t)$ in the complex plane. Note that $n_i(t)$ is in-phase with $s(t)$, whereas $n_q(t)$ is in quadrature with $s(t)$. Also recall from Section 5.4 that $n_i(t)$ and $n_q(t)$ have equal power. A coherent receiver derotates the received signal $s(t) + n(t)$ such that it is aligned with the real axis, and then interprets the signal as the magnitude of the in-phase component. After derotation $n_q(t)$ makes no contribution to this component of the signal, and thus can have no

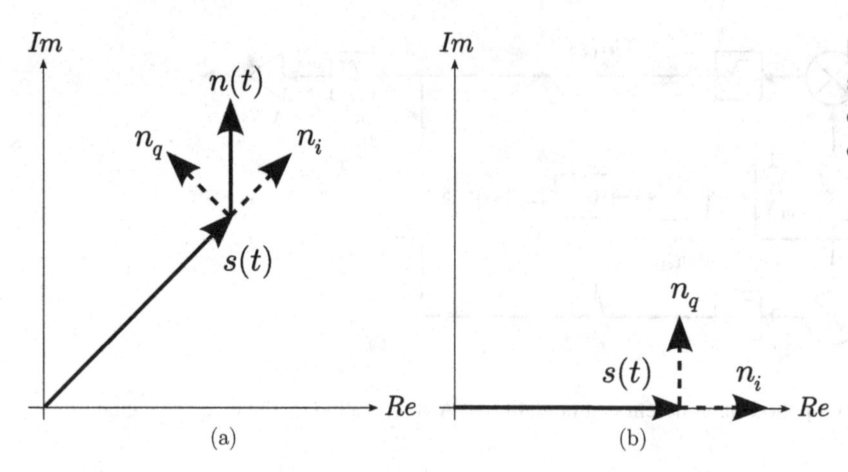

Figure 5.15. (a) Vectorial summation of $s(t)$ and $n(t)$ in the complex plane; (b) $s(t) + n(t)$ after derotation.

effect. Since only the in-phase component of the noise affects the estimated magnitude of the envelope, and since the power in the in-phase and quadrature components of the noise is equal, the output SNR resulting from coherent demodulation is greater than the input SNR by a factor of two for sufficiently large SNR.

In contrast, incoherent demodulation assumes the envelope is $|s(t) + n(t)|$; thus incoherent demodulation is always affected by both components of $n(t)$. Furthermore, the impact of the out-of-phase component of $n(t)$ becomes increasingly important with decreasing input SNR. We are led to the conclusion that if the input SNR is large – say $\geq$10 – then coherent demodulation of DSB-LC is just a little bit better than incoherent demodulation of DSB-LC; however, if the input SNR is small – say $\sim$1 – then coherent demodulation of DSB-LC results in output SNR which is roughly 3 dB greater than incoherent demodulation of DSB-LC.

A different, but related, question is how the sensitivity of DSB-SC compares to DSB-LC. If we employ coherent demodulation in both cases, then the ratio of output SNR to input SNR is the same in both cases, and one might be tempted to assume that they are equally sensitive. However, this conclusion neglects the fact that the carrier in DSB-LC contains no information; i.e., the "signal" is actually entirely contained in the sidebands. So, for example, when DSB-LC and DSB-SC signals arrive at a receiver with equal power and equal noise, the *effective* SNR of DSB-SC is already greater by the ratio of the total power to the power in the sidebands (only) of the DSB-LC signal. Therefore the effective sensitivity for DSB-SC is also greater by this factor – not because of the demodulation, but rather because DSB-SC uses power more efficiently. If we choose incoherent demodulation for the DSB-LC signal, then the sensitivity is degraded by another factor of two or so for the reasons described in the previous paragraph.

While these observations are true enough, precise estimation of output SNR as a function of input SNR is quite difficult owing to practical considerations. For example, coherent estimation depends on correctly and robustly estimating the carrier phase ϕ_p, which is impossible to do perfectly, and which degrades with decreasing SNR and increasing interference. Another complication is the prevalence of audio processing techniques such as preemphasis (see Section 5.8), which generally improve the ratio of output SNR to input SNR. For analog AM, it usually suffices to say that input SNR $\geq$ 10 or so is sufficient for reasonable audio quality, and that useable audio quality can be obtained with SNR as low as about 1 (0 dB) using DSB-SC with coherent demodulation, robust carrier phase estimation, and audio processing.

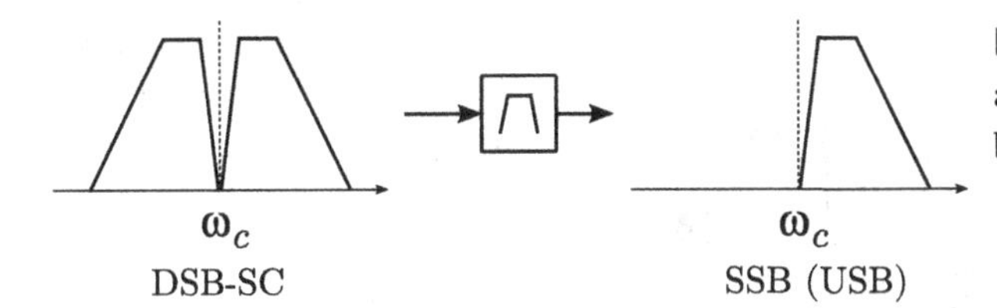

Figure 5.16. SSB interpreted as DSB-SC with a suppressed sideband. Here, generating USB by bandpass-filtering DSB-SC.

5.6 SINGLE SIDEBAND (SSB)

Referring back to Figure 5.5, we see that DSB AM signals (both the -LC and -SC versions) include two versions of the message spectrum: One version appearing at frequencies just above the carrier, and a second, spectrally-reversed version appearing below the carrier. The two versions are referred to as the upper and lower sidebands, respectively. The consequence of sending both sidebands is that the bandwidth of DSB AM is $B = 2M$.

The principle underlying single-sideband (SSB) modulation is to reduce the bandwidth by a factor of two by eliminating one of the redundant sidebands. Thus SSB may be interpreted as DSB-SC AM followed by filtering to pass only the upper or lower sideband, as illustrated in Figure 5.16. These two versions of SSB are known as USB and LSB, respectively.

Aside from reduced bandwidth, an advantage of SSB over DSB-SC or DSB-LC is *energy efficiency*. The term "energy efficiency," when applied to an analog modulation, refers to the SNR achieved at the output of a demodulator compared to the SNR delivered to the input of the demodulator. SSB has potentially twice the energy efficiency of DSB-SC, since all the transmit power in SSB lies within bandwidth W, and so the bandwidth of noise intercepted by a well-designed SSB receiver is also W; whereas for DSB-SC a noise bandwidth of $2W$ is intercepted, resulting in the effective input SNR being a factor of two less than the effective input SNR of SSB when the signal power is the same. This advantage is of course even greater with respect to DSB-LC, since the power in the carrier does not contribute to the output SNR.

SSB is generally preferred over DSB AM for modern analog voice communications, especially in the crowded HF spectrum where narrow bandwidth is a huge advantage. The sections that follow cover the basics of SSB. For additional details, the reader is referred to [70, Ch. 2].

5.6.1 Generation of SSB

SSB can be generated using the method implied in Figure 5.16, but this is not always practical as the required filter must have a very steep transition region in order to achieve sufficient suppression of the undesired sideband without distorting the desired sideband. Such filters are quite difficult to implement, as we shall see in Chapters 13 and 18. This motivates the Weaver and Hartley methods, which we shall now briefly describe.

The *Weaver method* is illustrated in Figures 5.17 and 5.18. We first convert $m(t)$ to complex baseband representation by shifting the spectrum of the message signal by $+W/2$ or $-W/2$, corresponding to LSB and USB respectively. This new signal is $m(t)e^{\pm j(W/2)t}$ in the time

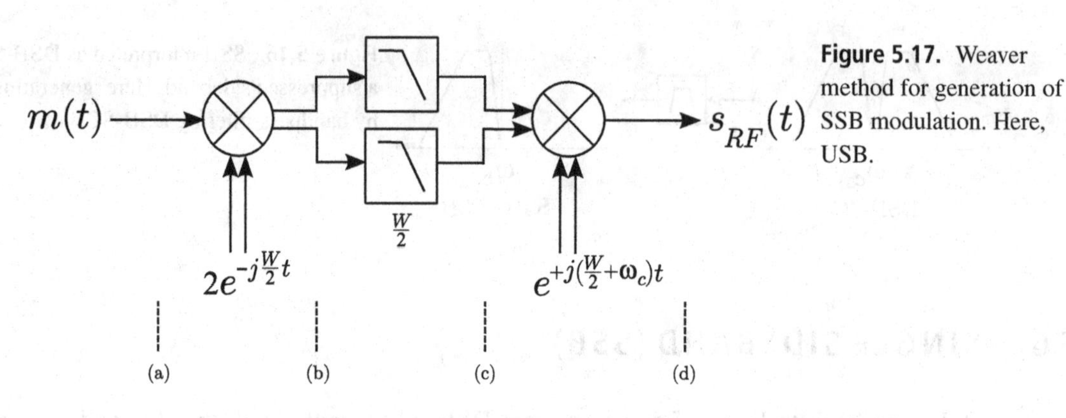

Figure 5.17. Weaver method for generation of SSB modulation. Here, USB.

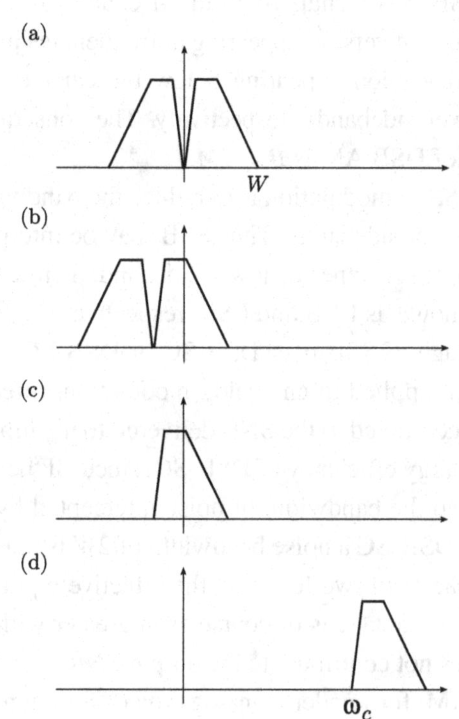

Figure 5.18. Spectra at various stages in the Weaver modulator of Figure 5.17. Here, USB.

domain and $M(f \pm W/2)$ in the frequency domain. The undesired sideband is then suppressed using a lowpass filter, which is practical because this filter can be implemented with relatively large fractional bandwidth. At this point we have the complex baseband representation of the desired SSB signal, so we may generate the associated real-valued signal using quadrature upconversion as described at the end of Section 5.3. As an aside, it turns out that the "Weaver modulator" has applications beyond generation of SSB: see Section 14.5.2.

The *Hartley method* is an ingenious method for suppressing one sideband by cancellation, as opposed to filtering. The method is illustrated in Figures 5.19 and 5.20. We first make a copy of $m(t)$ and apply a phase shift of $+\pi/2$ rad to the copy (equivalently, multiply by j). Note that a phase shift of $+\pi/2$ rad for positive frequencies corresponds to a phase shift of $-\pi/2$ rad for negative frequencies, since real-valued signals have conjugate-symmetric spectrum. The resulting spectrum is shown in Figure 5.20(b). Multiplying $m(t)$ by $\cos \omega_c t$ results in the spectrum shown in Figures 5.20(c); note the positive side of the spectrum is now centered

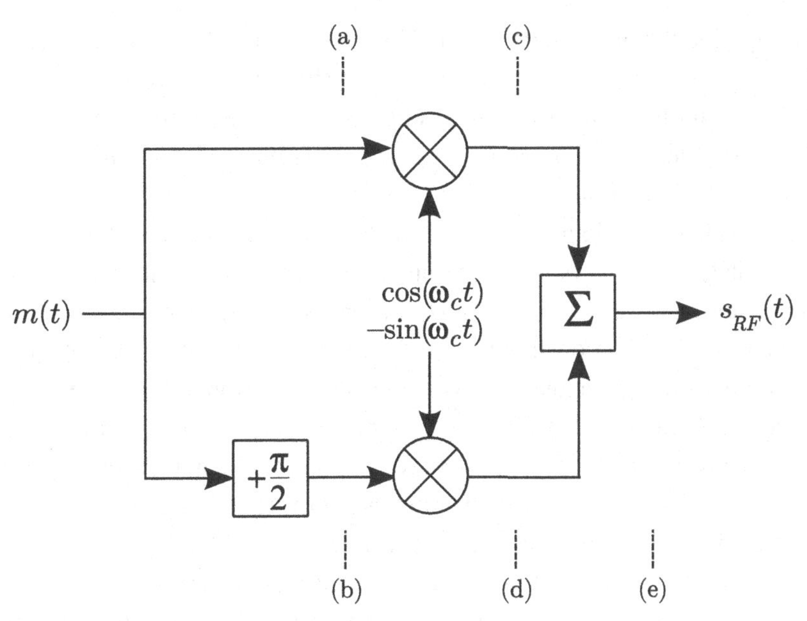

Figure 5.19. Hartley method for generation of SSB modulation; in this case, USB.

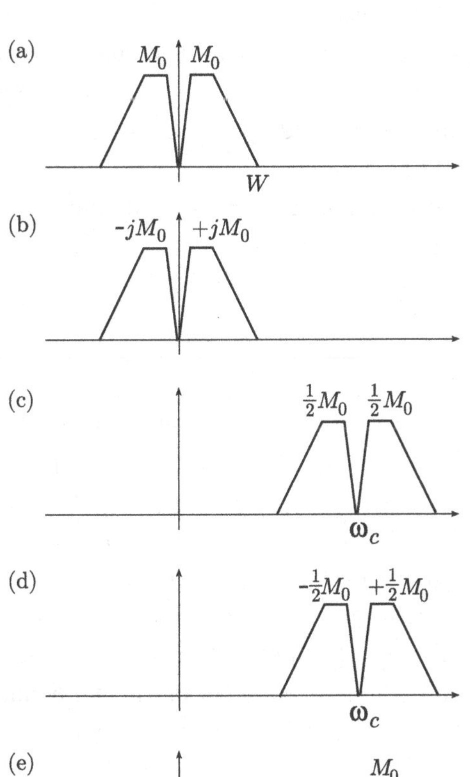

Figure 5.20. Spectra at various stages in the Hartley USB modulator of Figure 5.19.

around $+\omega_c$. Multiplying the phase-shifted signal by $-\sin\omega_c t$ results in the spectrum shown in Figure 5.20(d). This is because $-\sin\omega_c t$ imparts a phase shift of $-\pi/2$ rad relative to $+\cos\omega_c t$, which is equivalent to multiplication by $-j$. Adding the latter result to the former result, the lower sideband is cancelled whereas the upper sideband is preserved. A simple modification (see Problem 5.3) results in LSB as opposed to USB.

Hartley modulation has two big advantages over Weaver modulation, and one big disadvantage. The two advantages are: (1) only one modulating frequency is required, as opposed two for the Weaver method; and (2) cancellation of the undesired sideband is essentially perfect for Hartley, whereas suppression of the undesired sideband by the Weaver method is only as good as the selectivity of the baseband lowpass filter. The disadvantage of the Hartley method is that it is relatively difficult to apply a frequency-independent $\pi/2$ phase shift to a real-valued signal. Specifically, this phase shift can be achieved at one frequency exactly by introducing the appropriate delay, or by using a narrowband phase-shifting circuit (see, e.g., Figure 14.9). To achieve this over all frequencies requires a *Hilbert transform*. The Hilbert transform can be interpreted as an allpass filter with unit magnitude response and phase response equal to $+\pi/2$ or $-\pi/2$ for positive and negative frequencies, respectively. This filter has infinite impulse response, so any practical implementation is an approximation to the exact transform. On the other hand, a finite-length approximation can typically be made sufficiently good over the narrow range of frequencies occupied by the message signal; and this becomes particularly easy using digital signal processing techniques (more on this in Chapter 18). Thus, the inconvenience associated with the Hilbert transform is usually outweighed by the advantages of the Hartley method.

Like Weaver modulation, Hartley modulation has applications beyond generation of SSB: see Section 14.5.1.

5.6.2 SSB as a Quadrature Modulation

A second look at Figure 5.19 reveals something interesting: the Hartley modulator can be interpreted as a two-step process in which the first step is generation of a complex baseband signal, and the second step is quadrature upconversion (as in Figure 5.1). The complex baseband signal is $s(t) = s_i(t) + js_q(t)$ (as usual) with $s_i(t) = m(t)$ (as in AM) and $s_q(t) = \mathcal{H}\{m(t)\}$, where $\mathcal{H}\{\cdot\}$, is the Hilbert transform. This is in contrast to AM, for which $s_q(t) \equiv 0$. Thus, SSB signals occupy both dimensions in the real–imaginary plane. For this reason we refer to SSB as a *quadrature modulation*. Quadrature modulation turns out to be a recurring theme in bandwidth-efficient modulations, as we shall see in Chapter 6, and also has implications for demodulation.

5.6.3 Demodulation and Performance of SSB

Because the carrier is suppressed in traditional SSB, only coherent demodulation is possible. The methods described in Section 5.5.4 work equally well for SSB; this is because the received sideband is restored to its original frequency, whereas the missing sideband is "automatically" restored simply by virtue of taking the real part of the complex baseband signal (recall real-valued signals have conjugate-symmetric spectra).[3]

It is also possible to demodulate SSB using product detection *without* carrier phase estimation; i.e., as shown in Figure 5.11 except using an LO with essentially random phase.[4] The output in this case is obviously not properly derotated in this scheme, but this is of little consequence because – being a quadrature modulation – SSB occupies both dimensions of the signal space. The incorrect phase response imparted on the recovered message signal in this case has essentially no effect on the fidelity of the audio in voice communications applications. In applications where the phase of the message signal is important – e.g., $m(t)$ is phase-modulated data – then this approach is clearly not acceptable and carrier phase estimation is required.

5.6.4 Vestigial Sideband (VSB) Modulation

While SSB works well for voice signals, there are certain types of message signals for which traditional SSB becomes awkward to implement. Prominent among these are analog video signals, for which much important spectral content is concentrated at frequencies very close to DC. In SSB, this content lies within the transition regions of the filter used to suppress or cancel the undesired sideband. *Vestigial sideband* (VSB) modulation is a more practical approach for such signals. The difference between VSB and SSB is simple: The spectrum of the RF signal extends partially into the opposite sideband, as opposed to being limited to the primary sideband.

Analog VSB is becoming rare in modern communications systems for the simple reason that the message signals for which it was designed – such as analog video – are being replaced by digital versions. However – and perhaps ironically – we shall find in Chapter 6 (Section 6.17) that the VSB concept is central to the digital modulation scheme used by the North American broadcast television system.

5.6.5 Pilot-Assisted SSB and VSB

One of the principal difficulties in implementing high-performance SSB and VSB receivers is precise and robust determination of carrier frequency and phase. This is because the carrier is absent, and so its properties must be inferred from the signal. A compromise is to add an explicit carrier signal before transmission. This makes SSB/VSB essentially analogous to DSB-LC, but the motivation is quite different. First, the carrier is used solely for estimation of frequency and phase, and does not permit incoherent demodulation. Second, the ratio of power in the carrier to power in the sidebands is typically made relatively small; i.e., just large enough to facilitate the estimation of frequency and phase with sufficient accuracy. In this application, the carrier is more accurately referred to as a *pilot signal*, and this technique is referred to as *pilot-assisted*

[3] One obvious exception, just to be clear: Estimation of ϕ_p using the method implied by Equation (5.19) will *not* work for SSB, because SSB is a quadrature modulation extending in both dimensions in the complex plane, whereas the method of Equation (5.19) requires the modulation to be limited to one dimension.

[4] Amateur radio enthusiasts will be most familiar with this method.

modulation. Again, we'll see this technique reappear in our discussion of the North American digital broadcast television system in Section 6.17.

5.7 FREQUENCY MODULATION (FM)

The third major category of analog modulations remaining in common use in modern radio communications systems is frequency modulation (FM). As indicated in Table 5.1, FM is the representation of $m(t)$ as a variation in the frequency of the carrier. This single aspect of FM results in three advantages which have made FM immensely popular. First, FM may be generated by using non-linear (primarily "Class C") power amplifiers, which are simpler and more power-efficient than the linear (Class A or AB) amplifiers required for AM and SSB (more on this in Chapter 17). Second, FM benefits from relatively high immunity to additive noise and interference, since these affect primarily the magnitude and phase of the signal as opposed to the frequency. Third, there is no risk of fast fading being interpreted as part of the message signal, which makes FM somewhat better suited for use in terrestrial communications at VHF and above. However, these advantages come at the cost of significantly greater bandwidth, as we shall see below.

The following sections provide a brief but reasonably complete overview of FM in modern radio communications systems. Among all modulations FM is arguably the most difficult to analyze in a rigorous mathematical way, and so we shall sidestep all but the most essential aspects of that analysis here. For additional mathematical details some recommended references include [31] and [41, Ch. 8].

5.7.1 Characterization of FM

Since FM involves representing information as variations in frequency, a useful starting point is to determine the *instantaneous frequency* of the signal in Equation (5.1). Writing that equation here with $A(t) = A_c$ (a constant), we have

$$s_{RF}(t) = A_c \cos\left[\omega_c t + \phi(t)\right] \tag{5.20}$$

The instantaneous frequency f_i is the rate at which the argument of the cosine function varies with time:

$$f_i = \frac{1}{2\pi}\frac{d}{dt}\left[\omega_c t + \phi(t)\right] = f_c + \frac{1}{2\pi}\frac{d}{dt}\phi(t) \tag{5.21}$$

The factor of $1/2\pi$ emerges because the argument of the cosine function has units of radians, and therefore its derivative has units of rad/s, whereas we seek units of Hz (i.e., 1/s). For FM modulation we require the instantaneous frequency to vary around the center frequency in proportion to the message signal; i.e.,

$$f_i = f_c + k_f m(t) \quad \text{(FM)} \tag{5.22}$$

Setting the above two expressions equal, we find

$$\frac{1}{2\pi}\frac{d}{dt}\phi(t) = k_f m(t) \tag{5.23}$$

Solving for $\phi(t)$, we obtain

$$\phi(t) = 2\pi k_f \int_{-\infty}^{t} m(\tau)d\tau \tag{5.24}$$

Integrating from $-\infty$ is bit awkward. However, without loss of generality, we can begin integrating from any time t_0. To see this, note that the above integral can be separated into two integrals as follows:

$$\phi(t) = 2\pi k_f \int_{-\infty}^{t_0} m(\tau)d\tau + 2\pi k_f \int_{t_0}^{t} m(\tau)d\tau \tag{5.25}$$

The first integral does not depend on t, and so is a constant. For the signal represented by Equation (5.20), this constant will be indistinguishable from phase shifts imparted by the radio or imparted by propagation. Therefore this constant phase may be similarly ignored for the purposes of subsequent analysis here. Summarizing:

$$s_{RF}(t) = A_c \cos\left(\omega_c t + 2\pi k_f \int_{t_0}^{t} m(\tau)d\tau\right) \tag{5.26}$$

From this expression we may gain some insight into the spectrum and bandwidth of the FM signal. A useful approach is to consider the simple message signal $m(t) = A_m \cos\omega_m t$ where ω_m is constant. In this case we have:

$$s_{RF}(t) = A_c \cos\left(\omega_c t + \beta \sin\omega_m t + \theta_0\right) \tag{5.27}$$

where β and θ_0 are constants, with

$$\beta \equiv \frac{k_f A_m}{f_m} \tag{5.28}$$

referred to as the *FM modulation index*. This may be alternatively written

$$\beta = \frac{\Delta f}{f_m} \tag{5.29}$$

where $\Delta f \equiv k_f A_m$ is seen to be the *maximum instantaneous frequency deviation*. It can be shown that the spectrum associated with Equation (5.27) consists of discrete tones centered at ω_c with spacing ω_m. The number of tones and the rate at which they decrease with separation from the carrier is a function of β, which in turn is proportional to Δf.

Of course $m(t) = A_m \cos\omega_m t$ is not very representative of the audio signals for which FM is usually employed, and therefore Equation (5.27) is not quite representative of the usual RF signal. Simple expressions for the spectrum in the more general case are not available. However, it is possible to generalize the above results in a manner that is approximate, but nevertheless useful. The idea is that the modulation index may be defined more generally as

$$\beta \equiv \frac{\Delta f}{W} \tag{5.30}$$

and we may subsequently estimate the bandwidth of the modulation as

$$B \approx 2(\beta + 1)W \tag{5.31}$$

The above expression is known as *Carson's Rule*. Note this expression is consistent with the result suggested by Equation (5.27): B increases approximately linearly with β (or, equivalently, with k_f). However, this tells us nothing about the spectrum of the resulting signal.

To better understand the spectrum of FM signals, consider the role of β in Equation (5.27). If $\beta \ll 2\pi$, then $\beta \sin \omega_m t \ll 2\pi$ regardless of ω_m; whereas the carrier phase can vary over the full range $[0, 2\pi]$. Thus the spectrum of $s_{RF}(t)$ must be very narrow and the line spectrum described above must roll off very quickly with increasing distance from the carrier. If we further assume $\beta \ll 1$, then from Carson's Rule $B \approx 2W$; i.e., bandwidth depends only on the bandwidth of $m(t)$. One expects a qualitatively similar result if we replace the sinusoidal message signal with an arbitrary message signal having lowpass bandwidth W. The associated spectrum becomes a "smeared" version of the line spectrum, rolling off continuously but rapidly from the peak value at the carrier frequency. Modulation with $\beta \ll 2\pi$ is said to give rise to *narrowband FM* (NBFM).

FM modulation that is not obviously NBFM is considered *wideband FM* (WBFM). There is no clear dividing line between NBFM and WBFM in terms of β. However, as $\beta \gg 1$, Carson's Rule becomes $B \approx 2\beta W$; i.e., we observe approximately linear dependence of B on β. The spectrum of WBFM representation of audio typically also rolls off with increasing distance from a peak value at the carrier frequency, although typically not as smoothly as is the case for NBFM.

EXAMPLE 5.1

Broadcast FM stations transmit an audio signal with 15 kHz bandwidth using a maximum instantaneous frequency deviation of 75 kHz. Determine the modulation index, transmitted bandwidth, and classify as NBFM or WBFM.

Solution: Here $\Delta f = 75$ kHz and $W = 15$ kHz, so the modulation index $\beta = \Delta f / W = 5$. Transmitted bandwidth is estimated by Carson's Rule: $B \approx 2\,(\beta + 1)\,W = 180$ kHz. β is not $\ll 2\pi$ here, so this is considered WBFM.

In practice, broadcast FM channels are 200 kHz wide – slightly wider than necessary in order to lessen receiver selectivity requirements. Also, 10% of the maximum frequency deviation of a broadcast FM signal is given over to "subcarriers" located at frequencies above the primary 15 kHz audio lowpass channel. One of these subcarriers is used to transmit information required to generate the stereo audio signal.

EXAMPLE 5.2

LMR systems typically use FM designed to transmit voice signals having 3 kHz bandwidth within 12.5 kHz channels. Estimate the modulation index, maximum instantaneous frequency deviation, and classify as NBFM or WBFM.

Solution: Assuming $W = 3$ kHz for voice, we find from Carson's Rule that $\beta < B/2W - 1 = 1.08$ since B must be less than 12.5 kHz. Therefore $\Delta f = \beta W < 3.25$ kHz. This is considered narrowband FM because $\beta \ll 2\pi$.

In practice, Δf up to 5 kHz or so is tolerated as the spectrum of voice signals is not flat, but rather tends to be concentrated around 1 kHz.

5.7.2 Generation of FM

The two primary methods used to generate FM signals are known as the *direct method* and the *indirect method*, respectively. The *direct method* is as implied by Equation (5.22): $m(t)$ is used as the frequency control signal for a variable-frequency oscillator. This may be implemented in either the analog domain using a VCO or the digital domain using an NCO; for more details see Sections 16.6 or 16.9 respectively. The VCO technique is sometimes referred to as *reactance modulation*, which refers to the use of a varactor – a diode whose capacitance is a function of voltage – as the means to control the frequency of the VCO. The *indirect method* consists of first generating an NBFM signal, and then using frequency multiplication to increase the frequency deviation of the signal. Either method can be applied directly to the carrier, or alternatively implemented in complex baseband form and then converted to $s_{RF}(t)$ as a subsequent step.

5.7.3 Demodulation

There are two popular methods for demodulation of FM. One method is simply to differentiate $s_{RF}(t)$ (or its complex baseband representation) with respect to time, which yields a DSB-LC AM signal. For example, from Equation (5.26):

$$\frac{d}{dt}s_{RF}(t) = -A_c\left[\omega_c + 2\pi k_f m(t)\right]\sin\left(\omega_c t + 2\pi k_f \int_{-\infty}^{t} m(\tau)d\tau\right) \quad (5.32)$$

Now $m(t)$ may be recovered using any method suitable for DSB-LC AM, including envelope detection. The differentiation may be implemented in a number of ways. If the signal has been digitized, then the derivative is proportional to the difference between adjacent time samples. In the analog domain, essentially the same operation can be implemented using analog delay lines or by *slope detection*, including the venerable *Foster–Seeley discriminator*.

The second method employs a PLL, as shown in Figure 5.21. Here the phase of the FM-modulated signal is compared with the phase of a reference signal provided by a VCO (analog domain implementation) or NCO (digital domain implementation). The frequency of the VCO/NCO is determined by the magnitude of the error signal created by the phase comparator. Specifically, the VCO/NCO frequency is increased if the phase comparator determines that the input frequency is greater than the reference frequency, and the VCO/NCO frequency is decreased if the phase comparator determines that the input frequency is less than the reference frequency. Thus the phase comparator output is proportional to the instantaneous frequency deviation, and is therefore also proportional to $m(t)$. PLL design entails a number of considerations that must be addressed for this method to be successful: For example, how

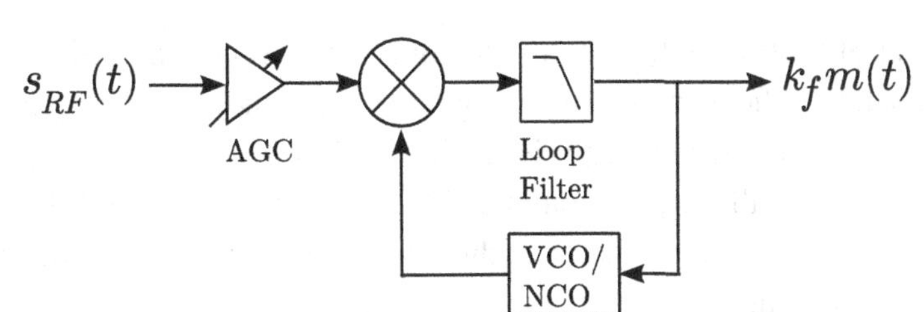

Figure 5.21. Phase-locked loop (PLL) method of FM demodulation. Here the VCO/NCO is generating a slightly-delayed estimate of the carrier and the multiplier is being used as the phase comparator.

exactly to measure phase difference, how fast to change frequency as a function of phase comparator output, and how to initialize the PLL and maintain "lock" in the presence of noise and interference. These issues are specific to PLL design and therefore somewhat beyond the scope of this chapter. However, they are addressed in Section 16.8.

Most FM demodulators implement some form of *fast AGC* and/or *limiting* to normalize the signal magnitude prior to demodulation. The basic idea is that the magnitude of the FM waveform contains no information, only propagation-induced variations and noise. Also, there is typically a limited range of magnitude over which analog detectors and analog-to-digital converters work most effectively. Fast AGC applies gain in inverse proportion to the dynamically-varying propagation gain; however, it is often not practical to have AGC vary at the rate required to accurately track this variation. (See Section 15.8 for more about this.) Limiting finishes the job by allowing a moderate amount of "clipping" of the waveform prior to demodulation.

An extreme version of this concept is so-called "hard limiting" followed by zero-crossing detection, which constitutes a third major class of FM demodulators. In this scheme the instantaneous frequency of the input signal is determined by converting zero-crossings of the modulated waveform into short pulses, and then integrating over the resulting pulse train to obtain a waveform whose magnitude is proportional to frequency.

5.7.4 Preemphasis

One unique aspect of FM is that the differentiation that is intrinsic to FM demodulation tends to exaggerate the impact of noise at frequencies farthest from the carrier. This is because FM demodulation converts large frequencies into large magnitudes. Thus, input noise at frequencies far from the carrier is converted to output noise in the message signal that is relatively strong, whereas input noise at frequencies close to the carrier, but having the same power spectral density, is converted to output noise in the message signal that is relatively weak. Ironically, the spectrum of audio signals tends to be concentrated around 1 kHz, with relatively less content at higher frequencies. Therefore it makes sense to modify the spectrum of audio signals before modulation such that more power is present at higher frequencies and less power is present at lower frequencies, as this equalizes the SNR as a function of frequency after the signal is demodulated. This technique is known as *preemphasis*, and the reciprocal operation at the demodulator is known as *deemphasis*. This process significantly improves the perceived quality of audio that has been transmitted using FM, and so is nearly universally employed.

5.7.5 Performance in Varying SNR; Threshold Effect

It is quite difficult to precisely estimate the output SNR of FM as a function of SNR. Not only does this relationship depend on the message signal characteristics and modulation parameters in a complex way, but also propagation characteristics and audio processing techniques play a role. Suffice it to say that FM typically requires input SNR on the order of 3–6 dB for intelligible voice, and perhaps more like 10–15 dB for quality audio. However, there are some other aspects of this issue that are worth knowing about.

One reason for the popularity of FM is that it achieves relatively high output SNR relative to input SNR when compared to AM. However, this occurs only when the input SNR is above a certain threshold, and below this threshold the ratio of output SNR to input SNR is significantly worse than that for AM. This characteristic of FM is known as *threshold effect*, and the critical SNR is sometimes referred to as the *capture threshold.* In audio applications, exceeding the capture threshold is perceived as *quieting*; i.e., an apparent reduction of background noise. The condition in which background noise seems to have been essentially completely suppressed – which is the demodulator operating fully in the capture region – is referred to as *full quieting*. No such effect is apparent in AM.

Threshold effect makes FM superior for audio communications applications in which the capture threshold – typically on the order of 10 dB – can reasonably be expected to be achieved. This is normally the case for FM broadcasting and LMR systems. Conversely, AM is a better choice for audio communications in which the SNR cannot reasonably be expected to be consistently greater than the FM capture threshold. This is more likely to be the case in low-power or long-range communications systems, especially those vulnerable to propagation uncertainties such as HF skywave communications.

This property of FM also applies to interference. A clas111gsic example pertains to broadcast FM, as follows: Because broadcast FM frequencies are reused, it is not uncommon for two stations to be individually strong enough to be received with sufficient SNR at a given location. However, the weaker station will be strongly suppressed in the audio output as long as the stronger station exceeds the capture threshold. It is rare to hear FM broadcast stations interfering with each other, as this requires both stations to be significantly below the FM capture threshold at the same time. This is in contrast to broadcast AM, in which multiple stations are frequently audible in the output.

5.8 TECHNIQUES FOR IMPROVING AUDIO

Analog modulations are used predominantly to transmit audio. Thus it is appropriate to consider techniques that can be used to improve the performance of these modulations specifically when transmitting audio. We have already noted one example: preemphasis in FM (Section 5.7.4). Preemphasis and other forms of equalization are routinely used in amplitude modulations as well, but for somewhat different reasons. These reasons include compensating for missing high-frequency content due to narrow bandwidth, and improving fidelity in the presence of non-ideal noise and interference.

Squelch is the silencing of the audio output of a receiver when there is no signal present, or when the input SNR drops below a specified threshold. Squelch is particularly important in LMR applications because it inhibits the noise ("static") that would otherwise be heard when no one is transmitting. Squelch is typically not implemented in receivers using amplitude modulation, because there is no obvious threshold SNR that can be universally considered to indicate the presence of an intelligible signal. In FM, however, the capture threshold serves this purpose. Thus, squelch in FM receivers is typically set such that audio is enabled only when the demodulator has achieved quieting or full quieting. Squelch is considered so desirable that

FM modulators in LMR systems typically embedded a tone in the audio signal that can be used by the demodulator to confirm the presence of an intelligible signal, providing an extremely robust squelch that requires no user adjustment. See [65, Section 9.3] for additional reading on this topic.

Noise blanking is the disabling of the audio output of a receiver when an interfering RF pulse is detected. As explained Section 4.5, such pulses are emitted by a variety of circuits and machines, and may also originate from systems which are deliberately producing pulses, such as HF-band radar.[5] Since amplitude modulations (including SSB) represent information as amplitude, these pulses contribute directly to the demodulated output, and may also interfere with the operation of the receiver's AGC. However, these pulses are also typically quite short – typically on the order of 1 μs – and occur with low duty cycle. A noise blanker disables the receiver for the duration of the pulse, which significantly improves the quality of the audio output relative to the result if the impulse is not blanked. This also prevents the pulse from disrupting the AGC. Noise-blanking circuits can be quite sophisticated; see [65, Section 9.2] and [70, Section 7.3] for additional reading on this topic.

Problems

5.1 (a) Verify Equation (5.10) by explicitly working through the integration. (b) Repeat for Equation (5.11). (c) Repeat for Equation (5.12).

5.2 A DSB-SC AM signal arrives with power P_{in} at the input of a coherent demodulator. The associated noise power at the input of the receiver is N_{in}. Then the modulation is changed to DSB-LC AM with 50% modulation, arriving with the same power. (a) What is the effective SNR (i.e., ratio of power associated with the message signal-to-noise power) at the input of the receiver in both cases? (b) What is the SNR at the output in both cases? (c) If incoherent demodulation is used for the DSB-LC AM signal, what is the SNR at the output?

5.3 Create figures similar to Figures 5.19 and 5.20 explaining the generation of LSB (as opposed to USB) using the Hartley method.

5.4 LMR in the USA is currently transitioning from 12.5 kHz channels to 6.25 kHz channels. Repeat Example 5.2 for this case.

5.5 Analog modulations commonly appearing in modern communications systems include DSB-LC, DSB-SC, SSB, NBFM, and WBFM. For each, indicate which of the following techniques is effective, and why or why not: (a) AGC, (b) limiting, (c) preemphasis, (d) squelch, (e) noise blanking.

[5] Recall from Section 3.8 that HF signals are capable of propagating great distances; therefore one need not be in close proximity to be affected by such systems.

6 Digital Modulation

6.1 INTRODUCTION

In Section 5.1 we defined modulation as the representation of information in a form suitable for transmission. *Digital* modulation might be defined as the representation of information as a sequence of *symbols* drawn from a *finite alphabet* of possible symbols. This definition is not completely satisfactory, since it would seem to include a variety of techniques that many practitioners would consider to be analog modulations; these include continuous wave (CW) signaling (the use of an interruptible carrier to send Morse code). In Section 5.1 we settled on the notion that digital modulation also requires the information to be transmitted to be represented in digital (not merely discrete-time) form at some point in the modulation process,[1] whereas this is not required in analog modulation.

6.1.1 Overview of a Digital Communications Link and Organization of this Chapter

With this definition in mind, Figure 6.1 shows an overview of a digital communications link. Here is the usual signal flow.

- The starting point of the link is *source coding*, which is the processing of data and/or digitized analog message signals – e.g., voice – in order to render the data in a form better-suited for digital modulation. A common goal in source coding is *compression*; that is, reduction of data rate so as to reduce the required bandwidth. Source coding is addressed in Section 6.2.
- Data are often encrypted for security. The principles of encryption are not specific to radio communications and so are not addressed further in this book.
- *Channel coding* is the processing of data with the goal of making them more robust to propagation channel impairments, including noise, multipath, and interference. Channel coding is addressed in Sections 6.13 (error control coding) and 6.14 (interleaving).
- At this point the data are processed by what is traditionally considered "modulation." The result is an analog signal representing a discrete-time finite-alphabet message, and which is suitable for frequency conversion and amplification. It is this form of the signal that is ultimately transmitted to the other end of the link. Modulation is addressed in Sections 6.3, 6.4, and portions of Sections 6.10–6.12, 6.18 and 6.19.

[1] "Discrete-time" means merely "changing only at specified times"; whereas "digital" implies quantization and reduction to binary as well.

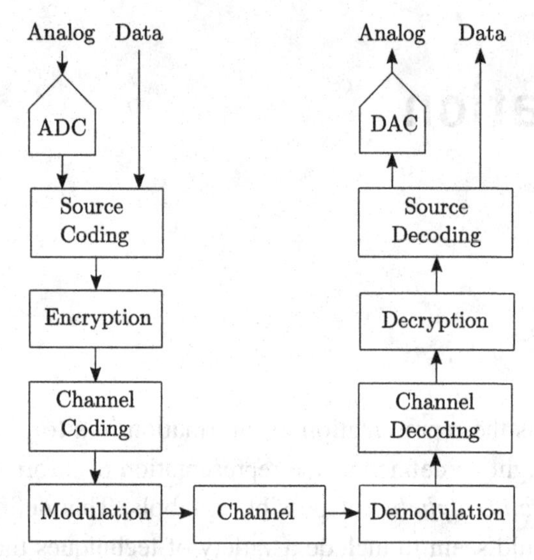

Figure 6.1. Overview of a digital communication link.

- At the receiver, the reciprocal operations are applied in reverse order: demodulation, channel decoding, decryption, source decoding, and (for analog signals) digital-to-analog conversion. Demodulation is addressed in Sections 6.4.1–6.9 and portions of Sections 6.10–6.12, 6.18 and 6.19.

It should be noted that modern digital communications systems cannot always be represented in precisely this structure. In particular, there are many ways in which source and channel coding might become commingled, and channel coding is intrinsic to some forms of modulation. Furthermore, a few key functions are glossed over in this depiction, including techniques used to mitigate flat fading (Section 6.14) and intersymbol interference (Section 6.15). Nevertheless Figure 6.1 facilitates a productive start to discussion of these functions.

6.1.2 Motivation for Digital Modulation

Since digital modulation is arguably more complex than analog modulation, we should be clear on the features of digital modulation that make it compelling in most scenarios. Primary among these are the aforementioned characteristics of *finite alphabet* and *discrete time representation*. A digital modulation consists of a finite set of *symbols* (the "alphabet"), which are sent at a specified rate (thus, discrete time representation). This reduces the task of demodulation to that of making the best possible guess as to the particular sequence of symbols that has been transmitted. For a modulation consisting of M unique symbols, one need only discriminate between M possible options per symbol period. This is in contrast to analog modulation, in which one must generate a continuous estimate from a continuum of possible sent waveforms. At moderate signal-to-noise ratios typical for radio communications systems, the discrete-time finite alphabet property typically results in superior performance. For voice communication in particular, it is possible to achieve fidelity which is limited primarily by the digitization and source coding as opposed to the noise and propagation characteristics of the RF channel. That said, there is no "free lunch" and so we shall see throughout this chapter the various measures required to achieve and sustain this advantage.

Other reasons for the popularity of digital modulation have relatively little to do with digital modulation *per se*, but rather with the fact that information is usually more conveniently represented in digital form. Again there is no particular reason why data cannot be transmitted by using analog modulation; however, opting for digital modulation allows a number of additional features to be brought to bear. These include the following.

- Lossless regeneration: If a digitally-modulated signal can be demodulated without error, then the signal can be re-modulated – "regenerated" – and sent through a second link with no degradation. In principle this process can be repeated indefinately. This is never the case with analog signals. In a system using analog modulation, each regeneration degrades the signal, however slightly, because analog demodulation cannot perfectly recover the original signal.
- Digital representation facilitates flexible management of the use of bandwidth by voice, data, and signaling. *Signaling* is data, separate from user data, required to operate a radio link. For example, each radio link in a modern cellular telecommunications system includes a nearly-continuous parallel stream of data that conveys information including identity, assignment to frequency and time slots, power control, and so on. If the user data are digital, then integration of signaling is straightforward, whereas this integration is awkward and limited in systems which are primarily analog.
- The same concept also applies to message information specifically: If the message will be represented as data, then the data rate provided by the radio link may be allocated in a flexible and dynamic way; e.g., entirely voice, voice+data, entirely data, and so on.
- Digital representation of information facilitates the use of a variety of signal processing techniques that have no counterpart in analog modulation. These include compression, encryption, and error control coding.

6.2 SOURCE CODING

Source coding is the processing of data and/or digitized analog message signals – e.g., voice – in order to render the data in a form better-suited for digital modulation. A simple taxonomy of source coding techniques includes *pulse code modulation*, *linear predictive coders*, and *multimedia coders*. Here we will briefly examine these schemes. For a more detailed discussion of source coding for voice signals, [61, Ch. 8] is a useful starting point.

The most elementary form of pulse code modulation (PCM) is essentially no more than rearranging the bits representing each sample output from an ADC to form a serial data stream. A common scheme for PCM encoding of voice is simply to lowpass-filter the voice signal to a little less than 4 kHz, and then to sample at 8 kHz with 8 bits/sample. This results in a 64 kb/s data stream.

Typically voice waveforms consist mostly of values that are much less than the mean, with short but important bursts having relatively large magnitude. However, ADCs are typically designed to encode values that are evenly distributed over the input magnitude range; i.e., assuming all encodable values occur at equal rate. Thus a voice-digitizing ADC that is able to encode the bursts without clipping uses, most of the time, only a few of the available levels.

Table 6.1. **Vocoders used in a few popular voice communications systems.**

Vocoder	Rate	Applications
Code Excited Linear Prediction (CELP)	0.8–13 kb/s	Cellular CDMA (older)
Enhanced Variable Rate Codec (EVRC)	0.8–8.55 kb/s	Cellular CDMA (newer)
Improved Multiband Excitation (IMBE)	7.2 kb/s	P.25 LMR (older)
Advanced Multiband Excitation (AMBE)	2–9.6 kb/s	P.25 LMR (newer); also Iridium, Inmarsat
Mixed Excitation Linear Prediction (MELP)	2.4 kb/s	US Military

To better utilize the range of values provided by the ADC, voice signals are typically subjected to *companding* before digitization. Companding is a non-linear scaling of the waveform magnitude such that small values are increased relative to large values. This process may be implemented in analog form as implied by the previous sentence; alternatively the waveform might first be encoded with a greater number of bits (e.g., 16) and then digitally transformed to an 8-bit waveform by companding. The predominant companding transfer functions are known as "μ-Law", used in North America and Japan, and "A-Law", used in Europe. The (wired) public switched telephone network (PSTN) uses primarily 64-kb/s PCM with companding, and therefore this level of fidelity, sometimes referred to as *toll quality*, is regarded as a benchmark against which other voice encoding schemes are compared.

Variants of PCM – in fact, each beginning with a PCM-encoded signal – include *differential PCM* (DPCM), *adaptive DPCM* (ADPCM), and *continuously-variable slope delta modulation* (CVSDM). The primary advantage of these schemes is a reduction in data rate by a factor of 2–4 (i.e., to 16–32 kb/s) with little or no compromise in voice quality, but some consequences in terms of robustness to bit errors.

The bandwidth available for radio communications systems is often so limited that even 16 kb/s may be considered unacceptable, and it is therefore reasonable to consider "less than toll quality" voice in return for a further reduction in data rate.[2] This is possible using *linear prediction* techniques, and the associated devices are commonly referred to as *vocoders*. The essential idea in these schemes is to identify a mathematical model for the voice signal which can be expressed as a small number of slowly-varying parameters, and then to send these values as opposed to sending the signal itself. Voice is reconstructed by using the received parameters with the assumed model to synthesize the waveform. The principal difference among source coders employing this principle is the model used to represent voice signals, and the method used to decompose voice into parameter sets. Vocoders used in nearly all modern mobile radio communications systems employing digital modulation employ some version of this scheme; a very brief list is shown in Table 6.1. Note that rates of 2.4 kb/s and less are possible, albeit with a tradeoff in voice quality.

[2] This sentence should not be interpreted as meaning that it is not possible to achieve toll quality at rates less than 16 kb/s. Instead, we are merely saying that lower rates are normally achieved by compromising on voice quality in some way.

A similar taxonomy can be defined for multimedia (e.g., video) information. Although a discussion of these techniques is beyond the scope of this book, it is useful to be aware of at least two commonly-used multimedia *codecs* (coder-decoder): These are the ITU-T H.262 MPEG-2 and H.264 MPEG-4 codecs. The MPEG-2 codec is used in the North American broadcast digital television standard (ATSC; see Section 6.17).

6.3 SINUSOIDAL CARRIER MODULATION, REDUX

In Section 5.2 we introduced the concept of sinusoidal carrier modulation as employed in analog modulation. The vast majority of digital modulation schemes employ sinusoidal carrier modulation as well. The principal difference is that digital sinusoidal carrier modulations represent information using *discrete-time* sequences of magnitude, phase, and frequency states. In this section and those that immediately follow, we shall address specifically *phase-amplitude* modulations. Beginning in Section 6.11 we shall consider digital modulations that use frequency.

As was the case for analog sinusoidal carrier modulation, digital sinusoidal carrier modulation is conveniently represented in complex baseband form (for a refresher, see Section 5.3) This representation is conveniently visualized in the complex plane, as in Figures 5.4 and 5.7. So, without further ado, Table 6.2 and Figure 6.2 show a "starter kit" of digital sinusoidal carrier modulations for us to consider. Here's a list of the acronyms:

- **OOK**: On-off keying
- **BPSK**: Binary phase-shift keying
- **QPSK**: Quadrature phase-shift keying
- ***M*-PSK**: *M*-ary phase-shift keying
- ***M*-ASK**: *M*-ary amplitude-shift keying
- ***M*-QAM**: *M*-ary quadrature amplitude modulation.

Table 6.2. **Representation of some common digital phase-amplitude modulations in the form $A_k \cos[\omega_c t + \phi_k]$. See also Figure 6.2.**

Modulation	A_k	ϕ_k	M	bits/symbol
OOK	$\{A_c \vert 0\}$	0	2	1
BPSK	A_c	$\{0 \vert \pi\}$	2	1
QPSK	A_c	$\left\{0 \vert \frac{\pi}{2} \vert \pi \vert \frac{3\pi}{2}\right\} + \frac{\pi}{4}$	4	2
M-PSK	A_c	$\left\{0 \vert \frac{2\pi}{M} \vert \cdots \vert \frac{2\pi}{M}(M-1)\right\} + \frac{\pi}{M}$	M	$\log_2 M$
M-ASK	Linear grid		M	$\log_2 M$
M-QAM	Square $\sqrt{M} \times \sqrt{M}$ grid		M	$\log_2 M$

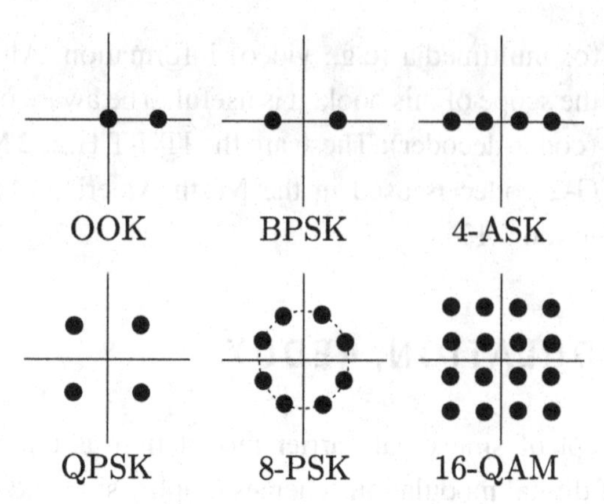

Figure 6.2. Constellation diagrams for the modulations shown in Table 6.2.

Note that k in Table 6.2 indexes the symbols (i.e., k is the discrete-time representation of t) and M is the number of symbols used in the modulation. The graphical representations of digital sinusoidal carrier modulations as shown in Figure 6.2 are known as *constellation diagrams*. Be forewarned that this set of modulations is hardly exhaustive, either in terms of possible values of M or as methods to digitally modulate the amplitude and phase of a sinusoidal carrier.

Note that M determines the number of bits that each symbol represents; for example, if $M = 2$ then each symbol can represent only one bit; namely 0 or 1. However, if $M = 4$, then it is possible to send two bits simultaneously; e.g., the four symbols represent 00, 01, 10, and 11. Pairs of bits associated with symbols in this manner are sometimes referred to as *dibits*. If $M = 8$, then it is possible to send three bits simultaneously. The eight possible sets of three bits are sometimes referred to as *tribits*. Extrapolating: a modulation having M symbols conveys $\log_2 M$ bits per symbol.

The assignment of bits to symbols in a modulation is typically done according to a principle commonly referred to as *Gray coding*. The concept is to assign bit sequences to symbols in such a way that the most likely symbol decision errors result in the smallest number of bit errors. This is illustrated for QPSK in Figure 6.3. Given symbols identified as A, B, C, and D with increasing phase, a straightforward assignment of dibits to symbols would be as follows: A: 00, B: 01, C: 10, D: 11. The problem with this is that for usable levels of SNR, small phase errors are more likely than large phase errors. So, for example, the most likely misdetections of symbol A are B and D (equally likely), and much less likely is C. If D is detected in lieu of A, then two bits are incorrect. A better scheme is to swap the assignments for symbols C and D. In this case, erroneously detecting B or D in lieu of A results in at most one bit error, and only the much less likely misdetection of C for A results in both bits being wrong. In fact, the latter assignment is readily verified to be be optimal in this respect for all four symbols, not just A.

Before moving on, it is worth pointing out a few popular QPSK variants and the reason why they are popular. One is *offset QPSK* (OQPSK). In OQPSK, the in-phase and quadrature components are offset in time by exactly $T/2$, where T is the symbol period. There is essentially no difference in the corresponding spectrum or fundamental performance of the modulation. However, this scheme eliminates the possibility that the magnitude of the complex baseband signal goes to zero when transitioning between symbols located on opposite sides of the constellation; such transitions occur by first transitioning to one of the adjacent symbols, and

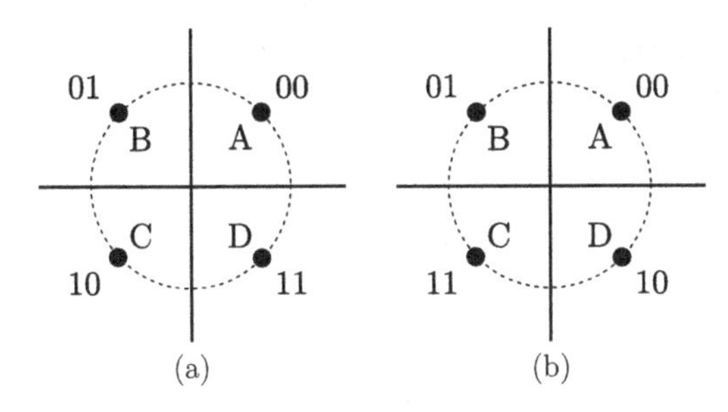

Figure 6.3. (a) An ill-advised assignment of dibits to symbols in QPSK. (b) A better assignment employing Gray coding.

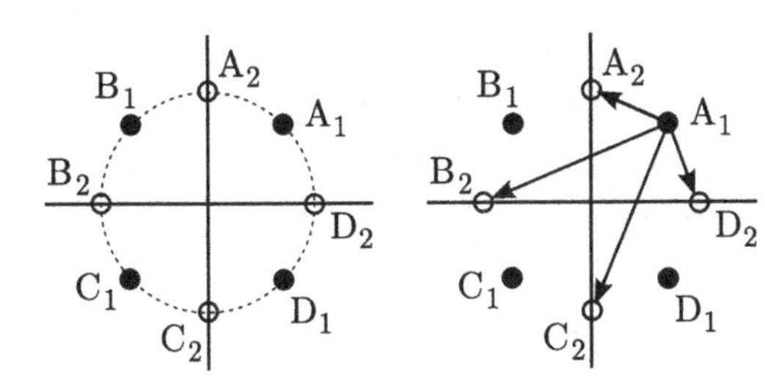

Figure 6.4. *Left*: The two QPSK constellations used in $\frac{\pi}{4}$-QPSK. *Right*: All possible transitions from A_1 to the next symbol.

then onward to the desired symbol. As a result, magnitude remains very nearly constant at all times, as opposed to intermittently going to zero and then back to full-scale within a symbol period. This in turn facilitates the use of non-linear (primarily "Class C") power amplifiers, which are simpler and more power-efficient than the linear (Class A or AB) amplifiers that would otherwise be required to accommodate the intermittent crossing of the origin of the complex plane. It is worth noting that this same issue provides a contraindication for the use of BPSK when high power is required, since with two symbols there is no alternative but to frequently cross the origin. This particular concern about power amplifiers will arise repeatedly throughout this chapter; see Chapter 17 for a more thorough discussion of this issue.

The other popular QPSK variant is $\frac{\pi}{4}$-QPSK. This modulation actually consists of two QPSK constellations, with one rotated by $\pi/4$ relative to the other. This is illustrated in Figure 6.4, with symbols A_1, B_1, C_1, and D_1 representing the first constellation, and symbols A_2, B_2, C_2, and D_2 representing the second constellation. This figure also illustrates the four possible transitions from symbol A_1. Note that none of these transitions crosses the origin; therefore we once again have a modulation that is suitable for less-linear power amplifiers. OQPSK has lower overall magnitude variation; however, $\frac{\pi}{4}$-QPSK has the property that it guarantees a symbol transition on every "tic" of the symbol clock. The latter turns out to be useful in symbol timing recovery (Section 6.16).

6.4 PULSE SHAPES AND BANDWIDTH

The bandwidth of a digital modulation is determined primarily – but not exclusively – by the symbol rate $1/T$. If the symbol rate is zero, then the modulation degenerates into an unmodulated carrier having essentially zero bandwidth. When the symbol rate is greater than zero, the transition between symbols gives rise to bandwidth proportional to symbol rate. To be more precise, it is necessary to specify how the transitions between symbols occur.

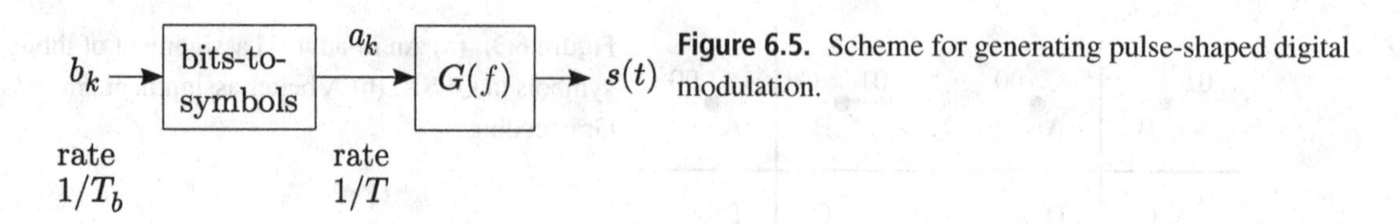

Figure 6.5. Scheme for generating pulse-shaped digital modulation.

6.4.1 Representation of Symbols as Pulses

It is convenient to interpret modulation symbols as *pulses*. The basic scheme is to define a "prototype" pulse $g(t)$ centered at time $t = 0$. The symbol centered at time $t = kT$ is then represented as $a_k g(t - kT)$, where a_k is $A_k e^{j\phi_k}$; that is, the complex baseband representation of the symbol from the constellation diagram. The transmit signal in complex baseband representation may then be interpreted as a sequence of pulses as follows:

$$s(t) = \sum_{k=-\infty}^{+\infty} a_k g(t - kT) \tag{6.1}$$

A scheme for generating $s(t)$ is shown in Figure 6.5. On the left we have a stream of bits b_k at rate $1/T_b$, which are subsequently mapped to modulation symbols a_k emerging at rate $1/T$. A convenient representation for this signal is in terms of the delta function as follows:

$$\sum_{k=-\infty}^{+\infty} a_k \delta(t - kT) \tag{6.2}$$

This is subsequently passed through a filter having impulse response equal to the desired pulse shape, or equivalently with frequency response $G(f)$ equal to the Fourier transform of $g(t)$. The output is the input convolved with $g(t)$:

$$s(t) = \left[\sum_{k=-\infty}^{+\infty} a_k \delta(t - kT) \right] * g(t) \tag{6.3}$$

which, by the properties of the delta function, is equal to Equation (6.1).

The simplest imaginable pulse waveform $g(t)$ is simply a rectangular pulse as shown in Figure 6.6. Such a pulse has constant value for an interval equal to T, and is identically zero at all other times. For a phase-amplitude modulation, this corresponds to holding the magnitude and phase of the carrier constant for the duration of the symbol, and then switching instantaneously to the magnitude and phase corresponding to the next symbol when the time comes. The spectrum of this pulse, as characterized by its Fourier transform, is given by

$$G(f) = \frac{\sin(\pi f T)}{\pi f T} \equiv \text{sinc}(fT) \tag{6.4}$$

and is also shown in Figure 6.6.[3] A problem with rectangular pulses is immediately apparent: They have unbounded bandwidth. In practical radio communications problems one typically

[3] A comment for advanced readers: Note we have implicitly assumed that the spectrum of $s(t)$ is the same as the spectrum of the pulse. This is true enough for most engineering applications, accounting for a constant of proportionality which is typically not of interest. However, this is *formally* true only if the autocorrelation of the sequence of a_ks is identically zero for all lags other than the zero lag; i.e., if the sequence is sufficiently random.

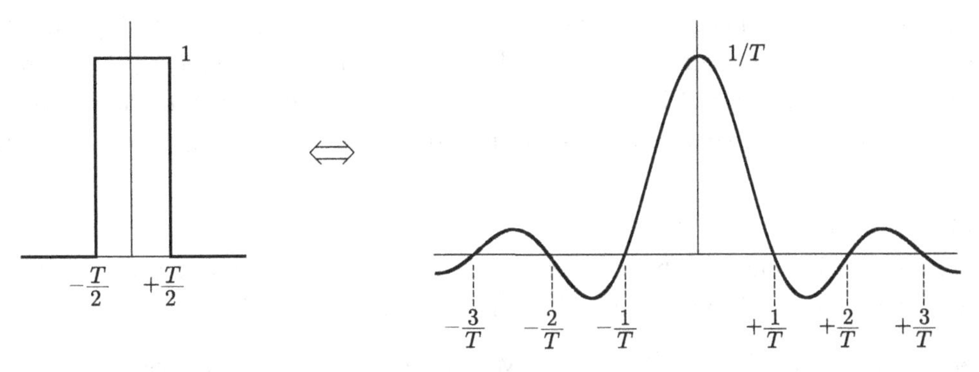

Figure 6.6. Rectangular pulse. *Left*: Time domain. *Right*: Frequency domain.

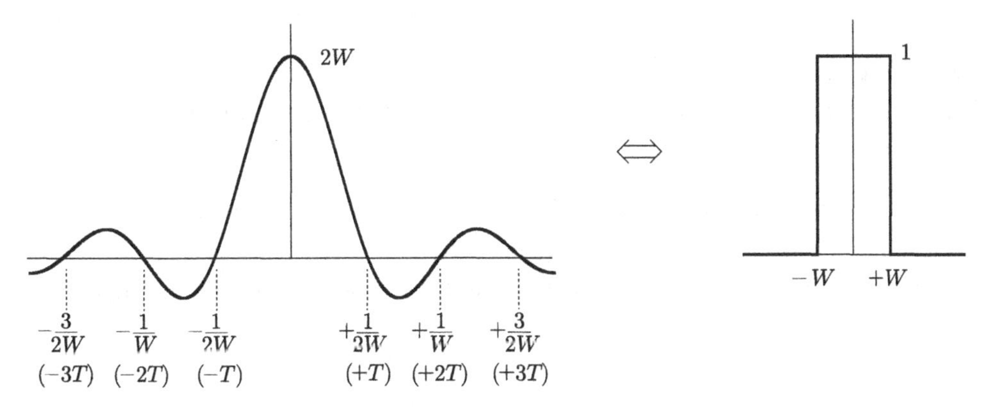

Figure 6.7. Sinc pulse. *Left*: Time domain. *Right*: Frequency domain.

faces stringent constraints on bandwidth and spectrum, so it would seem rectangular pulses may not be desirable.

6.4.2 Sinc Pulses and Intersymbol Interference

If one takes the position that it is the bandwidth, as opposed to the duration of the pulse, which should be strictly limited, then Figure 6.7 shows the way to go. Here the spectrum is required to have so-called "brick wall" response with lowpass bandwidth $W = B/2$, where B is the RF bandwidth since complex baseband representation is assumed. The associated pulse is – by the duality of the Fourier transform – the sinc waveform

$$g(t) = \text{sinc}\left(\frac{t}{T}\right) \tag{6.5}$$

As attractive as the sinc pulse is in terms of bandwidth, there is clearly some additional complexity associated with its use: In particular, each pulse is unbounded in the time domain such that each pulse would appear to interfere with all other pulses. This effect is known as *intersymbol interference* (ISI). A solution to ISI in this case lies in the fact that the value of sinc goes to zero at integer multiples of T in each direction from the defining *maximum effect time* (MET) $t = kT$ of the pulse waveform. Therefore if T is set equal to $1/B$, then the MET for each symbol occurs when the pulse waveform associated with every other symbol happens to be zero. This is commonly known as the *Nyquist criterion for zero ISI.*[4] Using sinc pulses with this particular relationship between B and T, we have that a_k is simply $y(kT)$; i.e. the symbol decisions are free from ISI. This subsequently constrains the symbol rate at which ISI-free transmission is possible to be equal to the non-zero bandwidth B of the sinc waveform.

EXAMPLE 6.1

What is the maximum symbol rate for which symbol detection may be ISI-free if the required RF bandwidth must be strictly bandlimited to 1 MHz bandwidth?

Solution: $B = 1$ MHz, so $1/T = B = 1$ MSy/s, where "Sy" is an abbreviation for "symbol".

EXAMPLE 6.2

What's the first-null bandwidth (FNBW) of a digital modulation that uses rectangular pulses having $T = 1$ μs?

Solution: From Figure 6.6, we see FNBW$= 2/T = 2$ MHz. For the same symbol rate obtained in Example 6.1, we see that ISI-free rectangular pulses require twice the bandwidth as measured between first nulls, and in fact greater because the bandwidth of the rectangular pulse is not strictly bandlimited.

6.4.3 Raised Cosine Pulses

It is apparent from Figure 6.7 that the consequence of strictly-bandlimited spectrum with "brick wall" transitions is unbounded impulse response. This makes clear the fundamental quandary of pulse design: It is impossible to define a pulse that is strictly limited in both time and frequency simultaneously. The situation is actually somewhat worse than this: For practical reasons it will be desirable to make the impulse response of the filter in Figure 6.5 finite, and so the associated pulse waveform will be truncated even if we intended it to be unbounded. These considerations lead one to consider pulse shapes which are intermediate between rectangular (strictly time-limited) and sinc (strictly frequency-limited).

One approach is to consider time-limited alternatives to the rectangular pulse; commonly-cited alternatives include the triangular pulse and the cosine (actually half-period cosine) pulse. From Fourier analysis the tradeoff is easily quantified: FNBW is $2/T$, $3/T$, and $4/T$ for the rectangular, cosine, and triangular pulse, respectively. Ironically, the level of the first sidelobe is about -13 dB, -22 dB, and -26 dB, respectively. Therefore bandwidth gained in the sense of FNBW is lost in the sense of increasing sidelobe level. In other words, simply choosing alternative shapes for a pulse that remains time-limited to an interval T does not accomplish much in terms of modifying bandwidth.

A far more useful approach begins with a spectrum that is strictly bandlimited but not having "brick wall" characteristics. This results in a waveform which remains unbounded, but with

[4] Not to be confused with Nyquist's other famous criterion; see Section 15.2.2.

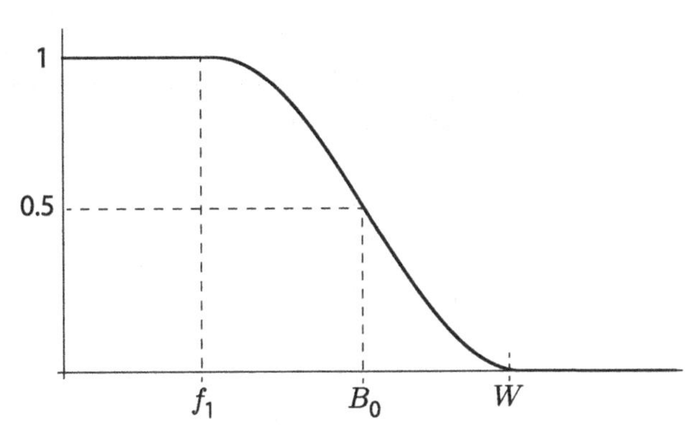

Figure 6.8. Specification of the spectrum $G(f)$ of the raised cosine pulse. Here, $r = 0.5$.

magnitude that is constrained within an envelope that decays much more quickly than that of the sinc function. When such a waveform is truncated by a finite impulse response filter, the resulting expansion of the spectrum is then much less serious.

By far the most popular scheme to accomplish this is known as the *raised cosine* spectrum. The raised cosine spectrum is shown in Figure 6.8 and is specified as follows: $G(f) = 1$ for $|f| \leq f_1$, $G(f) = 0$ for $|f| \geq W$, and

$$G(f) = \frac{1}{2} + \frac{1}{2} \cos\left[\frac{\pi}{2} \cdot \frac{|f| - f_1}{B_0 - f_1}\right], \quad f_1 \leq |f| \leq W \tag{6.6}$$

where the parameters B_0 and f_1 are related through a single parameter known as the *roll-off factor* r as follows

$$r \equiv \frac{B_0 - f_1}{B_0} \tag{6.7}$$

The roll-off factor is an expression of the width of the transition region, ranging from zero to 1; $r = 0$ ($f_1 = B_0$) corresponds to the sinc pulse (brick wall spectrum), whereas $r = 1$ ($f_1 = 0$) corresponds to a spectrum that has only transition, and no flat spectrum component. Values of r in the range 0.1 to 0.6 are typical.

The corresponding pulse waveform is

$$g(t) = 2B_0 \operatorname{sinc}(2B_0 t) \left[\frac{\cos 2\pi r B_0 t}{1 - (4rB_0 t)^2}\right] \tag{6.8}$$

As an example, the $r = 0.35$ waveform is shown in Figure 6.9. Choosing $T = 1/(2B_0)$ results in zeros spaced at integer multiples of T from the maximum effect point; therefore this waveform has the zero-ISI property of the sinc pulse but for a longer sample period. Specifically:

$$W = B_0 + (B_0 - f_1) = B_0 + rB_0 = (1 + r)\, B_0 = (1 + r)\, \frac{1}{2T} \tag{6.9}$$

Thus, one may also interpret r as representing the efficiency with which the available bandwidth $B = 2W$ is used, ranging from $r = 0$ (brick wall response) yielding $1/T = B$; to $r = 1$ yielding $1/T = B/2$; i.e., reduction of the symbol rate by a factor of 2.

Now let us return to the issue of the necessarily finite length of the pulse shaping filter. Comparing $r = 0$ and $r = 0.35$ in Figure 6.9, it is clear that the raised cosine pulse has a faster decay away from the MET. Figure 6.10 compares the spectrum that results when each of these pulses is truncated to an interval of only $4T$ centered around the MET. As expected, neither

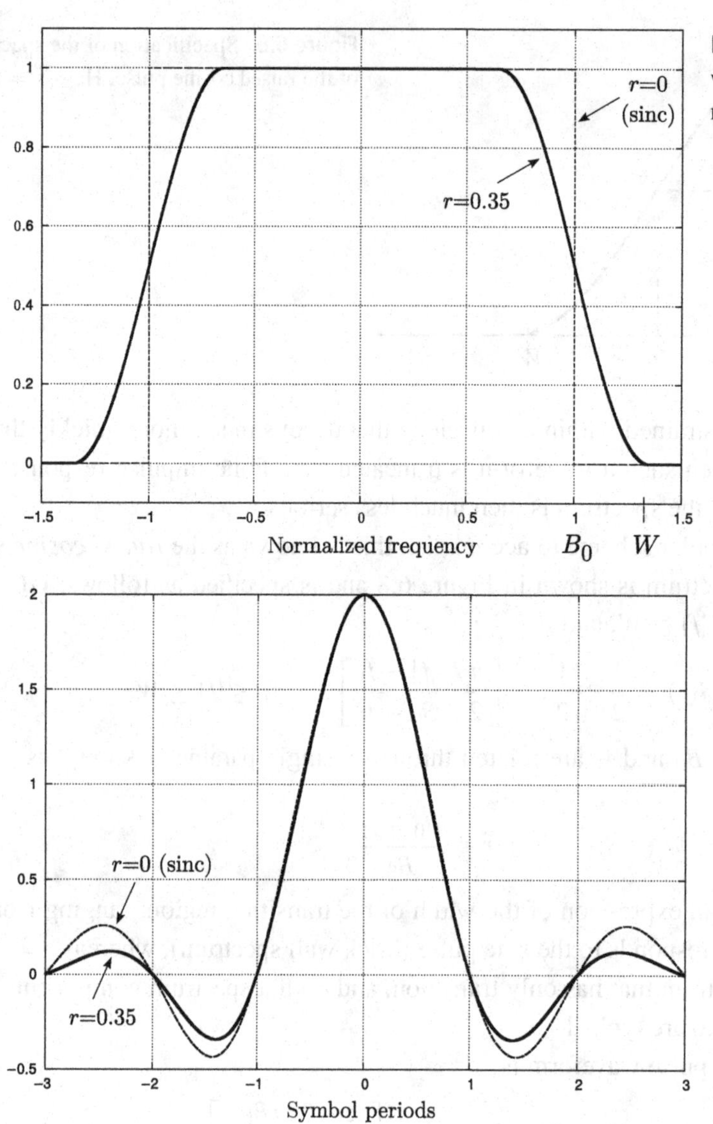

Figure 6.9. Raised cosine pulse waveform (*bottom*) and spectrum magnitude (*top*).

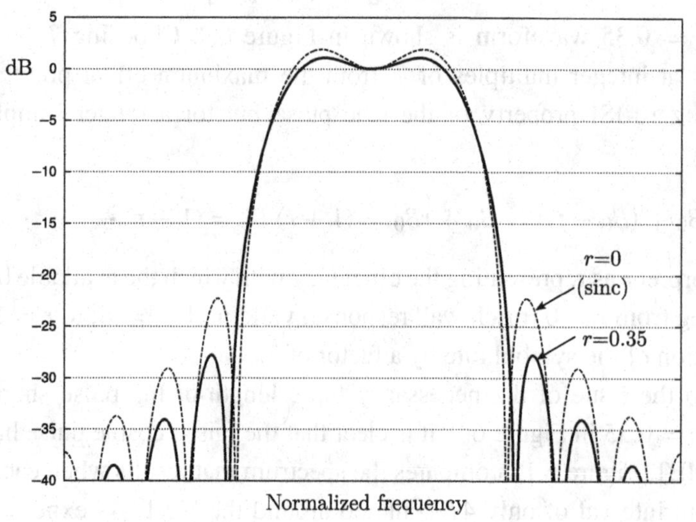

Figure 6.10. Comparison of the spectrum of the sinc ($r = 0$) and $r = 0.35$ raised cosine pulses after truncation to an interval of $\pm 2T$ centered around the MET.

spectrum is band-limited; however, the perturbation to the spectrum of the truncated $r = 0.35$ is much less that of the $r = 0$ pulse.

Before moving on, there is one additional thing to know about raised cosine pulse shaping: It is rarely used in quite the form presented above. This is because – as we shall see in Section 6.6.3 – receivers typically require some form of matched filtering, and matched filtering of raised cosine pulses destroys the zero-ISI property. Fortunately the situation is salvaged using a relatively simple modification that will leave intact all of the desirable properties of raised cosine pulse shaping described above.

6.4.4 Spectral Efficiency

We are now in a position to reconsider the question of bandwidth of phase-amplitude modulations, and the efficiency with which the available spectrum is used. It is already abundantly clear that bandwidth B is proportional to symbol rate T. A more precise relationship requires a more precise definition of bandwidth. Among the various possibilities, first null bandwidth (FNBW) – i.e., the spacing between the nulls bounding the center lobe of the spectrum – is certainly the easiest to work with. From the considerations of the previous sections, we have

$$\text{FNBW} = \frac{2}{T}, \quad \text{ISI-free rectangular pulses} \tag{6.10}$$

$$\text{FNBW} = (1+r)\frac{1}{T}, \quad \text{ISI-free raised cosine pulses} \tag{6.11}$$

We may now define *spectral efficiency* η_B in the following way:

$$\eta_B = \frac{1/T_b}{\text{FNBW}} = \frac{\log_2 M}{2}, \quad \text{ISI-free rectangular pulses} \tag{6.12}$$

$$\eta_B = \frac{1/T_b}{\text{FNBW}} = \frac{\log_2 M}{1+r}, \quad \text{ISI-free raised cosine pulses} \tag{6.13}$$

Note that the units of η_B are bps/Hz: η_B is the data rate that can be accommodated within a certain bandwidth using a particular combination of modulation (M) and pulse shaping (r). Table 6.3 shows some expected values of η_B for typical modulation parameters. It should

Table 6.3. **Spectral efficiency η_B for some typical modulation parameters.**

	η_B (bps/Hz)		
M	rectangular	$r = 1$	$r = 0$
2	0.5	0.5	1
4	1	1	2
8	1.5	1.5	3
16	2	2	4

be noted that because η_B depends on a particular definition of bandwidth, η_B is a somewhat arbitrary measure in the sense that other definitions of bandwidth would result in other, equally valid measures of spectral efficiency.

EXAMPLE 6.3

A transmitter sends 50 kb/s using QPSK with $r = 0.4$ raised cosine pulse shaping. Characterize the spectrum and spectral efficiency.

Solution: QPSK uses $M = 4$ symbols, and therefore sends two bits per symbol. The symbol rate $1/T$ is therefore 25 kSy/s. Since raised cosine pulse shaping is used, the spectrum is ideally strictly bandlimited to $(1 + r)/T = 35$ kHz; however, truncation of the pulse shaping filter will result in some leakage outside this bandwidth. The spectral efficiency is $\eta_B = \left(\log_2 4\right) / (1+r) = 1.43$ bps/Hz. It should be noted that if that quoted data rate includes in-band signaling and error control coding, then the spectral efficiency with respect to the user data will be less.

6.5 REPRESENTATIONS OF SIGNAL POWER, NOISE POWER, AND SNR IN DIGITAL MODULATIONS

An important consideration for any modulation is its performance with respect to the SNR provided to the demodulator. To most conveniently address this issue, it is useful to define some additional terminology that is specific to digital modulation.

To begin, let us assume we are working with signals in complex baseband representation. (For a review of this concept, see Section 5.3.) The concepts we are about to cover are in no way limited to this representation, but for clarity it is useful to narrow the scope.

6.5.1 Symbol Energy and Energy per Bit

It is preferred to express signal power in terms of *energy per bit*, E_b. Only in $M = 2$ modulations do symbols map one-to-one to bits, so let us first define the *symbol energy* E_m, where m refers to the mth symbol of a set of M symbols. For a phase-amplitude modulation in complex baseband representation

$$E_m = \int_{-\infty}^{+\infty} |a_m g(t)|^2 \, dt = |a_m|^2 \int_{-\infty}^{+\infty} |g(t)|^2 \, dt = A_m^2 E_g \tag{6.14}$$

where E_g is the energy of the pulse waveform $g(t)$, which is the same for all symbols. So, for example, for unit-magnitude rectangular pulses we have $E_m = A_m^2 T$. Next we define the *mean energy per symbol*, E_S, which – assuming all symbols are used with equal frequency – is simply the mean over the available symbols as follows:

$$E_S \equiv \frac{1}{M} \sum_{m=1}^{M} E_m \tag{6.15}$$

For modulations in which all symbols have constant magnitude, then $E_S = E_m$ for all m. For other modulations – in particular OOK, M-ASK with $M > 2$, and QAM with $M > 4$ – the symbol energies may be greater or smaller than E_S.

One very important use of E_S is that it facilitates "fair" comparisons of digital modulations. That is, when we compare modulations we specify equal E_S, so that no modulation appears to be better or worse simply because it includes symbols having more or less energy. Finally, we may define E_b as follows:

$$E_b = \frac{E_S}{\log_2 M} \tag{6.16}$$

So, for example, $E_b = E_S$ for $M = 2$ modulations, and is proportionally less than E_S for modulations with $M > 2$.

Note that only in the case of $M = 2$ is E_b a directly measurable "physical" quantity. That is, modulations send symbols, not bits, and only when $M = 2$ is there a one-to-one mapping between symbols and bits. Nevertheless, the concept of E_b proves useful for comparing the performance of modulations, even for $M > 2$.

6.5.2 The E_b/N_0 Concept

We now define E_b/N_0, a convenient metric for expressing SNR in digital communications systems. Let us begin with the denominator. It is preferred to express noise power in terms of the noise power spectral density, N_0. Recalling Section 5.4, the noise in the in-phase and quadrature components of a complex baseband signal is uncorrelated, with power in each component being equal to σ_{RF}^2. If the noise spectrum is white, then N_0 is simply this power divided by the noise-equivalent bandwidth B.

In later sections we shall see that the quantity E_b/N_0 emerges as a convenient description of the SNR determining the performance of digital modulation. The fact that this quantity is in fact a signal-to-noise ratio can be confirmed by first confirming that it is dimensionally correct: E_b has units of energy, which is typically expressed in joules (J). However, 1 J is equal to 1 W · s, since 1 W = 1 J/s. Since 1 W · s may be rewritten 1 W/Hz, we see E_b can also be interpreted as power spectral density, and therefore E_b/N_0 is unitless, as is expected for SNR.

However, E_b/N_0 is not *exactly* the same as the ratio of signal power to noise power, as is normally implied by the term "SNR." It is straightforward to determine the actual relationship between these quantities. First note that signal power can be expressed as E_b/T_b; i.e., energy per bit period. Then note that noise power is simply N_0B, as explained above. Thus we find:

$$\text{SNR} \equiv \frac{E_b/T_b}{N_0 B} = \frac{1}{T_b B} \cdot \frac{E_b}{N_0} \tag{6.17}$$

The above expression subtly demonstrates another reason why E_b/N_0 turns out to be useful in the analysis of digital modulations: E_b/N_0 describes the contribution of the modulation to SNR in a way which is independent of data rate $1/T_b$ and bandwidth B. Data rate and bandwidth can be varied in all kinds of ways without changing the modulation, so if one is really interested in comparing the *fundamental* performance of modulations independently of these additional factors, then E_b/N_0 is the way to go.

A final note before moving on: It is not uncommon to see E_b/N_0 referred to as "SNR per bit." This refers to the notion that, in principle, one receives and attempts to detect symbols; only later does one interpret symbols as collections of bits. Therefore E_b/N_0 can be viewed as a normalization of SNR to account for the impact on a bit-by-bit basis, as opposed to being limited to interpreting performance as the ability to detect symbols.

6.6 COHERENT DEMODULATION

This section introduces a common "recipe" for demodulation of all types of digital phase-amplitude modulations. Before we get to this somewhat specific approach, it is useful to have some idea of the scope of possible approaches and, in particular, what constitutes an *optimal* demodulator.

To make this discussion tractable, we will assume throughout this section that the modulation is *memoryless* and that the propagation channel is both *memoryless* and *time-invariant*. A memoryless modulation is one in which any given symbol is independent from all previous symbols. A memoryless propagation channel is one which is free of resolvable multipath; that is, the demodulator does not perceive additional delayed copies of symbols except possibly as flat fading. A time-invariant propagation channel is – for present purposes – one in which the change in amplitude and phase of the channel over a symbol period is negligible. None of these assumptions is necessarily reasonable or strictly necessary for practical work, and we will take steps to account for more general situations in Sections 6.14 and 6.15.

6.6.1 Optimal Demodulation

The optimum demodulator under these assumptions is known generally as the *correlation receiver*.[5] The correlation receiver concept is discussed in Section 6.6.4. In this section we will consider the most common implementation of the correlation receiver, known as the *matched filter receiver*. This particular architecture is shown in Figure 6.11. The signal flow in the matched filter receiver is as follows.

- Quadrature downconversion to complex baseband representation.
- Matched filtering (filter having frequency response $G^*(f)$).
- Calculation of a decision metric by sampling the matched filter output at the appropriate MET. This also entails determination of symbol timing.
- Carrier phase tracking, which may also be interpreted as derotation of the symbol constellation. This might be integrated into quadrature downconversion.
- AGC, with the primary purpose of maintaining constant mean symbol energy E_S. This too may be implemented as part of an earlier stage in the demodulator.

[5] Another note on nomenclature: Communication theorists commonly refer to demodulators as receivers, although clearly the intent is not to account for all of the functions that a complete radio receiver would normally provide.

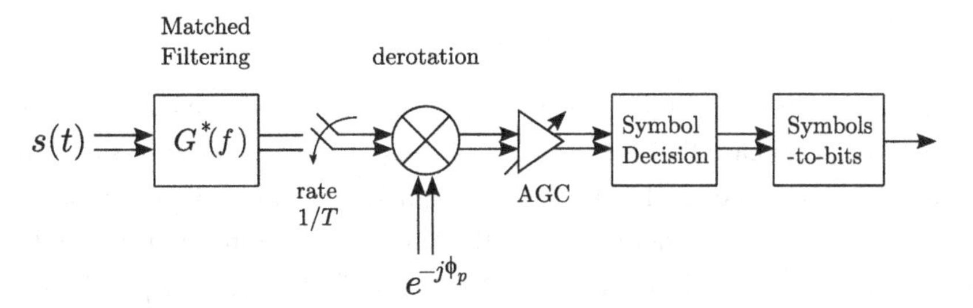

Figure 6.11. Coherent demodulation of digital phase-amplitude modulations by matched filtering in complex baseband representation.

- Symbol detection, consisting of selecting the most likely symbol given the decision metric.
- Conversion from symbol decisions to bits or bit sequences.

Not mentioned above but important to consider is carrier frequency tracking. This is often not an identifiable stage of processing in the demodulator, but may instead be integrated into quadrature downconversion or – when frequency errors can be assumed to be relatively small – simply "absorbed" into carrier phase tracking. Techniques for estimation of carrier frequency and phase, as well as symbol timing recovery, are addressed in Section 6.16.

6.6.2 Matched Filtering

Matched filtering can be interpreted as an optimal method to reduce a continuous waveform to a single value that best represents the symbol in effect at the time the filter output is sampled. Alternatively, matched filtering can be interpreted as a scheme that maximizes the SNR of the sampled output; i.e. of the decision metric. From this perspective, it is clear that a matched filter must take into account the symbol pulse waveform so as to emphasize portions of the spectrum where the power spectral density is high relative to the noise, and to deemphasize portions of the spectrum where the power spectral density is low relative to the noise. Therefore it should come as no surprise that the response of the optimal matched filter assuming spectrally-white noise is $G^*(f)$ (as shown in Figure 6.11), which is simply $G(f)$ in the vast majority of implementations in which the pulse waveform is real-valued.[6] Alternatively, the matched filter may be described in terms of its impulse response $\mathcal{F}^{-1}\{G^*(f)\} = g^*(-t)$, which is simply $g(t)$ in the vast majority of implementations in which $g(t)$ is both real-valued and time-symmetric.

6.6.3 Square Root Raised Cosine (SRRC) Matched Filtering

Comparing Figures 6.5 and 6.11 we note that we have a "pulse-matched" filter $G(f)$ in the modulator in order to set the bandwidth while enforcing the Nyquist criterion for ISI-free modulation, and essentially the same filter $G^*(f)$ in the demodulator in order to implement optimal matched filtering. At this point a problem becomes evident: The combined response of the filters in the modulator and demodulator is

[6] A rigorous proof is relatively straightforward, but a bit tedious. Since the result is not surprising, we shall not present the derivation here.

$$G(f)\,G^*(f) = |G(f)|^2 \tag{6.18}$$

For pulses which are strictly time-limited to duration $\leq T$, this poses no particular problem. For the more general and useful case that the duration of the pulse may not be so time-limited, the resulting response does not necessarily satisfy the Nyquist criterion for ISI-free modulation. In particular, it is simple enough to demonstrate that the ISI-free criterion is *not* satisfied for raised cosine pulses after passing through the associated matched filter. This is a problem because it is not enough to satisfy the ISI-free criterion in the transmit signal: One needs the result to be ISI-free where the symbol is to be detected; i.e., after the matched filter in the demodulator.

The fix is simple enough: We let $G(f)$ be not the raised cosine response, but rather the *square root* of the raised cosine response. To avoid ambiguity, let us define $G_{RC}(f)$ to be the particular choice of raised cosine response as given by Equation (6.6), and let us define $G_{SRRC}(t)$ to be the square root of this response. Now if we use $G(f) = G_{SRRC}(f)$ we have

$$G_{SRRC}(f)\,G^*_{SRRC}(f) = |G_{SRRC}(f)|^2 = |G_{RC}(f)| \tag{6.19}$$

which *does* satisfy the ISI-free criterion, and retains the ability to control the bandwidth-impulse response tradeoff (parameterized by roll-off factor r) as worked out in Section 6.4.3. For this reason, *square root raised cosine* (SRRC) pulse shaping emerges as the most popular scheme employed in modern digital communications systems.

6.6.4 The Correlation Receiver

As noted in Section 6.6.1, the matched filter receiver described above is actually a special case of the correlation receiver, which is applicable to *any* digital modulation. We will encounter a number of situations in the remainder of this chapter where the correlation receiver concept provides a useful framework for understanding modulation performance. So, although the details are beyond the scope of this book, it is nevertheless worthwhile to be familiar with the concept.

The basic principle of the correlation receiver is as follows: For each symbol to be received, one correlates the input with each of the M possible symbol waveforms. This results in M *decision metrics*. Each decision metric indicates the degree to which the input is correlated with the associated symbol waveform, and therefore the probability that the associated symbol is present increases with the decision metric. If E_m is not constant for all symbols in the constellation, the metrics are adjusted in proportion to the associated symbol energy to ensure fairness. Also if there is *a priori* knowledge that some symbols appear more often than others, then the metrics are adjusted to account for this as well. The decision metrics are then compared, and the "winner" is the symbol associated with the largest metric. This particular approach goes by various other names, some referring to statistical criteria such as *maximum a posteriori* (MAP) and *maximum likelihood* (ML; essentially MAP in which all symbols are assumed to be equally probable).

The design and analysis of correlation receivers is a advanced topic, and is quite useful for obtaining deep insight into problem of modulation design and analysis. A recommended starting point for interested readers is [59, Ch. 4].

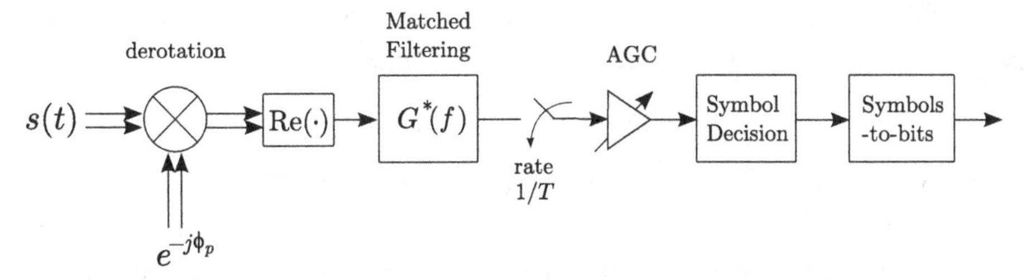

Figure 6.12. Demodulation of BPSK and OOK.

6.7 DEMODULATION OF BPSK AND OOK

BPSK and OOK are the two canonical $M = 2$ phase-amplitude modulations. Simplification relative to the general architecture of Figure 6.11 is possible because these modulations are "one-dimensional" in the sense that the quadrature component of the complex baseband representation is zero once derotation is applied. This allows the quadrature component of the decision metric – including the noise associated with that component – to be discarded without consequence.

The simplified architecture is shown in Figure 6.12. Moving derotation to precede matched filtering allows the quadrature component to be discarded before matched filtering, which eliminates unnecessary computation.

6.7.1 Optimal Demodulation of BPSK

Let us now consider the specific method for extracting a symbol decision from the detection metric d for BPSK. A simple and generic method to work this out is as follows. We first write the two possible symbols, as they appear at the input of the matched filter, as follows:

$$s_1(t) = -\sqrt{E_S}\, g(t) \tag{6.20}$$

$$s_2(t) = +\sqrt{E_S}\, g(t) \tag{6.21}$$

where – with no loss of generality – we have required $g(t)$ to have unit energy; i.e.,

$$\int_{-\infty}^{+\infty} |g(t)|^2\, dt = 1 \tag{6.22}$$

Presumably the probability of erroneously detecting $s_1(t)$ when $s_2(t)$ has been sent is the same as the probability of erroneously detecting $s_2(t)$ when $s_1(t)$ has been sent. We may therefore calculate the probability of symbol error P_S as the former, with confidence that the result applies in either case. So, let us assume symbol $s_1(t)$ has been sent.

The resulting decision metric is the output of the matched filter sampled at the MET, yielding

$$d = [s_1(t) + n(t)] * g^*(t)\, \Big|_{\text{MET}} \tag{6.23}$$

where $n(t)$ is the in-phase component of the noise waveform in complex baseband representation, prior to matched filtering. Separating this into two terms we have

$$d = s_1(t) * g^*(t)\Big|_{\text{MET}} + n(t) * g^*(t)\Big|_{\text{MET}} \tag{6.24}$$

The first term is simply $\sqrt{E_S}$ as a consequence of Equation (6.22). The second term is a real-valued random variable N which is Gaussian-distributed with zero mean. The variance of N, σ_N^2, is the power associated with $n(t)$ after matched filtering. Because the power spectral density (PSD) of $n(t)$ is expected to be uniform (white) over the bandwidth of the matched filter, a simple expression for σ_N^2 can be obtained by integration in the frequency domain as follows:

$$\sigma_N^2 = \int_{-\infty}^{\infty} \left|G^*(f)\right|^2 \frac{N_0}{2} df \tag{6.25}$$

where N_0 is the "one-sided" PSD of $n(t)$. By "one-sided" we mean the PSD defined so that the integral of N_0 over the real frequencies (only) accounts for all of the noise power. Since we are working in complex baseband representation, the total noise power is distributed between evenly among positive and negative frequencies; therefore the relevant PSD in this case is $N_0/2$. Evaluating the above expression:

$$\sigma_N^2 = \frac{N_0}{2}\int_{-\infty}^{\infty} \left|G^*(f)\right|^2 df = \frac{N_0}{2} B_N \tag{6.26}$$

Although not really relevant here, it is worth pointing out that B_N is the *noise equivalent bandwidth* of the matched filter. What is relevant is that $B_N \equiv 1$, since we defined $g(t)$ to have unit energy.[7] Therefore we have simply

$$\sigma_N^2 = \frac{N_0}{2} \tag{6.27}$$

If it bothers you that this expression has units of power on the left and units of PSD on the right, just keep that factor of B_N attached and remember that it is equal to 1 with respect to whatever units you decide to use for energy.

Now let us consider the probability distribution function (PDF) of the decision metric, which is shown in Figure 6.13. In the absence of noise, $d = -\sqrt{E_S}$. In the presence of noise, possible values of d are scattered around the nominal value according to a Gaussian distribution with variance $N_0/2$, as we established above. The same is true in the case where $s_2(t)$ has been sent, except the nominal value is $d = +\sqrt{E_S}$. It is evident that the following *decision rule* will minimize the probability of symbol error: Assume $s_1(t)$ was sent if $d < 0$, and assume $s_2(t)$ was sent if $d > 0$. In this case, the value of zero is considered the *decision threshold* d_{th}.

That completes the design of the demodulator. Now we consider the probability of symbol error. Following previous arguments, we may evaluate P_S as the probability of detecting $s_2(t)$ when $s_1(t)$ has been sent:

$$P_S = \int_0^{\infty} P_d(x)dx \tag{6.28}$$

[7] If it is not obvious why this is so, the quick answer is *Parseval's Theorem*: The integral of a magnitude-squared signal over all time is equal to the integral of the magnitude-squared spectrum over all frequencies.

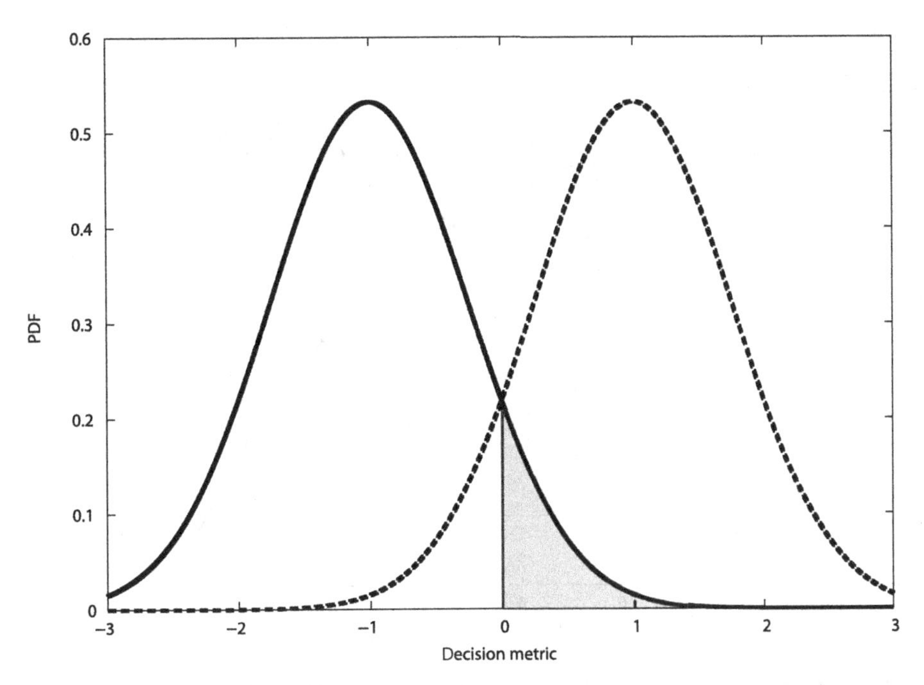

Figure 6.13. PDFs for BPSK symbols assuming zero-mean Gaussian noise. The horizontal axis is in units of $\sqrt{E_S}$. Here, $\sigma_N = 0.75\sqrt{E_S}$. The shaded area is the probability of symbol error when $s_1(t)$ has been sent.

where $P_d(x)$ is the PDF of d, as shown in Figure 6.13. This is somewhat more easily interpreted in terms the PDF of N:

$$P_S = \int_{\sqrt{E_S}}^{\infty} P_N(y)dy \tag{6.29}$$

This in turn may be written in terms of the cumulative distribution function (CDF) of N:

$$P_S = 1 - F_N\left(\sqrt{E_S}\right) \tag{6.30}$$

N is Gaussian-distributed with zero mean and standard deviation σ_N; the PDF in this case is known to be

$$F_N(y) = 1 - \frac{1}{2}\text{erfc}\left(\frac{y}{\sqrt{2}\,\sigma_N}\right) \tag{6.31}$$

where "erfc" is the *complementary error function* defined as

$$\text{erfc}(y) \equiv 1 - \frac{2}{\sqrt{\pi}}\int_0^y e^{-z^2}dz \tag{6.32}$$

Returning to Equation (6.30) and making the available substitutions we find:

$$P_S = \frac{1}{2}\text{erfc}\left(\sqrt{\frac{E_S}{N_0}}\right) \tag{6.33}$$

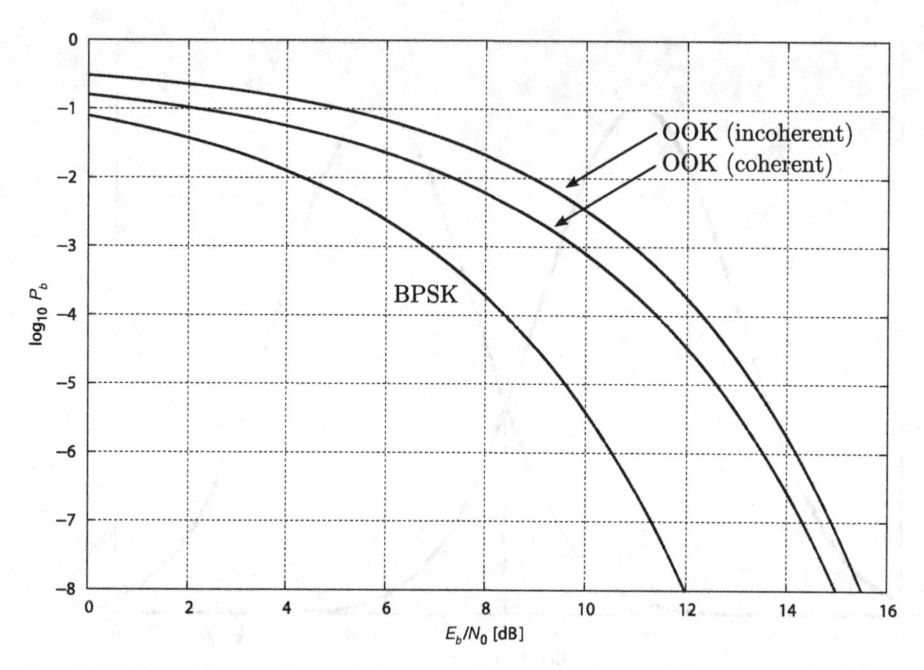

Figure 6.14. Probability of bit error P_b as a function of E_b/N_0 for BPSK and OOK.

While this is a perfectly reasonable answer, it is somewhat more common in communications systems analysis to express the result in terms of the "Q" function as opposed the complementary error function. The "Q" function is given by:

$$Q(y) \equiv \frac{1}{\sqrt{2\pi}} \int_{y}^{\infty} e^{-z^2/2} dz = \frac{1}{2} \text{erfc}\left(\frac{y}{\sqrt{2}}\right) \tag{6.34}$$

So in terms of the Q function we have:

$$P_S = Q\left(\sqrt{2\frac{E_S}{N_0}}\right) \tag{6.35}$$

For $M = 2$, symbols map one-to-one to bits. Therefore the probability of bit error P_b is identical to P_S. Also E_S is identically E_b. Thus we have

$$P_b = Q\left(\sqrt{2\frac{E_b}{N_0}}\right) \tag{6.36}$$

This result is plotted in Figure 6.14. Finally, note that P_b is also identical to the *bit error rate* (BER), which is the number of bits that are expected to be in error per bit transmitted.

6.7.2 Optimal Demodulation of OOK

Optimal demodulation and probability of error for OOK may be determined in a very similar fashion. The two symbols as they appear at the input of the matched filter are:

$$s_1(t) = 0 \tag{6.37}$$

$$s_2(t) = +\sqrt{2E_S}\, g(t) \tag{6.38}$$

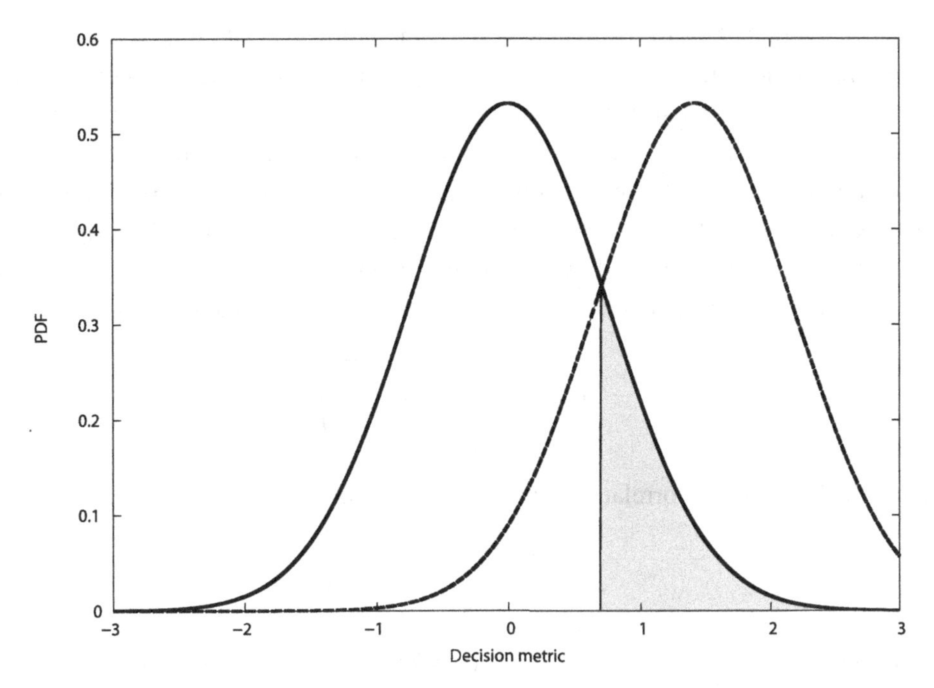

Figure 6.15. PDFs for OOK symbols assuming zero-mean Gaussian noise. The horizontal axis is in units of $\sqrt{E_S}$. Here, $\sigma_N = 0.75\sqrt{E_S}$. The shaded area is the probability of symbol error when $s_1(t)$ has been sent.

where $g(t)$ is again defined to have unit energy, as in the previous section. Note that the energy of $s_2(t)$ is $2E_S$ (*not* E_S): This is because E_S is the *mean* symbol energy and the energy of $s_1(t)$ is zero. Let us assume symbol $s_1(t)$ has been sent. The decision metric associated with this case is

$$d = n(t) * g^*(t) \Big|_{\text{MET}} \tag{6.39}$$

The PDF of the decision metric is shown in Figure 6.15. In the absence of noise, $d = 0$. In the presence of noise, possible values of d are scattered around zero according to a Gaussian distribution with variance $N_0/2$. The same is true in the case where $s_2(t)$ has been sent, except the nominal value is $d = +\sqrt{2E_S}$. It is evident that the following *decision rule* will minimize P_S: Assume $s_1(t)$ was sent if $d < d_{th} = \sqrt{2E_S}/2 = \sqrt{E_S/2}$, otherwise assume $s_2(t)$ was sent. In this case, $\sqrt{E_S/2}$ is the decision threshold.

Following previous arguments, we may evaluate P_S for OOK as the probability of detecting $s_2(t)$ when $s_1(t)$ has been sent:

$$P_S = \int_0^{\infty} P_d(x)dx = \int_{\sqrt{E_S/2}}^{\infty} P_N(y)dy = Q\left(\sqrt{\frac{E_S}{N_0}}\right) \tag{6.40}$$

Since $P_b = P_S$ and $E_S = E_b$, and we have

$$P_b = Q\left(\sqrt{\frac{E_b}{N_0}}\right) \tag{6.41}$$

This result is plotted with BPSK in Figure 6.14.

Comparing results for BPSK and OOK, we see that OOK is clearly inferior in terms of the probability of error delivered for a given E_b/N_0. From a system perspective, OOK requires

SNR twice that required by BPSK to deliver the same probability of error. Why should this be? A simple way to rationalize this result is in terms of the optimal correlation receiver described in the beginning of Section 6.6.4. BPSK symbols are *anti-correlated*; i.e., the correlation between symbols having unit energy is -1. OOK symbols, on the other hand, are merely *uncorrelated*; i.e., the correlation between symbols is zero. For either modulation, the correlation between the same symbols is $+1$. So for a given mean symbol energy, it is easier to discriminate between BPSK symbols than between OOK symbols. Although we shall not derive this here, it turns out that this principle applies to *any* $M = 2$ phase-amplitude modulation as follows:

$$P_b = Q\left(\sqrt{\frac{E_b}{N_0}(1-\rho)}\right) \tag{6.42}$$

where ρ is the normalized correlation between the symbols

$$\rho \equiv \frac{1}{E_S}\int_{-\infty}^{+\infty} s_1(t)s_2^*(t)dt \tag{6.43}$$

We previously established that BPSK and OOK have equal spectral efficiency (Section 6.4.4), and now we see that OOK has inferior error performance. For this reason, OOK is rarely used in modern communications systems. There are some important exceptions, falling into two broad categories. First, since OOK modulation might be easier to generate, OOK becomes particularly attractive if transmitter complexity is severely constrained. The second exception has to do with the fact that OOK – unlike BPSK – can be demodulated incoherently. This results in considerable simplification of the receiver and intrinsic robustness to carrier phase variation. We consider this possibility next.

6.7.3 Incoherent Demodulation of OOK

OOK is analogous to DSB-LC AM in that information is represented entirely as changes in carrier magnitude. This facilitates incoherent modulation in essentially the exact same manner described in Section 5.5.3. Just as OOK might be selected over BPSK when transmitter power/complexity is severely constrained, one might consider incoherent OOK over coherent OOK when *receiver* power/complexity is severely constrained.

The simplified demodulator architecture for incoherent OOK is shown in Figure 6.16. Compare this to Figures 6.12 and 5.8. The principal difference is that derotation to account for propagation-induced carrier phase is no longer required. Carrier frequency tracking is typically not required, either. This approach has some associated pros and cons that are important to consider. First, eliminating derotation considerably simplifies the receiver, since there is then no need to estimate carrier phase – recall from Section 5.5.4 (and looking forward to Section 6.16) that this is one of the principal headaches in coherent demodulation. At the same time, sacrificing derotation means the receiver becomes sensitive to both the in-phase and quadrature components of the noise – review Section 5.5.5 and Figure 5.15 to reacquaint yourself with that concept.

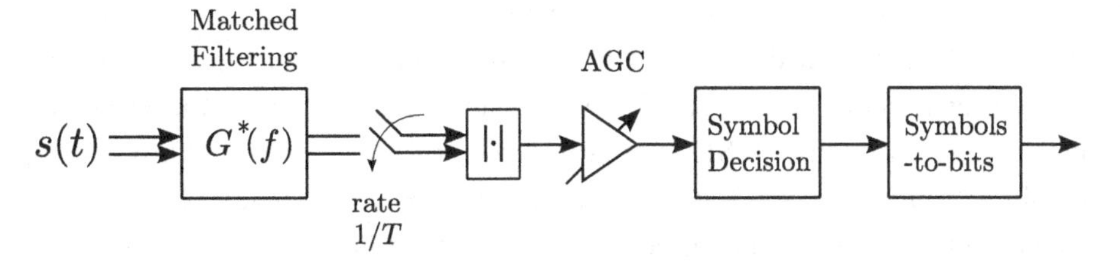

Figure 6.16. Incoherent demodulation of OOK. Note derotation is not required. As a practical matter, it is probably easier to calculate magnitude-squared ($|\cdot|^2$) as opposed to magnitude ($|\cdot|$) and then adjust the AGC or symbol decision threshold accordingly.

Derivation of the error performance of incoherently-detected OOK is straightforward, but beyond the scope of this book. The result is

$$P_b = \frac{1}{2}e^{-\frac{1}{2}E_b/N_0} \tag{6.44}$$

This is plotted in Figure 6.14. Note that incoherently-detected OOK is roughly 1 dB worse than coherent OOK. In many applications, this amount of degradation is worth the reduction in receiver complexity.

6.8 DEMODULATION OF QPSK

Coherent demodulation of QPSK requires the complete architecture of Figure 6.11, since information exists in both the in-phase and quadrature components of the complex baseband representation. Extrapolating from Section 6.7.1 it is not difficult to see that the optimal decision rule is simply to determine the quadrant in which the now-complex-valued decision metric d lies, and then declare the result to be the symbol associated with that quadrant. The associated derivation is straightforward, if a bit tedious; however, there is a simpler way to interpret QPSK that also yields some useful additional insight. Here it is: Gray-coded QPSK is simply two BPSK signals in phase quadrature.

To see this, note in Figure 6.3(b) that the in-phase component of the Gray-coded constellation encodes the low-order bit of the dibit, whereas the quadrature component of the constellation encodes the high-order bit. Thus, each symbol decision may actually be interpreted as two independent operations: Determination of the low-order bit, depending only whether $\mathcal{Re}\{d\}$ is $>$ or $<$ zero, and determination of the high-order bit, depending only whether $\mathcal{Im}\{d\}$ is $>$ or $<$ zero. Because each component of the complex baseband representation includes statistically-independent noise with equal variance, the probability of bit error in each of these operations is equal, and equal to that of BPSK. We are left to conclude that the probability of bit error for QPSK is the same as that for BPSK; i.e.,

$$P_b = Q\left(\sqrt{2\frac{E_b}{N_0}}\right) \tag{6.45}$$

This result has a rather important consequence: For a given symbol rate $1/T$ and mean symbol energy E_S, QPSK has twice the bit rate $1/T_b$ of BPSK, and yet has precisely the same probability of bit error. Since the spectrum is determined by the symbol rate, QPSK and BPSK have equal bandwidth even though QPSK has twice the bit rate. Recall also that QPSK is "power amplifier friendly" in the sense that it requires much smaller variation in magnitude over time. For these reasons, QPSK is preferred over BPSK in nearly all modern applications that are able to support the modest increase in complexity associated with the simultaneous processing of both in-phase and quadrature components of the baseband signal.

6.9 DEMODULATION OF HIGHER-ORDER PHASE-AMPLITUDE MODULATIONS

For the purposes of this book, *higher-order* modulations refer to modulations having $M > 2$. Techniques for the generation and demodulation of higher-order phase-amplitude modulations are straightforward extensions of techniques described for OOK, BPSK, and QPSK in previous sections. Determination of probability of symbol or bit error, on the other hand, is a considerably more difficult problem. This section briefly summarizes a few higher-order phase-amplitude modulations.

6.9.1 *M*-ASK

BPSK may be straightforwardly generalized to symmetric M-order amplitude shift keying (M-ASK). The constellation for symmetric M-ASK is shown in Figure 6.17. "Symmetric" refers to the fact that the constellation is centered on the origin of the complex plane; otherwise the modulation would include a power-wasting DC component as does DSB-LC AM and OOK.

Demodulation of M-ASK is straightforwardly accomplished by using the simplified architecture of Figure 6.12. In this case the decision rule can be summarized as follows: The symbol is determined to be one that minimizes distance between the real-valued decision metric d (or $\mathcal{R}e\{d\}$ depending on your interpretation) and the coordinate of the symbol along the real axis of the complex plane. An alternative but equivalent description of the decision rule uses the concepts of decision regions. In Figure 6.17 (4-ASK), for example, the rule might be expressed as follows: If d lies in Region I, then the symbol is assumed to be A; If d lies in Region II, then the symbol is assumed to be B; and so on.

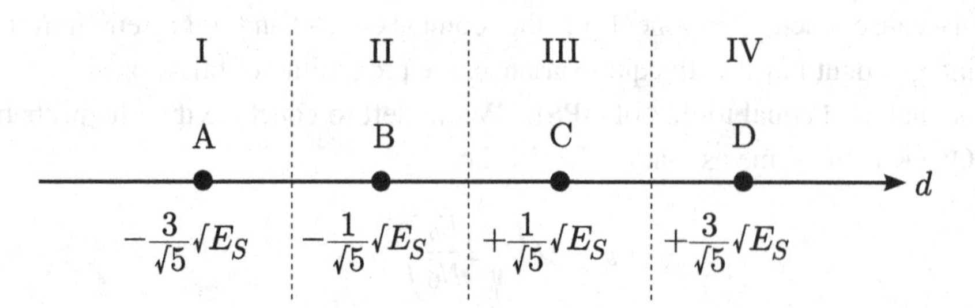

Figure 6.17. Symmetric M-ASK. In this case, $M = 4$. Capital letters index the symbols, whereas Roman numerals index the associated decision regions.

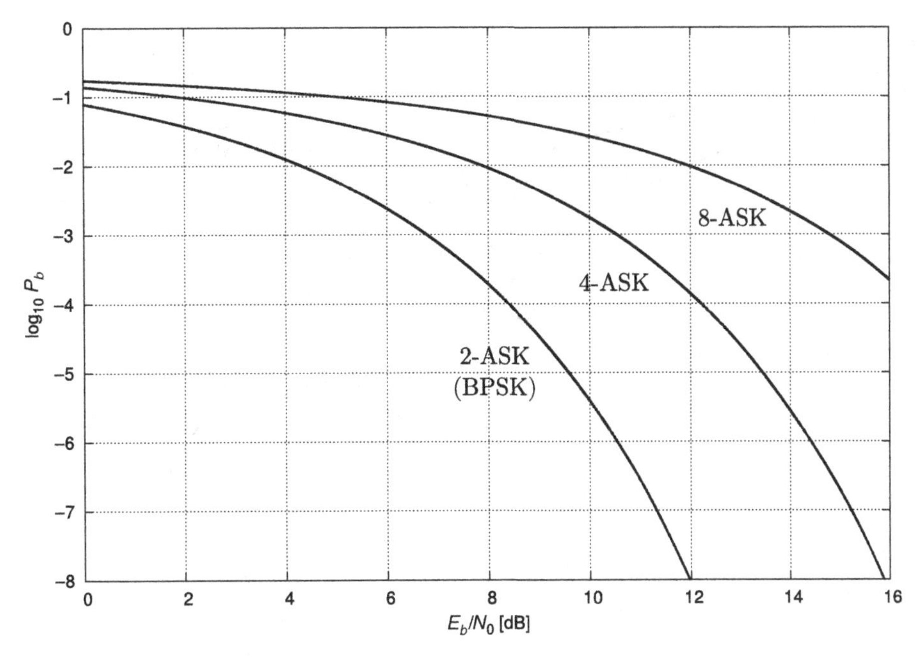

Figure 6.18. Probability of bit error as a function of E_b/N_0 for M-ASK.

The probability of bit error for symmetric M-ASK may be derived by a generalization of the derivation for BPSK. The result assuming Gray coding is

$$P_b = \frac{2\,(M-1)}{M \log_2 M}\, Q\left(\sqrt{\frac{6 \log_2 M}{M^2-1} \cdot \frac{E_b}{N_0}}\right) \tag{6.46}$$

For BPSK, which is symmetric 2-ASK, the above expression yields the expected result $P_b = Q\left(\sqrt{2E_b/N_0}\right)$. The result for M-ASK for other values of M is shown in Figure 6.18. Note that the E_b/N_0 required to achieve a particular BER increases with increasing M. This is because the separation between symbols within the constellation – relative to $\sqrt{E_S}$ – decreases with increasing M.

The primary drawback of M-ASK is that it requires many large variations in magnitude in symbol transitions. As pointed out above, this requires relatively higher-performance and less-efficient power amplifiers. A practical case where M-ASK does find application is in the North American broadcast television system; see Section 6.17.

6.9.2 *M*-QAM

M-QAM may be interpreted as a generalization of symmetric M-ASK in the same way that QPSK may be interpreted as a generalization of BPSK. Like all phase-amplitude modulations, demodulation of M-QAM is straightforwardly accomplished by using the architecture of Figure 6.11. In the particular case of M-QAM the decision rule can be summarized as follows: The symbol is determined to be one that minimizes the distance between d and the coordinate of the symbol the complex plane. Alternatively (and again, analogous to previous modulations

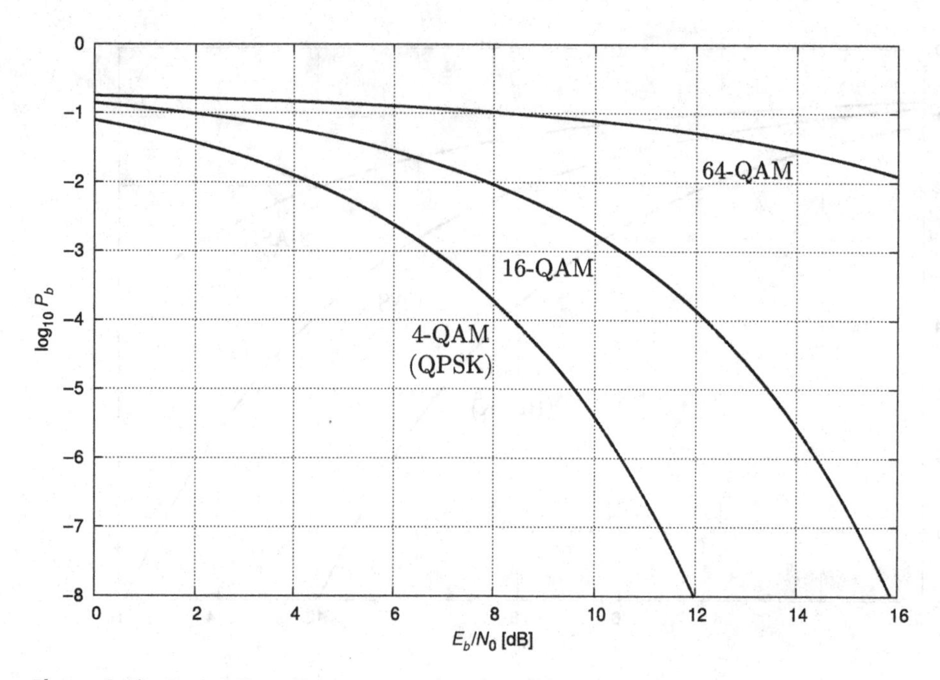

Figure 6.19. Probability of bit error as a function of E_b/N_0 for M-QAM.

addressed) the decision rule may be interpreted in terms of decision regions surrounding each symbol.

We found in Section 6.8 that QPSK could be interpreted as two BPSK constellations in phase quadrature, and therefore exhibited the same BER as BPSK for a given E_b/N_0. This analogy extends to M-QAM, which may be interpreted as $\sqrt{M}$ each $\sqrt{M}$-ASK constellations in phase quadrature. The probability of bit error for M-QAM is therefore Equation (6.46) with each instance of M replaced by $\sqrt{M}$:

$$P_b = \frac{2\left(\sqrt{M}-1\right)}{\sqrt{M}\log_2\sqrt{M}}\, Q\left(\sqrt{\frac{6\log_2\sqrt{M}}{M-1}\cdot\frac{E_b}{N_0}}\right) \tag{6.47}$$

The above result is readily verified to yield the expected result for QPSK, which is 4-QAM. The result for M-QAM for other values of M is shown in Figure 6.19. Here, as in the case of M-ASK, the effect of increasing M is to increase the E_b/N_0 required to maintain a specified BER. However, note that relatively larger values of M may be accommodated within a particular constraint on E_b/N_0 and BER. For this reason, M-QAM is far more popular than M-ASK when large M is desired.

6.9.3 *M*-PSK

Like all phase-amplitude modulations, demodulation of M-PSK is straightforwardly accomplished by using the architecture of Figure 6.11. However, because M-PSK encodes information entirely in phase, the nature of the decision rule is somewhat different: For M-PSK, the symbol is determined to be the one that minimizes the difference in phase between d and the symbol

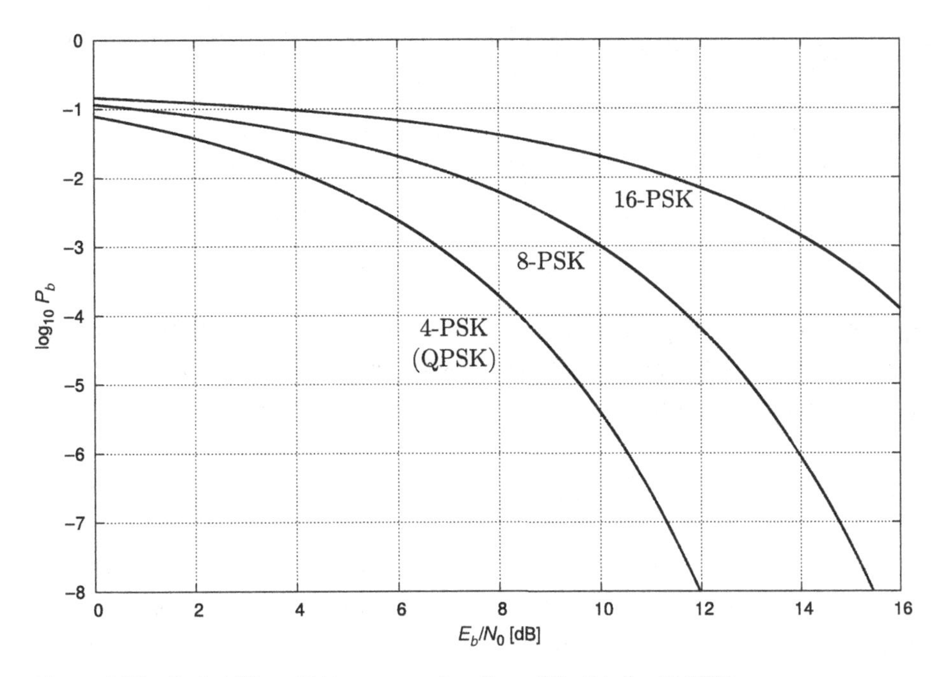

Figure 6.20. Probability of bit error as a function of E_b/N_0 for M-PSK.

the complex plane. Decision regions for M-PSK take the shape of wedges with a common point at the origin.

The probability of bit error for M-PSK is relatively difficult to work out; however, a usable approximation exists in the form of the following expression:

$$P_b \approx \frac{2}{\log_2 M} Q\left(\sqrt{2\frac{E_b}{N_0}\left(\log_2 M\right)} \cdot \sin\frac{\pi}{M}\right) \tag{6.48}$$

The above result is readily verified to yield the expected result for QPSK, which is 4-PSK. Interestingly the above result does *not* yield the expected result for BPSK, which is 2-PSK. (Problem 6.7 invites you to speculate on the reason for this apparent discrepancy.) The result for M-PSK for $M \geq 2$ is shown in Figure 6.20. As for previous modulations considered, the effect of increasing M is to increase the E_b/N_0 required to maintain a specified BER. Generally, M-PSK performs significantly worse than M-QAM. Nevertheless, M-PSK continues to find applications. The principal advantage of M-PSK over M-QAM is reduction in complexity, since accurate AGC is not required for M-PSK.

6.10 DIFFERENTIAL DETECTION

So far we have classified demodulators as *coherent* or, in the case of those modulations that permit it, *incoherent*. A third option for digital demodulation is *differential detection*. The fundamental idea in differential detection is to forgo detection of individual symbols and instead detect the *change* from one symbol to the next. Although generally applicable, the

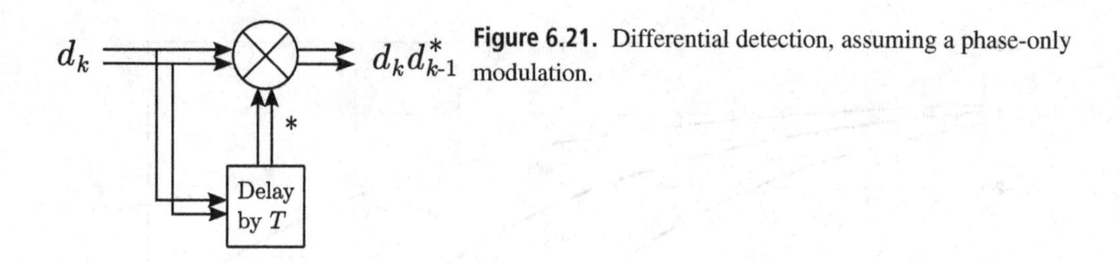

Figure 6.21. Differential detection, assuming a phase-only modulation.

technique is most often applied to phase-only modulations, for which there is a particular advantage in differential detection: Robustness to carrier phase variation and, to a limited extent, carrier frequency error.

6.10.1 Concept

The basic idea is illustrated in Figure 6.21. To begin, let us assume negligible error in carrier frequency, and let us further assume that the variation in the carrier phase ϕ_p over any two contiguous symbol periods is negligible. In this case, the decision metric d_k for the kth symbol, assuming perfect symbol timing, may be written

$$d_k = \sqrt{E_S}\, e^{j(\phi_k+\phi_p)} \tag{6.49}$$

After delay by T and conjugation, we have

$$d_{k-1}^* = \sqrt{E_S}\, e^{-j(\phi_{k-1}+\phi_p)} \tag{6.50}$$

So the output of the detector is

$$d_k d_{k-1}^* = E_S\, e^{j(\phi_k-\phi_{k-1})} \tag{6.51}$$

The magnitude of the result is irrelevant since we are assuming a phase-only modulation. The remarkable feature of the result is that the contribution of the carrier phase has been eliminated, so derotation is not required. On the other hand, we are left not with the absolute phase, but only the phase difference between the current and previous symbols.

This is useful only if the data are represented as phase differences, as opposed to absolute phases. Fortunately, this is quite easily accomplished using a scheme known as *differential coding*. The idea is easily demonstrated by example, using BPSK. Consider the following bit stream:

```
0 1 0 0 1 1 0 0 0 1 1 1 0 0 0
```

Assuming BPSK symbols `A` and `B`, we might normally represent these data as follows:

```
A B A A B B A A A B B B A A A
```

In differential coding, we would instead interpret a "`0`" to mean "use the previous symbol" and "`1`" to mean "change symbols". Let us assume the last symbol preceding the data considered here was `A`. Then we would have:

```
A B B B A B B B B A B A A A A
```

The data are recovered as follows: Working from left to right, we write "`1`" if the current symbol is different from the previous symbol, and "`0`" otherwise. This yields:

```
0 1 0 0 1 1 0 0 0 1 1 1 0 0 0
```

Thus, we have recovered the original data by considering only the change from symbol to symbol, as opposed to determining the symbols themselves. BPSK encoded in this way is sometimes referred to as *differential BPSK* (DBPSK). Similarly, differentially encoded QPSK, which is also quite common, is known as DQPSK.

It is worth noting that differential coding may be implemented regardless of whether the modulation is detected coherently or differentially. For this reason, it is common to encode data differentially, so that either method may be used. Differential coding has the additional advantage in that it eliminates the need to resolve the rotational ambiguity associated with all symmetric constellations: That is, the fact that the constellation looks the same after a certain amount of rotation. Using coherent demodulation without differential encoding, it is necessary to implement some additional method to resolve the ambiguity, such as a *training sequence*. A training sequence is a known sequence of symbols that are periodically inserted into the data.

Returning to our original motivation: The primary advantage of differential detection is that it eliminates the need for carrier phase estimation and derotation. Further: Since frequency is the derivative of phase with respect to time, differential detection is also robust to small errors in carrier frequency; all that is required is that the associated phase error accumulated over a symbol period is small compared to the minimum phase difference between symbols.

6.10.2 Performance

These nice features of differential detection come at the cost of error performance as a function of SNR, which is not quite as good as coherent demodulation. There are two ways to view this degradation of performance. First, note that any symbol that would have been detected erroneously by coherent detection could result in two successive errors in differential detection, since each symbol is used twice. Second, note that by combining two versions of the signal (one delayed) together, one is effectively doubling the noise associated with any one symbol decision. The probability of bit error for DBPSK is found to be[8]

$$P_b = \frac{1}{2}e^{-E_b/N_0} \tag{6.52}$$

This is compared to coherently-detected BPSK in Figure 6.22. The performance shortfall is noted to be about 1.5 dB for 1% BER; that is, E_b/N_0 must be about 1.5 dB greater for DBPSK in order to achieve the same BER as coherently-detected BPSK. For higher-order PSKs, the penalty tends to be closer to 3 dB.

[8] This is relatively difficult to derive. For details, see [59].

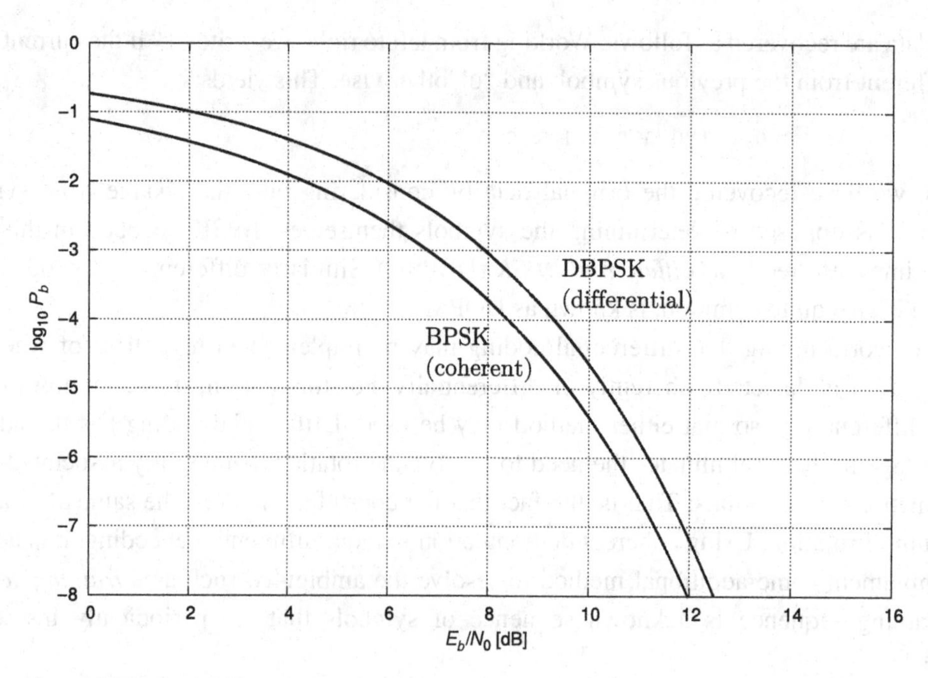

Figure 6.22. Probability of bit error for differentially-detected BPSK (DBPSK) compared to optimally-(coherently-) detected BPSK.

Differential-encoded and detected QPSK is common. The tradeoff with respect to coherent BPSK is simpler implementation with about twice the spectral efficiency, but with only about half the energy efficiency.

6.11 FREQUENCY-SHIFT KEYING (FSK)

Within the broad class of sinusoidal carrier modulations, the alternative to phase-amplitude modulation is frequency modulation. When used as a digital modulation, frequency modulation is typically referred to a frequency-shift keying (FSK). In modern communications systems, FSK fits two particular niches. First, FSK serves as a low-complexity alternative to phase-amplitude modulation when high robustness to channel fading is required. Second, FSK is useful when very high energy efficiency is required and spectral efficiency (i.e., bandwidth) is less important. These two properties should become evident as we progress through this section.

6.11.1 Concept

The modulation symbols in M-order FSK are simply M sinusoids with frequency separation $\Delta\omega = 2\pi\,\Delta f$. The symbols may be pulse-shaped in precisely the same way as symbols in phase-amplitude modulations. Although FSK symbols can be represented in complex baseband representation, they cannot be represented as fixed points in the complex plane;[9] therefore constellation diagrams for FSK modulations are not possible.

[9] Actually, they can if one is willing to redefine the meaning of the coordinates of the complex plane. This approach, known as *signal space representation*, is beyond the scope of this book.

The principal design parameters for an FSK modulation are therefore M, the pulse shape, and $\Delta\omega$. The principal consideration in selecting $\Delta\omega$ is a tradeoff between bandwidth, which favors minimizing $\Delta\omega$; and correlation between symbols, which favors maximizing $\Delta\omega$ as we shall soon see. Low correlation between symbols is desired because this reduces the probability of symbol decision error, as we have already noted.

Strategies for selecting $\Delta\omega$ can be demonstrated by the following simple example: Consider two FSK symbols as follows:

$$s_{RF,1}(t) = A_c \cos \omega_c t \tag{6.53}$$

$$s_{RF,2}(t) = A_c \cos \left([\omega_c + \Delta\omega] \, t \right) \tag{6.54}$$

Let these symbols be defined to be identically zero for $t < 0$ and $t > T$; i.e., we assume rectangular pulse shaping. The unnormalized correlation between symbols is

$$\int_0^T s_{RF,1}(\tau) s_{RF,2}(\tau) d\tau \tag{6.55}$$

Substitution of the symbol waveforms and application of a trigonometric identity yields

$$\frac{A_c^2}{2} \int_0^T \cos \left(\Delta\omega\tau \right) d\tau + \frac{A_c^2}{2} \int_0^T \cos \left([2\omega_c + \Delta\omega] \, \tau \right) d\tau \tag{6.56}$$

The second term is negligible as long as $2\omega_c\tau \gg 1/T$; this is a condition which is typically very well-met for radio communications. Using the definition of sinc from Equation (6.4) and $\Delta\omega = 2\pi \Delta f$ the first term is

$$\frac{A_c^2 T}{2} \text{sinc} \left(2\Delta f T \right) \tag{6.57}$$

Note that $A_c^2 T/2$ is the symbol energy; therefore the normalized correlation is simply

$$\rho \cong \text{sinc} \left(2\Delta f T \right) \tag{6.58}$$

This result is plotted in Figure 6.23. Two strategies for low intersymbol correlation become apparent. The first strategy is simply to make $\Delta f \gg 1/T$. This results in low intersymbol correlation independently of Δf. This approach has horrible spectral efficiency, but is convenient if precise frequency control is difficult; for example, in propagation channels subject to large Doppler variations. The second strategy is to select $2\Delta f T$ to be an integer, such that $\rho = 0$. In fact, this naturally facilitates a spectrally-efficient M-FSK for $M \geq 2$; just separate the symbol frequencies by $\Delta f = 1/\left(2T\right)$.

6.11.2 Minimum-Shift Keying (MSK)

Binary frequency-shift keying (2-FSK) with $\Delta f = 1/\left(2T\right)$ is referred to as *minimum-shift keying* (MSK), where "minimum" refers to the minimum frequency spacing that results in uncorrelated symbols.

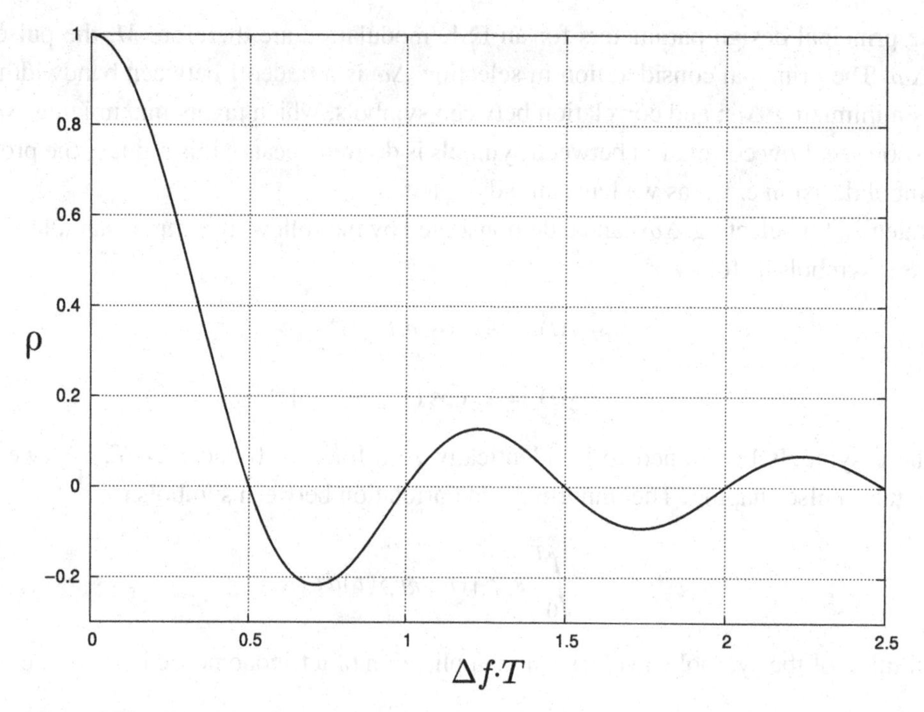

Figure 6.23. Correlation between FSK symbols as a function of normalized frequency spacing $\Delta f \cdot T$.

MSK has some very nice features from the perspective of the transmitter. If care is taken to make carrier phase continuous through symbol transitions, then MSK has constant magnitude (like FM), which facilitates the use of relatively efficient power amplifiers. Furthermore, MSK has spectral efficiency $\eta_B = 1.5/T_b$, which is superior to QPSK without pulse shaping ($1/T_b$) and almost as good as QPSK with optimal sinc pulse shaping ($2/T_b$). Note that the frequency shift of $\Delta f = 1/2T$ between symbols can alternatively be interpreted as a *phase* difference of $2\pi\Delta fT = \pi$ radians between symbols; therefore continuous-phase MSK with rectangular pulse shaping is "barely" an FSK and instead might be more accurately described as a phase-only modulation with sinusoidal pulse shaping. The gentle transitions between symbols results in spectrum with a slightly wider main lobe but with lower sidelobes. A particularly popular variant of continuous-phase MSK employs Gaussian pulse shaping, which further improves the spectral characteristics. This variant is known as *Gaussian MSK* (GMSK), which is the basis for the second-generation cellular telecommunications system known as GSM and also appears in DECT, DCS1800, DCS1900, and a variety of other systems.

6.11.3 Demodulation and Performance

Optimal demodulation of FSK is not easily described within the same context as optimal demodulation of phase-amplitude modulations, as in Figure 6.11. However, FSK demodulation is relatively easily implemented and analyzed in the context of the optimal correlation receiver (Section 6.6.4) The performance of coherently-detected BFSK with orthogonal symbols (i.e., $\rho = 0$) is given by Equation (6.42); i.e., the same as coherently-detected OOK, and not as good

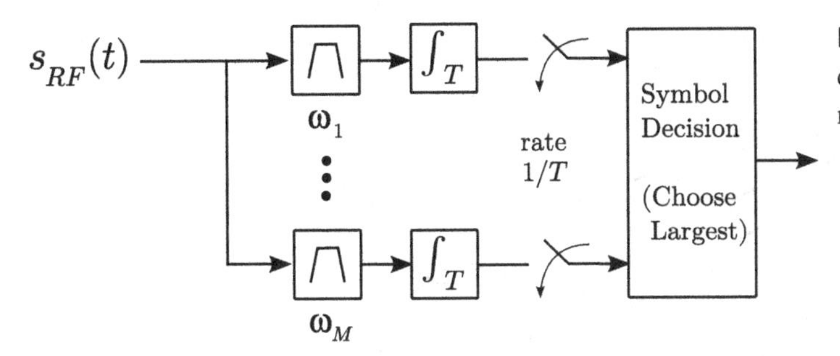

Figure 6.24. Bandpass incoherent demodulation of M-FSK with rectangular pulse shaping.

as BPSK or QPSK. The performance of optimally-demodulated continuous-phase MSK is the same as that of BPSK and QPSK; i.e., $P_b = Q\left(\sqrt{2E_b/N_0}\right)$.

A possible advantage of FSKs with sufficiently large Δf is that they may be incoherently demodulated, resulting in particularly simple demodulators which are robust to Doppler shift, carrier phase variation, and – if $\Delta f \geq$ the coherence bandwidth B_c – multipath fading. The scheme is as shown in Figure 6.24. The frequency channels representing each symbol are isolated using bandpass filters, and then the energy in each is evaluated once per symbol period. The channel displaying the largest energy is presumed to represent the current symbol. If significant differences in channel gain or noise exist from channel to channel, these must of course be equalized before the decision. The probability of bit error for BFSK detected in this manner is

$$P_b = \frac{1}{2}e^{-\frac{1}{2}E_b/N_0} \tag{6.59}$$

– note this is the same as for incoherent detection of OOK (Equation (6.44)), which is also an orthogonal modulation. For orthogonal FSK with $M > 2$, this can be generalized approximately to

$$P_b \leq \frac{M-1}{2\log_2 M}e^{-\frac{1}{2}(E_b/N_0)\log_2 M} \tag{6.60}$$

Results for a variety of M are shown in Figure 6.25. Here, we note a remarkable difference from all phase-amplitude modulations previously considered: Performance *improves* with increasing M. This can be explained as follows: The combination of Δf and bandpass filtering ensures that all symbols are utterly uncorrelated – i.e., *orthogonal* – to each other. Therefore, adding symbols does not increase the correlation between symbols, and instead reduces the overall probability of symbol error by increasing the frequency separation between symbols. Sadly, this performance is obtained at the cost of spectral efficiency which, compared to all phase-amplitude modulations, is truly atrocious.[10] Nevertheless, M-FSK is reasonable to consider when the best possible BER vs. E_b/N_0 performance is required and the associated penalty in spectral efficiency is acceptable.

[10] At the risk of getting ahead of ourselves, it could be noted that this weakness of M-FSK is largely mitigated using OFDM, in which all frequencies are used simultaneously; see Section 6.19.

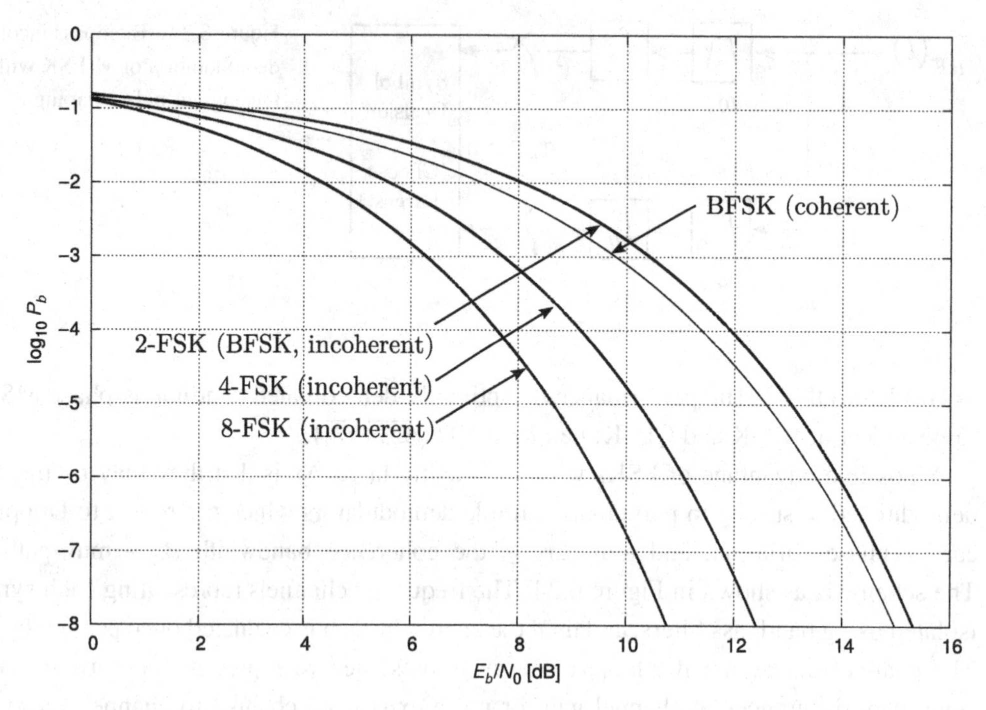

Figure 6.25. Probability of bit error for incoherently-detected orthogonal M-FSK compared to optimally-(coherently-) detected BFSK.

6.12 TRADEOFF BETWEEN SPECTRAL EFFICIENCY AND ENERGY EFFICIENCY

Among a broad range of practical considerations, two fundamental parameters inevitably arise when comparing modulations: These are *spectral efficiency* and *energy efficiency*. Spectral efficiency was introduced in Section 6.4.4, where it was defined in terms of the parameter η_B. Energy efficiency refers to the SNR (or E_b/N_0) required to achieve a specified BER (P_b) or symbol error rate (P_S). Having now considered a large number of modulations, we have considerable empirical evidence as to the nature of the spectral efficiency–energy efficiency tradeoff. However, as in all engineering problems, it is useful to be aware of the theoretical limits. What, if any, are the ultimate limits in performance? For example, what is the best possible spectral efficiency one may achieve for a given SNR, or what SNR is required to pass information at a rate greater than zero? In this case the limit is concisely quantified in terms of the principal result of the *Shannon–Hartley Theorem*, known as the *Shannon bound*:

$$C = \log_2(1 + \text{SNR}) \tag{6.61}$$

where C is the *normalized capacity of the channel*, which is unitless but customarily expressed as bits per second per Hz (bps/Hz). Normalized capacity is the fundamental rate of information that a given channel can support. By "support" we mean specifically the *reliable* rate of information transfer. This is not referring to a specified BER, but rather something more fundamental. For example, given bandwidth B and a certain amount of time Δt, the maximum number of bits that can be reliably transferred is equal to $CB\Delta t$.

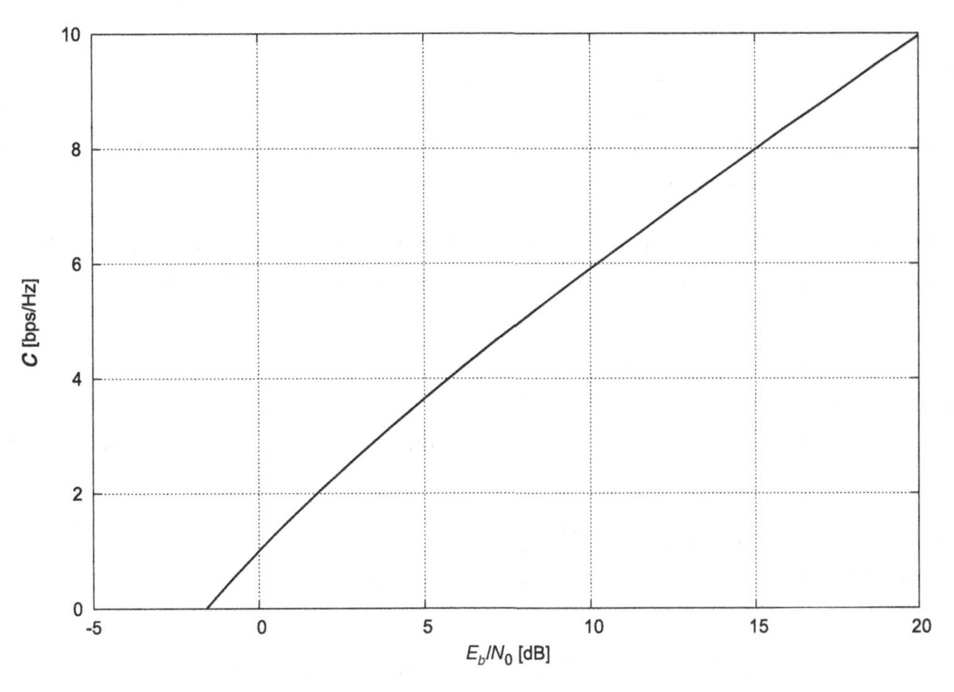

Figure 6.26. Maximum normalized capacity as a function of E_b/N_0.

Three useful insights can be gleaned from the above result. First: The direct – and unsurprising – interpretation is that $C > 0$ requires SNR>0; that is, you need non-zero signal power in order to communicate. Second: The normalized capacity depends *only* on SNR, and not on modulation or other technical details. The third piece of information is a bit more subtle, and may be obtained as follows. First, employing Equation (6.17), we note:

$$\text{SNR} = \frac{1}{T_b B} \cdot \frac{E_b}{N_0} \geq C \cdot \frac{E_b}{N_0} \tag{6.62}$$

since $1/T_b B$ is the actual rate (bits per second) per unit bandwidth.[11] Substitution into Equation (6.61) and solving for E_b/N_0 yields the following expression:

$$\frac{E_b}{N_0} \geq \frac{1}{C}\left(2^C - 1\right) \tag{6.63}$$

This result is plotted in Figure 6.26.

The Shannon bound reveals two useful pieces of information: First, it gives the minimum E_b/N_0 required to achieve a specified normalized capacity. Second, it reveals the minimum E_b/N_0 for information transfer, at *any* rate: This is obtained by taking the limit of Equation (6.63) as $C \to 0$, yielding

$$\frac{E_b}{N_0} \geq \ln 2 \cong -1.6 \text{ dB}, \text{ for reliable communication} \tag{6.64}$$

as can be verified from Figure 6.26.

[11] Information theorists may cringe at the sight of this particular substitution, since bandwidth B is subject to arbitrary definition and is not a fundamental quantity. In practical work, however, one is unlikely to approach sufficiently close to the bound for this particular issue to become significant.

EXAMPLE 6.4

The North American digital television system ATSC (see Section 6.17) is designed to work at a minimum SNR of about 15 dB in a bandwidth of 6 MHz. What is the theoretical maximum data rate under these conditions?

Solution: Here, SNR = 15 dB = 31.6 in linear units. So $C = \log_2(1 + 31.6) \cong 5.03$ bps/Hz, so 30.2 Mb/s in 6 MHz. For the record, ATSC achieves about 19.39 Mb/s.

EXAMPLE 6.5

In the above problem, what is the theoretical minimum E_b/N_0 required to sustain 30.2 Mb/s reliably?

Solution: $C \cong 5.03$ bps/Hz, so $E_b/N_0 \geq \left(2^C - 1\right)/C \cong 8.0$ dB. ATSC requires at least 7 dB greater SNR, despite operating at a significantly lower spectral efficiency.

Having established fundamental limits on communication, it is instructive to compare this with what is achieved by the modulations considered in earlier sections. This is done in an approximate way in Figure 6.27 by plotting the FNBW-based spectral efficiency η_B versus the E_b/N_0 required for 0.1% BER assuming optimal coherent detection. It is quite apparent that none of the modulations approach the Shannon bound. With respect to the 0.1% BER criterion, it appears E_b/N_0 in the range 5–16 dB is required for the modulations considered. To do significantly better, channel coding is required. This is considered next.

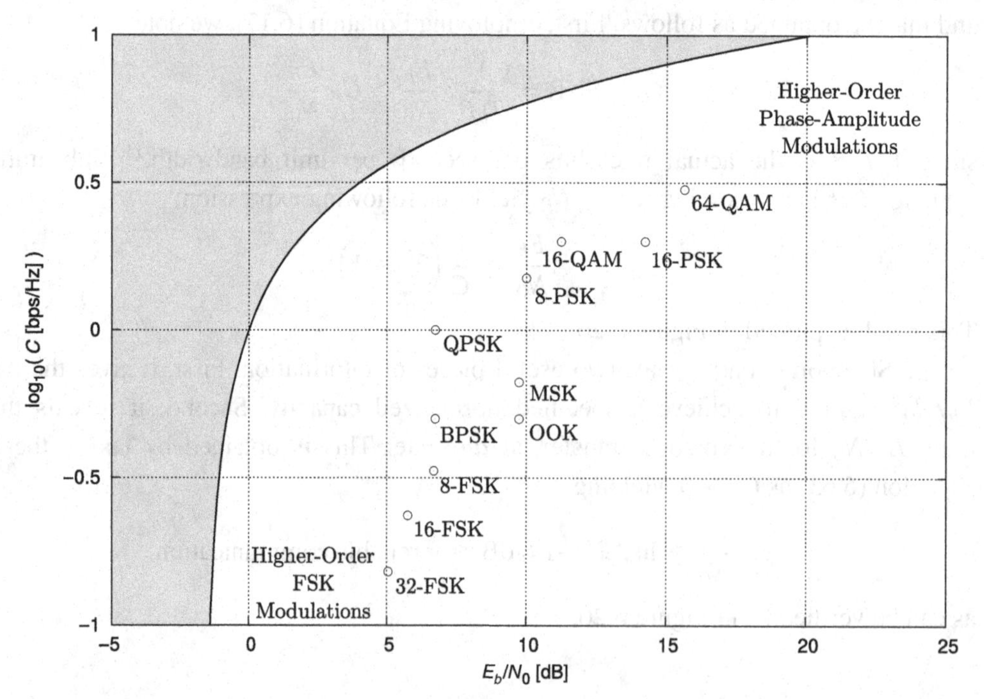

Figure 6.27. Performance of modulations relative to the Shannon bound. Each modulation is plotted at the E_b/N_0 for which it achieves 0.1% BER and taking η_B (FNBW basis, rectangular or $r = 1$ pulses) as C. Note spectral efficiency is now plotted in log scale.

6.13 CHANNEL CODING

We noted in the previous section that despite the wide range available in the tradeoff between spectral efficiency and energy efficiency, no modulation by itself approaches the Shannon bound. To do significantly better requires channel coding; in particular, error control coding. This section provides a brief introduction to the topic.

Channel codes selectively introduce redundant information into the transmitted data stream so that errors can be detected and corrected. It should be immediately apparent that this strategy simultaneously degrades spectral efficiency, since redundant bits by definition convey no information. The goal of channel coding design is to implement codes that significantly improve energy efficiency through error correction, without unacceptable degradation in spectral efficiency. This is nearly always possible, and as a result nearly all modern digital communications systems use some form of channel coding.

A variety of channel coding schemes are in common use. Perhaps the simplest to understand are block codes. In a block code, contiguous blocks of k bits are translated into contiguous blocks of $n > k$ bits, resulting in $n - k$ redundant bits. (Alternatively, the coder may operate on bytes or modulation symbols as opposed to bits.) One way to generate redundant data is to calculate *parity* or *checksum* data by analysis of the input data, and then to append this to the input data. An example of a commonly-used scheme that operates in this manner is *Reed–Solomon coding*.

More generally, an input block may be mapped to a longer output block referred to as a *codeword*. The goal in designing an effective block code is to maximize the minimum *Hamming distance* d_H, which is the number of bits in which two codewords differ. At the receiver, each codeword – now potentially containing errors – is compared with the original set of codewords and the one exhibiting the minimum Hamming distance is selected. This codeword is then replaced with the associated block of k bits. In practice, systematically comparing received codewords to valid codewords is computationally intractible, so codes must be designed with mathematical structure such that the minimum-distance codeword can be found without an exhaustive search over the possible codewords. A commonly-used scheme that operates in this manner is *Bose, Chaudhuri, and Hocquenghem (BCH) coding*.

It can be shown that a block code with a minimum Hamming distance of $d_{H,min}$ is able to detect any combination of up to $d_{H,min} - 1$ errors, and correct any combination of $t = \left(d_{H,min} - 1\right)/2$ errors. The quantity t is referred to as the *error correcting capability* of the code. The minimum Hamming distance is related to n/k, which indicates that for any given scheme, increasing n/k increases the "strength" of the code. The contraindication is that increasing n/k reduces spectral efficiency, and increases the computation burden associated with decoding, which can be formidable.

The ratio $r_c = k/n$ is referred to as the *code rate*, with smaller values indicating greater degrees of redundancy and thus greater potential for error correction. Simultaneously, spectral efficiency (using any measure of bandwidth) is multiplied (i.e., decreased) by r_c. Code rates for practical block codes may be as small as 1/4, with n typically in the range 20–1024.

The improvement in energy efficiency due to coding is referred to as *coding gain*. Coding gain is the reduction in E_b/N_0 required to achieve the same BER without coding. A code is said to be "strong" if the coding gain is relatively high (e.g., at least 1–2 dB) for sufficiently-large E_b/N_0.

Many codes fail below a threshold E_b/N_0 and, in that condition, actually degrade energy efficiency. In block codes, this reflects the fact that a threshold number of bits must be detected correctly in order for the associated codeword to be reliably identified. Thus it is important to choose a coding scheme that is appropriate for the SNR regime in which the system is expected to operate.

EXAMPLE 6.6

As a demonstration of what is possible, let us consider the use of the BCH (63,45) block code with coherently-demodulated QPSK.

Solution: "(63,45)" indicates $n = 63$ and $k = 45$; therefore $r_c = 0.714$ and the spectral efficiency is degraded by about 29% relative to uncoded coherently-demodulated QPSK. A (63,45) BCH block code used with QPSK reduces the E_b/N_0 required to achieve $P_b = 10^{-6}$ from about 11 dB (without coding) to about 8 dB; therefore the coding gain is about 3 dB with respect to $P_b = 10^{-6}$ and this code is considered to be quite strong. Unfortunately the code results in *increased* error rate for E_b/N_0 below about 6 dB; therefore this solution would not be suitable for weak signal applications.

After considering the above example, an appropriate question to ask might be: Is the modest improvement in energy efficiency (i.e., the 3 dB coding gain) worth the 29% penalty in spectral efficiency? Typically any modulation capable of providing comparably-high energy efficiency will have a much lower spectral efficiency. Generalizing, it is typical that a lower-order modulation such as QPSK with appropriate coding will operate closer to the Shannon bound than a higher-order modulation without coding. Generalizing further: Channel coding is necessary if one wishes to operate as close to the Shannon bound as possible.

An alternative form of block coding is low-density parity-check (LDPC) coding. LPDC codes are *very* long ($n = 20\,000$ is typical), but are able to yield performance within 1 dB of the Shannon bound even when suboptimal computationally-tractable decoding is used. However, not all applications can tolerate the latency associated with the relatively long gaps in time between the delivery of contiguous data blocks.

Alternatives to block coding include *convolutional coding*, trellis codes (*trellis-coded modulation*; TCM), and *turbo codes*. In convolutional coding, data pass through the coder in a continuous stream with effective k and n that are small relative to block coders. In TCM, convolutional coding and modulation are combined in a way that tends to result in increases in energy efficiency that are large relative to the degradation in spectral efficiency. In newer communications systems, traditional convolutional coding schemes have largely been replaced with turbo codes, which deliver very good performance (within 1 dB of the Shannon bound) using suboptimal computationally-tractable decoding.

6.14 COMMUNICATION IN CHANNELS WITH FLAT FADING

Section 3.5.3 introduced the problem of flat fading: rapidly-varying magnitude (as shown by example in Figure 3.12) and phase due to multipath. Flat fading greatly complicates the implementation of coherent demodulation, particularly for phase-amplitude modulations. Coherent demodulation is made more difficult due to the need to keep up with the rapidly-varying channel phase ϕ_p, so that the constellation remains properly derotated – more on this in Section 6.16. The rapidly-varying magnitude manifests as two problems: (1) Maintaining sufficiently accurate AGC, especially if the modulation encodes information in amplitude (e.g., M-ASK with $M > 2$, M-QAM with $M > 4$), and (2) intermittently low SNR, which leads to a host of problems including, but not limited to, symbol errors. In this section we focus on the latter.

6.14.1 Probability of Error in Flat Fading

For simplicity, let us focus on coherently-demodulated BPSK. The *instantaneous* probability of a symbol error in flat fading is

$$P_S = Q\left(\sqrt{2\beta}\right)\ , \text{ where } \beta \equiv \frac{E_b}{N_0}\,|\gamma|^2 \tag{6.65}$$

and γ is the complex-valued channel coefficient introduced in Section 3.5.3, which we view here as a random variable having Rayleigh-distributed magnitude with mean equal to 1, and uniformly-distributed phase (see Equation (3.67) and associated text). Therefore β is central χ^2-distributed with two degrees of freedom, and it can be shown (using a derivation beyond the scope of this book) that the *mean* probability of symbol error is

$$P_S = \frac{1}{2}\left(1 - \sqrt{\frac{\bar{\beta}}{1+\bar{\beta}}}\right) \tag{6.66}$$

where $\bar{\beta}$ is the mean value of β. For large SNR ($\bar{\beta} \gg 1$) this is well-approximated by

$$P_S \approx \frac{1}{4\bar{\beta}} \tag{6.67}$$

The resulting performance is illustrated in Figure 6.28. Note that performance is severely degraded relative to a non-fading channel having the same mean SNR. Channel (error control) coding alone usually does not offer much improvement, because such codes depend on errors being ergodic and randomly distributed over time, whereas the nature of the fading channel is to instead produce bursts of bit errors which tend to overwhelm the error-correcting capabilities of codes. The primary countermeasures in this situation are interleaving (next section) and space diversity (Section 6.14.3).

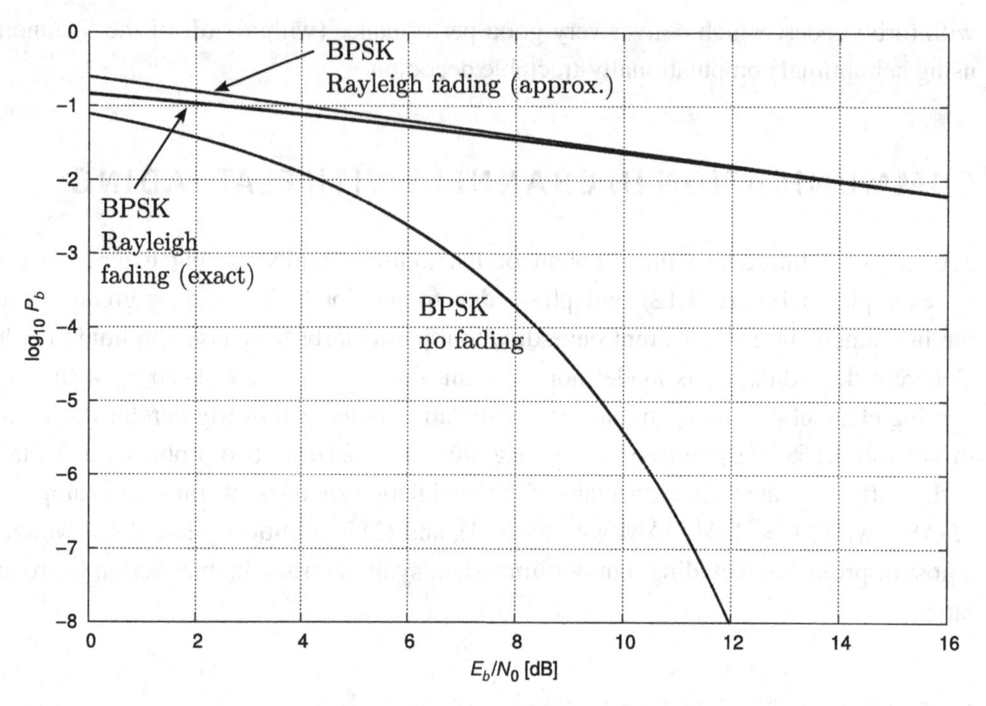

Figure 6.28. Performance of BPSK in a flat fading channel having Rayleigh-distributed magnitude.

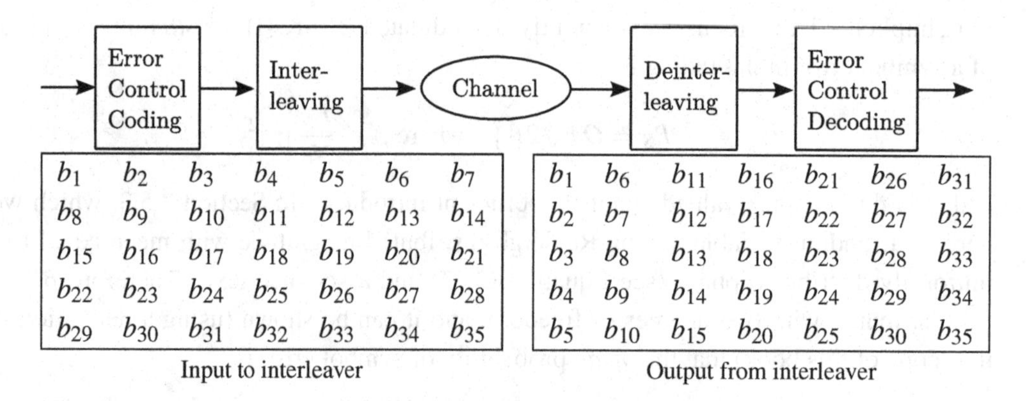

Figure 6.29. A simple example of interleaving having depth 5 and length 7.

6.14.2 Interleaving

Interleaving is systematic reordering of bits or symbols in the transmitter, so that burst errors become more evenly distributed when the received data are reordered in the receiver. Figure 6.29 shows an example. In this scheme, the interleaver is said to have "length" 7 and "depth" 5, and operates on blocks of $7 \times 5 = 35$ bits at a time. The first seven bits emerging from the interleaver are b_1, b_6, b_{11}, and so on up to b_{31}; skipping a number of bits equal to the depth of the interleaver each time. The process then repeats beginning with b_2 and ending with b_{32} and so on, until the block is exhausted. The reordered bits are subsequently modulated and transmitted.

Figure 6.30 shows how this scheme ameliorates a burst error. In the receiver, four contiguous bits, denoted with $\times$s, have been incorrectly detected due to a fading event. Once the

b_1	b_6	b_{11}	b_{16}	b_{21}	b_{26}	b_{31}
b_2	b_7	b_{12}	b_{17}	b_{22}	b_{27}	b_{32}
b_3	b_8	×	×	×	×	b_{33}
b_4	b_9	b_{14}	b_{19}	b_{24}	b_{29}	b_{34}
b_5	b_{10}	b_{15}	b_{20}	b_{25}	b_{30}	b_{35}

Input to deinterleaver

b_1	b_2	b_3	b_4	b_5	b_6	b_7
b_8	b_9	b_{10}	b_{11}	b_{12}	×	b_{14}
b_{15}	b_{16}	b_{17}	×	b_{19}	b_{20}	b_{21}
b_{22}	×	b_{24}	b_{25}	b_{26}	b_{27}	×
b_{29}	b_{30}	b_{31}	b_{32}	b_{33}	b_{34}	b_{35}

Output from deinterleaver

Figure 6.30. Amelioration of a burst error (×s) by deinterleaving.

de-interleaver reorders the bits, the incorrect bits are pseudo-randomly distributed, as one would expect for the constant-SNR channel for which coders are normally designed. Summarizing: Interleavers make error control coding robust to burst errors associated with fast fading.

Block interleavers combined with appropriate coding routinely provide coding gains of 5–10 dB in Rayleigh fading channels, and so routinely appear in modern mobile radio communications systems. The fundamental tradeoff in interleaver design is between block length and *latency*. A "deep" interleaver does a better job at randomizing the bit order, and the larger the block length the longer the error burst the interleaver can effectively disperse. From this perspective, an "ideal" (and utterly unrealizable) interleaver is one which is able to convert burst errors of any length into errors having the same occurance statistics as a constant-SNR channel.

On the other hand, the block length of the interleaver (35 in the example above) imposes a latency that may be difficult to accommodate in some applications. Communications systems typically organize data into frames having length on the order of milliseconds, which accommodates both carrier frequency/phase tracking and source coding. For example, if the data rate is 10 kb/s and a frame is 10 ms, then there are on the order of 100 bits per frame. An interleaver having length × depth equal to a sub-multiple of this number of bits is ideal from a latency perspective, whereas other lengths would delay the processing of frames with potentially deleterious effects to frame-level processing. This problem is especially serious in two-way voice communications, where latencies approaching a fraction of a second may be perceived by the user.

6.14.3 Space Diversity

The concept of space diversity as a countermeasure to flat fading was introduced in Section 3.5.4. Techniques that are predominantly used to improve the performance of digital communications systems vulnerable to flat fading include maximal ratio combining (MRC)[12] and space-time coding. In this section we consider the benefits of MRC. Space-time coding is an

[12] It is traditional to list *selection combining* and *equal-gain combining* as the first and second items in such a list, as these techniques have historically been important. However, selection and equal-gain combining are suboptimal (in contrast to MRC), and have value primarily because they are very simple to implement. In modern communications systems, and in particular with the modern trend toward implementation in digital baseband processing, it is possible to implement MRC with relatively little additional effort, so applications where suboptimal combining is appropriate are becoming rare. For additional information about these techniques specifically, [41, Ch. 10] is recommended.

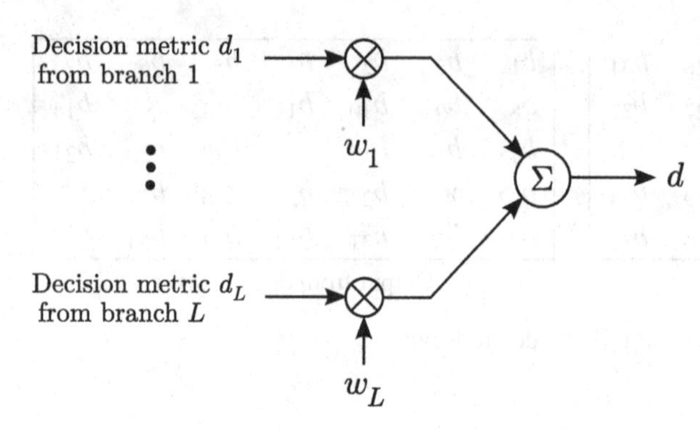

Figure 6.31. Diversity combining within a coherent receiver.

advanced topic beyond the scope of this book; for additional information, a suggested starting point is [59, Ch. 15].

The basic recipe for MRC applied to space diversity is shown in Figure 6.31. Each of the L antennas is equipped with a separate receiver and generates a decision metric d_l in the usual manner. These parallel processing streams are commonly referred to as *branches*; so we are here considering an L-branch diversity receiver. Because each antenna experiences a different channel propagation coefficient γ_l, the d_ls vary by this much in the absence of noise. So, for example, a decision metric associated with a branch experiencing a large value of $|\gamma_l|$ will have high SNR; whereas a decision metric associated with a branch experiencing a small value of $|\gamma_l|$ (e.g., in a deep fade) will have low SNR. It can be shown that the optimal decision metric d can be obtained from the branch decision metrics as follows:

$$d = \sum_{l=1}^{L} w_l d_l \text{ , where } w_l = \gamma_l^* \tag{6.68}$$

The resulting decision metric d is optimal in the sense that SNR is maximized.

The particular choice of $w_l = \gamma_l^*$ is what makes this MRC. The derivation is beyond the scope of this book; however, the result is possible to justify informally. First, note that multiplying each of the d_ls by γ_l^* tends to emphasize the contribution of the d_ls that are large, and tends to deemphasize the contribution of the d_ls that are small. Assuming the noise power is equal among the branches, this means the decision metrics having the largest SNR are emphasized, and decision metrics having low SNR are deemphasized. The effect of multiplying by the conjugate of γ_l is to equalize the phases of the branch statistics, so that they add in-phase. Thus the signal component of the decision metrics add coherently, whereas the noise component of the decision metrics – which are nominally uncorrelated between branches – add incoherently and the SNR is further increased.

Note that there is no particular difficulty in determining the coefficients w_l, since the γ_ls are normally required for coherent demodulation anyway. That is, the magnitude of γ_l is side information required for AGC, whereas the phase of γ_l (i.e., ϕ_p) is side information required for derotation. All that is required is to make these values available to the combiner.

It is possible to show that the probability of error for BPSK in Rayleigh flat fading after L-branch MRC is:

$$P_S = \frac{1}{2}\left[1 - \sqrt{\frac{\bar{\beta}}{1+\bar{\beta}}} \sum_{k=0}^{L-1} \binom{2k}{k} \frac{1}{4^k \left(1+\bar{\beta}\right)^k}\right] \tag{6.69}$$

and for high SNR ($\bar{\beta} \gg 1$):

$$P_S \approx \left(\frac{1}{4\bar{\beta}}\right)^L \binom{2L-1}{L} \tag{6.70}$$

where

$$\binom{n}{k} \equiv \frac{n!}{k!\,(n-k)\,!} \tag{6.71}$$

Note that Equations (6.69) and (6.70) are analogous to Equations (6.66) and (6.67). These equations assume perfectly uncorrelated fading between branches, which requires large spacings between antennas; see Section 3.5.4 for details about that. Equation (6.70) hints at the effectiveness of MRC: Comparing to Equation (6.67), we see the result improves exponentially with L. MRC performance is shown in its full glory in Figure 6.32; for example, simply increasing from $L = 1$ (single antenna) to $L = 2$ provides an effective gain of about 5 dB for 1% BER. The analogous gains for lower BER are much greater.

Before moving on, it is appropriate to point out an additional benefit of MRC. As described above, MRC is a diversity combining technique. However, in some propagation scenarios, scattering is very low. When this is the case, magnitude fading is mild relative to Rayleigh and signals arrive at the antenna array from a narrow range of angles. An example is the link between a base station antenna array on a high tower and a mobile station in a flat, rural environment. In this case, beamforming (see Figure 2.24 and associated text for a reminder) would be preferable to diversity combining. A marvelous feature of MRC – which can easily be verified from Equation (6.68) – is that when the situation calls for a beamformer, this is

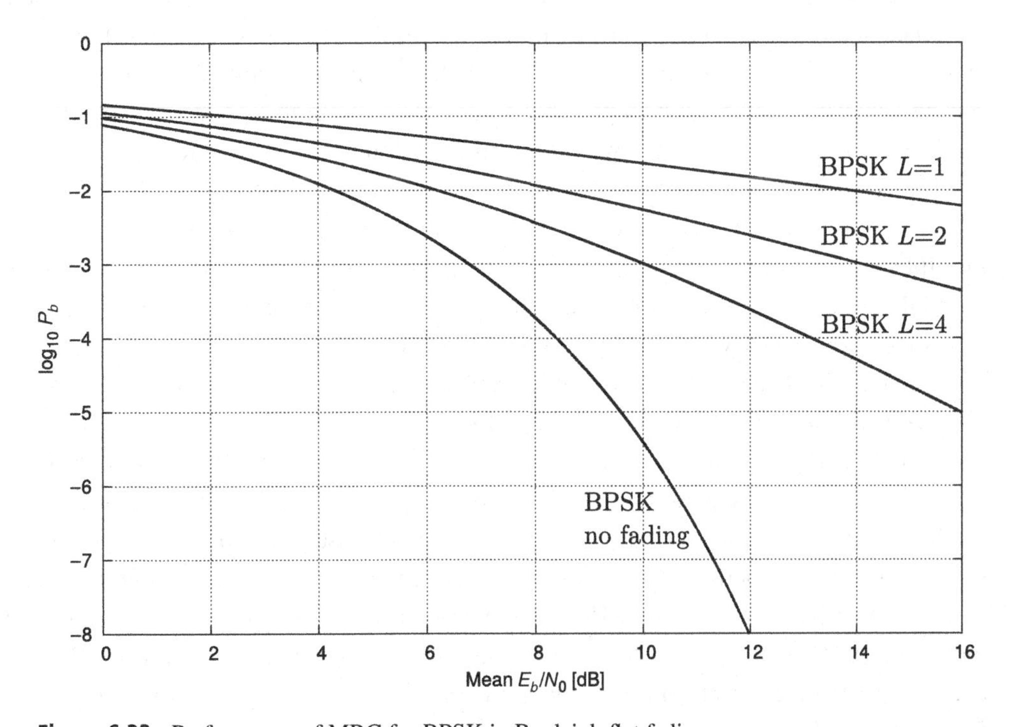

Figure 6.32. Performance of MRC for BPSK in Rayleigh flat fading.

what MRC provides. Said differently: MRC automatically provides the optimal combining coefficients, regardless of whether the situation calls for diversity combining or beamforming.

Finally, note that the above discussion has assumed widely separated antennas as the origin of the signals for each branch of the combiner. However, this is certainly not the only option. Two other possibilities including *polarization diversity* and *sector diversity*. In polarization diversity, the sources of signals for the branch receivers are orthogonal polarizations, as opposed to widely-separated co-polarized antennas. In sector diversity, the sources are antennas having patterns covering angular regions with low overlap. As far as MRC is concerned, these sources can be accommodated in any combination. So, for example, a cellular base station receiver might employ $L = 4$ MRC using two orthogonal polarizations from each of two widely-separated antennas, and if the fading between polarizations is uncorrelated in the same way as the fading between antennas, then the performance shown in Figure 6.32 will be achieved.

6.14.4 Multiple-Input Multiple-Output (MIMO)

The "input" and "output" in the term "MIMO" refer to the transmitter and receiver, respectively. Specifically, the term "MIMO" indicates the use of multiple antennas at both transmitter and receiver. However, this is not a sufficient definition. MIMO is a generalization of the concept of spatial diversity, and is an advanced topic well beyond the scope of this book (for additional information, a suggested starting point is [59, Ch. 15]). However, MIMO is sufficiently-well established in modern radio systems that it is useful to know at least the top-level concept. This is presented below.

All previous discussion of the propagation channel in this chapter has presumed the existence of a single channel. The transmitter inserts information into this channel, the receiver extracts information from this channel, and the Shannon bound limits the data rate that can be achieved. However, it is possible to create multiple independent channels between a transmitter and a receiver. Here we mean "independent" in the statistical sense; in particular, exhibiting uncorrelated fading between channels even after accounting for spatial diversity. These additional channels are created by using multiple antennas at the transmitter as well as multiple antennas at the receiver. MIMO is a generalization of spatial diversity in the sense that each additional transmit antenna nominally results in one additional independent channel that is available for diversity combining by the receive antenna array.

Effective MIMO requires considerable scattering in the propagation channel. For example, if there is no scattering and only direct "line of sight" propagation, then this scheme degenerates into simple beamforming at the transmitter and receiver, which yields only one independent channel, albeit with improved SNR. Even with copious scattering, a MIMO-enabled system might only yield one independent channel (in which case it is no different than traditional spatial diversity), or perhaps multiple channels in which only one channel yields SNR which is sufficient to be effectively used. However, many propagation channels can support multiple independent channels each with usable SNR. In this case the ultimate capacity that exists between receiver and transmitter is the sum capacity of the independent channels, since the Shannon bound applies to each independent channel separately. Thus MIMO systems achieve increased data rates when multiple independent channels are simultaneously available for the transmission of data.

6.15 COMMUNICATION IN CHANNELS WITH INTERSYMBOL INTERFERENCE

In Section 3.6.2 it was noted that the delay spread in the propagation channel could be large enough to decorrelate multipath components. When this happens, the receiver perceives multiple copies – or perhaps even a continuum – of transmit signals. This results in another form of ISI, distinct from that associated with pulse shape. So far in this chapter we have neglected this possibility; now we confront it.

6.15.1 Zero-Forcing Equalization

If the possibility of significant propagation-induced ISI exists, then it is reasonable to consider *equalization*. A possible approach to equalization is to determine the frequency response $H(\omega)$ of the propagation channel, and then to apply a filter $H^{-1}(\omega)$. This is known as *zero-forcing equalization* (ZFE). This is rarely done. Reasons include the following.

- The necessary filter has an impulse response that is at least as long as the impulse response of the propagation channel, and may be significantly degraded when truncated to a reasonable length.
- Noise enhancement: The signal components associated with low values of $H(\omega)$ have low SNR. When ZFE is applied, the noise associated with these components is enhanced because the associated frequencies receive relatively high gain.
- ZFE doesn't take into account the primary goal of optimizing BER, and therefore there is no assurance that it will do so.

Better alternatives to ZFE are maximum likelihood sequence estimation (MLSE), minimum mean square error equalization (MMSE) and decision feedback equalization (DFE). Although a detailed treatment of these topics is beyond the scope of this book ([59, Chs. 9 and 10] are recommended), it is useful for the radio systems engineer to be familiar with the pros and cons of these techniques and to be familiar with the associated terminology.

6.15.2 Maximum Likelihood Sequence Estimation

MLSE is the optimal scheme for demodulation in channels with significant ISI. The recipe for MLSE is shown in Figure 6.33. First, the pulse-matched filter $G^*(\omega)$ is replaced with the modified matched filter $H^*(\omega)$. The impulse response of this filter is derived from the convolution of the impulse response of the pulse shaping filter ($g(t)$) and the channel impulse response (CIR) $h_c(t)$; i.e.,

$$h(t) = g(t) * h_c(t) \tag{6.72}$$

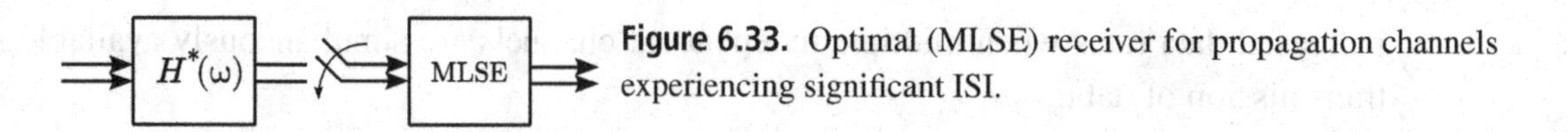

Figure 6.33. Optimal (MLSE) receiver for propagation channels experiencing significant ISI.

The receiver then compares a length-N sequence of decision metrics sampled from this filter to the set of all possible sequences of N symbols that may have actually been sent, accounting for the expected effect of the CIR. NT is nominally at least as long as the duration over which the CIR is significantly greater than zero. The sequence that minimizes the mean square error with respect to the observed sequence of decision metrics is declared the winner, and the associated most-recent symbol of the test sequence is declared the detected symbol.

One aspect of MLSE that should be apparent is that it requires an enormous amount of computation. To be optimal N must span the non-zero portion of the CIR, which might be tens to thousands of symbols long. Assuming the lowest-order ($M = 2$) modulation, the number of sequences that must be checked is 2^N. A direct implementation of this approach is usually not practical without significant compromises. Fortunately, the *Viterbi algorithm* comes to the rescue. This clever algorithm reorganizes the brute-force search over candidate symbol sequences, dramatically reducing computational burden. MLSE receivers using the Viterbi algorithm, typically implemented on an FPGA or DSP, are now practical and fairly common (see Sections 18.5–18.6). Nevertheless, the primary contraindication against the optimal MLSE receiver remains computational burden and the associated size, weight, power, and heat concerns.

Another aspect of MLSE that should be conspicuous is the need to know the CIR. This may prove to be a daunting challenge by itself. A typical method is to treat the CIR as a "system identification" problem, in which the "system" (the CIR) may be deduced by applying a known signal $x(t)$ and observing the response $y(t)$. In this case $x(t)$ is known as a *training sequence*. Obviously the training sequence must be sent often enough to effectively track the time variation of the CIR, and according to a schedule of which the receiver is aware. A consequence of the use of training signals is reduction in spectral efficiency, since symbols sent as training signals reduce the number of symbols per unit time available for user data. A less daunting but non-trivial additional problem is the design of the training sequence itself; i.e., how to select a sequence of symbols that results in the best possible estimation of the CIR. To the extent that the CIR is mis-estimated, and to the extent that the CIR becomes inaccurate due to time lapsed since the last measurement, the performance of MLSE is degraded.

6.15.3 Minimum Mean Square Error (MMSE) Equalization

When the complexity of MLSE becomes overwhelming, a reasonable fall-back position is MMSE equalization, as shown in Figure 6.34. In MMSE equalization, we revert to pulse-matched filtering ($G^*(\omega)$) and replace ML sequence estimation with a finite impulse response (FIR) filter. The nominal length of the FIR filter is equal to the length of the impulse response associated with the combined frequency response $\left[G(\omega)H_c(\omega)G^*(\omega)\right]^{-1}$; however, this is not critical, and MMSE equalizers may be much shorter and remain effective.

In MMSE equalization, the primary problem is how to select the impulse response of the FIR filter. The answer is to select filter coefficients representing the impulse response in such a

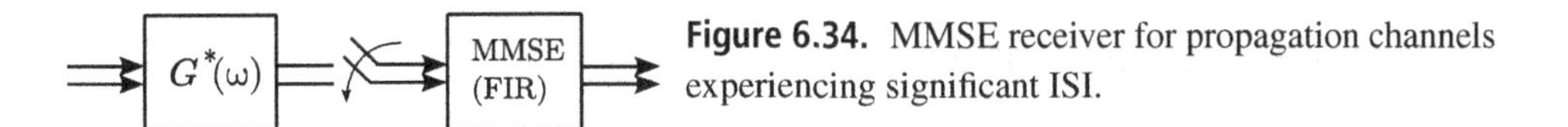

Figure 6.34. MMSE receiver for propagation channels experiencing significant ISI.

way as to minimize the mean square error between the detected sequence and the transmitted sequence. This, of course, raises the question of how the receiver is supposed to know the transmitted sequence in advance; the customary answer is a training sequence. This raises the same issues that were addressed with the use of training sequences in the previous section. In this case, however, the training sequence is not needed to determine CIR explicitly, but rather simply as a known input to "train" the equalizer. So, for example, the training sequence need not be as long as the CIR; but rather only as long as the equalizer.[13] MMSE is often referred to as a form of *adaptive equalization*.

MMSE equalization is a decidedly non-optimal approach to demodulation in channels with significant ISI. The primary reason for selecting MMSE equalization over MLSE is computational burden. The computational burden of MMSE is a combination of the burden of operating the FIR filter plus the burden of computing filter coefficients each time the training sequence arrives. Both of these operations are typically orders of magnitude less burdensome than MLSE, even when the Viterbi algorithm is employed.

Both MMSE equalization and MLSE are "one shot" techniques in the sense that once a solution is obtained from the training sequence, that solution is used until the next training sequence arrives, at which point the solution is discarded in favor of the new solution. A modification to the one-shot strategy facilitates an additional reduction in computational burden for MMSE. This modification employs alternative forms of MMSE known as *recursive least squares* (RLS) and *least mean squares* (LMS). In both RLS and LMS, the MMSE filter coefficients are estimated not in a "one shot" operation, but rather by updating a previous solution using the latest training sequence. The performance of these algorithms can be crudely characterized as follows: RLS generally converges to the MMSE solution as long as the latter does not vary too quickly, whereas LMS converges to the MMSE solution only if the channel varies sufficiently slowly *and* the channel is not too complicated. Generally, RLS should be avoided unless MMSE poses too large a computational burden, and LMS should be avoided unless RLS poses too large a computational burden.

6.16 CARRIER FREQUENCY, PHASE, AND SYMBOL TIMING

Much of the complexity of coherent demodulation is associated with determining the frequency and phase of the carrier, and the timing of symbols. Section 5.5.4 described techniques for dealing with the carrier from the perspective of analog modulation. Symbol timing recovery is unique to digital modulation, but is a similar problem.

[13] Not to confuse the issue, but it is worth pointing out that if both the training sequence and the FIR filter are at least as long as the non-zero portion of the CIR, then the impulse response determined by MMSE will end up looking a lot like the inverted CIR, without the noise enhancement problem associated with ZFE. This is one way to get the CIR estimate needed for MLSE.

6.16.1 Carrier Frequency Estimation

The usual procedure is to first obtain a usefully-accurate estimate of carrier frequency, and then to work out the carrier phase once the frequency is settled. In estimating carrier frequency, there are often actually two problems: *acquisition* and *tracking*. Frequency acquisition is the process of estimating frequency when there is no *a priori* estimate of the frequency; e.g., immediately after the receiver is turned on, upon the appearance of a new signal, and immediately following a severe outage. Frequency tracking is the process of updating the estimate of frequency given an earlier but recent estimate of frequency.

Two prominent counterexamples in which frequency acquisition is necessary are described below.

EXAMPLE 6.7

Iridium is a mobile telecommunications system using satellites. A particular challenge in the space-to-ground link is very large and variable Doppler shift, associated with the high speed of the satellite. To aid the receiver, transmissions are organized into 15-ms bursts, and each burst begins with 3-ms of unmodulated carrier. The carrier frequency and phase can be precisely estimated from this tone. The optimal frequency estimator consists of evaluating the following correlation (in complex baseband representation)

$$\int_{t_0}^{t_0+T_d} s(t)e^{-j\omega_c t}\,dt \tag{6.73}$$

where $s(t)$ is the tone waveform, t_0 is the tone start time, T_d is the tone duration (3 ms), and ω_c is the candidate carrier frequency. This correlation is recalculated while varying ω_c. The best possible estimate of ω_c is the value that maximizes the magnitude of the correlation. Subsequently the phase and amplitude of the correlation give the carrier phase and magnitude for derotation and AGC, respectively.

EXAMPLE 6.8

The ATSC broadcast digital television system (described in detail in Section 6.17) uses a continuous pilot tone embedded in the modulation which may be used for carrier frequency acquisition. In this case the motivation for the pilot tone is to facilitate low-cost receivers, which must be as simple as possible and are likely not to have very accurate internal frequency synthesis.

Once the carrier frequency has been acquired (or if explicit acquisition is not required), then all that remains is to track. A common carrier frequency tracking scheme is correlation testing performed on training sequences (analogous to the tone correlation test described for *Iridium* in Example 6.7). An alternative approach is *band-edge tracking*. In band-edge tracking, a filter is designed with frequency response $H_{BE}(\omega)$ having magnitude proportional to the magnitude of the first derivative of the spectrum $S(\omega)$ of the signal, as shown in Figure 6.35.

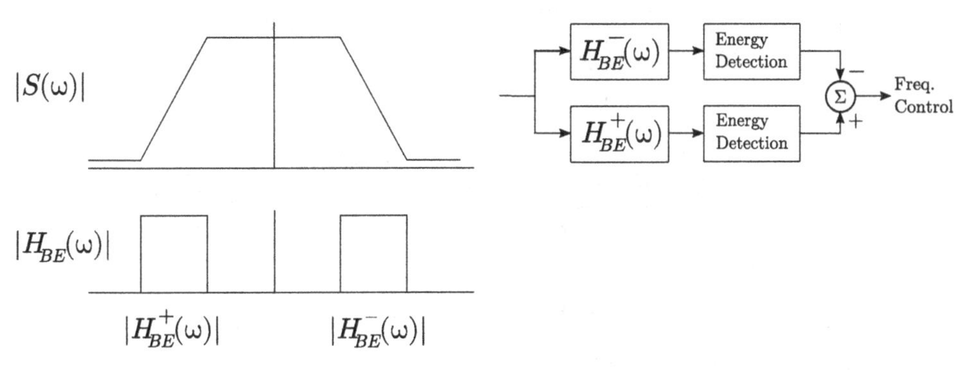

Figure 6.35. Band-edge scheme for tracking carrier frequency.

This filter may be decomposed into two filters $H_{BE}^{+}(\omega)$ and $H_{BE}^{-}(\omega)$, representing the low- and high-frequency band edges, respectively. When the frequency is correct, the spectrum is centered between the band-edge filters and $H_{BE}^{+}(\omega)$ and $H_{BE}^{-}(\omega)$ output equal power. Should the frequency drift high, then the power output from $H_{BE}^{-}(\omega)$ will decrease relative to the power in $H_{BE}^{+}(\omega)$. This difference in power can therefore be used as a signal which adjusts the assumed carrier frequency.

In some applications frequency estimation is not necessary. This might be because both transmitter and receiver have sufficiently good frequency accuracy and stability that any residual error can be absorbed into frequency tracking. In some cases frequency performance might be sufficiently good that it is not even necessary to track frequency, because the small error in frequency can be absorbed by phase tracking. This is possible because a sufficiently small frequency offset appears to a phase estimator as a continuous series of equal phase corrections.

6.16.2 Carrier Phase Estimation

This brings us to carrier phase tracking. A very common technique for carrier phase tracking is precisely the same open-loop scheme shown in Figure 5.12, which works equally well for all forms of phase-amplitude modulation. The Costas loop concept demonstrated in Figures 5.13 and 5.14 also applies to digital modulation, but modifications are often required. In particular, the Costas loop of Figure 5.13 works as-is for OOK, BPSK, and M-ASK, since these are "one-dimensional" modulations as is AM. To extend this to modulations that exist in both dimensions, all that is necessary is to use both components of the complex baseband signal that emerges from the multiplier in Figure 5.13.

6.16.3 Symbol Timing

Finally we consider symbol timing. By "symbol timing" we mean specifically the problem of determining the MET; i.e., when to sample the output of a matched filter in order to obtain a decision metric with maximum SNR. Just as carrier estimation involves two parameters (frequency and phase), symbol timing involves two parameters: rate and MET. Although in

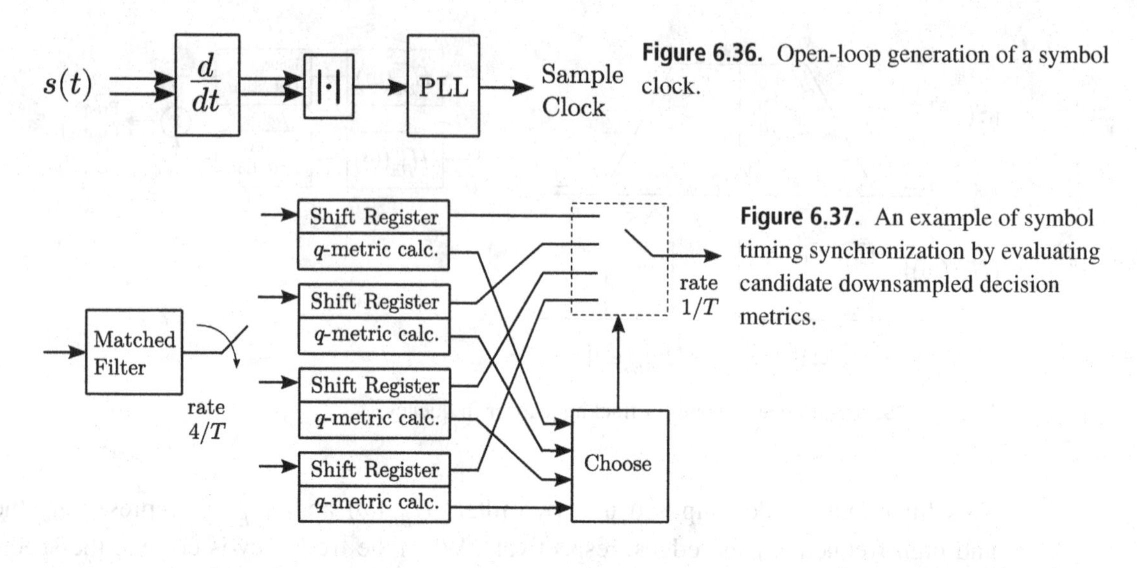

Figure 6.36. Open-loop generation of a symbol clock.

Figure 6.37. An example of symbol timing synchronization by evaluating candidate downsampled decision metrics.

principle it is possible to derive symbol rate from carrier frequency, this is not often done. Three somewhat more common approaches are described below.

A broad class of open-loop symbol clock-generating schemes is represented by the structure shown in Figure 6.36. In this scheme, the derivative is used to generate pulses associated with the symbol transitions. This occurs because the input $s(t)$ varies slowly between transitions, resulting in the derivative having small magnitude; whereas $s(t)$ varies relatively quickly during transitions, resulting in the derivative having relatively large magnitude. The pulses are rectified by the magnitude-squared operation, resulting in a train of narrow pulses at rate $2/T$. However, this signal cannot be used directly, for three reasons: (1) The rate is too fast by a factor of two; (2) some pulses are missing, namely those associated with adjacent symbols having the same value; and (3) the pulses themselves may not be sufficiently narrow to serve as an effective sample trigger. All three problems can be resolved using a phase-locked loop (PLL) which synchronizes an internally-generated symbol clock to the double-rate input signal generated by the above process. (For a primer on PLLs, see Section 16.8.) Note that the accuracy of this approach is limited by the degree to which $s(t)$ is oversampled with respect to $1/T$; generally a rate of at least $4/T$ or so is necessary for reasonable performance.

Another broad class of techniques involves oversampling the decision metric and then using some other criteria to determine how to downsample to the symbol rate. A representative scheme from this class is shown in Figure 6.37. In this example, the pulse-matched filter operates at four times the symbol rate, and the objective is to downsample by four. Therefore, there are four possible "phases" from which to select. To determine the best choice, a contiguous sequence of outputs associated with each candidate MET p is evaluated to determine a quality metric q_p, and the q_ps are compared to select the one closest to the exact MET. A simple quality metric suitable for phase-only modulations is to sum the magnitude-squared samples. In this scheme, the metric yields large values for the best estimate of the MET; whereas the other outputs represent samples taken during transitions from one symbol to the next, yielding smaller metrics. A more sophisticated and generally-applicable scheme is to compute the quality metric as the variance of the distances between each sample and the associated symbol decision. In this case, the variance is very small (nominally noise-dominated) for the best choice, and is larger (representing the timing bias) for the other choices. An advantage of this class of schemes is that explicit calculation of symbol rate or MET is avoided; the disadvantage is that one is limited to the P choices provided by oversampling by P.

When training sequences are available, correlation against the training sequence can be used to determine or refine the MET. The basic idea is to select t_0 to maximize:

$$\int_0^{T_d} s(\tau - t_0) s_t^*(\tau) d\tau \tag{6.74}$$

where $s_t(t)$ is the training sequence. Then t_0 corresponds to the start of the training sequence, which is also the MET for the first symbol in the training sequence. Again, oversampling relative to the symbol rate is necessary for this to be effective.

The three classes of schemes discussed above are hardly exhaustive, and the question of optimality has not been addressed. Nevertheless, these schemes are representative of the vast majority of practical applications.

6.17 ATSC: THE NORTH AMERICAN DIGITAL TELEVISION STANDARD

As an example of a synthesis of the techniques described in previous sections, this section presents a short overview of the North American broadcast digital television standard known as ATSC (so named after the defining organization, the Advanced Televisions Systems Committee).[14] ATSC is a system that is used to broadcast high-definition digital television in 6 MHz channels spanning the VHF and UHF frequency ranges [1]. An example of an ATSC signal is shown in Figure 6.38. Note that we have already had a few encounters with ATSC in previous sections; see Sections 6.2 (source coding; last paragraph), and 6.9.1 (M-ASK; last paragraph); also Examples 6.4–6.5 (performance with respect to the Shannon bound) and 6.8 (carrier frequency estimation).

6.17.1 Transmitter

Here is the transmission process for ATSC. (It may be useful to refer to Figure 6.1 while reading this section.)

(1) **Source coding.** The source coder for ATSC is the MPEG-2-TS ("transport stream") codec, which produces integrated video and audio at 19.39 Mb/s, organized as 188-byte packets. The first byte of the packet is a "sync" (synchronization) sequence which is removed, leaving 187-byte packets. The sync sequence is reintroduced in Step (5) below.

(2) **Randomization.** The packets are randomized by XOR-ing the stream with a pseudo-random sequence known to both the transmitter and receiver. The purpose of randomization

[14] The dominant schemes outside of North America are DVB-T (see also Example 6.10) and ISDB-T. For more on all three standards, see Section B.1.

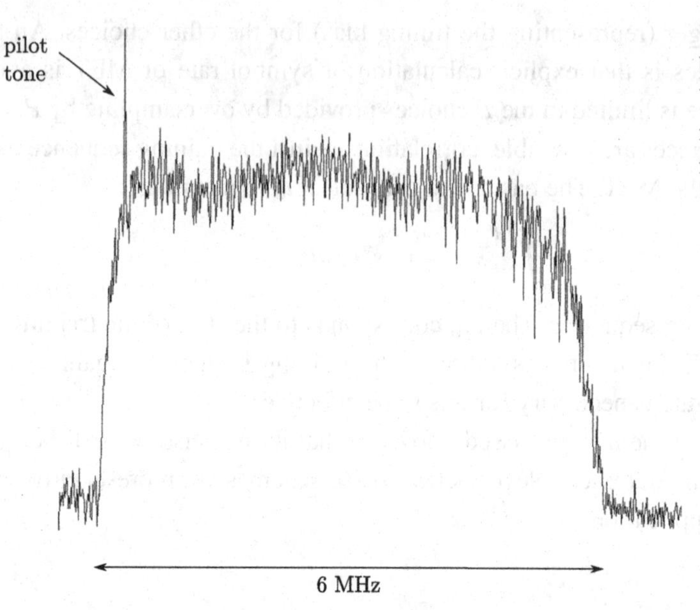

Figure 6.38. Spectrum of an ATSC broadcast signal (WPXR in Roanoke, VA, centered at 605 MHz). Actual measured spectrum, vertical scale in dB.

is to prevent data anomolies – e.g., long strings of zeros – from creating undesirable transients or spectral features in the transmitter output.

(3) **Error control coding.** Reed–Solomon (207,187)-byte forward error control (FEC) is applied. In this scheme, the 187-byte packet is analyzed to generate a 20-byte *check sequence*, which is appended to the data. This scheme has $t = 10$, so up to 10 byte errors can be detected and up to $t/2 = 5$ can be corrected.

(4) **Interleaving.** ATSC uses *convolutional byte interleaving*. This is essentially block interleaving done on a byte-by-byte basis, with 208-byte blocks interleaved to a depth of 52. Note that the interleaver block length is one byte longer than the FEC output packet; this causes the MPEG packet boundaries to "precess" within the span of the interleaver, which is intentional.

(5) **Trellis-coded modulation (TCM).** The MPEG sync byte is appended to the beginning of each 207-byte sequence. The resulting 208-byte sequence is known as a *segment*. The segment is converted to tribits (sequences of three bits representing the symbols in a $M = 8$ constellation) using rate-$\frac{2}{3}$ TCM. So, the number of bits representing the segment increases by a factor of $\frac{3}{2}$ due to convolutional coding, and the number of symbols is $\frac{1}{3}$ the number of bits. Therefore each segment is now represented by $208 \times 8 \times \frac{3}{2} \times \frac{1}{3} = 832$ symbols representing an $M = 8$ modulation, but not yet modulated.

(6) **Organization into fields.** Segments are organized into *fields*. A field consists of a *field sync* segment followed by 312 contiguous data segments. A field sync segment, like the data sync segments, consists of 832 symbols beginning with the MPEG-2 sync byte (now represented as four symbols). Next in the field sync segment is a 511-symbol pseudorandom sequence known as "PN511." PN511 is, among other things, a training sequence that may be used for carrier estimation, symbol timing recovery, channel estimation and/or equalizer training. The remainder of field sync segment contains

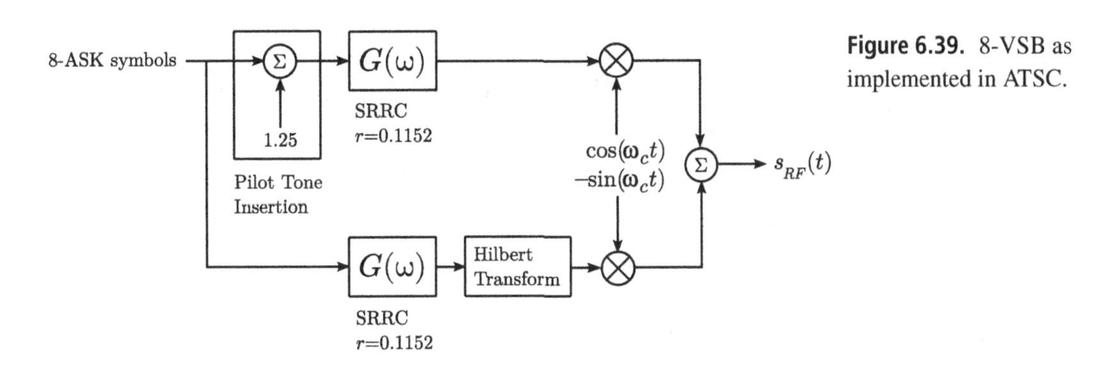

Figure 6.39. 8-VSB as implemented in ATSC.

additional (shorter) reference sequences and metadata indicating things like specific modes or advanced features in effect. The symbol rate at this point is 10.76 MSy/s.

(7) **8-VSB modulation.** 8-VSB is essentially symmetric 8-ASK (see Section 6.9.1) with a minor tweak and a major modification. Tribits representing modulation symbols are mapped to the 8-ASK modulation symbols

$$\{-7, -5, -3, -1, +1, +3, +5, +7\}$$

Now, the major modification: Because 8-ASK symbols are real-valued, the positive- and negative-frequency regions of the spectrum of the complex baseband representation are conjugate-symmetric. Therefore, it is sufficient to send only the positive frequencies, since the negative frequencies are redundant. Noting that this is essentially SSB modulation (Section 5.6.1), we may use the Hartley USB modulator (Figures 5.19 and 5.20). The scheme is summarized in Figure 6.39. This scheme uses SRRC pulse shaping with rolloff factor $r = 0.1152$. The broadband $\pi/2$ phase shift required to generate the quadrature branch is generated using the Hilbert transform, implemented using an FIR filter.[15] The minor tweak is the addition of a "pilot tone" to the in-phase branch of the modulator, having magnitude 1.25 at a frequency of 309 kHz. (The pilot tone is pointed out in Figure 6.38.) The relevance of this feature will be explained shortly.

A few characteristics of the ATSC transmit signal that bear notice: First, ATSC uses a very common channel coding strategy. The scheme consists of block error control coding, followed by interleaving to improve the performance of the block coder in the presence of burst errors, followed by convolutional coding (in this case, intrinsic to TCM). The interleaver is very deep, which allows the combination of block coding and interleaving to be robust to very long fades.

Second, note that the training sequence (assuming PN511 is used for this purpose) arrives once per field. There are 832×313 symbols per field, and the symbol rate is 10.76 MSy/s, so each field is about 24.2 ms long. Therefore, the receiver receives a new training sequence this often. Referring back to the discussion of channel coherence time in Section 3.6.3, we see this is a compromise: plenty fast for non-mobile applications (presumably accounting for most DTV use cases), but vulnerable in vehicular applications. To do better in this respect the field sync segment would need to be sent more often, which would make it impossible to send the same MPEG-2 rate in the same bandwidth.

[15] This scheme is approximate because the Hilbert transform has unbounded impulse response. But then again, so does the SRRC filter.

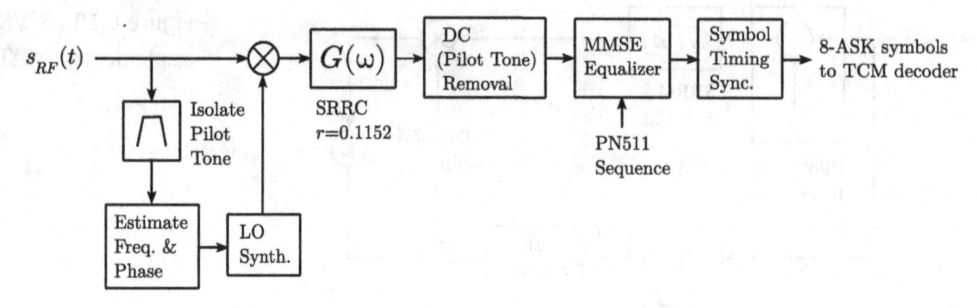

Figure 6.40. One possible scheme for demodulating an ATSC 8-VSB signal.

Finally, let us consider the spectrum and the spectral efficiency. Using $r = 0.1152$ SRRC pulse shaping, we expect a spectrum that is strictly bandlimited to

$$\text{FNBW} = (1+r)\frac{1}{T} = (1+0.1152)\frac{1}{10.76\text{ MSy/s}} \cong 12\text{ MHz}. \tag{6.75}$$

By retaining only the upper sideband, this has been reduced in half to 6 MHz. The spectral efficiency η_B using the FNBW criterion is therefore a respectable

$$(19.3\text{ Mb/s}) \;/\; (6\text{ MHz}) = 3.2\text{ bps/Hz},$$

despite the fact that only about 60% of the transmitted information is MPEG-2 data before coding.

6.17.2 Receiver

A reasonable starting point for demodulating ATSC is complex baseband representation $s(t) = s_i(t) + js_q(t)$ as shown in the architecture of Figure 6.11. Since 8-VSB is a "one-dimensional" modulation, the simplified architecture of Figure 6.12, which disposes of the unnecessary $s_q(t)$ component, could also work as a starting point. The scheme shown in Figure 6.40 is an even simpler version of this scheme, which is facilitated by the relative ease of carrier frequency and phase estimation in ATSC using the pilot tone. The idea is simply to isolate the pilot tone using a narrow bandpass filter, estimate its frequency and phase, and then use this information to generate a coherent local oscillator (LO) signal that can be used to convert $s_{RF}(t)$ to $s_i(t)$.[16] Also since the subsequent pulse-matched filtering has lowpass response, no additional filter is needed to reject the undesired sum-frequency product from the multiplier. This downconversion could be done in either the analog or digital domain.

Regardless of how $s_i(t)$ is obtained, most ATSC receivers follow a sequence of steps similar to those shown in Figure 6.40.

(1) SRRC pulse-matched filtering.
(2) The pilot tone is removed simply by subtracting the mean ("DC") value of $s_i(t)$.
(3) ISI mitigation is applied. In Figure 6.40, this is indicated as MMSE equalization. Alternatively, it could be MLSE with the appropriate change of matched filter response, or an alternative equalizer using RLS or DFE. In any event, the ISI mitigation is conveniently trained using the PN511 sequence, which arrives every 24.2 ms.

[16] Note this is very similar to the "product detection" scheme shown in Figure 5.11.

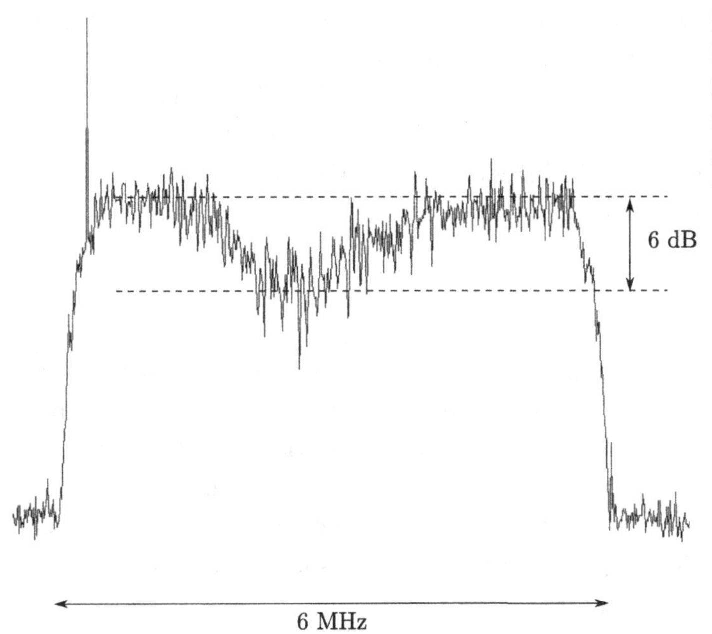

Figure 6.41. Same as Figure 6.38, but taken just a few minutes later. Strong time-resolved multipath is now evident, whereas only a slight amount of frequency-selective fading was evident previously.

(4) Symbol timing recovery can be implemented using any the methods described in Section 6.16.3.
(5) At this point the symbols are input to the TCM decoder and then on to deinterleaving, FEC decoding, and derandomization.

Finally, some form of AGC is required since the symbol decision regions in M-ASK depend on magnitude.

The particular procedure described above is certainly not the only valid approach. In addition to the various architectures that may be used to obtain $s_i(t)$ as described above, there are many alternative methods for acquiring/tracking the carrier and symbol timing.

Finally, it should be noted that ISI mitigation (e.g., equalization, in the above scheme) is not optional. To emphasize the point, Figure 6.41 shows the same signal shown in Figure 6.38 but just a few minutes later. The new spectrum implies the appearance of a strong time-resolved (i.e., ISI-generating) multipath component.[17]

6.18 DIRECT SEQUENCE SPREAD SPECTRUM (DSSS) AND CODE DIVISION MULTIPLE ACCESS (CDMA)

Direct sequence spread spectrum (DSSS) is a scheme in which bandwidth is expanded by replacing bits or modulation symbols with streams of modulation symbols at a much higher rate. What's the point in expanding bandwidth? The principal advantage is that this provides an alternative and somewhat simpler method to deal with propagation-induced ISI, which, as noted in Section 6.15, can be a formidable problem for traditional modulation. DSSS also

[17] One way to understand what is going on here is to compare Figure 6.41 with Figure 3.10, and review the associated text.

offers some simplification with respect to carrier estimation and symbol timing recovery, and allows multiple signals to coexist in the same bandwidth with low mutual interference. These attributes make DSSS quite popular, appearing in code division multiple access (CDMA) cellular telecommunication systems, the US GPS and Russian GLONASS satellite navigation systems, and in various modes of the IEEE 802.11 WLAN system. This section provides a brief overview of the DSSS concept and its applications.

6.18.1 Fundamentals

The principal characteristic of modulation in a DSSS system is *spreading*, illustrated in Figure 6.42. Spreading consists of replacing information units – typically bits, but sometimes modulation symbols – with long sequences of modulation symbols known as *codes*. The symbols within a code are referred to as *chips* in order to distinguish them from the original information units. Chips may be implemented using essentially any form of modulation already discussed, although BPSK and QPSK are popular. Pulse shaping is applied to chips as a means to shape the spectrum, just as in traditional modulations. Given symbol period T and chip period T_{chip}, the resulting bandwidth expansion is a factor of T/T_{chip}, which is also known as the *processing gain*.

Information can be recovered using the optimal correlation receiver described in Section 6.6.4, in which the pulse waveform in a traditional application is replaced by the spreading code here. This is shown in Figure 6.43. Of course the spreading code must be properly synchronized with the arriving signal for the correlation to be successful – more on this in a moment. This use of the correlation receiver is commonly referred to as *despreading*, as it collapses the transmit bandwidth of approximately $1/T_{chip}$ to the information bandwidth of approximately $1/T$.

As we shall see in a moment, DSSS is most attractive when the processing gain is large, so it is typical that a single DSSS signal occupies the bandwidth available for all users in a given segment of spectrum. In a CDMA system, all users use DSSS to occupy all available spectrum. In a traditional communications system, this would be a disaster. In a CDMA system, however,

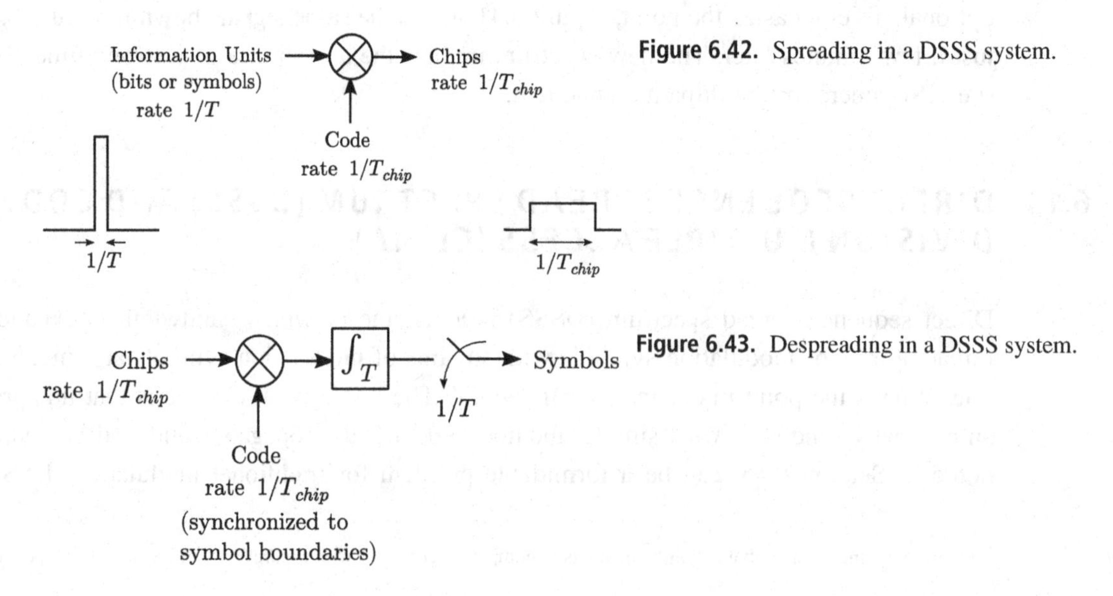

Figure 6.42. Spreading in a DSSS system.

Figure 6.43. Despreading in a DSSS system.

there is no problem as long as each user employs a unique spreading code, and the spreading codes are mutually uncorrelated. (This is precisely the idea being conveyed in Figure 1.5.) Under these conditions, the act of despreading strongly suppresses all user signals other than the particular one corresponding to the spreading code employed. In principle, the number of signals that can be accommodated in the same bandwidth in this manner is equal to the number of spreading codes that yield sufficiently low mutual correlation. This makes DSSS a multiple access technique[18] as well as a modulation; CDMA is essentially DSSS employed to allow multiple users to share the same bandwidth. In practice it turns out that the product of the number of users times information rate possible in a given bandwidth is about the same for DSSS as for traditional modulations, although there are some caveats and a rigorous analysis is beyond the scope of this book.

EXAMPLE 6.9

GPS. The US Global Positioning System employs a variety of DSSS modes, and multiple satellites share the same spectrum using CDMA. The simplest GPS DSSS mode is the Coarse/Acquisition (C/A) mode, in which each satellite sends information at 50 b/s spread by a 1023-bit spreading code repeating once per millisecond. Estimate the bandwidth and processing gain of the C/A signal.

Solution: The chip rate $1/T_{chip}$ is 1023 chips per millisecond, which is 1.023 MHz. The bandwidth is approximately equal to the chip rate, so the bandwidth of the C/A signal is approximately 1 MHz. The processing gain is T_{chip}/T where $1/T = 50$ Hz, so the processing gain is $20460 = 43.1$ dB. At this level of processing gain, all satellites above the horizon may be received simultaneously and separated using the correlation receiver architecture of Figure 6.43 with negligible mutual interference.

It is worth noting that one of the fundamental difficulties in DSSS is chip timing synchronization, which must be accurate to a fraction of T_{chip}. At first glance this appears to be a disadvantage. However, note that the spreading code can be interpreted as a training sequence that is continuously present in the signal. Thus, the techniques for carrier and symbol timing estimation utilizing training sequences described in Section 6.16 are all straightforwardly implementable in DSSS systems, using the spreading code as a continuously-available training signal.

6.18.2 Cellular CDMA

CDMA emerged in the 1990s as a strong contender for cellular telecommunications applications due to the particularly elegant manner in which it contends with multipath. As noted previously, the two primary problems created by multipath are fading and ISI. We noted in Section 3.6.2 that fading is associated with a coherence bandwidth B_c which

[18] see Section 1.3 for a reminder of this concept

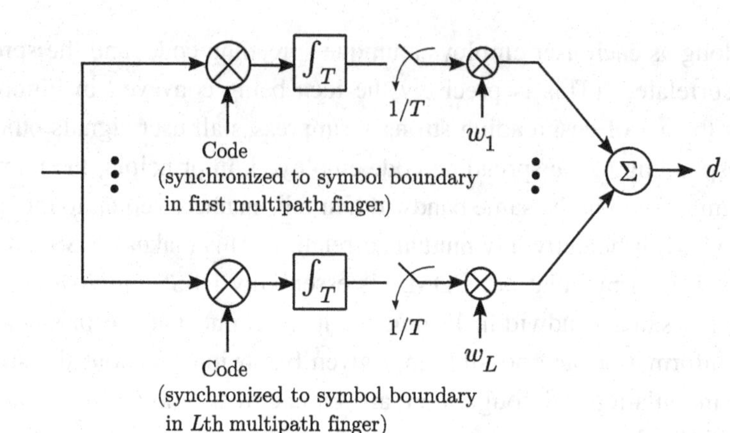

Figure 6.44. Rake receiver for exploiting resolvable ISI in DSSS communications systems.

(depending on one's definition of "coherence") ranges from about 10 kHz to 1 MHz in cellular telecommunications applications. Using DSSS, it is possible to expand the bandwidth of a communications signal to be many times B_c, making it quite unlikely for the entire bandwidth to be subject to a deep fade simultaneously. Thus, DSSS is intrinsically robust to flat fading if $1/T_{chip} \gg B_c$.

Similarly, DSSS is intrinsically robust to ISI if T_{chip} is much less than the delay spread σ_τ. This is because multipath-generated versions of the signal which arrive with delays significantly greater than T_{chip} will be suppressed by the correlation receiver of Figure 6.43. For example, a DSSS system with $1/T_{chip} = 10$ MHz will be robust to ISI associated with multipath components which are delayed by at least $T_{chip} = 100$ ns, which is typical for indoor scenarios and much less than most outdoor scenarios.

This remarkable property of DSSS can be exploited to actually *enhance* performance in ISI channels with sufficiently large delay spread. This is achieved by using a *rake receiver*, illustrated in Figure 6.44. A rake receiver consists of L branches, where each branch is a correlation receiver (identical to the one shown in Figure 6.43) synchronized to a different *finger* (resolvable multipath component) of the channel impulse response. The output of these branches are then combined using MRC as described in Section 6.14.3. In effect, the rake receiver decomposes the multipath channel into its constituent fingers, and then uses diversity combining to optimally combine the fingers. In summary, the management of ISI in DSSS systems is remarkably simple compared to the analogous processing (optimally, MLSE; suboptimally, equalization) required to make traditional modulations robust to ISI.

With this plethora of attractive features, one could fairly ask why it is that all cellular communications systems don't migrate to DSSS. There are essentially two reasons. First, DSSS entails processing data at a rate on the order of $1/T_{chip}$, whereas traditional modulation requires processing data at a much lower rate on the order of $1/T$. If the advantages of DSSS are not required – e.g., if multipath can be effectively mitigated using a combination of diversity and equalization – then the increase in computation burden associated with DSSS may not be justified.[19]

[19] This was essentially the crux of the debate throughout the 1990s about whether CDMA or GSM was preferable as the foundation for future cellular telecommunications systems. This debate has never been resolved and both concepts continue to find applications today.

The second contraindication for DSSS is when insufficient bandwidth exists to make DSSS worthwhile. DSSS is most effective when $1/T_{chip} \gg B_c$ and $T_{chip} \ll \sigma_\tau$, meaning occupied bandwidths of at least 1 MHz or so in most applications. This is one reason why DSSS is rarely considered for systems operating below 300 MHz: Contiguous bandwidth necessary to realize the advantages of DSSS is simply not available.

6.19 ORTHOGONAL FREQUENCY DIVISION MULTIPLEXING

Orthogonal frequency division multiplexing (OFDM) is another alternative to traditional modulation techniques. Like DSSS, it may alternatively be viewed as either an ISI countermeasure or a multiple access technique.

6.19.1 Concept

The basic recipe for OFDM is as follows: Demultiplex the symbol stream into N parallel channels, and assign each channel to a center frequency according to a grid with spacing equal to Δf. Now consider the correlation between a symbol $s_1(t)$ assigned to center frequency ω_1 and a symbol $s_2(t)$ assigned to center frequency ω_2:

$$\int_0^T \left[s_1(t)e^{j\omega_1 t}\right]\left[s_2(t)e^{j\omega_2 t}\right]^* dt \tag{6.76}$$

For simplicity (and without loss of generality) let us assume rectangular pulse shaping, so that $s_1(t)$ and $s_2(t)$ are constant over the duration of the symbol. Then we have

$$s_1 s_2^* \int_0^T e^{j\Delta\omega t} dt = \frac{s_1 s_2^*}{j\Delta\omega}\left(e^{j\Delta\omega T} - 1\right) \tag{6.77}$$

where $\Delta\omega \equiv \omega_1 - \omega_2$. Note that the correlation goes to zero when $\Delta\omega T = 2\pi n$, where n is any integer. Therefore channels spaced by

$$\Delta f = \frac{n}{T} \tag{6.78}$$

are uncorrelated, and so any number of channels with a spacing of $1/T$ are mutually uncorrelated. Furthermore, there is no smaller spacing that has this property. At the receiver, the data may be recovered simply by performing the reverse operation, and no channel will interfere with any other channel.

OFDM is a canonical example of a *multicarrier modulation*; i.e., a modulation in which multiple carriers are employed simultaneously and to synergistic effect.

Note that there is no processing gain associated with OFDM; that is, regardless of whether one employs OFDM or not, the transmit bandwidth is $1/T$; the only difference is whether you have one channel at that rate, or N adjacent channels operating at a rate of $1/NT$. So why bother? If you can make N sufficiently large that $1/NT$ is much less than the coherence bandwidth B_c, then each individual channel can experience only flat fading, and is essentially immune to ISI. If you can manage to fill a bandwidth much greater than B_c (i.e., $1/T \gg B_c$) under the previous condition, then it becomes very unlikely that all channels can be in a deep fade simultaneously. In effect, we have achieved some of the ISI-counteracting benefits of equalization without actually doing equalization. Alternatively, one might view OFDM as *frequency diversity*; i.e., diversity in which the "branches" are independently-fading frequency channels.

6.19.2 Implementation

Organizing data into N orthogonal carriers entails considerably complexity if N is large. However, the ISI-robust property of OFDM isn't really significant unless N is large. To make OFDM really attractive from a systems perspective, one needs a low-complexity scheme to generate the carriers. As it turns out, we get this on a silver platter using the discrete Fourier transform (DFT) through its reduced complexity equivalent, the fast Fourier transform (FFT). An N-point DFT decomposes a single time-domain signal sampled at a rate of $1/T$ into N orthogonal channels having spacing of $1/NT$ – exactly what we need for the receiver. For the transmitter, we simply use the DFT in reverse; i.e., the inverse DFT (IDFT). The DFT and IDFT have computational burden proportional to N^2, which is reduced to $N \log_2 N$ by replacing these functions with the equivalent FFT and IFFT. This reduction in complexity makes OFDM very attractive, so it is rare to see OFDM described as using any approach other than the IFFT/FFT even though the latter is not strictly necessary.

A generic block diagram of an OFDM system is shown in Figure 6.45. Note that a *cyclic prefix* is included. The purpose of a cyclic prefix is to prevent temporally-resolvable multipath from disrupting the orthogonality of the subcarriers. A cyclic prefix is simply an extension of the OFDM symbols by a time at least as large as the delay spread. This creates a "guard time" that ensures that the FFT in the receiver perceives only one symbol per channel on each execution of the FFT. This results in a minor degradation of spectral efficiency, but is important for robust operation.

OFDM tends to appear in applications in which is ISI is likely to be a severe problem, but in which the available bandwidth limits the processing gain that can be achieved using DSSS. Common applications include digital television systems (see example below), some WLAN systems including IEEE 802.11a and -g (see Section B.4.2), and 4G cellular telecommunications standards such as LTE (see Section B.3.3).

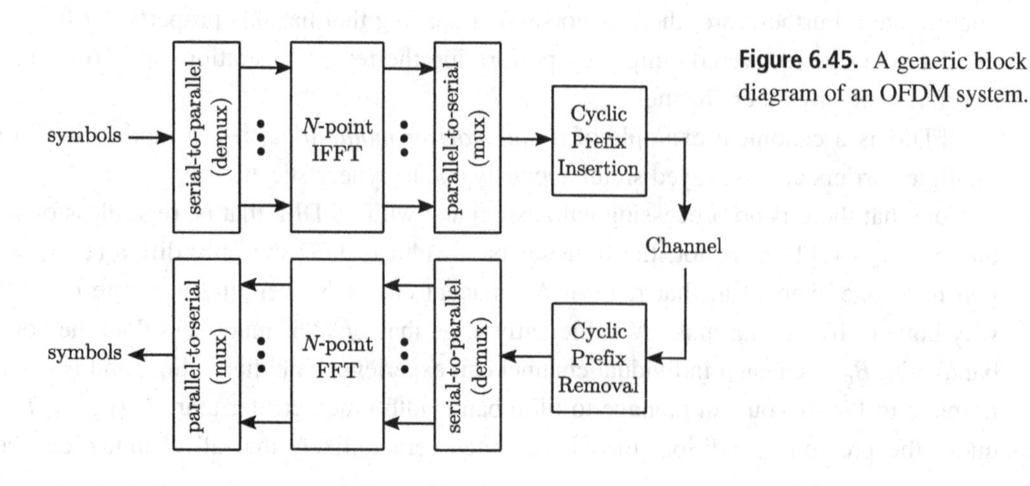

Figure 6.45. A generic block diagram of an OFDM system.

EXAMPLE 6.10

DVB-T. In Eurpope, the predominant system for broadcast television is *digital video broadcasting – terrestrial* (DVB-T), which in contrast to the North American ATSC system (Section 6.17) uses OFDM.[20] DVB-T operates in 6 MHz channels (similar to ATSC) with $N = 2048$ or 8192 subcarriers and QPSK, 16QAM, or 64QAM.

OFDM has two primary disadvantages over competing schemes. One is that OFDM has intrinsically large peak-to-average power ratio (PAPR), which is due to the potential for combinations of subcarriers to intermittently add in phase. Because the carriers are at different frequencies, this condition is never persistent; however, the large number of carriers in practical OFDM systems results in this happening quite frequently. Occasionally, all N carriers will add in phase; thus the large PAPR. This requires highly linear power amplifiers, which tend to be inefficient relative to power amplifiers that designed for constant magnitude or nearly-constant magnitude modulations. This is a relatively small issue in broadcast applications (since there is only one transmitter), but is a significant disadvantage in mobile and personal telecommunications applications. For this very reason, LTE cellular systems (Section B.3.3) use OFDM on the downlink, in which the power amplifiers are at the base station, but use single-carrier modulation on the uplink, in which the power amplifier is in a mobile device.

The second disadvantage of OFDM is sensitivity to Doppler spread. Doppler must be precisely tracked in order to keep subcarriers centered in FFT bins. Further, since Doppler spread is frequency-dependent, higher-frequency subcarriers are shifted by a greater amount than lower-frequency subcarriers, so no single frequency adjustment is optimal. This problem can be corrected using channelizers other than the Fourier transform, which unfortunately do not have low-complexity equivalent implementations analogous to the FFT. Combined with the PAPR issue, this tends to limit the use of OFDM in mobile applications and in particular vehicular applications.

Problems

6.1 Figures 6.3(a) and (b) show "straightforward" and Gray-coded assignments of bit groups to symbols for QPSK. Repeat for 8-PSK.

6.2 Develop an expression analogous to Equation (6.17) that relates SNR to E_S/N_0; i.e., the SNR per symbol.

6.3 Reproduce the BPSK result from Figure 6.14, except now plot the result with SNR as opposed to E_b/N_0 as the independent variable. Assume a bit rate of 10 kb/s using SRRC pulse shaping with $r = 0.22$.

6.4 Confirm that the symbol coordinates shown in Figure 6.17 yield mean symbol energy E_S.

[20] The debate between single-carrier 8-VSB and OFDM during the development of digital television – never fully resolved – starkly demonstrates that there is rarely an obvious optimal design for modern radio communications systems.

6.5 Work out expressions for the probability of *symbol* error as a function of E_S/N_0 for: (a) M-ASK, starting with Equation (6.46); (b) M-QAM, starting with Equation (6.47); (c) M-PSK, starting with Equation (6.48).

6.6 What E_b/N_0 is required to achieve a BER of 10^{-6} using (a) incoherently-detected OOK, (b) coherently-detected OOK, (c) BPSK, (d) QPSK, (e) 8-ASK, (f) 16-QAM, (g) 32-PSK.

6.7 Speculate on why Equation (6.48) appears to give an incorrect result for $M = 2$ (BPSK).

6.8 You are to transmit 100 kb/s using optimum sinc pulse shaping over a non-fading, ISI-free channel having SNR 10 dB. Do not employ pulse shaping or error control coding. Which of the following modulations yields the best BER, assuming coherent demodulation–OOK, BPSK, QPSK, 8-ASK, 16-QAM, 32-PSK?

6.9 What is the theoretical maximum reliable data rate that can be sustained in a 1 MHz channel if the SNR is 10 dB?

6.10 In the previous problem, what is the theoretical minimum E_b/N_0 required to sustain the theoretical maximum reliable data rate?

6.11 What E_b/N_0 is required to reliably transfer one bit in 1 Hz?

6.12 Following up Example 6.6, characterize the spectral efficiency and energy efficiency with and without coding.

6.13 A 30 kb/s BPSK radio link using SRRC pulse shaping with $r = 0.22$ is operating at SNR $\approx$ 20 dB and experiencing Rayleigh flat fading. What is the BER? By how much is the BER expected to improve if the transmit power is doubled?

6.14 In the above problem (at the original power level), evaluate BER if we employ space diversity with MRC, using two and four antennas. Assume perfectly uncorrelated fading at the four antennas. Which leads to the bigger improvement: doubling transmitter power or implementing space diversity?

6.15 What is the largest carrier frequency error that can be accommodated by a coherent QPSK demodulator that has only carrier phase tracking, and no explicit compensation for carrier frequency error? Assume one carrier phase update per symbol period, and assume the carrier phase tracking is otherwise perfect.

7 Radio Link Analysis

7.1 INTRODUCTION

In previous chapters, we considered four of the five primary aspects of a radio link identified in Chapter 1: antennas (Chapter 2), propagation (Chapter 3), noise (Chapter 4), and modulation (Chapters 5 and 6). The remaining aspect is the radio itself, and is the topic of much of the rest of this book. However, Chapters 1–6 provide sufficient background to analyze radio links, which is the topic of this chapter. Analysis of radio links in a few representative applications will reveal some of the principal requirements for radios; in particular, transmit power and receive sensitivity. This information will serve as useful guidance in the study of radio design in subsequent chapters.

However, the analysis of radio links is, by itself, an important topic. The general idea is this: Given information about the antennas, propagation, noise, and modulation, plus additional details including transmit power and receiver sensitivity, one is able to quantify the performance of the link. For digital modulations, one is able to determine bit error rate (BER) as a function of these parameters. In general, one is able to determine if the link performance is acceptable, and the changes or combinations of changes that are needed to improve performance.

The organization of this chapter is as follows. In Section 7.2, we modify the Friis transmission equation to account for losses associated with antenna efficiency, antenna impedance mismatch, and mismatch between the polarization of the receive antenna and the incident wave. The concepts of EIRP and ERP are presented in Section 7.3. In Section 7.4 we consider noise, which facilitates calculation of the signal-to-noise ratio at the input of the receiver's demodulator and subsequently E_b/N_0. This leads to the concept of sensitivity, which is the topic of Section 7.5. Section 7.6 introduces the link budget as a useful method for managing the analysis of sensitivity. Sections 7.7–7.10 serve the dual purpose of demonstrating link budget analysis in a variety of applications, and illustrating requirements for transmit power and receive sensitivity as they emerge in these applications. This chapter concludes with a summary of these requirements in Section 7.11.

7.2 FRIIS TRANSMISSION EQUATION REVISITED

In Chapter 3 we derived the Friis transmission equation (Equation (3.7))

$$P_R = P_T G_T L_p^{-1} G_R \tag{7.1}$$

which relates the power delivered to a receiver, P_R, to the power provided by a transmitter to its antenna, P_T. G_T and G_R are the gains of the transmit and receive antennas, respectively, and L_p is the path loss. Path loss was defined in Chapter 3 as the ratio of power transmit to power received for unit antenna gains ($G_T = G_R = 1$). The free space path loss L_{p0} was defined as L_p for the special case of unimpeded spherical spreading of the radio wave along the path from transmitter to receiver, and is given by Equation (3.6):

$$L_{p0}(R) = \left(\frac{\lambda}{4\pi R}\right)^{-2} \tag{7.2}$$

Generally, we found that path loss is well-modeled as variation around a mean path loss $\overline{L_p}(R)$, where the variation is due to fading, primarily associated with multipath. We found $\overline{L_p}(R)$ is typically well-modeled as

$$\overline{L_p}(R) = \overline{L_p}(R_b)\left(\frac{R_b}{R}\right)^{-n} e^{+2\alpha R}, \quad R > R_b \tag{7.3}$$

where n is the path loss exponent; R_b is the breakpoint distance; and α is the coefficient of attenuation, accounting for lossy media including the possibility of atmospheric losses and rain fade above a few GHz (Sections 3.7.2–3.7.4).

Implicit in Equation (7.1) are the following assumptions: (1) any impedance mismatch between transmitter and transmit antenna, and receive antenna and receiver, is negligible; and (2) mismatch between the polarization of the incident wave and the receive antenna is negligible. However, there are many commonly-encountered problems in which one or more of these assumptions is not reasonable (including examples appearing later in this chapter). Furthermore, we have the ever-present possibility of ambiguity in how radiation efficiency and other losses are taken into account, so it is useful to be explicit about that. We may accommodate all of these issues using an expanded version of Equation (7.1):

$$P_R = P_T\, \eta_T\, \epsilon_T\, G_T\, L_p^{-1}\, \epsilon_p\, G_R\, \epsilon_R\, \eta_R \tag{7.4}$$

where the additional factors are explained below.

- η_T is the fraction of P_T that is delivered to the antenna, and which is less than 1 due to impedance mismatch. If we define P_T as "available power" as in Section 2.6.1, then η_T is transducer power gain (TPG) and is given by Equation (2.47) in general, or by Equation (2.48) if the imaginary part of the impedance of the transmitter or antenna (or both) is zero.
- ϵ_T accounts for transmit-side losses due to dissipation (as opposed to reflection or transmission). This includes the radiation efficiency ϵ_{rad} of the antenna, but might also include other dissipative losses such as those due to electromagnetic coupling into the ground, prevalent especially at VHF and below.
- $\epsilon_R \cdot \eta_R$ accounts for the analogous receive-side losses.
- ϵ_p represents power loss due to mismatch between transmit and receive antennas, ranging from 1 (matched polarizations) to 0 (orthogonal polarizations).

Note that when the factors are defined as indicated above, G_T and G_R represent the *directivity* of the antennas as opposed to the *gain* of the antennas. It is important to be aware of this, as it is also common to associate ϵ_{rad} with G_T and G_R, as opposed to ϵ_T and ϵ_R as we have done above. It is also common to associate part of ϵ_{rad} to antenna gain and part to dissipative loss, or simply to leave ϵ_{rad} as a separate factor. These forms are all correct in a sense, and one should always take care to note the particular "accounting scheme" that is being used in any particular application.

EXAMPLE 7.1

Expanding on Example 3.1: A satellite transmits 5 W into a lossless and well-matched antenna having a directivity of 6 dBi. The antenna polarization is right circular. The center frequency is 1.55 GHz. The mobile, which is 800 km directly below the satellite, uses a horizontally-oriented linearly-polarized antenna with about 2 dBi directivity in the direction of the satellite. The receive antenna has a VSWR of 2:1 and is 70% efficient. What is the path loss and what is the power delivered to the receiver?

Solution: For 1.55 GHz, $\lambda = 19.4$ cm. Since atmospheric losses are negligible at L-band, we can assume free space path loss. Therefore $L_p = L_{p0} = 154.3$ dB from Equation (7.2). From the problem statement we have $P_T = 5$ W, $G_T = 6$ dBi, $\epsilon_T\eta_T \approx 1$, $G_R = 2$ dBi, and $\epsilon_R = 0.7$. Receive VSWR of 2:1 corresponds to $|\Gamma_R| = 1/3$; therefore $\eta_R = 1 - |\Gamma|^2 = 0.889$.[1] Also $\epsilon_p = 0.5$ since a linearly-polarized antenna captures only half the power in circularly-polarized wave (Section 2.5). Using Equation (7.4), we find $P_R = 3.61 \times 10^{-15}$ W, which is -144.4 dBW or -114.4 dBm. In Example 7.4 we'll find out this is typically plenty of power for a usable data link.

7.3 EFFECTIVE RADIATED POWER (EIRP AND ERP)

When a radio link consists of one radio transmitting to multiple receivers, or to a receiver whose location is not fixed, it is usually desired to specify the power available to receivers over the entire region of interest. For example, broadcasters aim to deliver power density [W/m^2] equal to or greater than some threshold over their coverage area. Regional regulatory agencies (such as the FCC in the USA) aim to limit the power density generated by a broadcaster, so as to make it possible to reuse the frequency in some other geographical region. Equation (7.4) indicates that the power density generated within some region is proportional to $P_T\eta_T\epsilon_T G_T$, and that there are no other relevant factors under the control of the broadcaster. Thus, it is not sufficient for broadcasters or the regulatory agencies simply to specify P_T, because if they do, then they must also specify the product $\eta_T\epsilon_T G_T$ to ensure that the power density is constrained as intended.

A simpler approach is to specify the product $P_T\eta_T\epsilon_T G_T$, which results in an unambiguous specification without arbitrarily constraining P_T, G_T, and the associated efficiency and

[1] This assumes the VSWR is with respect to a standard real-valued impedance such as 50 Ω or 75 Ω, which is essentially always the case. Thus, Equation (2.48) may be used.

mismatch losses. In terms of Equation (7.4), we define *effective isotropically radiated power* (EIRP) as this product:[2]

$$\text{EIRP} = P_T \, \eta_T \, \epsilon_T \, G_T \tag{7.5}$$

EIRP can be more formally defined as the power applied to an efficient ($\epsilon_T = 1$), conjugate-matched ($\eta_T = 1$), and isotropic ($G_T = 1$) antenna that results in the same power density at some distant point that would result from applying the *actual* power to the *actual* antenna, including efficiency and mismatch losses. In other words, EIRP is really a description of power density, but is quantified in terms of power delivered through a standard "reference" antenna.

EXAMPLE 7.2

An L-band satellite mobile data communication service provider currently promises an EIRP of 55 dBW to its subscribers using a transmitter equipped with a dish-type reflector antenna. For the next generation of satellites, the antenna must fit into half the area occupied by the antenna on the current generation of satellites. What does this imply for the new satellite's transmit power requirement?

Solution: Directivity is proportional to area (Equation (2.64)), so a 50% reduction in area amounts to a 50% reduction in directivity. Since EIRP is proportional to $P_T G_T$, P_T must increase by a factor of 2.

EIRP is certainly not the only way to specify the effective power of a transmitter. An alternative is "effective radiated power" (ERP) with respect to a specified antenna directivity, G_0. ERP is formally defined as the power applied to an efficient, conjugate-matched antenna *having directivity* G_0 that results in the same power density at some distant point that would result from applying the actual power to the actual antenna, including efficiency and mismatch losses. For example, a transmitter using an antenna with directivity $G_T = G_0$ has ERP $= P_T \eta_T \epsilon_T$. An advantage of this scheme is that when $G_T = G_0$, ERP can be physically measured as the power delivered to the antenna terminals. ERP defined with respect to the peak directivity of an ideal half-wave dipole (2.15 dBi) is convenient in applications in which antenna directivity is specified in dBd – that is, dB relative to the maximum directivity of a dipole – as opposed to dBi. Here is an example:

EXAMPLE 7.3

In the USA, the FM broadcasters of the 88–108 MHz band are typically limited to 100 kW ERP with respect to a half-wave dipole. Consider an FM transmitter using a vertical array of the type discussed in Example 2.7, which achieves 8 dBi. Assume this system suffers 3 dB loss in cables; the cables being considered part of the antenna system in this case. If this broadcaster achieves 100 kW ERP, what is the associated transmitter output power?

[2] We'll have a lot more to say about the product $\eta_T \, \epsilon_T$ in Section 12.3.2.

Solution: From the problem statement, $G_T/G_0 \cong 8$ dBi $- 2$ dBi $= 6$ dB $\cong 3.99$ and $\epsilon_T = -3$ dB $\cong 0.50$. VSWR is not mentioned, so we will assume $\eta_T \approx 1$. In this case ERP $= P_T\eta_T\epsilon_T G_T/G_0$, so $P_T = (100\text{ kW}) / (1 \cdot 0.50 \cdot 3.99) \cong 50.1$ kW. Note that this would be the power measured at the output of the transmitter and at the input to the (lossy) cables leading to the dipole array.

7.4 SIGNAL-TO-NOISE RATIO AT THE INPUT OF A DETECTOR

The Friis transmission equation gives the power P_R delivered to a receiver. However, the quality of the output of a receiver depends less on P_R than on the *signal-to-noise ratio* (SNR) delivered to the input of the detector (typically a demodulator) within the receiver.

The situation is illustrated in Figure 7.1. Here we have partitioned the receiver into two parts: The detector, and the signal path preceding the detector, characterized by gain G_{RX} and

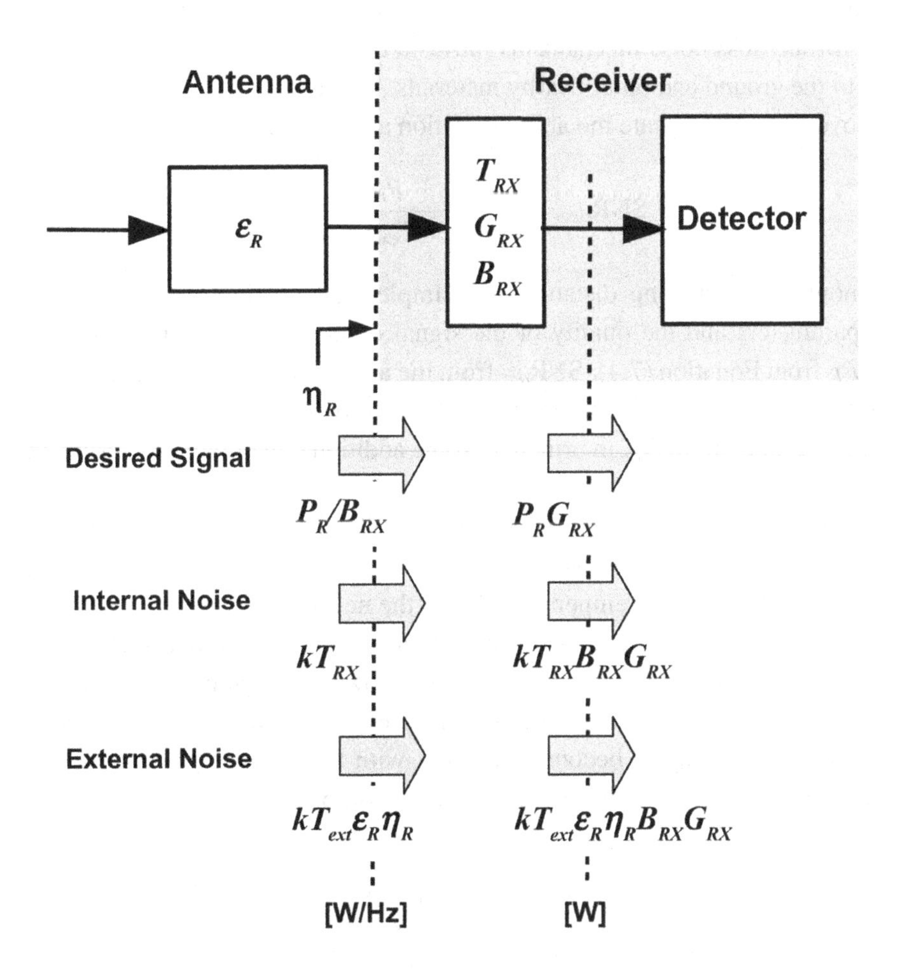

Figure 7.1. System model for determining the signal to noise ratio at the input to a receiver's detector. Quantities indicated by the left column of block arrows are expressed as power spectral densities referred to the input of the receiver; on the right as power delivered to the detector.

input-referred noise temperature T_{RX}. The signal power associated with P_R, which is delivered to the detector, is $P_R G_{RX}$, assuming the receiver bandwidth B_{RX} is at least as large as the signal bandwidth B. Since T_{RX} is input-referred, the receiver's contribution to the noise at the input of the detector is $kT_{RX}B_{RX}G_{RX}$. Finally there is the possibility that external noise is significant. Using the concept of antenna temperature T_A to represent this external noise, the external noise delivered to the detector is $kT_A\epsilon_R\eta_R B_{RX}G_{RX}$. Thus the signal-to-noise ratio at the input to the detector is

$$\text{SNR}_{det} = \frac{P_R}{k\left[T_{RX} + T_A\epsilon_R\eta_R\right]B_{RX}} \tag{7.6}$$

A potential ambiguity arises in the definition of T_A when the antenna exists in the proximity of – and therefore may be electromagnetically coupled to – lossy materials, or when potentially-lossy feedlines are defined as being part of the antenna. The ambiguity in these cases is that it may not be clear whether T_A accounts for these circumstances, or whether T_A is intended to be a value which exists independently of loss/noise mechanisms local to the antenna. To avoid potential confusion, we define the *external noise* T_{ext} to be the antenna temperature prior to accounting for any loss/noise mechanisms intrinsic to the antenna, including electromagnetic coupling into the ground and other nearby materials. This choice preserves the definition of ϵ_R as given above. Thus, we restate the above equation as follows:

$$\text{SNR}_{det} = \frac{P_R}{k\left[T_{RX} + T_{ext}\epsilon_R\eta_R\right]B_{RX}} \tag{7.7}$$

We are now within striking distance of a simple mathematical relationship between the radio link parameters and the quality of the signal output by the receiver. For example, we can obtain P_R from Equation (7.4), SNR_{det} from the above equation, and then knowing SNR_{det} and the modulation, we could compute the relevant metric of quality (e.g., BER) at the output of the receiver. While this is OK in principle, some additional manipulation results in a simpler equation and a new, useful concept: *system temperature*.

The system temperature T_{sys} can be defined as follows. Imagine a second version of our system in which there is miraculously no noise; i.e., $T_{ext} = T_{RX} = 0$. Now allow only T_{ext} to become non-zero. The system temperature T_{sys} is the new (higher) value of T_{ext} for which the noise power delivered to the detector is the same as that in the original system. An alternate definition of T_{sys} is as follows: kT_{sys} [W/Hz] is the power spectral density of a desired signal received in the same manner as kT_{ext}, but which results in $\text{SNR}_{det} = 1$. These definitions are equivalent, and a good way to become proficient with the concept of system temperature is to make sure you understand both of them. Application of either definition to Equation (7.7) yields

$$T_{sys} = T_{ext} + \frac{T_{RX}}{\epsilon_R\eta_R} \tag{7.8}$$

Note that T_{sys} is simply T_{ext}, plus T_{RX} referred to the "input" of the antenna.

It should be noted that in many problems we can assume T_{sys} to be dominated by either the T_{ext} term or the T_{RX} term of Equation (7.8), in which case the other can be ignored. In particular it is frequently assumed that $T_{RX}/\epsilon_R\eta_R \gg T_{ext}$ such that $T_{sys} \approx T_{RX}/\epsilon_R\eta_R$.[3] However, there are many practical situations in which this assumption becomes invalid; see Section 7.9 for an example. In these cases, it might be that $T_{sys} \approx T_{ext}$, or it might be important to account for both T_A and T_{RX}.

The T_{sys} parameter can be used to combine and simplify Equations (7.4) and (7.7) as follows. First, we use Equation (7.8) to rewrite Equation (7.7) as

$$\text{SNR}_{det} = \frac{P_R}{kT_{sys}\epsilon_R\eta_R B_{RX}} \tag{7.9}$$

Then we eliminate P_R by substitution of Equation (7.4):

$$\text{SNR}_{det} = \frac{P_T\eta_T\epsilon_T G_T L_p^{-1}\epsilon_p G_R}{kT_{sys}B_{RX}} \tag{7.10}$$

This is the desired expression. However, when digital modulations are employed, E_b/N_0 is a more suitable metric than SNR_{det} (see Section 6.5.2 for a quick review of the E_b/N_0 concept). We may obtain this as follows. First, note that the bit energy E_b at the input to the detector is

$$E_b = P_R G_{RX} T_b = P_R G_{RX}/r_b \tag{7.11}$$

where $r_b = 1/T_b$ is the bit rate. Similarly, the noise power spectral density N_0 at the input to the detector is

$$N_0 = kT_{sys}\epsilon_R\eta_R G_{RX} \tag{7.12}$$

thus we find

$$\frac{E_b}{N_0} = \frac{P_T\eta_T\epsilon_T G_T L_p^{-1}\epsilon_p G_R}{kT_{sys}r_b} \tag{7.13}$$

These equations will serve as the starting point for the development of link budgets beginning in Section 7.6.

Before moving on, it should be noted that the noise represented by T_{sys} is sometimes more conveniently represented in terms of a noise figure. Recalling Chapter 4, we may define a system noise figure F_{sys} with respect to a reference temperature T_0 as follows:

$$F_{sys} = 1 + \frac{T_{sys}}{T_0} \tag{7.14}$$

where T_0 is typically chosen to be 290 K (see Section 4.4.1 for a reminder about this). Thus one may replace T_{sys} with $T_0\left(F_{sys} - 1\right)$ in Equations (7.10) and (7.13) if desired.

EXAMPLE 7.4

In Example 7.1, we determined the power delivered to a receiver. Assume the antenna temperature is 100 K. If the receiver noise figure is 4 dB and the receiver bandwidth is 30 kHz, then what is the signal to noise ratio at the input of the detector? For what range of bit rates do we achieve $E_b/N_0 \geq 10$ dB?

[3] Often the assumption is not stated; T_{ext} will simply not be considered. If the only identified sources of noise are receiver noise figure and "reference noise" T_0, then this assumption is in effect.

Solution: $F_{RX} = 4$ dB, so $T_{RX} = T_0(F_{RX} - 1) = (290\text{ K})\ (2.51 - 1) = 438$ K. It is stated that $T_A = 100$ K; taking this to mean $T_{ext} = T_A$; and $\epsilon_R = 0.7$ and $\eta_R = 0.889$ (from Example 7.1) we have $T_{sys} = T_{ext} + T_{RX}/\epsilon_R\eta_R = 805$ K. Using Equation (7.9) with quantities computed in Example 7.1, we find $\text{SNR}_{det} = 17.5 = 12.4$ dB. This is the answer to the first question.

The second question can be addressed as follows: Comparing Equations (7.10) and (7.13), note

$$\frac{E_b}{N_0} = \text{SNR}_{det}\frac{B_{RX}}{r_b} \tag{7.15}$$

so here we seek bit rates for which $10 \geq (17.5)\ (30\text{ kHz})\ /r_b$. This condition is met for $r_b \leq$ 52.7 kb/s.

7.5 SENSITIVITY AND G/T

The sensitivity of a receiving system is the power density of an incident radio wave that is required to achieve a specified SNR_{det} or E_b/N_0 at the input of the receiver's detector. However, there are many alternative definitions of sensitivity, so we shall refer to this particular definition as *formal sensitivity*. To say a radio is more or less sensitive is to mean that the radio requires lower or higher incident power density to achieve a specified level of output signal quality. The concept of sensitivity plays an important role in the design and analysis of radio links.

Different applications employ different definitions of sensitivity, many not entirely consistent with the formal definition offered above. A very common way to define sensitivity is to specify the value of P_R (that is, the power at the input of the receiver, neglecting any antenna-related losses) required to achieve a specified level of output signal quality. For example, the US TIA-603 standard for land mobile radio (LMR) systems using narrowband FM specifies that the receiver should produce an output audio SINAD[4] of at least 12 dB when a properly-modulated audio test tone is applied at a level (P_R) of -116 dBm. The antenna and associated losses are not considered, and the result depends in part on the details of the modulation.

Schemes that exclude the antenna and associated losses are popular because they are easy to measure. (The most popular of these will be addressed in Section 11.6.) The drawback of such schemes is that not all antennas have the same directivity, efficiency, or VSWR, so two radios having the same "P_R-based" sensitivity can exhibit significantly different formal sensitivity when employed in a link. In applications such as LMR this is not considered a serious problem, because mobile antenna technology options are very limited, so all manufacturers of LMR radios typically end up building radios with similar G_R, ϵ_R, and η_R, and any two radios used in the same location will experience a similar T_A. Technology options for receiver design, on the other hand, are many and vary over time, so differences in the formal sensitivities of mobile radios tend to be attributable to differences in receiver noise figure (or linearity, as we shall see in Chapter 11), which *is* taken into account in receiver-only sensitivity testing.

In other applications, sensitivity may be specified in terms of T_{sys} (or, occasionally, F_{sys}). We encountered system temperature T_{sys} in the previous section. Compared to a "receiver–only"

[4] SINAD is the ratio of desired signal to the sum of noise, interference and distortion.

sensitivity specification, T_{sys} has the advantage of taking into account T_{ext} (that is, T_A), ϵ_R, and η_R, but it does not take into account the role of the receive antenna directivity G_R in improving signal-to-noise ratio.

A T_{sys}-based sensitivity metric that does account for G_R is "G/T" (spoken as "G over T"). This metric emerges from Equation (7.10); note that the contribution of the receiving system to the signal-to-noise ratio at the input of its detector is proportional to G_R/T_{sys}. The ratio G_R/T_{sys} is commonly known as G/T, and frequently referred to as a "figure of merit," because it facilitates "apples-to-apples" comparison of the formal sensitivity of systems with different hardware characteristics.

The G/T characterization of sensitivity is most useful when both G_R and T_{sys} are separate parameters of interest under the control of the designer of the receive end of the link, and both parameters can be known to a useful degree of accuracy. An example is a ground station receiving signals transmit from a satellite. In this scenario, P_T is typically severely constrained due to limitations in power, space, and mechanical factors in the space environment. Thus, there is particularly high incentive to maximize G_R and minimize T_{sys}. Consistently high G_R can be achieved by using an electrically-large reflector antenna which tracks the transmitting satellite. However, such antennas are mechanically complex and expensive to construct and operate, so there is strong incentive to make G_R only as large as is necessary to achieve acceptable sensitivity. Minimizing T_{sys} means reducing the noise figure of the receiver, which may not be technically possible, or might require compromises in other system specifications, such as linearity. Such systems are typically analyzed and specified in terms of G/T because this metric concisely represents the chosen tradeoff between antenna performance and receiver noise figure, as well as indicating the formal sensitivity.

EXAMPLE 7.5

What is G/T for the mobile receiving system considered in Examples 7.1 and 7.4?

Solution: From Example 7.1 (problem statement), $G_R = 2$ dBi. From Example 7.4, $T_{sys} =$ 805 K. Therefore G/T $= G_R/T_{sys} = 0.00\,197\ \mathrm{K}^{-1} = -27.1\ \mathrm{dB(K^{-1})}$. Note it is fairly common to use the informal notation "dB/K" in lieu of the proper but cumbersome "dB(K^{-1})".

This section has provided an overview of sensitivity that is sufficient to tackle the remaining topics in this chapter. There is much more to consider, but for that we require concepts from Chapters 8–11. We shall return to this topic when we consider the topic of antenna–receiver integration in Chapter 12.

7.6 LINK BUDGET

A link budget is simply a calculation of SNR$_{det}$ or E_b/N_0 for a radio link, typically using expressions from Section 7.4, with all values separately and explicitly stated. Frequently, SNR$_{det}$ or E_b/N_0 is specified, and the purpose of the calculation is to determine the values of the

Table 7.1. **Interpretation of Equation (7.13) as a link budget. The last column indicates whether the calculated value is added or subtracted.**

Parameter	Unit	Calculation	+/−
Transmit Power	dBW	$10\log_{10} P_T$	+
Transmit Mismatch Efficiency	dB	$10\log_{10} \eta_T$	+
Transmit Efficiency	dB	$10\log_{10} \epsilon_T$	+
Transmit Antenna Directivity	dBi	$10\log_{10} G_T$	+
Path Loss	dB	$10\log_{10} L_p$	−
Polarization Match Efficiency	dB	$10\log_{10} \epsilon_p$	+
Receive Antenna Directivity	dBi	$10\log_{10} G_R$	+
Boltzmann's Constant	dB(J/K)	$10\log_{10} k$	−
System Temperature	dB(K)	$10\log_{10} T_{sys}$	−
Bit Rate	dB(b/s)	$10\log_{10} r_b$	−
E_b/N_0	dB	(sum)	

other parameters necessary to "close the link" – that is, to meet or exceed the specification. It is traditional to express the calculation in logarithmic form, in which the factors are expressed in decibels and added (if they appear in the numerator) or subtracted (if they appear in the denominator). To illustrate this approach, consider Equation (7.13), and let us assume P_T is in units of W. After taking the base-10 logarithm of each side, the right side is expanded into the sum of terms, and then each term is multiplied by 10 to obtain expressions in decibels. Table 7.1 shows the result in the customary form.

There is no one "correct" way to construct a link budget. For example, we noted in Section 7.4 that kT_{sys} could be replaced by $kT_0(F_{sys} - 1)$ if we are inclined to use noise factors in lieu of noise temperatures. In this case, the lines "Boltzmann's Constant" and "System Temperature" could be replaced as follows:

Parameter	Unit	Calculation	+/−
Reference Noise	dB(W/Hz)	$10\log_{10} kT_0$	−
Receiver Sensitivity	dB	$10\log_{10}(F_{sys} - 1)$	−

In the "Reference Noise" plus "Receiver Sensitivity" approach, T_0 is usually specified to be 290 K. Then the constant $kT_0 = 4.00 \times 10^{-21}$ W/Hz, more commonly expressed as −204.0 dB(W/Hz) or −174.0 dB(mW/Hz).[5] It is often useful to identify kT_0 separately, since it is not a design parameter and cannot be changed. The minimum value of F_{sys} is 1 (0 dB), meaning there is no noise, and therefore $E_b/N_0 \to \infty$ in this case. $F_{sys} = 2$ (3.0 dB) means simply that $T_{sys} = T_0$. Generalizing, $F_{sys}-1$ is the factor by which T_{sys} is greater than T_0.

It should be noted that T_{ext}, T_{RX}, ϵ_R, and η_R do not appear explicitly in either the kT_{sys} or $kT_0\left(F_{sys}-1\right)$ approaches; instead, they are "embedded" in the definitions of T_{sys} and F_{sys}, respectively (see e.g. Equation (7.8)). In some applications it may be desirable to show some of these factors explicitly, as separate lines. This is possible in the T_{sys} approach if $T_{ext} \ll T_{RX}/\epsilon_R\eta_R$.[6] In that case we can substitute the following for the k and T_{sys} lines (using Equation (7.8)):

Parameter	Unit	Calculation	+/−
Receive Efficiency	dB	$10\log_{10}\epsilon_R$	+
Receive Mismatch Efficiency	dB	$10\log_{10}\eta_R$	+
Receiver Noise	dB(W/Hz)	$10\log_{10}kT_{RX}$	−

Using the relationship $T_{RX} = T_0\,(F_{RX}-1)$ we have, under the same assumption:

Parameter	Unit	Calculation	+/−
Receive Efficiency	dB	$10\log_{10}\epsilon_R$	+
Receive Mismatch Efficiency	dB	$10\log_{10}\eta_R$	+
Reference Noise	dB(W/Hz)	$10\log_{10}kT_0$	−
Receiver Noise	dB	$10\log_{10}(F_{RX}-1)$	−

If it is also true that $F_{RX} \gg 1$, then we have

Parameter	Unit	Calculation	+/−
Receive Efficiency	dB	$10\log_{10}\epsilon_R$	+
Receive Mismatch Efficiency	dB	$10\log_{10}\eta_R$	+
Reference Noise	dB(W/Hz)	$10\log_{10}kT_0$	−
Receiver Noise	dB	$10\log_{10}F_{RX}$	−

[5] It is common to see this factor identified in link budgets as "thermal noise" or "noise floor." Both are misleading, as the value T_0 is arbitrary and is specified only because it is intrinsic to the definition of noise figure; see Section 4.4.

[6] As demonstrated in Example 7.4 and Section 7.9, this assumption is not always valid, so be careful to check first!

Now, an important point: This last form of the link budget is not only the $F_{RX} \gg 1$ approximation, but also turns out to be *exact* in the special case that $T_{ext} = T_0$.[7] This fact tends to generate a lot of confusion, since it is easy to forget that this form of the link budget accounts for the *in situ* sensitivity only in the special case that $T_{ext} = T_0$, which is often not true – not even approximately. Otherwise, it is valid only if F_{RX} is quite large, which is not a reasonable assumption for many receivers.

Link budgets are used in many different ways, depending on the application and the intent of the analysis. Also, it is common to include additional factors in link budgets to account for particular concerns. Two common additions are "link margin" (demonstrated in Section 7.7) and "fade margin" (demonstrated in Section 7.9). In the following sections we demonstrate some common uses of link budgets, including these additional factors.

7.7 ANALYSIS OF A 6 GHZ WIRELESS BACKHAUL; LINK MARGIN

The base stations and switches in a cellular telecommunications system are sometimes interconnected using fixed point-to-point data links operating at frequencies of about 6 GHz. This application is commonly referred to as "backhaul." Wireless backhaul links are typically implemented using reflector antennas mounted high above the ground, using the same towers used to mount the cellular antennas. In a well-designed backhaul link the propagation is nearly "free space" because the power is concentrated in a narrow beam formed by the transmit antenna. Here are some design parameters for the particular system we would like to analyze:

- Transmit power is 10 mW.
- Transmit and receive antennas are identical 0.5 m diameter circular paraboloidal reflector systems. Aperture efficiency is 60% and losses due to antenna efficiency and impedance mismatch are negligible.
- The link range is 30 km.
- The system noise figure is 11 dB with respect to $T_0 = 290$ K, which includes losses in cables between the antennas and ground-mounted equipment. The contribution from noise external to the antenna/receiver system is negligible in comparison.
- 45 Mb/s QPSK modulation with BER $\leq 10^{-6}$, without forward error correction, is specified.

Can we close the link? The first step is determine the E_b/N_0 requirement. Recall from Chapter 6 that the BER for optimally-detected QPSK is $Q\left(\sqrt{2E_b/N_0}\right)$ (Equation (6.45)). From this we find the required $E_b/N_0 \geq 10.5$ dB. Before constructing a link budget, note that the question can be answered by using Equation (7.13) directly:

- From the problem statement, $P_T = 0.01$ W.
- The directivity of the transmit and receive antennas can be computed by using Equation (2.65) with $\epsilon_{rad} = 1$, $\epsilon_a = 0.6$, $\lambda = c/f \cong 5$ cm, and $A_{phys} = \pi\ (0.5\ \text{m}/2)^2 = 0.196\ \text{m}^2$. This gives $G_T = G_R = 592 = 27.7$ dBi.

[7] If this is not already apparent, see the step-by-step derivation in Section 11.6 leading to Equation (11.37).

- From the problem statement we assume $\epsilon_T = \eta_T = 1$ and $\epsilon_R = \eta_R = 1$.
- Path loss in this case can be assumed to be equal to the free space value (L_{p0}).
- Presumably the polarizations are aligned, so $\epsilon_p \approx 1$.
- $T_{sys} = T_0(F_{sys} - 1)$ and $F_{sys} = 11$ dB with respect to $T_0 = 290$ K, so $T_{sys} = 3361$ K.
- The bit rate $r_b = 45 \times 10^6$ b/s

Applying Equation (7.13), we obtain $E_b/N_0 = 14.7$ dB. This is greater than the required E_b/N_0, so the link is closed.

Now we approach the problem using a link budget. First we populate the entries in the link budget as shown in Table 7.2. Note P_T is expressed in dBm (as opposed to dBW), which is typical when dealing with powers less than a few watts. Consequently, the reference noise power spectral density is expressed as -174 dB(mW/Hz).

Notice that the achieved E_b/N_0 is not explicitly shown in this link budget; shown instead are the required E_b/N_0 and a "link margin" equal to the ratio of the calculated E_b/N_0 to the required E_b/N_0. The sum of these (in dB) is equal to the expected 14.7 dB. The concept of link margin is useful as a measure of the ability of the link to accommodate unanticipated losses while remaining closed. This link can accommodate 4.2 dB of additional losses before failing the E_b/N_0 requirement.

In this application, a common cause for additional loss is misalignment of the antennas, either as a result of inaccuracy in installation, or due to effects such as wind loading. Since each

Table 7.2. **Link budget for the 6 GHz wireless backhaul of Section 7.7.**

Parameter	Value	+/−	Comments
Transmit Power	10 dBm	+	10 mW
Transmit Mismatch Efficiency	0 dB	+	
Transmit Efficiency	0 dB	+	
Transmit Antenna Directivity	27.7 dBi	+	see text
Path Loss	137.6 dB	−	$R = 30$ km; see text
Pol. Mismatch Efficiency	0 dB	+	
Receive Antenna Directivity	27.7 dBi	+	see text
Reference Noise	−174 dB(mW/Hz)	−	$T_0 = 290$ K
Receiver Sensitivity	10.6 dB	−	$F_{sys} = 11$ dB
Bit Rate	76.5 dB(Hz)	−	45 Mb/s
Required E_b/N_0	10.5 dB		BER=10^{-6} for QPSK
Link Margin	4.2 dB		

antenna is 10λ across, the HPBW is only about 6°. If each antenna is mispointed by just 3°, then the power delivered to the receiver will be reduced by 3 dB. Since the link margin is 4.2 dB, this very plausible problem will not cause the link to open. Similarly, small amounts of impedance mismatch, additional loss, and polarization misalignment will not cause the link to open. Link margins are included in link budgets for most applications, and for some applications can be very large, especially if the reliability of the link must be very high.

7.8 ANALYSIS OF A PCS-BAND CELLULAR DOWNLINK

Mobile cellular telecommunications systems consist of fixed base stations operated by a service provider, and mobile phones operated by the users (see Figure 1.6 and associated text). A base station typically includes one or more tower-mounted antennas intended to serve all mobiles in a predetermined geographical area known as a "cell". Partitioning of the service area into cells is essential due to regulatory constraints on transmit power (which limit range) and the need to reuse frequency channels due to the limited availability of spectrum.

The principal metrics of performance for contemporary cellular systems are data rate and availability. Availability can be defined as the fraction of a cell over which a specified data rate can be supported. Cellular service providers strive to offer data rates comparable to contemporary Wi-Fi systems, which are on the order of 10 Mb/s. The bandwidth of cellular channels is on the order of 10 MHz (significantly larger for some systems), so such rates are possible in principle. However, 10 Mb/s is quite difficult to achieve on a typical cellular radio link with a high degree of availability. The highest rates realized in operational systems using the latest technology ("4G") systems is about 10 Mb/s, but with low availability. Most systems are able to sustain rates of only 0.1–1 Mb/s for availability approaching 100%. In this section, we will use link budget analysis to understand why this is so.

Let us consider the design parameters for a hypothetical PCS-band cellular downlink. The PCS band is spectrum around 1900 MHz allocated to the operation of mobile cellular telecommunications systems in the USA. The path from the base station to mobile, usually referred to as the "downlink," is of particular interest because this link typically requires the higher data rate. Each PCS system follows a technical standard that specifies specific services, data rates, and modulations and coding schemes (see Appendix B). However, it is possible to understand generally what is possible without limiting our attention to a specific standard.

Here are parameters for a hypothetical PCS downlink (all typical).

- Maximum available base station transmit power: 10 W. This is constrained by at least two factors: (1) Necessity to avoid interference to nearby cells which must reuse frequencies due to the limited availability of spectrum, and (2) the technical challenge (hence, cost) of power amplifiers with sufficient linearity to accommodate complex modulations and constrain "regrowth" of spectrum into adjacent frequency channels. (More on this in Section 17.4.)
- Base station transmit antenna directivity of 15 dBd over the entire cell, with negligible losses due to efficiency and impedance mismatch. This can be achieved using a vertical column of patch-type elements which are combined to make a pattern which is broad in azimuth (to

cover the sector) and narrow in elevation. It is typically not practical to significantly increase directivity, since this would require narrowing the beam in azimuth (leading to reduced ERP at the azimuthal edges of the cell) or narrowing the beam in elevation (leading to reduced ERP close to the base station).

- Propagation is characterized by Equation (7.3) with $\overline{L_p}(R_b) = 128.3$ dB, $R_b = 1$ km, $n = 3.2$, and $\alpha \approx 0$, with Rayleigh fading.[8]
- Typical PCS base station antennas employ dual orthogonal polarizations, and base stations are typically able to near-optimally combine these polarizations, so we will assume that there is no loss due to polarization mismatch; i.e., $\epsilon_p \approx 1$.
- The mobile's antenna will be assumed to have an approximately omnidirectional pattern with maximum directivity of 0 dBi, a VSWR of about 2:1, and is about 50% efficient.
- The receiver noise figure is 7 dB with respect to $T_0 = 290$ K.
- External noise will be assumed to be negligible compared to internal noise.

Before constructing the link budget, we should determine how we will account for noise. A receiver noise figure of $F_{RX} = 7$ dB with respect to $T_0 = 290$ K corresponds to a receiver noise temperature $T_{RX} \cong 1163$ K. From the problem statement we have $\eta_R \approx 0.5$, $\Gamma_R \approx 1/3$, and T_{ext} negligible; thus

$$T_{sys} = T_{ext} + \frac{T_{RX}}{\epsilon_R \eta_R} \approx 2618 \text{ K} \tag{7.16}$$

Table 7.3 shows a link budget. To facilitate our analysis, it is structured somewhat differently from the previous examples: We keep R and r_b as free parameters, and express the result in terms of those parameters in the bottom line. Also, path loss is now represented as *two* entries in the link budget: These are obtained by taking $10\log_{10}$ of both sides of Equation (7.3) and arranging the right side as follows

$$10\log_{10}\overline{L_p}(R) = 10\log_{10}\left[\overline{L_p}(R_b)\left(\frac{1}{R_b}\right)^n\right] + 10n\log_{10}R \tag{7.17}$$

Now, as long as we are careful to use the same units for R and R_b, we can use any distance units we like. Using km, the first term is simply $\overline{L_p}(R_b)$ in dB, and the second term is simply a "correction" to obtain $\overline{L_p}(R)$ for $R > R_b$.

Another feature of this link budget is that bit rate is represented so as to make Mb/s (as opposed to b/s) the "natural" unit of data rate:

$$10\log_{10} r_b = 10\log_{10}\left[\frac{r_b}{\text{Mb/s}}\left(10^6 \text{ b/s}\right)\right] = 60 \text{ dB(b/s)} + 10\log_{10}\frac{r_b}{\text{Mb/s}} \tag{7.18}$$

Because we are almost certain to have fading but have not yet accounted for this possibility, we identify the result as the mean E_b/N_0 and obtain:

$$\overline{E_b/N_0} = 33.2 \text{ dB} - 32\log_{10}(R/\text{km}) - 10\log_{10}\left(r_b/(\text{Mb/s})\right) \tag{7.19}$$

[8] These are pretty typical parameters. I obtained them by fitting Equation (7.3) to the COST231-Hata suburban model (Appendix A.3) with base station height 100 m and mobile height 2 m. The agreement is to within 0.2 dB for R from 1 km to 10 km.

Table 7.3. **Analysis of the hypothetical PCS downlink of Section 7.8.**

Parameter	Value	+/−	Comments
Transmit Power	40.0 dBm	+	10 W
Transmit Mismatch Efficiency	0.0 dB	+	
Transmit Efficiency	0.0 dB	+	
Transmit Antenna Directivity	17.1 dBi	+	15 dBd
Path Loss at 1 km	128.3 dB	−	
Path Loss Dependence on R	$32 \log_{10}(R/\text{km})$ dB	−	$n = 3.2$
Receive Antenna Directivity	0.0 dBi	+	
Boltzmann's Constant	−198.6 dB(mJ/K)	−	$k = 1.38 \times 10^{-23}$ J/K
System Temperature	34.2 dB(K)	−	$T_{sys} \approx 2618$ K $\eta_R = 50\%$ VSWR = 2 $F_{RX} = 7$ dB Negligible T_{ext}
Bit Rate	60 dB(b/s) $+10 \log_{10}(r_b/(\text{Mb/s}))$	−	
$\overline{E_b/N_0}$	33.2 dB $-32 \log_{10}(R/\text{km})$ $-10 \log_{10}(r_b/(\text{Mb/s}))$		Mean over fading

For example: For $R = 1$ km and $r_b = 1$ Mb/s, the last two terms are zero and we have simply $\overline{E_b/N_0} = 33.2$ dB. That is a pretty high value (even in the presence of fading), so the situation initially appears promising. However, it is also clear from the above equation that each order-of-magnitude increase in R "costs" 32 dB, and each order-of-magnitude increase in r_b "costs" 10 dB. So, for example, if the base station needs to support R up to 10 km and r_b up to 10 Mb/s, $\overline{E_b/N_0}$ sinks to −8.8 dB, which would be unacceptable even if there were no fading to consider. It is also clear that there is no readily available method to significantly improve this situation, since it is not reasonable to expect to be able to scrounge more than a few dB from any parameter appearing in the link budget other than R or r_b.

Figure 7.2 shows the relationship between r_b, $\overline{E_b/N_0}$, and R under the assumptions made above. This provides a broader perspective on the situation. Note that any modulation +

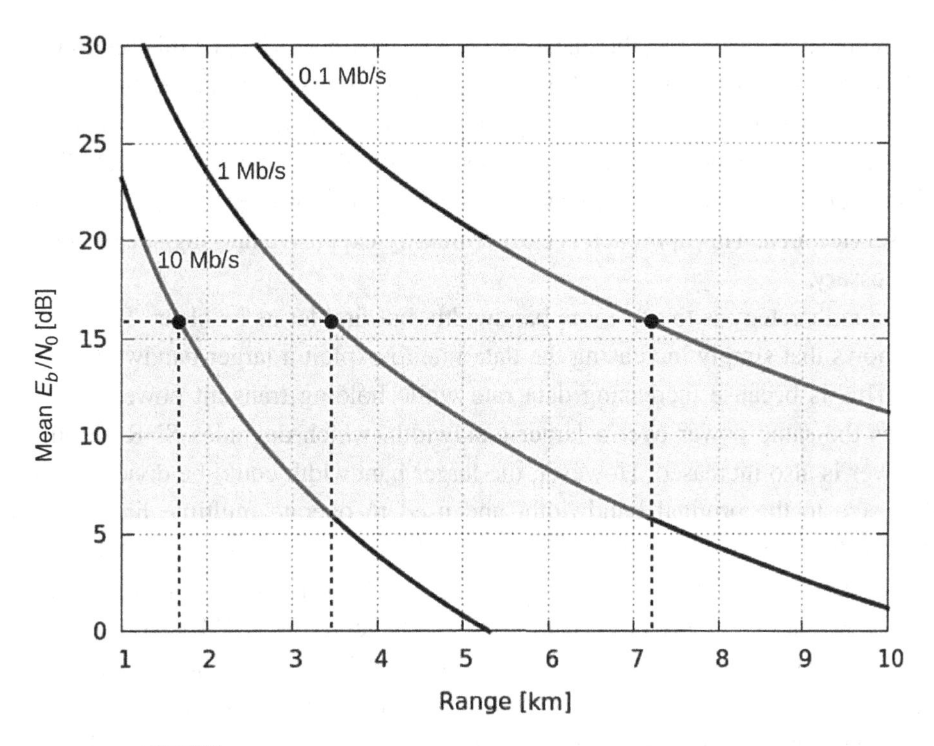

Figure 7.2. $\overline{E_b/N_0}$ vs. range (R) for $r_b = 0.1$, 1, and 10 Mb/s in the hypothetical PCS system of Section 7.8. The shaded area indicates the range of $\overline{E_b/N_0}$ that might reasonably be proposed as an operating point for this system. The horizontal dashed line is the $\overline{E_b/N_0}$ required for QPSK to achieve BER= 10^{-2} in Rayleigh fading without channel coding or interleaving.

channel coding scheme capable of delivering data rates greater than $\approx$ 0.1 Mb/s must be able to work at $\overline{E_b/N_0}$ which is quite low. For example, a scheme yielding 10 Mb/s at ranges greater than about 5.3 km must be able to deliver satisfactory BER for $\overline{E_b/N_0} < 0$ dB. Recall from Chapter 6 that this is not likely to happen – see e.g. Figure 6.26 and the associated text.

The horizontal line at $\overline{E_b/N_0} \approx 16$ dB in Figure 7.2 represents a plausible operating point, corresponding to BER $\approx 10^{-2}$ for coherently-detected QPSK in Rayleigh fading, without channel coding. Although 10^{-2} would probably not be an acceptable BER, this could be improved to an acceptable level using an appropriate channel coding scheme (Section 6.13). However, the channel coding scheme would also degrade spectral efficiency by a factor equal to the code rate, which could be as low as $1/4$.

Returning to Figure 7.2, we see this approach would limit the availability of 10 Mb/s service to significantly less than 1.6 km at best. One additional observation from Figure 7.2: The reason $\overline{E_b/N_0} = 6$ dB is identified as a lower bound is that modulations with channel coding tend to perform *worse* than modulations without channel coding below some threshold level of $\overline{E_b/N_0}$ (Section 6.13), and that threshold is typically below 6 dB. Since channel coding is essential in this application, $\overline{E_b/N_0} = 6$ dB is a practical lower limit, and we should be prepared to increase this to achieve an acceptable level of reliability. Using the $\overline{E_b/N_0} = 6$ dB threshold implies 10 Mb/s service will be limited to about 3.3 km and that any data rate covering the entire 10 km range of interest will be significantly less than 1 Mb/s.

We now have an answer to the question posed at the beginning of this section: "Why is it so hard for PCS to offer 10 Mb/s?". So what would it take to do significantly better? There are essentially two possible strategies. One strategy is to simply make the cells smaller, so that R can be smaller. This is practical in some cases (resulting in so-called "nanocells," "picocells," and "femtocells"); however, this dramatically increases the number of cells required to cover a defined service area. This approach is extraordinarily expensive and logistically onerous – but often necessary.

The second strategy is to use more bandwidth, but first let us be clear: The above analysis clearly shows that simply increasing the data rate to exploit a larger bandwidth would not be helpful. This is because increasing data rate while holding transmit power constant simply distributes the same power over a larger bandwidth, which degrades SNR because the total noise power is also increased. However, the larger bandwidth could be divided into subbands equal in size to the original bandwidth, and used to operate multiple links ("carriers") in parallel. With N carriers running in parallel, the data rate available to a mobile that is able to simultaneously access those carriers increases by N. This is referred to as *carrier aggregation*.

A different but essentially equivalent approach is adopt the CDMA strategy, in which the data are distributed across a larger bandwidth using different spreading codes, as opposed to different carriers. The downside in both schemes is that capacity (the number of users that can be simultaneously supported at a given level grade of service) is reduced linearly with increasing data rate provided to users. Since the number of users of such systems is continuously increasing, and additional spectrum is not readily available, increased carrier aggregation (in either the multicarrier or CDMA forms) is challenging to implement in PCS cellular systems. This analysis should also provide some insight as to why spectrum is considered so valuable.

7.9 ANALYSIS OF AN HF-BAND NVIS DATA LINK; FADE MARGIN

In this section we consider a long-range portable HF-band "near-vertical incidence skywave" (NVIS) data link. In an NVIS system, the transmitter uses an antenna with intended maximum directivity toward the zenith. The signal reflects from the ionosphere and returns to Earth, resulting in a circular coverage area with an effective radius up to hundreds of kilometers. Under the right conditions this propagation mechanism can have path loss similar to that of free space, so link closure is possible with transmit power that is very low compared with higher-frequency systems. Examples of applications in which NVIS systems are used include military operations in large, undeveloped areas; emergency communications following natural disasters which disrupt existing means of communications; and communications from remote sensors used in geophysical and biological data collection.

Here are the parameters for the particular system we wish to consider:

- Transmit frequency is 8 MHz, and a single channel having 3 kHz bandwidth is dedicated to this application.
- Transmit power: To be determined; our goal is to specify the minimum transmit power that meets the system requirements.

- Both transmit and receive antennas are nominally horizontally-mounted half-wave dipoles. However, the antennas are likely to be misshapen due to the "field expedient" (temporary, perhaps improvised) nature of their installation. As a result, G_R and G_T could be anywhere between 0 dBi and 8 dBi. The dipoles are impedance-matched using automatic antenna tuners (more on those in Section 12.6), so $\eta_T \approx \eta_R \approx 1$. If no ground screen is used, $\epsilon_T \approx \epsilon_R \approx 0.5$ due to ground loss (as to why the ground emerges as a loss mechanism, see Section 3.3.2).
- The distance between radios could be up to 200 km. Since the altitude of the ionosphere's reflecting (F) layer varies between 130 km and 400 km (Section 3.8), so the maximum path length R is about $2\sqrt{400^2 + (200/2)^2} = 825$ km.
- NVIS path loss is highly variable. However, a reasonable starting point is to assume free space propagation over the path length and perfectly efficient reflection, plus 20 dB to account for D layer absorption during the day (Section 3.8). In addition, an 18 dB *fade margin* is specified, which in principle provides about 98% availability assuming the fading is Rayleigh-distributed (Figure 3.13 and associated text).
- The ionosphere rotates and scatters the polarization of the signal. It is reasonable to assume that the scattering is sufficient to result in equal amounts of power in any two orthogonal polarizations. Therefore, $\epsilon_p \approx 0.5$ regardless of the azimuthal orientations of the transmit and receive antennas.
- NVIS receive sites tend to be located in unpopulated areas with negligible anthropogenic noise, so we shall assume this. However, this does not mean that T_{ext} is negligible, as is explained below.
- BPSK modulation with BER $\leq 10^{-3}$ is specified. For simplicity, we shall assume there is no channel coding.

Note that three important parameters have not been identified above: receiver noise temperature, transmit power, and data rate. Let us begin by considering receiver noise figure F_{RX}. An initial reaction might be simply to specify F_{RX} to be the minimum value available from state-of-the-art technology, which would then minimize the system noise figure and maximize the data rate for the specified BER. However, minimizing noise figure has some severe consequences for other aspects of performance. In Chapters 10 and 11 we will find that there is typically a tradeoff between noise figure and linearity; i.e., improving one typically happens to the detriment of the other. The principal problem with linearity in this particular application is that we can expect to receive multiple strong signals closely spaced with the signal of interest. This is because this portion of the radio spectrum is very crowded, and because skywave propagation allows transmitters to be received over global distances with relatively low path loss (Section 3.8). So, one should be careful not to over-specify F_{RX}; otherwise we risk forcing a receiver designer to unnecessarily give up some valuable linearity.

A better way to specify F_{RX} in this application is as follows. Since $\eta_R \approx 1$ due to the use of an antenna tuner, Equation (7.8) becomes:

$$T_{sys} \approx T_{ext} + \frac{T_{RX}}{\epsilon_R} \tag{7.20}$$

where T_{ext} quantifies of the external (environmental) noise captured by the antenna.[9] At 8 MHz, the smallest T_{ext} can be is $\sim 10^5$ K, attributable to Galactic noise (see Equation (4.29) and associated text). That is a pretty big value, so it is worth considering if the receiver noise temperature even matters. For example, 10^5 K $> 10 \times T_{RX}/\epsilon_R$ when $T_{RX} < 5000$ K. So: if $T_{RX} < 5000$ K, then T_{sys} is dominated by T_{ext} in the sense that T_{RX} contributes at most 10% to T_{sys}. Summarizing:

$$T_{sys} \approx T_{ext} \text{ if } T_{RX} < 5000 \text{ K} \tag{7.21}$$

This is useful information because the receiver noise figure $F_{RX} = 1 + T_{RX}/T_0 \approx 12.6$ dB (with respect to $T_0 = 290$ K) for $T_{RX} = 5000$ K. 12.6 dB is a relatively high noise figure, and there does not appear to be much benefit to making F_{RX} smaller, since we have just shown that there will be very little effect on T_{sys} if we do so. This condition is known as *external noise dominance*: The sensitivity of the receiver is determined to a good approximation only by external noise, which is beyond the control of the engineering team. In a sense, this represents the theoretical best possible sensitivity, and therefore makes further improvements to F_{RX} essentially irrelevant.

We now construct a link budget based on Equation (7.13). Recall from Chapter 6 that the BER for optimally-detected BPSK is $Q\left(\sqrt{2E_b/N_0}\right)$. From this we find the required $E_b/N_0 \geq$ 6.7 dB. The link budget is shown in Table 7.4 (constructed following a scheme similar to that employed in the previous section) and indicates we can satisfy this requirement with a link margin equal to

$$19.1 \text{ dB} + 10 \log_{10} P_T - 10 \log_{10} r_b \tag{7.22}$$

In other words, we have a link margin of 19.1 dB for $P_T = 1$ W and $r_b = 1$ b/s.

The data rate is constrained by the 3 kHz specified bandwidth, so let us consider that first. BPSK with optimal pulse shaping has a spectral efficiency of 1 bps/Hz, so $r_b \leq 3$ kb/s. For $r_b = 3$ kb/s, the link margin becomes -15.7 dB $+ 10 \log_{10} P_T$. Thus the link is closed for the maximum data rate supported by the bandwidth if $P_T > 36.9$ W. Any increase in transmitter power beyond 36.9 W increases the link margin – not a bad idea, as noted below – but cannot be used to increase the data rate. Of course, the data rate will be reduced by any overhead incurred in channel coding.

It should be noted that there are some good reasons why the necessary transmit power might be significantly less than or greater than the value of 36.9 W that we have derived. At night, D-layer absorption is negligible and thus transmit power can be reduced by 20 dB to about 369 mW. Similarly, recall our assessment of antenna directivity in the link budget is pretty pessimistic. If the directivities of the transmit and receive antennas turn out to be 8 dBi as opposed to 0 dBi, then the required daytime transmit power is reduced by 16 dB to just 920 mW. On the other hand, it should be noted that there are some good reasons to consider transmit power of 100 W to as high as 1 kW in this application. These include atmospheric noise "eruptions" (typically associated with thunderstorms and solar activity; Section 4.5.2) and anthropogenic interference (Section 4.5.3), which can increase T_{ext}, and therefore T_{sys}, by orders of magnitude. Also, NVIS propagation is vulnerable to anomalous

[9] Also, do not forget the choice made earlier to define T_{ext} as the antenna temperature T_A with ϵ_R factored out.

Table 7.4. **Link budget for the HF data link of Section 7.9.**

Parameter	Value	+/−	Comments
Transmit Power	$10\log_{10} P_T$ dBW	+	
Transmit Mismatch Efficiency	0 dB	+	tuner in use
Transmit Efficiency	−3 dB	+	50% ground loss
Transmit Antenna Directivity	0.0 dBi	+	worst case; see text
Free Space Path Loss	108.8 dB	−	$R = 825$ km
D-layer Absorption	20 dB	−	Daytime
Fade Margin	18 dB	−	est. 98% availability
Polarization Match Efficiency	−3 dB	+	
Receive Antenna Directivity	0.0 dBi	+	worst case; see text
Boltzmann's constant	−228.6 dB(J/K)	−	$k = 1.38 \times 10^{-23}$ J/K
System Temperature	50.0 dB(K)	−	$T_{sys} \approx T_{ext} = 10^5$ K $\epsilon_R = 50\%$, $\eta_R \approx 1$ $F_{RX} \leq 12.6$ dB
Bit Rate	$10\log_{10} r_b$ dB(b/s)	−	
E_b/N_0	6.7 dB		BER=10^{-3} for BPSK
Link Margin	19.1 dB $+10\log_{10} P_T$ $-10\log_{10} r_b$		

but nevertheless common ionospheric scattering, which can temporarily increase path loss by orders of magnitude. In general, increased transmit power is useful for improving the reliability of the link. Thus, 100 W to 1 kW is justifiable on that basis.

A caveat before moving on: As noted above, HF propagation and noise conditions are extremely variable. In the scenario considered in this problem, the irreducible environmental noise T_{ext} will typically be significantly higher than assumed here, allowing F_{RX} to be be considerably higher while maintaining external noise dominance. Also, the propagation conditions will frequently be very good, temporarily obviating the need for a large fade margin. Since higher noise figure implies improved linearity, there is incentive to allow noise figure to be variable so as to optimally accommodate both the pessimistic situation presumed by the link budget, and more favorable typical conditions.

A second consequence of the variability of HF propagation and noise conditions is that link budgets typically cannot be used to design specific links, as is common practice at higher frequencies. At HF the utility of link budget analysis is limited to understanding the link in a generic way. This is nevertheless useful: In this application, for example, link budget analysis provided a means to synthesize reasonable hardware requirements and to confirm the feasibility of a link capable of the maximum data rate supported by the available bandwidth.

7.10 ANALYSIS OF A KU-BAND DIRECT BROADCAST SATELLITE SYSTEM

Direct broadcast satellite (DBS) is a popular method for distribution of television channels to homes. A DBS link consists of a broadcast satellite in geosynchronous orbit, and a home-user receiving system consisting of a dish (reflector) antenna assembly and a receiver. Because the satellite is in geosynchronous orbit, its position in the sky remains fixed, eliminating the need for the dish assembly to move. Dish assemblies consist of a dish with a *low noise block* (LNB) feed system. The LNB is a single assembly containing the feed horn, the first stages of gain and amplification, and a downconverter. The purpose of the downconverter is to shift the center frequency of the signal from Ku-band ($\approx$ 12 GHz) to L-band (approximately 1–2 GHz); see Chapters 14 and 15 for more on this technique. The *reason* for the shift to L-band is that this results in lower loss when the signal is sent by cable from the dish assembly (typically on the roof) to the receiver (typically near a television), which subsequently mitigates the contribution of the cable loss to the system temperature – see Example 4.6 for an example of how that can happen.

In this section, we wish to consider how to specify the performance of the dish assembly, and then consider the range of dish sizes and LNB noise figures that is viable for this application. As we have done in previous sections, let us begin by itemizing what is known in this problem. All of the following values are typical.

- The satellite is in geosynchronous orbit, so the range from satellite to home dish assembly is approximately 36 000 km.
- The satellite broadcast EIRP is at least 55 dBW per channel.
- The center frequency is 12 GHz.
- Free space propagation can be assumed, with two modifications. First, about 0.5 dB attenuation is expected due to absorption by the atmosphere. Second, the link is vulnerable to severe fading when it rains. A 7 dB fade margin to account for rain is assumed sufficient for 99.9% availability; that is, it is assumed the loss due to rain is less than 7 dB at least 99.9% of the time.
- Linear polarization is used. It is difficult to align polarizations with high precision, so we shall assume 0.4 dB loss due to polarization mismatch. You might ask why not instead use circular polarization, which would obviate the need for precise alignment. The answer in this case is that horns tend to yield intrinsically linear polarization, and synthesis of

a circular polarization from linear polarizations (as described in Section 2.5) requires RF hybrid devices whose intrinsic losses would significantly impact the satellite's power budget and the receiver's system temperature. Since neither end of the link moves very much relative to the other, linear polarization works well in this application and yields slightly better performance from a link budget perspective.

- The antenna temperature varies with pointing relative to zenith, which depends on the geographical location of the user. Anthropogenic contributions are typically negligible. A worst case value of $T_{ext} = 100$ K will be assumed.
- The design of the LNB+cable scheme is such that the signal-to-noise ratio at the receiver input is approximately equal to the signal-to-noise ratio at the LNB output. (Again, this is the "trick" demonstrated in Example 4.6.)
- The nominal channel data rate is 10 Mb/s, and $E_b/N_0 \geq 8$ dB is desired at the receiver input.
- A link margin of 3 dB is deemed adequate to account for manufacturing tolerances and likely errors associated with installation and aging. The dominant source of error is expected to be dish assembly pointing error.

First, we wish to determine the required G/T for the dish assembly. We wish to specify G/T in particular, as opposed to dish directivity and LNB noise figure separately. This is for two reasons. First, G/T conveniently accounts for all factors associated with the dish assembly in one figure of merit. Second, dish assemblies are typically purchased from manufacturers who are separate from the DBS service provider. The G/T specification allows the manufacturer maximum flexibility in design. For example, a manufacturer that is particularly skilled in LNB design might be able to meet the G/T specification with a smaller-than-expected dish, whereas a manufacturer skilled in dish design might be able to achieve a lower overall cost by improving aperture efficiency, allowing a noisier but less-expensive LNB to be used. By specifying G/T as opposed to dish size or LNB noise figure, all of these options remain open.

We now construct a link budget. As before, we begin with Equation (7.13), but some modifications are in order. Since transmit power and receive sensitivity are quantified in terms of EIRP and G/T respectively, we use Equation (7.5) and G/T=G_R/T_{sys} to obtain

$$\frac{E_b}{N_0} = \frac{(\text{EIRP})\, L_p^{-1} \epsilon_p\, (\text{G/T})}{k r_b} \tag{7.23}$$

The above equation is the basis for the link budget shown in Table 7.5. We find that the required G/T for the dish assembly is 11.2 K^{-1} = 10.5 dB(K^{-1}).

Now we consider the range of possibilities available to the dish assembly manufacturer. If the dish has a circular aperture then the dish assembly can be described in terms of the dish diameter D, the aperture efficiency ϵ_a, and the system temperature T_{sys}. From Equation (7.8), the system temperature depends on T_{ext} and the LNB noise temperature T_{RX}. If we assume that $\epsilon_R \approx \eta_R \approx 1$, then $T_{sys} \approx T_{ext} + T_{RX}$. The G/T specification can now be written in terms of the remaining parameters as follows:

$$\frac{\epsilon_a \pi^2 (D/\lambda)^2}{T_A + T_{RX}} \geq 11.2\ \text{K}^{-1} \tag{7.24}$$

Table 7.5. **Link budget for the Ku-band DBS satellite link of Section 7.10.**

Parameter	Value	+/−	Comments
Satellite EIRP	55 dBW	+	
Free Space Path Loss	205.2 dB	−	Geosynchronous Orbit
Atmospheric Absorption	0.5 dB	−	
Rain Fade Margin	7.0 dB	−	99.9% Availability
Polarization Match Efficiency	−0.4 dB	+	
Ground Station G/T	10.5 dB(K^{-1})	+	
Boltzmann's Constant	−228.6 dB(K)	−	
Bit Rate	70.0 dB(bps)	−	10 Mb/s
Required E_b/N_0	8.0 dB		
Link Margin	3.0 dB		

where the numerator is the directivity of a circular reflector (Equation (2.64)). We might conservatively assume an aperture efficiency of 50% (Section 2.10.2), then $\epsilon_a = 0.5$. Earlier in this section, a planning value of T_{ext} was indicated to be 100 K. The only two remaining parameters are D and T_{RX}, and the G/T specification bounds one once the other is set.

Figure 7.3 shows the resulting tradeoff between D and T_{RX}, which is expressed in the figure as noise figure (F_{RX}) with respect to 290 K. Note that D must be $\approx$ 0.4 m or larger, and that the penalty for increasing noise figure is a significant increase in the required dish size. Many contemporary dish assemblies have noise figures of about 1 dB and dish diameters of about 0.5 m, which corresponds to a point on the curve shown in Figure 7.3.

7.11 SPECIFICATION OF RADIOS AND THE PATH FORWARD

In the introduction to this chapter two goals were stated: Presenting techniques for the analysis of radio links, and quantifying typical specifications for radios. Table 7.6 summarizes the latter. From the examples presented in this chapter, we have seen transmit power range from levels on the order of 10 mW to levels on the order of 1 kW, depending quite a bit on the application. These values are representative of the vast majority of applications one is likely to encounter. We have seen useful receiver noise figures also vary quite a bit depending on application, ranging from ~1 dB (DBS LNB) to ~12 dB or higher (HF NVIS). Here too, these values are typical of the vast majority of applications one is likely to encounter. This information, combined with the theory of Chapters 2–6, provides the necessary background to confront the problem of radio design in a properly well-informed manner.

Table 7.6. **Comparison of the radio links considered in this chapter, sorted by frequency. For more systems, see Appendix B.**

Application	Sec.	Freq.	P_T	F_{RX}	r_b	R
DBS[(1)]	7.10	12 GHz		≤ 1 dB	10 Mb/s	36,000 km
Backhaul[(2)]	7.7	6 GHz	10 mW	8.0 dB	45 Mb/s	30 km
PCS Downlink	7.8	1.9 GHz	10 W	7.0 dB	10 Mb/s	≈ 2 km
					1 Mb/s	≈ 3 km
					0.1 Mb/s	≈ 7 km
HF NVIS[(3)]	7.9	8 MHz	10 W -1 kW	≥ 12.6 dB	3.0 kb/s	200 km

[(1)] For DBS, only EIRP was indicated, so P_T is not available.
[(2)] For backhaul, $F_{sys} = 11$ dB. F_{RX} value shown accounts for 3 dB cable loss.
[(3)] For HF NVIS, data rate was constrained primarily by 3 kHz channel bandwidth, and indicated R is distance between radios (i.e., not path length).

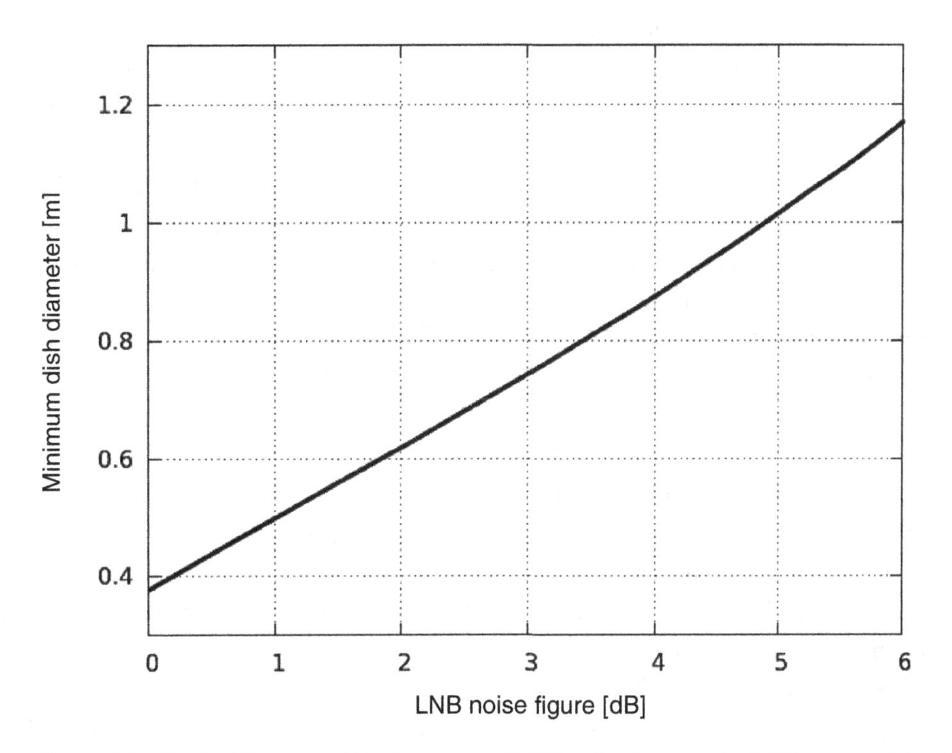

Figure 7.3. Tradeoff between dish diameter and LNB noise figure in the DBS system of Section 7.10.

Beginning with the next chapter, we consider the elements of radio design. At various points we will need to know something about the goals of the design; in other words, the system requirements. Suggestions will typically be offered at these points, perhaps referring to an example presented in this chapter. It is useful to keep in mind that the simple methods demonstrated in this chapter are always applicable, so one can often derive a requirement, or propose a requirement by making a few reasonable assumptions.

Problems

7.1 A land mobile radio base station transmits 50 W into a lossless and well-matched antenna having a directivity of 9 dBi. The antenna polarization is vertical. The center frequency is 155 MHz. The propagation is well modeled as free space up to 500 m, and $n = 4$ beyond that range. The receiving mobile is 5 km distant, using a vertically-polarized antenna with about 4 dBi directivity in the direction of the base station. The receive antenna has a VSWR of 1.5:1 and is connected to the radio using a cable with 2 dB loss. What is the path loss and the power delivered to the receiver?

7.2 Broadcast emission limits are sometimes specified in μV/m at the receiver location, as opposed to EIRP in W. If the 200 MHz emission limit 20 km from the broadcast antenna is 7.2 μV/m, and free space propagation is assumed, what is the associated EIRP?

7.3 Show that Equation (7.13) is dimensionally correct. That is, show that the units of all quantities in the equation are mutually consistent.

7.4 In Section 7.6 it is noted that if T_{ext} is negligible and $F_{rx} \gg 1$, it is possible to construct a link budget in which η_R, $\left(1 - |\Gamma_R|^2\right)$, and F_{RX} appear as separate lines. Beginning with Equation (7.13), derive this link budget.

7.5 We noted at the end of Section 7.6 that if T_{ext} is negligible and $F_{rx} \gg 1$, then it is reasonable to represent kT_{sys} in the link budget as two lines, one for kT_0 and one for F_{RX}. Show that this condition does (or does not) apply in the case of the wireless backhaul link of Section 7.7. What is the error if we use this approximation?

7.6 In Example 7.4, it was found that bit rates up to 52.7 kb/s could be supported. What combinations of modulation and pulse shaping would accommodate this rate?

7.7 In the PCS system of Section 7.8, the receiver noise figure F_{RX} was assumed to be 7 dB. Recreate Table 7.3 and Figure 7.2 assuming $F_{RX} = 1$ dB. Comment on the availability of 10 Mb/s service with this revision.

7.8 HF NVIS can also be used for analog voice communications. For the HF NVIS system of Section 7.9, recreate the link budget (Table 7.4) assuming SSB voice with a required predetection signal-to-noise ratio of 6 dB, and assume fixed transmitter power of 100 W.

7.9 In the DBS system of Section 7.10, we considered only EIRP, not transmit power or transmit antenna directivity. If the transmit antenna aperture is circular and the transmit radiation efficiency is 70%, plot the relationship between aperture diameter and transmit power required to achieve the stated EIRP.

7.10 Create a plot of maximum data rate as a function of range for a 60 GHz data link having the following characteristics: Transmit power 1 mW, receive and transmit use identical phased arrays with radiation efficiency 40%, VSWR 1.5:1, and directivity 8 dBi, negligible polarization mismatch loss and external noise, receiver noise figure 4 dB. Assume BPSK without channel coding, BER $\leq 10^{-5}$.

8 Two-Port Concepts

8.1 INTRODUCTION

The electronic circuits that comprise modern radios are complex, typically consisting of discrete components numbering in the hundreds to thousands. Analysis and design of a radio entirely at the component level, while technically possible, is usually not efficient and is rarely done. Instead, radios are viewed as consisting of blocks, where each block is a circuit which performs a function that can be readily understood without detailed knowledge of the consistent circuitry.

Two such blocks are filters and amplifiers. A filter may consist of many capacitors, inductors, or other structures, but to one using a previously-designed filter it is usually just frequency response and interface impedance that are of interest – a detailed schematic is not required. Similarly, an amplifier may consist of transistors plus circuitry for biasing and impedance matching, but a useful block diagram-level description might require just two numbers: gain and port impedance.

Two-port theory is the tool that facilitates representation of complex circuits as simpler building blocks. A two-port is shown in Figure 8.1. As the name suggests, it is a device consisting of two ports. Each port has an associated voltage and current, defined as shown in the figure. Assuming the two-port consists entirely of linear, time-invariant, and causal devices, the port voltages and currents can be expressed as linear combinations of other voltages and currents at the same and other ports. Following from this idea there are four common schemes for describing relationships among the port voltages and currents: impedance ("Z") parameters:

$$V_1 = Z_{11}I_1 + Z_{12}I_2 \tag{8.1}$$

$$V_2 = Z_{21}I_1 + Z_{22}I_2 \tag{8.2}$$

admittance ("Y") parameters:

$$I_1 = Y_{11}V_1 + Y_{12}V_2 \tag{8.3}$$

$$I_2 = Y_{21}V_1 + Y_{22}V_2 \tag{8.4}$$

hybrid ("H") parameters:

$$V_1 = H_{11}I_1 + H_{12}V_2 \tag{8.5}$$

$$I_2 = H_{21}I_1 + H_{22}V_2 \tag{8.6}$$

and scattering (s) parameters, which are introduced in Section 8.2. Other two-port parameter schemes include inverse hybrid parameters, $ABCD$ parameters, and transmission ("t") parameters, which appear in Section 8.6. In this book, s-parameters will be used primarily.

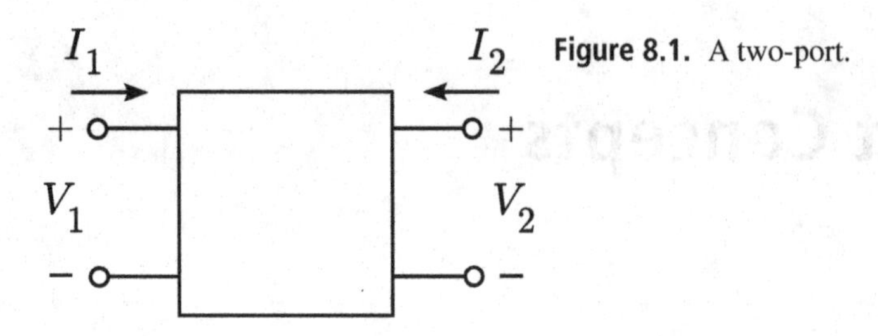

Figure 8.1. A two-port.

The objectives of this chapter are to introduce two-port theory, and explain how this theory is employed to understand the gain, input and output impedances, and stability of two-ports and cascades of two-ports. The organization of this chapter is as follows: Section 8.2 introduces *s*-parameters. Sections 8.3 and 8.4 summarize the intrinsic and embedded properties, respectively, of two-ports. Section 8.5 addresses the issue of stability and the determination of two-port gain given the possibility of unstable operation. Section 8.6 introduces *t*-parameters and describes their use in determining the *s*-parameters of systems consisting of cascaded two-ports. Finally, Section 8.7 introduces differential circuits, describes their applications, and describes how they can be accommodated within the framework for analysis and design established in earlier sections.

8.2 *s*-PARAMETERS

In this section we introduce the two-port single-ended *s*-parameters (Section 8.2.1) and derive the *s*-parameters for a few important two-ports, including series and shunt impedances (Section 8.2.2) and transmission lines (Section 8.2.3). We conclude with a summary of techniques for obtaining *s*-parameters for other two-ports (Section 8.2.4).

8.2.1 Derivation of *s*-Parameters

Consider a two-port device that is accessed using transmission lines having characteristic impedance Z_0, as shown in Figure 8.2. As with any transmission line, Z_0 is real-valued.[1] A voltage wave $V_1^i(d)$ travels toward the input ("1") port: d indicates distance from the associated port along the transmission line; i.e., increasing with increasing distance from the associated port. Another voltage wave $V_1^o(d)$ travels away from the input port. From transmission line theory, the total voltage and current at any point on the input side transmission line is

$$V_1(d) = V_1^i(d) + V_1^o(d) \tag{8.7}$$

$$I_1(d) = \left[V_1^i(d) - V_1^o(d)\right]/Z_0 \tag{8.8}$$

(The minus sign in the above equation is correct, and follows from elementary transmission line wave theory.) These equations can be manipulated to obtain expressions for $V_1^i(d)$ and $V_1^o(d)$ in terms of the total voltage and current:

[1] This is not strictly necessary; however, the presentation in this book assumes this and the equations for the general complex-valued case are different. A discussion of the general case is presented in [37].

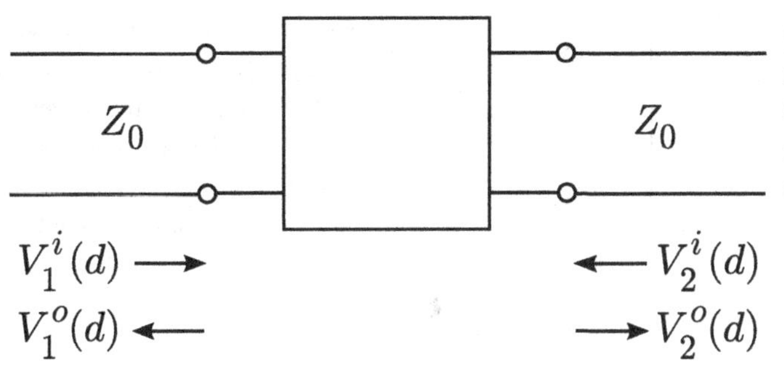

Figure 8.2. Voltage waves incident and reflected at the terminals of a two-port; d is defined to be zero at the port and increases with increasing distance from the port.

$$V_1^i(d) = \left[V_1(d) + Z_0 I_1(d)\right]/2 \tag{8.9}$$

$$V_1^o(d) = \left[V_1(d) - Z_0 I_1(d)\right]/2 \tag{8.10}$$

Finally, we define quantities a_1 and b_1 in terms of $V_1^i(d)$ and $V_1^o(d)$ respectively, as follows:

$$a_1 \equiv \frac{V_1^i(0^+)}{\sqrt{Z_0}} = \frac{V_1(0^+) + Z_0 I_1(0^+)}{2\sqrt{Z_0}} \tag{8.11}$$

$$b_1 \equiv \frac{V_1^o(0^+)}{\sqrt{Z_0}} = \frac{V_1(0^+) - Z_0 I_1(0^+)}{2\sqrt{Z_0}} \tag{8.12}$$

where $d = 0^+$ indicates that the quantity is measured just outside the two-port and on the transmission line, where the characteristic impedance is Z_0. Note that a_1 and b_1 are associated with the inbound and outbound waves, respectively, and have units of $\mathrm{V}/\sqrt{\Omega}$; i.e., square root of power.

On the output ("2") side, a voltage wave $V_2^i(d)$ travels toward the output, a voltage wave $V_2^o(d)$ travels away from the output, and the total voltage and current at any point on the output side transmission line are $V_2(d)$ and $I_2(d)$, respectively. Following the same procedure previously applied to the input side, we obtain

$$a_2 \equiv \frac{V_2^i(0^+)}{\sqrt{Z_0}} = \frac{V_2(0^+) + Z_0 I_2(0^+)}{2\sqrt{Z_0}} \tag{8.13}$$

$$b_2 \equiv \frac{V_2^o(0^+)}{\sqrt{Z_0}} = \frac{V_2(0^+) - Z_0 I_2(0^+)}{2\sqrt{Z_0}} \tag{8.14}$$

Since $V_1(d)$, $I_1(d)$, $V_2(d)$, and $I_2(d)$ are linearly related (that is, any one can be represented as a linear combination of the others), and since a_1, b_1, a_2, and b_2 are linear combinations of $V_1(0^+)$, $I_1(0^+)$, $V_2(0^+)$, and $I_2(0^+)$, then a_1, b_1, a_2, and b_2 must be linearly related. Thus:

$$b_1 = s_{11}a_1 + s_{12}a_2 \tag{8.15}$$

$$b_2 = s_{21}a_1 + s_{22}a_2 \tag{8.16}$$

for some set of complex-valued coefficients s_{11}, s_{12}, s_{21}, and s_{22}. These are the *scattering parameters* or simply "s-parameters." Equations (8.15) and (8.16) can be interpreted as shown in Figure 8.3: Each outbound wave is a linear combination of the two inbound waves, and the s-parameters are simply the combining coefficients that apply at the ports.

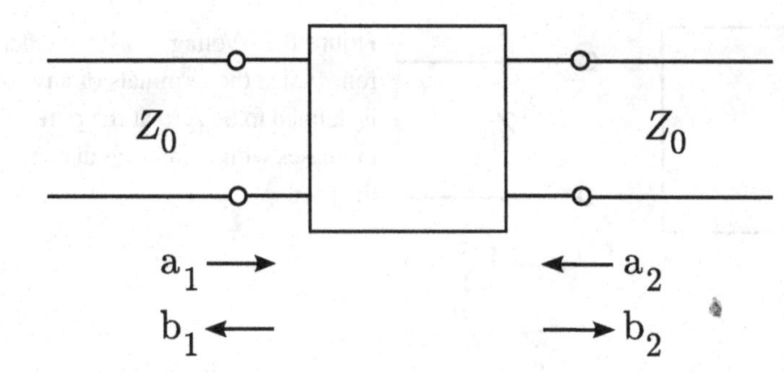

Figure 8.3. Power waves used in the *s*-parameter description of a two-port. Note the values of the power waves are specified to be the values exterior to the two-port; that is, on the transmission line side of the terminals of the two-port.

The reader may fairly note that the above derivation of *s*-parameters seems a bit convoluted, and wonder why *z* or *y* parameters might not be a better way to characterize two-ports. There are two reasons why *s*-parameters are usually preferred for RF work. First, *s*-parameters are relatively easy to measure, because they are always ratios of voltages. For example:

$$s_{21} = \left.\frac{b_2}{a_1}\right|_{a_2=0} = \frac{V_2^o(0^+)}{V_1^i(0^+)} \tag{8.17}$$

in other words, to measure s_{21} you launch a signal into the input port and compare it to the signal that emerges from the output port. For radio frequency devices, this is pretty easy compared to measuring voltages and currents simultaneously, while also shorting or open-circuiting ports. In fact, the latter is often not even an option, because some devices become unstable or otherwise inoperable when open- or short-circuited.

The second reason *s*-parameters are worth the trouble is that each of the *s*-parameters have a useful physical interpretation. For example, we see above that s_{21} is the voltage transfer coefficient from port 1 to port 2. Similarly, s_{12} is the "reverse" (port 2 to port 1) voltage transfer coefficient. Examining s_{11}:

$$s_{11} = \left.\frac{b_1}{a_1}\right|_{a_2=0} = \frac{V_1^o(0^+)/\sqrt{Z_0}}{V_1^i(0^+)/\sqrt{Z_0}} = \frac{V_1^o(0^+)}{V_1^i(0^+)} \tag{8.18}$$

We see s_{11} is the voltage reflection coefficient, typically assigned the symbol "Γ" with the appropriate subscript. Similarly, s_{22} is the voltage reflection coefficient for a wave arriving at the output from a transmission line with characteristic impedance Z_0.

Now for an important point: The quantity Z_0 is part of the definition of the *s*-parameters, and in that sense is referred to as the *reference impedance*. In other words, the values of the *s*-parameters of a two-port depend on Z_0. Z_0 is nearly always chosen to be 50 Ω, so this point is easily forgotten. However, it is reasonable and occasionally useful to choose Z_0 to have some other value, so one should always be careful to check this. The policy in this book shall be that $Z_0 = 50\ \Omega$ unless indicated otherwise.

8.2.2 *s*-Parameters for Series and Shunt Impedances

Let us now work out the *s*-parameters for some simple but important two-ports. Figure 8.4 shows perhaps the two simplest possible two-ports: A series impedance and a shunt impedance. In each case, the internal impedance is identified simply as *Z*. Let us consider the series

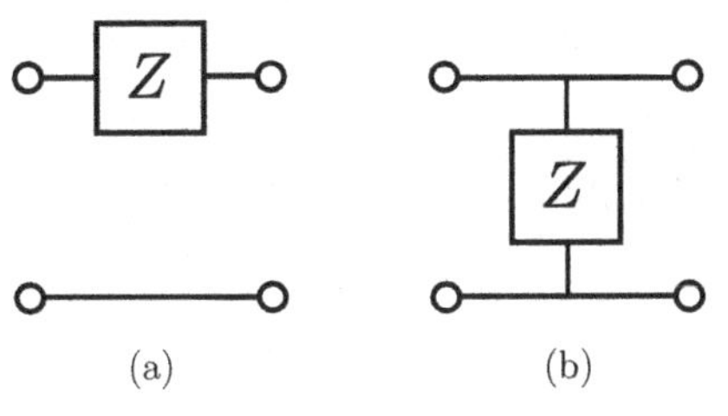

Figure 8.4. Two-ports consisting of a (a) series impedance and (b) shunt impedance.

impedance first: s_{11} is simply the reflection coefficient at the input when the source and load impedances are Z_0. Let Z_{in} be the impedance looking into port 1. Then:

$$s_{11} = \frac{Z_{in} - Z_0}{Z_{in} + Z_0} = \frac{(Z + Z_0) - Z_0}{(Z + Z_0) + Z_0} = \frac{Z}{Z + 2Z_0} \tag{8.19}$$

and from the symmetry of the problem it is apparent that

$$s_{22} = s_{11} = \frac{Z}{Z + 2Z_0} \tag{8.20}$$

The forward voltage transfer coefficient s_{21} can be determined as follows: First, we replace the wave incident on port 1 with the following Thévenin equivalent circuit: A voltage source $2V_1^i(0^+)$ in series with a discrete impedance Z_0. This is shown in Figure 8.5. This equivalent circuit can be confirmed by replacing everything to the right of port 1 with another discrete impedance Z_0 and noting that the voltage across the load is $V_1^i(0^+)$. Upon substitution of the equivalent circuit, we have a voltage divider in which $V_2^o(0^+)$ is given by

$$V_2^o(0^+) = 2V_1^i(0^+)\frac{Z_0}{Z_0 + Z + Z_0} = V_1^i(0^+)\frac{2Z_0}{Z + 2Z_0}, \text{ thus:} \tag{8.21}$$

$$s_{21} = \left.\frac{b_2}{a_1}\right|_{a_2=0} = \frac{V_2^o(0^+)}{V_1^i(0^+)} = \frac{2Z_0}{Z + 2Z_0} \tag{8.22}$$

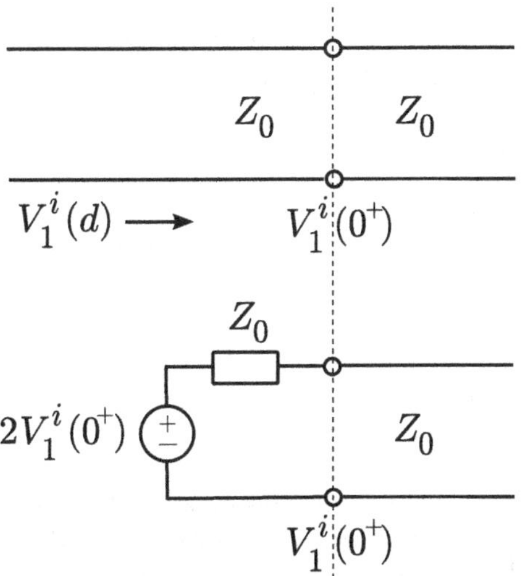

Figure 8.5. A voltage wave on an uninterrupted transmission line (*top*) replaced by a Thévenin equivalent circuit at $d = 0^+$ (*bottom*).

and from the symmetry of the problem it is apparent that

$$s_{12} = s_{21} = \frac{2Z_0}{Z + 2Z_0} \tag{8.23}$$

EXAMPLE 8.1

What are the s-parameters for a two-port in which (a) the output terminals are connected to the input terminals ($Z = 0$), (b) the output terminals are disconnected from the input terminals ($Z \to \infty$), and (c) the output terminals are connected to the input terminals by a series resistor $Z = Z_0$?

Solution: In each case, we can use Equations (8.20) and (8.23). (a) $Z = 0$, so $s_{11} = s_{22} = 0$ and $s_{21} = s_{12} = 1$. (b) $Z \to \infty$, so $s_{11} = s_{22} = 1$ and $s_{21} = s_{12} = 0$. (c) $Z = Z_0$, so $s_{11} = s_{22} = 1/3$ and $s_{21} = s_{12} = 2/3$.

The s-parameters for a *parallel* impedance Z are provided below, and their derivation is left as an exercise for the student (Problem 8.1):

$$s_{11} = s_{22} = \frac{-Z_0}{2Z + Z_0} \tag{8.24}$$

$$s_{21} = s_{12} = \frac{2Z}{2Z + Z_0} \tag{8.25}$$

8.2.3 s-Parameters for Transmission Lines

It is useful to know the s-parameters for a section of transmission line having length l and characteristic impedance equal to the s-parameter reference impedance Z_0. This is shown in Figure 8.6. In this case there will be no reflection, because the input and output impedances are matched. Thus, $s_{11} = s_{22} = 0$. The other two s-parameters can be obtained from the voltage transfer functions, which can be worked out from transmission line theory. First, let us represent the propagation constant of the transmission line comprising the two-port using the symbol $\gamma = \alpha + j\beta$.[2] The real part of γ is α, the attenuation constant, having units of

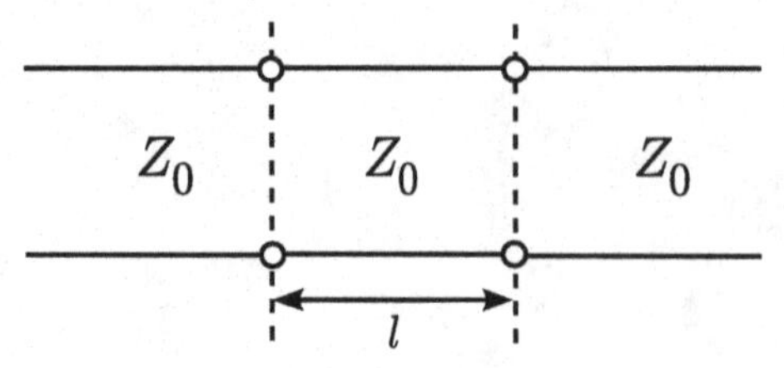

Figure 8.6. A matched transmission line represented as a two-port.

[2] As is common in this field, some of these variables are being reused with different meanings. Here is a reminder if you might have lost track. The "γ" appearing here has no relationship to the fading envelope $\gamma(t)$ in Chapter 6. The "α" appearing here *is* the same one appearing in Chapter 7. The "β" appearing here *is* the same one appearing in Chapters 2 and 3, but has no relation to the quantities appearing in Chapters 5 and 6.

$$s_{11} = \frac{(z_{11}-1)(z_{22}+1)-z_{12}z_{21}}{(z_{11}+1)(z_{11}+1)-z_{12}z_{21}} \quad z_{11} = \frac{(1+s_{11})(1-s_{22})+s_{12}s_{21}}{(1-s_{11})(1-s_{22})-s_{12}s_{21}}$$

$$s_{12} = \frac{2z_{12}}{(z_{11}+1)(z_{22}+1)-z_{12}z_{21}} \quad z_{12} = \frac{2s_{12}}{(1-s_{11})(1-s_{22})-s_{12}s_{21}}$$

$$s_{21} = \frac{2z_{21}}{(z_{11}+1)(z_{22}+1)-z_{12}z_{21}} \quad z_{21} = \frac{2s_{21}}{(1-s_{11})(1-s_{22})-s_{12}s_{21}}$$

$$s_{22} = \frac{(z_{11}+1)(z_{22}-1)-z_{12}z_{21}}{(z_{11}+1)(z_{11}+1)-z_{12}z_{21}} \quad z_{22} = \frac{(1+s_{22})(1-s_{11})+s_{12}s_{21}}{(1-s_{11})(1-s_{22})-s_{12}s_{21}}$$

Figure 8.7. Conversion between z-parameters and s-parameters: z-parameters are normalized to Z_0, so multiply by Z_0 to get Z-parameters in impedance units.

1/length. The imaginary part of γ is the wavenumber β, having units of rad/length. Now for a forward-traveling wave:

$$V_2^o(0^+) = V_1^i(0^+)e^{-\gamma l}\text{ , so } s_{21} = e^{-\gamma l} \tag{8.26}$$

and for a reverse-traveling wave

$$V_1^o(0^+) = V_2^i(0^+)e^{-\gamma l}\text{ , so } s_{12} = e^{-\gamma l} \tag{8.27}$$

8.2.4 s-Parameters for Other Two-Ports

For other two-ports, methods for computing s-parameters include analysis (as above); simulation, in the case of more complicated circuits; and measurement. In Section 8.6 we will develop a fourth method in which the s-parameters of complex two-ports can be determined from the s-parameters of simpler series-connected two-ports. Finally, it is possible to convert between s-parameters and Z, Y, or H-parameters. These conversions are shown in Figures 8.7, 8.8, and 8.9, respectively. Note that the first step in the scheme to obtain s-parameters from Z, Y, or H parameters (whichever is of interest) is to first obtain the normalized (unitless) z, y, or h-parameters, respectively.[3] Similarly, the scheme to obtain Z, Y, or H-parameters from s-parameters requires "denormalizing" from unitless quantities in the final step.

8.3 INTRINSIC PROPERTIES OF TWO-PORTS

We now consider some metrics that are commonly used to characterize two-ports.

Forward & Reverse Power Gain. The forward power gain of a two-port is defined as the ratio of the power delivered to an output load equal to Z_0 (i.e., equal to the reference impedance) to the power applied to the input from a source impedance equal to Z_0. Because s_{21} is a voltage ratio defined with respect to input and output impedances equal to Z_0, the forward power gain

[3] Note the use of lower case to distinguish the normalized two-port parameters from the unnormalized parameters. The s-parameters are normalized "as is."

$$s_{11} = \frac{(1-y_{11})(1+y_{22})+y_{12}y_{21}}{(1+y_{11})(1+y_{22})-y_{12}y_{21}} \quad y_{11} = \frac{(1+s_{22})(1-s_{11})+s_{12}s_{21}}{(1+s_{11})(1+s_{22})-s_{12}s_{21}}$$

$$s_{12} = \frac{-2y_{12}}{(1+y_{11})(1+y_{22})-y_{12}y_{21}} \quad y_{12} = \frac{-2s_{12}}{(1+s_{11})(1+s_{22})-s_{12}s_{21}}$$

$$s_{21} = \frac{-2y_{21}}{(1+y_{11})(1+y_{22})-y_{12}y_{21}} \quad y_{21} = \frac{-2s_{21}}{(1+s_{11})(1+s_{22})-s_{12}s_{21}}$$

$$s_{22} = \frac{(1+y_{11})(1-y_{22})+y_{12}y_{21}}{(1+y_{11})(1+y_{22})-y_{12}y_{21}} \quad y_{22} = \frac{(1+s_{11})(1-s_{22})+s_{12}s_{21}}{(1+s_{11})(1+s_{22})-s_{12}s_{21}}$$

Figure 8.8. Conversion between y-parameters and s-parameters: y-parameters are normalized to $Y_0 = 1/Z_0$, so divide by Y_0 to get Y-parameters in admittance units.

$$s_{11} = \frac{(h_{11}-1)(y_{22}+1)-h_{12}h_{21}}{(h_{11}+1)(h_{22}+1)-h_{12}h_{21}} \quad h_{11} = \frac{(1+s_{11})(1+s_{22})-s_{12}s_{21}}{(1-s_{11})(1+s_{22})+s_{12}s_{21}}$$

$$s_{12} = \frac{2h_{12}}{(h_{11}+1)(h_{22}+1)-h_{12}h_{21}} \quad h_{12} = \frac{2s_{12}}{(1-s_{11})(1+s_{22})+s_{12}s_{21}}$$

$$s_{21} = \frac{-2h_{21}}{(h_{11}+1)(h_{22}+1)-h_{12}h_{21}} \quad h_{21} = \frac{-2s_{21}}{(1-s_{11})(1+s_{22})+s_{12}s_{21}}$$

$$s_{22} = \frac{(1+h_{11})(1-h_{22})+h_{12}h_{21}}{(h_{11}+1)(h_{22}+1)-h_{12}h_{21}} \quad h_{22} = \frac{(1-s_{22})(1-s_{11})-s_{12}s_{21}}{(1-s_{11})(1+s_{22})+s_{12}s_{21}}$$

Figure 8.9. Conversion between h-parameters and s-parameters: h-parameters are normalized to Z_0, so $H_{11} = h_{11}Z_0$, $H_{12} = h_{12}$, $H_{21} = h_{21}$, $H_{22} = h_{22}/Z_0$.

is $|s_{21}|^2$, or $20 \log_{10} |s_{21}|$ in dB. Similarly the reverse power gain of a two-port is defined as the ratio of the power delivered to an input load equal to Z_0 to the power applied to the output from a source impedance equal to Z_0; i.e., $|s_{12}|^2$, or $20 \log_{10} |s_{12}|$ in dB. For passive devices, the power gain is less than or equal to 1. For active devices, the power gain may be greater than, equal to, or less than 1.

Insertion Loss (IL) is the ratio of power applied to the input of a two-port to the power delivered to a load attached to the output, when the source and load impedances are Z_0. In other words, IL is the reciprocal of forward power gain, and is ≥ 1 for passive two-ports. In terms of s-parameters, IL= $|s_{21}|^{-2}$ or $-20 \log_{10} |s_{21}|$ in dB. Although forward power gain and IL report essentially the same information, forward power gain is more commonly used to describe active two-ports whereas IL is more commonly used to describe passive two-ports.

Reverse Isolation is analogous to IL, but applies in the reverse direction: It is the ratio of power applied to the output of a two-port to the power delivered to a load attached to the input, when the source and load impedances are Z_0. Thus, it is the reciprocal of reverse power gain and is ≥ 1 for passive two-ports. In terms of s-parameters, reverse isolation is $|s_{12}|^{-2}$ or $-20 \log_{10} |s_{12}|$ in dB.

Input and Output Return Loss (RL) are defined as the ratio of power applied to the port to the power returning from the port, when the source and load impedances are Z_0. Applying s-parameter definitions, we see input RL is $|s_{11}|^{-2}$ or $-20\log_{10}|s_{11}|$ in dB, and output RL is $|s_{22}|^{-2}$ or $-20\log_{10}|s_{22}|$ in dB. Note passive devices always have RL ≥ 1 (> 0 dB).[4]

Input and Output Impedance. The impedance Z_1 looking into the input of a two-port whose output is terminated into Z_0 can be determined by first noting

$$s_{11} = \Gamma_1 = \frac{Z_1 - Z_0}{Z_1 + Z_0} \tag{8.28}$$

and then solving for Z_1, yielding

$$Z_1 = Z_0 \frac{1 + s_{11}}{1 - s_{11}} \tag{8.29}$$

Similarly the output impedance Z_2 looking into the output of a two-port whose input is terminated into Z_0 is

$$Z_2 = Z_0 \frac{1 + s_{22}}{1 - s_{22}} \tag{8.30}$$

An important point about all of the above metrics: These are *intrinsic* characteristics of a two-port; they do not necessarily indicate what happens when you connect the two-port to other devices. For example: If you connect the output of a two-port to a load Z_0, then the magnitude of the reflection coefficient with respect to a source impedance Z_0 is equal to the inverse square root of the input RL. However, if the source or load impedances are *not* equal to Z_0, then the input reflection coefficient is *not* simply related to the RL; instead, it is given by Equation (8.35) as explained in the next section.

EXAMPLE 8.2

What is the forward power gain, IL, reverse isolation, RL, and input impedance for a two-port in which (a) the output terminals are directly connected to the input terminals, (b) the output terminals are disconnected from the input terminals, and (c) the output terminals are connected to the input terminals by a series resistance equal to Z_0?

Solution: We worked out the s-parameters for these three cases in Example 8.1, so we need only to apply the definitions:

Two-Port	$\lvert s_{21}\rvert^2$	IL	Rev. Iso.	RL	Z_1
(a)	0 dB	0 dB	0 dB	$+\infty$ dB	Z_0
(b)	$-\infty$ dB	$+\infty$ dB	$+\infty$ dB	0 dB	∞
(c)	-3.5 dB	3.5 dB	3.5 dB	9.5 dB	$2Z_0$

[4] One commonly finds RL reported as a gain, so that passive devices have values are reported to have RL less than 1 (<0 dB). This practice can usually be inferred from context.

We continue with some additional ways in which two-ports are characterized.

Reciprocity. Passive two-ports will usually be *reciprocal*, meaning their behavior does not depend on which port is considered to be the "input" and which port is considered to be the "output." Consequently, $s_{mn} = s_{nm}$. All two-ports considered so far in this chapter have been reciprocal. Two-ports containing active devices are typically not reciprocal, nor are passive two-ports containing non-reciprocal devices or materials.

Lossless vs. Lossy. Passive two-ports can be classified as "lossless" or "lossy." Since a_1 and a_2 represent power flowing into the two-port, and b_1 and b_2 represent power flowing out of the two-port, the principle of conservation of power dictates that a lossless two-port has $|b_1|^2 + |b_2|^2 = |a_1|^2 + |a_2|^2$, and a lossy two-port has $|b_1|^2 + |b_2|^2 < |a_1|^2 + |a_2|^2$. The lossless/lossy distinction is usually not meaningful for a two-port containing active devices.

Unilateral vs. Bilateral. Usually, we prefer power within the two-port to flow only in one direction; i.e., only from the input ("1") port to the output ("2") port. This condition implies $s_{12} = 0$. A two-port having this property is referred to as "unilateral"; otherwise, it is classified as "bilateral." This distinction will become an issue when we consider gain and stability of embedded two-ports in Section 8.5, and in Chapter 10 when we consider amplifier design.

8.4 PROPERTIES OF EMBEDDED TWO-PORTS

We now consider the behavior of two-ports embedded in a larger circuit. From the perspective of the two-port, the rest of the circuit presents an impedance to the input and/or output ports that is possibly different from the reference impedance Z_0. When embedded, the input impedance (looking into port "1" of the two-port) and output impedance (looking into port "2" of the two-port) change, with an associated change in reflection coefficient, gain, and so on.

8.4.1 Reflection Coefficient for Embedded Two-Ports

A common problem is to determine the reflection coefficient Γ_{in} at the input of a two-port with respect to the reference impedance Z_0, when the output is terminated into an impedance $Z_L \neq Z_0$. This is shown in Figure 8.10 (top). First, by definition:

$$\Gamma_{in} = \frac{b_1}{a_1} = \frac{s_{11}a_1 + s_{12}a_2}{a_1} = s_{11} + s_{12}\frac{a_2}{a_1} \tag{8.31}$$

Let Γ_L be the reflection coefficient looking into Z_L, with respect to Z_0; i.e.,

$$\Gamma_L = \frac{Z_L - Z_0}{Z_L + Z_0} \tag{8.32}$$

Note $\Gamma_L \equiv a_2/b_2$. We can use this to find a_2/a_1 in terms of the s-parameters and Γ_L as follows:

$$\frac{a_2}{a_1} = \frac{\Gamma_L b_2}{a_1} = \Gamma_L \frac{s_{21}a_1 + s_{22}a_2}{a_1} = s_{21}\Gamma_L + s_{22}\Gamma_L\frac{a_2}{a_1} \tag{8.33}$$

Solving for a_2/a_1:

$$\frac{a_2}{a_1} = \frac{s_{21}\Gamma_L}{1 - s_{22}\Gamma_L} \tag{8.34}$$

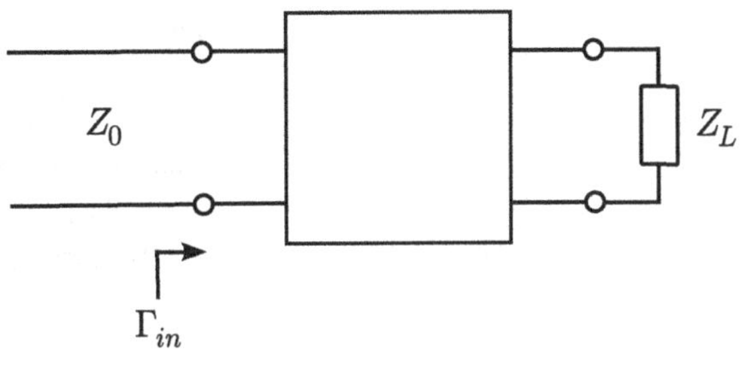

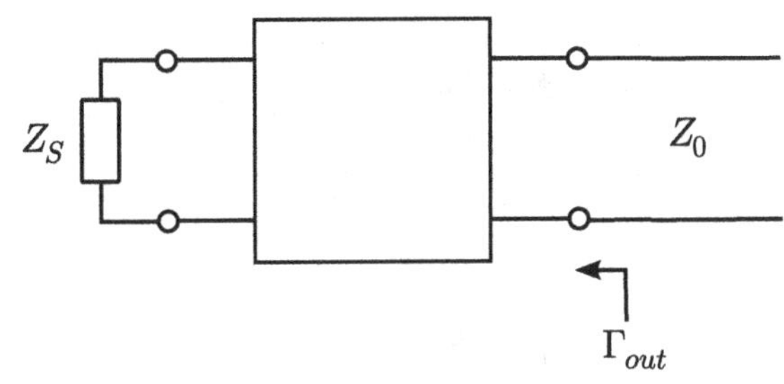

Figure 8.10. Definitions of the embedded reflection coefficients Γ_{in} *(top)* and Γ_{out} *(bottom)*.

Substituting this expression into Equation (8.31):

$$\Gamma_{in} = s_{11} + \frac{s_{12}s_{21}\Gamma_L}{1 - s_{22}\Gamma_L} \tag{8.35}$$

EXAMPLE 8.3

In Examples 8.1 and 8.2, we considered a two-port consisting of a series resistor connected between the input and output terminals. What is the reflection coefficient at the input if the source impedance is Z_0 and the load impedance is (a) Z_0, (b) 0, (c) ∞?

Solution: We determined in Example 8.1 that $s_{11} = s_{22} = 1/3$ and $s_{12} = s_{21} = 2/3$. (a) By definition, $\Gamma_{in} = \Gamma_1 = s_{11} = 1/3$. However, we can also use Equation (8.35) with $Z_L = Z_0$, which yields $\Gamma_L = 0$ so again $\Gamma_{in} = 1/3$. (b) When $Z_L = 0$, $\Gamma_L = -1$, so from Equation (8.35), $\Gamma_{in} = 0$. (c) $\Gamma_L = 1$, so $\Gamma_{in} = 1$.

Similarly, we can find the reflection coefficient Γ_{out} at the output of a two-port with respect to the reference impedance Z_0, when the input is terminated into an impedance $Z_S \neq Z_0$. The derivation is left as an exercise for the student (Problem 8.7). The result is

$$\Gamma_{out} = s_{22} + \frac{s_{12}s_{21}\Gamma_S}{1 - s_{11}\Gamma_S} \tag{8.36}$$

where

$$\Gamma_S = \frac{Z_S - Z_0}{Z_S + Z_0} \tag{8.37}$$

8.4.2 Transducer Power Gain (TPG)

Another common problem is to determine the contribution of a two-port to the gain of a system in which it is embedded. This is the intrinsic forward gain (defined in the previous section)

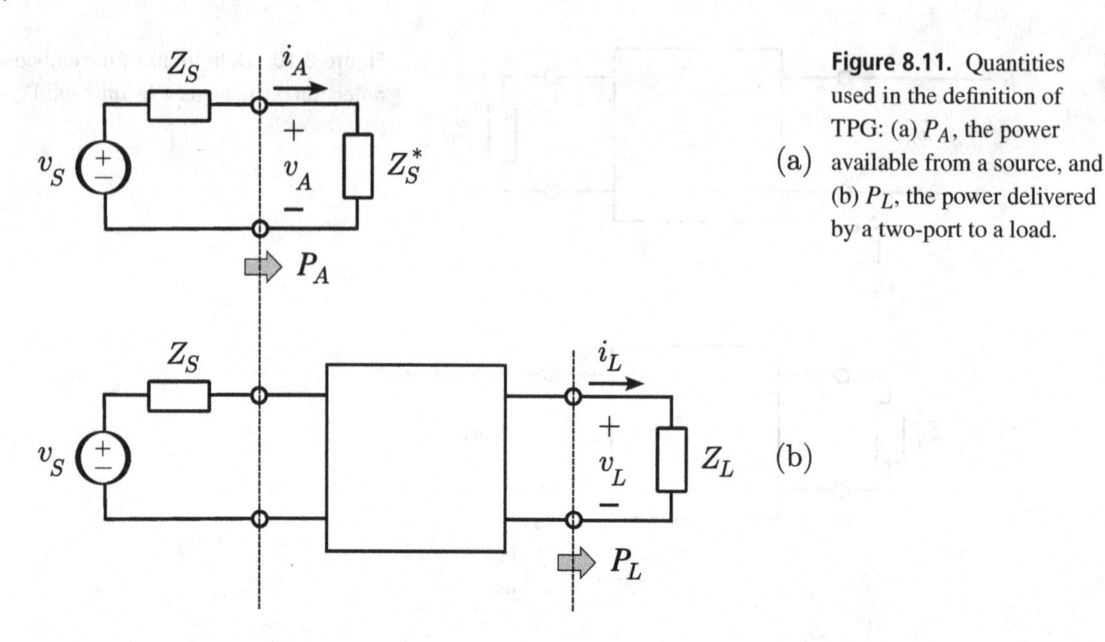

Figure 8.11. Quantities used in the definition of TPG: (a) P_A, the power available from a source, and (b) P_L, the power delivered by a two-port to a load.

only if $Z_S = Z_L = Z_0$. If either Z_S or Z_L are different from Z_0, then the situation is not so straightforward. A particularly useful definition that applies in this case is *transducer power gain*, also known as TPG or G_T, defined as follows:

$$G_T \equiv \frac{\text{Power delivered by the two-port to a load impedance } Z_L}{\text{Power delivered by the source to a conjugate-matched load}} \tag{8.38}$$

This is a relatively complex definition. To make sense of it, let us first consider the numerator and denominator separately.

The denominator in Equation (8.38) is the definition of *available power*, P_A. Figure 8.11(a) illustrates the concept. For simplicity, let us refer to all circuitry connected to the input of the two-port as "the source." A Thévenin equivalent circuit for the source consists of a source voltage v_S in series with the source impedance Z_S. The source delivers maximum power to the load when the impedance of the load is Z_S^*; i.e., a "conjugate match." P_A is the power delivered to the load in this scenario, which is readily determined to be

$$P_A = \text{Re}\left\{v_A i_A^*\right\} = \text{Re}\left\{\left[v_S \frac{Z_S^*}{Z_S + Z_S^*}\right]\left[\frac{v_S}{Z_S + Z_S^*}\right]^*\right\} = \frac{|v_S|^2}{4R_S} \tag{8.39}$$

where we have assumed v_S is in root mean square (rms) as opposed to peak units. Similarly the numerator in Equation (8.38) is given the symbol P_L and is simply

$$P_L = \text{Re}\left\{v_L i_L^*\right\} \tag{8.40}$$

as indicated in Figure 8.11(b).

EXAMPLE 8.4

A two-port consists of a series impedance equal to the reference impedance $Z_0 = 50\ \Omega$. This two-port is driven by a source with open-circuit voltage 1 V (rms) and impedance Z_S. The output is terminated into a load Z_L. Let $Z_S = Z_L = Z_0$. Compute the power available from the source, power delivered to the load, and from these quantities compute TPG.

Solution: The power available from the source is $P_A = |v_S|^2/(4R_S) = 5$ mW. The power delivered to the load is $P_L = \text{Re}\{v_L i_L^*\}$. From elementary circuit analysis, $v_L = 1/3$ V and $i_L = 20/3$ mA, so $P_L = 2.22$ mW. Applying the definition of TPG: $G_T \equiv P_L/P_A = 0.444 = -3.5$ dB.

EXAMPLE 8.5

Repeat Example 8.4, now with $Z_L = 100\ \Omega$.

Solution: The power available from the source is still $P_A = 5$ mW, since nothing has changed on the input side. The power delivered to the load increases to $P_L = 2.5$ mW. Applying the definition of TPG: $G_T \equiv P_L/P_A = 0.5 = -3$ dB. Note that TPG has increased even though the available power and the two-port remain the same. It is the increase in the load impedance that has resulted in the increase in delivered power and TPG. This underscores the fact that TPG is an embedded – not intrinsic – property of the two-port.

It is possible to calculate TPG directly from s-parameters and source and load reflection coefficients, as follows:

$$G_T = \frac{\left(1 - |\Gamma_S|^2\right)|s_{21}|^2\left(1 - |\Gamma_L|^2\right)}{|(1 - s_{11}\Gamma_S)(1 - s_{22}\Gamma_L) - s_{12}s_{21}\Gamma_S\Gamma_L|^2} \tag{8.41}$$

The derivation of this expression is beyond the scope of this book; a suggested reference is [20, Sec. 2.6]. We can, however, confirm that the above equation behaves as expected in cases where the outcome is already known: For example, when the input and output ports are terminated into impedances equal to Z_0. In this case, $\Gamma_S = \Gamma_L = 0$ and we obtain $G_T = |s_{21}|^2$. Thus we find TPG is equal to intrinsic forward gain in this case, as expected. We can also repeat the previously-worked examples using this new expression:

EXAMPLE 8.6

Repeat Example 8.5, now using Equation (8.41).

Solution: As before $Z_S = Z_0$ so $\Gamma_S = 0$. However, now $Z_L \neq Z_0$, and so we find Γ_L is non-zero; in fact it is $1/3$. Applying Equation (8.41), we find the TPG is $1/2 = -3$ dB, in agreement with the result of Example 8.5.

Problems 8.10 and 8.11 offer additional practice with the concepts of available power, delivered power, and TPG.

It might now appear that we are able to compute the transducer power gain for any embedded two-port given the s-parameters, Z_S, and Z_L. However, examination of the denominator of Equation (8.41) reveals that there may be combinations of these parameters for which G_T becomes infinite, which implies that there is output power even when there is no input power. Although this might initially seem to be very wrong, in fact this is an apt description of an oscillator, as we shall see in Chapter 16. Assuming the intent is not to create an oscillator, the conditions under which G_T might become variable or unbounded are to be avoided. The relevant issue is *stability*, which is considered next.

8.5 STABILITY AND GAIN

We now consider the issue of two-port stability, and the implications for the usable gain of active two-ports. After an introduction to the concept of stability in Section 8.5.1, we consider methods for determining the conditions under which a two-port may be considered stable (Section 8.5.2). Section 8.5.3 introduces simultaneous conjugate matching, which maximizes the TPG of a stable two-port. In Section 8.5.4 we consider the question of how to estimate the maximum TPG of an active two-port, taking into account the potential for instability.

8.5.1 Instability and Oscillation

Stability is the absence of instability. Instability is a property of some two-ports which are *active* and have *internal feedback*. The concept is best demonstrated by example. Consider a voltage amplifier circuit. The voltage amplifier produces an output voltage which is larger than the input voltage. If there is perfect isolation between output and input, the amplifier will be stable. If on the other hand some fraction of the voltage at the output makes its way to the input, then there is feedback, but not necessarily instability. In fact, properly implemented "negative" feedback is often desirable in voltage amplifiers. Instability occurs when the voltage feedback creates a "vicious cycle" in which the modified input voltage results in a significant increase in the output voltage, which increases the feedback voltage, which further increases the output voltage, and so on. In this scenario, the apparent voltage gain of the amplifier is continuously increasing, even though the input voltage has remained constant.

From the principle of conservation of power it is clear that this continuous increase is not sustainable, since the power available from the power supply is presumably bounded. Instead, the characteristics of the amplifier must change so as to stop the unbounded escalation of the output voltage. This "change in characteristics" might be as simple (and dramatic) as the self-destruction of the amplifier. Typically, however, what happens is that the intrinsic gain of the amplifier decreases with increasing output so that the combined input-plus-feedback voltage results in a sustainable output voltage. This non-linear behavior is known as *gain compression* and is addressed further in Section 11.2.3. A second phenomenon associated with instability is *oscillation*, which for our purposes here can be defined as a periodically-varying output in response to a constant (or zero) input. While oscillation has applications (see Chapter 16), it is not desirable in most active two ports, and is especially undesirable in amplifiers.

In practice, all practical active two-ports have some degree of internal feedback and therefore are potentially unstable. Even if this internal feedback is negligible, supporting circuitry (e.g., bias circuitry in the case of transistors) and reflection from impedance mismatches provide paths for feedback that can lead to instability. This raises the question: Is it possible to characterize the stability of an embedded two-port from its s-parameters and the source and load impedances? The answer is yes, and this is the topic of the next section.

8.5.2 Determination of Stability

The instability of an embedded two-port can be recognized from outside the two-port as a condition in which incident power waves result in "reflected" power waves with greater magnitude; i.e., $|\Gamma_{in}| > 1$ or $|\Gamma_{out}| > 1$. The term "reflected" is in quotes here because the power flowing in the opposite direction is not simply a fraction of the incident power, but rather a sum of simply-reflected power plus power entering the two-port some other way; i.e., from a power supply or through the input. Thus we can write the conditions for stability in terms of Equations (8.35) and (8.36) as follows:

$$|\Gamma_{in}| = \left| s_{11} + \frac{s_{12}s_{21}\Gamma_L}{1 - s_{22}\Gamma_L} \right| < 1 \tag{8.42}$$

$$|\Gamma_{out}| = \left| s_{22} + \frac{s_{12}s_{21}\Gamma_S}{1 - s_{11}\Gamma_S} \right| < 1 \tag{8.43}$$

We should also be clear that the conditions assume that the source and load are themselves well-behaved, with $|\Gamma_S| < 1$ and $|\Gamma_L| < 1$.

If the above conditions are satisfied for all possible Γ_S and Γ_L with magnitudes less than 1, then the two-port is said to be *unconditionally stable*. If the above conditions are satisfied only for some combinations of Γ_S and Γ_L, then the two-port is said to be *conditionally stable*. Unconditionally stable two-ports are highly-valued in practical design, as the task of impedance matching is unencumbered by concern that the match may make the two-port unstable. However, conditionally stable two-ports are hard to avoid in many circumstances; some techniques for dealing with this possibility in amplifier design are considered in Chapter 10.

It is also important to note that the stability of a two-port may depend on frequency. It it quite common for a two-port to be stable at one frequency and unstable at a different frequency. Furthermore: If some device within the two-port begins to oscillate because it has been stimulated by a signal at a frequency at which the two-port is unstable, then the two-port's performance may be affected at all frequencies. Therefore it is important to consider the two-port's stability over a frequency range that contains all possible input signals that might exist; not just the signals for which the two-port is intended.

We now consider the question of how to determine the stability of a two-port. There are four ways to go about this: (1) sample testing, (2) exhaustive ("brute force") testing, (3) stability circles, and (4) computing a stability metric. We now consider each of these in turn.

Sample Testing. If we wish to test for stability given the s-parameters with Γ_S and Γ_L (or, equivalently, with Z_S and Z_L), then we can simply substitute these values into Equations (8.42) and (8.43) and see if the inequalities are satisfied. If they are, then the two-port is stable for that choice of Γ_S and Γ_L, and the two-port must be either conditionally or unconditionally stable at that frequency.

Exhaustive ("Brute Force") Testing. If we do not have a specific Γ_S and Γ_L in mind and instead wish to know all values resulting in a stable embedded two-port, then we can systematically test all possible combinations of Γ_S and Γ_L. This is not quite as bad as it sounds, as long as you are content to let a computer do it. The reason this approach is tractable is because we have already constrained $|\Gamma_S| < 1$ and $|\Gamma_L| < 1$; that is, all allowed values of these reflection coefficients lie inside a circle of radius 1 on the real–imaginary plane. The astute reader will observe that this is essentially the famous "Smith chart." However, there is no specific need to invoke the Smith chart or associated methods for the work at hand.

EXAMPLE 8.7

Table 8.1 shows the s-parameters of the Avago AT-41511 transistor in common-emitter configuration, biased at $V_{CE} = 2.7$ V and $I_C = 5$ mA, for three frequencies. Perform an exhaustive search to determine all values of source and load impedance for which this transistor is unstable at 1 GHz.

Table 8.1. **Some s-parameters for the Avago AT-41511 transistor in common-emitter configuration, biased at $V_{CE} = 2.7$ V and $I_C = 5$ mA, from the datasheet. Magnitudes are in linear units.**

Freq. GHz	s_{11}	s_{21}	s_{12}	s_{22}
1.0	$0.48\angle-149°$	$5.19\angle+89°$	$0.073\angle+43°$	$0.49\angle-39°$
1.5	$0.46\angle-176°$	$3.61\angle+72°$	$0.087\angle+44°$	$0.44\angle-43°$
2.0	$0.46\angle+162°$	$2.774\angle+59°$	$0.103\angle+45°$	$0.42\angle-47°$

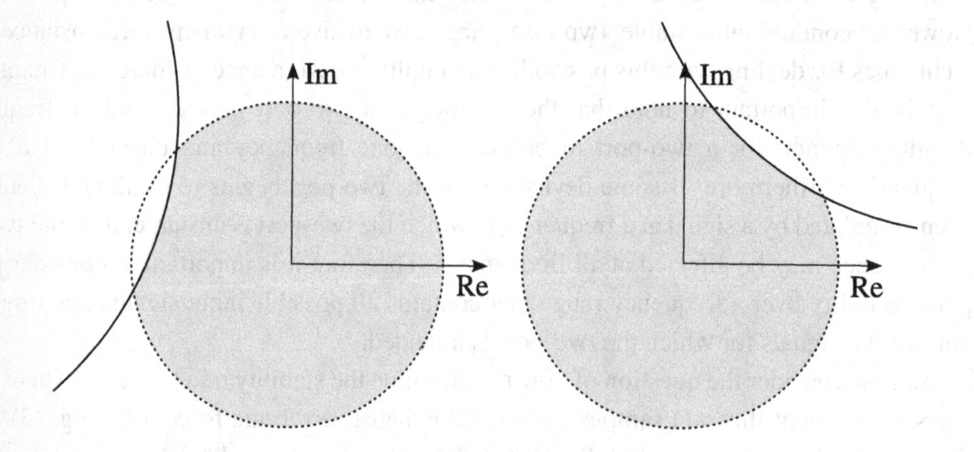

Figure 8.12. Shaded areas indicate values of Γ_S (*left*) and Γ_L (*right*) that yield a stable embedded two-port. The unit circle is indicated by the dashed line. The arcs indicate where an input or output reflection coefficient becomes 1. In this example, the two-port is the Avago AT-41511 transistor at 1 GHz; see Table 8.1.

Solution: First, realize that we can save ourselves a lot of work by doing the search over Γ_S and Γ_L, since these parameters map one-to-one with Z_S and Z_L but have the advantage of being constrained to have magnitude less than 1 – i.e., fall within the unit circle. Figure 8.12 shows the regions within the unit circle for which Γ_S satisfies Equation (8.42) and for which Γ_L satisfies Equation (8.43). As long as both inequalities are satisfied over some part of the unit circle, the embedded two-port is at least conditionally stable; we see that this is true in this example.

If both conditions are satisfied over the entire unit circle, then the two-port is unconditionally stable; this is *not* true in this example. Therefore the AT-41511 is conditionally stable at 1 GHz, and the values of source and load impedance leading to instability correspond to the unshaded regions in Figure 8.12.

Stability Circle Method. The above approach is not particularly elegant or convenient. However, note from the example in Figure 8.12 that the boundaries between regions associated with stability and instability appear as arcs in the real–imaginary plane. In fact, they are circles: These circles are known as *stability circles*, formally defined as the locus of points for which $|\Gamma_{in}| = 1$ with varying Γ_L, or $|\Gamma_{out}| = 1$ with varying Γ_S. Expressions for the centers and radii of these circles can be derived directly from Equations (8.42) and (8.43). We will not present that derivation here, as it is tedious and there is not much insight to be gained in the process.[5] Before presenting the result, we make the following definition:

$$\Delta \equiv s_{11}s_{22} - s_{12}s_{21} \tag{8.44}$$

Now the center of the $|\Gamma_{in}| = 1$ stability circle – also known as the *output stability circle* – is:

$$C_L = \frac{\left(s_{22} - \Delta s_{11}^*\right)^*}{|s_{22}|^2 - |\Delta|^2} \tag{8.45}$$

and the radius is

$$r_L = \left| \frac{s_{12}s_{21}}{|s_{22}|^2 - |\Delta|^2} \right| \tag{8.46}$$

Similarly, the center and radius of the $|\Gamma_{out}| = 1$ stability circle – also known as the *input stability circle* – is:

$$C_S = \frac{\left(s_{11} - \Delta s_{22}^*\right)^*}{|s_{11}|^2 - |\Delta|^2} \tag{8.47}$$

and the radius is

$$r_S = \left| \frac{s_{12}s_{21}}{|s_{11}|^2 - |\Delta|^2} \right| \tag{8.48}$$

Note that the stability circles indicate only the boundary between stable and unstable regions. To determine whether the interior or exterior of the circle corresponds to stable operation, one must test at least one point inside or outside the circle. This can be done using the "sample testing" method as explained above.

EXAMPLE 8.8

For the Avago AT-41511 transistor at 1 GHz (i.e., the same problem considered in Example 8.7), determine the input and output stability circles.

Solution: The center and radius of the input stability circle are $3.10\angle + 162^\circ$ and 2.25 respectively. The center and radius of the output stability circle are $2.98\angle + 51^\circ$ and 2.13 respectively. These circles are illustrated in Figure 8.13.

[5] A complete derivation can be found in [20, App. B].

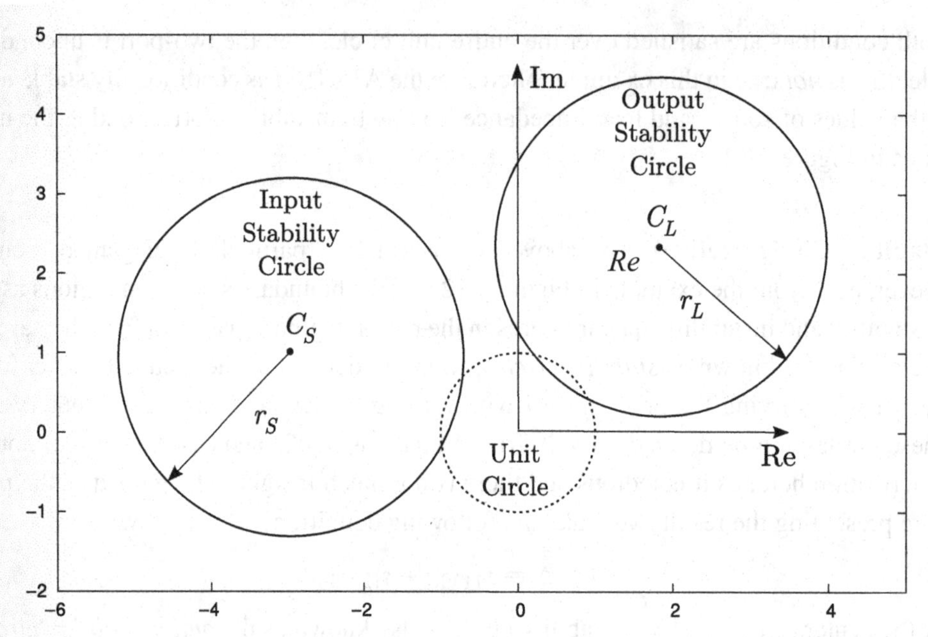

Figure 8.13. Finding values of Γ_S and Γ_L that correspond to a stable embedded two-port using the stability circle method. This example uses the same two-port and frequency considered in Figure 8.12.

Metrics for Unconditional Stability. If our interest is limited to knowing whether or not a two-port is unconditionally stable, then the above approaches are more work than is necessary. Beginning with Equations (8.42) and (8.43) plus the constraints $|\Gamma_S| < 1$ and $|\Gamma_L| < 1$, it is possible to derive simple tests for unconditional stability. One such test is:

$$|\Delta| < 1 \text{ and } K > 1 \text{ for unconditional stability} \tag{8.49}$$

where

$$K \equiv \frac{1 + |\Delta|^2 - |s_{11}|^2 - |s_{22}|^2}{2\,|s_{12}s_{21}|} \tag{8.50}$$

The value K is known as the *Rollett stability factor* [66]. It is worth noting that unconditional stability corresponds to stability circles that do not intersect the unit circle in the real–imaginary plane; in fact this is an alternative approach that can be used to derive the Rollett stability factor.

EXAMPLE 8.9

For the Avago AT-41511 transistor at 1 GHz (i.e., the same problem considered in Figure 8.12 and Example 8.8), test for unconditional stability. Repeat at 2 GHz and calculate the stability circles.

Solution: At 1 GHz we find $|\Delta| = 0.250$ and $K = 0.781$, indicating that this two-port is *not* unconditionally stable, as expected. At 2 GHz we find $|\Delta| = 0.103$ and $K = 1.09$. The transistor is therefore unconditionally stable at 2 GHz. The center and radius of the 2 GHz input stability circle are $2.47\angle - 159^\circ$ and 1.72 respectively. The center and radius of the 2 GHz output stability circle are $2.78\angle + 50^\circ$ and 1.42 respectively. Since $|C_S| - r_S > 1$ and $|C_L| - r_L > 1$, neither stability circle overlaps the unit circle. This confirms our assessment of unconditional stability at 2 GHz.

It should be noted that the Rollett stability factor is not the only metric of stability in common practice: A prominent alternative is the *Edwards–Sinsky metric* [17]:

$$\mu \equiv \frac{1-|s_{11}|^2}{\left|s_{22}-s_{11}^*\Delta\right|+|s_{12}s_{21}|} > 1 \quad \text{for unconditional stability} \tag{8.51}$$

The principal advantage of the Edwards–Sinsky criterion is that it requires computation of only one metric (μ), whereas the Rollett criterion requires two (Δ and then K). On the other hand, K appears as a parameter in other calculations (as we shall soon see), so both sets of criteria continue to be used in common practice. (Problem 8.14 gives you the opportunity to experiment with the Edwards–Sinsky metric, following up Example 8.9.)

8.5.3 Simultaneous Conjugate Matching

Let us say a two-port is at least conditionally stable. What is the maximum possible TPG and for what values of Γ_S and Γ_L is it obtained? This turns out to be one of the fundamental problems in amplifier design, so we address this question in detail here.

If we temporarily ignore the possibility of instability, then G_T is maximized when the input and output ports are conjugate-matched; i.e., $Z_S = Z_{in}^*$ and $Z_L = Z_{out}^*$, or equivalently $\Gamma_S = \Gamma_{in}^*$ and $\Gamma_L = \Gamma_{out}^*$. These values can be substituted into Equations (8.35) and (8.36), yielding simultaneous equations for Γ_S and Γ_L in terms of the s-parameters of the embedded two-port. The result is that G_T is maximized when

$$\Gamma_S = \Gamma_{S,max} \equiv \frac{B_1 - \sqrt{B_1^2 - 4|M|^2}}{2M} \tag{8.52}$$

$$\Gamma_L = \Gamma_{L,max} \equiv \frac{B_2 - \sqrt{B_2^2 - 4|N|^2}}{2N} \tag{8.53}$$

where

$$B_1 \equiv 1 + |s_{11}|^2 - |s_{22}|^2 - |\Delta|^2 \tag{8.54}$$

$$B_2 \equiv 1 + |s_{22}|^2 - |s_{11}|^2 - |\Delta|^2 \tag{8.55}$$

$$M \equiv s_{11} - s_{22}^*\Delta \tag{8.56}$$

$$N \equiv s_{22} - s_{11}^*\Delta \tag{8.57}$$

Now, if the embedded two-port is stable for this choice of Γ_S and Γ_L (easily determined by sample testing), then we have a valid *simultaneous conjugate match*. In this case, the maximum value of G_T is obtained by substituting these values back into Equation (8.41); after some algebra it is possible to show that this maximum value of G_T – known as the *maximum available gain* (MAG) – is equal to

$$G_{T,max} \equiv \left|\frac{s_{21}}{s_{12}}\right| \left(K - \sqrt{K^2-1}\right) \tag{8.58}$$

Alternatively, it could be that the values of Γ_S and Γ_L obtained by simultaneous conjugate matching lead to unstable operation. If this is found to be the case, then the above expression for $G_{T,max}$ is not valid and one must resort to other methods to identify the maximum possible TPG. This accounts for much of the effort in the design of transistor amplifiers, as we shall see in Chapter 10.

EXAMPLE 8.10

For the Avago AT-41511 transistor at 2 GHz (previously considered in Example 8.9), determine maximum TPG. If a stable simultaneous conjugate match solution exists, give Γ_S and Γ_L.

Solution: We determined in Example 8.9 that the transistor is unconditionally stable at this frequency. Therefore the maximum TPG is equal to the MAG = 12.5 dB. Since the transistor is unconditionally stable, the simultaneous conjugate match solution is guaranteed to be stable, and is given by $\Gamma_S = 0.782\angle - 159^\circ$ and $\Gamma_L = 0.767\angle + 50^\circ$. As a check, we substitute these values of Γ_S and Γ_L into the general TPG equation (8.41) and obtain 12.5 dB, which is equal to the MAG calculated above.

Looking forward, note that we may not be content to leave the answer in the form of the reflection coefficients Γ_S and Γ_L, but rather would prefer to have answers in the form of impedances. This is easily done: Since

$$\Gamma_S = \frac{Z_S - Z_0}{Z_S + Z_0}$$

we may solve for Z_S to obtain

$$Z_S = Z_0 \frac{1 + \Gamma_S}{1 - \Gamma_S} \tag{8.59}$$

and similarly

$$Z_L = Z_0 \frac{1 + \Gamma_L}{1 - \Gamma_L} \tag{8.60}$$

Finally, we note that when the two-port is unilateral, i.e., when $s_{12} = 0$, $K \to \infty$ and Equation (8.58) becomes awkward to use even though the maximum TPG remains well-defined. A more convenient way to compute MAG for unilateral two-ports is Equation (10.50) (Section 10.4.2).

8.5.4 Maximum Stable Gain

If a two-port is only conditionally stable, and a stable simultaneous conjugate match is not possible, then which values of Γ_S and Γ_L maximize TPG? The short answer to this question is that there is no simple answer. We have seen above that given the s-parameters of the embedded two-port, it is possible to identify the values of Γ_S and Γ_L that result in stable operation. We can also test for the possibility of a stable simultaneous conjugate matching solution, which guarantees maximum TPG but only if the resulting values of Γ_S and Γ_L result in stable operation. This accounts for all unconditionally-stable two-ports and some conditionally-stable two-ports. For those conditionally-stable two-ports for which a simultaneous conjugate match is not stable, more effort is required to determine the maximum stable TPG.

A quite reasonable option is to test all sets of values of Γ_S and Γ_L for which the two-port is stable using Equation (8.41), with the goal of determining the choices that maximize G_T. Again, this is tractable because all reasonable choices for Γ_S and Γ_L have magnitudes ≤ 1 (i.e., lie within the unit circle), so there is a finite and simply-bounded domain within which all possibilities lie. (This same procedure can also be used to determine all possible combinations of Γ_S and Γ_L which yield a *specified* TPG.)

One is sometimes faced with the problem of having to select from among a number of possible devices for a particular application, in which case the procedure described in the previous paragraph may become impractical. In this situation, one desires an easy-to-calculate characterization of the maximum TPG, analogous to Equation (8.58). A suitable metric in this case is the *maximum stable gain* (MSG), derived as follows: First, we note that Equation (8.58) assumes that the embedded two-port is stable for $\Gamma_S = \Gamma_{in}^*$ and $\Gamma_L = \Gamma_{out}^*$. This in turn implies $K > 1$. Note that $K - \sqrt{K^2 - 1} = 1$ for $K = 1$, and $K - \sqrt{K^2 - 1} \rightarrow 0$ as $K \rightarrow \infty$. Therefore we estimate that

$$G_{T,max} < \left| \frac{s_{21}}{s_{12}} \right| \quad \text{(estimate)} \tag{8.61}$$

The quantity on the right is the MSG. Despite the name, MSG is *not* the maximum G_T of an embedded two-port, nor is it even the maximum stable TPG for a two-port. Instead, MSG is merely the best guess of the maximum stable TPG that is possible without a brute force search over all combinations of Γ_S and Γ_L that yield stable operation. Thus, as noted above, MSG is useful for comparing devices, but does not convey the same information as MAG, and is typically not equal to the maximum stable TPG.

EXAMPLE 8.11

For the Avago AT-41511 transistor at 1 GHz (previously considered in Examples 8.7–8.10), determine maximum TPG.

Solution: In Example 8.9 we found the transistor is not unconditionally stable at 1 GHz; therefore the maximum TPG is not equal to MAG. We estimate the maximum TPG instead using MSG, which is found to be 18.5 dB.

Another approach to this problem is a brute force search over Γ_S and Γ_L, seeking the maximum TPG over the domain of values for which the two-port is stable. Using a rectangular search grid with spacings of 0.033 in the real and imaginary plane, it is found that the largest TPG for which $|\Gamma_S|$, $|\Gamma_L|$, $|\Gamma_{in}|$ and $|\Gamma_{out}|$ are all <1 is approximately 23.8 dB. Thus, we see the estimate provided by MSG in this case is quite conservative.

Before moving on, an important caveat about the brute force solution in the above example: This solution is obtained when $\Gamma_S = 0.907\angle 144^\circ$ and $\Gamma_L = 0.898\angle 68^\circ$. Referring to Figure 8.13, note that these values are on the borders between stable and unstable regions. It is typical for the largest stable TPGs achieved for a conditionally-stable two-port to be located at the edge of regions of instability. In practical design, it would be quite risky to intentionally choose such a solution, since small errors in component values or changes due to temperature could cause the two-port to drift over the line into instability. Revising the search to build in some margin – e.g., to requiring $|\Gamma_{in}| < 0.95$ and $|\Gamma_{out}| < 0.95$, is a good idea. In this case, the revised maximum stable TPG is found to be 21.8 dB – still quite conservative relative to the MSG.

8.6 CASCADED TWO-PORTS

The previous sections of this chapter have demonstrated a variety of applications of two-port analysis. This analysis can be extended to a broader class of applications if it possible to combine series arrangements ("cascades") of two-ports into a single two-port. For example: We will see in Chapter 10 that transistor amplifiers are typically designed as a cascade of three two-ports (see Figure 10.12): an input matching circuit, the transistor itself, and an output matching circuit. Once the design is completed, however, it is useful to combine these three two-ports into a single amplifier two-port accounting for all three parts.

To consider how this might be done, let us begin by representing Equations (8.15) and (8.16) as a single matrix equation as follows:

$$\begin{bmatrix} b_1 \\ b_2 \end{bmatrix} = \begin{bmatrix} s_{11} & s_{12} \\ s_{21} & s_{22} \end{bmatrix} \begin{bmatrix} a_1 \\ a_2 \end{bmatrix} \tag{8.62}$$

Note the matrix of s-parameters relates the inbound power waves to the outbound power waves. For cascading two-ports, this is not quite what we need. What would be better is a form in which the power waves at the output were related to the power waves at the input by some matrix; i.e.,

$$\begin{bmatrix} b_1 \\ a_1 \end{bmatrix} = \begin{bmatrix} t_{11} & t_{12} \\ t_{21} & t_{22} \end{bmatrix} \begin{bmatrix} a_2 \\ b_2 \end{bmatrix} \tag{8.63}$$

The parameters in this matrix are known as transmission (t) parameters, and are related to the s-parameters as shown in Figure 8.14. Now given a two-port with s-parameters $s_{11}^{(1)}$, $s_{12}^{(1)}$, $s_{21}^{(1)}$, and $s_{22}^{(1)}$; and a second two-port with s-parameters $s_{11}^{(2)}$, $s_{12}^{(2)}$, $s_{21}^{(2)}$, and $s_{22}^{(2)}$, we have

$$\begin{bmatrix} b_1^{(1)} \\ a_1^{(1)} \end{bmatrix} = \begin{bmatrix} t_{11}^{(1)} & t_{12}^{(1)} \\ t_{21}^{(1)} & t_{22}^{(1)} \end{bmatrix} \begin{bmatrix} a_2^{(1)} \\ b_2^{(1)} \end{bmatrix} \tag{8.64}$$

and

$$\begin{bmatrix} b_1^{(2)} \\ a_1^{(2)} \end{bmatrix} = \begin{bmatrix} t_{11}^{(2)} & t_{12}^{(2)} \\ t_{21}^{(2)} & t_{22}^{(2)} \end{bmatrix} \begin{bmatrix} a_2^{(2)} \\ b_2^{(2)} \end{bmatrix} \tag{8.65}$$

$$\begin{aligned} t_{11} &= (s_{12}s_{21} - s_{11}s_{22})/s_{21} & s_{11} &= t_{12}/t_{22} \\ t_{12} &= s_{11}/s_{21} & s_{12} &= (t_{11}t_{22} - t_{12}t_{21})/t_{22} \\ t_{21} &= -s_{22}/s_{21} & s_{21} &= 1/t_{22} \\ t_{22} &= 1/s_{21} & s_{22} &= -t_{21}/t_{22} \end{aligned}$$

Figure 8.14. Conversion from s-parameters to t-parameters and vice versa.

If we connect these two-ports together, then $a_2^{(1)} = b_1^{(2)}$ and $b_2^{(1)} = a_1^{(2)}$, so

$$\begin{bmatrix} b_1^{(1)} \\ a_1^{(1)} \end{bmatrix} = \begin{bmatrix} t_{11}^{(1)} & t_{12}^{(1)} \\ t_{21}^{(1)} & t_{22}^{(1)} \end{bmatrix} \begin{bmatrix} t_{11}^{(2)} & t_{12}^{(2)} \\ t_{21}^{(2)} & t_{22}^{(2)} \end{bmatrix} \begin{bmatrix} a_2^{(2)} \\ b_2^{(2)} \end{bmatrix} \tag{8.66}$$

In other words, the t-parameters of the combined two-port are

$$\begin{bmatrix} t_{11} & t_{12} \\ t_{21} & t_{22} \end{bmatrix} = \begin{bmatrix} t_{11}^{(1)} & t_{12}^{(1)} \\ t_{21}^{(1)} & t_{22}^{(1)} \end{bmatrix} \begin{bmatrix} t_{11}^{(2)} & t_{12}^{(2)} \\ t_{21}^{(2)} & t_{22}^{(2)} \end{bmatrix} \tag{8.67}$$

So the scheme for obtaining the s-parameters for a cascade of two-ports is as follows: (1) Obtain the t parameters for each two-port in the cascade, (2) multiply the associated t-parameter matrices in order from input to output, and then (3) convert the resulting t parameters back to s parameters.

Now, an important note: While there is essentially ubiquitous agreement on the definition of s-parameters, others may define t parameters in different ways. Common alternative conventions for specifying t parameters involve swapping a and b in the input and output vectors, and swapping port 1 and port 2 in Equation (8.63). There is no one "correct" definition, so one must take care to be aware of the convention in effect.

EXAMPLE 8.12

The conversions between t and s parameters given in Figure 8.14 were provided without derivation. However, we can devise a simple check. Recall that the s-parameters for a parallel impedance Z are given in Equations (8.24) and (8.25). Calculate the s-parameters for a cascade of two parallel impedances, each equal to Z_0 (a) by using basic circuit theory and then (b) by using the s-parameter cascade approach.

Solution: (a) The parallel combination of two impedances, each equal to Z_0, is an impedance $Z_0Z_0/(Z_0 + Z_0) = Z_0/2$. From Equations (8.24) and (8.25), we find the s-parameters for a parallel impedance $Z = Z_0/2$ are $s_{11} = s_{22} = -1/2$ and $s_{12} = s_{21} = +1/2$.
(b) The s parameters for a parallel impedance $Z = Z_0$ are $s_{11} = s_{22} = -1/3$ and $s_{12} = s_{21} = +2/3$. The associated t parameters are $t_{11} = t_{21} = +1/2$, $t_{12} = -1/2$, and $t_{22} = +3/2$. The t parameters for a cascade of two such parallel impedances is $t_{11} = 0$, $t_{12} = -1$, $t_{21} = +1$, $t_{22} = +2$. Converting this back to s parameters, we find $s_{11} = s_{22} = -1/2$ and $s_{12} = s_{21} = +1/2$, in agreement with (a).

An important application of t parameters is the evaluation of TPG of cascaded two-ports when the port impedances are mismatched. When the impedances of the connecting ports are connected in a reflectionless match (note: not the same as conjugate match!) then TPG with respect to Z_0 is simply the product of the intrinsic forward power gains of the individual two-ports; i.e., $|s_{21}^{(1)}|^2|s_{21}^{(2)}|^2$. When the port impedances are mismatched such that there is reflection (including conjugate matching), then this is no longer true. Here is an example from transistor amplifier design.

EXAMPLE 8.13

As we shall find in Chapter 10, one of the principal headaches in amplifier design is ensuring the stability of transistors. One way to accomplish this is by *resistive loss stabilization* (Section 10.4.3) which can be accomplished using a resistor across the output; i.e., from collector to emitter in a common-emitter BJT design. The s-parameters of the parallel resistance are simple to calculate (Section 8.2.2), and from this we can calculate the s-parameters of a two-port cascade consisting of the transistor and the stabilization resistor. From the s-parameters of the cascade, it is possible to reevaluate stability (e.g., using the Rollett criterion) and to simultaneously determine the effect – usually deleterious – on gain. Although gain is reduced, the advantage in terms of improved stability is often worth the trouble.

So here's the problem: In Example 8.9 we examined the AT-41511 NPN BJT in common emitter configuration and found it to be potentially unstable at 1 GHz. Make this transistor unconditionally stable at 1 GHz by adding a resistor in parallel with the output.

Solution: The s-parameters for the transistor were given in the cited example. The s-parameters for the resistor two-port are given by (see Equations (8.24) and (8.25)):

$$s_{11} = s_{22} = \frac{-Z_0}{2R_{os} + Z_0}$$

$$s_{12} = s_{21} = \frac{2R_{os}}{2R_{os} + Z_0}$$

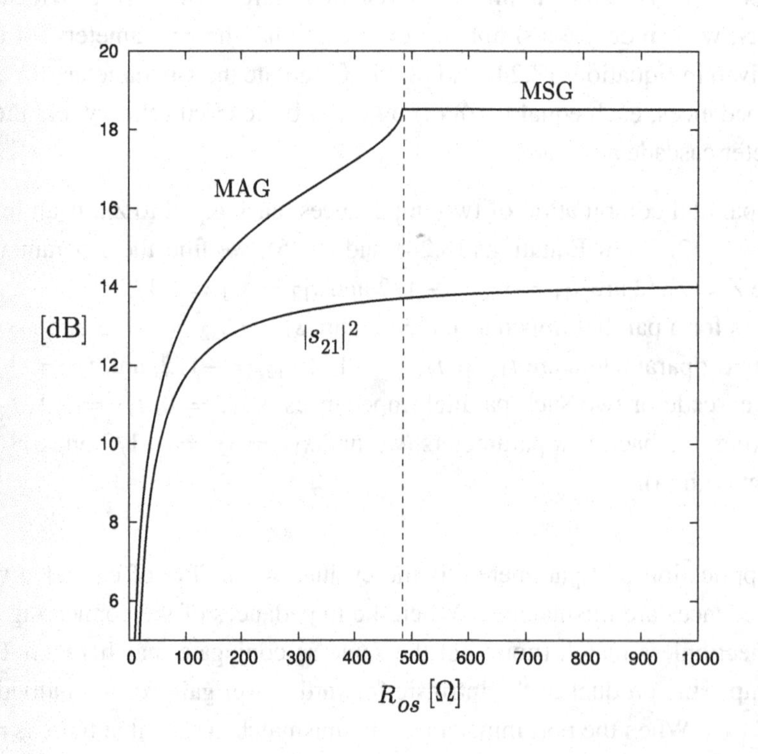

Figure 8.15. Gain as a function of the value of the stabilization resistor in Example 8.13.

where R_{os} is the value of the stabilization resistor. Since we would like to consider a range of values for R_{os}, calculation of the s-parameters for the transistor–resistor cascade is a good job for a computer program. The program steps through candidate values of R_{os}, and for each value does the following: (1) Computes the s-parameters of the resistor two-port, (2) computes the s-parameters of the transistor-resistor cascade, (3) computes the stability parameters $|\Delta|$ and K of the cascade.

Using the program, the cascade is found to be unconditionally stable ($|\Delta| < 1$, $K > 1$) for $R_{os} \leq 483\ \Omega$. So how small should R_{os} actually be? From the perspective of stability, there is not much to be gained once the threshold of unconditional stability is crossed, allowing of course ample margin to accommodate component tolerances. Furthermore, R_{os} dissipates increasing power as its value is reduced. The impact on gain is easily determined by having the computer program also calculate maximum TPG (either MSG or MAG, whichever is appropriate) and perhaps also intrinsic forward gain ($|s_{21}|^2$) for the cascade. The result is shown in Figure 8.15. As expected, gain decreases with decreasing R_{os}. Thus, the appropriate choice of R_{os} should account for the gain required from the amplifier.[6]

8.7 DIFFERENTIAL CIRCUITS

So far in this chapter we have made an implicit assumption that signals are *single-ended*: That is, voltages are defined as a potential differences with respect to a datum or "ground", as shown in Figures 8.16(a) and 8.17(a). In circuits that are designed to process single-ended signals, currents emerge from sources, are processed, and return to the source via ground (hence, the term "circuit"). But this is not the only possibility.[7]

A prominent alternative to single-ended design is *differential* design, as illustrated in Figures 8.16(b) and 8.17(b). In this scheme, sources and loads are arranged symmetrically around the datum, voltage signals are defined as the potential differences between symmetric nodes (as opposed to being defined with respect to the datum), and an explicit ground is not necessary. Independence from an explicit ground is perhaps the most important distinction between single-ended and differential circuits: In differential circuits, the current return path is an integral component of the signal, whereas in a single-ended circuit, currents are merely "sunk" at a ground node which is assumed to be at the same potential as the ground node nearest the source, often with no explicitly-defined return path. As a result, currents in a single-ended circuit can return to the source by *any* path with sufficiently low potential difference. If not carefully managed, these paths may be prone to interference pick-up and can lead to instability, resulting in some unpleasant surprises.

[6] One final consideration: Since R_{os} connects collector to emitter, R_{os} will change the operating point unless it is very large. One way to mitigate this problem is to place a DC blocking capacitor in series with R_{os}.

[7] If you read Chapter 2, then this would be a good time to review Section 2.6.2.

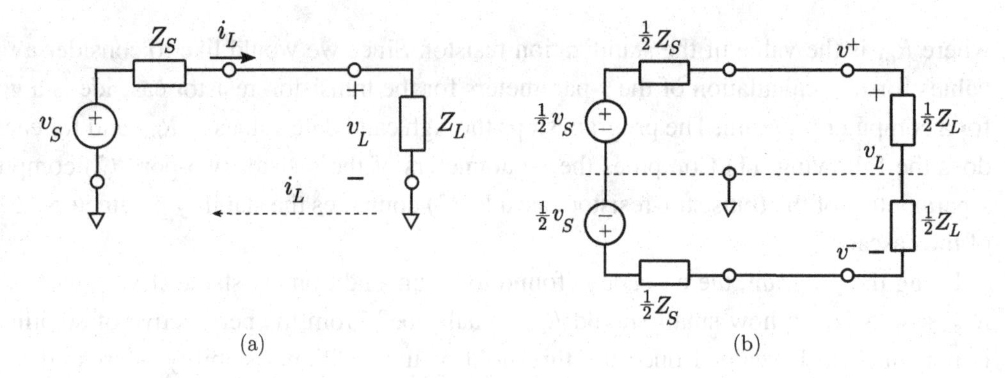

Figure 8.16. Single-ended (a) vs. differential (b) circuit topology. The loads in (b) can, of course, be combined into a single load Z_L.

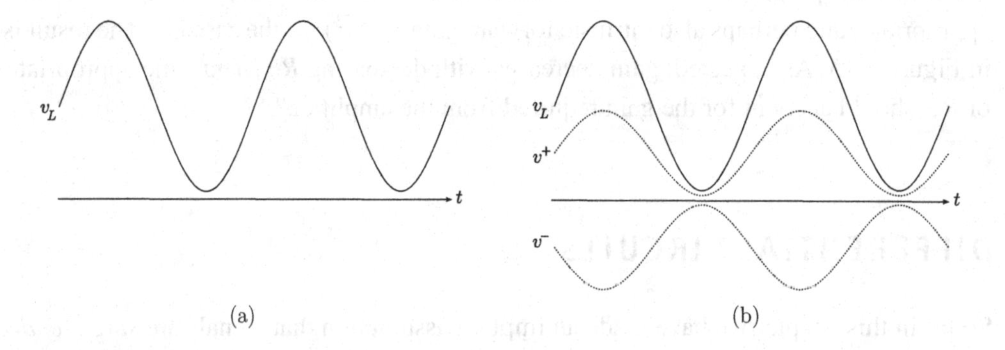

Figure 8.17. A DC-offset sinusoid represented as (a) a single-ended signal (v_L measured with respect to a physical datum), and (b) a differential signal ($v_L = v^+ - v^-$).

8.7.1 Applications of Differential Circuits

There are four principal applications for differential circuits in radio engineering. The first pertains to sources and loads which are inherently differential: prominent among these are dipole antennas. Examination of Figures 2.1 and 2.15 reveals that dipoles are inherently differential (in reception, as sources; in transmission, as loads), whereas monopoles are inherently single-ended, with the ground plane serving as both datum and ground.[8] We can expect that there will be some potential for mischief when we attempt to connect a dipole to a single-ended radio circuit without accounting for the differential nature of the dipole. The same distinction applies to transmission lines. In particular, parallel conductor transmission line is intrinsically differential, whereas coaxial transmission line (in which the shield is typically used as both datum and ground) is intrinsically single-ended. For convenience, the situation is summarized in Table 8.2.

[8] At this point it should be noted that the single-ended equivalent circuits shown in Figures 2.4 and 2.6 are technically correct for the monopole and other single-ended antennas, but should properly be shown as differential circuits when used to model dipoles and other differential-mode antennas. Nevertheless, there is no harm done in using the simpler single-ended models for the purposes of Chapter 2.

Table 8.2. **A summary of single-ended versus differential circuits.**

	Single-Ended	Differential
Also known as...	"Unbalanced"	"Balanced"
Datum	Explicit (usually ground)	Implicit
Antenna example	Monopole	Dipole
Transmission line example	Coaxial	Parallel conductor
Common mode immunity	Low	High

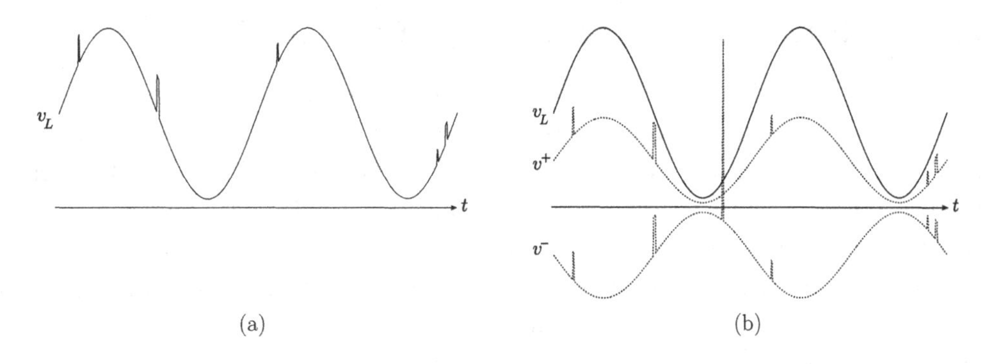

Figure 8.18. Following up the example from Figure 8.17: Effect of impulsive common-mode interference on single-ended vs. differential circuits.

Another common application for differential circuits is to improve immunity to *common mode interference*. This is most easily explained by example. Consider a circuit in close proximity to another circuit which is radiating electromagnetic interference (a common offender: switching-mode DC power supplies). This interference adds to the desired signal. However, the effect is quite different for single-ended and differential circuits, as illustrated in Figure 8.18. The single-ended circuit has no immunity to the interference. In the differential circuit, the common mode interference is effectively canceled in the final interpretation of the signal as the difference between the conductor potentials. The immunity offered by differential circuits to common-mode interference becomes essential in compact designs, and especially in designs in which disparate circuits must coexist on a single integrated circuit.

A third application for differential circuits is ability to handle higher power. All devices used in radio receive and transmit circuits are ultimately limited in the amount of power they can process. In both transmitters and receivers, the limits may be due to non-linearity; for example, the inability of an amplifier to deliver a specified gain once the signal power exceeds a certain threshold. In transmitters, the situation is sometimes further aggravated by heating due to large currents encountering the unavoidable intrinsic resistance of materials. In a differential circuit, power is distributed across two identical circuits, so in principle a differential circuit can handle twice the power at the same level of performance as a single-ended circuit using identical hardware.

A fourth application for differential circuits is transmission of signals in situations in which the current return path between single-ended circuits becomes ambiguous. This commonly occurs on printed circuit boards, in which the desired (or assumed) return path is intended to be through a ground layer, but in which other paths exist over which the impedance is sufficiently low to attract significant amounts of current. In particular: When the intended connection to a ground plane does not have sufficiently low impedance, a significant fraction of current may return via a highly undesirable path, such as through a power supply. This problem is largely mitigated when the signal is conveyed differentially because the current return path is intrinsic to the signal path.

8.7.2 Interfaces between Differential and Single-Ended Circuits

Since both single-ended and differential circuits are commonly used, one commonly encounters the problem of how to connect one to the other. A direct connection which simply ignores the difference between the two types of circuits is not necessarily out of the question: For example, dipole antennas are often connected directly to single-ended radio circuits. However, such connections typically yield degraded performance and often unintended consequences: In the case of dipoles, such connections result in modified distributions of current on the dipole, and the circuit itself radiates and receives as part of the dipole. (For more on this topic, see Sections 2.6.2 and 12.7.)

In general, an explicit conversion between signal modes is desirable. A device which performs this conversion is known as a *balun*. Perhaps the simplest device that can be used as balun is a transformer, as shown in Figure 8.19(a). In this scheme, the single-ended and differential ports are electrically isolated, connected only by the magnetic flux linking the coils. Thus the explicit datum required by the single-ended port can be implemented with no effect on the differential port. This particular scheme is often of interest since it can be used to simultaneously change impedance – more on that in Chapter 9.

Another way to use a transformer as a balun is as shown in Figure 8.19(b). In this scheme the transformer is wound so that the magnetic field associated with the current at the single-mode port excites a current in the other coil that flows in the opposite direction, yielding a differential current at the opposite port. Variations on this idea are the *Guanella* and *Ruthroff* baluns, which use multiple transformers to simultaneously perform impedance transformation.

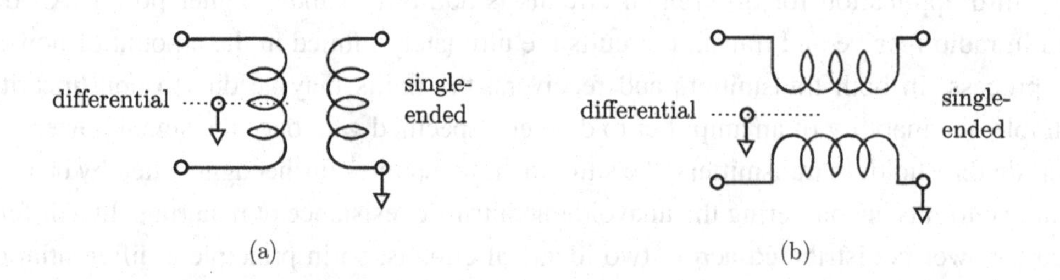

Figure 8.19. Two ways a transformer can be used as a balun.

The above schemes can also be implemented using transmission line techniques in lieu of transformers. Transmission lines yield a variety of additional strategies for converting between single-ended and differential signals; see Section 12.7 for more on this topic.

Finally, we note one other approach: The *LC balun* comprises of discrete inductors (Ls) and capacitors (Cs) arranged to implement the necessary phase shifts. A useful introduction to this class of devices can be found in [43, Sec. 15.3]. The advantage of this approach is that the balun may be very small. The disadvantages of this approach are (1) it is difficult to achieve large bandwidth and (2) discrete Ls and Cs of sufficient quality may be difficult to find for frequencies in the UHF band and above.

8.7.3 Analysis of Differential Circuits

Given the advantages and prevalence of differential circuitry in radio systems, it seems reasonable to expect that the single-ended s-parameter methods presented in this chapter could be extended to deal with differential two-ports, as well as to combinations of differential and single-ended two-ports. In fact, the necessary "mixed-mode" s-parameter analysis is a relatively recent development [7]. In practice, it is often possible to avoid the complexity of mixed-mode analysis. Differential circuits are frequently designed and analyzed first as single-ended circuits, and then converted to equivalent differential forms as a final step. Interfaces between single-ended and differential circuits, such as baluns, can often be accommodated by initially assuming ideal conversion between single-ended and differential signal modes. Explicit mixed-mode analysis becomes important when the conversion between single-ended and differential modes cannot reasonably be approximated as ideal. This is sometimes the case when the mode conversion is considered over very large bandwidths, and in integrated circuit implementations.

Problems

8.1 Derive expressions for the s-parameters of a parallel impedance Z. See Equations (8.24) and (8.25).

8.2 Determine the s-parameters for a lossless transmission line having characteristic impedance equal to the reference impedance, and with length equal to (a) one-fourth wavelength, (b) one-half wavelength, and (c) one wavelength.

8.3 Derive expressions for the s-parameters of a two-port consisting of a length-l section of transmission line with characteristic impedance Z_c and propagation constant $\gamma = \alpha + j\beta$. Do **not** assume $Z_c = Z_0$.

8.4 Determine expressions for the s-parameters of a unilateral amplifier with voltage gain A_V and which is reflectionless at input and output.

8.5 An isolator is a device that allows power to flow only from input to output. Determine the s-parameters for an ideal (lossless and reflectionless) isolator.

8.6 A 3-dB attenuator is a reflectionless, lossy, reciprocal device which reduces the power at the input by 3 dB when terminated into the reference impedance Z_0 on both ports. (a) Based solely on this much information, derive the *s*-parameters of this two-port. (b) This two-port can be implemented as a "pi"-configuration of resistors in which the series resistor is $0.35Z_0$ and the shunt resistors are $5.83Z_0$. Show that this two-port is reflectionless using traditional circuit theory, leaving Z_0 as a free parameter. (c) Continuing, show that this implementation is also a 3-dB attenuator. (d) Determine the TPG of this two-port when $Z_0 = 50\ \Omega$, using Equation (8.41). (e) Repeat (d) for this same circuit, but now for a load impedance of 100 Ω.

8.7 Derive Equation (8.36), which gives the reflection coefficient (with respect to Z_0) looking into the output of a two-port whose input is terminated into a source impedance Z_S.

8.8 Active two-ports are bilateral and can sometimes be approximated as unilateral. Can *passive* devices be unstable? Unilateral? Explain.

8.9 Find the simplest possible expression for TPG when the input and output impedances are matched to the reference impedance, assuming the two-port is stable. Under what conditions is this the maximum possible TPG?

8.10 Repeat Example 8.4, now for a shunt (parallel) impedance equal to $2Z_0$.

8.11 Repeat Examples 8.5 and 8.6, now for a series impedance equal to $2Z_0$.

8.12 Show that the TPG of a embedded unilateral two-port is maximized by setting $\Gamma_S = s_{11}^*$ and $\Gamma_L = s_{22}^*$. Assume unconditional stability.

8.13 For the Avago AT-41511 at 1.5 GHz: (a) Using the Rollett stability test, is the device unconditionally stable? (b) Determine the centers, radii, and stable regions of the stability circles. (c) Compute MAG or MSG as appropriate. (d) Repeat (a) using the Edwards–Sinsky criterion.

8.14 Repeat Example 8.9 using the Edwards–Sinsky criterion. (No need to compute stability circles this time.)

8.15 Calculate the *t*-parameters for each of the cables considered in Problem 8.2.

8.16 Determine the *s*-parameters for the series connection of the three cables considered in Problem 8.2. Do this first using the method of cascaded *t*-parameters. Then check your work by computing the answer starting with a single cable equal in length to the sum of the lengths of the three original cables.

8.17 A common problem when measuring the *s*-parameters of a two-port is accounting for the cables that connect the device to the instrument making the measurements. Consider a two-port consisting of a shunt impedance equal to $Z_0 = 50\ \Omega$. The input of the two-port is connected directly to the instrument, and the output connected to the instrument using a lossless coaxial cable having $\beta l = \pi$ rad. What *s*-parameters does the instrument measure? Explain and demonstrate a technique for recovering the two-port *s*-parameters from the combined two-port + cable *s*-parameters.

8.18 Repeat Example 8.13 (resistive loss stabilization), this time for a transistor at 500 MHz with the following *s*-parameters: $s_{11} = 0.63\angle{-148^\circ}$, $s_{21} = 8.26\angle 97^\circ$, $s_{12} = 0.06\angle 136^\circ$, and $s_{22} = 0.36\angle - 67^\circ$. Also, choose a value for a capacitor to be inserted in series with the stabilization resistor to preserve the DC bias.

9 Impedance Matching

9.1 INTRODUCTION

In Chapter 8 we noted the central role two-port theory plays in RF circuit design, and especially the role of reflection coefficients in determining the stability and gain of two-ports. Reflection coefficients were introduced as surrogates for source and load impedance (recall Equations (8.32) and (8.37)), facilitating the use of scattering parameters. In this chapter we consider *impedance matching*, which is the process of converting impedances (equivalently, embedded reflection coefficients) from a given value to some other value.

Applications for impedance matching include (1) maximizing power transfer between two-ports, (2) eliminating reflections (not the same thing!), and (3) converting from a given impedance to another impedance necessary for setting the stability and/or gain of a two-port. In each of these cases, the problem is essentially as shown in Figure 9.1: We wish to design a passive two-port that converts the output impedance of the source, $Z_S = R_S + jX_S$, into the desired input impedance for the load, $Z_L = R_L + jX_L$. Usually we would like to achieve this with a TPG as close to 1 as possible, and often there is a frequency span over which a minimum TPG which must be achieved. At the design frequency, the nominal embedded input impedance Z_{in} of the matching two-port is either Z_S^* for maximum power transfer or Z_S for reflectionless matching. Similarly, the nominal embedded output impedance Z_{out} of the matching two-port is either Z_L^* or Z_L at the design frequency.

The organization of this chapter is as follows. In Section 9.2 we consider some rudimentary approaches to impedance matching, including resistive matching, transformer matching, and reactance canceling. Section 9.3 introduces narrowband matching using two discrete reactances in an "L" configuration, which is useful both as a standalone technique as well as a building block for more sophisticated techniques. Section 9.4 considers the issue of bandwidth, and introduces the concept of quality factor to provide some insight into how the bandwidth of matching circuits is determined and can be manipulated. In Section 9.5 we combine the concepts of previous sections to develop impedance matching circuits having bandwidth which can be wider or narrower than that obtained by using discrete two-reactance matching. Section 9.6 explains how to obtain differential versions of the single-ended circuits developed in earlier sections. In Section 9.7 we introduce distributed matching devices, which use transmission lines in lieu of discrete reactances and thereby find application at higher frequencies, where discrete inductors and capacitors become difficult to use. The chapter concludes with a discussion of impedance inverters (Section 9.8), a class of impedance matching devices which have some useful special applications and can be implemented in either discrete or distributed form.

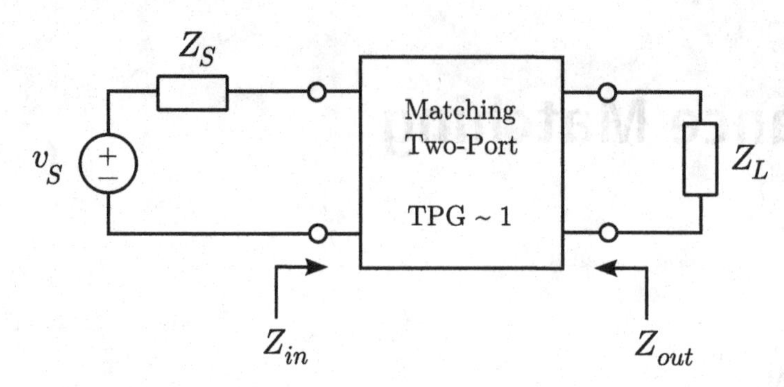

Figure 9.1. A summary of the impedance matching problem. $Z_S = R_S + jX_S$ and $Z_L = R_L + jX_L$. For conjugate matching, $Z_{in} = Z_S^*$ and $Z_{out} = Z_L^*$. For reflectionless matching, $Z_{in} = Z_S$ and $Z_{out} = Z_L$.

9.2 SOME PRELIMINARY IDEAS

Before getting started in earnest, let us consider the pros and cons of some simple methods that we might consider to address the problem posed in Figure 9.1.

Matching Real Impedances Using a Series or Parallel Resistance. Let us say $X_S = X_L = 0$ (i.e., the source and load impedances are real-valued), as shown in Figure 9.2. If $R_S > R_L$, a possible solution is a series resistance $R_S - R_L$, resulting in $Z_{in} = R_S$. This is shown in Figure 9.2(a). Similarly, if $R_S < R_L$ (Figure 9.2(b)), then a parallel resistance can be used to reduce R_L to R_S. These methods are rarely used because the matching two-port dissipates power and is therefore lossy with TPG much less than 1. On the other hand, this approach is worth keeping in mind because it has one extremely useful feature: It has infinite bandwidth.[1] In other words, if R_S and R_L are constant with frequency, then the associated matching two-port will have TPG which is constant with frequency.

Matching Real Impedances Using a Transformer. Let us say $X_S = X_L = 0$ and you are serious about achieving TPG≈1. In this case you might consider using a transformer, as shown in Figure 9.3. A transformer with a turns ratio of N has an impedance ratio of N^2, so the matching two-port could be a transformer with an input-to-output turns ratio of $\sqrt{R_S/R_L}$. Transformations to impedances that are both lower and higher are possible, and are reciprocal. Furthermore, this approach is nominally lossless, since no resistance is required for basic function (however, more on this in the next paragraph). Even better, transformers are capable of very large bandwidth. Yet another advantage of this approach has to do with single-ended to differential mode conversion: If you need to do this, you could use the same transformer

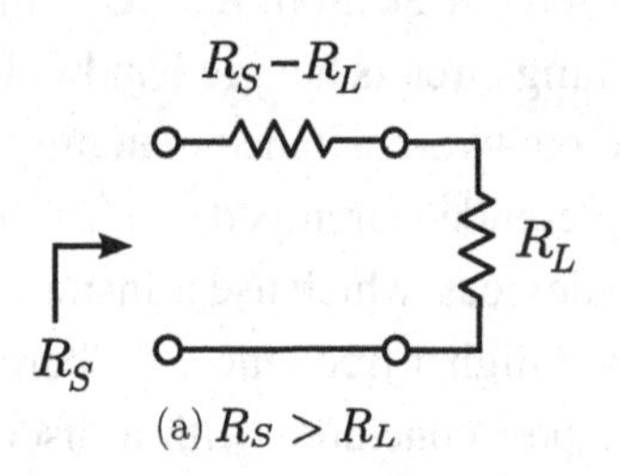

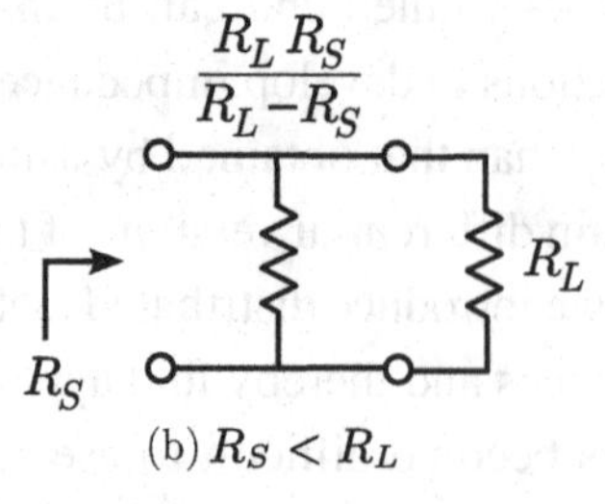

Figure 9.2. Lossy matching of real impedances using a series or parallel resistance. Note that these matching two-ports are not reciprocal; i.e., only the left-facing port is matched.

[1] Well, bandwidth at least as large as that of the resistors, anyway.

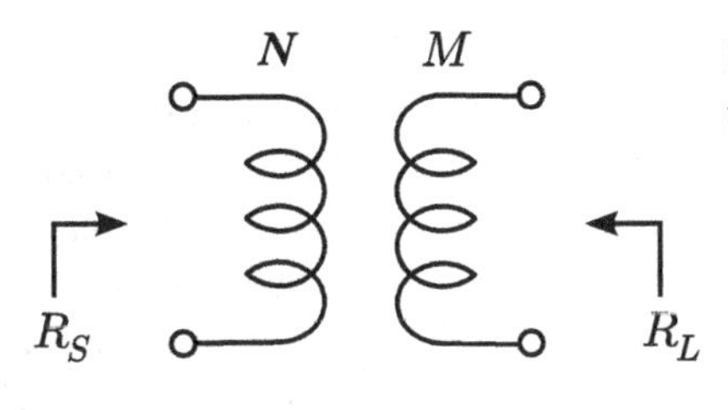

Figure 9.3. Matching real impedances using a transformer having a turns ratio of $N/M = \sqrt{R_S/R_L}$.

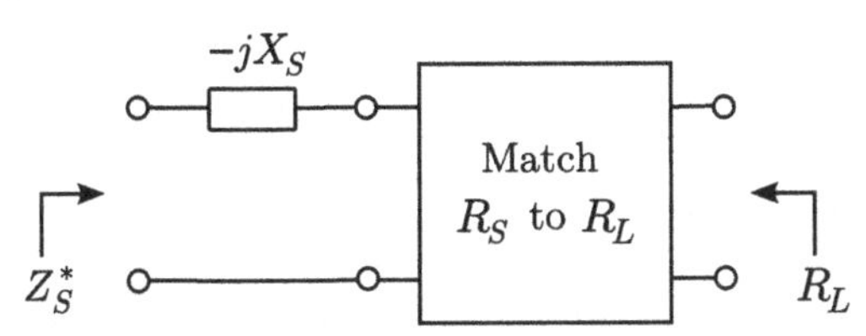

Figure 9.4. Matching a complex source impedance by first canceling the imaginary part of the source impedance.

to accomplish both the mode conversion (as explained in Section 8.7.2) and the impedance transformation simultaneously.[2]

The principal disadvantages of transformers in this approach are relatively large size, non-negligible loss (although not nearly as bad as resistive matching), and poor performance near DC (due to flux saturation) and above VHF (due to capacitive coupling within coils). Transformers are large in the sense that they require at least two coils and typically also a ferrite core in order to contain the magnetic flux that links the coils and to keep the magnetic flux from interacting with nearby devices. To the extent that the flux is not completely contained, or to which the ferrite core dissipates magnetic flux, the transformer will be lossy. Loss on the order of 0.5–1 dB is typical for transformers used in broadband RF applications.

Canceling Reactance. A third strategy worth considering is impedance matching in two steps. The conjugate matching version of this strategy is shown in Figure 9.4: First, a series reactance equal to $-X_S$ to cancel the imaginary part of Z_S, followed a "real-to-real" conversion of R_S to R_L. (Alternatively, the transformer can be used to transform Z_S to R_L plus some imaginary component, followed by a series reactance to cancel the imaginary-valued component.) This approach is particularly attractive if $R_S \cong R_L$; then all that is needed is a single reactance to cancel X_S and no other devices are necessary.

The primary disadvantage of this approach is that it has narrow bandwidth. This is because reactances nearly always increase with increasing frequency. This is demonstrated by example in Figure 9.5: The reactance of a capacitor ($-j/\omega C$) increases from $-\infty$ to 0 as frequency increases from 0 to ∞, whereas the reactance of an inductor ($+j\omega L$) increases from 0 to $+\infty$ over the same range. Therefore if you use a capacitor to cancel the reactance of an inductor, or vice versa, the sum reactance can be exactly zero only only at one frequency.

9.3 DISCRETE TWO-COMPONENT ("L") MATCHING

For reasons listed in the previous section, transformers (with or without canceling of source reactance) are often unacceptable as impedance matching devices. This is the case even

[2] It is worth knowing that similar tricks are possible using transmission lines, if narrow bandwidth is OK: See for example Figure 12.27 and associated text.

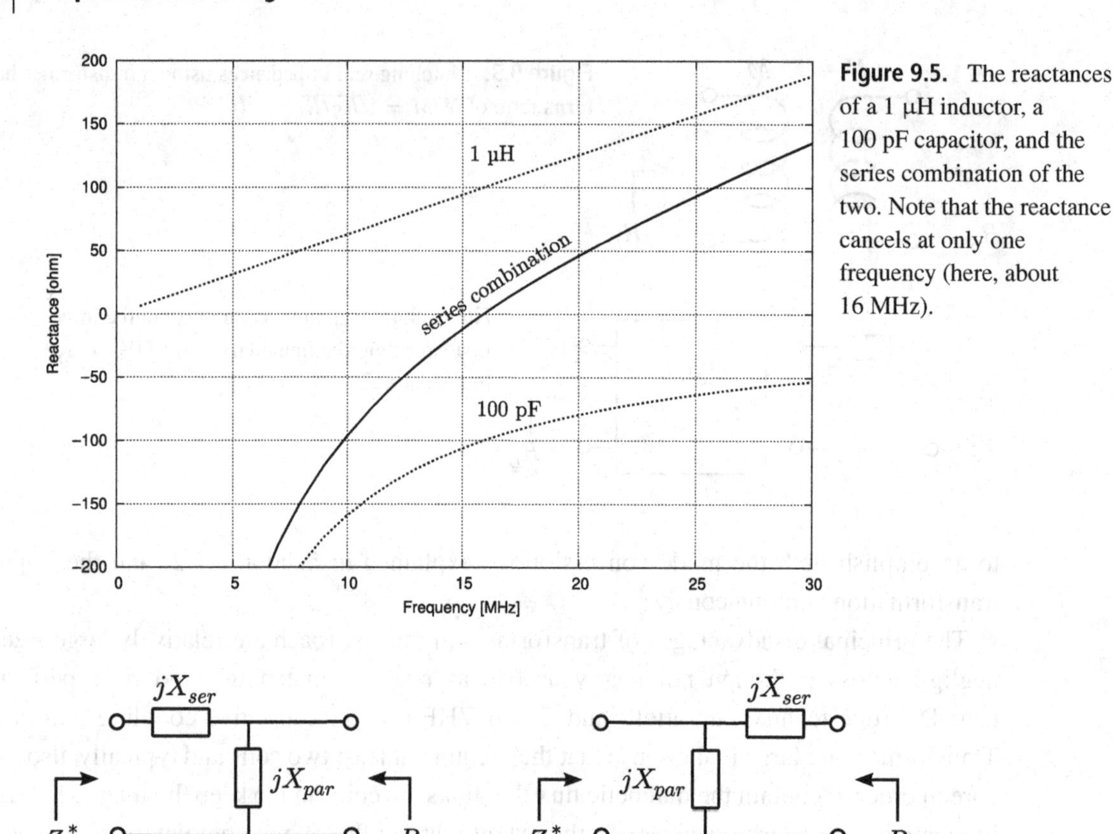

Figure 9.5. The reactances of a 1 µH inductor, a 100 pF capacitor, and the series combination of the two. Note that the reactance cancels at only one frequency (here, about 16 MHz).

Figure 9.6. Options for discrete two-component "L" matching.

though they achieve the desired impedance transformation with less loss than resistive matching, and even though they can deliver very large bandwidth for real-to-real valued impedance transformation. The alternative is to design two-port circuits that explicitly meet the requirements inferred by Figure 9.1 using combinations of discrete or distributed reactances. In this section, we describe the design and performance of the simplest of this class of two-ports: The discrete two-component match, also known as the "L-match".[3]

To simplify subsequent discussion, we shall assume $X_L = 0$; that is, the matching is from a complex- (or real-) valued impedance to a real-valued impedance. This is not a fundamental limitation of the L-match, but greatly simplifies the analysis and addresses the vast majority of practical applications. In cases where complex-to-complex impedance matching cannot be avoided, it is possible to use the complex-to-real L-match technique first to match the real part of the load impedance, and then to simply add a series reactance to match the imaginary part of the load impedance.

There are two possible L-match schemes, both shown in Figure 9.6. In the "series first" scheme, we select X_{par} such that $\text{Re}\left\{jX_{par} \parallel R_L\right\} = R_S$, and then use X_{ser} to match the remaining reactance. The resulting design equations for conjugate matching are:

[3] Note that the symbol "L" is used in this chapter to denote both the two-element matching circuit topology as well as inductance. Which should be clear from context.

$$X_{par} = \pm\sqrt{\frac{R_L^2 R_S}{R_L - R_S}} \tag{9.1}$$

$$X_{ser} = -X_S - \text{Im}\left\{\left(jX_{par}\right) \parallel R_L\right\} \tag{9.2}$$

Note that we only get valid solutions only for $R_L \geq R_S$; i.e., when the parallel reactance is adjacent to the larger resistance. Also, we see there are two possible solutions, associated with the plus and minus signs in Equation (9.1). To complete the solution, reactance is converted into capacitance and inductance in the usual way:

$$C = -\frac{1}{\omega X}\ , X < 0 \text{ and} \tag{9.3}$$

$$L = +\frac{X}{\omega}\ , X > 0\ . \tag{9.4}$$

In the "parallel first" scheme (Figure 9.6(b)), we select X_{par} such that
$\text{Re}\left\{jX_{par} \parallel Z_S\right\} = R_L$, and then use X_{ser} to cancel the remaining reactance. The resulting design equations are:

$$\left(R_L - R_S\right) X_{par}^2 + \left(2X_S R_L\right) X_{par} + R_L \left|Z_S\right|^2 = 0 \tag{9.5}$$

which is a quadratic equation that is easily solved for X_{par}, and

$$X_{ser} = -\text{Im}\left\{\left(jX_{par}\right) \parallel Z_S\right\} \tag{9.6}$$

EXAMPLE 9.1

A particular electrically-short monopole (see Section 2.8) has a self-impedance $Z_A = 1.2 - j450.3\ \Omega$ at 30 MHz. Identify all possible discrete two-component conjugate matches to 50 Ω.

Solution: Recall from Chapter 2 that a monopole can be modeled as a source using an equivalent circuit exactly in the form depicted in Figure 9.1. Thus, $Z_S = 1.2 - j450.3\ \Omega$ and $R_L = 50\ \Omega$. $R_L \geq R_A$, so there are two valid series-first solutions as well as two parallel-first solutions. All four solutions are shown in Figure 9.7.

If X_S is also zero (i.e., real-to-real matching) the method simplifies considerably. In this case Equations (9.1), (9.2), (9.5), and (9.6) reduce to a single pair of design equations that can be summarized as follows:

$$X_{par} = \pm\sqrt{\frac{R_{hi}^2 R_{lo}}{R_{hi} - R_{lo}}} \tag{9.7}$$

$$X_{ser} = -\text{Im}\left\{\left(jX_{par}\right) \parallel R_{hi}\right\} \tag{9.8}$$

where R_{hi} is the larger of R_S and R_L, R_{lo} is the smaller of R_S and R_L, and the parallel reactance (X_{par}) is adjacent to the larger resistance R_{hi}. Thus in the real-to-real case we always have two possible solutions, each of which are either series-first or parallel-first depending on which places the parallel reactance adjacent to the larger resistance.

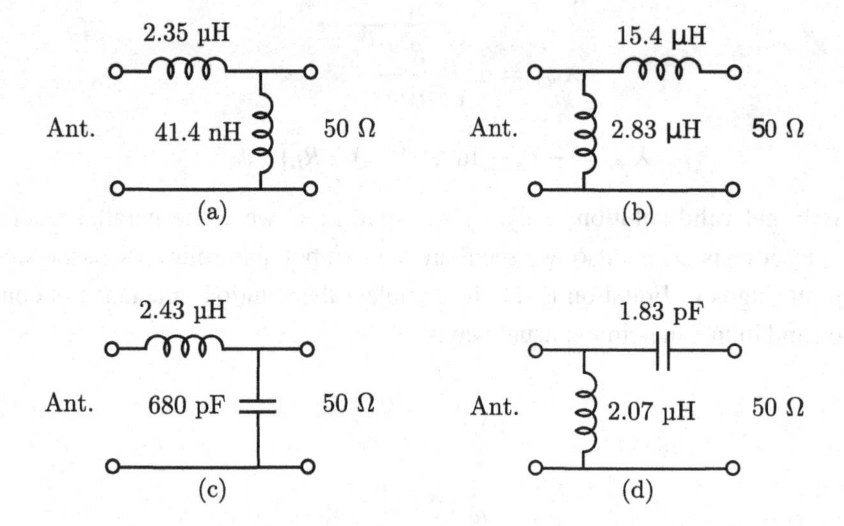

Figure 9.7. Options for conjugate matching of $1.2 - j450.3\ \Omega$ to 50 Ω at 30 MHz (Example 9.1).

Equations (9.7) and (9.8) can be simplified further using the following definition:

$$Q \equiv \sqrt{\frac{R_{hi}}{R_{lo}} - 1} \tag{9.9}$$

The parameter "Q" is known as the *quality factor*, but for the moment we consider this to be merely a convenient shorthand that allows Equation (9.7) to be written

$$X_{par} = \pm \frac{R_{hi}}{Q} \tag{9.10}$$

and, after a little algebra, Equation (9.8) is found to be

$$X_{ser} = \mp R_{lo} Q \tag{9.11}$$

where the upper and lower signs correspond to the two possible solutions. Note that X_{par} and X_{ser} have opposite sign; therefore the two possible matching circuit topologies are parallel inductor with series capacitor, or parallel capacitor with series inductor, with no other possibilities. Equations (9.9)–(9.11) constitute the simplest possible method for determining both discrete two-reactance solutions to a real-to-real impedance matching problem.

EXAMPLE 9.2

Identify all possible discrete two-reactance matches between 5000 Ω and 50 Ω for a frequency of 75 MHz.

Solution: Since this is a "real-to-real" matching problem, the Q-parameter method suggested above is available. Using $R_{hi} = 5000\ \Omega$ and $R_{lo} = 50\ \Omega$, we find $Q \cong 9.95$ and obtain the two solutions shown in Figure 9.8.

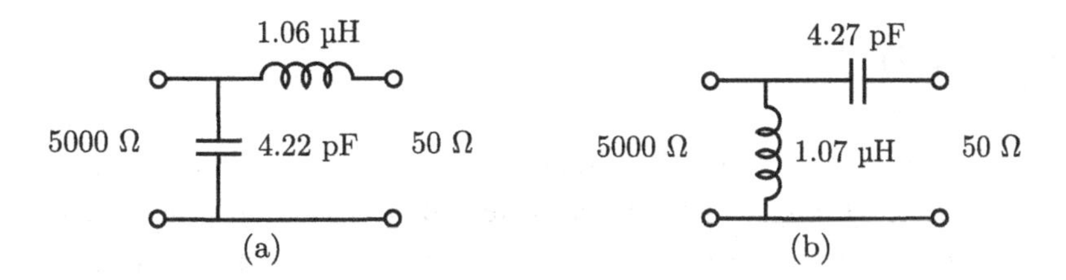

Figure 9.8. Options for matching 5000 Ω to 50 Ω at 75 MHz (Example 9.2).

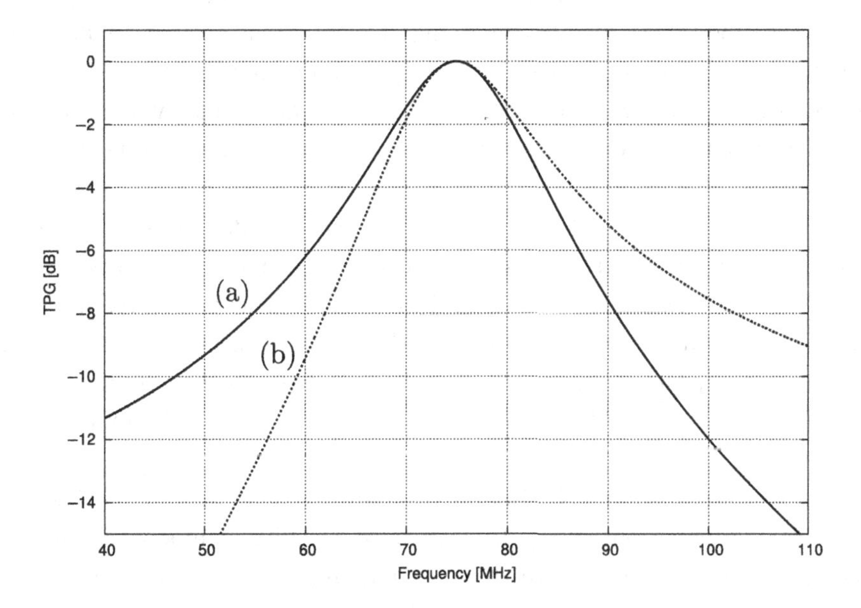

Figure 9.9. Frequency response of the matching circuits obtained in Example 9.2 (see also Figure 9.8). Curve (a) corresponds to the parallel capacitor solution, and therefore has lowpass response. Similarly curve (b) corresponds to the parallel inductor solution, and therefore has highpass response.

How should one choose from among the available matching solutions? In the case of real-to-real impedance matching, one distinction between the two solutions is that one has "highpass" frequency response and the other has "lowpass" frequency response. This is apparent from inspection of the circuits shown in Figure 9.8, and Figure 9.9 underscores the point.

As hinted at in Section 9.2, this simple "Q-parameter" technique can also be applied in the more general case in which the source impedance is complex-valued. This can be done by either first canceling X_S and then solving the remaining "real-to-real" problem, or by matching the real parts of the source and load impedances first and then canceling the remaining reactance. One of these approaches will always result in a design in which the reactance-canceling component is either in series or in parallel with another reactance, and thus can be combined to reduce the matching circuit to two components. In fact, this solution will always be one of the two or four solutions identified by the "direct" approach using Equations (9.1), (9.2), (9.5), and (9.6).

9.4 BANDWIDTH AND Q

We observed in the previous sections that impedance matching is inherently sensitive to frequency; i.e., it is quite simple to achieve a perfect match at a specified frequency, but the performance of the resulting design degrades at higher or lower frequencies. In other words, the matching techniques considered so far are *narrowband*.[4] Sometimes narrowband matching circuits are acceptable, and perhaps even preferred since they also provide a degree of selectivity. Often, however, the bandwidth obtained using narrowband matching techniques is too small. In some other cases, the bandwidth may be larger than desired. In this section we develop some theory pertaining to the bandwidth of matching circuits, and in Section 9.5 we apply this theory to develop matching circuits with larger or smaller bandwidth. In Chapter 13 we will take up this topic again, considering the problem of how to achieve a specific frequency response.

The first order of business should be to establish a rigorous definition of bandwidth. In fact, it is useful to have two definitions. The first and very common definition of bandwidth is *half-power bandwidth* B_{HP} (also known as "3 dB bandwidth"): This is the separation between frequencies at which the transducer power gain of the two-port has degraded by a factor of 2 relative to the nominal (peak) value. As an example, half-power bandwidth for Example 9.2 (Figure 9.9) is 15.2 MHz and 15.5 MHz for the parallel capacitor and parallel inductor versions, respectively.

For the purposes of this section, it is useful to employ a second definition. This definition assumes a passive two-port consisting of reactive or resistive components, driven by a sinusoidal source, and operating in steady-state conditions (i.e., any transient response has reduced to an insignificant level). Energy injected into such a two-port by the source can do one of two things: It can escape, or it can be stored. The two possible methods of energy escape are dissipation (conversion to heat through resistance) and transfer through the output. Energy storage is by containment in the electric field of capacitors and/or the magnetic field of inductors. With this perspective, we can define a bandwidth B_Q as follows:

$$B_Q \equiv \frac{U_{es}/T}{U_{st}} \tag{9.12}$$

where the numerator is the escaped energy U_{es} per period T of the sinusoidal stimulus; and U_{st} is the energy stored, which is remains constant given the steady-state assumption made above.

How is this a definition of bandwidth? First, note that it is dimensionally correct, having units of 1/time. Next, note that if the two-port is unable to store energy (e.g., perhaps because it contains no capacitors or inductors), then $U_{st} = 0$ and therefore $B_Q \to \infty$. This is precisely the expected behavior: For example, impedance matching using a resistive voltage divider has no energy storage and has bandwidth (by any definition) which is essentially infinite. If on the other hand $U_{st} > 0$ but $U_{es}/T = 0$ for $T > 0$, then there is no energy escape; only energy

[4] Here, we use the term "narrowband" to refer to circuits designed for nominal performance at a specific frequency, ignoring all other frequencies. Often, but not always, such circuits also have very narrow bandwidth. A counter-example of narrowband design leading to large bandwidth is Problem 9.5 (75 Ω to 50 Ω match).

storage. In this case, the two-port is "frozen" and there is no possibility of passing a time-varying signal through it: $B_Q \to 0$ and bandwidth (by any definition) is essentially zero. In the intermediate case in which both $U_{es} > 0$ and $U_{es}/T > 0$, the rate at which a voltage or current signal passing through the two-port can be varied (another way to say "bandwidth") increases with increasing U_{es}/T. Thus Equation (9.12) is clearly a definition of bandwidth.

This bandwidth is clearly not identical to half-power bandwidth, however, and in particular differs in that B_Q is an evaluation of bandwidth specific to a frequency $f = 1/T$. Said differently, $B_Q(f)$ is a assessment of bandwidth based on the change in TPG with changing frequency in the neighborhood of $f = 1/T$.

This concept turns out to be closely related to the concept of *resonance*. For the purposes of this section, resonance is a state in which all reactances internal to a lossless two-port cancel, such that the input and output impedances become completely real-valued. In particular: For two-component discrete "L" matches between real valued impedances, there is precisely one resonance, and it occurs at the design frequency. f_0. Let this frequency be expressed as the angular frequency $\omega_0 = 2\pi f_0$, having units of rad/s.

The ratio of ω_0 to $B_Q(f_0)$ is a unitless quantity known as the *quality factor Q*:

$$Q \equiv \frac{\omega_0}{B_Q(f_0)} = \omega_0 \frac{U_{st}}{U_{es} f_0} = 2\pi \frac{U_{st}}{U_{es}/\text{cycle}} \tag{9.13}$$

The quality factor quantifies the steady-state energy storage of a two-port, relative to the energy escaped in one period ("cycle") of the sinusoidal stimulus, evaluated specifically at the indicated frequency. Two-ports that efficiently store energy have $Q \gg 2\pi$; whereas $Q < 2\pi$ means the two-port loses more energy per cycle than it is able to store. Because U_{es}/T has units of power and because steady-state conditions are assumed, Equation (9.13) is sometimes expressed as

$$Q = \omega_0 \frac{U_{st}}{P_{es}} \tag{9.14}$$

where $P_{es} \equiv U_{es} f_0$ is the time-average power escaping the two-port when operated at the indicated frequency.

The concept of quality factor has multiple applications in radio engineering. An important one is the characterization of the lossiness of practical inductors and cavity resonators. However, Q plays a distinct and important role in analysis and design of impedance matching circuits, as we shall now explain.

Let us begin by working out the Q for the matching circuit shown in Figure 9.8(a). First, recall that the capacitor and inductor in an LC circuit share the same stored energy, transferring all energy from inductor to capacitor and back, once per period. Therefore we can determine Q based on the energy storage in the series inductor only, and know that it is the same value obtained if we considered the parallel capacitor only, or the entire circuit at once. Further, we note that the inductor is in series with the load, so the same current I_L flows through both. Therefore $U_{st} = |I_L|^2 L$ (assuming RMS units), $P_{es} = |I_L|^2 R_L$, and we find

$$Q = \omega_0 \frac{|I_L|^2 L}{|I_L|^2 R_L} = \frac{\omega_0 L}{R_L} = \frac{X_{ser}}{R_L} \tag{9.15}$$

Remarkably, we have independently derived the "+"-signed case of Equation (9.11), in which Q was simply a parameter we defined to simplify the equations. The Q parameter of Section 9.3 is in fact the quality factor Q defined in Equation (9.13). Furthermore, according to Equation (9.9), Q depends only on R_{hi}/R_{lo} (that is, the ratio of the port impedances), and not directly on the reactances used in the two-port.

As an example, let us reexamine the solution to Example 9.2 (Figure 9.9). Here, $Q \cong 9.95$, so $B_Q = \omega_0/Q \cong 47.4$ MHz. This is quite different from B_{HP}, which we found to be 15.2 MHz and 15.5 MHz for the two possible solutions; this underscores the fact that these two definitions of bandwidth are measuring different things. Here one finds B_Q is closer to the 10-dB bandwidth. For sufficiently large Q (i.e., $Q \gg 2\pi$) one finds

$$B_{\text{HP}} \cong \frac{B_Q}{2} = \frac{\pi f_0}{Q} \quad \text{for } Q \gg 2\pi \tag{9.16}$$

However, in this example Q is not much greater than 2π, and therefore the estimate $B_{\text{HP}} \cong B_Q/2 = 23.7$ MHz is not very accurate. Nevertheless, these relationships are useful for making estimates of the bandwidth of matching circuits.

Since $B_Q \propto Q^{-1}$, it must also be true that the bandwidth of the matching two-port is inversely proportional to R_{hi}/R_{lo}. In other words, bandwidth is large if the impedance ratio is small (perhaps obvious since bandwidth should be infinite if the ratio is 1) and bandwidth is small if the impedance ratio is large.

With these insights in mind, we are now prepared to address the problem of how to increase the bandwidth of a matching two-port: The key is to decrease the Q of the two-port. Decreased Q can be achieved by increasing internal loss (since this increases U_{es}), or by decreasing U_{st}. Both methods have applications, but usually the latter is preferred since we usually prefer TPG ≈ 1 over the bandwidth of interest.

9.5 MODIFYING BANDWIDTH USING HIGHER-ORDER CIRCUITS

Section 9.3 presented the simplest possible scheme for lossless impedance matching, using two reactances in an "L" configuration. In Section 9.4 we found the half-power bandwidth of such a match was limited: This bandwidth is often found to be insufficient, especially when the impedance ratio (R_{hi}/R_{lo}) is large, resulting in high Q. When the impedance ratio is low, Q is low and bandwidth may be quite large. In some applications where this occurs, it is desirable to reduce bandwidth. A good example of this appears in Chapter 10, in which reducing the bandwidth of matching circuits used in transistor amplifiers can reduce the likelihood of undesirable oscillation. Thus, we find there are a variety of reasons why we might desire matching two-ports with bandwidth either larger or smaller than that provided by the two-reactance matches of the type considered in Section 9.3.

9.5.1 Increasing Bandwidth using Cascades of Two-Reactance Matching Circuits

Increasing or decreasing the bandwidth of a matching two-port can be accomplished with little or no reduction in nominal TPG using higher-order circuits; that is, using circuits containing additional discrete reactances. A simple but effective method is to implement the change in impedance in smaller steps, with each step implemented using a discrete two-reactance two-port of the type described in the previous section. This results in a cascade of two-reactance "L" two-ports having the so-called "ladder" topology. Typically the intermediate impedance R_{mid} is selected to be the geometrical mean of R_{hi} and R_{low}; i.e., $\sqrt{R_{hi}R_{lo}}$, which results in the two stages having the same Q. Because each stage implements a smaller change in impedance than a single two-port implementing the entire change at once, the former has smaller Q and thus larger bandwidth. Generally the overall bandwidth of a cascade of impedance transformations arranged in this way will also be greater than that of a single impedance transformation from the initial to final value.

EXAMPLE 9.3

Recall in Example 9.2 we designed matching circuits that transformed $R_{hi} = 5000\ \Omega$ to $R_{lo} = 50\ \Omega$ at 75 MHz using two discrete reactances in an "L" configuration. Design a two-stage "L" match with increased bandwidth.

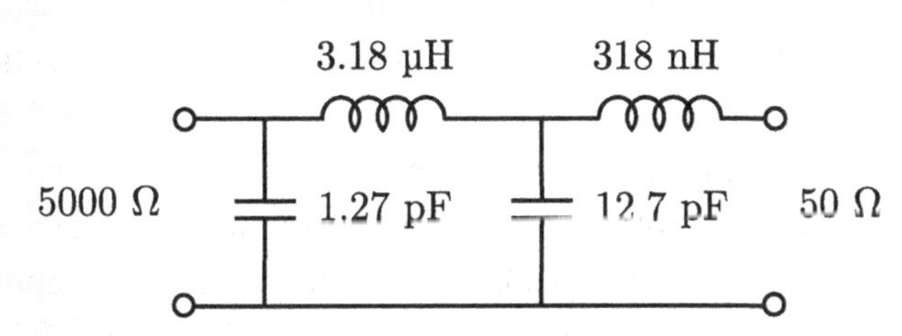

Figure 9.10. An "L-L" design for matching for 5000 Ω to 50 Ω at 75 MHz with increased bandwidth (Example 9.3).

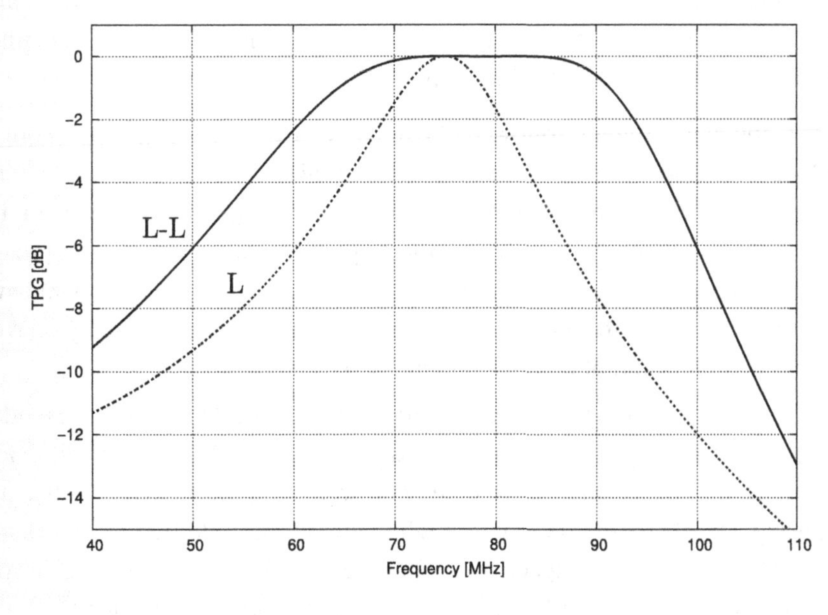

Figure 9.11. Frequency response of the "L-L" match shown in Figure 9.10, compared to the "L" match of Figure 9.8(a).

Solution: The first stage will reduce 5000 Ω to an intermediate impedance R_{mid}, and the second state will reduce R_{mid} to 50 Ω. We set $R_{mid} = \sqrt{R_{hi}R_{lo}} = 500\ \Omega$, resulting in $Q = \sqrt{R_{hi}/R_{mid} - 1} = 3$ for the first stage and $Q = \sqrt{R_{mid}/R_{lo} - 1} = 3$ for the second stage. Thus each stage has much lower Q than the original value of 9.95 for the single-stage design. Each stage is now designed using the technique outlined in Section 9.3. Thus, there are two possibilities for the first stage (i.e., parallel capacitor + series inductor, and vice-versa) and two possibilities for the second stage. In either case, the parallel reactance will appear on the side closest to the larger impedance. Figure 9.10 shows the implementation in which both stages use the parallel capacitance configuration, and Figure 9.11 shows the associated frequency response.

It is clear from the above example that bandwidth is considerably increased. In general it is difficult to accurately predict the increase that is achieved. A very rough estimate can be computed as follows: Note both matching stages are nominally lossless and have the same Q. The escaped power is entirely by power transfer as opposed to loss, so P_{es} is equal for both stages and is equal to P_{es} for the cascade. Thus we estimate Q, the overall quality factor of the cascade, to be equal to $Q_1 + Q_2$, the sum of the quality factors of the first and second stages considered individually. In the above example, $Q_1 + Q_2 = 3 + 3 = 6$, which is much less than 9.95 for the single-stage match, so the increased bandwidth is predicted. Accurately estimating the increased bandwidth is another matter. If we use the rule-of-thumb $B_{\text{HP}} \cong B_Q/2$ with $Q = 6$, we obtain $B_{\text{HP}} \cong 39.3$ MHz, whereas Figure 9.10 indicates $B_{\text{HP}} = 37.2$ MHz. The apparent agreement here is a bit of a fluke: Normally $B_Q/2$ overestimates B_{HP} when Q is not $\gg 2\pi$; however, B_{HP} itself is larger than might otherwise be expected because the two-stage match has four reactances as opposed to two, with a second resonance at 82.9 MHz. This is apparent in Figure 9.11 as the asymmetry in the frequency response around the design value of 75 MHz; in fact, the TPG goes to 1 (0 dB) at both 75 MHz and 82.9 MHz. This second resonance has further increased the B_{HP}, and is not taken into account by the simple "$Q_1 + Q_2$" calculation.

Bandwidth can be further increased in four ways:

- The "divide and conquer" strategy followed above can be extended to three or more stages, resulting in longer cascades of "L" type circuits matching smaller impedance ratios. In principle there is no limit to the bandwidth that can be achieved in this way, since in the limit as the number of stages increases to infinity, the impedance ratio per stage approaches 1, and therefore each stage can have essentially zero Q and essentially infinite bandwidth. In practice, the number of stages that can be effectively implemented is limited by unavoidable dissipative losses and precision of component values, as well as size and cost.
- Additional resonances can be employed in order to extend the bandwidth. We see evidence of this effectiveness of this strategy in Example 9.3, as noted above. Generalizing this idea, we could attempt to design the circuit with specific constraints on the frequency response: For example, requiring TPG = 1 (or some other high value) at multiple frequencies distributed over the desired passband. An impedance matching circuit which achieves a specified frequency response is essentially a filter. Filter design is addressed in Chapter 13.
- Bandwidth can always be increased by introducing dissipative loss. This is because dissipative loss increases P_{es} without necessarily affecting U_{st}, thereby decreasing Q. The obvious disadvantage is that the passband TPG is degraded. Odd though it may seem, this is often

a reasonable tradeoff: For example, this technique is commonly used to increase the usable bandwidth of antennas embedded in handheld devices; see e.g. Example 12.11 (Chapter 12).

- Combinations of the above approaches are frequently employed; for example, a three-stage "divide and conquer" design with additional loss to increase the bandwidth by some additional amount that could not otherwise be achieved due to limitations in the available reactances.

It is prudent to make one additional point since we are on the topic of increasing bandwidth: As in the "divide and conquer" strategy, there is a fundamental limit on the maximum TPG that can be achieved over a specified frequency range using a given number of reactances. The most commonly-cited expression of this limit is the *Fano–Bode relation* [18] (more on this in Section 12.5). However, this limit is quite difficult to apply to practical problems. Here there is a "no free lunch" principle in action: For any given number of reactances employed in a matching circuit, the maximum TPG that can be uniformly achieved over a specified bandwidth is fundamentally constrained, no matter how cleverly the circuit is designed.

9.5.2 Decreasing Bandwidth Using "Pi" and "T" Circuits

At the beginning of Section 9.5.1 we considered how the bandwidth of a matching circuits could be increased by adding additional sections, with an intermediate impedance R_{mid} between R_{lo} and R_{hi}. Similarly, the bandwidth of an impedance match can be *decreased* using a two-stage cascade of "L" type matching circuits in which R_{mid} is selected to be either lower than R_{lo} or higher than R_{hi}. These two strategies are illustrated in Figure 9.12. In the first strategy, R_{mid} is chosen to be less than R_{lo}. This results in two "L" matches in which the series reactances appear in series with each other, and therefore can be combined into a single series reactance. This results in the "Pi" topology (so-named for its resemblance to the Greek letter π). Because R_{hi}/R_{mid} and R_{lo}/R_{mid} are both greater than R_{hi}/R_{lo}, the Q of the Pi circuit is greater than the Q of a single-stage match. Therefore the bandwidth of the Pi circuit is less, and the degree to which it is less depends on how much lower R_{mid} is than R_{lo} and R_{hi}.

Alternatively, R_{mid} can be chosen to be greater than R_{hi}. This results in two "L" matches in which the parallel reactances appear in parallel with each other, and therefore can be combined

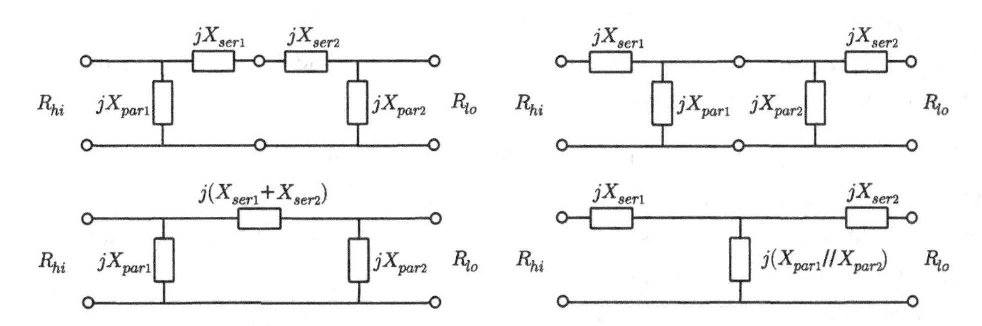

Figure 9.12. *Left*: Development of a "Pi" circuit using back-to-back parallel-first "L" circuits. *Right*: Development of a "T" circuit using back-to-back series-first "L" circuits.

into a single parallel reactance. This results in the "T" topology. Because R_{mid}/R_{hi} and R_{mid}/R_{lo} are both greater than R_{hi}/R_{lo}, the Q of the T circuit is greater than the Q of a single-stage match. Therefore the bandwidth of the T circuit is less, and the degree to which it is less depends on how much greater R_{mid} is than R_{lo} and R_{hi}.

The choice between a Pi configuration and a T configuration for a reduced-bandwidth impedance match depends primarily on the requirements of interfacing circuits; e.g., whether DC signals should be shorted to ground or simply blocked, and the ability to absorb components in adjacent stages into the design so as to decrease overall parts count.

We noted in the previous section that the Q of a multistage matching circuit can be estimated as the sum of the Q of the stages, if all Qs are sufficiently high. Similarly, the bandwidth of the Pi and T circuits can be roughly estimated as the sum of quality factors of the original two "L" match sections. This provides an approximate method for specified-bandwidth design: Bandwidth implies overall Q, which in turn constrains R_{mid} to one of two possible values; one for Pi variant and one for the T variant. For example for the Pi circuit we have

$$\sqrt{\frac{R_{hi}}{R_{mid}} - 1} + \sqrt{\frac{R_{lo}}{R_{mid}} - 1} \cong Q \tag{9.17}$$

for $Q \gg 1$. Assuming the individual sections also have sufficiently high Q, we can further simplify as follows:

$$\sqrt{\frac{R_{hi}}{R_{mid}}} + \sqrt{\frac{R_{lo}}{R_{mid}}} \cong Q \tag{9.18}$$

Now solving for R_{mid}:

$$R_{mid} \cong \frac{\left(\sqrt{R_{hi}} + \sqrt{R_{lo}}\right)^2}{Q^2} \tag{9.19}$$

From this information the two L circuits can be designed and then combined into a single circuit. In practice some "tuning" of reactances will be required to achieve a specified bandwidth precisely.

EXAMPLE 9.4

Recall in Example 9.2 we designed matching circuits that transformed $R_{hi} = 5000\ \Omega$ to $R_{lo} = 50\ \Omega$ at 75 MHz using two discrete reactances in an "L" configuration. This match had a half-power bandwidth of roughly 15 MHz. Then in Example 9.3 we expanded the bandwidth to about 37 MHz using a two-section "L–L" technique. Now, let us design a Pi matching circuit which reduces the bandwidth to about 7 MHz.

Solution: The first stage will reduce 5000 Ω to an intermediate impedance R_{mid} which is less than both R_{hi} and R_{lo}. The required overall Q is estimated to be $\omega_0/2B_{\text{HP}} = 33.7$. Using Equation (9.19), R_{mid} should be about 5.34 Ω. Noting that we already have several approximations at work here, there is no harm in rounding this value to 5 Ω. Thus we choose $Q = \sqrt{R_{hi}/R_{mid} - 1} = 31.6$ for the first stage and $Q = \sqrt{R_{lo}/R_{mid} - 1} = 3$ for the second stage. Each stage is now designed using the technique outlined in Section 9.3. As usual, there are two possibilities for the first stage (i.e., parallel capacitor + series inductor, and vice versa) and two possibilities for the second stage. In keeping with previous examples, let us arbitrarily choose the parallel-capacitor realization for both stages. Figure 9.13 shows the resulting design (after combining the series reactances), and Figure 9.14 shows the resulting frequency response.

The resulting half-power bandwidth is 4.4 MHz, whereas we were aiming for 7 MHz. Much of the error can be attributed to the low Q of the second section; recall that the approximation $B_{HP} \cong B_Q/2$ is accurate only for $Q \gg 2\pi$. The half-power bandwidth can be adjusted to meet the 7 MHz specification by modifying R_{mid}, calculating the new response, and iterating as necessary.

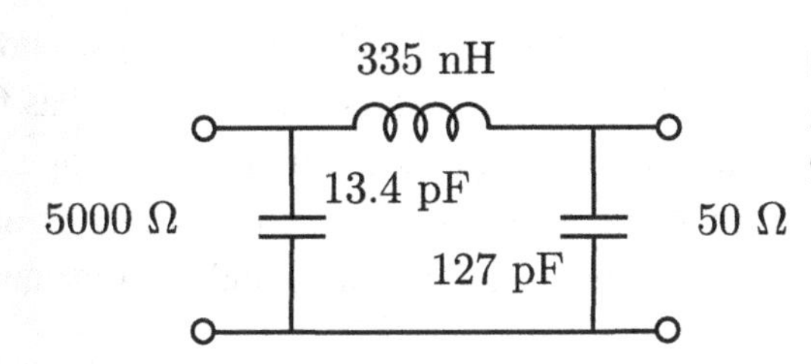

Figure 9.13. A Pi design for matching for 5000 Ω to 50 Ω at 75 MHz with decreased bandwidth (Example 9.4).

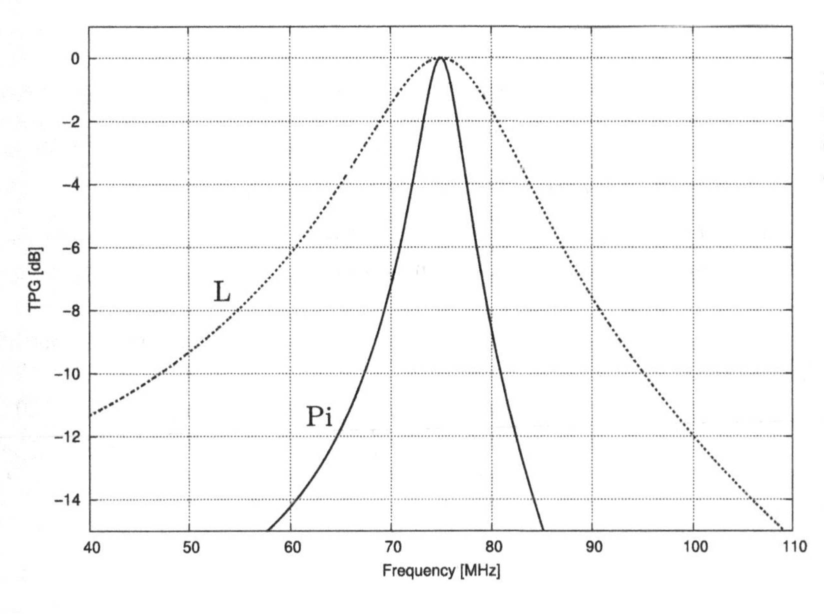

Figure 9.14. Frequency response of the "Pi" match shown in Figure 9.13, compared to the "L" match of Figure 9.8(a).

9.5.3 Other Considerations and Variants

There may be other considerations when designing matching circuits as described in the previous sections. We have already noted the intrinsic lowpass/highpass behavior of various matching circuits, and suggested that this might influence the selection of a circuit. However, recall from elementary electronics that the delta-wye transformation[5] may be used to transform any T-type impedance matching circuit into a Pi-type circuit, and vice versa. The circuits will

[5] This transformation goes by many other names, including Δ-Y, Π-Y, star-mesh, etc.

be equivalent at the frequency assumed for the transformation, but will behave differently away from this frequency. There are a number of reasons why such a transformation may be desirable: For example, it may yield more favorable component values, it may avoid a DC short circuit or provide DC blocking at the input or output, or may position a component in such a way that the component may be "absorbed" into the adjacent stage, thereby reducing the parts count.

9.6 IMPEDANCE MATCHING FOR DIFFERENTIAL CIRCUITS

The techniques developed in previous sections, although presented as single-ended, are straightforwardly modified for use in differential matching circuits. One method is as follows: First, design a single-ended circuit which matches one-half the desired source impedance to one-half the desired load impedance. Next, combine the resulting circuit with a mirror image of itself using a common datum. Finally, eliminate the common datum and combine any series reactances.

EXAMPLE 9.5

Design a circuit that matches a 10 kΩ differential source impedance to a 100 Ω differential load impedance at 75 MHz. Use the differential version of the "L" match technique.

Solution: First, we identify the "prototype" single-ended matching circuit: It is 5000 Ω (single-ended) to 50 Ω (single-ended). This is precisely the problem we solved in Example 9.2. Figure 9.15 shows the corresponding differential matching circuits.

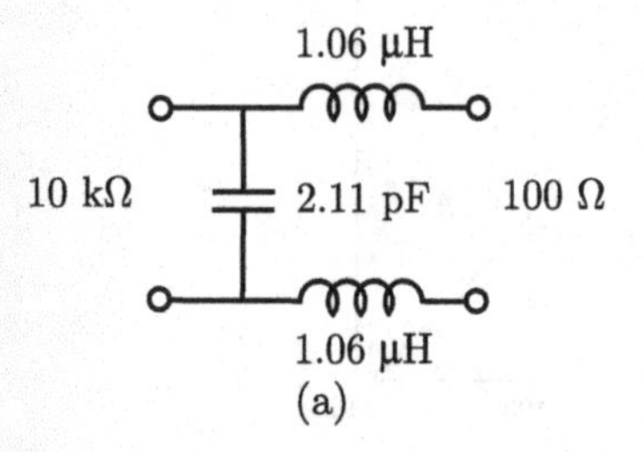

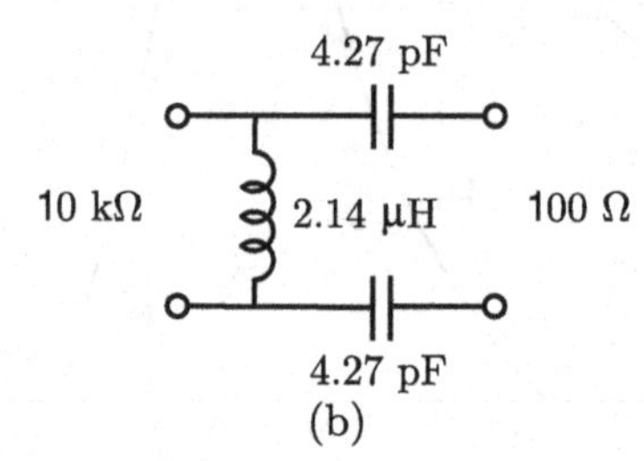

Figure 9.15. Matching a 10 kΩ differential source impedance to a 100 Ω differential load impedance at 75 MHz (Example 9.5).

9.7 DISTRIBUTED MATCHING STRUCTURES

In previous sections of this chapter we have considered lossless impedance matching circuits using discrete capacitors and inductors. This approach is generally applicable for frequencies up to a few hundred MHz, and to higher frequencies if the non-ideal behaviors of these components can be properly taken into account. Common examples of non-ideal behaviors that become important at higher frequencies include: The capacitance between turns in the coils used in inductors, the inductance associated with the leads of capacitors, and in both cases the resistance of materials may become significant relative to the nominal reactance. Although it

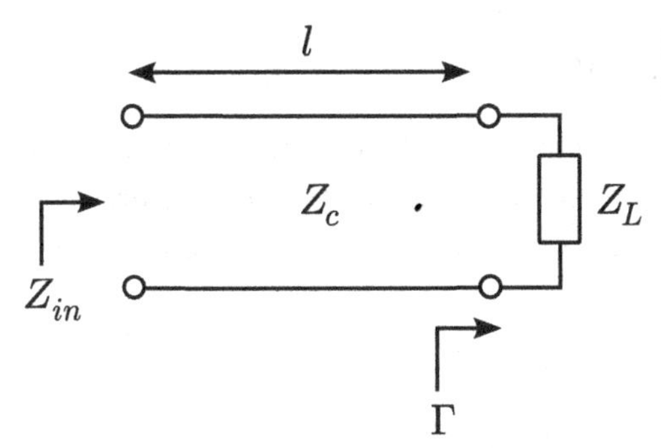

Figure 9.16. A transmission line having characteristic impedance Z_c terminated by a load impedance Z_L.

is possible to compensate for these effects well into the SHF regime, sufficient accuracy and repeatability is typically quite difficult to achieve.

Fortunately, there is an alternative to discrete reactances that works well at higher frequencies: transmission lines. The essential concept is shown in Figure 9.16: When the output of a transmission line having length l and *characteristic impedance* Z_c is terminated into a load impedance Z_L, the input impedance Z_{in} is given by

$$Z_{in} = Z_c \frac{1 + \Gamma e^{-j4\pi l/\lambda}}{1 - \Gamma e^{-j4\pi l/\lambda}} \tag{9.20}$$

where λ is wavelength in the transmission line, and Γ is the reflection coefficient

$$\Gamma \equiv \frac{Z_L - Z_c}{Z_L + Z_c} \tag{9.21}$$

There are two ways this impedance transformation can be used: directly, as a means to transform the load impedance to some other impedance; or alternatively the transmission line can be open- or short-circuited (i.e., Z_L can be set to ∞ or 0, respectively), resulting in a one-port device that is purely reactive and can be used in the same manner as the ideal discrete inductors and capacitors assumed in previous sections. In Section 9.7.3 we consider single-stub matching, which uses both techniques and is able to transform a complex-valued impedance into any other complex-valued impedance. In Section 9.7.4 we consider quarter-wave matching, which uses only the first technique and is limited to converting real-valued impedances into other real-valued impedances. However, we begin with a brief description of the characteristic impedance and wavelength of practical transmission lines (Section 9.7.1) and impedance properties of open- and short-circuited transmission lines (Section 9.7.2).

Before moving on, note that we will sometimes find it convenient to work in terms of admittance (Y) as opposed to impedance (Z). Generally $Y = 1/Z$, and we use the variables "G" and "B" to represent the real and imaginary parts of Y, respectively; i.e., $Y = G + jB$.[6] In particular it will be convenient to use Equation (9.20) in the form

$$Y_{in}(l/\lambda) = Y_c \frac{1 - \Gamma e^{-j4\pi l/\lambda}}{1 + \Gamma e^{-j4\pi l/\lambda}} = G_{in} + jB_{in} \tag{9.22}$$

where $Y_c = 1/Z_c$.

[6] The fact that the variables G and B are also commonly used for gain and bandwidth is admittedly awkward, but is customary and not normally a source of confusion.

9.7.1 Properties of Practical Transmission Lines

Any lossless transmission line can be completely characterized in terms of Z_c and λ. The characteristic impedance Z_c is determined by the geometry and materials used in the transmission line, and can usually be assumed to be independent of frequency. The wavelength is given by

$$\lambda = \frac{v_p}{f} \tag{9.23}$$

where v_p is the *phase velocity* in the material through which the signal-bearing electromagnetic fields propagate, and f is frequency. Note that v_p itself is

$$v_p = \frac{c}{\sqrt{\epsilon_{\text{eff}}}} \tag{9.24}$$

where c is the speed of light in free space and ϵ_{eff} is the *relative permittivity* of the material through which the signal-bearing electromagnetic fields propagate. For typical pre-manufactured transmission lines, such as coaxial cable, Z_c and v_p are specified. For example, RG-58/U coaxial cable has $Z_c \cong 50\ \Omega$ and $v_p \cong 0.67c$.

An extremely important class of transmission line for modern radio engineering work is the *microstrip line*, illustrated in Figure 9.17. The line is usually implemented as part of a printed circuit board (PCB), facilitating use with discrete components on the same PCB. The microstrip line can be viewed as a modified parallel plate transmission line in which one conductor is the PCB ground plane and the other conductor is a trace atop the dielectric slab. The effective permittivity for computing phase velocity is less than the relative permittivity of the dielectric material, because signal-bearing electromagnetic fields extend beyond the sides of the trace and into the space above the dielectric. An approximation that is commonly used for ϵ_{eff} in lieu of an exact value (which may be difficult to determine) is

$$\epsilon_{\text{eff}} \approx \frac{\epsilon_r + 1}{2} \tag{9.25}$$

i.e., the average of the relative permittivity of the dielectric (ϵ_r) and that of air (1).

Similarly, there is no exact expression for the characteristic impedance of a microstrip line, and instead an approximation must be employed. The recommended reference on the

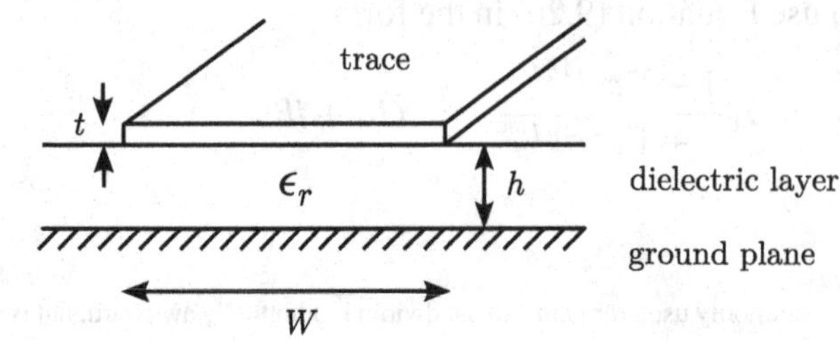

Figure 9.17. Microstrip line, shown in cross section.

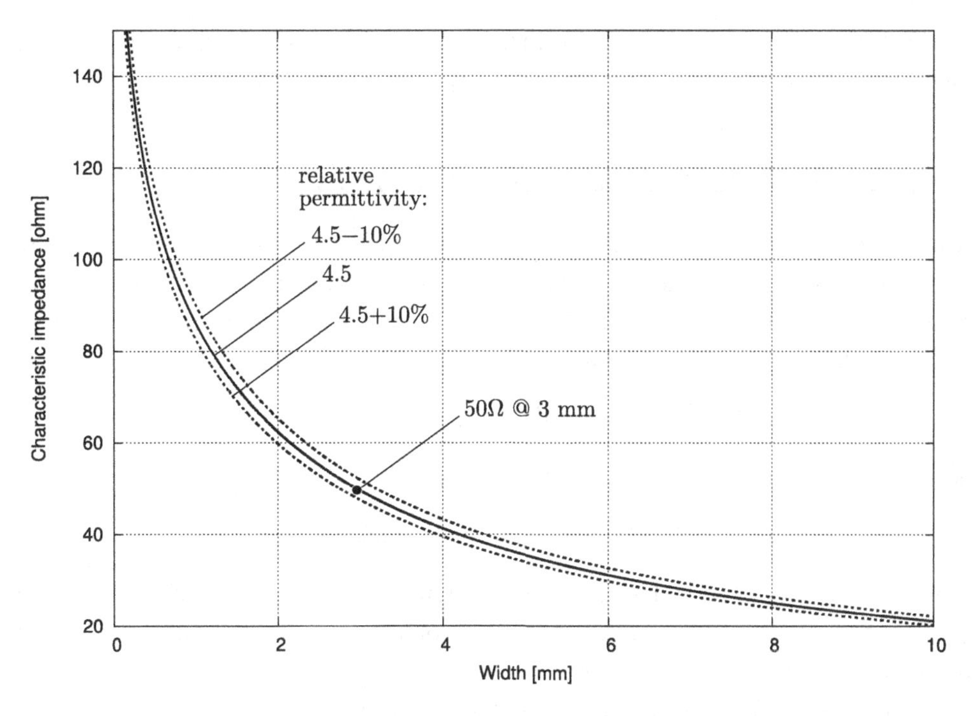

Figure 9.18. Relationship between characteristic impedance Z_c and width W for FR4 ($h = 1.575$ mm, nominal $\epsilon_r = 4.5$) using Equation (9.26). Also shown are the characteristic impedances computed assuming relative permittivity which is 10% greater and 10% less than the nominal value.

characteristic impedance of microstrip transmission lines is Wheeler 1977 [84], which offers the following expression for the characteristic impedance of a microstrip line:[7]

$$Z_c \cong \frac{42.4\ \Omega}{\sqrt{\epsilon_r + 1}} \ln\left[1 + \left(\frac{4h}{W'}\right)\left(\Phi + \sqrt{\Phi^2 + \frac{1 + 1/\epsilon_r}{2}\pi^2}\right)\right] \tag{9.26}$$

where

$$\Phi \equiv \frac{14 + 8/\epsilon_r}{11}\left(\frac{4h}{W'}\right) \tag{9.27}$$

and W' is W adjusted to account for the thickness t of the microstrip line. Typically $t \ll W$ and $t \ll h$, for which $W' \approx W$. Other approximations commonly employed in the design and analysis of microstrip lines are limited in the range of h/W for which they are valid, and can usually be shown to be special cases or approximations of Equation (9.26).

Perhaps the world's most common PCB material is FR4, which consists of a $h = 1.575$ mm-thick fiberglass epoxy dielectric layer having $\epsilon_r \cong 4.5$. Figure 9.18 shows the width W required to achieve a specified characteristic impedance Z_c for FR4, using Equation (9.26). Note that $Z_c = 50\ \Omega$ is obtained when $W \cong 3$ mm. Estimates for other characteristic impedances can be obtained by reading values from Figure 9.18, or by trial and error using Equation (9.26). More accurate approximations are typically not useful because the error associated with

[7] This appears in [84] as Equation 10.

manufacturing variances – specifically, in the value of the relative permittivity of FR4 dielectric material – is at a comparable level and typically cannot be avoided.

9.7.2 Impedance of Single-Port Transmission Line Stubs

Earlier we noted the possibility of using a transmission line to emulate a discrete reactance. The simplest possible way to do this is to create a "stub" consisting of a given length of transmission line which is either open- or short-circuited on one end. For the open-circuited case we use Equations (9.20) and (9.21) to obtain

$$Z_{OC}(l/\lambda) \equiv Z_{in}|_{Z_L \to \infty} = -jZ_c \cot(2\pi l/\lambda) \tag{9.28}$$

and in the short-circuited case we obtain

$$Z_{SC}(l/\lambda) \equiv Z_{in}|_{Z_L \to 0} = +jZ_c \tan(2\pi l/\lambda) \tag{9.29}$$

Note that for small l/λ, Z_{OC} has a negative imaginary-valued impedance and therefore can be used to cancel a positive reactance, and in other applications in which a discrete capacitor might otherwise be used. Similarly Z_{SC} has a positive imaginary-valued impedance and therefore can be used to cancel a negative reactance, and in other applications in which a discrete inductor might otherwise be used. Note, however, that the change in impedance with frequency of a stub is different from the change in impedance with frequency of an inductor or capacitor; therefore any equivalence between a stub and a discrete reactance applies only at a single frequency. Nevertheless, the stub is an extremely useful tool, as it is a practical way to generate any value of imaginary impedance and is not limited by non-ideal effects in the same manner as discrete capacitors and inductors.

As noted previously, it is often convenient to work in terms of admittances as opposed to impedances. Here are Equations (9.28) and (9.29) written in terms of admittance, and specifically in terms of the imaginary component of the admittance:

$$B_{OC}(l/\lambda) = +Y_c \tan(2\pi l/\lambda) \tag{9.30}$$

$$B_{SC}(l/\lambda) = -Y_c \cot(2\pi l/\lambda) \tag{9.31}$$

9.7.3 Single-Stub Matching

The single-stub matching technique, illustrated in Figure 9.19, allows any complex-valued impedance Z_L to be transformed to any complex-valued impedance Z_{in}, represented in the figure as the source admittance $Y_{in} = 1/Z_{in}$. The essential idea is to use a transmission line to match the real part of Y_S, and then to use a stub attached in parallel to match the transmission line + stub combination to the imaginary part of Y_S. We prefer to work with admittances as opposed to impedances on the source side because this greatly simplifies the math: the admittance of the parallel combination of the transmission line and the stub is simply the sum of the associated admittances, whereas the impedance of the parallel combination of the transmission line and the stub is given by a much more complex expression.

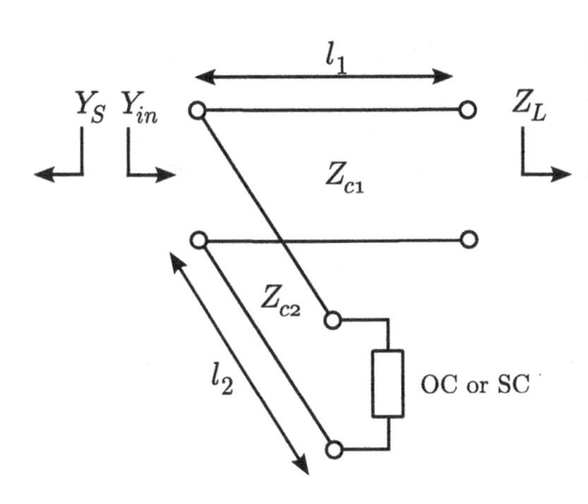

Figure 9.19. Single-stub technique for matching a load impedance Z_L to a source impedance $Z_S = 1/Y_S$. As in other matching circuits, $Y_{in} = Y_S^*$ for a conjugate match and $Y_{in} = Y_S$ for a reflectionless match.

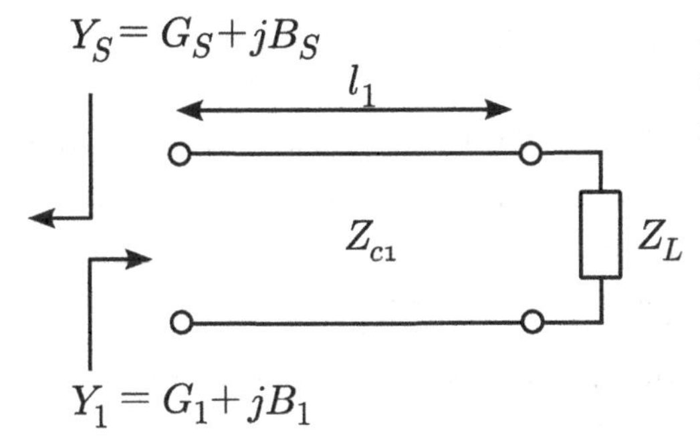

Figure 9.20. Design of the transmission line section of the single-stub match of Figure 9.19: l_1 is selected to make $G_1 = G_S$.

A separate consideration is what Y_{in} should be. As always, the choices are $Y_{in} = Y_S$ for reflectionless matching and $Y_{in} = Y_S^*$ for conjugate matching. If X_S (the imaginary part of Z_S) is zero, then $B_S = 0$ and there is no distinction between reflectionless and conjugate matching.

The single-stub design procedure is as follows.

(1) **Select a transmission line** and determine its characteristic impedance Z_{c1}. Z_{c1} is commonly, but not necessarily, chosen to be equal to either the desired real component of the transformed load impedance, or the reference impedance for s-parameters employed in the design. Thus, $Z_{c1} = 50\ \Omega$ is quite common. Similarly, select a transmission line type for the stub and determine its characteristic impedance Z_{c2}. The transmission line and transmission line stub are commonly, but not necessarily, chosen to be identical transmission line types, differing only in length and termination; thus, $Z_{c2} = Z_{c1}$ typically.

(2) **Design the primary (in-line) transmission line section.** This is illustrated in Figure 9.20. Note that Y_1, the input admittance of the transmission line, is given by Equation (9.22). Use this equation to find the smallest value of l_1/λ that makes G_1 (the real part of Y_1) equal to G_S (the real part of Y_S). The resulting imaginary value of the admittance (B_1) is unimportant at this step. This step can be done graphically using a Smith chart; however, accuracy of three or more significant figures is possible with just a few rounds of trial and error with Equation (9.22) using a short computer program or a simple spreadsheet.

(3) **Determine the required stub input susceptance B_{stub}.** For a conjugate match, the imaginary part of the source admittance B_S should be equal to $-(B_1 + B_{stub})$, where jB_{stub} is the input admittance of the stub. Therefore $B_{stub} = -B_S - B_1$ for a conjugate match. For

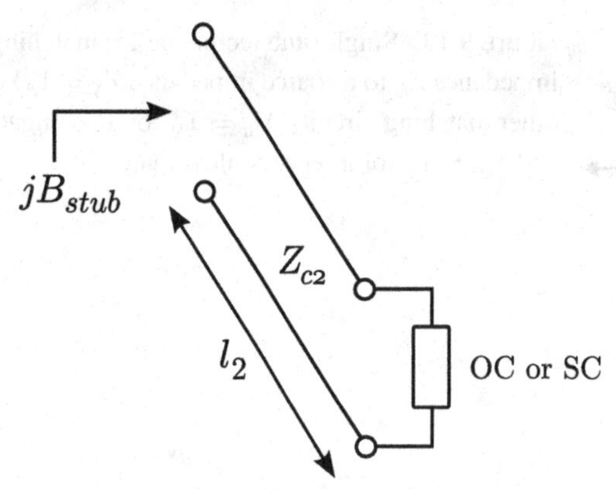

Figure 9.21. Design of the stub section of the single-stub match of Figure 9.19: l_2 is selected to make $B_{stub} = -B_S - B_1$ for a conjugate match or $B_{stub} = +B_S - B_1$ for a reflectionless match.

a reflectionless match, B_S should be equal to $+(B_1+B_{stub})$, and therefore $B_{stub} = +B_S - B_1$ in this case.

(4) **Design the stub.** This is illustrated in Figure 9.21. Using Equations (9.30) and (9.31), find a stub that has an input admittance jB_{stub} as selected in the previous step. Since stubs can be either open- or short-circuited, you should check both possibilities. Typically – but not always – the shorter stub is desired (more on this below).

We should add a final step: Check your work. It is pretty easy to make a mistake in the design process, but such mistakes are easily found simply by computing Y_{in} for the candidate design, and confirming that it is the correct value.

EXAMPLE 9.6

Design a single-stub match that transforms $Z_L = 5.1 + j7.5\ \Omega$ to $Z_S = 50\ \Omega$ at 2 GHz. Assume microstrip implementation on FR4 and use lines with a characteristic impedance of 50 Ω.

Solution: The smart play in such problems is to first work out an "implementation-independent" solution in which lengths are expressed in units of wavelengths; and then to work out the details of the specific implementation. Following the procedure outlined above:

(1) $Z_{c1} = Z_{c2} = 50\ \Omega$ is specified in the problem statement.

(2) Using Equation (9.21), $\Gamma = (Z - Z_{c1}) / (Z + Z_{c1}) = -0.782 + j0.242$. Now we seek the value of l_1 for which G_1 equals $Y_S = 1/Z_S = 0.02$ mho. Using a trial-and-error approach, one may test values of l_1 starting with zero and increasing 0.001λ at a time until G_1 is as close as possible to G_S. We find $l_1 = 0.024\lambda$. The resulting value of Y_1 is $0.0202 - j0.0573$ mho, so $B_1 = -0.0573$ mho.

(3) In this problem Z_S is real-valued, so $B_S = 0$ and there is no difference between a conjugate match and a reflectionless match. Thus, $B_{stub} = -B_1 = +0.0573$ mho.

(4) Now we design both open- and short-circuited stubs and see which is shorter. In the open-circuited case we use Equation (9.30) and find $l_2 = 0.197\lambda$. In the short-circuited case we find $l_2 = 0.447\lambda$. The open-circuited stub is shorter, so this is the configuration that is selected.

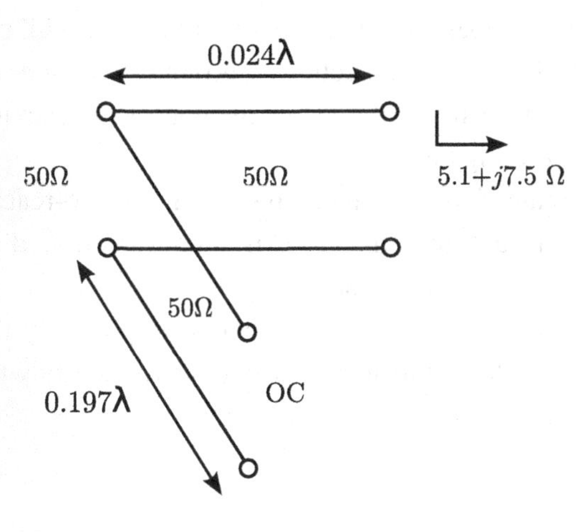

Figure 9.22. Implementation-independent solution to the single-stub matching problem of Example 9.6.

At this point we have the implementation-independent solution shown in Figure 9.22.

The problem statement specifies a solution in FR4 microstrip line. Since FR4 has a specified height $h = 1.575$ mm and relative permittivity $\epsilon_r = 4.5$, we are left to specify the trace width W, and the line lengths l_1 and l_2 must be specified in terms of physical lengths (e.g., meters) as opposed to wavelengths. We noted in Section 9.7.1 (see Figure 9.18) that $W \cong 3$ mm yields a characteristic impedance of 50 Ω. To determine the physical lengths, we first calculate the wavelength in the transmission line as follows:

$$\lambda = \frac{v_p}{f} = \frac{c/\sqrt{\epsilon_{\text{eff}}}}{f} \approx \frac{c/\sqrt{\frac{1}{2}(\epsilon_r + 1)}}{f} = 90.4 \text{ mm} \tag{9.32}$$

where we have used Equations (9.23), (9.24), and (9.25) in turn. A sketch of this microstrip implementation is shown in Figure 9.23.

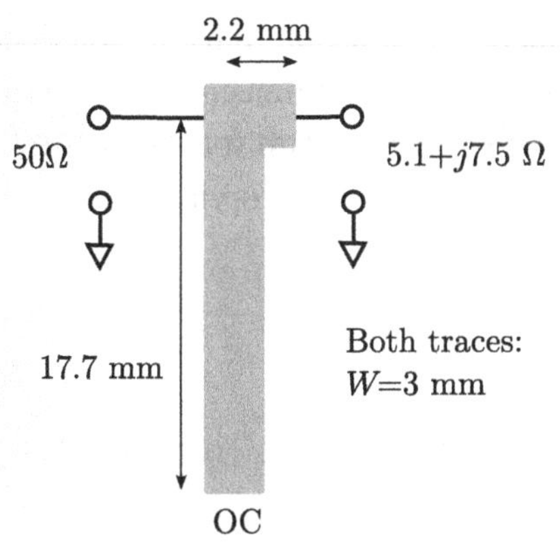

Figure 9.23. Solution to the single-stub matching problem of Example 9.6 as microstrip on FR4.

While it is normally desired to minimize the length of the stub, this may not always be the preferred solution. For example, if the shortest stub requires a short-circuit termination, then any DC signal encountering this matching structure will experience a short-circuit to ground through the stub. This problem is commonly encountered in the design of transistor amplifiers, in which DC power supplies are often commingled with the RF signals being processed. In this case, there are two solutions: Either a longer open-circuited stub could be selected, or a DC-blocking capacitor could be used in lieu of a direct connection to ground. The latter solution allows the shorter stub to be used, but introduces the problem that the non-ideal behavior of the capacitor may be significant and difficult to mitigate.

It is useful to note that the single stub match has a lot in common with the two-reactance L-match considered earlier in this chapter. In fact, in some cases it is convenient and possible to replace the transmission line stub with a discrete reactance. Furthermore, it is possible to cascade two single-stub matches to increase or decrease bandwidth in precisely the same manner as it is possible to cascade two L-match circuits to increase or decrease bandwidth; this technique is referred to as *double-stub tuning*.

9.7.4 Quarter-Wave Matching

An alternative to stub matching is *quarter-wave matching*. Here is the concept: The input impedance for a section of transmission line which has length $l = \lambda/4$ is found from Equation (9.20) to have the remarkably simple form:

$$Z_{in} = \frac{Z_c^2}{Z_L} \tag{9.33}$$

Solving for the characteristic impedance:

$$Z_c = \sqrt{Z_{in} Z_L} \tag{9.34}$$

In other words, we can match Z_L to a specified impedance by using a quarter-wave section of transmission line having this characteristic impedance. Note that this works only if the product $Z_{in}Z_L$ is real-valued, because the characteristic impedance of useful transmission lines must be real-valued. Also, this approach requires that we be able to design transmission lines to have the necessary characteristic impedance; however, this is straightforward for microstrip lines since their characteristic impedance can be selecting by choosing the appropriate trace width, which is easy to do on a printed circuit board.

EXAMPLE 9.7

Design a quarter-wave match that transforms $Z = 23.7\ \Omega$ to $Z_0 = 50\ \Omega$ at 2 GHz. Assume microstrip implementation on FR4.

Solution: First, we confirm that the product of the impedances to be matched is real-valued. Since both impedances are real-valued, the technique is applicable. The required characteristic impedance is $\sqrt{50\ \Omega \cdot 23.7\ \Omega} = 34.4\ \Omega$. Referring to Figure 9.18 (or directly from Equation (9.26) if desired), we find $W = 5.3$ mm. We already worked out that $\lambda = 90.4$ mm in Example 9.6; thus the quarter-wave matching section in this problem is $\lambda/4 = 22.6$ mm long.

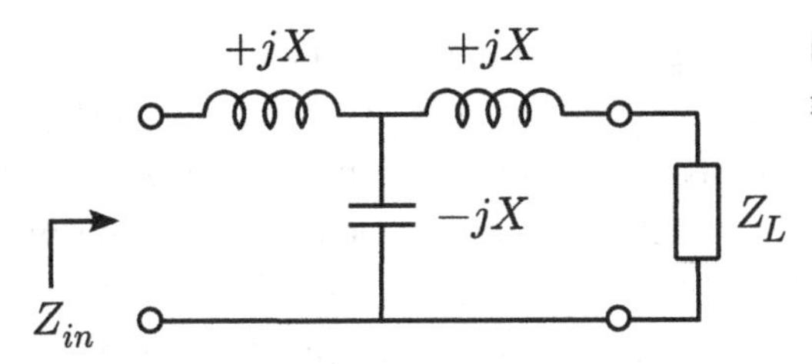

Figure 9.24. An impedance inverter using discrete reactances.

9.8 IMPEDANCE INVERSION

The quarter-wave match belongs to a class of impedance matching devices known as *impedance inverters*. An impedance inverter is a two-port device for which the input impedance is proportional to the reciprocal of the load impedance. For the quarter-wave match, this is clear from Equation (9.33). Impedance inverters can also be constructed from discrete reactances; for example, the input impedance of the Pi circuit shown in Figure 9.24 is

$$Z_{in} = \frac{X^2}{Z_L} \tag{9.35}$$

which is seen to have the same form as Equation (9.33). (Proof of this is left as an exercise for the reader; see Problem 9.17.)

Impedance inverters have a number of applications beyond impedance matching. Before moving on, it is worthwhile to identify at least two of these applications. One application is the replacement of inductors with capacitors, using an impedance inverter to change the sign of the capacitor's reactance. This is useful at frequencies in the UHF band and above, where (1) inductors become very difficult to use whereas capacitors are somewhat easier to work with, and (2) quarter-wave transmission lines are sufficiently small to fit on PCBs.

A second application of impedance inversion is injection of DC power into RF circuits. As mentioned earlier, and as we shall see many times in Chapter 10, it is often necessary for DC power to share paths used by RF signals. However, it is also important to keep RF signals out of power supply circuits. At lower frequencies, this isolation is possible using a series inductor, which passes DC while blocking RF. However, suitable inductors become difficult to find at RF frequencies in the VHF band and higher. An alternative is to use an impedance inverter between the power supply, which has low output impedance, and the point of power injection. The quarter-wave transmission line transforms the low output impedance of the power supply into a high AC impedance, so that RF is blocked while DC passes freely.

Problems

9.1 Derive the design equations (Equations (9.1) and (9.2)) for the "series-first" discrete two-component L-match.

9.2 Derive the design equations (Equations (9.5) and (9.6)) for the "parallel-first" discrete two-component L-match.

9.3 Derive Equations (9.7) and (9.8) beginning with Equations (9.1)–(9.6).

9.4 The output matching network for a 150 MHz transistor amplifier must present an impedance of $82.4 + j64.5\ \Omega$ to the transistor, and 50 Ω to the next device in the signal chain. Identify all possible discrete two-reactance circuits that accomplish this with a TPG of 0 dB at 150 MHz.

9.5 Design all possible discrete two-reactance matches between 75 Ω and 50 Ω for a frequency of 30 MHz. Plot frequency response (TPG vs. frequency) for each solution. Summarize any differences in performance and identify any other considerations that might lead the designer to choose one of these solutions over the others.

9.6 Figure 9.10 shows one solution for the increased bandwidth L-L match problem of Example 9.3. Design all other possible L-L solutions to this problem.

9.7 Derive the version of Equation (9.19) that applies to T-match circuits.

9.8 Verify the differential matching circuit of Figure 9.15(a) using elementary circuit theory. You can do this by showing that the input impedance is 10 kΩ when the load impedance is 100 Ω.

9.9 A certain dipole antenna has a differential self-impedance of $5 - j150\ \Omega$ at 150 MHz. Design a circuit consisting of discrete reactances that matches this antenna to a 75 Ω differential impedance. An additional requirement of this design is that the matching circuit must not present a DC short to the 75 Ω termination.

9.10 Design a single-stub match that conjugate-matches a source impedance of $105.4 - j24.6\ \Omega$ to a load impedance of 50 Ω at 1.5 GHz. Assume characteristic impedances of 50 Ω throughout and leave your answer in terms of wavelengths.

9.11 Complete the design started in the previous problem assuming implementation on an FR4 PCB.

9.12 Design a single-stub match that conjugate-matches a source impedance of 50 Ω to a load impedance of $33.9 + j17.6\ \Omega$ at 1.5 GHz. Assume characteristic impedances of 50 Ω throughout and leave your answer in terms of wavelengths.

9.13 Complete the design started in the previous problem assuming implementation on an FR4 PCB.

9.14 Match a 150 Ω source impedance to a 50 Ω load impedance using a quarter-wavelength section of transmission line at 2.45 GHz. Assume microstrip implementation on an FR4 PCB.

9.15 A patch antenna has a self-impedance of $35 - j35\ \Omega$ at 2.45 GHz. Match this impedance to 50 Ω using a short section of 50 Ω transmission line followed by a quarter-wave matching section.

9.16 Complete the design started in the previous problem assuming implementation on an FR4 PCB.

9.17 Using elementary circuit theory, prove that the input impedance of the impedance inverter shown in Figure 9.24 is given by Equation (9.35).

10 Amplifiers

10.1 INTRODUCTION

The primary purpose of most amplifiers used in radio frequency systems is to increase signal magnitude. This is necessary in receivers because propagation path loss results in a signal arriving with voltage or current which is far too small to be digitized or directly demodulated. Amplification appears in transmitters when it is convenient to synthesize a signal at low power and amplify it, as opposed to synthesizing signals directly at the intended transmit power; the latter is difficult for all but very simple analog modulations.

This chapter is organized as follows. Section 10.2 introduces the transistor, its use as a gain device, and provides a brief summary of transistor types and device technologies. The design of narrowband single-transistor amplifiers is introduced in four parts: biasing (Section 10.3), small-signal design to meet gain requirements (Section 10.4), designing for noise figure (Section 10.5), and designing for voltage standing wave ratio (Section 10.6). Section 10.7 presents a design example intended to demonstrate the principles of amplifier design as laid out in the previous sections. Section 10.8 introduces the concepts of multistage, differential, and broadband transistor amplifiers. Section 10.9 describes the implementation of transistor amplifiers as integrated circuits.

10.2 TRANSISTORS AS AMPLIFIERS

The fundamental building block of modern radio frequency amplifiers is the transistor.[1] A transistor is a three-terminal semiconductor device in which the current between two terminals can be manipulated using the voltage or current at a third terminal. The application to RF design is most easily explained in the context of a particular device – for our purposes, the silicon (Si) *bipolar junction transistor* (BJT) is a good place to start.

10.2.1 Bipolar Transistors

A BJT is depicted in Figure 10.1. The three terminals of a BJT are known as the *base*, the *collector*, and the *emitter*. The associated terminal currents and voltages are as defined in Figure 10.1. The Si BJT comes in two varieties, known as "NPN" and "PNP," referring to the

[1] Full disclosure: Very-high-power applications continue to use devices such as traveling wave tubes and klystrons, and there continue to be a few – albeit somewhat eclectic – applications for vacuum tubes.

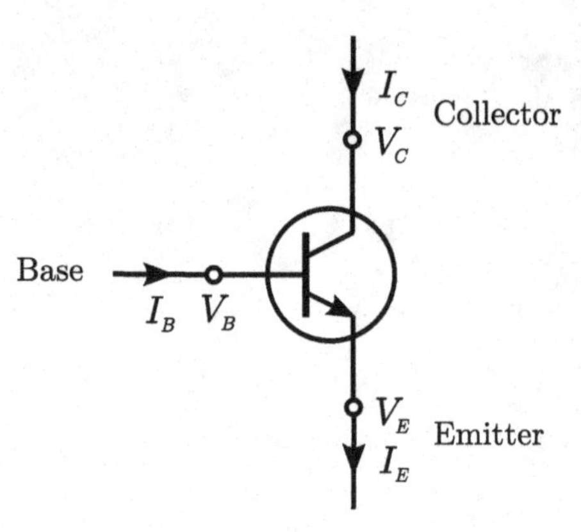

Figure 10.1. Circuit schematic depiction of an NPN BJT. The PNP variant is indicated by changing the direction of the arrow on the emitter.

geometry of semiconductor materials within the device. The NPN variant is somewhat more common in RF applications and is specifically considered here (and shown in Figure 10.1); the PNP variant is relatively easily understood once the NPN variant is mastered.

The properties of the semiconductor materials and the physical principles at work are sometimes useful to know but somewhat beyond the scope of this book. (A useful reference for additional details on all types of transistors considered in this chapter is [48, Ch. 2].) For our immediate purposes, it suffices to know that the BJT can be used as an amplifier when operating in the *forward-active mode*. This mode is achieved by constraining $V_{CE} \equiv V_C - V_E$ to be greater than about 0.3 V, which results in the base–emitter junction behaving as a diode with $V_{BE} \equiv V_B - V_E \approx 0.65$ V when appropriately biased. Under these conditions, there are two useful descriptions of the behavior of the transistor. The first is the following approximate version of the *Ebers–Moll model*:

$$I_C \approx I_{C0} e^{V_{BE}/V_T} \tag{10.1}$$

where I_{C0} is a device constant and V_T is the *thermal voltage*

$$V_T \equiv \frac{kT}{q} \cong 25 \text{ mV at room temperature} \tag{10.2}$$

where k is Boltzmann's constant, T is physical temperature, and q is the charge of an electron (1.6×10^{-19} coulomb). The primary application of this model will come in Chapter 11 (Section 11.2.2), when we consider linearity issues.

For amplifier design, a somewhat more useful model is the equivalent circuit of Figure 10.2. Note $I_C = \beta I_B$, where β, also known as the *DC current gain* and sometimes by the symbol h_{FE}, is a constant value on the order of 10 to a few hundred. Thus, small changes in I_B cause large and proportional changes in I_C, assuming a power supply exists to provide the necessary collector current. The specific value of β depends primarily on V_{CE} and I_C; these two quantities together are known as the *operating point* of the transistor.[2]

An important step in using a BJT as an amplifier is to select an operating point that results in forward-active mode and is able to support the necessary variation around V_{CE} ("signal swing") for a specified supply voltage V_{CC}. Then small variations around I_B result in proportionally larger variations around the selected value of I_C. Note that the increase in power is not provided

[2] The operating point is also sometimes referred to as the *quiescent point* or simply the "Q point."

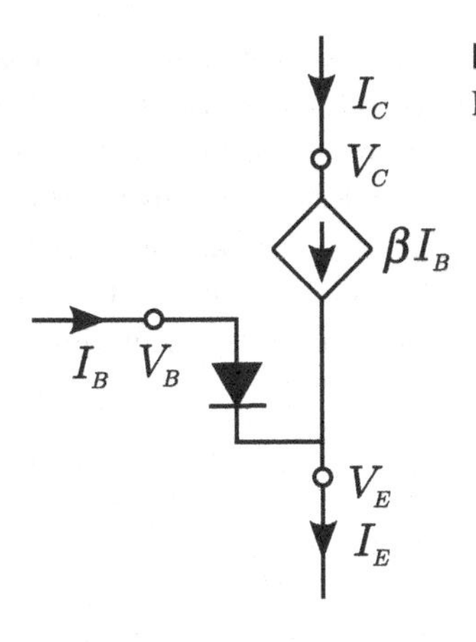

Figure 10.2. Equivalent circuit for forward-active region operation of the NPN BJT.

by the transistor, but rather by the power supply; the transistor merely moderates the flow of collector current.

Finally, to get this scheme to work, we need to interface an input signal to the transistor and extract an output signal from the transistor. It is unlikely that the transistor will have the appropriate input and output impedances as-is, so impedance matching will typically be necessary. Also, because the transistor is being used as an active device, stability must be considered.

In summary, the process of transistor amplifier design boils down to two parts: *biasing*, which sets the operating point and is addressed in more detail in Section 10.3; and impedance matching with stability constraints, addressed in more detail in Section 10.4.

Si BJTs are but one category from an expansive family of qualitatively similar devices, which we shall refer to as *bipolar* transistors. Whereas the maximum useful frequency of Si BJTs is on the order of a few GHz, frequencies up to about 10 GHz are possible by replacing Si with various combinations of alternative semiconductor materials to form what is known as a *heterojunction bipolar transistor* (HBT). HBTs are typically identified by their constituent materials, which may include gallium arsenide (GaAs), silicon geranium (SiGe), indium phosphide (InP), indium gallium phosphide (InGaP), and indium GaAs nitride (InGaAsN). Bipolar transistors may also be implemented directly in complementary metal–oxide semiconductor (CMOS) devices using manufacturing processes such as BiCMOS, enabling the compact implementation of large and complex radio frequency circuits as single integrated circuit (IC) devices (more on this in Section 10.9).

The primary driver for alternative transistor technologies is the quest for improved gain, noise figure, and linearity at ever-increasing frequencies. The maximum usable frequency of a transistor depends on the application. A common and useful characterization is *unity-gain bandwidth*, f_T. The concept is that the realizable gain inevitably decreases with frequency, and eventually becomes unity; i.e., no gain. For bipolar transistors in amplifier applications, the decline in gain with frequency is due to the capacitance of the base-emitter junction, and "realizable gain" is most conveniently defined as the "short circuit gain" $\Delta I_C/\Delta I_B$; i.e., the value of β applicable to small excursions around a specified operating operating point. In this terminology, Si BJTs have f_T in the range 100 MHz – 4 GHz, whereas HBTs have f_T up to about 100 GHz.

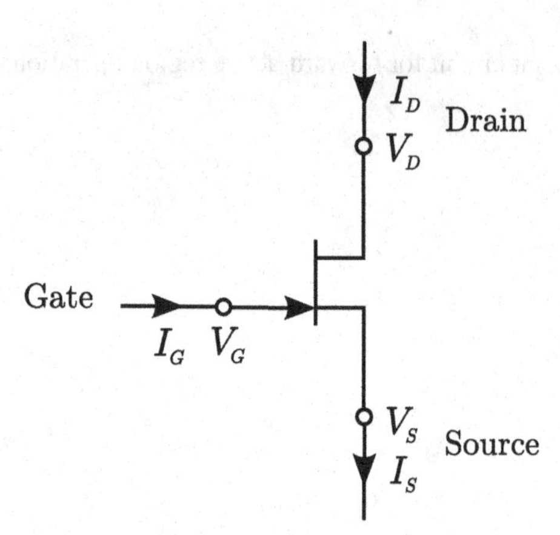

Figure 10.3. Circuit schematic depiction of an N-channel FET. Some N-channel FET varieties are indicated with the arrow on the source terminal facing outward (analogous to a PNP bipolar symbol). P-channel variants are indicated by reversing the direction of the arrow on the gate, or moving the arrow from source to drain. Some transistors of both varieties – in particular, MOSFETs – are sometimes shown with a fourth ("body") terminal.

10.2.2 Field Effect Transistors

A distinct and equally important category of transistors is known as *field effect transistors* (FETs). A circuit schematic depiction of a FET is shown in Figure 10.3. The three terminals are known as the *gate*, *drain*, and *source*; these are analogous to the base, collector, and emitter, respectively, of the bipolar transistor. A fourth terminal, known variously as *body* or *base*, is sometimes identified but is typically important only in integrated circuit applications.

A big difference between FETs and bipolar transistors is that I_G (analogous to I_B for bipolar transistors) is often small enough that it may be neglected, with the result that $I_S \cong I_D$. When used as an RF amplifier, a FET is essentially a *voltage*-controlled current source. The drain current I_D is controlled by the gate-to-source voltage V_{GS} according to the following relationship:

$$I_D = I_{DSS}\left[1 - \frac{V_{GS}}{V_P}\right]^2 \tag{10.3}$$

where I_{DSS} and V_P are device-specific parameters normally specified in the data sheet.

In terms of performance relative to bipolar transistors, FETs dominate above about 4 GHz, and appear in applications up to 100 GHz and beyond. Below 4 GHz, the primary advantage of FETs over bipolar transistors is low noise figure, although there are copious applications for both technologies throughout the MF–SHF bands.

As with bipolar transistors, FETs constitute a broad class of devices. Although a detailed discussion is beyond the scope of this book, it is useful to know something about the family tree of FET devices so that they may be recognized when encountered. First, FETs come in two varieties: *depletion-mode* devices, and *enhancement-mode* devices. The primary distinction is that depletion-mode FETs require V_{GS} to be negative for use as an amplifier, which affects the design of bias circuitry. FETs also come in "N-channel" and "P-channel" variants, which are analogous to NPN and PNP for bipolar transistors. Many types of FETs have appeared over the years as a result of the never-ending quest to improve gain, noise figure, and linearity at ever-increasing frequencies. Modern applications are dominated by *metal–oxide semiconductor FETs* (MOSFETs), *metal–epitaxy–semiconductor FETs* (MESFETs), and *high electron mobility transistors* (HEMTs). MOSFETs appear primarily in applications at UHF and below, and include the increasingly popular category of *laterally-diffused MOS* (LDMOS) devices. MESFETs are commonly constructed from GaAs, known as GaAsFETs, and increasingly from gallium nitride (GaN) and silicon carbide (SiC). HEMTs are heterojunction devices, analogous to HBTs, utilizing AlGaAs on GaAs; AlGaAs–InGaAs layers on GaAs, known as *pseudomorphic HEMTs* (PHEMTs); and "lattice-matched" AlInAs–GaInAs layers on InP. Also notable are the *modulation doped FET* (MODFET), and the venerable *junction FET* (JFET), which was the first popular FET technology but is now largely deprecated in favor of newer variants.

10.2.3 Designing with Transistors

Given the dizzying array of bipolar and FET device technologies, it might seem that a broadly-applicable design methodology is not on the cards. In practice, the situation is not quite that bad. The process of designing amplifiers from transistors inevitably boils down to the problems of biasing and small-signal design, as explained above. The particulars of biasing of bipolars and FETs may be somewhat different, but the considerations and methods are, in a broad sense, the same. Once the operating point is set, the transistor may be reduced to a two-port model (e.g., *s*-parameters), which is essentially technology-independent; thus the design process from that point proceeds in essentially the same way regardless of the device family in use.

10.3 BIASING OF TRANSISTOR AMPLIFIERS

Biasing involves designing a circuit around the transistor that results in DC operation at the desired operating point, and which facilitates interfacing of the AC (in our case, RF) signal. The operating point determines the small-signal two-port behavior – e.g., the *s*-parameters – so once biased, the circuit can be reduced to a two-port model for subsequent analysis of RF behavior and the design of impedance matching networks, as described in Section 10.4.

10.3.1 Bipolar Transistors

The most common (but certainly not only) topology for an RF amplifier using an NPN bipolar transistor is the *common emitter* configuration. In this configuration, the input signal is applied to the base, the output signal is extracted from the collector, and the emitter serves as a datum. For bipolar transistors, the common emitter configuration results in high current gain and high output impedance, which is often what is needed in an amplifier. For simplicity, we will assume common emitter configuration throughout this section, and defer discussion of other configurations to Section 10.3.3.

Once the configuration is settled, there are two possible strategies for selecting the operating point. The classical and rigorous strategy is based on an analysis of the *characteristic curves* of the transistor. For a bipolar transistor in common emitter configuration, these curves are simply plots of I_C as a function of V_{CE} for various values of I_B. The trick is to choose a combination of I_C and V_{CE} that yields desirable small-signal characteristics and allows sufficient "swing" in the output signal with respect to the collector supply voltage V_{CC} while keeping the transistor firmly biased in the forward-active mode (see e.g. [72, Ch. 5]). This strategy is not always practical because such data are typically not available; especially for UHF and above. The simpler and more common strategy is simply to refer to the manufacturer's datasheet for a recommended operating point. For RF transistors of modern vintage, manufacturers typically document a set of nominal operating points in advance, providing also the associated *s*-parameters, noise figure parameters, and other potentially useful information. Even for older transistors, the trend is provide data that allow one to quickly grasp the range of reasonable operating points and the implications of various choices within that range, in lieu of characteristic curves. Alternatively, a sufficiently detailed equivalent circuit ("device model") for the transistor can be used in simulation to determine a suitable bias point.

EXAMPLE 10.1

The Avago AT-41511 (previously considered in Chapter 8) is a modern NPN Si BJT. Seek out the datasheets for this device and determine some reasonable operating points.

Solution: The datasheet for this device is readily available on the internet, through the manufacturer. You'll find that this device is very well-characterized for $I_C = 5$ mA and 25 mA, and $V_{CE} = 2.7$ V and 5 V; i.e., four operating points. The datasheet provides *s*-parameters and noise figure characterization data for these operating points. This is certainly not to say that other operating points are not possible or recommended; however, this represents a pretty good set of options for our initial design. For many applications, one of these would probably be sufficient without modification.

EXAMPLE 10.2

An example of an older NPN Si BJT still appearing in radios is the 2N3904, offered by a variety of manufacturers (and under a variety of different model names, associated with different packaging options). Seek out the datasheet and determine some reasonable operating points.

Solution: The datasheets for this device are readily available on the internet. Different manufacturers will provide the data in somewhat different formats. Nevertheless, you probably will not find specific operating point recommendations. You probably will see a lot of data for I_C in the range 0.1–100 mA and V_{CE} in the range 1–10 V, which indicate relevant tradeoffs and can be used to narrow down the range of possibilities. In the absence of any particular constraint, it should be apparent that a pretty good starting point for the 2N3904 is $I_C = 10$ mA and $V_C = 5$ V, as the device is typically well-characterized for operating points near those values. You will most likely not find *s*-parameter or noise figure characterization data provided as part of the datasheet. These data are typically available elsewhere (e.g., as data files intended for computer-aided design software), or can be extracted from an analysis of a device model (e.g., a SPICE netlist).

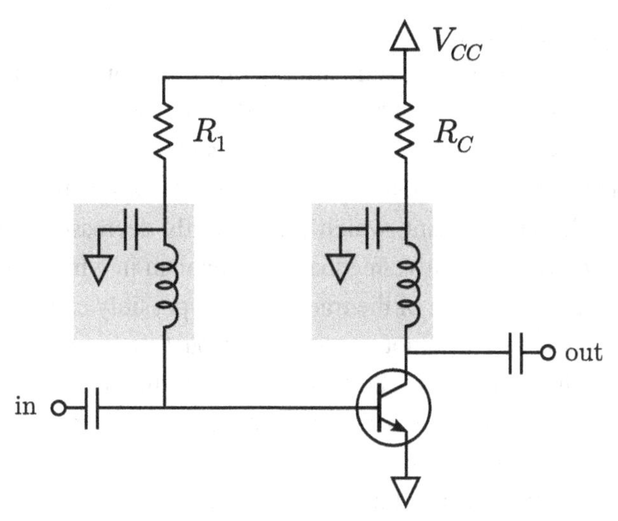

Figure 10.4. Fixed biasing of a bipolar transistor in common emitter configuration (not recommended). Shaded components are associated with RF/DC isolation.

Once the operating point is decided, a circuit that sets those values can be designed. For bipolar transistors, there are a variety of techniques, described below.

Fixed Bias. Fixed bias for common emitter bipolar transistors is shown in Figure 10.4. First, note some elements that tend to appear in all bias schemes: Capacitors at the input and output terminals isolate the bias (DC) voltages and currents from the RF input and output. The inductor–capacitor "L" circuits appearing within the shaded boxes in Figure 10.4 form lowpass filters intended to keep RF signals out of the power supply. Any of the inductors or capacitors shown in this figure may or may not actually be necessary, depending on the specific application. We include them here for two reasons: (1) To account for the possibility that they may be necessary; and (2) we will find later that these components may need to be considered in the small-signal design, and may in fact turn out to be useful for impedance matching. For purposes of our discussion of biasing, it is safe to assume all capacitors are open circuits and all inductors are short circuits.

The idea behind fixed bias is pretty simple: First set I_C using R_1, and then set V_{CE} using R_C. To see how this works, first apply Kirchoff's Voltage Law (KVL)[3] to the path from V_{CC} through the base to ground:

$$V_{CC} - I_B R_1 - V_{BE} = 0 \tag{10.4}$$

In forward-active mode, $I_B = I_C/\beta$. Making this substitution and solving for I_C we find

$$I_C \cong \beta \left(\frac{V_{CC} - V_{BE}}{R_1} \right) \tag{10.5}$$

In forward-active mode $V_{BE} \approx 0.65$, so I_C is determined by β, V_{CC}, and R_1. Also we see

$$V_{CE} = V_{CC} - I_C R_C \tag{10.6}$$

and so V_{CE} is determined by R_C once I_C has been set.

Despite its apparent elegance, fixed bias is usually a bad idea. A big problem is *thermal stability*. The problem arises as follows: V_{BE} varies somewhat as a function of physical

[3] That is, the sum of voltages along a closed path equals zero.

temperature; as a rough rule of thumb, V_{BE} decreases by about 2 mV for each 1 °C increase in temperature. Electronic devices increase in temperature after being turned on, so what happens is this: V_{BE} decreases slightly, so I_B increases slightly. $I_C = \beta I_B$, and β is a large value, so I_C increases by a lot. This is already bad, because the transistor is straying from its intended operating point. However, what also happens is that the increase in I_C causes the temperature of the transistor to increase an additional amount, which subsequently decreases V_{BE} further. At this point, we have *thermal runaway*: a positive feedback mechanism in which I_C is driven to ever larger values, increasing the temperature of the transistor and possibly causing damage.

Another problem with fixed bias is that it is not particularly robust to variations in device characteristics. In particular, variations in β translate to variations in the operating point and subsequently in the *s*-parameters. This is a concern because values of β do vary considerably, even between apparently identical devices.

Self Bias. A simple remedy to both thermal runaway and robustness to variation in β is *collector feedback stabilization*, illustrated in Figure 10.5. In this simple modification to fixed bias, the collector is used as the supply for the base bias. Now when I_C begins to increase due to increasing temperature, V_C decreases, and so the increase in I_B due to decreasing V_{BE} is mitigated.

This scheme also reduces the dependence of the operating point on β. To see how that part works, first note that the voltage drop across R_C is $(I_B + I_C)R_C$. However, $I_C \gg I_B$, because β is large. So the voltage drop across R_C can be approximated as $I_C R_C$. Now invoking KVL:

$$V_{CC} - I_C R_C - I_B R_1 - V_{BE} \cong 0 \tag{10.7}$$

Making the substitution $I_B = I_C/\beta$ and solving for I_C, we find:

$$I_C \cong \frac{V_{CC} - V_{BE}}{R_C + R_1/\beta} \tag{10.8}$$

We can make this result insensitive to β by making $R_C \gg R_1/\beta$. Then:

$$I_C \approx \frac{V_{CC} - V_{BE}}{R_C} \text{ for } R_C \gg \frac{R_1}{\beta} \tag{10.9}$$

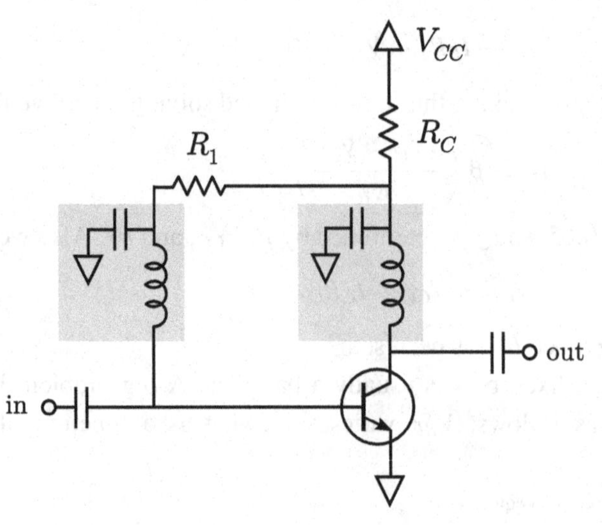

Figure 10.5. Collector feedback stabilization, a self-biasing technique here applied to a bipolar transistor in common emitter configuration.

You might be tempted to simply set $R_1 = 0$. Two pretty good reasons not to do that are current limiting and power consumption. Concerning power consumption: The smaller R_1 is, the more current is dissipated, and without much to show for it. Thus, R_1 is chosen according to a tradeoff between sensitivity to β and temperature on one hand, and power consumption on the other.

Summarizing: The scheme is to use R_C to set I_C, and to use R_1 to make sure $R_C \gg R_1/\beta$, but not by too much. Note that you cannot set V_{CE} independently in this scheme, since

$$V_{CE} = V_{CC} - (I_C + I_B) R_C \tag{10.10}$$

and we have run out of free parameters in the design. Nevertheless, the simplicity of this scheme makes it very attractive. The idea of using one part of the bias circuit to regulate another part is known generally as *self bias*. This idea shows up in a variety of other schemes, as we shall see.

EXAMPLE 10.3

An Si NPN BJT has a DC current gain in the 60–130 range and is to be biased to $I_C = 5$ mA using an 8 V supply. Design a bias circuit employing collector feedback stabilization.

Solution: The relevant schematic is Figure 10.5. Since we are dealing with an Si BJT, $V_{BE} \cong 0.65$ V. The design should satisfy Equation (10.9), so we have $R_C \cong (V_{CC} - V_{BE})/I_C = 1.47$ kΩ. It is also required that $R_1 \ll \beta R_C$. The smallest value of β is 60, so we require $R_1 \ll 88.2$ kΩ. Now R_1 should be small enough to make the operating point robust to changes in β, but selecting R_1 to be too small is detrimental to power consumption. One way to make a decision is to work out the results for a few possibilities over a range of candidate values; say 10 kΩ, 1 kΩ, and 100 Ω. It is best to do this using exact expressions. We have the following exact relationship from KVL:

$$V_{CC} - (I_B + I_C) R_C - I_B R_1 - V_{BE} = 0 \tag{10.11}$$

Using $I_C = \beta I_B$ to eliminate I_B and solving for I_C, we obtain

$$I_C = \frac{V_{CC} - V_{BE}}{(R_C + R_1)/\beta + R_C} \tag{10.12}$$

We can obtain V_{CE} exactly from Equation (10.10). First, let us consider $\beta = 60$:

For $R_1 = 10$ kΩ: $I_C = 4.42$ mA, $V_{CE} = 1.39$ V; Total current 4.50 mA

For $R_1 = 1$ kΩ: $I_C = 4.86$ mA, $V_{CE} = 0.73$ V; Total current 4.94 mA

For $R_1 = 100$ Ω: $I_C = 4.91$ mA, $V_{CE} = 0.66$ V; Total current 4.99 mA

If we increase β to 130, we find:

For $R_1 = 10$ kΩ: $I_C = 4.72$ mA, $V_{CE} = 1.01$ V; Total current 4.75 mA

For $R_1 = 1$ kΩ: $I_C = 4.94$ mA, $V_{CE} = 0.69$ V; Total current 4.97 mA

For $R_1 = 100$ Ω: $I_C = 4.96$ mA, $V_{CE} = 0.65$ V; Total current 5.00 mA

As expected, the benefit of reducing R_1 is that I_C is closer to the design value and less dependent on β. The disadvantage of reducing R_1 is significantly increased power consumption, as represented above by the total current. Since total power is V_{CC} times total current, the penalty for reducing R_1 from 10 kΩ to 100 Ω is ≅11% increase in power consumption for $\beta = 60$ and ≅5% increase in power consumption for $\beta = 130$. With this information and knowledge of the system requirements, the designer can now make an informed choice for R_1, and then may adjust the value of R_C slightly to achieve the specified collector current more precisely.

A typical amplifier output impedance is 50 Ω. If that were specified for this circuit, the inductor and capacitor on the collector supply would probably not be required, since R_C by itself would provide a very high impedance to RF relative to the specified 50 Ω. If the specified output impedance were higher, it might be necessary to use these reactances to increase the impedance into the power supply relative to the output impedance.

A similar consideration applies to the inductor and capacitor on the base supply: If the input impedance of the amplifier is specified to be 50 Ω, then these would probably not be required for $R_1 = 10$ kΩ, but would probably be important for $R_1 = 100$ Ω.

The necessity of the DC-blocking capacitors at input and output depends on the adjacent circuits. For sure, these capacitors are necessary if the DC impedance looking out the input or output is small, or if the adjacent circuits assert a DC voltage at the input or output.

Emitter Degeneration. Sometimes you just cannot do without independent control of V_{CE} and I_C in setting the operating point. In this case you might consider *emitter degeneration*, illustrated in Figure 10.6. In this scheme, you start with a fixed bias, but add a resistor R_E to the emitter. Common emitter small-signal design requires the emitter to be grounded at RF, so a bypass capacitor is added in parallel to R_E.

To see why this is effective, consider KVL to the path from V_{CC} through R_1 to ground:

$$V_{CC} - I_B R_1 - V_{BE} - I_E R_E = 0 \tag{10.13}$$

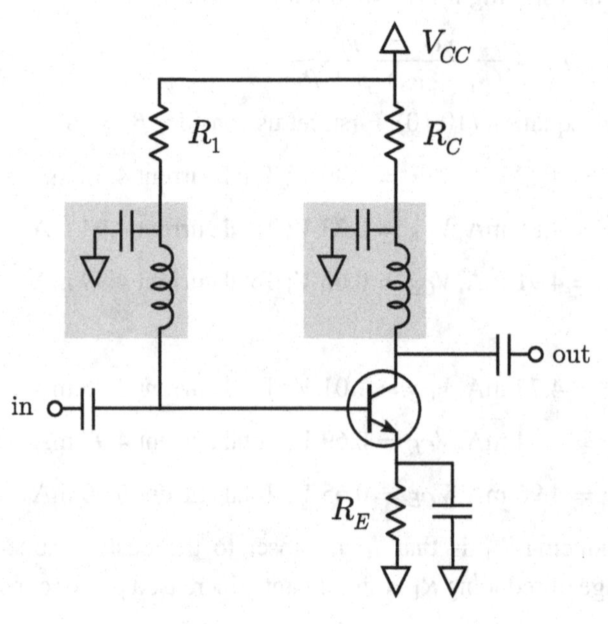

Figure 10.6. Emitter degeneration for a bipolar transistor in common emitter configuration.

Noting $I_E = I_B + I_C = (\beta + 1)I_B$, this equation may be solved for I_E:

$$I_E = \frac{V_{CC} - V_{BE}}{R_E + R_1/(\beta + 1)} \tag{10.14}$$

The above expression is a pretty good approximation to I_C, since $I_E \approx I_C$ when β is large. Thus:

$$I_C \approx \frac{V_{CC} - V_{BE}}{R_E} \text{ for } R_E \gg \frac{R_1}{\beta + 1} \tag{10.15}$$

In other words, the bias is unaffected by β to the extent that the condition on the right above is met.

A design strategy is as follows: Applying KVL to the path from V_{CC} through R_C to ground:

$$V_{CC} - I_C R_C - V_{CE} - I_E R_E = 0 \tag{10.16}$$

After substituting $I_E = I_C\,(\beta + 1)\,/\beta$, the above equation may be rearranged as follows:

$$R_C + R_E\frac{\beta + 1}{\beta} = \frac{V_{CC} - V_{CE}}{I_C} \tag{10.17}$$

The quantity on the right is a constant determined by the desired (or specified) bias conditions. Also, $(\beta + 1)\,/\beta \approx 1$, so we see that the sum $R_C + R_E$ must be approximately equal to (actually, slightly less than) this constant. We also need $R_E \gg R_1\,(\beta + 1)$, so an effective strategy is to set R_E to the largest standard value less than $(V_{CC} - V_{CE})\,/I_C$ and solve for R_C from the above equation. This leaves R_1, for which a suitable range of values can now be determined by using Equation (10.13):

$$R_1 = \frac{V_{CC} - V_{BE} - I_E R_E}{I_B} = \frac{V_{CC} - V_{BE} - I_C R_E\,(\beta + 1)\,/\beta}{I_C/\beta} \tag{10.18}$$

Finally, check to make sure the condition $R_E \gg R_1/\,(\beta + 1)$ is satisfied.

The result is a design in which robustness to temperature variation is achieved: When I_C increases due to increasing temperature, V_E increases, which increases V_B, subsequently reducing the voltage drop across R_1 and reducing I_B; just as in collector feedback stabilization.

Voltage Divider Bias with Emitter Degeneration. This scheme offers all the benefits of emitter degeneration and adds another degree of freedom to improve the tradeoff between robust biasing and power consumption. In this scheme, illustrated in Figure 10.7, you add a resistor R_2 between base and ground, forming a voltage divider which constrains the base voltage. To analyze this approach, first replace the voltage divider with its Thévenin equivalent circuit: a single resistor $R_B \equiv R_1 \parallel R_2$ between a supply voltage $V_{BB} \equiv V_{CC}R_2/(R_1 + R_2)$ and the base. KVL applied through this path to ground is

$$V_{BB} - I_B R_B - V_{BE} - I_E R_E = 0 \tag{10.19}$$

Noting $I_C = \beta I_B$ and $I_C \approx I_E$, we solve for I_C:

$$I_C \approx \frac{V_{BB} - V_{BE}}{R_E + R_B/\beta} \tag{10.20}$$

and so

$$I_C \approx \frac{V_{BB} - V_{BE}}{R_E} \text{ for } R_E \gg \frac{R_B}{\beta} \tag{10.21}$$

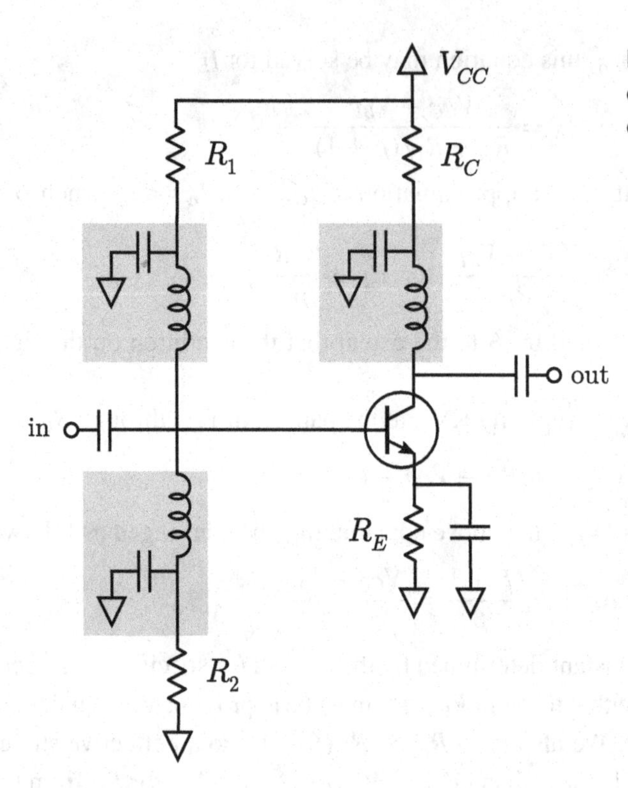

Figure 10.7. Voltage divider bias with emitter degeneration for a bipolar transistor in common emitter configuration.

which looks a lot like what we ended up with for emitter degeneration without the base voltage divider (see above). A big advantage over emitter degeneration alone has to do with power consumption: Neglecting I_B, the current consumed by the voltage divider to stabilize the bias with respect to β is inversely proportional to $R_1 + R_2$; however, the degree of stabilization is inversely related to $R_1 \parallel R_2$, which is less (perhaps much less) than $R_1 + R_2$. Therefore the degree of stabilization achieved for a given level of power consumption is considerably improved. A useful rule of thumb for an initial selection of R_1 and R_2 is to set $V_{CC}/(R_1 + R_2)$, the current through the voltage divider neglecting I_B, to be small relative to I_E; say $I_E/10$. This sets $R_1 + R_2$; then $V_{BB} = V_{CC}R_2/(R_1 + R_2)$ can be solved for R_2, leaving $R_1 = (R_1 + R_2) - R_2$. Finally, one should check the solution using KVL without approximation; it is not uncommon for the large number of approximations to accumulate into large errors.

EXAMPLE 10.4

Using base voltage divider bias with emitter degeneration, bias an NPN bipolar transistor to $I_C = 1$ mA and $V_{CE} = 5$ V using a supply voltage of +12 V. Assume β is in the range 80–125.

Solution: Since $I_E \approx I_C$, we find $R_C + R_E \approx (V_{CC} - V_{CE})/I_C = 7$ kΩ. It seems we can allocate this resistance between R_C and R_E however we'd like; let us arbitrarily choose $R_C = R_E = 3.5$ kΩ since that at least simplifies the bill of materials. From Equation (10.20) (or from inspection, since I_B should be very small) we find $V_{BB} \approx I_C R_E + V_{BE} = 4.15$ V. The base voltage divider (R_1 and R_2) must be such that $V_{BB} = 4.15$ V and the current flowing through R_2 is small compared to I_E. Using the "$I_E/10$" rule-of-thumb and again neglecting the base current, we choose $R_1 + R_2 \approx V_{CC}/(I_E/10) = 120$ kΩ. This fixes R_2 since $V_{BB} = V_{CC}R_2/(R_1 + R_2)$; we obtain $R_2 = 41.5$ kΩ and subsequently $R_1 = (R_1 + R_2) - R_2 = 78.5$ kΩ.

With this much approximation, it is wise to check the solution. We may obtain an exact expression for I_C from Equation (10.19) by substituting $I_B = I_C/\beta$ and $I_E = I_C(\beta + 1)/\beta$:

$$I_C = \frac{V_{BB} - V_{BE}}{R_B/\beta + R_E\beta/(\beta + 1)} \quad (10.22)$$

Similarly:

$$V_{CE} = V_{CC} - I_C R_C - I_C \frac{\beta}{\beta + 1} R_E \quad (10.23)$$

Now evaluating for $R_C = R_E = 3.5$ kΩ, $R_1 = 78.5$ kΩ, $R_2 = 41.5$ kΩ, and the minimum and maximum β:

For $\beta = 80$: $I_C = 0.988$ mA, $V_{CE} = 5.04$ V

For $\beta = 125$: $I_C = 0.992$ mA, $V_{CE} = 5.03$ V

These values are within a few percent of the specified operating point values, so our approximations were reasonable. It is worth noting that we did not deliberately consider the "$R_E \gg R_B/\beta$" requirement (Equation (10.21)); nevertheless, that requirement has been satisfied.

Self Bias with Base Voltage Divider. The primary drawback in emitter degeneration is potential difficulty in making R_E "disappear" at the frequency of operation, and without residual reactance to complicate the small-signal design. A safer alternative at UHF and above may be instead to "upgrade" the self-bias scheme with an additional resistor to create a base voltage

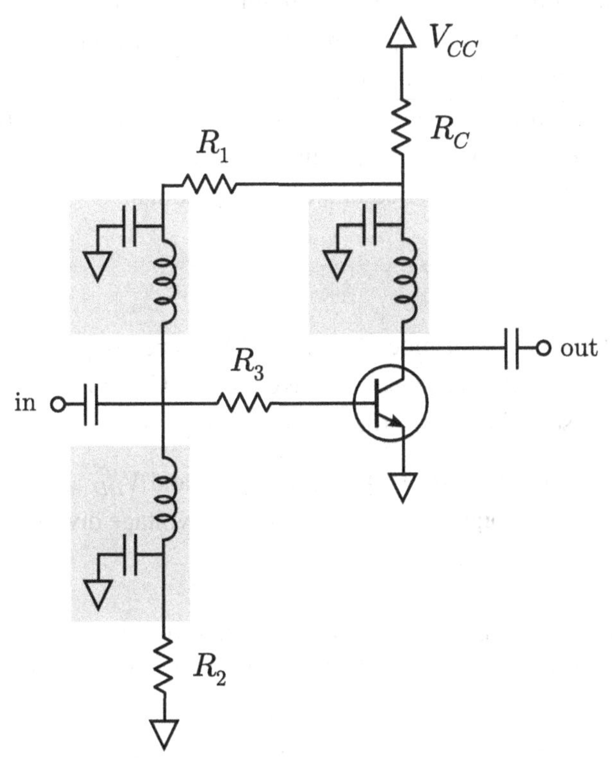

Figure 10.8. Self bias with base voltage divider. R_3 is optional and will be part of small-signal circuit; this is sometimes useful in stabilizing the transistor (see Section 10.4.3).

divider and thereby one additional degree of freedom. This is illustrated in Figure 10.8, with one additional tweak: A resistor at the base terminal for even better control of the tradeoff between stabilization and power consumption. The analysis and design of this scheme is very similar to those of the previous schemes.

Active Bias. In some applications, there is no combination of passive biasing schemes – by which we mean any of the preceding techniques – which provides an acceptable level of robustness to variations in temperature and β. In some designs it may be difficult to keep the operating point from shifting around in response to other things which are going on in the circuit, leading to fluctuations in V_{CC}, for example. In these cases, the alternative is to design and implement current and voltage sources around the transistor that fix the desired operating point, with feedback so as to "clamp" I_C and V_{CE} to their nominal values. This is known as *active bias*. The additional sources typically consist of transistors themselves, so it is not uncommon to encounter amplifier circuits which consist of several transistors, but in which only one is actually doing the work of amplification.

10.3.2 FETs

The FET version of the common emitter configuration is *common source* configuration: the input is AC-coupled to the gate, the output is AC-coupled from the drain, and the trick is to find an operating point (V_{DS}, I_D) that yields the desired gain with sufficient output signal swing relative to the supply voltage V_{DD}. Then V_{GS} can be determined by using Equation (10.3) (or read from the datasheet), and a bias circuit is designed to set these values with an eye toward thermal stability and minimizing sensitivity to manufacturing variations. As with bipolar transistors, manufacturers typically provide guidance as to nominal operating points.

For N-channel enhancement-mode devices, V_{GS} is positive and the techniques of Section 10.3.1 apply; in fact, biasing is considerably simplified by the fact that the gate current can be neglected, both because it is small and because it does not have much effect on the bias (remember: FETs are essentially *voltage*-controlled current sources). V_{GS} may also be positive for depletion-mode devices (N- and P-channel) at certain operating points.

EXAMPLE 10.5

The ATF-54143 is an N-channel enhancement-mode PHEMT. When operating at $I_D = 60$ mA and $V_{DS} = 3$ V, $V_{GS} = +0.6$ V. The worst-case gate leakage current (I_{GSS}) is specified to be 200 μA. Design a common source self-bias circuit for a supply voltage of +5 V.

Solution: The schematic is shown in Figure 10.9.[4] Here, both V_{GS} and V_{DD} are positive and the self-bias circuit is quite similar to the bipolar "self-bias with base voltage divider" circuit of

[4] Here we have drawn the circuit to emphasize the similarity to the bipolar self-bias scheme. However, it is common practice to draw schematics for FETs with all bias circuitry *below* the transistor. If a FET bias circuit seems confusing, it may help to redraw the circuit.

Figure 10.8. Note that a scheme analogous to the simpler self-bias circuit of Figure 10.5 would not work, since $I_G \approx 0$. The current I_1 through the voltage divider consisting of R_1 and R_2 should be as small as possible for low power consumption. However, I_1 must be large enough that variations in the gate current – however small – do not affect the bias. A typical tradeoff is to set $I_1 = 10 I_{GSS}$, which guarantees that any changes in the gate current will have a negligible effect on the bias. So, we will set $I_1 = 2$ mA. Thus, $R_1 \approx (V_{DS} - V_{GS})/I_1 = 1.2$ kΩ and $R_2 \approx V_{GS}/I_1 = 300$ Ω. Finally, $R_D \approx (V_{DD} - V_{DS}) / (I_D + I_1) = 32.3$ Ω.

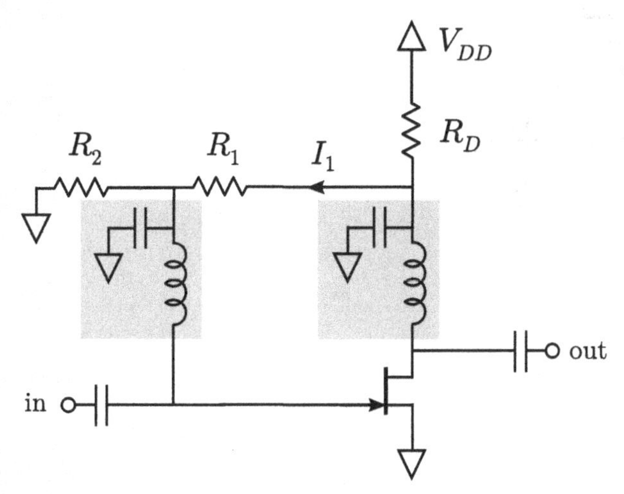

Figure 10.9. Self-biasing a FET with $V_{GS} > 0$ (see Example 10.5).

Note that R_D and R_2 have particularly low values compared with what we have seen in previous examples. If the RF input and output impedances have typical values; i.e., 50 Ω or greater, then the bias inductors and capacitors will be required – not optional – to keep RF out of the power supply.

V_{GS} is negative for P-channel enhancement-mode devices, as well as depletion-mode devices (N- and P-channel) at certain operating points. This makes biasing slightly more complicated. If the circuit already has positive and negative supplies, then the principles of the previous section still apply, with the minor difference that the gate should be tied to the negative supply. Often a negative voltage supply is not available. In this case the source terminal can be tied to a positive supply voltage V_{SS} that is less than V_{DD}, while the gate is tied to ground, causing V_{GS} to be negative. If dual positive supply voltages are not available either, then consider the self-biasing scheme shown in Figure 10.10. The trick here is that V_G is essentially tied to ground, since negligible current flows through R_1; whereas V_S will be positive due to the flow of current through R_S. Thus, V_{GS} will be negative. Although this facilitates single-supply operation, this scheme has the same drawback as emitter degeneration in bipolar transistor designs; namely that the bypassing of R_S necessary to AC-ground the source terminal for small-signal design may be difficult to implement at high frequencies.

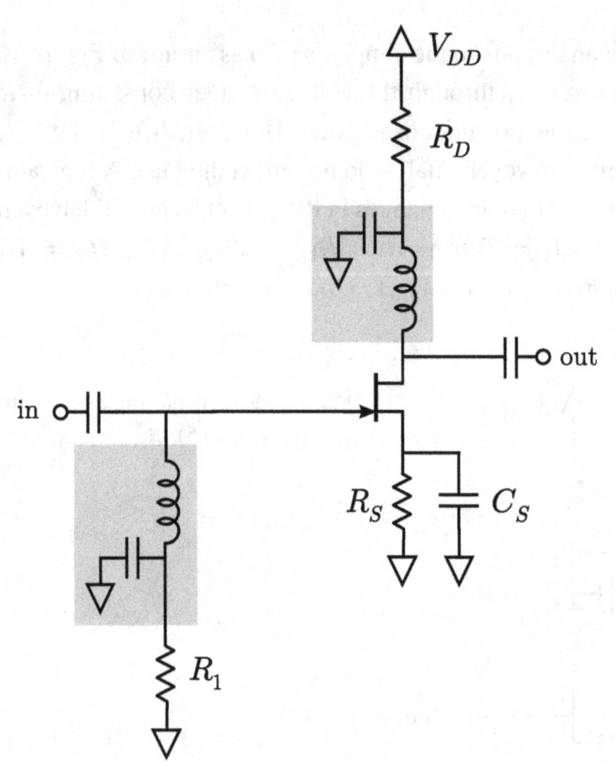

Figure 10.10. Self-biasing a FET with $V_{GS} < 0$ (see Example 10.6).

EXAMPLE 10.6

The ATF-34143 is an N-channel depletion-mode PHEMT. We seek an operating point of $I_D =$ 20.8 mA at $V_{DS} = 2$ V, with $V_{GS} = -0.25$ V. Design a common source self-bias circuit for a supply voltage of +2.5 V.

Solution: Here, V_{GS} is negative, and we have only a positive-voltage supply. Therefore, we set $V_G \approx 0$ by connecting it to ground through R_1. The value of R_1 is not critical since there is negligible current flowing through it.[5] Now any current flowing through R_S will make V_{GS} negative, as desired. Applying KVL while neglecting I_G, we find

$$V_{DD} - R_D I_D - V_{DS} - R_S I_D \cong 0 \tag{10.24}$$

so $R_D + R_S \cong (V_{DD} - V_{DS})/I_D = 24\ \Omega$. Since $V_S = I_D R_S$, $V_{GS} = -I_D R_S$ and therefore $R_S = 12\ \Omega$. Now $R_D = 24\ \Omega - R_S = 12\ \Omega$, which completes the design. As in the previous example, the low values of R_S and R_D will require the use of the bias inductors and capacitors to keep RF out of the power supply.

[5] In this case the manufacturer recommends 50 Ω, which optimizes low-frequency stability for this particular transistor.

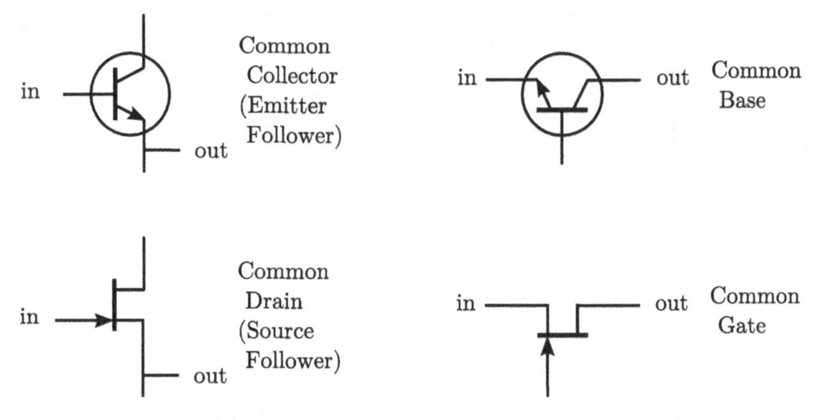

Figure 10.11. Alternatives to common emitter and common source configurations for NPN (bipolar) and N-channel (FET) transistors.

10.3.3 Beyond Common Emitter and Common Source

As explained above, the common emitter configuration (for bipolar transistors) and common source configuration (for FETs) account for most of the amplifier designs seen in radio circuits. However, transistors are three-terminal devices, and so there are two other configurations for attaching input and output signals, shown in Figure 10.11. If you need *low* output impedance as opposed to high output impedance, then the better choice is the common collector (bipolar) and common drain (FET) configurations. A bipolar transistor in common collector configuration is also known as an *emitter follower*. These configurations tend to show up in applications where the output needs to drive a low-impedance load with a lot of current – the classic example is a *cable driver*, which is a low-gain amplifier whose output drives a long section of coaxial transmission line. If you require low *input* impedance and are OK with relatively low gain, then the better choice is the common base (bipolar) and common gate (FET) configurations. A typical application for these configurations is in some low-noise amplifiers that are directly attached to resonant antennas with low resistive impedance. In this application, low gain is an advantage because the lack of preselection (see Section 15.4) makes the amplifier vulnerable to compression due to out-of-band signals unless the gain is low.

10.4 DESIGNING FOR GAIN

Once an operating point is selected, the single-transistor amplifier design problem can be addressed as a "small-signal" (AC) two-port synthesis problem at the frequency or frequencies of interest. The basic topology is shown in Figure 10.12, and consists of three two-ports: An *input matching network* (IMN), a two-port representation of the transistor as determined by the selected configuration (e.g., common emitter) and operating point, and an *output matching network* (OMN). As in any small-signal circuit analysis, DC supply voltages are interpreted as ground, and reactances which are not effectively open-circuited or short-circuited at the frequency of interest may be so represented. In particular, bias circuit reactances which are not

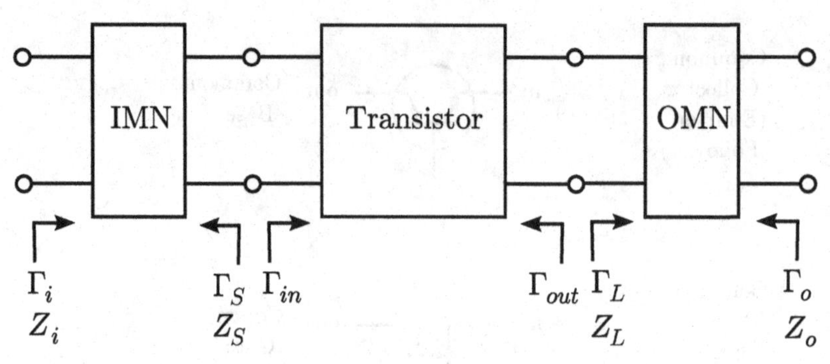

Figure 10.12. Small-signal amplifier design problem.

effectively open- or short-circuited at the relevant frequencies must be addressed as part of the IMN or OMN.

Entering the small-signal design problem, one has in mind goals for gain (by which we typically mean TPG), noise figure, center frequency, bandwidth, input impedance, and output impedance. The problem is therefore to identify the IMN output impedance Z_S and OMN input impedance Z_L that best satisfy the goals while ensuring stable operation. In this section we introduce methods suitable for meeting gain requirements; in Section 10.5 we will consider a simple modification to these techniques that facilitates optimization of noise figure, as well as joint specification of gain and noise figure. Fair warning: It is assumed you have mastered the theory presented in Chapters 8 and 9.

10.4.1 Bilateral Design to Meet a Gain Requirement

A principal decision entering the design process concerns gain. The top-level choices are that one might seek to maximize gain, or alternatively one might seek to achieve a specified gain. If the goal is to maximize gain at a specified frequency, then the first thing to consider is whether simultaneous conjugate matching (Section 8.5.3) yields a stable solution. If it does, then all you need to do is design an IMN that yields the necessary Z_S, an OMN that yields the necessary Z_L, and then do the bias circuit integration.

EXAMPLE 10.7

In Example 10.3, we self-biased an Si NPN BJT in common emitter configuration. Continuing: Design an amplifier that delivers maximum gain at 25 MHz, and which has 50 Ω input and output impedances. For the configuration and operating point established in Example 10.3, the s-parameters with respect to $Z_0 = 50\ \Omega$ at 25 MHz are as follows: $s_{11} = 0.3\angle 160^\circ$, $s_{12} = 0.03\angle 62^\circ$, $s_{21} = 6.1\angle 65^\circ$, $s_{22} = 0.40\angle -38^\circ$.

Solution: We begin by assessing the stability of the transistor. In this case we have

$$\Delta = s_{11}s_{22} - s_{12}s_{21} = 0.0643\angle -43.6^\circ \tag{10.25}$$

$$K = \frac{1 + |\Delta|^2 - |s_{11}|^2 - |s_{22}|^2}{2|s_{12}s_{21}|} = 2.06 \tag{10.26}$$

Since $K > 1$ and $|\Delta| < 1$, the transistor is unconditionally stable. Therefore, simultaneous conjugate matching should yield a stable design. Next, let us anticipate what gain we can anticipate. Since the transistor is unconditionally stable, we have

$$\text{max TPG} = \text{MAG} = \left|\frac{s_{21}}{s_{12}}\right| \left(K - \sqrt{K^2 - 1}\right) = 17.2 \text{ dB} \tag{10.27}$$

This will be useful later as a check of the completed design. The source and load reflection coefficients for simultaneous conjugate matching are, from Equations (8.52)–(8.57):

$$\Gamma_S = \Gamma_{S,max} \equiv \frac{B_1 - \sqrt{B_1^2 - 4|M|^2}}{2M} = 0.410 \angle -161.1^\circ \tag{10.28}$$

$$\Gamma_L = \Gamma_{L,max} \equiv \frac{B_2 - \sqrt{B_2^2 - 4|N|^2}}{2N} = 0.485 \angle 37.3^\circ \tag{10.29}$$

This result can be checked by computing the TPG using Equation (8.41) with the above reflection coefficients. One finds $G_T = 17.2$ dB, which is equal to the MSG as computed above. Next we determine the source and load impedances. Using Equations (8.59) and (8.60):

$$Z_S = Z_0 \frac{1 + \Gamma_S}{1 - \Gamma_S} = 21.4 - j6.8 \ \Omega \tag{10.30}$$

$$Z_L = Z_0 \frac{1 + \Gamma_L}{1 - \Gamma_L} = 82.4 + j63.5 \ \Omega \tag{10.31}$$

Now we have what we need to design the IMN and OMN.

IMN Design. The IMN must have $Z_i = 50\ \Omega$ and $Z_S = 21.4 - j6.8\ \Omega$. Performance is specified at only one frequency (25 MHz), so presumably a narrowband match is adequate. Recalling Chapter 9, narrowband options are discrete two-component "L" matching and single-stub matching. A wavelength at 25 MHz is very large, and there should be no particular difficulty in finding suitable inductors and capacitors at 25 MHz, so the discrete "L" match is the best option. In terms of Figure 9.6, the transistor is the complex impedance on the left, and so "Z_S^*" in the figure corresponds to "Z_S" in the present problem. Similarly, "R_L" in the figure is $R_i = 50\ \Omega$ in the present problem. Recalling that the parallel reactance should be adjacent to the larger of the terminating impedances, we choose the "series-first" topology of Figure 9.6(a). In terms of the present problem, Equations (9.1) and (9.2) give us:

$$X_{par} = \pm\sqrt{\frac{R_i^2 R_S}{R_i - R_S}} = \pm 43.2\ \Omega \tag{10.32}$$

$$X_{ser} = +X_S - \mathcal{Im}\left\{\left(jX_{par}\right) || R_i\right\} = -31.6\ \Omega, +17.9\ \Omega \tag{10.33}$$

(note "$+X_S$" in the above equation as opposed to "$-X_S$" in Equation (9.2) due to the "Z_S^* vs. Z_S" issue noted above). The first value given for X_{ser} corresponds to the "+" solution for X_{par} and the second value corresponds to the "−" solution for X_{par}. The positive-valued reactances are inductors and the negative-valued reactances are capacitors. One uses the expressions

$$X = 2\pi f L \tag{10.34}$$

$$X = -\frac{1}{2\pi f C} \tag{10.35}$$

where $f = 25$ MHz to obtain the associated values for inductors and capacitors, respectively. The two resulting solutions for the IMN are as shown in Figure 10.13.

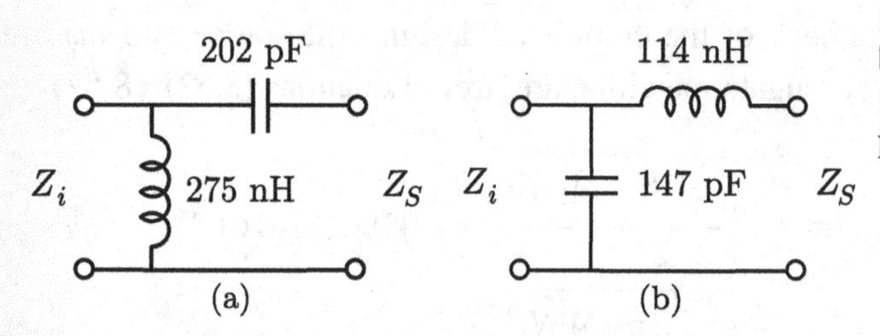

Figure 10.13. IMN candidates for Example 10.7. Impedances indicated are those looking *into* the labeled port.

Now: Which IMN candidate to choose? One consideration should be component values; that is, you may not want to choose a design that requires values that are smaller or larger than are commonly available for the specified frequency range, or maybe are awkward to generate from standard values. A second consideration is frequency response: Note that in this case one option exhibits a highpass response, whereas the other is lowpass. An additional consideration is integration with the bias circuitry. The problem statement specifies the self-bias design from Example 10.3, which uses the circuit prototype shown in Figure 10.5. Note that the bias circuit includes an optional capacitor in series with the input and an optional inductor in the base bias supply. If either IMN candidate had that same topology, that might be a good choice since the total number of components might be reduced. In the present case, it seems no such sharing of components is possible, and we have no particular requirements for component values or frequency response. In the completed schematic of Figure 10.15, we have arbitrarily chosen the parallel-capacitor option.

OMN Design. The OMN must present $Z_o = R_o = 50\ \Omega$ to the output and $Z_L = 82.4 + j63.5\ \Omega$ to the transistor. Noting that the parallel reactance should be adjacent to the larger of the terminating resistances, we now choose the "parallel-first" topology of Figure 9.6(b). In terms of the present problem, Equations (9.5) and (9.6) give us

$$(R_o - R_L)\, X_{par}^2 - (2X_L R_o)\, X_{par} + R_o\, |Z_L|^2 = 0 \tag{10.36}$$

(again, note the complex impedance is conjugate with respect to the original equations, as noted in the IMN design discussion, above)

$$X_{ser} = -\text{Im}\left\{ \left(jX_{par}\right) || Z_L^* \right\} \tag{10.37}$$

There are two valid solutions for X_{par}, namely -260.1Ω and $+64.2\Omega$. The corresponding solutions for X_{ser} are $+63.8\Omega$ and -63.8Ω, respectively. The associated OMN circuits are as shown in Figure 10.14. Since we have no particular requirements for standard values or frequency response, let us move directly to the issue of bias circuit integration. In this case, the parallel-inductor OMN (Figure 10.14(b)) is quite attractive, as the OMN components correspond topographically to reactances in Figure 10.5.

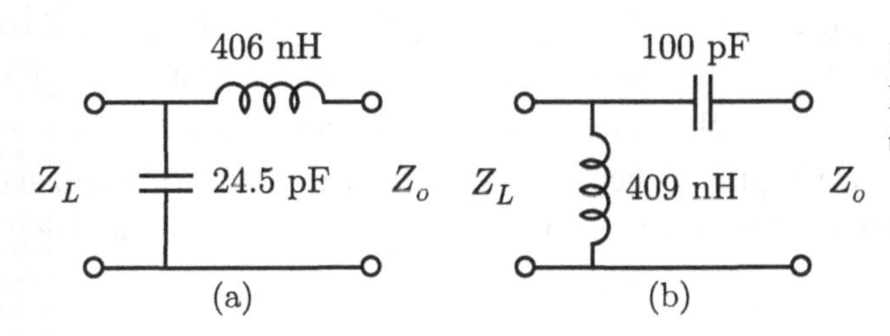

Figure 10.14. OMN candidates for Example 10.7. Impedances indicated are those looking *into* the labeled port.

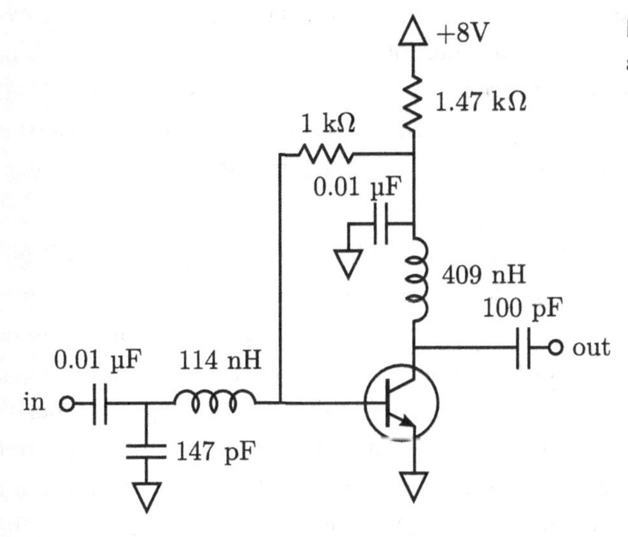

Figure 10.15. Completed maximum-gain amplifier (Example 10.7).

Bias Circuit Integration. The OMN after bias circuit integration is shown in the final schematic, Figure 10.15. The 100 pF capacitor does double-duty as OMN component and DC-blocking output capacitor. The 409 nH inductor does double-duty as OMN component and collector bias inductor.[6] The value of the bypass capacitor on the collector bias path was set to 0.01 µF so as to present a low impedance (just $-j0.6\ \Omega$) at 25 MHz, effectively grounding the 409 nH inductor so that it works properly as the OMN inductor while simultaneously providing additional isolation between the RF and the collector bias supply. Note that the impedance of the inductor is only about 64 Ω at 25 MHz, so it is not very effective as an RF choke. However, when combined with the adjacent bypass capacitor and large-valued bias resistors, the isolation is effective.

To block DC at the input, we have used another 0.01 µF capacitor, selected so as to not significantly change the nominal input impedance of 50 Ω at 25 MHz. For the base bias, we have chosen the "middle-of-the-road" value for R_1: 1 kΩ (see Example 10.3). This value is large enough that it allows us to eliminate the base bias inductor and capacitor, since the impedance looking into the base bias supply will be much greater than the impedance looking into the transistor.

[6] This is sometimes referred to as a *bias tee* configuration. A "bias tee" is a device that is used to combine or separate RF and DC signals appearing on the same signal path.

It is worth noting that conjugate matches tend to have relatively high Q, and thus tend to have bandwidth narrower than that which would be achieved using the same design modified for lower gain. This gives us one possible motivation for designing for lower gain.

If a gain or bandwidth specification indicates a need for less-than-maximum gain, or if simultaneous conjugate matching does not yield a stable solution, then here are some options you might consider:

- **Brute Force Search Method.** Impedances can be represented as reflection coefficients, and reflection coefficients must have magnitude less than or equal to 1 to ensure stability (Section 8.5.2). So, it is reasonable to consider a brute force search for all possible combinations of Γ_S (corresponding to Z_S) and Γ_L (corresponding to Z_L) for which the scheme of Figure 10.12 yields a stable design with the specified TPG. Simultaneously, one can evaluate stability for each combination of Γ_S and Γ_L considered and discard unstable solutions along the way (recall we did exactly this in Section 8.5.2).
- **Available Gain Method.** The brute force approach forces you to spend a lot of time looking at designs that fall far from the mark, and the method does not provide much insight into issues that might otherwise be exploited to improve the design. The available gain method provides an alternative approach. In this approach, one specifies a conjugate match at the transistor output, which then reduces the problem to that of selecting a Γ_S that yields the specified TPG. As we shall see shortly, all such solutions correspond to a circle in the real–imaginary plane. Recall that regions of stable operation are bordered by circles in the real–imaginary plane; thus, we may obtain a visual representation of all stable designs that meet a gain requirement. This is quite useful, especially since some solutions are likely to result in IMNs that are easier to implement than others.
- **Operating Gain Method.** Unlike the brute force approach, the available gain approach does not show all possible solutions; it reveals only those for which the output is conjugate-matched. Additional (but not all remaining) solutions can be found using the operating gain approach: One specifies a conjugate match at the *input*, which then reduces the problem to that of selecting a Γ_L from those which yield the specified TPG. Again, it turns out that all such solutions correspond to a circle in the real–imaginary plane.

The available gain and operating gain methods will be explained in a moment, but first a reminder: The transistor can be characterized in advance as being either unconditionally stable, or not. If the transistor is unconditionally stable, then no further consideration of stability is required: Any solution obtained using any of the methods above will result in a stable design. If on the other hand the transistor is not unconditionally stable, then there is no guarantee that any of above methods will yield a stable design for the specified gain. If that turns out to be the case, and there is no alternative to the selected transistor, then something has to be done to improve the stability of the transistor. Possibilities include resistive loss stabilization and neutralization. These are discussed in Section 10.4.3.

The available gain and operating gain methods are quite similar. Let us consider the available gain method first. *Available gain* G_A is formally defined as the ratio of the power delivered by the two-port (in this case, the transistor) into a conjugate-matched load, to the power available from the source. Alternatively, you can think of it this way: Arbitrarily select Z_S, and then conjugate-match the output. The resulting TPG, which is less than or equal to the MAG, is the available gain for that value of Z_S. Evaluating Equation (8.41) for $\Gamma_L = \Gamma_{out}^*$, one finds:[7]

$$G_A = \frac{1 - |\Gamma_S|^2}{|1 - s_{11}\Gamma_S|^2} |s_{21}|^2 \frac{1}{1 - |\Gamma_{out}|^2} \tag{10.38}$$

where Γ_{out} is (as shown in Figure 10.12) the embedded output reflection coefficient of the two-port representing the transistor. For convenience we define the *normalized available gain* g_A as the product of the first and third factors of the above equation, such that

$$G_A = g_A |s_{21}|^2 \tag{10.39}$$

All values of Γ_S for which g_A is a constant lie along a circle in the real-imaginary plane. The center and radius of an *available gain circle* are [20]:

$$C_A = \frac{g_A \left(s_{11} - \Delta s_{22}^*\right)^*}{1 + g_A \left(|s_{11}|^2 - |\Delta|^2\right)} \tag{10.40}$$

$$r_A = \frac{\left[1 - 2K |s_{12}s_{21}| g_A + |s_{12}s_{21}|^2 g_A^2\right]^{1/2}}{\left|1 + g_A \left(|s_{11}|^2 - |\Delta|^2\right)\right|} \tag{10.41}$$

The following example demonstrates the use of available gain circles for specified-gain amplifier design.

EXAMPLE 10.8

Using the available gain circle method, design a narrowband RF amplifier that achieves 14.1 dB gain at 1.5 GHz. The input and output impedances of the amplifier are to be 50 Ω. The transistor to be used is bipolar NPN in common emitter configuration. The operating point has been decided, but the bias circuit is not to be implemented at this time. For the selected operating point, the s-parameters with respect to 50 Ω are as follows: $s_{11} = 0.8\angle 120°$, $s_{12} = 0.02\angle 62°$, $s_{21} = 4.0\angle 60°$, $s_{22} = 0.2\angle -30°$.

Solution: We begin by assessing the stability of the transistor. In this case we have $\Delta = 0.101 \angle 65.3°$ and $K = 2.06$. Since $K > 1$ and $|\Delta| < 1$, the transistor is unconditionally stable. Therefore, any solution obtained using by the available gain method is guaranteed to be stable. Next, let us be certain that the transistor can deliver the specified gain. Since the transistor is unconditionally stable, The maximum TPG is equal to the MAG, which we find to be 17.1 dB. This is greater than the specified 14.1 dB, so there should be no difficulty in meeting the specification.

The first step in the available gain circle method is to compute the specified TPG normalized to the intrinsic forward gain:

$$g_A = \frac{G_A}{|s_{21}|^2} = \frac{25.7}{4.00^2} = 1.606 \tag{10.42}$$

[7] This is a somewhat difficult derivation. For details, the reader is referred to [20, Sec. 2.6].

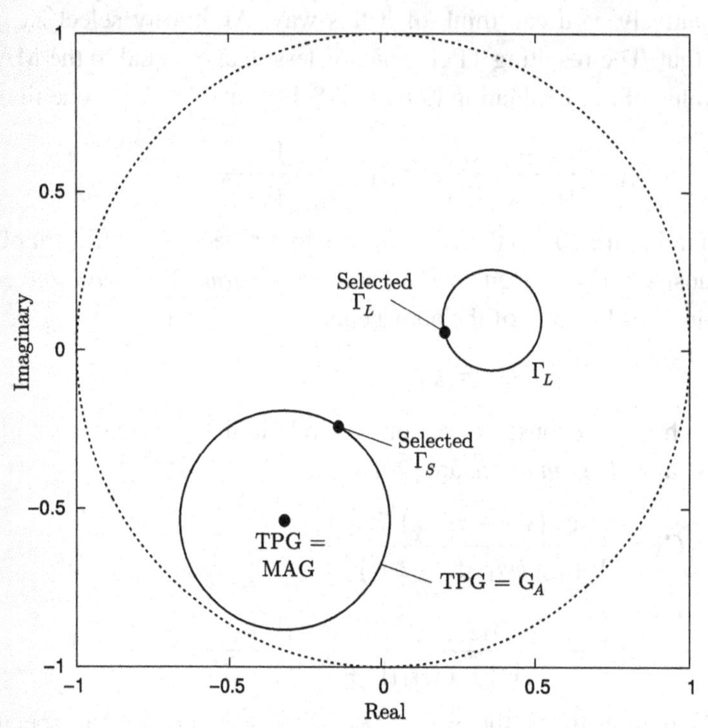

Figure 10.16. Example 10.8: The available gain circle is shown lower left. The circle in the upper right are the Γ_Ls associated with each Γ_S on the available gain circle.

The center C_A and radius r_A of the available gain circle are calculated using Equations (10.40) and (10.41), yielding $0.624\angle-121^\circ$ and 0.346, respectively. This circle is shown graphically in Figure 10.16. Any point along this circle corresponds to a value of Γ_S that delivers the specified gain when the output is conjugate-matched. How do we choose? First, consider that each possible Γ_S represents a different IMN, so you might choose a Γ_S that leads to a particularly convenient IMN design.[8] If there is no particular concern about the IMN design, then a simple-to-calculate but arbitrary approach is simply to choose the value of Γ_S having the minimum magnitude. The minimum-magnitude value of Γ_S is easy to calculate, since it corresponds to the point on the available gain circle which is closest to the origin. Thus:

$$\Gamma_S = (|C_A| - r_A)\frac{C_A}{|C_A|} \quad \text{for min. } |\Gamma_S| \tag{10.43}$$

which in this example is $0.278\angle-121^\circ$. Now we conjugate-match the output for the selected Γ_S:

$$\Gamma_L = \Gamma_{out}^* = \left[s_{22} + \frac{s_{12}s_{21}\Gamma_S}{1 - s_{11}\Gamma_S}\right]^* = 0.225\angle 26.2^\circ \tag{10.44}$$

where we have used Equation (8.36). The associated source and load impedances using Equations (8.59) and (8.60) are $Z_S = 33.9 - j17.6\ \Omega$ and $Z_L = 73.5 + j15.3\ \Omega$, respectively. At this point, it is a good idea to check Γ_S and Γ_L by confirming that Equation (8.41), with these reflection coefficients, yields the specified gain (it does).

[8] This is where knowing your way around a Smith Chart really pays dividends. Alternatively, you might consider calculating the Z_Ss that correspond to the Γ_Ss on the available gain circle, and look for attractive IMN candidates on that basis.

IMN Design. The IMN needs to transform 50 Ω to 33.9 − *j*17.6 Ω, with no particular constraint on bandwidth. At 1.5 GHz, suitable capacitors and inductors may be difficult to find. A wavelength at 1.5 GHz is about 10 cm in FR4, so single-stub matching is not exactly compact, but let us assume this is satisfactory for the present exercise. Given the specified input impedance, a natural choice for the characteristic impedance of the stripline is $Z_c = 50\ \Omega$. By using the procedure explained in Section 9.7.3, we obtain a design in which the primary line has length 0.020λ, with an open-circuited stub on the input side having length 0.083λ.[9]

OMN Design. Repeating the above procedure for the OMN, we obtain a design in which the primary line has length 0.107λ, with an short-circuited stub on the output side having length 0.181λ.

Bias Circuit Integration. The completed design is shown in Figure 10.17. Per the problem statement, the bias circuit is not indicated. However, a few points worth mentioning. First: The stub in the OMN is short-circuited, and you must be careful that this doesn't result in the collector being shorted to ground at DC. In this schematic, we have solved this problem by specifying a 1 nF capacitor between the collector and the OMN, which blocks DC while contributing only about −*j*0.1 Ω to the OMN input impedance at 1.5 GHz. Note that in this design the DC output impedance of the amplifier will be zero; if this is a problem then a second DC blocking capacitor could be added between the stub and the output, or a single blocking capacitor might be used at the end of the stub. Finally, a 1 nF capacitor is also indicated for the input to block the base bias.

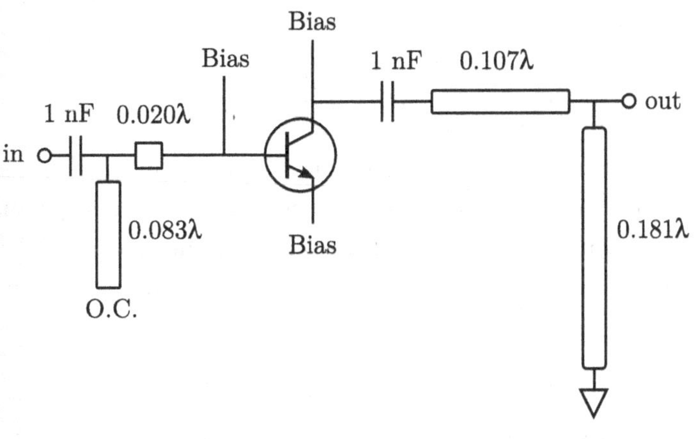

Figure 10.17. Completed design for Example 10.8. All stripline has characteristic impedance equal to Z_0.

As discussed above, an alternative to the available gain circle method is the operating gain circle method. *Operating gain* G_P is formally defined as the ratio of the power delivered to the load, to the power delivered to the two-port by a conjugate-matched source. Think of it this way: Arbitrarily select Γ_L, then select Γ_S to achieve a conjugate match; then G_P is the TPG, and $G_P \leq$ MAG. Evaluating Equation (8.41) for $\Gamma_S = \Gamma_{in}^*$, one finds [20]:

[9] Note you do not have to perform the design procedure to check this result; instead you can calculate the embedded impedances looking into the IMN and OMN and confirm they have the desired values.

$$G_P = \frac{1}{1-|\Gamma_{in}|^2}\,|s_{21}|^2\,\frac{1-|\Gamma_L|^2}{|1-s_{22}\Gamma_L|^2} \tag{10.45}$$

where Γ_{in} is (as shown in Figure 10.12) the embedded input reflection coefficient of the two-port representing the active device. We define the *normalized operating gain* g_P as the first and third factors of the above equation, such that

$$G_P = g_P\,|s_{21}|^2 \tag{10.46}$$

As in the available gain method, all values of Γ_L for which g_P is a constant lie along a circle in the real–imaginary plane, with center and radius given by [20]:

$$C_P = \frac{g_P\left(s_{22}-\Delta s_{11}^*\right)^*}{1+g_P\left(|s_{22}|^2-|\Delta|^2\right)} \tag{10.47}$$

$$r_P = \frac{\left[1-2K\,|s_{12}s_{21}|\,g_P+|s_{12}s_{21}|^2\,g_P^2\right]^{1/2}}{\left|1+g_P\left(|s_{22}|^2-|\Delta|^2\right)\right|} \tag{10.48}$$

The method of design is essentially the same as the available gain approach except you first select Γ_L using a gain circle, and then choose Γ_S to achieve a conjugate match.

10.4.2 Unilateral Design to Meet a Gain Requirement

Recall that a unilateral two-port is one for which s_{12} (the intrinsic reverse voltage gain) is zero. For a unilateral two-port, G_T becomes

$$G_{TU} \equiv \frac{1-|\Gamma_S|^2}{|1-s_{11}\Gamma_S|^2}\,|s_{21}|^2\,\frac{1-|\Gamma_L|^2}{|1-s_{22}\Gamma_L|^2} \tag{10.49}$$

which is simply Equation (8.41) with s_{12} set to zero. This form is useful because it isolates the contribution of the input match in the first factor and the contribution of the output match in the third factor. In particular, simultaneous conjugate matching simplifies to $\Gamma_S = s_{11}^*$ and $\Gamma_L = s_{22}^*$,[10] and substituting these values into Equation (10.49) we obtain the *maximum unilateral transducer power gain*

$$G_{TU,max} \equiv \frac{1}{1-|s_{11}|^2}\,|s_{21}|^2\,\frac{1}{1-|s_{22}|^2} \tag{10.50}$$

The form of Equation (10.50) facilitates a straightforward solution to specified-gain problems. Because the contributions of source and load matching are represented as separate factors in Equation (10.49), we may consider the terminations separately – literally, "unilaterally" – without consideration of the termination of the other port. This approach is known as *unilateral design*. This is in contrast to the methods of Section 10.4.1 in which the termination of both ports must be considered jointly, and which therefore is referred to as *bilateral design*.

The procedure for unilateral specified gain design consists of choosing Γ_S and Γ_L such that G_{TU} (in Equation (10.49)) equals the specified gain. There is a continuum of solutions,

[10] Note the same result can be obtained by solving for the values of Γ_S and Γ_L that maximize Equation (10.49).

representing different allocations of gain between the input match (first factor in Equation (10.49)) and the output match (third factor in Equation (10.49)). For this reason it is convenient to consider Equation (10.49) explicitly in terms of these factors; i.e.

$$G_{TU} = G_1 \left|s_{21}\right|^2 G_2 \tag{10.51}$$

The procedure is essentially the same for input and output. For input, the gain factor is

$$G_1 \equiv \frac{1 - |\Gamma_S|^2}{|1 - s_{11}\Gamma_S|^2} \tag{10.52}$$

For a given choice of G_1, the problem is to find values of Γ_S that satisfy the above equation. One option is simply to do a brute force search, which is especially easy since we need search over just one complex-valued parameter. However, there is once again a more elegant method based on the observation that the values of Γ_S that satisfy Equation (10.52) for a given G_1 correspond to a circle in the real–imaginary plane [20]. The center of this circle lies at

$$C_1 = \frac{g_1 s_{11}^*}{1 - \left|s_{11}\right|^2 (1 - g_1)} \tag{10.53}$$

where

$$g_1 \equiv G_1 \left(1 - \left|s_{11}\right|^2\right) \tag{10.54}$$

The radius of this circle is

$$r_1 = \frac{\sqrt{1 - g_1}\left(1 - \left|s_{11}\right|^2\right)}{1 - \left|s_{11}\right|^2 (1 - g_1)} \tag{10.55}$$

For the *output* match, the contribution to G_{TU} is a factor of

$$G_2 \equiv \frac{1 - |\Gamma_L|^2}{|1 - s_{22}\Gamma_L|^2} \tag{10.56}$$

The gain circles are given by the same expressions (Equations (10.53)–(10.55)) with each "1" replaced by "2" in the subscripts.

We will consider an example in a moment, but first we should consider the practical reality that no transistor is exactly unilateral. Nevertheless, the simplicity of unilateral design – relative to the bilateral methods of Section 10.4.1 – is so compelling that it is frequently of interest to assume unilateral behavior even if the transistor is not exactly unilateral. For this reason, it is useful to have some way to assess the error incurred in making the unilateral assumption. This is provided by the *unilateral figure of merit*, defined as

$$u \equiv \frac{\left|s_{11}s_{12}s_{21}s_{22}\right|}{\left|\left(1 - \left|s_{11}\right|^2\right)\left(1 - \left|s_{22}\right|^2\right)\right|} \tag{10.57}$$

For an exactly unilateral two-port, $u = 0$. To the degree that a two-port is not unilateral, u increases. Using this figure of merit, the error in TPG associated with the unilateral assumption is bounded as follows:

$$\frac{1}{|1+u|^2} < \frac{G_T}{G_{TU}} < \frac{1}{|1-u|^2} \tag{10.58}$$

This is demonstrated in the first step of the unilateral design example that follows.

EXAMPLE 10.9

Repeat Example 10.8, except now using the unilateral gain circles method, seeking a gain of 15.2 dB.

Solution: We previously identified that the transistor was unconditionally stable, so any solution from the unilateral gain circles method is guaranteed to yield a stable design. We previously noted that the MAG was 17.1 dB, so the specified gain of 15.2 dB is not out of the question. Next we consider the expected accuracy of the unilateral assumption in this case. From Equation (10.57), we find $u = 0.037$. From Equation (10.58), we find the associated range of TPG error is about ± 0.3 dB; we'll assume this is satisfactory. Using Equation (10.50) we find

$$G_{TU,max} \equiv \frac{1}{1-|s_{11}|^2}|s_{21}|^2\frac{1}{1-|s_{22}|^2} = 4.4 + 12.0 + 0.2 \text{ dB} = 16.6 \text{ dB}$$

The question is now where we should shed $16.6 - 15.2 = 1.4$ dB of gain. The options are reducing G_1 from 4.4 dB, reducing G_2 from 0.2 dB, or some combination of the two. One way to choose would be to figure out if there is any choice that yields a particularly attractive IMN and OMN design. For simplicity in this example, let us arbitrarily choose to address the input match only. Therefore, we choose for the gain associated with the input match to be reduced from 4.4 dB to 3.0 dB.

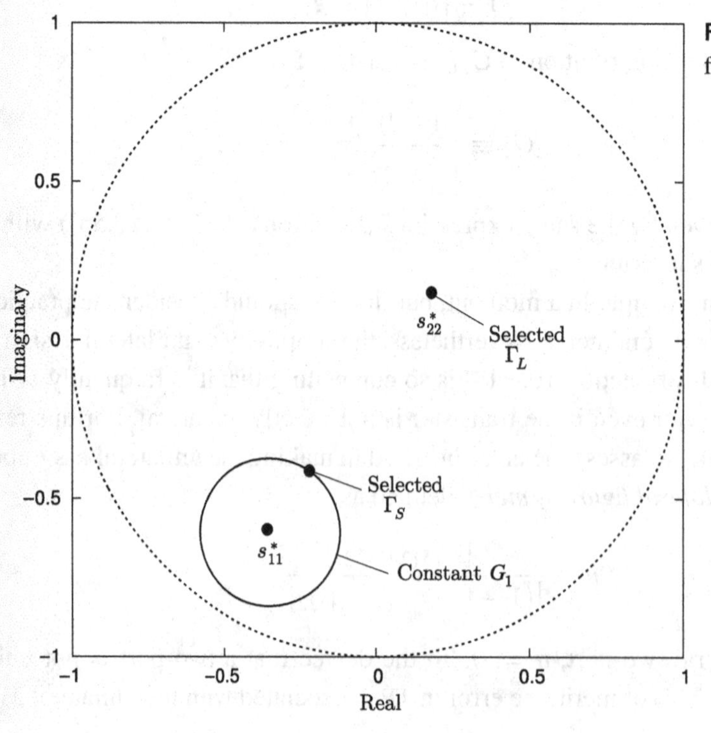

Figure 10.18. Unilateral gain circle for Example 10.9.

Next, we determine the source and load impedances. For the output, we have $\Gamma_L = s_{22}^*$. The associated load impedance is $Z_L = 69.2 + j14.4\ \Omega$. The range of possible Γ_S corresponds to a unilateral gain circle. First we find the normalized gain: From Equation (10.54), $g_1 = 0.715$. Then using Equations (10.53) and (10.55) we find the center and radius of the Γ_S unilateral gain circle to be $C_1 = 0.701\angle - 120°$ and $r_1 = 0.234$, respectively. This circle is shown graphically in Figure 10.18.

As in the previous example, we will arbitrarily choose the value of Γ_S that has the smallest magnitude (analogous to the calculation shown in Equation (10.43)): This is $\Gamma_S = 0.468\angle - 120°$. The associated source impedance is $Z_S = 23.1 - j24.0\ \Omega$.

Now it is a good idea to check that the values of Γ_S and Γ_L that we have obtained are sufficiently close to the gain specification. Using Equation (8.41) (exact TPG), we obtain $G_T =$ 15.3 dB, which is 0.1 dB above the specification. This is well inside the range of error indicated by the unilateral figure of merit.

IMN Design. The IMN is implemented as a single-stub match with a primary line of length 0.003λ, with an open-circuited stub on the input side having length 0.129λ.

OMN Design. The OMN is implemented as a single-stub match with a primary line of length 0.100λ, with a short-circuited stub on the output side having length 0.188λ.

Bias Circuit Integration. The only difference from the previous example (available gain, Figure 10.17) is the lengths of transmission lines.

10.4.3 Taming Unruly Transistors: Unilateralization and Stabilization

If the unilateral figure of merit is worse than you can bear, but you are determined to proceed with unilateral design anyway, then *unilateralization* is a possible course of action. Unilateralization is the addition of circuitry that results in s_{12} being reduced to zero, or at least to a negligible level. A common method of unilateralization is *neutralization*, in which a passive feedback path is added, connecting output to input through a reactive impedance. This is always a bit tricky, since the DC bias circuit is usually in the way, and it may be difficult to implement the necessary reactance over the necessary frequency range without creating instability or otherwise degrading the performance of the transistor.

It is not uncommon to find oneself in the position of having to use a transistor which is unstable in any design that would otherwise meet amplifier design requirements. In this case, *resistive loss stabilization* may be an option. The fundamental concept is pretty simple: Instability happens when internal reflections get out of control, so add loss to keep that from happening. The loss can be added in the form of a resistor in series or in parallel with either the output or the input, or in some combination of resistors. The s-parameters of series and parallel resistances are simple to determine (Section 8.2.2), and determining the s-parameters of a two-port cascade consisting a transistor and stabilization resistors is also a straightforward problem (Section 8.6). From the s-parameters of the cascade, it is possible to reevaluate stability (e.g., using the Rollett criterion) and to simultaneously determine the effect – usually deleterious – on gain. Although gain usually suffers, the advantage in terms of improved stability and bandwidth (since the Q is usually reduced in the process) is often worth the trouble. See Example 8.13 (Chapter 8) and Section 10.7.4 for examples.

10.5 DESIGNING FOR NOISE FIGURE

In the previous section, we considered the design of transistor amplifiers to meet a specification on gain. In many applications, noise figure is equally important or may be the primary consideration. The process of design to meet noise figure specifications is surprisingly similar to the process of design to meet gain specifications. We continue to assume that the amplifier can be described in terms of the IMN-transistor-OMN paradigm of Figure 10.12, and we assume that the IMN and OMN are lossless. Then the noise figure F of the amplifier is given by the following expression [20]:

$$F = F_{min} + \frac{4r_n \left|\Gamma_S - \Gamma_{opt}\right|^2}{\left(1 - |\Gamma_S|^2\right)\left|1 + \Gamma_{opt}\right|^2} \tag{10.59}$$

where Γ_{opt} is the value of Γ_S that minimizes F, F_{min} is that minimum value of F, and r_n is the *normalized noise resistance*, a unitless quantity. The normalized noise resistance is the *noise resistance* R_n (units of Ω), another intrinsic characteristic of the transistor, divided by the reference impedance Z_0; i.e.,

$$r_n = \frac{R_n}{Z_0} \tag{10.60}$$

Like s-parameters, Γ_{opt}, F_{min}, and r_n are characteristics of the transistor that depend on the configuration (e.g., common emitter), bias, and frequency.

Given these parameters, designing for minimum noise figure is easy: Set Γ_S to Γ_{opt}; then F will be equal to F_{min}. As long as the transistor is stable, Γ_L has no impact on F: This is because the output mismatch loss reduces signal and noise by the same factor, and so does not affect the output signal to noise ratio. Therefore, Γ_L can be used to adjust the gain once the noise figure has been minimized using Γ_S.

As it turns out, Γ_{opt} is often not close to the value of Γ_S required for maximum gain. As a result, it is usually not possible to design an amplifier that simultaneously maximizes gain and minimizes noise figure.

Let us now consider an example to demonstrate these concepts:

EXAMPLE 10.10

Design an LNA for 3.8 GHz using the NXP BFU725F/N1 SiGe NPN BJT. Minimize noise figure. Maximize gain to the extent that noise figure is not compromised. The input and output impedances are to be 50 Ω. Consider, but do not design the bias circuit. At the specified operating point and frequency, the s-parameters with respect to 50 Ω are $s_{11} = 0.513\angle{-158.8^\circ}$, $s_{12} = 0.063\angle 24.1^\circ$, $s_{21} = 8.308\angle 70.8^\circ$, $s_{22} = 0.322\angle -92.2^\circ$ and the noise parameters are $\Gamma_{opt} = 0.214\angle 68.8^\circ$, $F_{min} = 0.73$ dB, and $r_n = 0.0970$.

Solution: We begin as always by assessing the stability of the transistor. In this case we have $\Delta = 0.365 \angle -91.4^\circ$ and $K = 0.732$. Since $K < 1$, the transistor is only conditionally stable. Therefore we will need to check the stability of whatever solution we develop. To minimize noise figure, we choose $\Gamma_S = \Gamma_{opt}$. To maximize TPG subject to the constraint on Γ_S, we conjugate-match the output; i.e., we set $\Gamma_L = \Gamma_{out}^*$, where Γ_{out} is the embedded reflection coefficient looking into the output of the transistor. Using Equation (8.36) we obtain:

$$\Gamma_L = \Gamma_{out}^* = \left[s_{22} + \frac{s_{12}s_{21}\Gamma_S}{1 - s_{11}\Gamma_S} \right]^* = 0.302\angle 112.4^\circ$$

Now let us check if this solution is stable. It is apparent from the above result that $|\Gamma_{out}| < 1$. Using Equation (8.35) we find:

$$\Gamma_{in} = s_{11} + \frac{s_{12}s_{21}\Gamma_L}{1 - s_{22}\Gamma_L} = 0.685\angle -156.7^\circ$$

Since $|\Gamma_{in}| < 1$ as well, this design is stable.

The noise figure achieved is $F = F_{min} = 0.73$ dB. The TPG can be determined by using Equation (8.41) (exact TPG); we find $G_T = 18.5$ dB. The source and load impedances are $Z_S = 53.5 + j22.4\ \Omega$ and $Z_L = 34.4 + j21.1\ \Omega$, respectively.

IMN Design. Given that this is a narrowband SHF-band design, single-stub matching is advisable. The IMN has a primary line of length 0.047λ, with an short-circuited stub on the input side having length 0.184λ.

OMN Design. The OMN is implemented as a single-stub match with a primary line of length 0.194λ, with a open-circuited stub on the output side having length 0.090λ.

Bias Circuit Integration. The completed design (minus the bias circuit) is shown in Figure 10.19. Note that the short-circuited stub on the input must be either bypassed at DC using a blocking capacitor, or the base bias must be isolated from the IMN using a blocking capacitor. The latter scheme is used in this design. The value of both DC blocking capacitors is selected to be 100 pF, which is $-j0.4\ \Omega$ at 3.8 GHz. A larger capacitor would present a smaller RF impedance, but at somewhat greater risk of difficulties due to internal inductance and resistance.

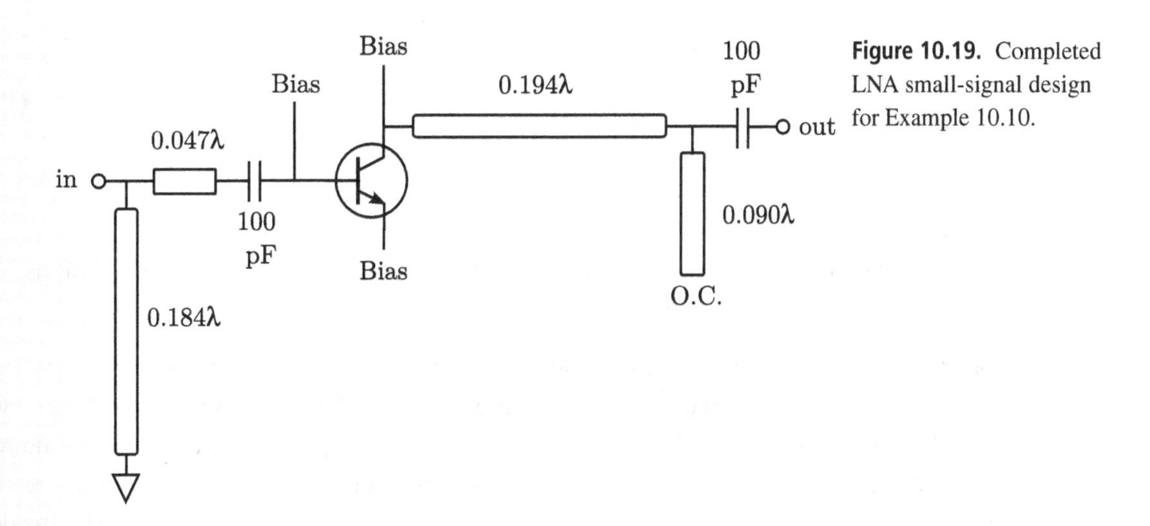

Figure 10.19. Completed LNA small-signal design for Example 10.10.

Recall that resistors generate noise (Section 4.2). This is an LNA, so a direct connection between bias resistors and RF signal paths should be avoided if possible. The use of lowpass LC circuits between bias resistors and the RF path, as previously discussed, is one option. However, at 3.8 GHz suitable inductors will be difficult to find. Another possibility is to employ a quarter-wave impedance inverter to apply the bias, as described in Section 9.8.

One final note before moving on: We will find in Example 10.12 that this design has very high input VSWR, which is a common consequence of unconstrained optimization of noise figure. We will deal with that issue in Example 10.13.

Often, the design problem is not simply to minimize noise figure, but rather to jointly optimize gain and noise figure. If the transistor is not unconditionally stable, then this optimization needs to also account for the limited domain of stable solutions. All this can be achieved using a brute force method: That is, one may search over the domain of reflection coefficients, evaluating stability, gain, and noise figure at each point, and then select from the subset of stable solutions. Alternatively, one may use stability circles combined with one of the "gain circles" methods (i.e., available gain, operating gain, or unilateral gain) to identify stable solutions, and then evaluate noise figure at each point on a gain circle within the stable domain.

However, as it turns out, all values of Γ_S for which Equation (10.59) yields a specified noise figure lie – you guessed it – on a circle in the real–imaginary plane. Therefore amplifiers may be designed using stability circles to identify stable domains, gain circles (by whatever method) to identify constant-gain contours, and noise figure circles to identify constant-noise figure contours; all this yielding a convenient "map" from which reflection coefficients may be chosen in an informed manner and with relatively little effort.

The center C_F and radius r_F of a circle of constant noise figure F are [20]:

$$C_F = \frac{\Gamma_{opt}}{1 + N_F} \tag{10.61}$$

$$r_F = \frac{\sqrt{N_F^2 + N_F\left(1 - |\Gamma_{opt}|^2\right)}}{1 + N_F} \tag{10.62}$$

where

$$N_F \equiv \frac{F - F_{min}}{4r_n}\left|1 + \Gamma_{opt}\right|^2 \tag{10.63}$$

EXAMPLE 10.11

Let us reexamine the problem posed in Example 10.10, this time considering solutions which tradeoff between noise figure and gain.

Solution: As always, we first consider stability, depicted in Figure 10.20. The input and output stability circles are determined by using Equations (8.45)–(8.48), yielding $C_L = 16.8\angle{-80.3^\circ}$, $r_L = 17.5$, $C_S = 4.8\angle 162.6$, and $r_S = 4.0$. It is found that $|\Gamma_{in}| < 1$ for $\Gamma_L = C_L$, therefore points inside the input stability circle (and inside the unit circle) are stable. Also, it is found that $|\Gamma_{out}| > 1$ for $\Gamma_S = C_S$, therefore points *outside* the output stability circle (and inside the unit circle) are stable.

For noise figure, we are concerned only with the source reflection coefficient. Referring to Figure 10.21, $F = F_{min}$ is achieved for $\Gamma_S = \Gamma_{opt} = 0.214\angle 68.8^\circ$; this is plotted first. All other realizable noise figures can only be greater than F_{min}. We consider a few representative values: In this case, $F = 1$ dB, 2 dB, and 3 dB. Using Equations (10.61) and (10.62), we find

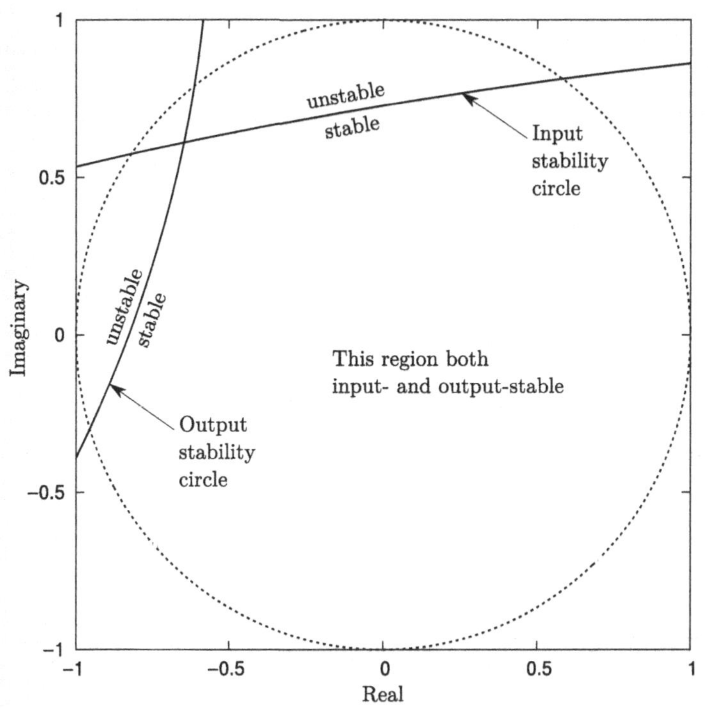

Figure 10.20. Stability circles for the LNA design problem of Example 10.11.

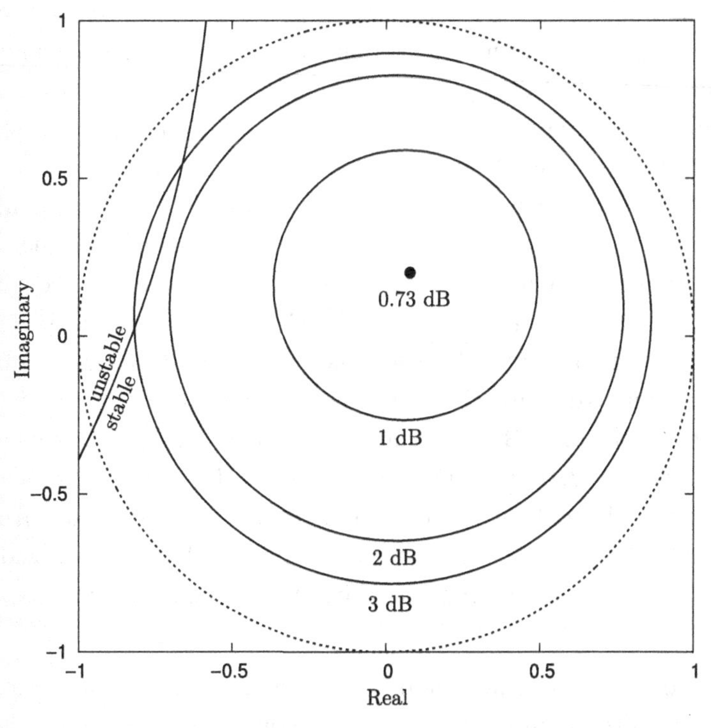

Figure 10.21. Noise figure circles for the LNA design problem of Example 10.11. Also shown is the output stability circle.

the associated values of C_F and r_F are $0.173\angle 68.8°$ and 0.428, $0.095\angle 68.8°$ and 0.737, and $0.061\angle 68.8°$ and 0.840, respectively. These are plotted on Figure 10.21. Note all solutions for noise figures of 1 dB and 2 dB are stable, but that some solutions for noise figure of 3 dB are unstable.

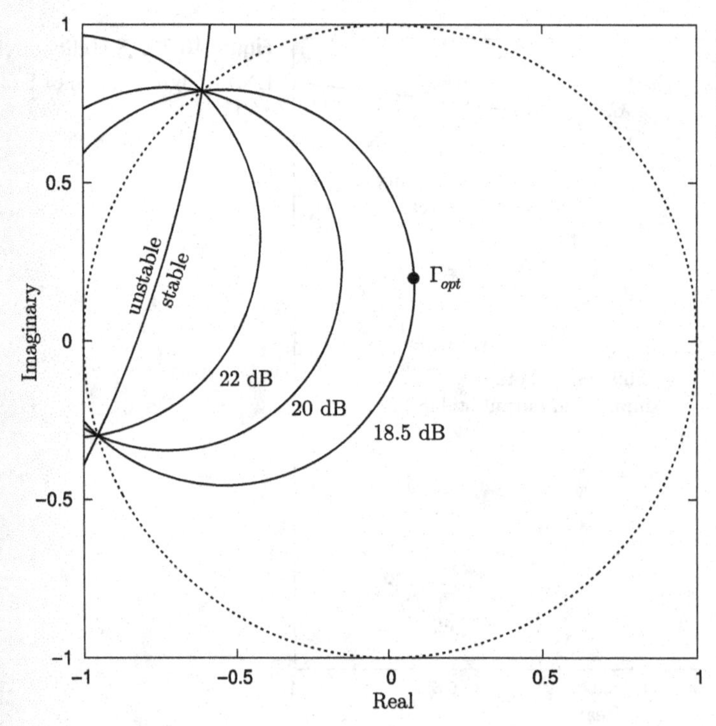

Figure 10.22. Available gain circles for the LNA design problem of Example 10.11. Also shown is the output stability circle and Γ_{opt}.

Next we consider gain. Since noise figure depends only on Γ_S, it is convenient to analyze gain using available gain circles. We determined in Example 10.10 that the TPG at the optimum noise figure was 18.5 dB, so let us declare that to be the available gain G_A and compute the associated available gain circle. The result is $C_A = 0.565\angle 162.6°$ and $r_A = 0.625$, and is plotted in Figure 10.22. Note that this circle intersects Γ_{opt}, as expected, and extends outside the unit circle. In other words, there are other choices for Γ_S which yield 18.5 dB, but all of them result in greater noise figure and some of them are potentially unstable. Two other available gain circles are shown, corresponding to $G_A = 20.0$ dB and 22.0 dB. Note that the gain circles contract toward the unstable region with increasing gain. Comparing to Figure 10.21, we see that the minimum noise figure associated with each of these circles also increases with increasing gain; for example, it is impossible to achieve $F = 1$ dB for $G_A \geq 22$ dB. Furthermore, it is clear that G_A much greater than 22 dB or so requires Γ_S to encroach on the output stability circle, which might be risky considering component tolerances. Thus Figures 10.20 and 10.21 provide a convenient "map" of the tradeoff between stability, gain, and noise figure, from which Γ_S may be decided in an informed manner. Of course, the input stability (associated with Γ_L and shown in Figure 10.21) should also be considered before making a final decision.

Simple modifications of the procedure demonstrated in the above example permit an analogous tradeoff analysis of solutions generated using the operating gain and unilateral gain methods.

10.6 DESIGNING FOR VSWR

Voltage standing wave ratio (VSWR) is a measure of the degree to which power is reflected from an interface, as opposed to being transmitted through the interface. Total transmission corresponds to VSWR = 1, whereas total reflection corresponds to VSWR$\rightarrow\infty$. Reflection is normally undesirable as this leads to frequency-domain ripple and, in the case of high-power systems, has the potential to cause damage. On the other hand, reflection is created by impedance mismatch, which is the very thing that we manipulate in amplifier design in order to set gain and noise figure. Thus, we have yet another tradeoff: low VSWR versus essentially everything else. Usually VSWR is regarded as a constraint in a "constrained optimization" statement of an amplifier design problem. For example, an LNA design problem might be articulated as follows: "Minimize noise figure subject to the constraint that the input and output VSWR shall not be greater than 2." (We'll consider this very problem in Example 10.13.)

VSWR is an embedded property of a two-port. Let us begin by considering the problem of how to calculate the input VSWR, VSWR_i. In terms of Figure 10.12 this is simply

$$\text{VSWR}_i = \frac{1+|\Gamma_i|}{1-|\Gamma_i|} \tag{10.64}$$

One way to perform this calculation is as follows: First calculate Γ_{in} from Γ_L and the s-parameters of the transistor using Equation (8.35), then calculate Γ_i from Γ_{in} and the s-parameters of the IMN using Equation (8.35) (again), and then finally apply Equation (10.64) to obtain VSWR_i. Since one is more likely to have Γ_S than the s-parameters of the IMN, it is usually more convenient to use the following expression [20, Sec. 2.8]:

$$|\Gamma_i| = \left|\frac{\Gamma_{in} - \Gamma_S^*}{1-\Gamma_{in}\Gamma_S}\right| \tag{10.65}$$

The corresponding equations for output VSWR are

$$\text{VSWR}_o = \frac{1+|\Gamma_o|}{1-|\Gamma_o|} \tag{10.66}$$

$$|\Gamma_o| = \left|\frac{\Gamma_{out} - \Gamma_L^*}{1-\Gamma_{out}\Gamma_L}\right| \tag{10.67}$$

EXAMPLE 10.12

Determine the input and output VSWR for the LNA designed in Example 10.10.

Solution: In the design problem the output was conjugate matched; i.e., $\Gamma_L = \Gamma_{out}^*$. From Equation (10.67) we find $|\Gamma_o| = 0$, and therefore $\text{VSWR}_o = 1$.[11] Using the values of Γ_S and Γ_{in} determined in Example 10.10, Equation (10.65) yields $|\Gamma_i| = 0.706$. Then using Equation (10.64), we find the input VSWR is 5.8.

[11] This may appear to contradict our previous understanding from Chapter 9 that conjugate matching does not necessarily minimize reflection. There is no contradiction: For example, if the OMN is a single-stub matching circuit, then there will be a standing wave within the OMN, but not at the output of the amplifier. The same is true at the input port.

The input VSWR in the above example is pretty high. Typically, VSWR is specified to be in the range 1–3. In that light, let us consider an example of a commonly-encountered design problem.

EXAMPLE 10.13

Revisiting Examples 10.11 and 10.12: Find Γ_S and Γ_L that minimize noise figure subject to the constraint that input VSWR is 2 or better and output VSWR is 2.5 or better. Of course it is also required that the transistor be stable for the selected reflection coefficients.

Solution: There are several ways to approach this problem. Here we consider a simple and direct brute force approach: We search over a grid of values of Γ_S and Γ_L for which both $|\Gamma_S| < 1$ and $|\Gamma_L| < 1$. This is a four-dimensional search since both reflection coefficients have real and imaginary components. For each trial value of Γ_S and Γ_L, the procedure is: (1) check stability using the criteria $|\Gamma_{in}| < 1$ and $|\Gamma_{out}| < 1$; (2) if stable, check to see if VSWR_i and VSWR_o meet requirements; (3) if VSWR requirements are met, then calculate F; If F is the lowest value encountered so far, then this choice of Γ_S and Γ_L is noted as the best solution so far. For a grid spacing of 0.05 in the real and imaginary coordinates, the "winner" is $\Gamma_S = 0.335\angle+153.4°$ and $\Gamma_L = 0.150\angle+90.0°$, for which $\text{VSWR}_i = 2.0$, $\text{VSWR}_o = 2.4$, $F = 0.9$ dB, and TPG $= 20.1$ dB.

The above example demonstrates why one might want to consider VSWR simultaneously with stability, gain, and noise figure in the amplifier design process. Thus it is sometimes useful to have a technique for mapping VSWR in the complex plane, as a function of Γ_S and Γ_L. As the reader might be prone to guess at this point, it turns out that contours of constant VSWR correspond to circles in the real–imaginary plane. *VSWR circles* are somewhat similar to stability circles, gain circles, and noise figure circles, but are a little bit trickier to use. For additional information, the reader is referred to [20, Sec. 3.8].

10.7 DESIGN EXAMPLE: A UHF-BAND LNA

Examples presented so far in this chapter have served primarily to illustrate a particular concept. Let us now pursue a design to conclusion so as to illustrate the various issues that arise in practical work. In this section we will design and demonstrate a UHF-band LNA based on the NXP BFG540W Si BJT. This example is based on an existing design (documented in the manufacturer's application note [57]), and is presented here as a beginning-to-end design problem.[12]

Here is the problem statement: Design a low-power LNA using the BFG540W in common emitter configuration, optimized for 400 MHz. In this case, "low power" should be interpreted as drawing no more than 10 mA from a +3V supply. The noise figure should be minimized, the

[12] Alternatively you could view this section as an explanation of the design presented in [57]. Here's a good exercise for students: Read the application note first, and see if you can figure out how and why the design works. Chances are you will encounter some mysteries. Then read this section to confirm your analysis and hopefully resolve the mysterious aspects.

Figure 10.23. The completed UHF-band LNA based on the NXP BFG540W Si BJT. See Figure 10.31 for schematic. The transistor is the rectangular four-terminal device near the center of the board. RF and DC input on left, RF output on right. The ruler along the bottom is marked in inches (top) and centimeters (bottom).

input and output interfaces should be 50 Ω single ended, and the associated VSWRs should be less than 2:1. A gain of about 15 dB is desired; 15 ± 2 dB would be acceptable.

Now, where to begin? A review of the datasheet is always a good idea. In this case the manufacturer's web site provides a datasheet, some application notes, *s*-parameters and noise figure parameters for a few operating points and frequencies, and an equivalent circuit suitable for analysis using SPICE. The datasheet indicates that the transistor is provided in a SOT343 package (visible in the completed design shown in Figure 10.23). In this four-terminal package, two of the pins are associated with the emitter. Although not mentioned in the datasheet, there is a reason for this seemingly odd packaging, which we now address.

10.7.1 Inductive Degeneration

Perusing the available application notes (including [57]) one finds that the manufacturer recommends using the SOT343 package's dual emitter pins to add inductance between emitter and ground. This is commonly known as *inductive degeneration* (being somewhat analogous to "emitter degeneration") and is sometimes referred to as *emitter induction* or *source induction* for bipolar transistors and FETs respectively. While a detailed explanation is beyond the scope of this book,[13] it suffices to say that adding about 0.5 nH to the emitter (or source) tends to reduce F_{min} while making Γ_{opt} closer to the value of Γ_S associated with conjugate matching. Thus, a lower noise figure can be achieved for a particular gain, providing some additional flexibility in the design process. The effects on the *s*-parameters of the transistor are normally quite small: There is typically a small reduction in realizable gain, and some improvement in stability from the negative feedback-like action of the degenerating inductor.

As in emitter degeneration the exact value of the degenerating inductance is not critical, but it is a good idea to follow the suggestions provided by the manufacturer unless you know better. In this case the manufacturer recommends using using short-circuited transmission line stubs to present an equivalent inductance of about 0.6 nH to each of the emitter pins (see Section 9.7.2 for a reminder about why this works), amounting to about 0.3 nH total inductance since these

[13] Excellent explanations can be found in [63, Sec. 5.3.4] and [35]. The latter includes some useful practical examples.

stubs appear in parallel at the emitter. The stubs can be seen in Figure 10.23. Transmission line stubs are definitely the way to go here, because it would be difficult to find a suitable discrete inductor of this very small value that did not also introduce a comparable amount of detrimental capacitance at this frequency. Furthermore the stub's impedance turns out to be almost proportional to frequency (as is that of an ideal discrete inductor) when used this way – to see this just consider the behavior of Equation (9.29) when $l \ll \lambda$, as it is here.

10.7.2 Selecting an Operating Point and Establishing RF Design Parameters

To get started in earnest on the RF design, we would like to have the relevant s-parameters as well as the noise parameter Γ_{opt}, and perhaps also F_{min} and r_n. These depend on the operating point of the transistor, so we should first establish that. If we take the given supply voltage of +3 V as V_{CC}, then a good initial guess for V_{CE} might be 2/3 of that value; i.e., $V_{CE} = 2$ V. The power supply current is specified to be less than 10 mA, so the collector current I_C will also end up being less than 10 mA.

A review of the datasheet confirms that the operating point $V_{CE} = 2$ V and $I_C \leq 10$ mA is a reasonable choice. However, we have two problems once we start looking for s-parameters and noise parameters. First, the manufacturer provides these only for higher values of V_{CE} than we intend to use – namely 4 V and 8 V – and for these, parameters are provided only for $I_C = 4$ mA and 10 mA. The second problem is that the parameters provided by the manufacturer are for the transistor without the emitter inductance that we decided to add in Section 10.7.1.

There are at least four potential ways to navigate this issue. First, you could ask the manufacturer to provide the parameters you need, or try to secure them from a third party. Second, you could try to extract the parameters from a SPICE model, which the manufacturer does provide. Third, you could measure the parameters for yourself, assuming you have the facilities to do this properly. Finally, you could attempt a best-guess design based on the best available information and making some educated guesses along the way. Each of these approaches has some merit, and some disadvantages.

For this example, here is the way we will go: First, experience suggests that the amount of inductance we have added to the emitter will have only a small effect on the s-parameters, so we will assume the manufacturer's unmodified s-parameters are a reasonable estimate of the actual s-parameters.[14] The effect of the emitter inductance on the noise performance is more difficult to anticipate, and this is supposed to be a *low-noise* amplifier, so we will simply state the relevant noise parameters here (Figure 10.24) and note that these could could have come from any of the first three methods mentioned in the previous paragraph.

So: Which particular set of the manufacturer's unmodified s-parameters to use? Since the problem statement indicates maximum supply current of 10 mA, the $I_C = 10$ mA parameters seem like a reasonable choice, assuming we do not end up with a scheme in which too much current ends up somewhere besides the collector. Also if you compare the available s-parameters for $I_C = 4$ mA to those for higher currents, you note the latter are quite similar to each other, and significantly different from the former. Comparing the s-parameters for

[14] You could also confirm this by using the manufacturer's SPICE model, if you would like to do so.

freq. MHz	s_{11}	s_{21}	s_{12}	s_{22}
400	$0.592\angle -138°$	$9.650\angle +104.5°$	$0.054\angle +47°$	$0.375\angle -60°$

freq. MHz	F_{min} dB	Γ_{opt}	r_n
400	1.0	$0.392\angle +113°$	0.100

Figure 10.24. Presumed s-parameters and actual noise parameters for the BFG540W with inductive degeneration in the UHF-band LNA example. $Z_0 = 50\ \Omega$.

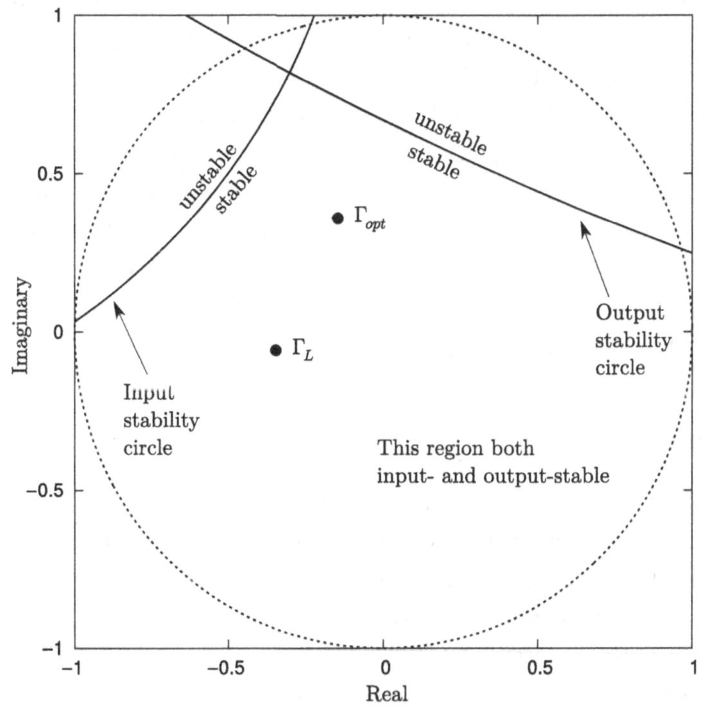

Figure 10.25. Stability circles for the UHF-band LNA example. Also shown are Γ_{opt}, and Γ_L for an output conjugate match when $\Gamma_S = \Gamma_{opt}$.

$V_{CE} = 4$ V to those for 8 V, both at 10 mA, we find there is very little difference; so we preliminarily assume that the s-parameters for $V_{CE} = 4$ V and $I_C = 10$ mA are the best guess for our preferred operating point ($V_{CE} = 2$ V and I_C a little less than 10 mA). These parameters are given in Figure 10.24.

10.7.3 Transistor Characterization

As always we should assess the stability of the transistor before getting too far along in the design. In this case we have $\Delta = 0.306\angle -36°$ and $K = 0.578$ at 400 MHz. Since $K < 1$, the transistor is only conditionally stable. The associated stability circles are shown in Figure 10.25. Γ_{opt} is also shown, along with the value of Γ_L required for a maximum gain (output conjugate-matched) design when $\Gamma_S = \Gamma_{opt}$. We see that this particular solution appears to be stable. The associated TPG is 19.6 dB, which is considerably more than what has been specified. So, it seems likely that a minimum noise figure design with TPG in the range 15 ± 2 dB is within reach.

One also finds the VSWR looking into the amplifier input ($VSWR_i$ as defined in Section 10.6) for this particular design – i.e., with conjugate-matched output – is 16.1, which is far higher than the specified maximum value of 2. So, we should be looking for ways to reduce the gain that also improve the input VSWR.

Going forward, two possible strategies might be considered. First, we could hunt for output matches (i.e., values of Γ_L) that reduce TPG, and hope that one of these is simultaneously stable and yields sufficiently low input and output VSWR. Not only are we not assured that such a solution exists, but furthermore recall that our *s*-parameters are at best approximate, so we face an elevated risk of ending up with an unsatisfactory solution in practice even if the present analysis indicates an acceptable solution. A better approach in this case is probably resistive loss stabilization. (For a refresher on this topic, see Section 10.4.3 and Example 8.13.) If we add a resistor at the transistor output, we may well be able to simultaneously achieve unconditional stability, reduce the TPG, and improve VSWR all at once. A resistor applied to the output has no effect on the noise figure of the transistor, because this depends entirely on the *input* match; and as long as the transistor's gain is sufficiently high, the noise contributed by a resistor at the output will be negligible. So, we proceed to "condition" the transistor by employing resistive loss stabilization at the output.

10.7.4 Transistor Output Conditioning

Going the route of resistive loss stabilization at the output, we must first decide whether the resistor should be in series or in parallel with the transistor output. A resistor in parallel with the output will provide an alternative path for collector bias current, and thereby increase power consumption. While this could be addressed by adding a DC-blocking capacitor in series with the stabilization resistor, we choose instead to proceed with a resistor in series with the output, as this ameliorates the issue without additional components.

Pursuing the series resistor idea, Figure 10.26 shows the TPG for the transistor–resistor cascade, calculated using the same procedure described in Example 8.13. We find that resistance greater than about 90 Ω is sufficient to make the transistor unconditionally stable, and still yields far more gain than we need. However, we also see that the input VSWR, while improved, remains very high.

Since we have gain to spare, it is worth considering how we might further improve the input VSWR before we take action to reduce gain and proceed to OMN design. It is clear from Figure 10.26 that no value of the stabilization resistor will reduce input VSWR to an acceptable level, so we need to introduce some imaginary-valued impedance at the transistor output. If we restrict ourselves to discrete components, we must initially choose between inductors and capacitors. In the UHF band capacitors are generally easier to work with, as inductors tend to exhibit significantly more onerous non-ideal behaviors in this frequency regime. So, we have just two options to consider: A capacitor in parallel with the transistor output, or a capacitor in series with the transistor output. In either case the same method of analysis can be used, except now with three stages as opposed to two. Analyzing these options is a straightforward extension of

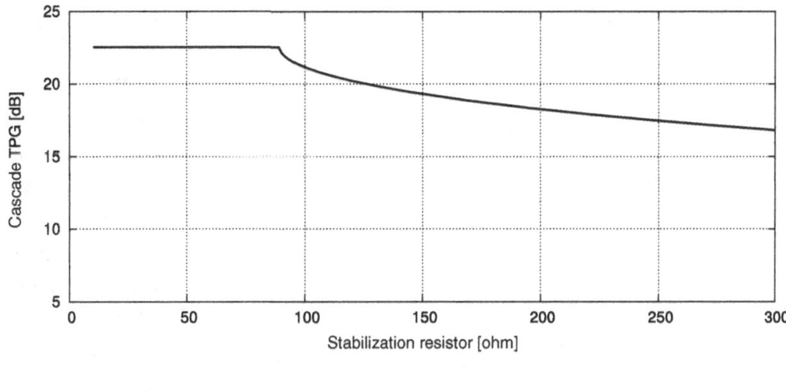

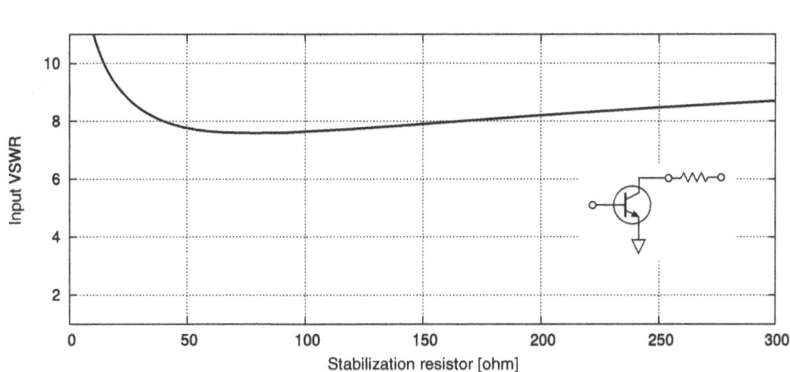

Figure 10.26. Analysis of TPG (top panel) and input VSWR (bottom panel) for the proposed transistor in cascade with an output series resistor of the indicated value for the UHF-band LNA example. The cascade schematic shown in inset of bottom panel. TPG and VSWR are computed for the minimum noise figure input termination and conjugate-matched output termination. The border between unconditional and conditional stability corresponds to the discontinuity in the TPG curve at $\cong 90\ \Omega$.

what we have previously done: The proposed cascade of two-ports is now (1) common-emitter transistor, (2) a capacitor which is either in series or in parallel, and (3) the series stabilization resistor. One finds that adding a 4.7 pF capacitor in parallel yields the nice result shown in Figure 10.27: The input VSWR is dramatically reduced, the minimum resistance needed for unconditional stability is reduced, and we still have plenty of gain to work with. The input VSWR is still not reduced all the way to the required maximum of 2 in this tentative solution; however, keep in mind that the VSWR will shift again as we adjust the gain using the output match, and furthermore recall the s-parameters are not expected to be very accurate. Settling for $\text{VSWR}_i < 3$ at this point will serve our purposes sufficiently well. Going forward, we will choose a stabilization resistance of 22 Ω because it is a standard value that is safely larger than that required to achieve unconditional stability (about 12 Ω).

10.7.5 IMN Design

The IMN must have $Z_i = 50\ \Omega$ (by specification) and $Z_S = 29.0 + j24.8\ \Omega$ (which corresponds to $\Gamma_S = \Gamma_{opt}$, since we wish to minimize noise figure). Performance is specified at only one frequency (400 MHz), so presumably a narrowband match is adequate. Using the method of Section 9.3, we find there are four possible "L"-type two-reactance matching circuits. Looking ahead for a moment to bias circuit integration, recall that the high-pass "bias tee" configuration of a shunt inductor and series capacitor is quite convenient, and it turns out that we have one of those. The selected IMN is shown in the preliminary schematic presented in Figure 10.28.

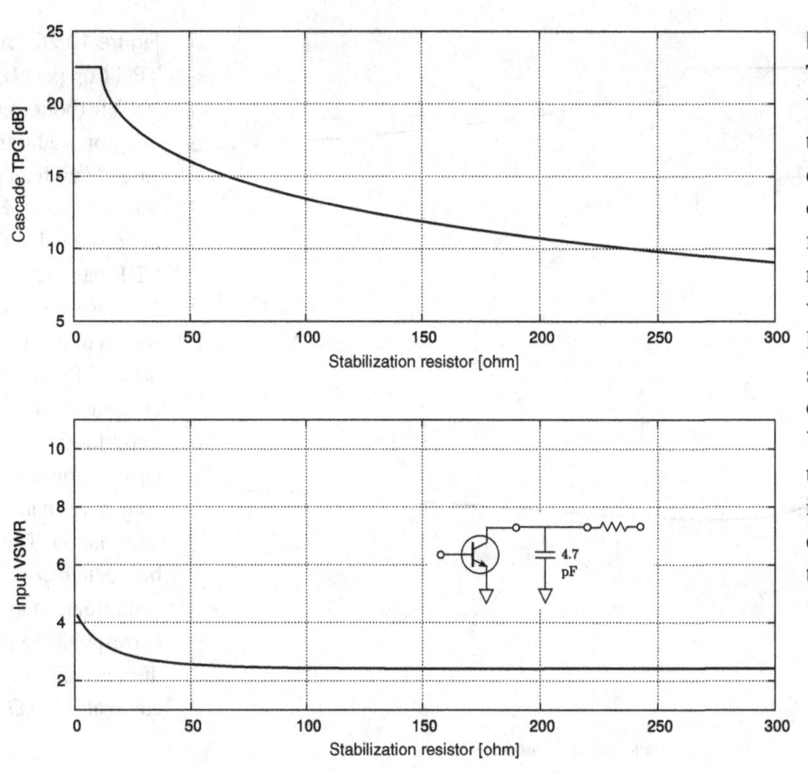

Figure 10.27. Analysis of TPG (top panel) and input VSWR (bottom panel) for the proposed transistor in cascade with an 4.7 pF output capacitor in parallel, followed by a series resistor of the indicated value, for the UHF-band LNA example. The cascade schematic is shown in inset of bottom panel. TPG and VSWR are computed for the minimum noise figure input termination and conjugate-matched output termination.

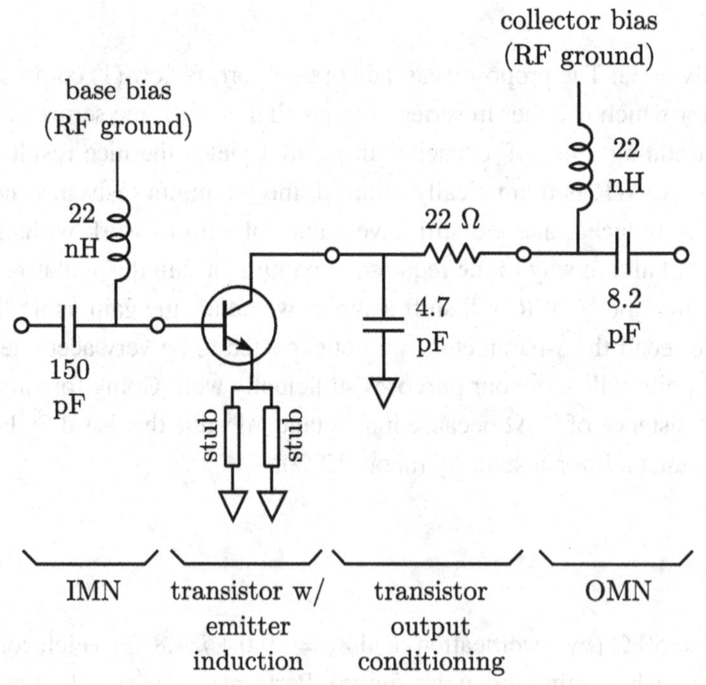

Figure 10.28. Preliminary (RF) schematic, including IMN, transistor, output conditioning stage, and OMN.

The reader may fairly wonder why in this case we pounce on a solution that involves a discrete inductor, whereas we have deliberately avoided these components at least twice previously. First, 22 nH amounts to a reactance of about 55 Ω at 400 MHz, which will be useful as part of the strategy for keeping RF out of the DC power supply. Because the inductance is relatively large, the intrinsic capacitance is relatively less important; this is in contrast to the problem we faced when adding emitter inductance. Finally, the discrete inductor will be dramatically smaller than the stub that would be required if we used the transmission line method. In summary: We use a discrete inductor here because it helps with RF-DC isolation and because we can find a suitable component of the required value at 400 MHz.

10.7.6 OMN Design

The straightforward procedure for OMN design is to first determine the desired value of Γ_L and then to design an OMN that delivers that value of Γ_L. In the present situation there is no obvious and preferred strategy for choosing Γ_L. Neither the available gain nor operating gain methods are available, since no ports are constrained to be conjugate-matched. Since we do not have a exact gain requirement, there is no particular benefit in following the unilateral design strategy, which requires making additional assumptions anyway. A completely reasonable approach would be to perform a brute force search over all possible values of Γ_L, seeking to minimize VSWR (both input and output) with TPG of about 15 dB as requested. However, recall that we are working with s-parameters that are approximate, since they do not account for the emitter induction in our design. So, there is no guarantee that an exhaustive brute force search will yield a result which is more reasonable than some simpler approach.

Taking all this into account, let us try a simpler approach. We note that it would be convenient to replicate the bias tee topology of the IMN in the OMN. (This is the scheme already unveiled in Figure 10.28.) Furthermore, let us consider if it is possible to use the same value (22 nH) inductor used in the IMN circuit, so as to simplify the parts list. Is there a series capacitance that works in this case? The answer is quite easy to work out since it is easy to calculate the TPG and VSWR of the amplifier as a function of the series capacitance, should we go this route. The result is shown in Figure 10.29. We see that a value of about 8 pF minimizes the input VSWR, reduces output VSWR below 2, and puts the TPG at about ≈ 17 dB, close to the high end of the specified range. This is very encouraging, since 8.2 pF is a standard value.

How concerned should we be that the input VSWR, while minimized, is still about 3; i.e., significantly greater than the specified value of 2? Once again, recall we are using s-parameters that are only approximate, since we are not accounting for emitter induction and are using s-parameters for a slightly different operating point than we intend to use. It is plausible – although not assured – that minimizing the input VSWR here results in input VSWR less than 2 in the final design. The selected IMN is shown in the preliminary schematic presented in Figure 10.28.

10.7.7 Bias Scheme

In Section 10.7.2 we decided to aim for $V_{CE} = 2$ V and $I_C < 10$ mA. As explained in Section 10.3.1, the simplest reasonable way to do this is with self bias. The appropriate DC equivalent circuit is shown in Figure 10.30. Note that this is quite similar to the previously-presented version of self bias (see Figure 10.5) with the complicating factor that we have an additional

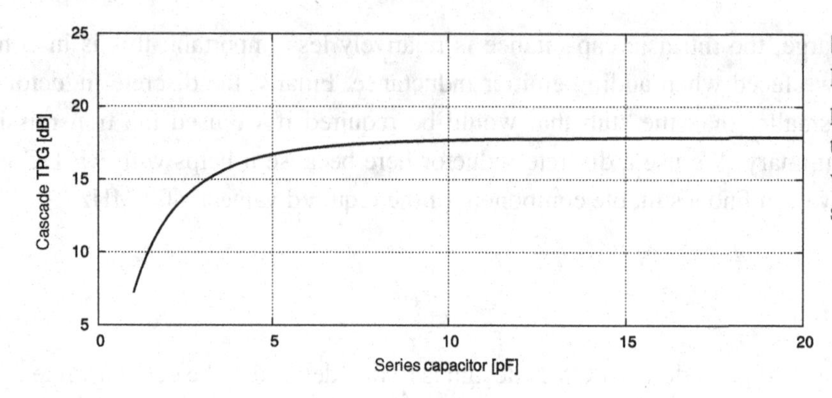

Figure 10.29. Performance of the preliminary design using the proposed OMN, varying the value of the series capacitor.

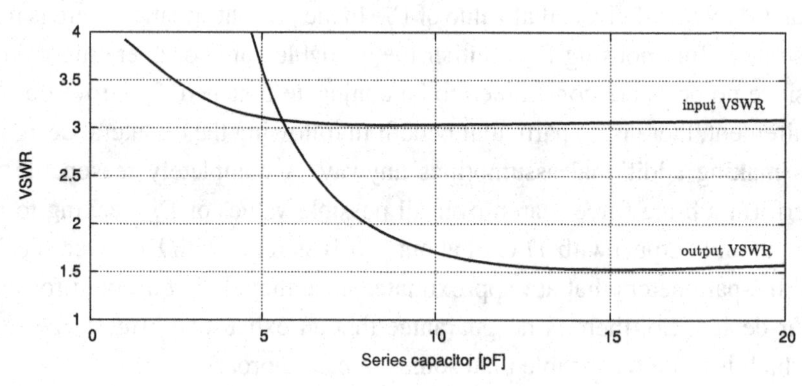

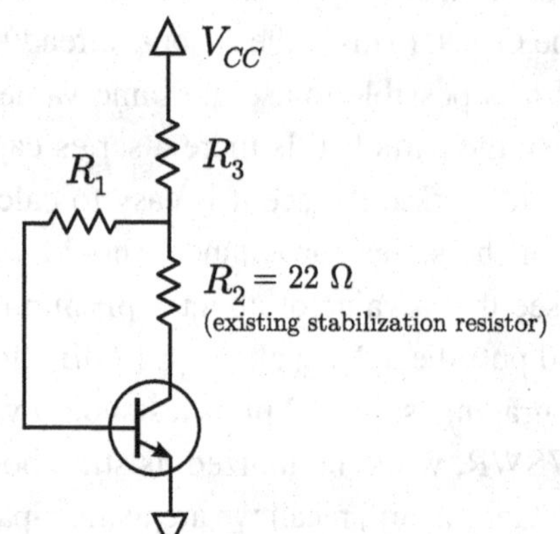

Figure 10.30. Preliminary design of a self-bias circuit for the UHF-band LNA example; DC equivalent circuit shown. As explained in the text, we end up setting $R_3 = 100\ \Omega$ and $R_2 = 22\ \text{k}\Omega$.

resistor: Namely the 22 Ω stabilization resistor (R_2 as shown in Figure 10.30), which lies unavoidably in the path of the collector bias current. Thus, the analysis and design are slightly different than as explained in Section 10.3.1. Here we go:

An appropriate value for the resistor R_3 (as shown in Figure 10.30) may be determined as follows. Applying KVL to the path from V_{CC} through the collector to ground, we have:

$$V_{CC} - (I_C + I_B)R_3 - I_C R_2 - V_{CE} = 0 \tag{10.68}$$

Making the substitution $I_B = I_C/\beta$ and solving for I_C, we find

$$I_C = \frac{V_{CC} - V_{CE}}{R_3/\beta + R_3 + R_2} \tag{10.69}$$

We can make this result insensitive to β by making $R_2 + R_3 \gg R_3/\beta$, which is quite easy since β is on the order of 100 or so for this transistor (as it is for any Si BJT). Choosing $R_3 = 100\ \Omega$ yields $R_2 + R_3 = 122\ \Omega$ compared to $R_3/\beta \sim 1$, and subsequently

$$I_C \approx \frac{V_{CC} - V_{CE}}{R_3 + R_2} = 8.2 \text{ mA} \tag{10.70}$$

The supply current will be only about 1% greater since $\beta \sim 100$, and so this appears to meet the 10 mA specification with a comfortable margin.

Since the ratio of the currents I_C and I_B is already constrained to be β, and since both currents "source" from the same point (i.e., from R_3), the value of the resistor R_2 is not critical. A reasonable range of values can be determined by applying KVL to the path from V_{CC} through the base to ground:

$$V_{CC} - (I_C + I_B) R_3 - I_B R_1 - V_{BE} = 0 \tag{10.71}$$

Making the substitution $I_B = I_C/\beta$ and solving for R_1, we find

$$R_1 = (V_{CC} - V_{BE}) \frac{\beta}{I_C} - (\beta + 1) R_3 \tag{10.72}$$

Assuming $V_{BE} = 0.65$ V and $I_C = 8$ mA, we obtain values of 19 kΩ to 48 kΩ for β from 100 to 250, respectively. In the design we adopt 22 kΩ as one of the standard values falling within this range.

10.7.8 Bias Circuit Integration

Figure 10.31 shows the final schematic, in which the bias circuit of Figure 10.30 is integrated with the RF design of Figure 10.28.

Note the use of bypass capacitors to provide the necessary RF ground on the supply side of the 22 nH inductors. The primary requirement on these capacitors is that they be large enough

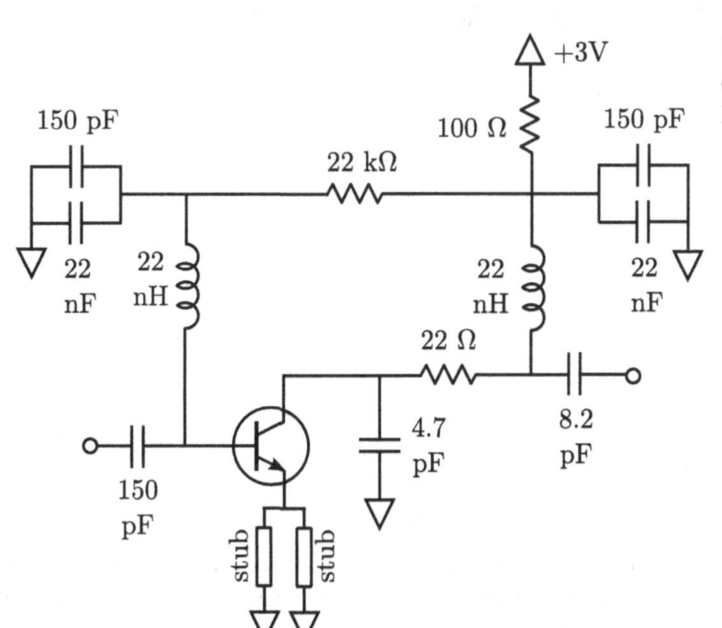

Figure 10.31. Schematic of the completed UHF-band LNA shown in Figure 10.23.

to provide a low-impedance path to ground at RF. The design from [57] uses 150 pF and 22 nF capacitors in parallel for each bypass, which in principle is equivalent to a single 22.150 pF capacitor having an impedance of $1/j\omega C = -j0.02\ \Omega$ at 400 MHz – plenty small.

To the uninitiated the 150 pF capacitors may seem superfluous. The additional 150 pF capacitors are warranted because physical capacitors have frequency-dependent behaviors. In particular, it is easy to find a 150 pF capacitor that works well at 400 MHz, but it is much harder to find a 22 nF capacitor that works well at that frequency.

So then why bother with the 22 nF capacitors? This is because these capacitors are doing double-duty: They also serve to filter interference generated by the power supply, which occurs at frequencies below the usual radio bands. By using both 22 nF and 150 pF capacitors, we ensure effective bypassing of both low-frequency power supply transients as well as 400 MHz RF signals that get by the inductors.

Finally: Why the value 150 pF specifically, for the RF bypass capacitors? This is to simplify the parts list: We have already specified a 150 pF capacitor for the IMN; repeating that value in the bias circuit reduces the number of different parts that the purchasers and technicians must contend with.

10.7.9 Measured Results

Figure 10.23 shows the completed LNA, which has been implemented in surface mount technology on FR4 substrate, and using SMA-type coaxial input and output connectors. The performance of this prototype was measured with the following results.

Noise Figure. The 400 MHz noise figure was determined be about 1.0 dB, meeting the 1.3 dB requirement and close to F_{min} (as anticipated since we endeavored to make $\Gamma_S = \Gamma_{opt}$).

Gain. Figure 10.32 shows that the 400 MHz gain is about 15.5 dB, which is well within the 15 ± 2 dB range specified. It is typical that gain is slightly less than expected due to various losses and other non-idealities associated with practical implementation; however, here the

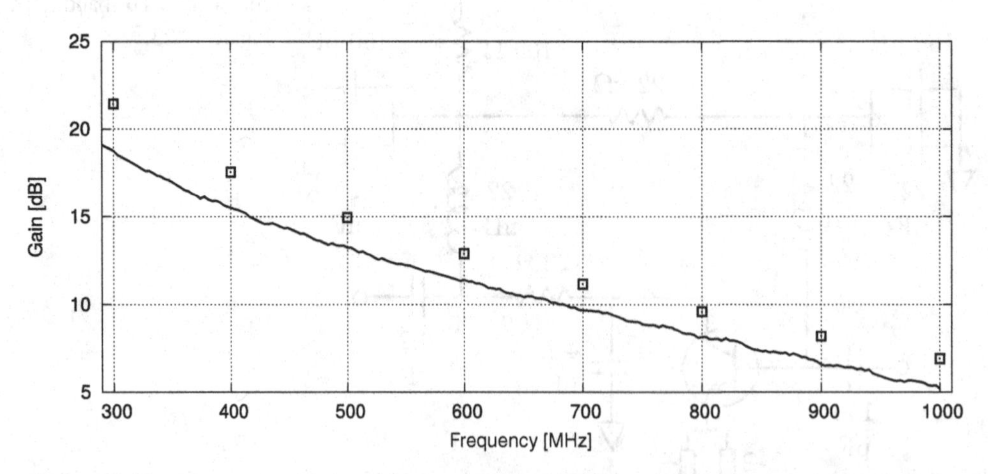

Figure 10.32. Measured gain ($|s_{21}|$) of the completed UHF-band LNA. Markers indicate the result predicted using the presumed *s*-parameters of the transistor ($V_{CE} = 4$ V, $I_C = 10$ mA, no emitter induction) for the frequencies at which these data are available.

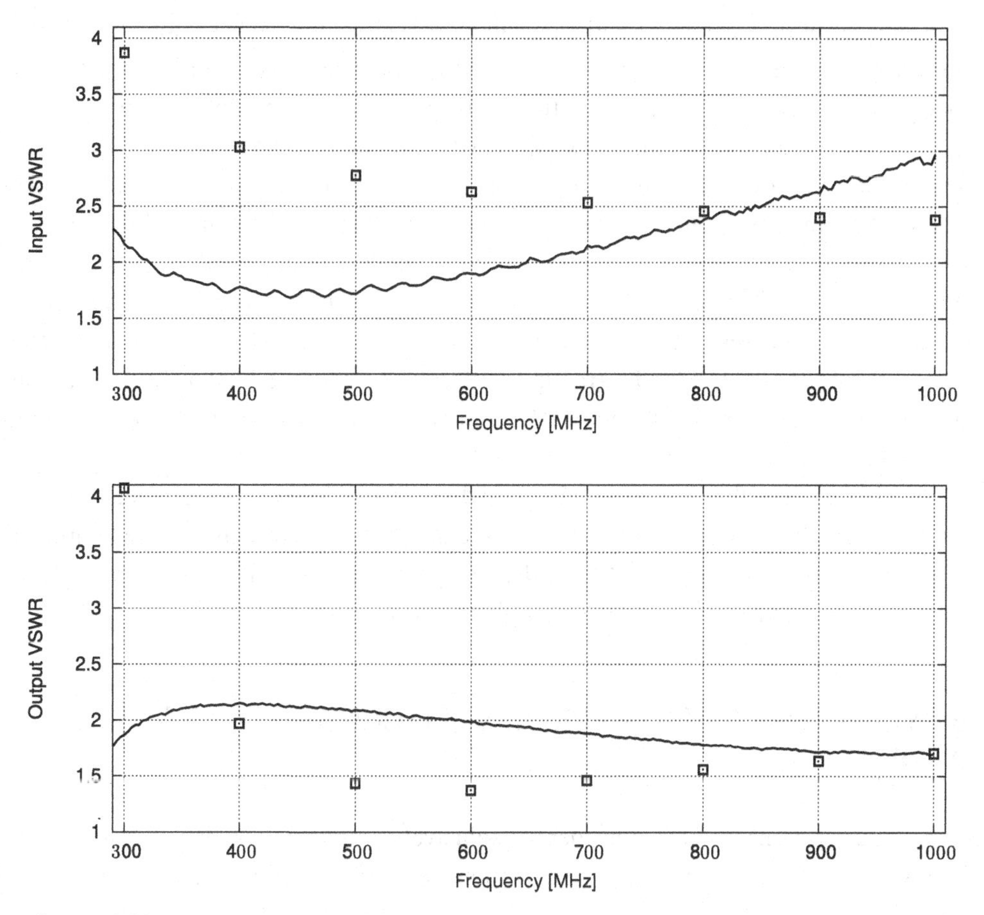

Figure 10.33. Measured input VSWR (*top*) and output VSWR (*bottom*) of the completed UHF-band LNA. The markers indicate the results predicted using the presumed *s*-parameters of the transistor ($V_{CE} = 4$ V, $I_C = 10$ mA, no emitter induction) for the frequencies at which these data are available.

difference is probably attributable mainly to the issue that our *s*-parameters do not account for emitter degeneration, and it is known that emitter degeneration tends to decrease gain somewhat. It is encouraging that the frequency response closely follows the response we would predict using the provided *s*-parameters at all frequencies (i.e., not just at 400 MHz).

Input VSWR. Figure 10.33 (top panel) shows that the input VSWR at 400 MHz is about 1.7, which is less than the specified value of 2 and much less than the value of 3 predicted using the presumed *s*-parameters. Thus the educated guess we made at the end of Section 10.7.6 has turned out to be correct, even though the predicted VSWR frequency response is not very accurate.

Output VSWR. Figure 10.33 (bottom panel) shows that the output VSWR at 400 MHz is about 2.1, which is slightly greater than the specified value of 2.

Power Consumption. The circuit draws 8.0 mA at +3 V, meeting the 10 mA requirement and being very close to the value of 8.2 mA predicted by the analysis of Section 10.7.7.

Summarizing, the prototype meets all design requirements except that the output VSWR is slightly too high. This could easily be addressed by slightly increasing the value of the 8.2 pF OMN capacitor – to see why this is so, take another look at Figure 10.29.

At this point we can also consider one additional issue that we have not considered explicitly until now: Figures 10.32 and 10.33 indicate the impedance bandwidth of the amplifier. Note that gain and VSWR are only weak functions of bandwidth and do not appear to be "peaked" at 400 MHz, even though the IMN and OMN are "narrowband" designs. How can this be? The primary reason is that the two-port parameters of the transistor vary independently with frequency, which tends to "unravel" the selectivity that the IMN and OMN might otherwise provide. If our intent was to achieve a particular bandwidth or selectivity, we could have taken this into account.[15] We also obtain some "response leveling" from the loss introduced by the 22 Ω stabilization resistor, which acts to reduce the overall Q of the circuit, which in turn increases impedance bandwidth (reminder: Section 9.4). In the final analysis, the usable range of frequencies is roughly 300–600 MHz, limited primarily by input VSWR and, of course, by noise figure.

Finally, we note that there are other equally valid alternative design methodologies (many identified in this process of working this example) that we might have employed, and a multitude of equally-good or better designs that would result. It is an excellent learning experience to repeat this example employing some of those methodologies.

10.8 BEYOND THE SINGLE-TRANSISTOR NARROWBAND AMPLIFIER

In previous sections we have considered exclusively amplifiers consisting of a single transistor, and which are designed to meet specifications at a single frequency. Let us now consider the wider range of possibilities.

At the end of the previous section (i.e., the UHF LNA design example) we considered for the first time the bandwidth of amplifiers. As in that example, it is typical that amplifiers designed using narrowband techniques are usable over bandwidths on the order of 50% of the carrier frequency. Modification of bandwidth typically requires at least that the designer accounts for the frequency variation of the transistor's two-port parameters. Greater selectivity can be achieved by also increasing the selectivity of the IMN and OMN, or by cascading the amplifier with one or more filters. Significantly greater bandwidth is typically not possible with IMNs and OMNs designed using narrowband techniques; in this case resistive matching may be required.

Bandwidth, gain, and other characteristics of amplifiers can be further modified using architectures employing cascades of transistors. To be clear, amplifiers consisting of a single discrete transistor are common in modern radio design; two examples include the low-noise amplifiers (LNAs) used in many receivers, and the power amplifiers used in many transmitters (more on that beginning in Section 17.4). However, there are many applications where a single amplifier may consist of two or more transistors in a cascade. One reason this happens is that any one transistor is limited in the gain it may achieve, regardless of the device technology and operating point. Typical maximum gains for a single-transistor amplifier are

[15] By the way, this is yet another task in which knowing your way around a Smith Chart really pays dividends.

in the neighborhood of 10–25 dB. However, some applications require "gain blocks" that are specified to deliver gain much greater than this, and there may be other compelling reasons to organize an amplifier as a cascade of single-transistor amplifier stages. The design of amplifiers consisting of transistor cascades is not much different from the design of single-transistor amplifiers. Typically, the topology consists of an IMN at the input, an OMN at the output, and *interstage matching networks* (ISMNs) between transistors. An ISMN can be viewed as a cascade of an OMN from the previous stage with the IMN of the next stage, each matched to an interstage impedance R_{mid}. How does one choose R_{mid}? The answer to that question lies in recognizing that this is precisely the scenario we considered in Section 9.5 ("Modifying Bandwidth Using Higher-Order Circuits"), where we used R_{mid} to control bandwidth. Summarizing: the multistage amplifier design problem decomposes to a cascade of single-transistor (IMN–transistor–OMN) amplifiers, and one uses the ISMNs constructed from adjacent OMN–IMN cascades to control the frequency response. In practice this is complicated by the fact that the s-parameters and noise figure parameters vary with frequency.

Another multi-transistor amplifier topology is the *differential amplifier*. Without pointing it out, all previous work in this chapter has assumed single-ended signals. Recalling the nature of differential signals (Section 8.7), it is obvious that single-ended amplifiers are not appropriate for amplifying differential signals. However, just as single-ended matching circuits may be easily re-fashioned into differential matching circuits (Section 9.6), single-ended amplifiers may be re-fashioned as differential amplifiers. A differential amplifier will always consist of pairs of transistors, symmetrically arranged around a physical or virtual datum.

10.9 IC IMPLEMENTATION

As in all areas of electronics, there is a unrelenting trend toward consolidation of RF devices and circuits into integrated circuits (ICs). For amplifiers, this trend plays out in a number of forms.

One form is the *microwave monolithic integrated circuit* (MMIC). MMIC amplifiers are useful because they implement what might ordinarily be a large and complex multi-transistor amplifier as a single compact device, and mostly (but not entirely) relieving the designer of the challenges of amplifier design. A MMIC amplifier can be regarded as a single two-port device, typically unconditionally stable, in which the biasing is greatly simplified. Furthermore, many MMIC amplifiers are designed to have 50 Ω port impedances and only a very weak dependence on frequency, so one does not necessarily need to consider or even know the s-parameters of a MMIC amplifier. As an example, Figure 10.34 shows the Mini-Circuits GALI-74, which is an InGaP HBT (bipolar) device that provides about 23 dB gain from nearly DC to $>$ 1 GHz, is unconditionally stable, requires only a single bias supply via the output, and has a maximum dimension of just 4.5 mm. The input and output terminations are very close to 50 Ω over the entire specified frequency range, so matching is required only if the VSWR is not already satisfactorily low. A wide variety of devices such as this are offered by a large number of manufacturers. The primary disadvantage of a MMIC is reduced ability to fine-tune characteristics such as gain, noise figure, and power consumption.

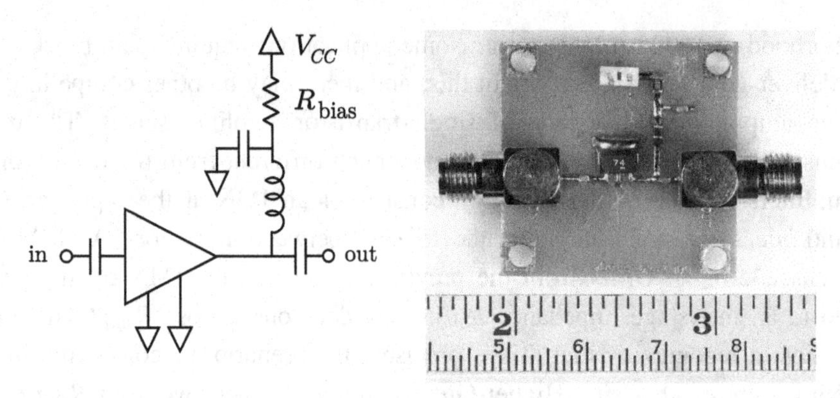

Figure 10.34. The GALI-74 MMIC. *Left:* Typical application circuit (R_{bias} sets I_C); *right:* As implemented on a PCB. The ruler along the bottom is marked in inches (top) and centimeters (bottom).

A second aspect of the trend toward increasing IC implementation of radio frequency amplifiers is CMOS integration. Radio frequency amplifiers may be implemented using the same CMOS process technology used for other analog and digital circuits, facilitating integration of amplifiers, analog control circuits, and digital control logic on the same IC. This has become commonplace using MOSFET transistors for frequencies up to the SHF band. A few of the challenges associated with CMOS implementation of amplifiers are poor repeatability and low realizable precision of device parameters in CMOS, and difficulty in implementing inductances and transmission line structures. For more information on this topic, [39] and [63] are recommended.

Before continuing, it is worth noting one radio frequency amplifier technology for which the concepts introduced in previous sections are largely *not* relevant: *operational amplifiers*, also commonly referred to as "opamps". An opamp is a three-terminal device in which the voltage at the output terminal is equal to the voltage difference across the input terminals, and for which the current into the input terminals is negligibly small. These characteristics make opamps particularly useful as *buffers*; i.e., as devices that isolate particularly low or high impedances, or devices with variable impedance, from parts of the circuit which work nominally at an utterly-fixed, moderate impedance. Modern opamps are typically effective into the UHF range, and commonly appear in modern radios as the final stage before a demodulator or analog-to-digital converter, and as the first stage following a modulator or digital-to-analog converter.

Problems

10.1 An NPN bipolar transistor is to be biased to a collector current of 10 mA using an 12 V supply, with input and output impedance equal to 75 Ω. The DC current gain is believed to be roughly 100. (a) Design a bias circuit using collector feedback stabilization, including schematic. (b) Estimate the operating point (I_C,V_{CE}) as accurately as you can. (c) Estimate power consumption as accurately as you can.

10.2 Repeat Problem 10.1, but now: (a) Design a bias circuit using emitter degeneration (without base voltage divider), including schematic. Set V_{CE} to 2 V. (b) Estimate the actual operating point (I_C,V_{CE}) as accurately as you can. (c) Estimate power consumption as accurately as you can.

10.3 Repeat Problem 10.1, but now: (a) Design a bias circuit using emitter degeneration with a base voltage divider, including schematic. Set V_{CE} to 8 V. (b) Estimate the resulting actual operating point (I_C,V_{CE}) as accurately as you can. (c) Estimate power consumption as accurately as you can.

10.4 An NPN bipolar transistor is to be biased to a collector current of 2 mA and collector–emitter voltage of 8 V, with input and output impedance equal to 50 Ω. The DC current gain is believed to be roughly 100. (a) Design a bias circuit using emitter degeneration with a base voltage divider, including schematic. (b) Estimate the resulting operating point (I_C,V_{CE}) as accurately as you can. (c) Estimate power consumption as accurately as you can.

10.5 Repeat Problem 10.3, but now using self bias with a base voltage divider.

10.6 An N-channel enhancement-mode MOSFET in common source configuration is to be biased for $I_D = 5$ mA and $V_{DS} = 10$ V, with $V_{GS} = +3.2$ V. The worst-case gate leakage current (I_{GSS}) is specified to be 100 μA. The drain supply voltage is +18 V. (a) Design the bias circuit using the self-biasing scheme. (b) Design the bias circuit as a gate voltage divider with source degeneration.

10.7 An N-channel depletion-mode PHEMT in common source configuration is to be biased for $I_D = 30$ mA at $V_{DS} = 4$ V, with $V_{GS} = -1.5$ V. The drain supply voltage is +6 V. Design the bias circuit using the self-biasing scheme.

10.8 Repeat Example 10.7, but now for 150 MHz. Assume the same *s*-parameters.

10.9 Design a maximum-gain amplifier for 2 GHz using an NPN Si BJT. The input and output impedances are to be 50 Ω. The transistor is to be biased for $I_C = 10$ mA and $V_{CE} = 8V$, using a +12 V supply. The *s*-parameters at this frequency and bias point are $s_{11} = 0.61\angle 165°$, $s_{21} = 3.72\angle 59°$, $s_{12} = 0.05\angle 42°$, and $s_{22} = 0.45\angle -48°$. (a) Assess stability. (b) Using only the *s*-parameters, determine the maximum possible gain. (c) Determine the IMN output impedance Z_S. (d) Determine the OMN input impedance Z_L. (e) Confirm this choice of Z_S and Z_L is consistent with the result of part (b). (f) Design the IMN using the single-stub matching technique. Lines may be specified in wavelengths. (g) Repeat (f) for the OMN. (h) Design a bias circuit of the type shown in Figure 10.7 (voltage divider bias with emitter degeneration). (i) Perform bias circuit integration and show the completed schematic.

10.10 Use the *available gain circle* method to design an amplifier for 9 dB gain at 1 GHz using an NPN Si BJT. The input and output impedances are to be 50 Ω. As in Example 10.8 consider the bias circuit, but do not design the bias circuit. The *s*-parameters at this frequency and bias point are $s_{11} = 0.28\angle -58°$, $s_{21} = 2.10\angle 65°$, $s_{12} = 0.08\angle 92°$, and $s_{22} = 0.80\angle -30°$. (a) Assess stability. (b) Using only the *s*-parameters, determine the maximum possible gain. (c) Plot the available gain circles for Γ_S and Γ_L. Choose the minimum-magnitude value of Γ_S and determine the IMN output impedance Z_S.

(d) Determine the OMN input impedance Z_L. (e) Confirm that this choice of Z_S and Z_L delivers the specified gain (use Equation (8.41)) and is consistent with the result of part (b). (f) Design the IMN using the single-stub matching technique. Lines may be specified in wavelengths. (g) Repeat (f) for the OMN. (h) As in Figure 10.17, show a completed schematic that accounts for any issues associated with bias circuit integration.

10.11 Repeat Problem 10.10 for 50 MHz and using discrete capacitors and inductors.

10.12 Repeat Problem 10.10 using the *operating gain circle* method, making the necessary modifications to steps (c) and (d).

10.13 Use the unilateral gain circles method to design an amplifier for 15 dB gain at 1 GHz using an NPN Si BJT. The input and output impedances are to be 50 Ω. As in Example 10.8 consider the bias circuit, but do not design the bias circuit. The s-parameters at this frequency and bias point are $s_{11} = 0.73\angle 175°$, $s_{21} = 4.45\angle 65°$, $s_{12} = 0.05\angle 77°$, and $s_{22} = 0.21\angle -80°$. (a) Assess stability. (b) Using only the s-parameters, determine the maximum possible gain. Repeat, making the unilateral assumption. (c) Compute the unilateral figure of merit and the range of error associated with the unilateral assumption. (d) Maximize the unilateral contribution of the output match, and plot the unilateral gain circle associated with input match necessary to meet the gain requirement. Choose the minimum-magnitude value of Γ_S and determine the IMN output impedance Z_S. (e) Determine the OMN input impedance Z_L. (f) Confirm that this choice of Z_S and Z_L delivers the specified gain (use Equation (8.41)) and is consistent with the results of parts (b) and (c). (g) Design the IMN using the single-stub matching technique. Lines may be specified in wavelengths. (h) Repeat (g) for the OMN. (i) As in Figure 10.17, show a completed schematic that accounts for any issues associated with bias circuit integration.

10.14 Repeat Problem 10.13 for a gain of 14 dB at 300 MHz, assuming the same s-parameters. Also, use discrete capacitors and inductors for the matching circuits.

10.15 Design an amplifier for minimum noise figure at 150 MHz using an NPN Si BJT. Optimize gain subject to the noise figure constraint. The input and output impedances are to be 75 Ω. Consider the bias circuit, but do not design the bias circuit. The s-parameters and noise parameters at this frequency and operating point are (Z_0 = 50 Ω): $s_{11} = 0.35\angle 165°$, $s_{21} = 5.90\angle 66°$, $s_{12} = 0.035\angle 58°$, $s_{22} = 0.46\angle -31°$, and $\Gamma_{opt} = 0.68\angle 142°$. (a) Assess stability and maximum possible gain. (b) Design the IMN using discrete inductors and capacitors. (c) Determine Γ_L and the resulting transducer power gain. (d) Design the OMN using discrete inductors and capacitors. (e) Show a completed schematic which accounts for any issues associated with bias circuit integration. (This problem may be continued as Problem 10.17.)

10.16 An LNA is to be designed for 2.4 GHz using an NPN Si BJT. The input and output impedances are to be 50 Ω. The s-parameters and noise parameters are (Z_0 = 50 Ω): $s_{11} = 0.60\angle 146°$, $s_{21} = 1.97\angle 32°$, $s_{12} = 0.08\angle 62°$, $s_{22} = 0.52\angle -63°$, $\Gamma_{opt} = 0.45\angle -150°$, F_{min} = 2 dB, and r_n = 0.2. (a) Assess stability and maximum possible gain. (b) Calculate and plot noise figure circles for F_1 = 2.5 dB and 3 dB. (c) Calculate and plot the available gain circle corresponding to the minimum noise figure case.

(d) Repeat (c) for available gain which is greater by 1 dB. (e) Continuing part (d), what is the minimum noise figure that can be achieved for this value of available gain? What is the corresponding value of Γ_S? (f) Continuing part (e), what is the corresponding value of Γ_L?

10.17 Determine the input and output VSWR for the LNA designed in Problem 10.15.

10.18 Repeat Example 10.13, this time with the constraint that input VSWR should be ≤ 1.5 and output VSWR should be ≤ 4. Report Γ_S, Γ_L, $VSWR_i$, $VSWR_o$ and TPG. Suggest an addition to the amplifier that can be used to decrease the output VSWR without degrading noise figure.

11 Linearity, Multistage Analysis, and Dynamic Range

11.1 INTRODUCTION

Previous chapters have described two-port devices in terms of three aspects: impedance, gain, and internal noise. In this chapter, we introduce a fourth aspect: linearity.

A system is linear if its behavior is independent of its input. This also implies that the output is proportional to the input, although the constant of proportionality may be a function of frequency. A brief review of the characteristics of transistors (in particular, Equations (10.1) and (10.3)) reveals that transistor amplifiers cannot be formally linear; one may only achieve an approximation to linearity in the vicinity of the operating point. (Passive devices are also intrinsically non-linear, although the approximation is typically much better.) Sections 11.2 and 11.3 formally introduce the concept of linearity, the consequences of non-linearity, and the various metrics that are used to quantify these effects in RF systems.

The next major topic in this chapter is the analysis of the gain, internal noise, and linearity of RF systems consisting of multiple two-ports arranged in cascade. The characteristics of a cascade of two-ports may be readily determined from the gain, internal noise, and linearity of the individual two-ports (the "stages") if the stages can be assumed to be impedance-matched. We first consider the linearity of cascaded two-ports in Section 11.4. Section 11.5 explains the technique of "stage/cascade analysis" and demonstrates its application in the design of practical RF systems. Section 11.6 explains the concepts of MDS and noise floor, both alternative descriptions of the sensitivity of an RF system.

The third major topic of this chapter is dynamic range (Section 11.7). Broadly speaking, dynamic range refers to the ability of a system to properly process very strong signals without unacceptably degrading sensitivity to weak signals.

11.2 CHARACTERIZATION OF LINEARITY

The term *linearity*, as it is used in modern radio engineering, has several possible interpretations; some leading to informal definitions that can be confusing to the uninitiated. Let us begin with a formal definition and then proceed to some of the more common interpretations in the context of radio engineering.

11.2.1 Linearity as Independence of Response

A formal definition of linearity is as follows.[1] Consider a system consisting of a single two-port, or consisting of a two-port cascade that may be described as a single two-port. Such a system may be characterized in terms of its impulse response $h(t)$, which is its output when the input is the Dirac delta function $\delta(t)$.[2] When a signal $x_1(t)$ is applied to this system, the output is given by

$$y_1(t) = x_1(t) * h(t)$$

where "$*$" denotes convolution. In a separate experiment, a signal $x_2(t)$ is applied to the same system, resulting in output

$$y_2(t) = x_2(t) * h(t)$$

The system $h(t)$ is formally linear if, for any constants a_1 and a_2:

$$[a_1x_1(t) + a_2x_2(t)] * h(t) = a_1y_1(t) + a_2y_2(t) \tag{11.1}$$

where $x_1(t)$, $y_1(t)$, $x_2(t)$, and $y_2(t)$ also satisfy the previous equations. This might also be referred to as the "scaling and superposition" definition of linearity: We say a system is linear if its response to inputs one at a time tells us everything we need to know about the system's response to linear combinations of those inputs. The frequency domain version of the "scaling and superposition" definition of linearity is

$$[a_1X_1(f) + a_2X_2(f)]\,H(f) = a_1Y_1(t) + a_2Y_2(t) \tag{11.2}$$

where the upper-case quantities are Fourier transforms of the lower-case quantities, and, in particular, $H(f)$ is the frequency response of the system.

Now, how does this apply to an RF system? $x_1(t)$ and $x_2(t)$, as well as $y_1(t)$ and $y_2(t)$, may be either voltages or currents, and $h(t)$ can be a description of any two-port, including cascades of two-ports. Equations (11.1) and (11.2) are essentially statements that the system is not affected by the input. A case where this is trivially true is the "null" system $h(t) = \delta(t)$ (equivalently, $H(f) = 1$), which is just a pair of wires connecting the input terminals to the output terminals. In this case it is easy to verify that Equations (11.1) and (11.2) apply, and it is also clear that $h(t)$ does not depend on $x(t)$. An ideal amplifier can be modeled as $h(t) = g\delta(t)$ where g is a constant (equivalently, $H(f) = g$); again it is clear that this is formally linear, and the response of the system does not depend on the input.

For any practical amplifier, g cannot be completely independent of the input. For an example, just look back to Section 10.3, where it is noted that the gain of a transistor amplifier varies as a function of the operating point. When the input current strays too far from the operating point, the output current becomes significantly distorted by the change in β over the resulting output signal swing. In effect, g – not just $y(t)$ – becomes a function of $x(t)$. It is only in a relatively small range around the operating point that the apparent gain of the amplifier is approximately independent of the input to the amplifier. In this region Equation (11.1)

[1] Hopefully this is familiar from an undergraduate course in signals and systems; if not, no worries.

[2] Reminder: $\delta(t)$ is defined as follows: $\delta(t) = 0$ for $t \neq 0$, and $\int_{-\infty}^{+\infty} \delta(t)dt = 1$.

holds – albeit approximately – and in that region we may reasonably refer to the amplifier as "linear." Generalizing, it is common to refer to a two-port device as linear if its transfer function is independent – at least approximately – of its input. Conversely, a two-port is said to be *non-linear* if this condition cannot reasonably be assumed.

11.2.2 Linearity of Systems with Memoryless Polynomial Response

A memoryless system is one in which the output is determined primarily by the present value of the input, or by a past input after a fixed delay. In a memoryless system, the output does not depend significantly on any other past values of the input. Many two-ports encountered in radio engineering are either electrically small or dispersionless, and thus may be assumed to be memoryless. Exceptions include certain power amplifiers (more on those beginning in Section 17.4).

Any system that can be described as a memoryless two-port has a response that can, in principle, be described by the following infinite sum:

$$y(t) = a_0 + a_1 x(t) + a_2 x^2(t) + a_3 x^3(t) + \cdots \tag{11.3}$$

where $x(t)$ is the input signal (voltage or current as appropriate), $y(t)$ is the output signal (again, voltage or current, as appropriate), and the a_ns are constants. This relationship may be viewed as a Taylor series representation, which provides a method for computing the coefficients. In this form, the system is seen to be formally linear when all coefficients except a_1 are zero; in this case, the gain is a_1. If $a_0 \neq 0$, then the system cannot satisfy the "scaling and superposition" definition of linearity. However, the a_0 term is a DC component, which is easily filtered out (e.g., using a blocking capacitor), or perhaps simply ignored. Thus if $a_1 \neq 0$ and $a_n = 0$ for $n \geq 2$, the system is typically considered linear from an RF engineering perspective, even though such a system might not be formally linear.

EXAMPLE 11.1

In Chapter 10 (Equation (10.1)) we found that the relationship between collector current I_C and base voltage V_B for a bipolar transistor could be written in the form $I_C = I_{C0} e^{V_{BE}/V_T}$ where I_0 and V_T are constants. Develop a memoryless polynomial model for the common emitter amplifier.

Solution: In a common emitter amplifier, the base voltage is the input and the collector current is the output.[3] Let us express the base voltage as $V_{BE} + v_B$; i.e., as the sum of the quiescent base-emitter voltage and the small-signal (RF) input voltage. Then the response equation may be modified as follows:

$$I_C = I_{C0} e^{(V_{BE}+v_B)/V_T} = \left[I_{C0} e^{V_{BE}/V_T} \right] e^{v_B/V_T} \tag{11.4}$$

[3] This might initially appear to be a contradiction relative to our preference in Chapter 10 to interpret the BJT as a linear current-controlled current source (i.e., $I_C = \beta I_B$), but that interpretation assumes that the transistor is being biased such that it behaves linearly over the desired range of signal magnitudes. What we are considering here is a bit more general.

The factor in brackets is a constant that represents the operating point; for convenience let us assign this factor the symbol I_0. The remaining factor may be represented using a Taylor series expansion around $v_B = 0$:

$$e^{v_B/V_T} = 1 + \frac{v_B}{V_T} + \frac{1}{2!}\left(\frac{v_B}{V_T}\right)^2 + \frac{1}{3!}\left(\frac{v_B}{V_T}\right)^3 + \cdots \tag{11.5}$$

Thus we have

$$I_C = I_0 + I_0\frac{v_B}{V_T} + \frac{I_0}{2!}\left(\frac{v_B}{V_T}\right)^2 + \frac{I_0}{3!}\left(\frac{v_B}{V_T}\right)^3 + \cdots \tag{11.6}$$

Which we may rewrite as

$$I_C = a_0 + a_1 v_B + a_2 v_B^2 + a_3 v_B^3 + \cdots \tag{11.7}$$

where

$$a_n \equiv \frac{1}{n!}\frac{I_0}{V_T^n} = \frac{1}{n!}\frac{I_{C0}e^{V_{BE}/V_T}}{V_T^n} \tag{11.8}$$

Note a_0 in this case is simply the quiescent collector current I_0, so we may rewrite the result in terms of the small-signal output i_C of the common emitter amplifier simply by omitting this first term; that is,

$$i_C = a_1 v_B + a_2 v_B^2 + a_3 v_B^3 + \cdots \tag{11.9}$$

Problem 11.1 gives you the opportunity to demonstrate this for a common source (FET) amplifier. In fact, Taylor series representation facilitates the reduction of essentially any practical two-port to polynomial form, provided it is memoryless.

For most practical systems, the relationship between output and input is approximately linear over some range. As a result, the higher-order coefficients – that is, a_n for $n \geq 2$ – tend to be small relative to a_1 and tend to decrease in magnitude with increasing n. If the input signal magnitude is appropriately constrained, their effect may be negligible. As input signal magnitude increases, and as these higher-order coefficients become significantly different from zero, non-linear effects become increasingly important. To quantify the effect, we consider the simplest imaginable non-trivial input signal: The sinusoid $x(t) = A\cos\omega t$. We apply this to a system for which $a_n = 0$ for $n \geq 4$ and DC terms are considered non-consequential; i.e.,

$$y(t) \equiv a_1 x(t) + a_2 x^2(t) + a_3 x^3(t) \tag{11.10}$$

After some algebra and disposal of DC terms we have

$$y(t) = \left[a_1 + \frac{3a_3}{4}A^2\right]A\cos\omega t + \left[\frac{a_2}{2}A\right]A\cos 2\omega t + \left[\frac{a_3}{4}A^2\right]A\cos 3\omega t \tag{11.11}$$

(Problem 11.2 offers the opportunity to confirm this on your own.) Note that the quantities in square brackets are essentially gain factors; for comparison, the output of a formally linear system $a_1 x(t)$ would be simply $a_1 A\cos\omega t$. We find that when the system is not formally linear (i.e., when a_2 and a_3 are non-zero), two things happen: The apparent gain of the system is different, and additional products appear at two and three times the frequency of the input.

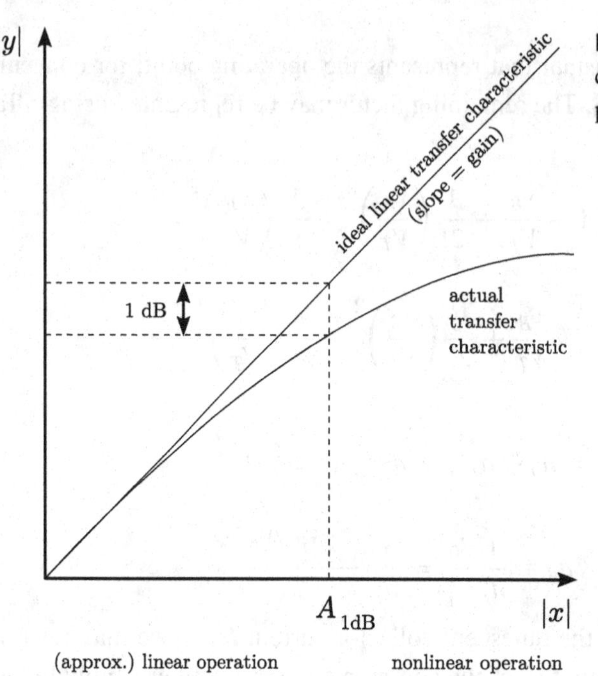

Figure 11.1. Transfer functions for linear and compressive systems, and 1 dB compression point.

Let us first consider the additional spectral content. The double-frequency term is referred to as the *second-order harmonic*, associated with a_2; and the triple-frequency term is referred to as the *third-order harmonic*, associated with a_3. The remaining term, known as the *fundamental* and associated with a_1, represents the nominally linear output and is located at the input frequency.

11.2.3 Gain Compression

Turning our attention to the fundamental, note that the apparent gain is $a_1+(3a_3/4)A^2$, whereas the nominal linear gain is simply a_1. Practical devices are *compressive*, meaning they have apparent gain which decreases with increasing input magnitude, as illustrated in Figure 11.1.[4] From this we can infer that a_1 and a_3 have opposite sign. This phenomenon is referred to as *gain compression* or simply *compression*.

A useful characterization of a compressive system is the *1 dB compression point*. This metric may be expressed with respect to the input or the output. The *input 1 dB compression point* A_{1dB} is defined as the magnitude of $x(t)$ for which the apparent gain of the fundamental is 1 dB less than the nominal linear gain. For the system model and stimulus considered above, this condition may be expressed mathematically as

$$20\log_{10}\left|a_1+\frac{3a_3}{4}A_{1dB}^2\right| = 20\log_{10}|a_1| - 1\text{ dB} \tag{11.12}$$

[4] Note that "small-signal" models employed for RF design are typically *not* compressive; however, the resulting amplifiers certainly *are* compressive. The issue is simply that the small-signal model does not capture the "large signal" behavior that leads to compression. See Problem 11.3 for an example.

which may be rewritten

$$\left|a_1 + \frac{3a_3}{4}A_{1dB}^2\right| \cong |0.891a_1| \tag{11.13}$$

By inspection it is apparent that A_{1dB} must have the form $\sqrt{c\,|a_1/a_3|}$. Making this substitution and solving for c, we obtain

$$A_{1dB} \cong \sqrt{0.145\left|\frac{a_1}{a_3}\right|} \tag{11.14}$$

It should be no surprise that to make A_{1dB} large, the ratio a_1/a_3 must be large. A little surprising is that A_{1dB} does not depend on a_2.

This metric may also be expressed with respect to the output. The *output 1 dB compression point* is defined as the magnitude of the output for which the apparent gain of the fundamental is 1 dB less than the nominal linear gain. This is simply $|a_1|A_{1dB}$; i.e., the input and output 1 dB compression points are related by the magnitude of the nominal linear gain a_1.

As formulated above, the 1 dB compression point is expressed in terms of either voltage or current. The 1 dB compression point may also be expressed in terms of power, since all three quantities (voltage, current, and power) are related by impedance. Power units are typically more convenient for measurement, and manufacturer's datasheets typically use power quantities. The output 1 dB compression point in power units is typically represented by the symbol P_{1dB}, P_1, or OP_1 ("O" for output). The input 1 dB compression is sometimes represented by the symbol IP_1, and is related to the output 1 dB compression point by the power gain; i.e., $\mathrm{OP}_1 = G \cdot \mathrm{IP}_1$. Datasheets frequently do not make the distinction between the input- and output-referred quantities, in which case it is typically the output-referred quantity that is indicated; however, it is always a good idea to check.

Because common emitter bipolar transistor amplifiers figure prominently in RF system design, it is useful to be aware of some rules-of-thumb applicable to these devices. Because bipolar transistors are essentially controlled current sources, the linearity depends most directly on collector bias current I_C. A common-emitter amplifier with a collector bias current of 20 mA and which is terminated into 50Ω will typically have an output 1 dB compression point of about +10 dBm. Generally, decreasing the collector bias current or decreasing load impedance tends to decrease P_1.

FET amplifiers used for as small-signal amplifiers typically have lower P_1 when designed for optimal or near-optimal noise figure, which is one reason why bipolar transistors continue to be used in some low-noise applications at UHF and above.

11.2.4 Intermodulation; Third-Order Intermodulation

In the previous section we assumed a single sinusoid at the input. A slightly more complicated input is two sinusoids:

$$x(t) = A_1 \cos \omega_1 t + A_2 \cos \omega_2 t \tag{11.15}$$

Following the procedure of the previous section, we apply this input to the system described by Equation (11.10). As in the previous section, it is useful to fully expand the result and then

Table 11.1. **Output products resulting from the application of a two-tone signal having frequencies of ω_1 and ω_2 to a memoryless polynomial non-linearity of third order with $a_0 = 0$. Products are listed in order by decreasing frequency assuming ω_2 is slightly greater than ω_1.**

Magnitude	Frequency	Identification
$(1/4)a_3A_2^3$	$3\omega_2$	IM_3 (third harmonic)
$(1/4)a_3A_1^3$	$3\omega_1$	IM_3 (third harmonic)
$(3/4)a_3A_1A_2^2$	$2\omega_2+\omega_1$	IM_3
$(3/4)a_3A_1^2A_2$	$2\omega_1+\omega_2$	IM_3
$(1/2)a_2A_2^2$	$2\omega_2$	IM_2 (second harmonic)
$(1/2)a_2A_1^2$	$2\omega_1$	IM_2 (second harmonic)
$a_2A_1A_2$	$\omega_2+\omega_1$	IM_2 (sum frequency)
$(3/4)a_3A_1^2A_2$	$2\omega_2-\omega_1$	IM_3
$a_1A_2+(3/4)a_3A_2^3+(3/2)a_3A_1^2A_2$	ω_2	Fundamental (desired)
$a_1A_1+(3/4)a_3A_1^3+(3/2)a_3A_1A_2^2$	ω_1	Fundamental (desired)
$(3/4)a_3A_1A_2^2$	$2\omega_1-\omega_2$	IM_3
$a_2A_1A_2$	$\omega_2-\omega_1$	IM_2 ("beat")
$(1/2)a_2\left(A_1^2+A_2^2\right)$	0	IM_2 (DC)

gather terms by frequency. The algebra is straightforward, but a bit tedious. The results are summarized in Table 11.1. Note that the output products include not only the fundamentals and their harmonics, but now also additional products whose frequencies depend on both ω_1 and ω_2. These additional products are referred to as *intermodulation*. Intermodulation products associated with a_2 are known as second-order intermodulation (IM_2) and those associated with a_3 are known as third-order intermodulation (IM_3).

As noted in Table 11.1 the fundamental (desired) output for each tone is modified by not only a_3 (as in the single-tone case), but now also by a_2 and by the magnitude of the other tone. Among all of the remaining products, the two of greatest concern are usually the IM_3 products at $2\omega_1-\omega_2$ and $2\omega_2-\omega_1$. This is because the separation between each of these products and the nearest fundamental is equal to the separation between the fundamentals; e.g., $(2\omega_2-\omega_1)-\omega_2=\omega_2-\omega_1$. If both fundamentals appear in the same passband, then the IM_3 products are likely to be either also in the passband, or close enough to present a filter design challenge.

Not only are the IM_3 products potentially difficult to filter out, but they are also potentially quite large. From Table 11.1 we see that the magnitude of an IM_3 product for equal-magnitude

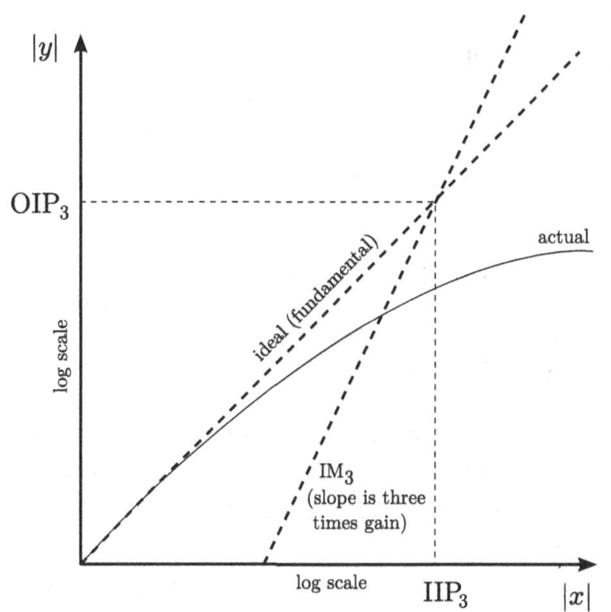

Figure 11.2. Third-order intercept point concept. Note by plotting the magnitude of the IM_3 in log–log form, the factor A^3 becomes $3 \log A$; thus the slope is three times the linear gain $A \rightarrow \log A$.

fundamentals (e.g., $A_1 = A_2 = A$) is $3a_3A^3/4$; that is, each IM_3 product is proportional to A^3, whereas the nominal (linear) fundamental output is proportional to A. In other words: When A increases by a factor of α, the IM_3 products increase by a factor of α^3. For this reason, the vulnerability of system to IM_3 is often characterized using a quantity known as the *third-order intercept point* (IP_3), illustrated in Figure 11.2. The concept is as follows: The IM_3 products increase more quickly with increasing $|A|$ than do the fundamentals themselves; therefore for some value of $|A|$ the magnitudes of the IM_3 products should in principle be equal to the magnitudes of the ideal (linear) fundamental products. The value of $|A|$ at which this occurs is the input IP_3 (IIP_3), and the associated value of $|a_1A|$ is the output IP_3 (OIP_3).

Note that IP_3 is *not* the value of $|A|$ for which the magnitude of the IM_3 is equal to the actual (e.g., compressed) magnitude of the fundamental output. Therefore, IP_3 cannot be measured "directly" in the sense of increasing $|A|$ until the IM_3 products are equal to the fundamentals and taking that value of $|A|$ to be the IIP_3. (Not only is this not consistent with the definition, but it is potentially dangerous to the device under test!) Nevertheless IP_3 may be easily (and safely) measured as follows: Using relatively low input levels, one measures the nominal linear gain as the slope of the output-to-input power curve, and also the IM_3 for the same input levels. When plotted on a log–log plot, these curves may be linearly-extrapolated to the point where they meet, as shown in Figure 11.2.

As is the case for 1 dB compression point, IP_3 is commonly expressed in power units, $OIP_3 = G \cdot IIP_3$, and datasheets usually report the output-referred quantity (OIP_3). The IIP_3 of typical receivers ranges from about -30 dBm to $+10$ dBm, depending on the application. For example, a typical IIP_3 specification for a cellular GSM receiver is -18 dBm, and a typical specification for a cellular CDMA receiver is -13 dBm. Linearity specifications for receivers in HF and LMR applications tend to be much higher.

For the system model of Equation (11.10) with the two-tone input of Equation (11.15), we are able to calculate IIP_3 in terms of the model parameters: It is simply the solution to

$$\left|\frac{3a_3}{4}A_{\mathrm{IP}_3}^3\right| = |a_1 A_{\mathrm{IP}_3}| \tag{11.16}$$

By inspection we see that

$$A_{\mathrm{IP}_3} = \sqrt{\frac{4}{3}\left|\frac{a_1}{a_3}\right|} \tag{11.17}$$

Comparing Equations (11.14) and (11.17), we find the following:

$$\frac{A_{\mathrm{IP}_3}}{A_{1\mathrm{dB}}} \cong 3.03 \cong 9.6 \text{ dB} \tag{11.18}$$

Remarkably, it turns out that the third-order intercept point is approximately 10 dB above the 1 dB compression point, independently of the model parameters (i.e., the a_ns) or the magnitude of the applied tones. This observation provides a rough estimate of IP_3 from P_1 and vice versa, knowing nothing else about the device.

In practice one finds Equation (11.18) is not a bad guess, but caveats apply. Passive non-linear devices are often well-modeled as third-order compressive, in which case Equation (11.18) is a good estimate. For active devices, this is typically not a safe assumption. To demonstrate, Figure 11.3 shows how this works out for modern commercially-available MMIC amplifiers: We find that Equation (11.18) underestimates IP_3 given $\mathrm{P}_{1\mathrm{dB}}$ by somewhere between 1 dB and 5 dB on average, with the largest error occurring for the largest values of IP_3. From this we can conclude that the linearity of commercially-available MMIC amplifiers

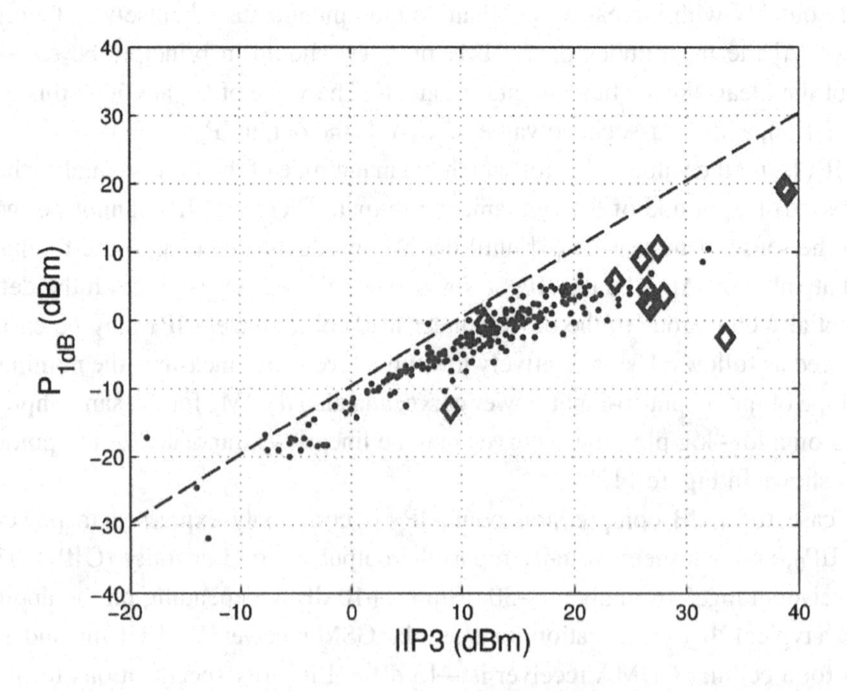

Figure 11.3. Relationship between input $P_{1\mathrm{dB}}$ and IIP_3 from a survey of 271 commercially-available MMIC amplifiers at 50 MHz operated at the recommended bias conditions, determined from datasheets. The dashed line shows the expected relationship from Equation (11.18). Dots indicate single-ended amplifiers and diamonds indicate differential amplifiers.

is not always well-modeled as third-order compressive, so Equation (11.18) is, at best, a rough bound for these devices. See [11] for more on this topic.

Similar behavior is seen in most RF transistor amplifiers, for which IP_3 is commonly 12–14 dB greater than P_{1dB} [11]. Other types of amplifers may exhibit greater differences; in particular operational amplifiers ("op-amps"), for which there is essentially no consistent relationship between IP_3 and P_{1dB}. The proper interpretation and pitfalls of system characterization in terms of P_{1dB} and IP_3 are a continuing topic of research [13].

11.2.5 Second-Order Intermodulation

It was noted in the previous section that, among the many products indicated in Table 11.1, the IM_3 products at $2\omega_2 - \omega_1$ and $2\omega_1 - \omega_2$ were of particular concern. Another product of particular concern is the IM_2 product at $\omega_2 - \omega_1$. For historical reasons, this frequency is sometimes referred to as the *beat frequency*, and the associated IM_2 product is thus known as the *beat product*. The beat product is often a concern because it occurs relatively close to DC, which (as we shall see in Chapter 15) is where the final stages of analog signal processing (i.e., digitization or demodulation) are often performed. In such receivers the beat product is prone to interfere with the desired signal in the final stages of the receiver. The beat product occurs at a relatively low frequency because the beat frequency is equal to the separation between the tones, which is generally small relative to the center frequency of the passband. Like IM_3, the beat product tends to be relatively strong, therefore it may be difficult to suppress this product sufficiently using analog filtering – more on that in later chapters.

From Table 11.1 we see that the magnitude of the beat and sum products for equal-magnitude fundamentals is $a_2 A^2/2$; that is, these products are proportional to A^2, whereas the nominal (linear) fundamental output is proportional to A. Therefore, when A increases by a factor of α, the beat product increases by a factor of α^2. For this reason, the vulnerability of system to the beat product is often characterized using a quantity known as the *second-order intercept point* (IP_2), illustrated in Figure 11.4. The concept is exactly analogous to IP_3: The input IP_2 (IIP_2) is the value for which $|A_1| = |A_2| = |A|$ results in a product having magnitude equal to the ideal (linear) fundamental product(s).[5] The associated value of $|a_1 A|$ is the output IP_2 (OIP_2). As in the case of IP_3, IP_2 is *not* the value of $|A|$ for which the magnitude of the IM_2 is equal to the actual (e.g., compressed) magnitude of the fundamental output; nevertheless IP_2 may be conveniently measured using the same low-level extrapolation technique described in the previous section (and as implied by Figure 11.4).

For the system model of Equation (11.10) with the two-tone input of Equation (11.15), we are able to calculate IIP_2 in terms of the model parameters: It is the solution to

$$\left|a_2 A^2_{IP_3}\right| = \left|a_1 A_{IP_2}\right| \tag{11.19}$$

which is simply

$$A_{IP_2} = \left|\frac{a_1}{a_2}\right| \tag{11.20}$$

[5] Note that this is the usual definition, based on two-tone input. It is also possible to define a second-order intercept point using the second harmonic resulting from a single tone excitation, but the value will be different by a factor of 3 dB.

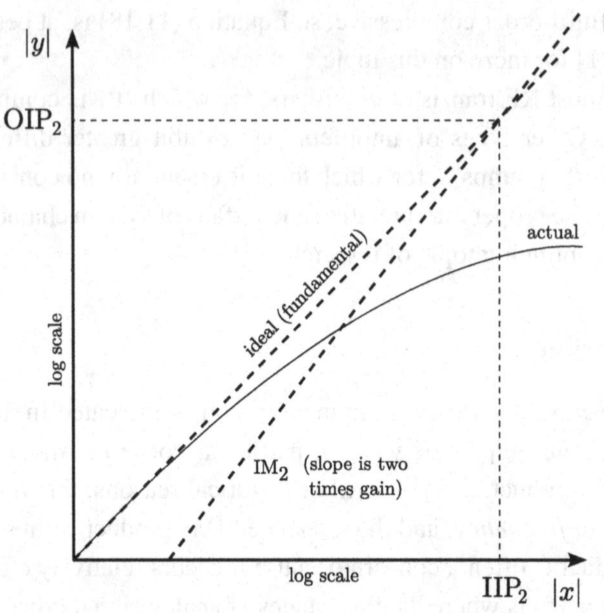

Figure 11.4. Second-order intercept point.

As is the case for P_1 and IP_3, IP_2 is commonly expressed in power units, $OIP_2 = G \cdot IIP_2$, and datasheets frequently report the output-referred quantity (OIP_2) without indicating the distinction between the input- and output-referred quantities. Because the beat product increases much more slowly with increasing $|A|$ than does IM_3, one normally finds that $IP_2 \gg IP_3$.

Figure 11.5 shows the relationship between IP_2 and IP_3 found in the same survey of commercially-available MMIC amplifiers described in Section 11.2.4 (Figure 11.3). One finds that IP_2 for single-ended devices tends to about twice the IP_3 in dB; e.g., $IP_3 = 10$ dBm implies $IP_2 \approx 20$ dB. However, this relationship is utterly empirical, not necessarily accurate, and likely to be different for different device technologies. Differential devices tend to do better in IM_2, as expected, but the advantage is difficult to anticipate in a general way.

11.2.6 AM–PM Conversion

In previous sections we have considered only constant-power sinusoids. When the power in a signal varies, the non-linearity of a system may manifest in more complex ways. Primary among these is AM–PM conversion. The concept can be explained as follows: Consider an amplitude-modulated (AM) signal applied to a two-port. The two-port may be arbitrarily simple or complex, but let us assume for the sake of illustration that it consists of at least one capacitor, and let us assume that the signal creates sufficient voltage across the capacitor that non-linear effects are apparent. Recalling that non-linearity is fundamentally the dependence of response on input, this means that the capacitance C of the capacitor is a function of the applied voltage, and that this variation in capacitance is significant. The impedance of a capacitor is $-j/\omega C$; therefore the reactance of this capacitor varies with the input signal magnitude, which is varying due to the modulation. Since there are also real-valued impedances in the system, and since these impedances are either not varying significantly or are varying independently of

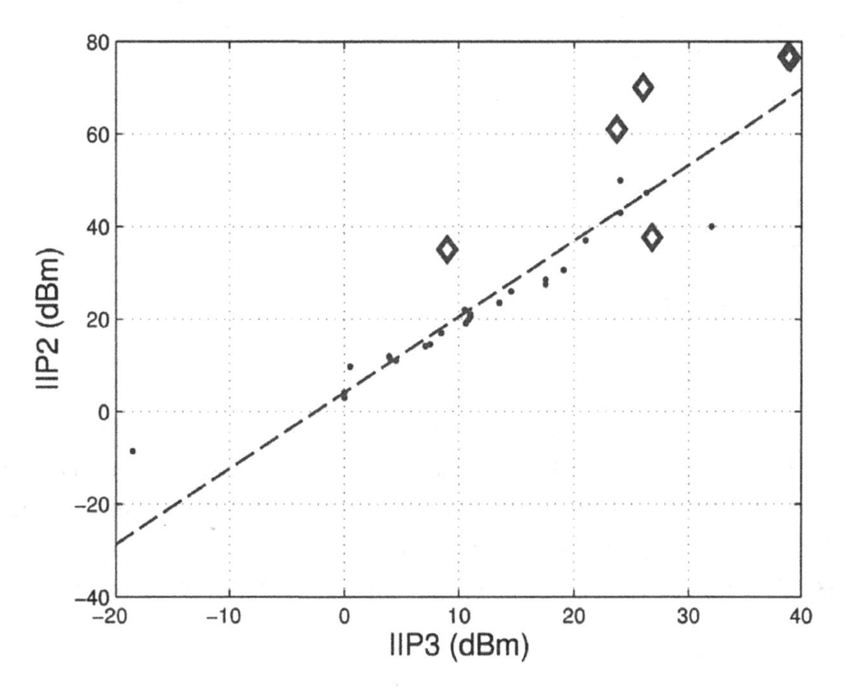

Figure 11.5. Relationship between IIP_2 and IIP_3 for the subset of MMIC amplifiers in the survey (see caption of Figure 11.3) for which IIP_2 is also reported. The dashed line shows the best linear-in-dB fit to the data. Dots indicate single-ended amplifiers and diamonds indicate differential amplifiers.

C, the phase response of the system must be time-varying. Thus, the non-linear capacitance is essentially converting the amplitude-modulated input signal into a signal which also contains phase modulation (PM). AM-PM conversion can also be attributed to inductors (also reactances), transistors (which exhibit capacitive terminal-to-terminal impedances), and gain compression relative to independently-non-linear reactances – all that is needed for the system phase response to become a function of the input magnitude.

AM–PM conversion is typically not a concern for AM signals, since the erroneous PM component does not affect the information, which is represented in the amplitude. AM–PM conversion is also typically not of much concern in systems using FM or PSK, since these signals have constant modulus. However, digital modulations employing sinusoidal carrier modulation with non-constant modulus – most notably, QAM – really suffer, as the PM component significantly distorts the constellation. This is one reason why amplifier design for QAM is considerably more demanding than amplifier design for AM, FM, and PSK, assuming a roughly equal power requirement.

11.3 LINEARITY OF DIFFERENTIAL DEVICES

So far in this chapter, we have implicitly assumed single-ended signals and devices. However, we know from previous sections (8.7, 9.6, and 10.8) that differential circuits may in some cases be preferred. In the context of linearity, let us consider one additional advantage of differential circuits: Inherently good second-order linearity. Figure 11.6 illustrates a differential

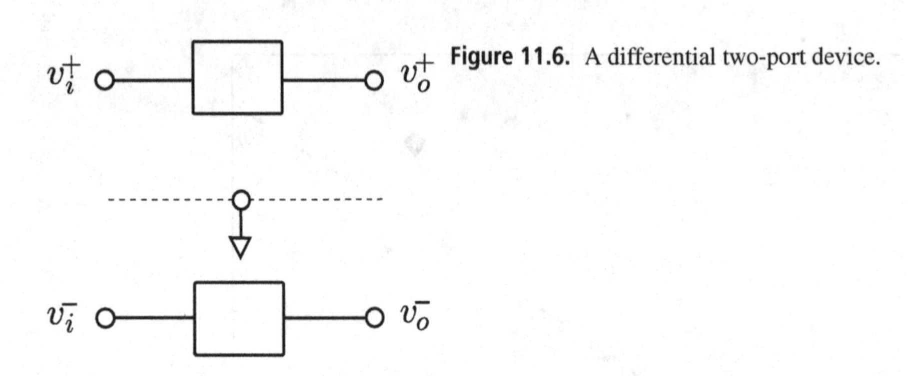

Figure 11.6. A differential two-port device.

two-port device, consisting (as usual) of a symmetric joining of two identical single-ended devices through a virtual ground which is not accessible directly from the terminals. The input terminal pins are $v_i^+(t)$ and $v_i^-(t)$, and the differential input signal is defined in the usual way as $v_i(t) = v_i^+(t) - v_i^-(t)$. Similarly, the output terminal pins are $v_o^+(t)$ and $v_o^-(t)$, and the differential output signal is defined as $v_o(t) = v_o^+(t) - v_o^-(t)$. In terms of the memoryless polynomial system model, we have

$$v_o^+(t) = \sum_{n=0}^{\infty} a_n \left[v_i^+(t)\right]^n \tag{11.21}$$

$$v_o^-(t) = \sum_{n=0}^{\infty} a_n \left[v_i^-(t)\right]^n \tag{11.22}$$

However, in a properly implemented differential system we have $v_i^-(t) = -v_i^+(t)$; therefore the preceding equation may be rewritten

$$v_o^-(t) = \sum_{n=0}^{\infty} a_n (-1)^n \left[v_i^+(t)\right]^n \tag{11.23}$$

Now the output may be written

$$v_o(t) = v_o^+(t) - v_o^-(t) = \sum_{n=1,3,5,\ldots}^{\infty} 2a_n \left[v_i^+(t)\right]^n \tag{11.24}$$

In other words, all even-order terms cancel. In particular, the second-order terms – i.e., those associated with a_2, including the troublesome "beat product" – vanish. For this reason, differential circuits are preferred when second-order linearity is important.

In practice, even-order products are not completely cancelled. This is because of inevitable imbalance in the signal (i.e., $v_i^-(t)$ not exactly equal to $-v_i^+(t)$) or the system. Nevertheless, the achieved level of suppression of even-order products, combined with the other advantages of differential implementation discussed previously, is often sufficient to justify the effort.

11.4 LINEARITY OF CASCADED DEVICES

In previous chapters we have noted the value of representing complex RF systems as cascades of two-port devices. The same principle can be applied to the analysis of linearity. Consider the

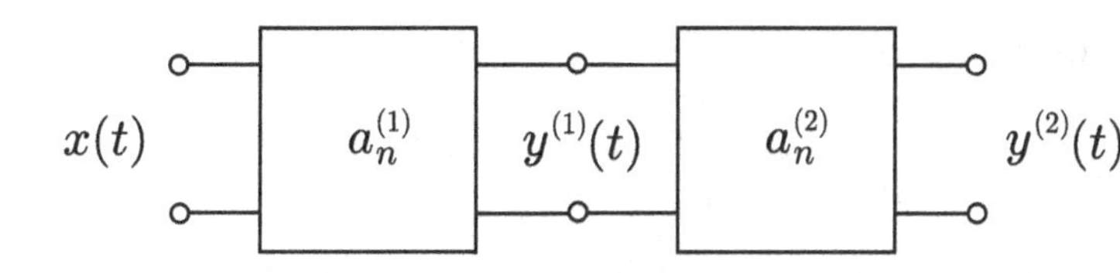

Figure 11.7. A two-stage cascade to be analyzed in terms of third-order intercept point.

two-stage cascade of two-ports shown in Figure 11.7. Following Section 11.2.2, the first stage is described using the voltage transfer characteristic

$$y^{(1)}(t) = a_1^{(1)}x(t) + a_2^{(1)}x^2(t) + a_3^{(1)}x^3(t) \tag{11.25}$$

where the superscript ("(1)") is used to denote stage order, and we will ignore the zero-order non-linearity ($a_0^{(1)}$) and non-linear terms of order 4 or greater. Similarly, the second stage is described using the voltage transfer characteristic

$$y^{(2)}(t) = a_1^{(2)}\left[y^{(1)}(t)\right] + a_2^{(2)}\left[y^{(1)}(t)\right]^2 + a_3^{(2)}\left[y^{(1)}(t)\right]^3 \tag{11.26}$$

The input IP3 may now be found using the procedure described in Section 11.2.4, and subject to the same assumptions and caveats. To make the problem tractable, we assume perfect impedance matching between stages. Then the solution is straightforward, but tedious.[6] The result is:

$$A_{\mathrm{IP}_3} = \frac{2}{\sqrt{3}}\left|\frac{a_1^{(1)}a_1^{(2)}}{a_3^{(1)}a_1^{(2)} + 2a_1^{(1)}a_2^{(1)}a_2^{(2)} + \left[a_1^{(1)}\right]^3 a_3^{(2)}}\right|^{1/2} \tag{11.27}$$

From this, the input-referred 1 dB compression point may be found using Equation (11.18). A calculation for 3 or more stages is possible by simply repeating the above calculation iteratively; that is, first calculating the result for stages 1 and 2, then for the combination of stages 1 and 2 with stage 3, and so on.

In practice, the above approach is rarely used, both because it is tedious and because it depends directly on model parameters (that is, the $a_n^{(m)}$s) that are rarely available. However, the above calculation can be approximated using the stage IP3s directly:

$$\frac{1}{\left[A_{\mathrm{IP}_3}\right]^2} \approx \frac{1}{\left[A_{\mathrm{IP}_3}^{(1)}\right]^2} + \frac{\left[a_1^{(1)}\right]^2}{\left[A_{\mathrm{IP}_3}^{(2)}\right]^2} \tag{11.28}$$

In fact, this result may be extended to any number of stages:

$$\frac{1}{\left[A_{\mathrm{IP}_3}\right]^2} \approx \frac{1}{\left[A_{\mathrm{IP}_3}^{(1)}\right]^2} + \frac{\left[a_1^{(1)}\right]^2}{\left[A_{\mathrm{IP}_3}^{(2)}\right]^2} + \frac{\left[a_1^{(1)}\right]^2\left[a_1^{(2)}\right]^2}{\left[A_{\mathrm{IP}_3}^{(3)}\right]^2} + \cdots \tag{11.29}$$

and since power is proportional to the square of voltage (or current), we have

$$\frac{1}{\mathrm{IIP}_3} \approx \frac{1}{\mathrm{IIP}_3^{(1)}} + \frac{G_1}{\mathrm{IIP}_3^{(2)}} + \frac{G_1G_2}{\mathrm{IIP}_3^{(3)}} + \cdots \tag{11.30}$$

[6] More complete derivations of this and subsequent results in this section can be found in [39, Sec. 19.2.2] and [63, Sec. 2.2.5].

Finally, the same expression applies to IP2; that is:

$$\frac{1}{\mathrm{IIP}_2} \approx \frac{1}{\mathrm{IIP}_2^{(1)}} + \frac{G_1}{\mathrm{IIP}_2^{(2)}} + \frac{G_1 G_2}{\mathrm{IIP}_2^{(3)}} + \cdots \tag{11.31}$$

Equations (11.30) and (11.31) convey some rather useful insights. First, consider a two-port cascade in which all the gains are unity; i.e., $G_n = 1$. Note that the resulting IP3 (or IP2) cannot be greater than the maximum stage IP3 (or IP2).[7] For G_ns greater than 1, we see that the situation is potentially worse – increasing stage gain tends to reduce cascade IP3 (or IP2). Since radios tend to consist of multiple stages containing a lot of gain, the consequence of Equations (11.30) and (11.31) is that radio design involves a tradeoff between gain and linearity; and in particular a tradeoff between gain and vulnerability to second- and third-order intermodulation products.

EXAMPLE 11.2

A system consists of two amplifiers, which are perfectly impedance matched at the frequency of interest. The gain and OIP_3 of the first amplifier are 15 dB and 15 dBm respectively. The gain and OIP_3 of the second amplifier are 15 dB and 20 dBm respectively. What is the IIP_3 of this system? What is the OIP_3 of this system?

Solution: From Equation (11.30), we have

$$\frac{1}{\mathrm{IIP}_3} \approx \frac{1}{\mathrm{IIP}_3^{(1)}} + \frac{G_1}{\mathrm{IIP}_3^{(2)}} \tag{11.32}$$

$G_1 = 31.6$ and $\mathrm{OIP}_3^{(1)} = 31.6$ mW, so $\mathrm{IIP}_3^{(1)} = 1$ mW. $G_2 = 31.6$ and $\mathrm{OIP}_3^{(2)} = 100$ mW, so $\mathrm{IIP}_3^{(2)} = 3.16$ mW. Applying the above equation, we find the cascade IIP_3 is 0.091 mW $= -10.4$ dBm. The cascade OIP_3 is greater by the cascade gain, which is $15 + 15 = 30$ dB. Therefore, the cascade OIP_3 is $+19.6$ dBm.

11.5 STAGE/CASCADE ANALYSIS; SIGNIFICANCE OF STAGE ORDER

In Section 4.4.2, we worked out a procedure for determining the sensitivity of a cascade of two-ports given the sensitivity (specifically, noise figure or equivalent temperature) of the constituent stages. Similarly, in earlier sections of this chapter we have worked out procedures for determining the linearity of cascades of two-ports given the linearity (specifically, IIP_3 and IIP_2) of the constituent stages. Under the same assumptions the cascade gain is simply the product of the stage gains; that is

$$G = G_1\, G_2\, G_3\, \ldots \tag{11.33}$$

[7] To see this, it may be useful to note the direct mathematical analogy with the equation for the resistance of resistors in parallel.

Stage/cascade analysis typically consists of constructing a table of stage gain, linearity, and sensitivity and the corresponding quantities for the associated cascade. This type of analysis is known by many other names, and exists in many different variants and presentations. The approach presented here is sufficiently generic that it should apply to most other forms that the reader is likely to encounter. The purpose of this analysis is to determine the impact of individual stages on the overall performance, thereby providing insight into how best to arrange the stages, and how to specify the stage characteristics when that flexibility exists. The concept is most easily demonstrated by example.

EXAMPLE 11.3

An LMR base station has a receive path consisting of a mast-mounted antenna separated by a long cable having 10 dB loss at the frequency of interest. The receiver consists of two parts: A front end including the antenna, cable, and two amplifiers; and a back end consisting of the rest of the receiver. The front end is specified to have a gain of 35 dB. To accomplish this, two amplifiers are provided. Amplifier 1 has 15 dB gain, +5 dBm OIP_3, and 4 dB noise figure. Amplifier 2 has 30 dB gain, +10 dB OIP_3, and 8 dB noise figure. The noise figure of the front end is required to be 7 dB or better. Assume perfect impedance matching, and either amplifier may be installed at the antenna side or receiver side of the cable. Design the front end.

Solution: This is a classic problem in radio engineering: You have multiple components and you need to arrange them to achieve acceptable overall performance. Since this is a receiver problem, the relevant performance considerations are sensitivity and linearity. In this application, linearity is commonly characterized in terms of IIP_3, whereas sensitivity is commonly characterized in terms of noise figure or system temperature. The linearity of the amplifiers is specified in terms of OIP_3 as opposed to IIP_3, but this poses no particular difficulty since IIP_3 is related to OIP_3 by the gain; thus we find $IIP_3 = -10$ dBm and -20 dBm for Amplifiers 1 and 2 respectively.

The noise figure of Amplifier 2 is larger than the noise figure specified for the front end, so there will be no solution in which Amplifier 2 precedes Amplifier 1 in the cascade. This leaves three possibilities: (1) The cable comes first, and both amplifiers are on the receiver side; (2) Amplifier 1 is mast-mounted, followed by the cable, followed by Amplifier 2, and (3) Amplifiers 1 and 2 are both mast-mounted, followed by the cable. Option 1 is worth considering only if the noise figure of the cable is less than the specified noise figure of the front end. If we assume the physical temperature of the cable is equal to the noise figure reference temperature T_0, then the noise figure of the cable is simply the reciprocal of the gain of the cable. (If this is not clear, see Equation (4.19) and associated text.) The IEEE-standard reference temperature $T_0 = 290$ K is a reasonable guess for the physical temperature of the cable, so we will take the noise figure of the cable to be 10 dB. This is greater than 7 dB, so Option 1 is out.

To choose between Options 2 and 3, it is useful to prepare a stage-cascade analysis for each option, as shown in Tables 11.2 and 11.3. Comparing the results, a familiar dilemma is apparent: Option 2 offers better linearity (IIP_3), but Option 3 offers better sensitivity (F). If we interpret the noise figure requirement from the problem statement as a constraint (as opposed to noise figure as being something to be optimized), then Option 1 is preferred since it meets the noise figure specification while optimizing the remaining parameter; i.e., cascade IIP_3. In practice, a decision would probably be made by considering the cascade IIP_3 and F findings from both options in the context of other system-level considerations, perhaps as part of an iterative system design process.

Note that Tables 11.2 and 11.3 also show the stage and cascade equivalent noise temperatures. Specifically, these are the input-referred equivalent noise temperatures. What value is there in this? There are two reasons why this is potentially useful. First, recall that we assumed the physical temperature of the cable was equal to the IEEE standard reference temperature for noise figure, T_0. We noted in Example 4.3 that this is a pretty good approximation over a broad range of physical temperatures. However, if the cable is at a temperature *not* close to T_0 (perhaps it is coiled up in a box with hot electronics), it is a little more straightforward to do the sensitivity computation in terms of noise temperature, and then convert the result back to noise figure. A second benefit of considering noise temperature is that the contribution of external noise is easily considered. Here is an example:

Table 11.2. **Stage-cascade analysis of Option 2 from Example 11.3. Stage values are in italics. Values for the complete (three-stage) cascade are in boldface. Equivalent temperatures are input-referred.**

	G [dB]	IIP_3 [dBm]	*F* [dB]	*G* [dB]	IIP_3 [dBm]	*F* [dB]	*T* [K]	*T* [K]
Amp 1	*15.0*	*−10.0*	*4.0*	15.0	−10.0	4.0	*438.4*	438.4
Cable	*−10.0*	*40.0*	*10.0*	5.0	−10.0	4.5	*2610.0*	521.0
Amp 2	*30.0*	*−20.0*	*8.0*	**35.0**	**−25.1**	**6.5**	*1539.8*	**1007.9**

Table 11.3. **Stage-cascade analysis of Option 3 from Example 11.3. Stage values are in italics. Values for the complete (three-stage) cascade are in boldface. Equivalent temperatures are input-referred.**

	G [dB]	IIP_3 [dBm]	*F* [dB]	*G* [dB]	IIP_3 [dBm]	*F* [dB]	*T* [K]	*T* [K]
Amp 1	*15.0*	*−10.0*	*4.0*	15.0	−10.0	4.0	*438.4*	438.4
Amp 2	*30.0*	*−20.0*	*8.0*	45.0	−35.0	4.3	*1539.8*	487.1
Cable	*−10.0*	*40.0*	*10.0*	**35.0**	**−35.0**	**4.3**	*2610.0*	**487.2**

EXAMPLE 11.4

Continuing Example 11.3, consider the contribution of external noise to the sensitivity of front end Options 1 and 2. Work this out for an antenna temperature of 400 K.

Solution: First, recall from Section 4.5 that antenna temperature T_A is a description of the amount of external noise delivered from the antenna, and that 400 K is pretty typical for mobile radio applications in the VHF–UHF bands. Recall also from Section 7.4 that a convenient way to characterize the overall sensitivity, accounting for external noise, is system temperature. The system temperature T_{sys} is simply the sum of T_A and the input-referred equivalent noise temperature of the front end, which was found to be 1008 K and 487 K for Options 1 and 2 respectively. Thus T_{sys} is 1408 K for Option 1 and 887 K for Option 2. With these values we could now proceed with radio link analysis as described in Section 7.6.

The above example can also be solved beginning with the cascade noise figure. However, this approach requires more work because noise figure is expressed with respect to a pre-determined input noise temperature; namely T_0, which is typically taken to be 290 K. On the other hand, if T_A just happens to be $\approx T_0$, then the noise figure approach is actually easier: The noise power spectral density delivered at the output of the cascade is simply kT_0FG (where F and G are the cascade values), so T_{sys} (being input-referred) is simply T_0F. Because this is such a simple calculation, and because it is frequently difficult to determine the actual antenna temperature, it is very common to assume $T_{sys} = T_0F$; however, this leads to potentially very large error unless $T_A \cong T_0$, or $T_A \ll T_0$ (so that external noise may be neglected).

11.6 OTHER COMMON CHARACTERIZATIONS OF SENSITIVITY

It was stated in Chapter 1 that we would find that receiver design was essentially a three-way tradeoff between sensitivity, dynamic range, and bandwidth. We are now almost ready to quantify the first two elements of this tradeoff. First, it is useful to make two additional definitions, both related to sensitivity. These are *minimum discernible signal* and *noise floor.*

11.6.1 Minimum Discernible Signal (MDS): Concept and Zero-Input-Noise Expressions

In Section 7.5 we defined "formal sensitivity" as the incident power density of a radio wave required to achieve a specified signal-to-noise ratio (SNR) at the input of a detector or demodulator. We also noted that there are many alternative definitions that are more or less appropriate depending on the application. Let us now consider one of the alternative definitions, known as *minimum discernible signal* (MDS).[8] The MDS of a two-port (including a cascade of two-ports) is the signal power applied to the input which results in a specified SNR at the

[8] This is sometimes written *minimum detectable signal.*

output, assuming perfect impedance matching at input and output, and assuming the bandwidth of the two-port is at least as large as the bandwidth of the signal. The specified SNR is typically 1, and this will be our definition, but the reader should be aware that it is not uncommon to define MDS to correspond to an SNR of 2, 3 dB (which is not exactly 2), 10 dB, or other values. The reader should also note that MDS is sometimes defined so as to account for a non-zero noise power applied to the input – more on that in the next section. Summarizing: We shall initially use a definition that assumes SNR = 1 and no noise entering from the input.

MDS is useful primarily because it can be easily measured in a laboratory, since the effect of antennas and variable external noise need not be considered. Here is an example:

EXAMPLE 11.5

A signal generator is applied to the input of a two-port cascade, and a power meter is attached to the output of the cascade. It is found that an RF tone at −127 dBm at the input results in a reading of +5 dBm on the power meter. When the tone is turned off, the power meter reads +2 dBm. What is the MDS? Assume the signal generator contributes negligible noise.

Solution: When the tone is turned off, the power meter is indicating the output-referred internal noise of the cascade. Let us call this quantity N. When the tone is turned on, the power meter is reading $S + N$ where S is the output-referred signal power.[9] The ratio of the two readings is 3 dB $\cong$ 2; therefore $(S + N)/N \cong 2$. Note $(S + N)/N = S/N + 1 = \text{SNR} + 1$, so we find that SNR = 1. Therefore MDS is (in this case) equal to the applied power, −127 dBm.

One can also calculate MDS from the cascade parameters. The output-referred internal noise power is $N = kT_{eq}BG$, where B is bandwidth and G is cascade gain. The output-referred signal power when the input power is equal to the MDS is $S = \text{MDS} \cdot G$. If the input power is equal to the MDS, then the output SNR is equal to 1; i.e., $kT_{eq}BG = \text{MDS} \cdot G$. Thus:

$$\text{MDS} = kT_{eq}B \tag{11.34}$$

This may also be expressed in terms of the cascade noise factor F:

$$\text{MDS} = kT_0 \left(F - 1\right) B \tag{11.35}$$

where, as usual, T_0 is the noise temperature used to define noise figure, and this is usually, but not always, chosen to be the IEEE standard value of 290 K.

11.6.2 Minimum Discernible Signal (MDS): Non-Zero-Input-Noise Expressions

Equation (11.34) indicates that the zero-input-noise definition of MDS is really just another way of indicating the equivalent noise temperature of the receiver. This is not exactly the same as sensitivity, since sensitivity is also affected by external noise applied to the input, which we assume to be zero in the previous section. The definition of MDS may be modified to account for this by changing Equation (11.34) as follows:

[9] This of course assumes that the cascade remains utterly linear when the tone is turned on; i.e., N does not change once the signal is applied.

$$\text{MDS} = kT_{sys}B = k\left(T_{in} + T_{eq}\right)B \tag{11.36}$$

where T_{in} is the noise temperature applied at the input; for example, $T_{in} = T_A$ if the input is connected to an antenna.

This raises the question of what value to use for T_{in}, since this is now intrinsic to the definition. One approach is to specify a value of T_{in} that corresponds to a value which is expected or typical when the two-port – which we might as well call a "receiver" in this case – is in operation. However, it is very common to arbitrarily assume $T_{in} = T_0$ for this purpose, as noted in the last paragraph of Section 11.5. Of course it is pretty uncommon for T_{in} to *actually* be equal to T_0 – even approximately – so this form of the definition is no less arbitrary than the zero-input-noise ($T_{in} = 0$) definition.

One nice feature of the $T_{in} = T_0$ definition is a convenient simplification of the math, since in this case we have

$$\text{MDS} = k\left(T_0 + T_{eq}\right)B = kT_0B + kT_{eq}B = kT_0B + kT_0\left(F-1\right)B$$

which reduces to

$$\text{MDS} = kT_0FB \quad \text{assuming} \quad T_{in} = T_0 \tag{11.37}$$

This equation is readily expressible in a popular decibel form:

$$\text{MDS [dBm]} \cong -174 \text{ dBm/Hz} + F \text{ [dB]} + 10\log_{10} B \tag{11.38}$$

where "dBm/Hz" is shorthand for "dB(mW/Hz)". It should also be noted that this definition of MDS is usually approximately equal to the zero-input-noise definition if output-referred external noise is small compared to output-referred internal noise (i.e., F is very large), since in that case it doesn't matter what you assume about the external noise.

Be careful to avoid the following pitfall: Given F you assume Equation (11.37) gives you the *in situ* sensitivity of the receiver. This is clearly not true, since again the noise input to the receiver may be (probably is!) very different from −174 dBm/Hz. Recall we noted precisely this type of error when discussing link budgets in Chapter 7 (see Section 7.6). In summary, MDS defined in this manner is not really useful for evaluating *in situ* sensitivity; instead its utility is in relating internal noise characteristics to sensitivity in laboratory conditions and other situations where it is convenient to specify a "standard" external noise condition.

EXAMPLE 11.6

The bandwidth and noise figure of a receiver are 30 kHz and 5 dB, respectively. (a) What is the MDS assuming the zero-input-noise definition? (b) What is the MDS assuming the $T_{in} = T_0$ definition?

Solution: Note $B = 30$ kHz and $F = 5$ dB $= 3.2$. For (a), we have from Equation (11.35) that MDS = 2.60×10^{-16} W = −125.9 dBm. For (b), we have from Equation (11.37) that MDS = 3.80×10^{-16} W = −124.2 dBm. Note that the receiver is no less sensitive; it is simply that we have defined the input noise condition differently. In case (b) we may also use Equation (11.38); then we find as expected:

$$\text{MDS} \cong -174 \text{ dBm/Hz} + 5 \text{ dB} + 44.8 \text{ dB (Hz)} = -124.2 \text{ dBm}$$

EXAMPLE 11.7

In Example 11.5, assume the bandwidth is 30 kHz. What is the noise figure?

Solution: Assuming the noise figure was defined with respect to $T_0 = 290$ K, then Equation (11.35) gives us

$$F = \frac{\text{MDS} + kT_0B}{kT_0B} \tag{11.39}$$

MDS is -127 dBm, and $kT_0B = -129.2$ dBm, so we have

$$F = \frac{\text{MDS} + kT_0B}{kT_0B} = 2.66 = 4.2 \text{ dB} \tag{11.40}$$

We must not use Equation (11.37) to solve this problem, since its assumption about the actual *in situ* input noise is incorrect.

11.6.3 Noise Floor

Very often the bandwidth of the signal (or signals) of interest is less than the bandwidth of the receiver, resulting in a situation such as the one shown in Figure 11.8. We have seen this above: The signal is a tone (taken to mean that the signal bandwidth is smaller than the spectral resolution), yet the bandwidth is 30 kHz. This realization leads us to the concept of a *noise floor*. The noise floor is best defined as the average power spectral density of the noise in the central region of the passband, where the noise power spectral density is relatively flat. So, in Example 11.7, this is $kT_{eq}G = kT_0(F-1)G$ referred to output, which is $kT_0(F-1) = -171.8$ dBm/Hz referred to input. An alternative and very common description of the noise floor is in terms of

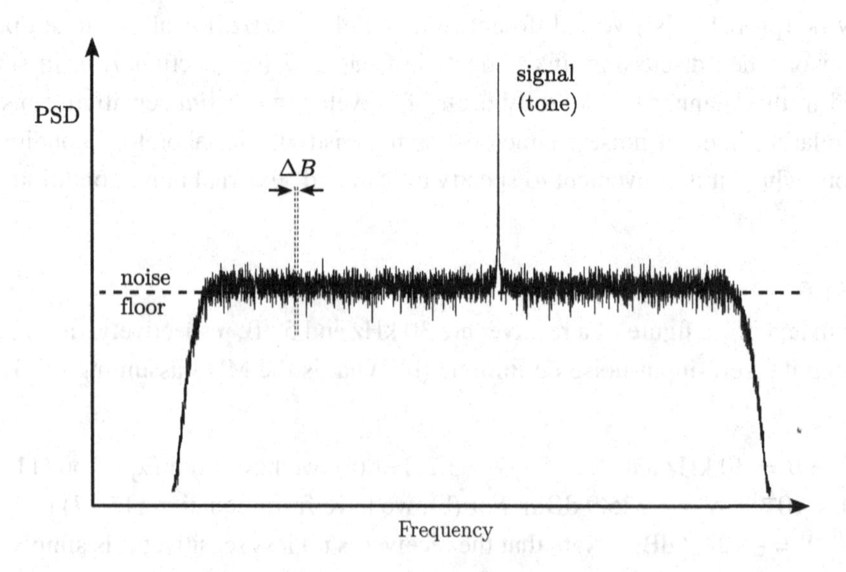

Figure 11.8. Noise floor. Note that the apparent level of the noise floor depends on the resolution bandwidth (ΔB), whereas the apparent level of the signal is independent of ΔB as long as the signal bandwidth is less than ΔB.

power within a specified *resolution bandwidth*. Resolution bandwidth is simply the bandwidth $\Delta B < B$ within which the power is measured.

EXAMPLE 11.8

Returning to Example 11.7, let us say the gain is 50 dB. (a) What is the output-referred noise floor in terms of power spectral density? (b) Let us say we terminate the input into a very cold matched load, and observe the output using an ideal (noiseless) spectrum analyzer with resolution bandwidth set to 1 kHz. What does the spectrum analyzer indicate as the noise floor?

Solution: The output-referred noise floor is $kT_{eq}G = kT_0(F-1)G = -171.8$ dBm/Hz + 50 dB = -121.8 dBm/Hz, which is the answer to (a). Terminating the input into a "very cold matched load" suggests that the applied external noise is negligible compared to the internal noise, so the spectrum analyzer is measuring only internal noise. Since the noise floor power spectral density is -121.8 dBm/Hz, and $\Delta B = 1$ kHz, the spectrum analyzer shows a trace at -121.8 dBm/Hz $+ 10\log_{10}\Delta B = -91.8$ dBm.

The above examples demonstrate that MDS and noise floor are closely-related concepts that have multiple definitions. To further complicate the issue, the terms "MDS" and "noise floor" are often used interchangeably. In practical problems one should always take great care to understand precisely how these terms are being used.

11.7 DYNAMIC RANGE

Finally, we consider the concept of *dynamic range*. In radio systems engineering, dynamic range can be broadly defined as the ratio of the powers of the strongest signal and the weakest signal that can be simultaneously accommodated without unacceptable distortion or loss of sensitivity. This definition is simultaneously complicated and too vague for practical use, but it conveys the essential concept.

To understand the concept in a more useful way, first consider a single signal. As its power increases, the non-linearity of the cascade becomes apparent, resulting in significant gain compression and generation of harmonics. Now consider a second, weaker signal at a different frequency, but within the passband. The degradation of the system due to the presence of the large signal results in diminished ability to receive the weaker signal.

This manifests in a number of ways: For example, if the gain is reduced but the output-referred internal noise remains the same (or increases), then MDS increases. It is said that the strong signal has *desensitized* the system. In this case, we could define dynamic range as the ratio of the power of the strong signal to the resulting MDS. To make this a measurable quantity, we specify that the desensitizing signal should be at a power level equal to the input 1 dB compression point. We now have the most common definition of *blocking dynamic range*

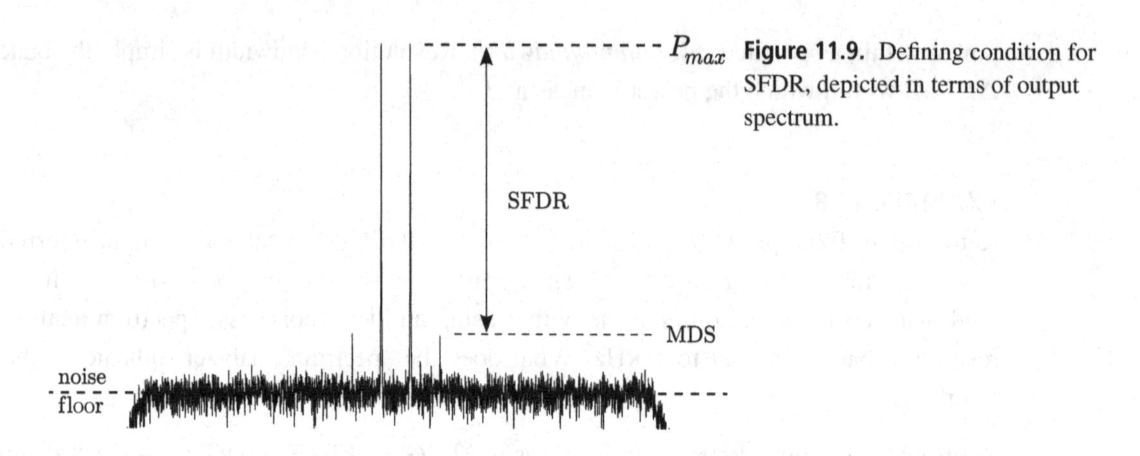

Figure 11.9. Defining condition for SFDR, depicted in terms of output spectrum.

(BDR). Note that BDR depends on both the linearity and the sensitivity of the system. The advantage of knowing the BDR is that is quantifies the loss of sensitivity associated with a signal that results in "just barely non-linear" operation of the system, and it is defined in a way that facilitates measurement and comparison among systems.

There are many other ways to define dynamic range. Another common definition is *two-tone spurious-free dynamic range* (2T-SFDR, or just SFDR), illustrated in Figure 11.9. Consider two tones, both within the passband of a system and separated by a span which is much less than one-third of the bandwidth of the system. The power of the tones is equal, and is increased until the close-in third-order intermodulation products at $2\omega_2 - \omega_1$ and $2\omega_1 - \omega_2$ (both within the passband) are apparent. At some per-tone power P_{max}, the power of each intermodulation product becomes equal to the MDS; in other words, the IM3 has become as strong as the weakest detectable (in the MDS sense) signal. SFDR is then defined as the ratio of P_{max} to MDS. SFDR is favored as a metric because the definition reflects a scenario that is commonly encountered: Two tones result in IM3 (the "spurious" signals alluded to in the term "SFDR") which becomes strong enough to be falsely detected as legitimate signals. If the IM3 is strong enough to be detected, then it is strong enough to mask legitimate signals at comparable levels; thus detectable IM3 can be interpreted as a form of desensitization.

As it turns out, SFDR is quite simple to calculate in terms of existing linearity and sensitivity metrics. This is demonstrated in Figure 11.10, which shows output power as a function of input power. Recall that the power of each IM3 tone increases at a rate of three times the linear gain (G). Also, by definition, the output power of each IM3 tone is equal to the output power of the fundamental when the input power of the fundamental is equal to IIP_3. Expressing all quantities in terms of dB power, we see that P_{max} is less than the IIP_3 by $(IIP_3 - MDS)/3$. Therefore we have

$$\begin{aligned} \text{SFDR [dB]} &\equiv P_{max} \text{ [dBm]} - \text{MDS [dBm]} \\ &= \frac{2}{3} \left(\text{IIP}_3 \text{ [dBm]} - \text{MDS [dBm]}\right) \end{aligned} \tag{11.41}$$

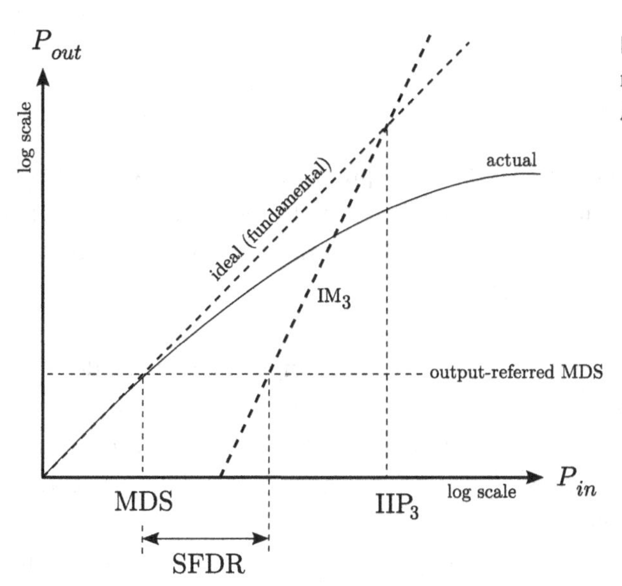

Figure 11.10. SFDR in terms of the relationship between output-referred power P_{out} and input-referred power P_{in}.

EXAMPLE 11.9

The IIP_3 and MDS of an cascade are +10 dBm and −110 dBm, respectively. What is the SFDR?

Solution: Using Equation (11.41), one obtains SFDR = $\frac{2}{3}$ (+10 − [−110]) = 80 dB.

EXAMPLE 11.10

Continuing the previous example: When measuring the SFDR, what is the input power of each applied tone when the IM3 is equal to the MDS?

Solution: Since the power of each IM3 product is equal to the MDS, the input power of each tone is $P_{max} = \text{IIP}_3 - (\text{IIP}_3 - \text{MDS})/3$ [dB], which is −30 dBm.

EXAMPLE 11.11

Continuing the previous examples: Estimate the BDR.

Solution: BDR is defined as $P_1 - \text{MDS}$ [dB]. If we employ the rule of thumb that $\text{IIP}_3 - P_1 =$ 9.6 dB (Following Equation (11.18) and associated caveats), then we may estimate BDR to be $\approx +10 - 9.6 - (-110) = 110.4$ dB.

Problems

11.1 Demonstrate that a common source (FET) amplifier can be expressed in the polynomial form of Equation (11.3), and find the coefficients. In this case, the answer will be an expression for the small-signal drain current in terms of the small-signal gate-to-source voltage.

11.2 Derive Equation (11.11).

11.3 Show that the transistor model considered in Example 11.1 either does or does not exhibit gain compression.

11.4 Does an ideal FET in common source configuration experience gain compression? (Hint: See Problem 11.1). Does a practical common source amplifier experience gain compression? Why or why not?

11.5 Develop an expression for the input-referred second-order intercept point, analogous to Equation (11.17).

11.6 Repeat Example 11.2 but now for the following configurations. In all configurations, the attenuator (if used) has 3 dB loss and $OIP_3 = +100$ dBm. (a) Amplifier 2 followed by Amplifier 1 (no attenuator); (b) attenuator, Amplifier 1, Amplifier 2; (c) Amplifier 1, attenuator, Amplifier 2; (d) Amplifier 1, Amplifier 2, attenuator.

11.7 Generate a table in the form shown in Tables 11.2 and 11.3 for Option 1 in Example 11.3

11.8 Repeat Example 11.3, but now assume that Amplifier 2 is required to come first in the cascade (for example, it might be physically integrated into the antenna). Identify the options, and for each option produce a table in the form shown in Tables 11.2 and 11.3. Comment on the merits of each option.

11.9 In Example 11.5, the output power is −17 dBm when the signal generator is on, and −19 dBm when the signal generator is off. What is the MDS for SNR=1?

11.10 Make a sketch similar to Figure 11.8 showing the situation when the input signal power is (a) equal to MDS and (b) equal to 3 dB over MDS. Assume the criterion for MDS is SNR = 1, and assume the bandwidth of the signal is much less than the resolution bandwidth. Sketch as accurately as you can, indicating the relevant levels.

11.11 A manufacturer reports the blocking dynamic range of a LNB (see Section 7.10) to be 60 dB. The gain and OIP_3 of the LNB is found to be 50 dB and −5 dBm, respectively. Estimate the MDS, and disclose any assumptions associated with this estimate.

11.12 Continuing the previous problem, estimate the two-tone SFDR.

12 Antenna Integration

12.1 INTRODUCTION

Chapter 2 provided an introduction to antenna fundamentals in which antennas were described as transducers of guided electrical signals to unguided electromagnetic waves, and vice versa. In Chapter 4, receive antennas were viewed as receptors of environmental noise, resulting in sensitivity-limiting electrical noise characterized in terms of antenna temperature. Intervening chapters have introduced methodology for analysis and design of radios which accounts for effective radiated power and sensitivity (Chapter 7), impedance matching (Chapter 9), and noise within a receiver (Chapter 11). We are now in a position to confront some antenna issues that were bypassed in Chapter 2, namely: How to most effectively integrate an antenna with radio electronics so as to achieve system-level design goals. In this chapter we address the topic of antenna integration so as to close this gap in our review of radio electronics, before moving on to the topics of receiver and transmitter design (Chapters 15 and 17, respectively).

The organization of this chapter is as follows. In Section 12.2 we consider the manner in which an antenna contributes to the SNR achieved by a receiver, thereby limiting sensitivity. In Section 12.3 we consider VSWR and efficiency of power transfer in transmit operation. These considerations are closely related to the issue of impedance matching between antenna and transceiver, which is addressed in Section 12.4. The trend toward increasingly small antennas is considered, first in terms of the impedance-matching challenge, and then (in Section 12.5) in terms of practical limits on an antenna size. Section 12.6 gives a brief introduction to antenna tuners, a resurgent technology that facilitates improvements in the performance of radios that must cover large tuning ranges using antennas which are narrowband or electrically-small. This chapter concludes with an introduction to baluns (Section 12.7), which are needed to accomplish the common tasks of interfacing single-ended antennas to differential transceivers, and differential antennas to single-ended transceivers.

12.2 RECEIVE PERFORMANCE

Let us now quantify the role an antenna plays in the sensitivity of a radio receiver. This issue was previously addressed in Sections 7.4 and 7.5, and a review of those sections is recommended before engaging with this section. We begin by elaborating on the concepts presented in those sections and consider the implications for the design of antennas and the transmission lines and matching networks used to interface them to receivers.

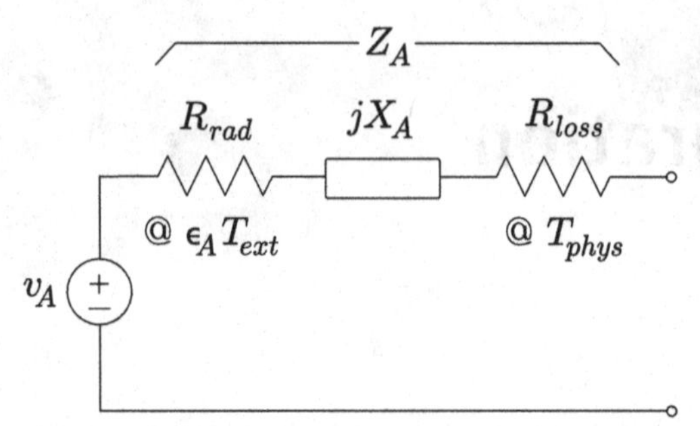

Figure 12.1. Model for receiving antenna. The temperature for calculation of noise contributions is indicated next to each resistance.

12.2.1 Antenna Receive Model, Revisited

Figure 12.1 shows a model for a receive antenna that accounts for all previous considerations, including the signal-of-interest, external noise, and inefficiencies and noise associated with internal and external losses. In Figure 12.1, v_A is the open-circuit voltage in response to the signal-of-interest $\mathbf{E}^i$, represented as an electric field intensity; i.e., a vector having units of V/m. As in Section 2.3.1,

$$v_A = \mathbf{E}^i \cdot \mathbf{l}_e \tag{12.1}$$

where $\mathbf{l}_e$ is the vector effective length, which describes the directivity and polarization of the antenna. The antenna self-impedance Z_A is composed of the radiation resistance R_{rad}, reactance X_A, and ohmic loss R_{loss}. External noise is represented as the thermal noise generated by R_{rad} when at a temperature equal to the antenna temperature. To avoid ambiguity, the antenna temperature is written $\epsilon_A T_{ext}$, where T_{ext} is the antenna temperature (T_A) for the antenna in isolation from its immediate surroundings – in particular, the ground – and ϵ_A accounts for the reduction in T_{ext} due to dissipative and absorptive mechanisms, such as ground loss, which are separate from losses internal to the antenna. With respect to the definitions of Chapter 7 (specifically, Equation (7.4)):

$$\epsilon_R = \epsilon_A \epsilon_{rad} \tag{12.2}$$

where ϵ_{rad} is the radiation efficiency, as in Chapter 2. Separately, internal noise is represented as the thermal noise generated by R_{loss} with T_{phys} equal to the physical temperature of the antenna.

In Figure 12.2, the model of Figure 12.1 is rearranged to show external and internal noise as the explicit series voltage sources v_{ext} and v_{int}, respectively; and so now Z_A consists of "ideal" (non-noise-generating) resistances. Following Section 4.2, the rms magnitudes of the new voltage sources are given by

$$\sigma_{ext} = \sqrt{4k\epsilon_A T_{ext} R_{rad} B}\ \text{, and} \tag{12.3}$$

$$\sigma_{int} = \sqrt{4kT_{phys} R_{loss} B} \tag{12.4}$$

respectively.

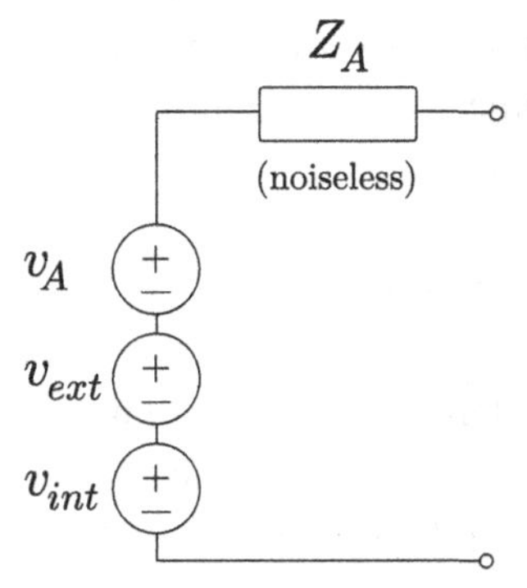

Figure 12.2. Model for receiving antenna, revised from Figure 12.1 to show external and internal noise as explicit series voltage sources.

From this model we may easily derive two quantities of subsequent interest. First is the power P_A available from the antenna. From the usual definition of available power – i.e., power delivered to a conjugate-matched load – we immediately find:[1]

$$P_A = \frac{\sigma_{tot}^2}{4R_A} \tag{12.5}$$

where $R_A = R_{rad} + R_{loss}$ (as usual) and σ_{tot}^2 is the variance of the sum of the three voltages. Because the signal-of-interest, the external noise, and the internal noise are expected to be utterly uncorrelated, we expect the variance of the sum of voltages to be equal to the sum of the individual variances; i.e.,

$$\sigma_{tot}^2 = \sigma_A^2 + \sigma_{ext}^2 + \sigma_{int}^2 \tag{12.6}$$

where σ_A^2 is the variance of v_A. A useful expression for σ_A^2 in terms of $\mathbf{E}^i$ may be obtained by direct application of the definition of variance:

$$\sigma_A^2 \equiv \lim_{\tau\to\infty} \frac{1}{\tau} \int_t^{t+\tau} \left|\mathbf{E}^i \cdot \mathbf{l}_e\right|^2 dt \tag{12.7}$$

At this point it is useful to separate the signal of interest into a scalar component representing the waveform of interest, and a vector component representing polarization. Thus we make the definition $\mathbf{E}^i \equiv E^i\hat{\mathbf{e}}^i$, with $\hat{\mathbf{e}}^i$ being a unit vector representing the polarization of $\mathbf{E}^i$. Further, we assume that both $\hat{\mathbf{e}}^i$ and $\mathbf{l}_e$ vary slowly with respect to E^i. Under these assumptions, the above equation becomes

$$\sigma_A^2 = \sigma_S^2 \left|\hat{\mathbf{e}}^i \cdot \mathbf{l}_e\right|^2 \tag{12.8}$$

where

$$\sigma_S^2 \equiv \lim_{\tau\to\infty} \frac{1}{\tau} \int_t^{t+\tau} \left|E^i\right|^2 dt \tag{12.9}$$

Now returning to Equation (12.5), we have

$$P_A = \frac{\sigma_S^2 \left|\hat{\mathbf{e}}^i \cdot \mathbf{l}_e\right|^2}{4R_A} + \frac{k\epsilon_A T_{ext} R_{rad} B}{R_A} + \frac{kT_{phys} R_{loss} B}{R_A} \tag{12.10}$$

This is the power available from the antenna, conveniently partitioned into separate terms for the signal-of-interest, external noise, and internal noise.

[1] See Equations (4.2) and (8.39) and associated text for a refresher.

Related to the available power is the *available signal-to-noise ratio*, which is simply the ratio of the first term of the above equation to the sum of the second and third terms:

$$\text{SNR}_A = \frac{\sigma_S^2 \left|\hat{\mathbf{e}}^i \cdot \mathbf{l}_e\right|^2}{4k\left(\epsilon_A T_{ext} R_{rad} + T_{phys} R_{loss}\right) B} \tag{12.11}$$

The significance of this quantity is that it is an upper bound on the SNR that can be achieved by a radio receiver, since any processing subsequent to the antenna affects the signal delivered by the antenna and the noise delivered by the antenna in precisely the same way. This is an *upper bound* on SNR because the receiver can only increase the total noise delivered to the receiver's detector or digitizer.

EXAMPLE 12.1

We wish to evaluate the receive-mode performance of a particular antenna to be used in an urban UHF-band LMR radio system operating in a wide band around 453 MHz using 12.5 kHz channels. The antenna is marketed as a quarter-wave monopole, and is found to have length of 15.7 cm and a radius of 1 mm. The LMR system can be expected to deliver a field strength of at least 10 μV/m (RMS) to the antenna. Estimate available signal power and available SNR for this antenna.

Solution: The available *signal* power is the first term of Equation (12.10). Polarization is not specified in the problem statement. Assuming the antenna is co-polarized with the arriving signal, then

$$\sigma_S^2 \left|\hat{\mathbf{e}}^i \cdot \mathbf{l}_e\right|^2 = \sigma_S^2 l_e^2$$

where $\sigma_S^2 = (10\ \mu\text{V/m})^2$.

There are two approaches that might be used to estimate l_e. If directivity measurements or specifications are provided, then those could be converted to effective aperture by using Equation (2.44) and then subsequently converted to effective length by using Equation (2.39). Alternatively we could assume the antenna is well-modeled as an ideal electrically-thin monopole over an infinite flat perfectly-conducting ground plane. Since the current distribution for the quarter-wave monopole is identical to that of either arm of the half-wave dipole, the current distribution in the present problem is given by Equation (2.49) with $h = L/2 = 15.7$ cm and z measures distance along the monopole with $z = 0$ being the base and feedpoint. This current model can then be used to find the effective length at any frequency using Equation (2.31), by numerical integration if desired. The effective length calculated using this approach is shown in Figure 12.3.

The antenna impedance is also required. Again, this could be obtained from measurements; however, here the equivalent circuit method described in Section 2.7.5 proves to be useful. This model was specified for dipoles, but monopole impedance may be easily obtained using image theory (reminder: Section 2.8.1); one simply divides the inductance and resistance by 2 and multiplies the capacitances by 2. The equivalent circuit obtained in this case is shown in Figure 12.4. Impedance calculated using this model is shown in Figure 12.5.

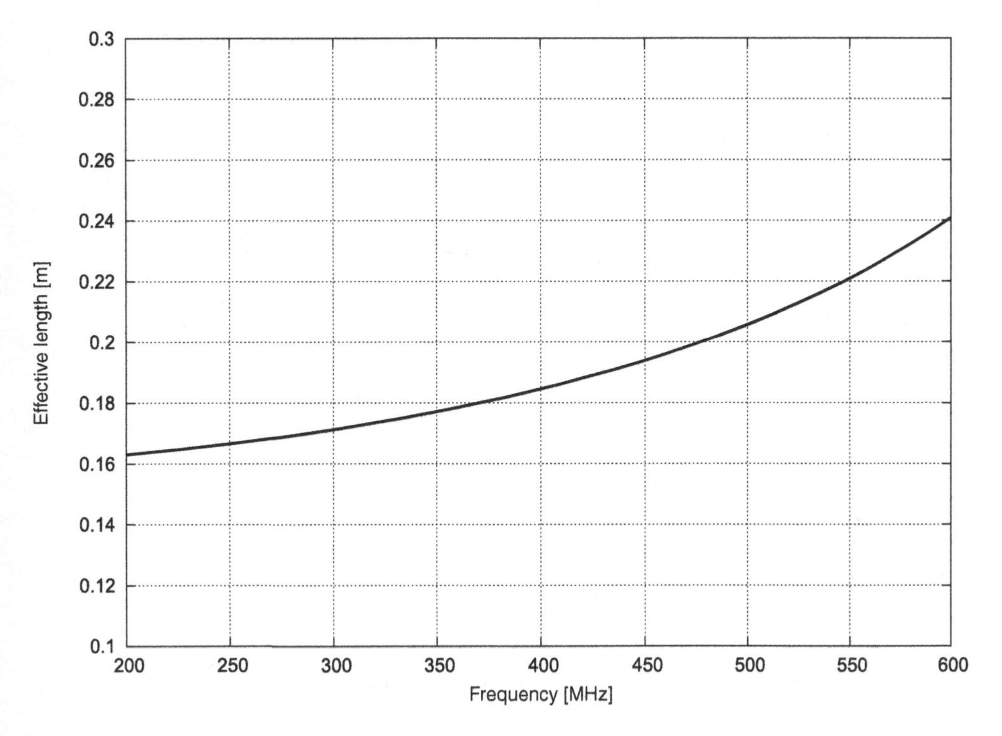

Figure 12.3. Estimated effective length for the antenna in Example 12.1. Note that the expected value of $\lambda/\pi \simeq 20$ cm is obtained for $h = \lambda/4$, which is at $\simeq 478$ MHz.

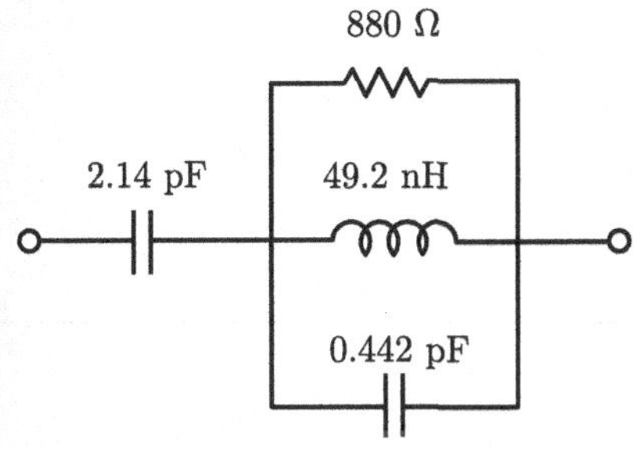

Figure 12.4. Equivalent circuit model for the impedance Z_A used for the antenna in Example 12.1.

Putting these elements together, we may now evaluate the available signal power $\sigma_S^2 l_e^2/4R_A$. The result is shown in Figure 12.6. This is the power associated with the signal of interest and that the antenna delivers to a load which is conjugate-matched over the range of frequencies considered. Note that the available power declines steadily with increasing frequency, even though effective length is increasing over the same frequency range.

Available SNR is calculated using Equation (12.11) with the reduced expression $\sigma_S^2 l_e^2$ in the numerator. Now some additional parameters must be determined. Here they are:

- $\epsilon_A = 1$ is assumed. As discussed in Chapter 2, ground loss is normally negligible above VHF. No other relevant mechanism is identified in the problem statement.
- T_{ext} is estimated using the "city" noise model from Table 4.2. Formally this model is not specified for use above 250 MHz; however, it is a reasonable guess in the absence of more specific information. This model gives a plausible value of $T_{ext} \cong 610$ K at 453 MHz, and exhibits the typical spectral dependence.

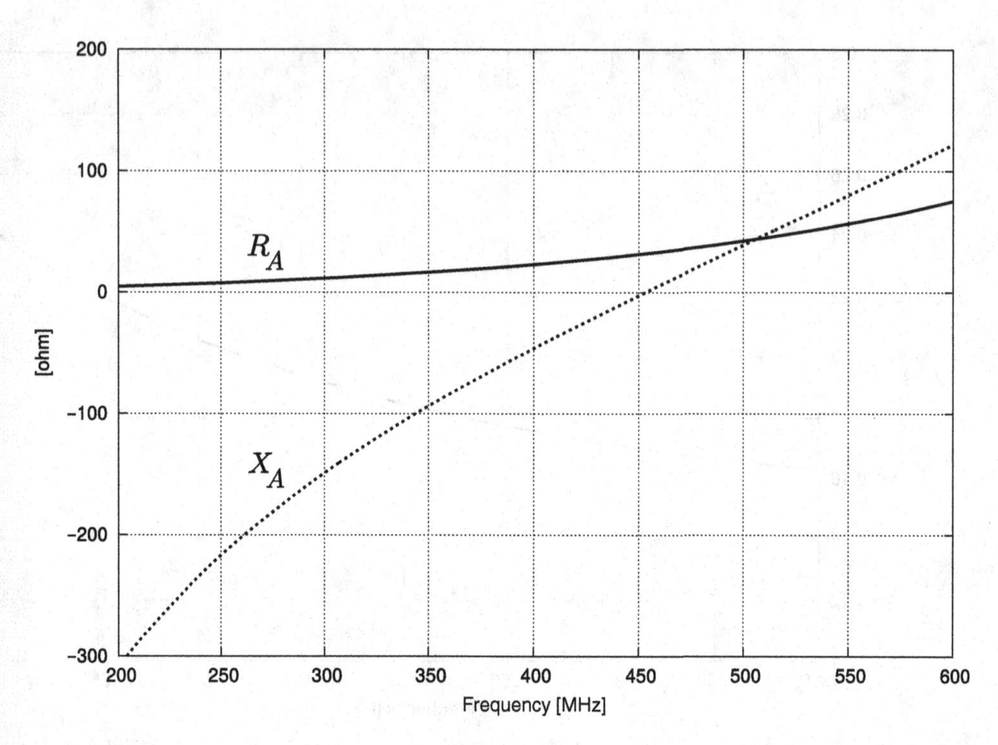

Figure 12.5. Impedance Z_A determined for the antenna in Example 12.1. The impedance when $h = \lambda/4$ ($\cong$ 478 MHz) is $37.0 + j20.7\ \Omega$, which is very close to the theoretical value for an infinitesimally-thin quarter-wave monopole (Equation (2.59)).

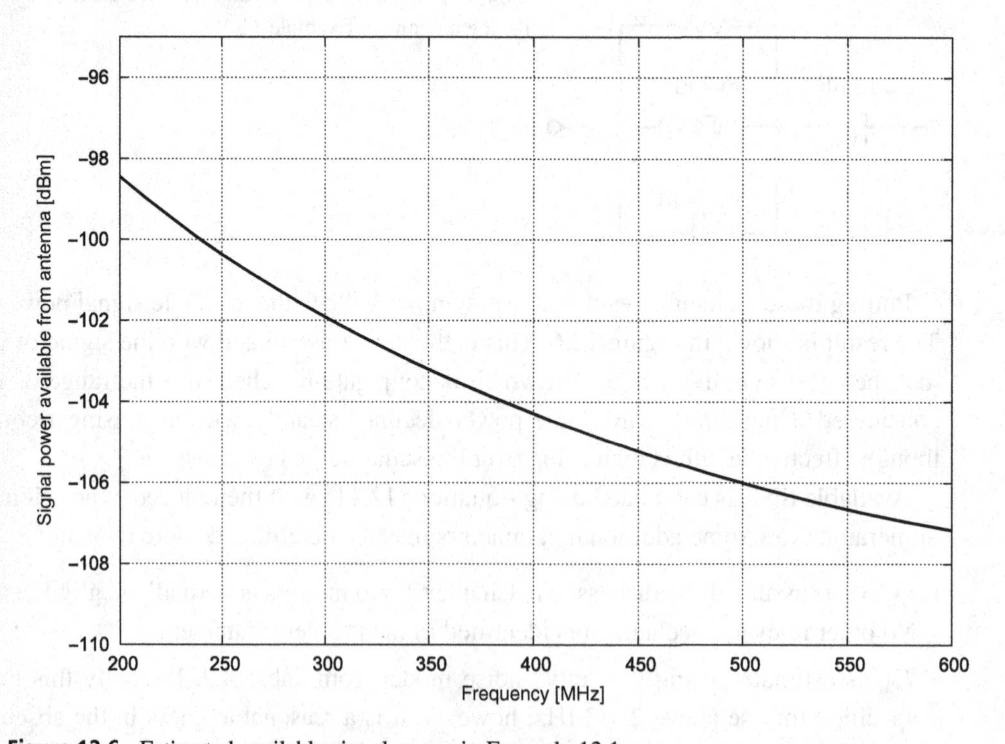

Figure 12.6. Estimated available signal power in Example 12.1.

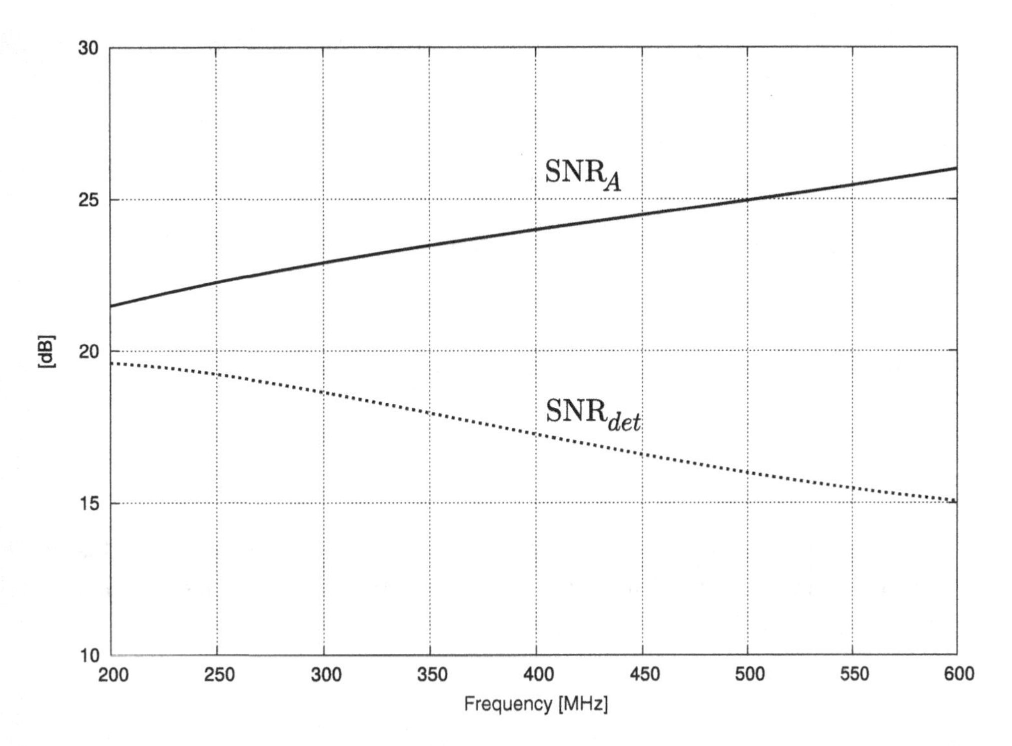

Figure 12.7. Estimated available SNR (SNR_A; *solid/top*) in Example 12.1. Also, estimated SNR delivered to the detector/digitizer in Example 12.3 (SNR_{det} (*dashed/bottom*) for a receiver having noise figure of 10 dB which is conjugate-matched to its antenna).

- R_{loss} is assumed to be negligible. This is reasonable because R_{loss} can be expected to be $\ll 1\ \Omega$ assuming typical construction techniques, whereas R_{rad} is seen to be $\gg 1\ \Omega$ in Figure 12.5.
- Because R_{loss} is assumed to be negligible, T_{phys} is irrelevant.
- $B = 12.5$ kHz as identified in the problem statement.

The resulting available SNR is shown in Figure 12.7. This is the SNR that the antenna delivers to a receiver which is conjugate-matched over the range of frequencies considered, and is therefore the maximum possible SNR that can be achieved under the conditions assumed above. $SNR_A \cong 24.5$ dB is achieved at 453 MHz, so this is the largest possible SNR that can be delivered to the digitizer or detector.

Note in the above example that SNR_A increases with increasing frequency. This is not unusual, but is not quite what the receiver sees since it is not reasonable to expect a conjugate match over such a large bandwidth.

12.2.2 Signal Power Delivered by an Antenna to a Receiver

The power delivered by the antenna to the receiver will be quite different from the power available from the antenna, unless the receiver happens to be conjugate-matched to the antenna

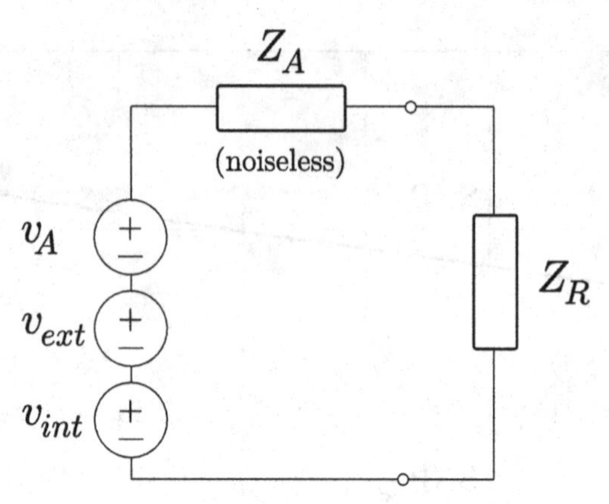

Figure 12.8. Model for receiving antenna interfaced to a receiver represented by a load impedance Z_R.

at all frequencies of interest. Figure 12.8 shows a model of the new situation, in which the receiver is modeled as a load impedance $Z_R \equiv R_R + jX_R$. A quick bit of circuit analysis reveals that the power delivered to the receiver is

$$P_R = \sigma_{tot}^2 \frac{R_R}{|Z_A + Z_R|^2} \tag{12.12}$$

Furthermore, we obtain the expected result in the special case that $Z_R = Z_A^*$; i.e., that P_R is maximized, and becomes equal to P_A (Equation (12.5)).

It is possible to express the efficiency of power transfer in the form of a transducer power gain (TPG) G_T using Equation (2.47):

$$G_T \equiv \frac{P_R}{P_A} = \frac{4R_A R_R}{|Z_A + Z_R|^2} \tag{12.13}$$

Therefore the signal power delivered to a receiver, accounting for impedance mismatch between antenna and receiver, is

$$P_R = \frac{\sigma_S^2 \left|\hat{\mathbf{e}}^i \cdot \mathbf{l}_e\right|^2}{4R_A} G_T \tag{12.14}$$

We further note that Equation (12.13) can often be written in terms of the reflection coefficient Γ_A looking into the antenna from the receiver: First, note Γ_A may be calculated as

$$\Gamma_A \equiv \frac{Z_A - Z_R}{Z_A + Z_R} \tag{12.15}$$

From Equation (2.48) and associated text, Equation (12.13) now simplifies to the following expression:

$$G_T = 1 - |\Gamma_A|^2 \quad \text{if } X_A = 0,\ X_R = 0, \text{ or both} \tag{12.16}$$

Quite often $X_A X_R \ll R_A R_R$, in which case $G_T \approx 1 - |\Gamma_A|^2$ is a good approximation even if the reactances are not exactly zero.

EXAMPLE 12.2

Continuing Example 12.1, calculate the signal power and noise power (separately) delivered to the receiver assuming that the receiver's input impedance is 50 Ω and independent of frequency.

Solution: Here, $Z_R = R_R = 50\ \Omega$. We return to the expression for P_A given in Equation (12.10). Note that the first term is available signal power, whereas the last two terms represent the available noise power. Therefore the delivered signal power is

$$\frac{\sigma_S^2 l_e^2}{4R_A} \cdot \frac{4R_A R_R}{|Z_A + Z_R|^2}$$

assuming the incident field is co-polarized with the antenna, and the delivered noise power is

$$\left(\frac{k\epsilon_A T_{ext} R_{rad} B}{R_A} + \frac{kT_{phys} R_{loss} B}{R_A} \right) \cdot \frac{4R_A R_R}{|Z_A + Z_R|^2}$$

The result is shown in Figure 12.9. Note that the difference between power delivered (Figure 12.9) and power available (Figure 12.6) is simply G_T, which is an expression of the impedance mismatch. Here the mismatch has essentially filtered the result leaving a maximum response where the match is best, and with suppressed response away from this peak.

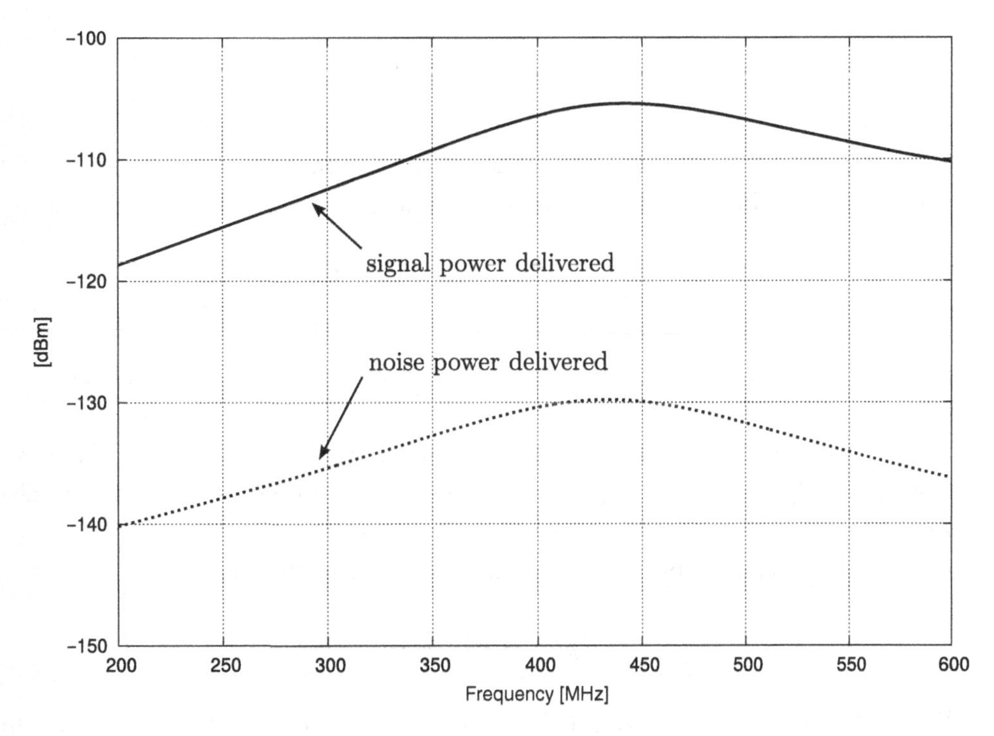

Figure 12.9. Signal power (*solid/top*) and noise power (*dashed/bottom*) delivered to the receiver in Example 12.2. The effect of impedance mismatch may be seen by comparison to Figure 12.6.

12.2.3 SNR Delivered to the Digitizer or Detector Assuming Conjugate Matching

The sensitivity of a receiver depends on the SNR delivered to the detector or digitizer of the receiver. This in turn depends on the noise that is contributed internally by the receiver.

A general expression for SNR delivered to the detector or digitizer in this case is tricky. This is because (1) noise generated internally by the receiver may depend on the impedance mismatch at the input; and (2) noise generated by the receiver may propagate in either direction; that is, toward the detector/digitizer or through the impedance mismatch and out the antenna. To address this problem, we will first consider two simpler questions: (1) What is the SNR realized by the receiver assuming conjugate matching (this section); and (2) what is the actual SNR when the two-port noise parameters of the receiver are known (next section)?

If the antenna is conjugate-matched to the receiver, then the input-referred noise delivered to the digitizer or detector of the receiver is simply the sum of the receiver's input-referred noise power kT_RB plus the available noise power from the antenna. The associated signal power is simply the available signal power from the antenna. The SNR realized by the receiver in this condition is an upper bound on the effective SNR, in the sense that impedance mismatch can only reduce the ratio of signal power to receiver-generated noise power. Using Equation (12.10) we find that the total noise power delivered to the detector or digitizer, referred to the input of the receiver, is

$$\frac{k\epsilon_A T_{ext} R_{rad} B}{R_A} + \frac{kT_{phys} R_{loss} B}{R_A} + kT_R B$$

whereas the associated signal power is

$$\frac{\sigma_S^2 \left|\hat{\mathbf{e}}^i \cdot \mathbf{l}_e\right|^2}{4R_A}$$

Therefore the SNR delivered to the detector/digitizer, SNR_{det}, is the ratio of the above expressions. Writing this ratio in a form similar to Equation (12.11), one obtains:

$$\text{SNR}_{det} = \frac{\sigma_S^2 \left|\hat{\mathbf{e}}^i \cdot \mathbf{l}_e\right|^2}{4k\left(\epsilon_A T_{ext} R_{rad} + T_{phys} R_{loss} + T_R R_A\right) B} \tag{12.17}$$

Note that the difference between this result and the available SNR (Equation (12.11)) is the term $T_R R_A$ in the denominator.

This result tells us something about how to go about specifying the receiver noise temperature T_R, should the opportunity to do so arise. Of course the optimal value of T_R is zero, but it is useful to know that there is a value which is small enough to make additional reduction ineffectual. From Equation (12.17) we see the criterion is

$$T_R \ll \frac{\epsilon_A T_{ext} R_{rad} + T_{phys} R_{loss}}{R_A} \tag{12.18}$$

In other words, the sensitivity of the conjugate-matched antenna–receiver system becomes approximately equal to the bound imposed by the antenna (SNR_A, Equation (12.11)) if T_R satisfies the above criterion. On the other hand if T_R is much greater than the right side of the above equation, then sensitivity is dominated by internal noise, reduction in T_R is productive, and external noise and losses intrinsic to the antenna are unimportant.

Finally, keep in mind that we are still assuming conjugate matching between antenna and receiver. The more general case will be considered in the next section.

EXAMPLE 12.3

Continuing Example 12.2: The receiver has a noise figure of 10 dB, which is approximately independent of frequency over the range of interest. What is the SNR delivered to the receiver's detector/digitizer?

Solution: $F = 10$ dB corresponds to $T_R = 3190$ K assuming $T_0 = 290$ K. All that remains is to apply Equation (12.17). The result is shown alongside SNR_A in Figure 12.7. As expected, accounting for the internally-generated noise of the receiver significantly degrades the realizable SNR – we are now down to an upper bound of 16.6 dB SNR at 453 MHz, not yet accounting for impedance mismatch.

EXAMPLE 12.4

In the previous example, what receiver noise figure should be specified if the goal is for the noise generated internally by the receiver to be no more than the noise delivered by the antenna, as measured at the input to the receiver's detector/digitizer?

Solution: Adapting Equation (12.18):

$$T_R = \frac{\epsilon_A T_{ext} R_{rad} + T_{phys} R_{loss}}{R_A} \tag{12.19}$$

Assuming the desired performance is specified at 453 MHz, $T_R = 610$ K. This corresponds to a noise figure of 4.9 dB.

12.2.4 SNR Delivered to the Digitizer or Detector when Two-Port Noise Parameters are Available

Section 10.5 introduced the two-port noise parameters F_{min}, Γ_{opt}, and r_n, which can be used to determine the two-port noise figure F via Equation (10.59) (repeated here):

$$F = F_{min} + \frac{4r_n \left|\Gamma_S - \Gamma_{opt}\right|^2}{\left(1 - |\Gamma_S|^2\right) \left|1 + \Gamma_{opt}\right|^2} \tag{12.20}$$

This provides a means to avoid both problems identified at the beginning of Section 12.2.3: One may use the above expression to determine the input-referred equivalent noise temperature under mismatched conditions, and also know that the internal noise computed in this manner is directed entirely toward the output. Figure 12.10 demonstrates the concept. This is simply the antenna model of Figure 12.2 terminated into the small-signal amplifier model of Figure 10.12. In this case, Γ_S is given by the embedded input impedance looking into the IMN from the gain device, now including the antenna. Adapting Equation (8.36):

$$\Gamma_S = s_{22} + \frac{s_{12} s_{21} \Gamma_A^{(s)}}{1 - s_{11} \Gamma_A^{(s)}} \tag{12.21}$$

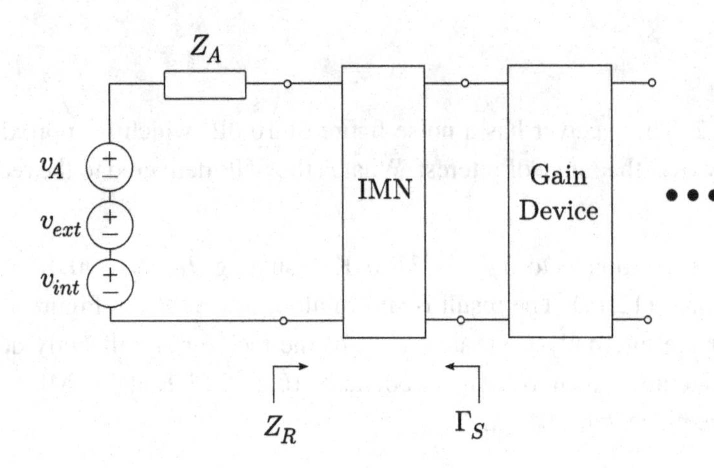

Figure 12.10. Model for a receiving antenna driving an amplifier using the IMN–Transistor–OMN model of Chapter 10 (Figure 10.12).

where the s-parameters are those of the IMN, and $\Gamma_A^{(s)}$ (different from Γ_A!) is defined as

$$\Gamma_A^{(s)} \equiv \frac{Z_A - Z_0}{Z_A + Z_0} \tag{12.22}$$

where Z_0 is the reference impedance for which the s-parameters are determined. Equation (12.20) may now be used with Equation (12.21) to obtain F, from which the desired value of T_R can be obtained.

SNR$_{det}$ may now be calculated using the signal and noise powers referred to the input of the receiver (i.e, delivered to the IMN). The noise power is

$$\left(\frac{k\epsilon_A T_{ext} R_{rad} B}{R_A} + \frac{k T_{phys} R_{loss} B}{R_A}\right) G_T + k T_R B$$

where G_T is the ratio of power delivered to the receiver to power available from the antenna, determined for example using Equation (12.13). (If not already known, the receiver input impedance Z_R needed in Equation (12.13) may be determined by using the two-port s-parameter techniques explained in Chapter 8.) The associated signal power is

$$\frac{\sigma_S^2 \left|\hat{\mathbf{e}}^i \cdot \mathbf{l}_e\right|^2}{4R_A} G_T$$

Finally we obtain

$$\text{SNR}_{det} = \frac{\sigma_S^2 \left|\hat{\mathbf{e}}^i \cdot \mathbf{l}_e\right|^2}{4k\left(\epsilon_A T_{ext} R_{rad} + T_{phys} R_{loss} + T_R R_A / G_T\right) B} \tag{12.23}$$

The "stopping criterion" for minimization of T_R, analogous to Equation (12.18), is in this case

$$\frac{T_R}{G_T} \ll \frac{\epsilon_A T_{ext} R_{rad} + T_{phys} R_{loss}}{R_A} \tag{12.24}$$

The difference is the factor of G_T in the denominator, which appears now because we account for the power loss due to impedance mismatch as opposed to assuming delivered power equals available power.

12.3 TRANSMIT PERFORMANCE

There are three primary considerations for antenna integration in transmit operation:

- VSWR should be low, so that reflected power does not degrade the performance of the transmitter or damage the radio;
- the efficiency of power transfer from transmitter to radio wave should be high, so as not to waste transmit power; and
- the directivity of the antenna should be such that radiated power is oriented primarily in the desired direction or directions, to the maximum extent possible.

Directivity of antennas was sufficiently well addressed in Chapter 2, although we shall circle back to that issue in Section 12.3.2.

12.3.1 VSWR

Let us begin with VSWR. Figure 12.11 shows an appropriate model for a transmitter driving an antenna. For generality we accommodate the possibility of feedlines, baluns, etc., using a single two-port between the transmitter and the antenna. In terms of the model, the transmit VSWR is given by

$$\text{VSWR} = \frac{1 + |\Gamma_T|}{1 - |\Gamma_T|} \tag{12.25}$$

where Γ_T is the voltage reflection coefficient looking into the two-port from the transmitter:

$$\Gamma_T = \frac{Z_{A,T} - Z_T}{Z_{A,T} + Z_T} \tag{12.26}$$

where $Z_{A,T}$ and Z_T are the impedances looking into the two-port and into the transmitter output, respectively. Note that $Z_{A,T}$ is the input impedance for the two-port when terminated on the other side by the antenna. This quantity may be obtained using s-parameter techniques as follows:

$$Z_{A,T} = Z_0 \frac{1 + \Gamma_{A,T}}{1 - \Gamma_{A,T}} \tag{12.27}$$

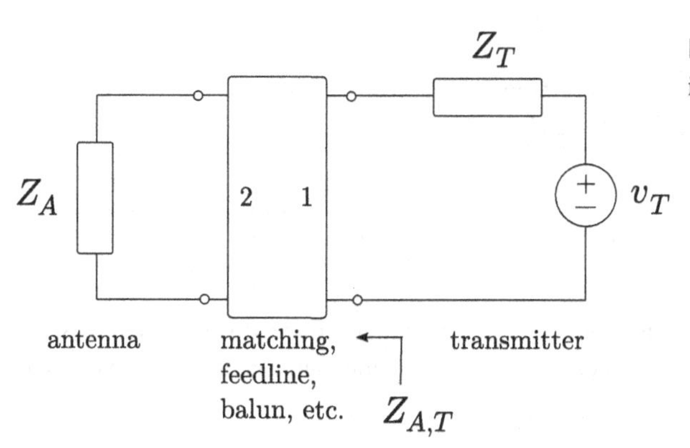

Figure 12.11. Model for a transmitter interfaced to an antenna.

where

$$\Gamma_{A,T} = s_{11} + \frac{s_{12}s_{21}\Gamma_A^{(s)}}{1 - s_{22}\Gamma_A^{(s)}} \tag{12.28}$$

Be careful to note the indexing of ports 1 and 2 in Figure 12.11.

EXAMPLE 12.5

Returning once again to the UHF quarter-wave monopole of Example 12.1: Now determine the VSWR in transmit operation assuming direct connection to a transmitter having output impedance 50 Ω.

Solution: Here, $Z_T = 50\ \Omega$ and $Z_{A,T} = Z_A$ since antenna and transmitter are directly connected. (Alternatively, or as a check, one may look at this as a transmitter connected to an antenna through a two-port having $s_{21} = s_{12} = 1$ and $s_{11} = s_{22} = 0$.) For Z_A we use the equivalent circuit model from Example 12.1, shown in Figure 12.4. The result is shown in Figure 12.12.

Two comments on this result: First, note that the nominal VSWR of 1 is not achieved, since Z_A is never exactly equal to 50 Ω. Second, note that the vertical scale has been limited to VSWR ≤ 3, which is often considered a maximum safe value for transmitter operation. Under this criterion, the useable bandwidth of the antenna–transmitter system is about 110 MHz – quite limited relative to the usable bandwidth achieved in receive operation (using the SNR_{det} criterion). This is typical.

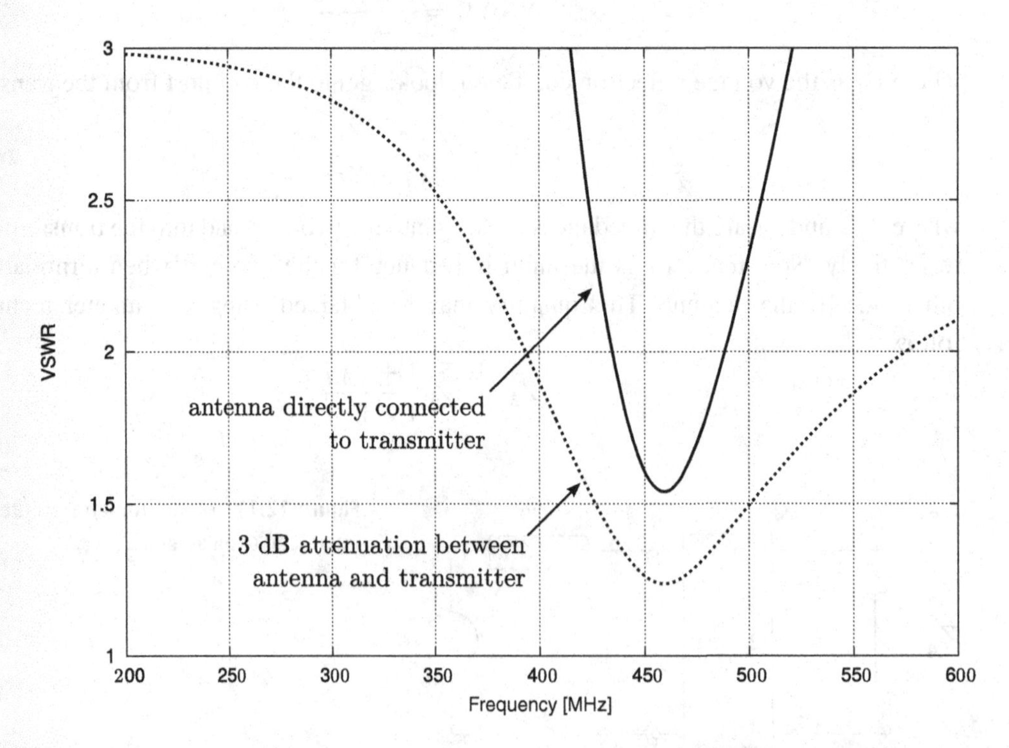

Figure 12.12. VSWR in transmit operation in Example 12.5 (direct connection between transmitter and antenna; *top/solid*) and Example 12.6 (3 dB loss between transmitter and antenna; *bottom/dashed*).

EXAMPLE 12.6

Let us repeat Example 12.5, but now let us account for some loss between transmitter and antenna. The loss could be caused by a long coaxial cable, or might be deliberately introduced for other reasons (soon to be evident). Let us assume a 3 dB attenuator, which can be modeled using the two-port parameters $s_{21} = s_{12} = 1/\sqrt{2}$, $s_{11} = s_{22} = 0$ with $Z_0 = 50\ \Omega$.

Solution: The result is shown in Figure 12.12. Note that the loss is beneficial in the sense that the VSWR at 453 MHz has improved significantly, and the bandwidth over which VSWR ≤ 3 has expanded dramatically. This positive outcome is attributable to the fact that the lossy two-port dissipates reflected power that would otherwise degrade VSWR. For this reason it is fairly common to introduce some loss in practical systems. Of course there is a price to be paid in terms of sensitivity; and in terms of transmit efficiency, as demonstrated in the next section.

12.3.2 Transmit Efficiency

Transmit efficiency is defined here as the ratio of power successfully transferred to a radio wave, to the power P_T available from the transmitter. From Figure 12.11 it is seen that the ratio of power delivered to the antenna to P_T is equal to the TPG of the interface two-port. By definition (see Section 7.2) this TPG is equal to η_T.[2] However, not all of the power delivered to the antenna is radiated; some is dissipated in R_{loss}, and some fraction is lost via ϵ_A; e.g., through dissipation into the ground. Thus we find that the transmit efficiency is given by

$$\eta_T \epsilon_T = \text{TPG} \cdot \epsilon_{rad} \epsilon_A \tag{12.29}$$

where ϵ_T is analogous to ϵ_R, as defined in Section 7.2. Now using Equation (8.41) for the TPG of the interface two-port, we find the transmit efficiency may be calculated as:

$$\eta_T \epsilon_T = \frac{\left(1 - \left|\Gamma_T^{(s)}\right|^2\right) |s_{21}|^2 \left(1 - \left|\Gamma_A^{(s)}\right|^2\right)}{\left|\left(1 - s_{11}\Gamma_T^{(s)}\right)\left(1 - s_{22}\Gamma_A^{(s)}\right) - s_{12}s_{21}\Gamma_T^{(s)}\Gamma_A^{(s)}\right|^2} \cdot \epsilon_{rad}\ \epsilon_A \tag{12.30}$$

Where $\Gamma_T^{(s)}$ (different from Γ_T!) is defined similarly to $\Gamma_A^{(s)}$ (Equation (12.22)):

$$\Gamma_T^{(s)} \equiv \frac{Z_T - Z_0}{Z_T + Z_0} \tag{12.31}$$

[2] We prefer to use "η_T" as opposed to "G_T" here to reduce the possibility of confusion with the antenna transmit directivity, which (unfortunately) is also indicated by the symbol "G_T".

and the s-parameters are those of the interface two-port, with ports indexed as shown in Figure 12.11.

EXAMPLE 12.7

What is the transmit efficiency in Example 12.5, in which the antenna is directly connected to the transmitter?

Solution: Reaffirming assumptions made much earlier, $\epsilon_{rad} \approx 1$ since R_{loss} has been assumed to be negligible, and $\epsilon_A \approx 1$ since ground (vicinity) loss has been assumed to be negligible. Thus the transmit efficiency $\eta_T \epsilon_T$ is approximately equal to the TPG of the two-port that interfaces the transmitter to the antenna. There is no interface two-port; however, the TPG is less than 1 because of the impedance mismatch between transmitter output and antenna input. (Alternatively, you can think of the interface two-port as being simply two short wires connecting the transmitter to the antenna.) There are a number of shortcuts available to calculate this TPG, but for the purposes of demonstration let us straightforwardly attempt Equation (12.30). If we choose $Z_0 = 50\ \Omega$ for the s-parameter reference impedance, then $\Gamma_T^{(s)} = 0$ because $Z_T = 50\ \Omega$. Since we have only a "null" two-port in this example, $s_{21} = s_{12} = 1$ and $s_{11} = s_{22} = 0$. Therefore in this example we have

$$\eta_T \epsilon_T \approx \eta_T = 1 - \left|\Gamma_A^{(s)}\right|^2 = 1 - |\Gamma_A|^2$$

which could have been anticipated (see Equation (2.48)).

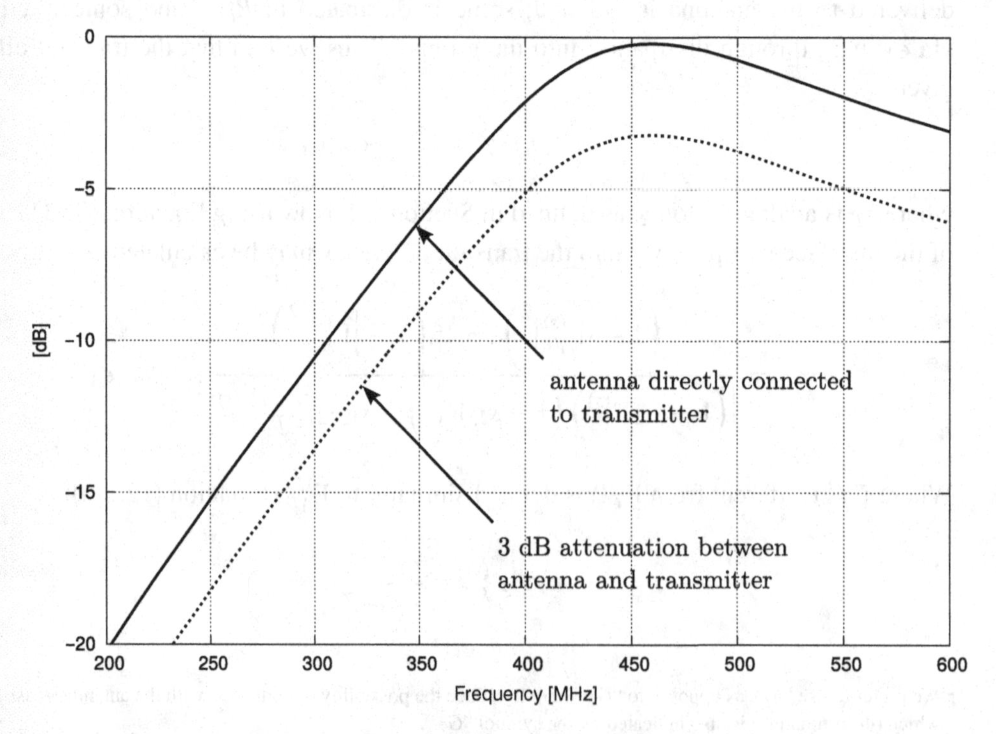

Figure 12.13. Transmit efficiency $\eta_T \epsilon_T$ in Example 12.7 (direct connection between transmitter and antenna; *top/solid*) and Example 12.8 (3 dB loss between transmitter and antenna; *bottom/dashed*).

The result is shown in Figure 12.13. As expected, the transmit efficiency is quite close to 0 dB where the impedance match is good (near 453 MHz) and degrades elsewhere, representing power reflected from the antenna and presumably absorbed by the transmitter. The bandwidth over which at least one-half of the available transmit power (P_T) is successfully converted into radio emission is about 200 MHz – greater than the VSWR ≤ 3 bandwidth calculated in Example 12.5.

EXAMPLE 12.8

What is the transmit efficiency in Example 12.7 when the interface two-port is a well-matched 3 dB attenuator?

Solution: The only difference from Example 12.7 is that $s_{21} = s_{12} = 1/\sqrt{2}$, representing the 3 dB attenuation. Therefore

$$\eta_T \epsilon_T \approx \eta_T = |s_{21}|^2 \left(1 - \left|\Gamma_A^{(s)}\right|^2\right) = \frac{1}{2}\left(1 - |\Gamma_A|^2\right)$$

i.e., the transmit efficiency is uniformly reduced by 3 dB. This is shown in Figure 12.13.

As pointed out in Section 7.3, one often cares less about *total power* radiated, and instead is more concerned with *power density* delivered to the location(s) of the receivers. The solution was to specify EIRP (alternatively, ERP) as opposed to total power. In the context of the present analysis, EIRP is simply

$$\text{EIRP} = P_T \, \eta_T \epsilon_T \, G_T \tag{12.32}$$

where (note well!) G_T in the above equation is not TPG, but rather the directivity of the transmit antenna, per Section 7.2. Moreover G_T is *not* antenna *gain*, since we've already accounted for ϵ_{rad}.

EXAMPLE 12.9

What is the EIRP in Example 12.5, in which the antenna is directly connected to the transmitter? Assume the transmit power is 1 W.

Solution: Here EIRP $= P_T \eta_T \epsilon_T G_T$, where $P_T = 1$ W, $\eta_T \epsilon_T$ is the same as determined in Example 12.7, and G_T is directivity. Since the antenna in question is a monopole being used in an LMR system, the direction of interest is presumably toward the horizon. Directivity in this case can be straightforwardly calculated from the effective length and antenna impedance by combining Equations (2.39) and (2.44), yielding

$$G_T = \frac{4\pi}{\lambda^2} \cdot \frac{\eta l_e^2}{4R_{rad}} = \pi \frac{\eta}{R_{rad}} \left(\frac{l_e}{\lambda}\right)^2$$

Recall we figured out l_e in Example 12.1 (Figure 12.3).

The result is shown in Figure 12.14. Note that the peak EIRP is about 3 W, which is 1 W (reduced slightly by the transmit efficiency) times the directivity of a quarter-wave monopole, which is about 5 dB = 3.2. This does not imply power added; this merely means that an isotropic antenna would require $P_T \approx 3$ W to produce the same power density toward the horizon as this antenna does when $P_T = 1$ W.

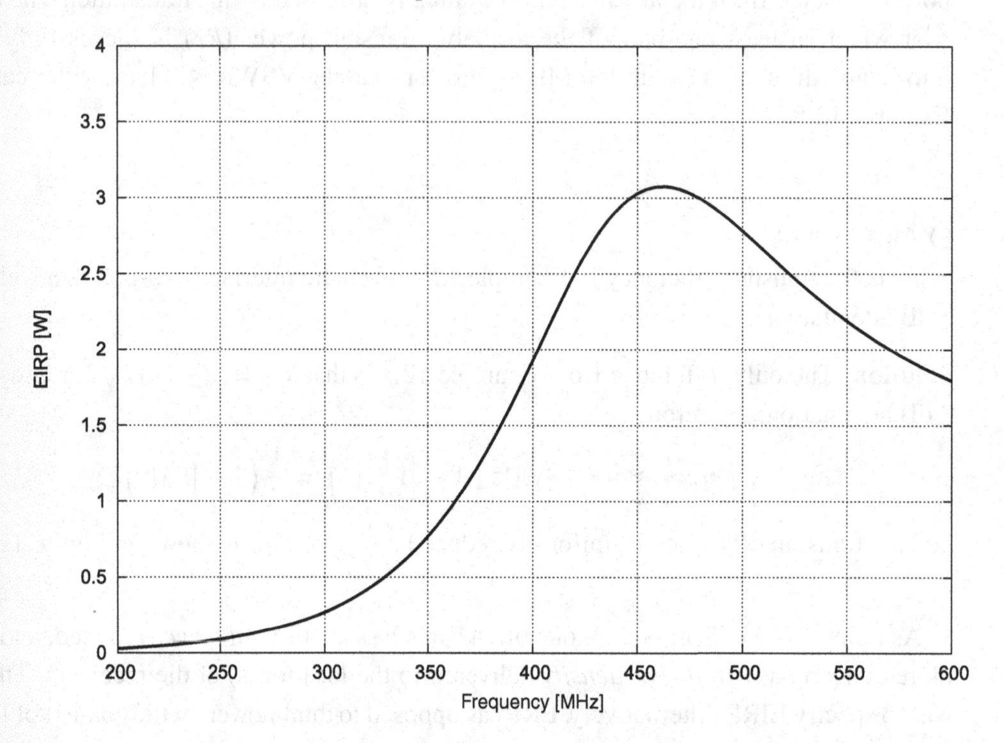

Figure 12.14. EIRP in Example 12.9 (direct connection between transmitter and antenna) assuming 1 W available transmit power.

12.4 ANTENNA–TRANSCEIVER IMPEDANCE MATCHING

From the perspective of impedance matching, antennas may be very broadly classified as either resonant or non-resonant, with electrically-small being a sub-category of the latter that is important enough to warrant a separate discussion. Here, we refer primarily to the manner in which the antenna is used, as opposed to some property intrinsic to the antenna. This particular categorization is useful because impedance-matching strategies depend primarily on this distinction.

12.4.1 Fractional Bandwidth Concept

Before considering the possibilities, it is useful to define the concept of *fractional bandwidth*. Fractional bandwidth is simply bandwidth B (by whatever definition) divided by a "nominal frequency" f_n within that bandwidth; i.e.,

$$\text{fractional bandwidth} = \frac{B}{f_n} \tag{12.33}$$

Note that f_n might be the center frequency within this bandwidth, or it might be some other frequency of particular relevance within the bandwidth; the meaning must typically be inferred from context. Often, f_n is defined to be the geometrical mean of the frequencies defining the edges of the relevant bandwidth; i.e.,

$$f_n \equiv \sqrt{f_1 f_2} \tag{12.34}$$

where f_1 and f_2 define the bandwidth such that $B = f_2 - f_2$. Sometimes f_n is defined as the arithmetic mean instead:

$$f_n \equiv \frac{1}{2}(f_1 + f_2) \tag{12.35}$$

The difference for small fractional bandwidths is usually negligible; however, for larger fractional bandwidths the former (geometric mean) definition is more common, and is adopted here.

By any definition, fractional bandwidth is B expressed as a fraction of f_n; so, for example, a device with a bandwidth of 1 MHz centered at 100 MHz is said to have a fractional bandwidth of 0.01 = 1%, in which f_n has been chosen to be the center frequency.

As we shall see below and in subsequent chapters, fractional bandwidth is a useful way to think about bandwidth in RF system engineering problems; this is because many design issues depend primary on fractional bandwidth as opposed to the absolute bandwidth B.

Note that for analog systems, the concept of fractional bandwidth makes sense only for systems with bandpass response, and does not necessarily make sense for systems with lowpass or highpass response. For example, a lowpass filter has infinite fractional bandwidth if f_n is chosen to be zero, and 100% fractional bandwidth if f_N is chosen to be the cutoff frequency. We will see in Chapter 18 that there is no such ambiguity for digital systems, because we define f_n to be equal to the Nyquist bandwidth; i.e., one-half the sample rate.

12.4.2 Resonant Antennas

A resonant antenna is defined here as an antenna which is operated primarily at or near a frequency of resonance; i.e., with $X_A \sim 0$, or small relative to R_{rad}. This condition leads to relatively easy impedance matching, since feedlines and many radios are designed to have interface impedances with negligible reactance. Such antennas may be either "narrowband" or "wideband." "Narrowband" normally implies small reactance but only over a "small" fractional bandwidth; normally considered to be <20% or so. Prominent examples include dipoles (1%–10%) and patches (a few percent). "Wideband" implies small reactance over a fractional bandwidth ≥20% or so. A prominent class of wideband resonant antennas is the horn, capable of >50% bandwidth.

12.4.3 Non-Resonant Broadband Antennas

A non-resonant antenna is defined as an antenna that is operated over a frequency range over which X_A does not remain small relative to R_A. This condition leads to relatively difficult impedance matching, again because most feedlines and radios are designed to have interface impedances with negligible reactance. Within this category there are actually two important categories: Electrically-small antennas, discussed in the next section; and broadband antennas. For the moment, we shall consider the discriminating characteristic being whether there is *any* frequency in or below the range of interest at which X_A is small; if so, it is broadband; otherwise it is electrically-small.

According to the above definitions, examples of broadband antennas include resonant antennas used over frequency ranges larger than those over which they may be considered resonant, as well as a few others including discones and log-periodic beam antennas. Effective impedance matching is typically not possible with these antennas because of their variable impedance. This results in VSWR that is typically too high for transmit without the use of an antenna tuner. Antenna tuners are discussed in Section 12.6.

Non-resonant broadband antennas (without antenna tuners) commonly appear in receive-only communications applications such as broadband scanning and spectrum monitoring. The impedance matching strategy in such applications turns out to be ironically simple: Typically, the best answer is to do nothing. Recall from Chapter 9 that impedance matching – with certain caveats – tends to reduce bandwidth, which is usually the opposite of what is intended in these applications. How can such antennas possibly be useful if they remain poorly matched? The answer is captured in Equation (12.24) and associated text. The message of that equation is that if the ratio of internal noise to interface TPG is sufficiently small, then reasonable sensitivity may be obtained even in the presence of an atrocious impedance mismatch.

12.4.4 Electrically-Small Antennas

An electrically-small antenna (ESA) is one which is much smaller than a wavelength, resulting in reactance which is huge and negative throughout the frequency range of interest – see Figure 2.5 and the associated text for an example. If the maximum dimension available for an antenna is less than $\lambda/4$ or so at the highest frequency of interest, the result will inevitably be electrically-small. To demonstrate the ubiquitous nature of ESAs, here are a few examples.

- Any MF/HF-band radio (including AM broadcast receivers) intended for mobile operation will inevitably require an ESA. This is because 1λ at 30 MHz is 10 m, so if the antenna is not at least 2.5 m or so it will operate as an ESA.
- Any cell phone intended to operate in the 850 MHz band without an extendable antenna (now considered unacceptable!) will require an ESA. This is because 1λ at 850 MHz is about 35 cm, so a resonant antenna requires roughly 9 cm or more. This is comparable to the largest dimension of some cell phones; however, this space must obviously be shared with other components that are not compatible with antenna operation. The maximum linear dimension available for an antenna in a modern cell phone is typically less than 5 cm, and sometimes as little as 2 cm.[3]
- Any antenna for a 2.4 GHz WiFi application (minimum $\lambda \approx 12$ cm) that must fit inside a volume with maximum dimension smaller than about 3 cm will be electrically-short. WiFi-equipped wireless sensor/actuator devices increasingly fall in this size regime.

[3] An informative review of antenna design for modern cell phones can be found in [67].

These are just a few examples. The broader view is that there is essentially no radio application in which a smaller antenna would not be welcome for reasons pertaining to size, weight, form factor, cost, convenience, and so on.

The principal difficulty in impedance-matching ESAs is that the resulting system has narrow impedance bandwidth. This occurs for two reasons. First, we noted in Section 9.4 that requiring TPG $= 1$ for a match between impedances with a large magnitude difference resulted in a high-Q circuit, which has narrow bandwidth. The second reason is that ESAs have capacitive reactance (again, see Figure 2.5 for an example). To cancel this reactance precisely over a broad range of frequencies requires negative capacitance – something nature does not provide. This problem can be (and often is) addressed using impedance inverters (Section 9.8); however, passive impedance inverters are themselves narrowband, so the extent to which this approach helps is limited.

EXAMPLE 12.10

Let us consider impedance matching of a particular antenna to be used in a mobile phone operating in the 850 MHz band. For this particular system, the handheld transmits at frequencies between 824 MHz and 849 MHz, and receives at frequencies between 869 MHz and 894 MHz. So that we may work with a realistic antenna impedance $Z_A(f)$, let us assume the antenna can be modeled as a monopole over a infinite ground plane so that the equivalent circuit model described in Section 2.7.5 can once again be used. A monopole having length 2 cm and radius 0.4 mm is assumed.[4] Assume the receiver presents an input impedance of 50 Ω to the antenna. Characterize the performance of the antenna without impedance matching, and then with a two-reactance "L" match yielding TPG $= 1$ (0 dB) at the center of the transmit band.

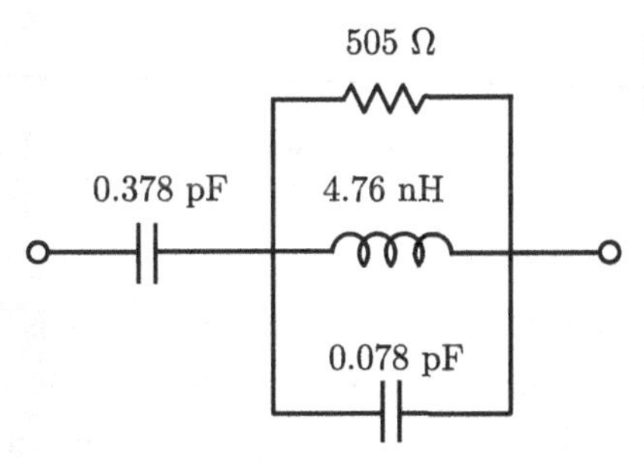

Figure 12.15. Equivalent circuit model for the impedance Z_A assumed for the cellular handheld antenna in Example 12.10.

Solution: The equivalent circuit for the antenna impedance is obtained in the same manner as in Example 12.1 (i.e., based on the model described in [77]) and is shown in Figure 12.15. The associated impedance is shown in Figure 12.16. As expected, R_A is small relative to 50 Ω, whereas X_A is large and negative. Without matching, VSWR > 1000 over the entire frequency range of interest, and so some matching is required if the antenna is to be useful for transmit.

Here, we will attempt a TPG $= 1$ match in the center of the transmit band, which is 836.9 MHz. Using the technique of Section 9.3, four possible "L" matches using discrete

[4] To be clear, this model is entirely bogus for the purposes of determining directivity in this case, since an antenna integrated into a handheld phone is geometrically quite different from a monopole in free space over a ground plane. Nevertheless the associated equivalent circuit model for $Z_A(f)$ is representative of this class of devices.

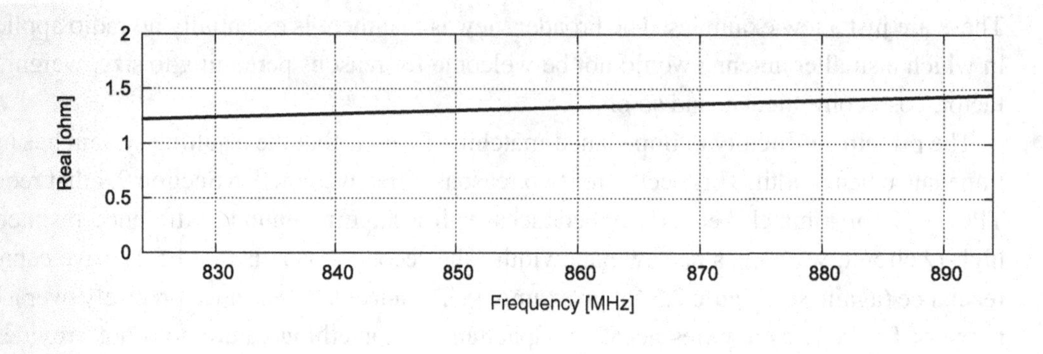

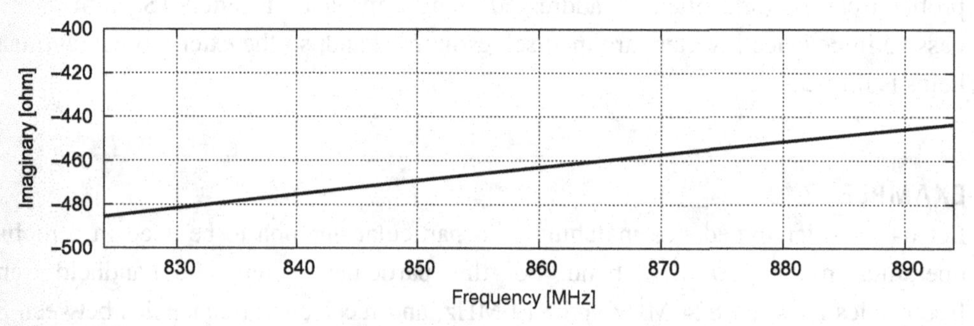

Figure 12.16. Impedance Z_A determined for the cellular handheld antenna in Example 12.10.

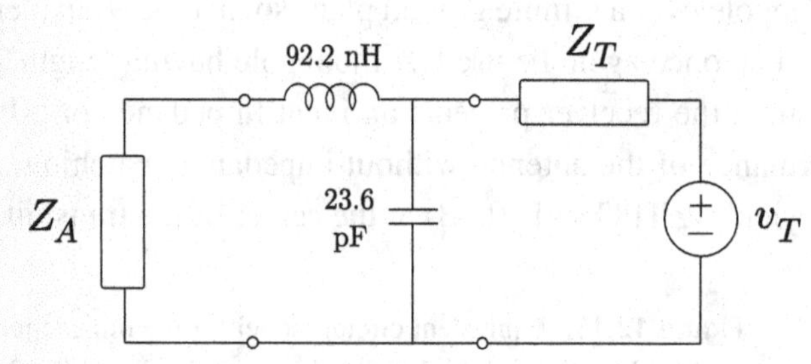

Figure 12.17. Proposed impedance matching scheme for the cellular handheld antenna in Example 12.10.

reactances can be identified.[5] The choice made in this case is shown in Figure 12.17, made primarily on the basis of realizability of component values.

The resulting VSWR is shown in Figure 12.18. Here the problem common to all ESAs is plainly evident: A very good match is possible at the frequency for which the match is designed, but the associated impedance bandwidth is very narrow. Here, VSWR < 3 is exceeded even within the 824–849 MHz transmit band. This design would therefore be considered unacceptable in most cases.

How does one proceed when faced with the problem identified in the above example? Certainly it is possible to use a different matching circuit that offers increased bandwidth; here the techniques described in Section 9.5.1 are applicable. However, in practice one encounters two problems: First a practical problem, then eventually a theoretical limitation. The practical problem is that a match with significantly larger bandwidth will require significantly larger numbers of components, and will be increasingly sensitive to the values of those components.

[5] This problem is quite similar to Example 9.1, in case you need a reminder.

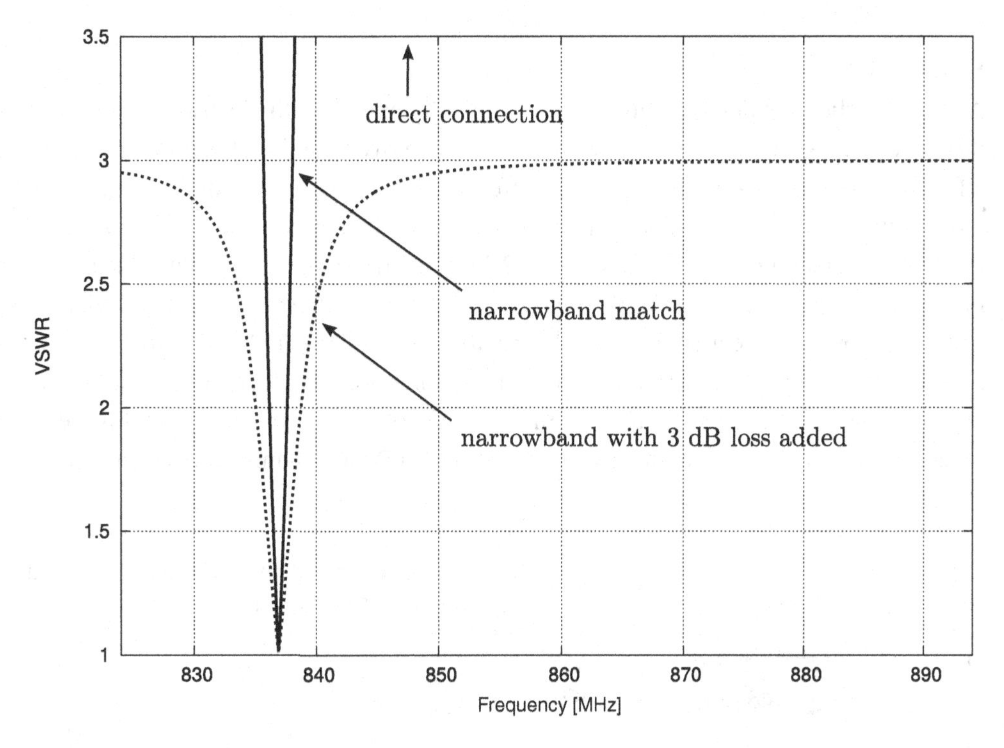

Figure 12.18. VSWR for various matching schemes in the cellular handheld antenna in Example 12.10 (no loss; *top/solid*) and Example 12.11 (3 dB loss added; *bottom/dashed*).

The theoretical limitation is known as the *Fano bound* – we will address that issue in Section 12.5.

If the VSWR bandwidth must be much larger, and the antenna cannot be made larger, then there is really only one option: One must reduce the TPG of the matching circuit. Reducing TPG reduces Q, which subsequently increases bandwidth. There are two ways to go about reducing TPG: (1) impedance *mismatching* and (2) increasing loss. (Recall we have already seen the positive effect of loss on VSWR in Example 12.6.) The particular combination of these tactics depends on the application. In receive-only applications, impedance mismatching makes a lot more sense, since sensitivity is not necessarily significantly degraded by impedance mismatch. In applications where transmit is also required, then only a relatively small amount of mismatch can be accommodated before the minimum VSWR becomes unacceptable. Any remaining TPG reduction required to achieve the desired bandwidth must then be implemented using loss.

EXAMPLE 12.11

Returning to Example 12.10: How much loss is required between the transmitter and the antenna to achieve VSWR < 3 across the entire transmit band?

Solution: Here we may use the same analysis technique employed in Example 12.8. In fact, it turns out that a 3 dB attenuator does the trick. The result is shown alongside the zero-loss result in Figure 12.18.

EXAMPLE 12.12

As noted earlier, handheld cellular phones typically end up with electrically-small antennas. At the same time, such devices are expected to operate over fractional bandwidths on the order of a few percent, and often times at a few different center frequencies due to regional differences and the increased use of widely-separated frequency bands (e.g., $\sim$ 850 MHz and $\sim$ 2 GHz). As demonstrated in Examples 12.10 and 12.11, it is often necessary to introduce loss in order to meet VSWR specifications. For this reason, the transmit efficiency of these devices is typically very low. However, the associated loss is not always implemented as "in line" attenuation, as in Example 12.11. An alternative approach is to intentionally degrade the radiation efficiency (ϵ_{rad}) of the antenna. For example, the Geo GC688 handset shown in Figure 12.19 uses a resistive foam ground plane as an intentional loss mechanism, resulting in ϵ_{rad} of just 5% [67].

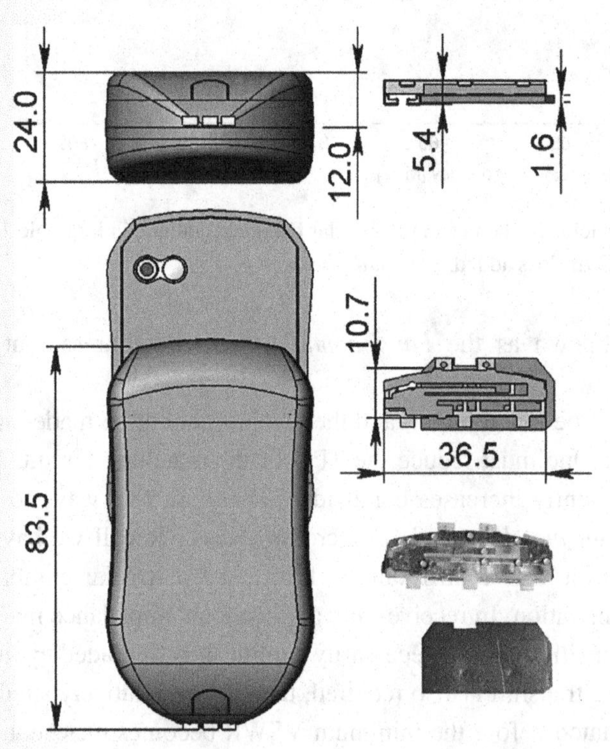

Figure 12.19. The antenna used in the Geo GC688 mobile phone, shown here, uses a multiband PIFA with a lossy ground plane to achieve acceptable VSWR in the GSM 850, 900, 1800, and 1900 MHz bands. All dimensions are in mm. ©2012 IEEE. Reprinted, with permission, from [67].

12.5 HOW SMALL CAN AN ANTENNA BE?

The discussion of ESAs in the previous section leads naturally to the question of limits: Specifically, how small can an antenna be, and still be useable? Of course, an answer to this question requires that one clearly specify what constitutes "useable," and the answer to that question is clearly application-specific. Nevertheless, some additional guidance is possible.

One way to answer this question is to identify the minimum Q (hence, bandwidth) that can be achieved from any antenna that fits entirely within a specified volume. The classical formulation of this problem is to specify a sphere of radius b which completely contains the candidate antenna. A commonly-cited estimate of the minimum Q in this case is from [51]:

$$Q \approx \left(2\pi \frac{b}{\lambda}\right)^{-3} + \left(2\pi \frac{b}{\lambda}\right)^{-1} \tag{12.36}$$

The associated impedance bandwidth is $B_Q \equiv 2\pi f_0/Q$ ($B_{HP} \approx \pi f_0/Q$) for $Q \gg 2\pi$, per Equation (9.16) (Section 9.4). Here, $f_0 = c/\lambda$. There are a variety of alternative versions and refinements of this bound, but the differences are typically not important because to achieve the bound requires that the antenna fill the available volume. Since applications for sphere-shaped antennas are rare, practical antennas typically fall far short of any version of the bound. For additional reading on this aspect of antenna theory, [76, Sec. 11.5.1] and [80] are recommended.

EXAMPLE 12.13

Evaluate the antenna from cellular handset antenna from Example 12.10 in terms of the bound of Equation (12.36).

Solution: In this case we may use image theory (reminder: Figure 2.15) to obtain $b = h = 2$ cm. Using Equation (12.36) at 836.5 MHz gives $Q \approx 26.1$, which implies a half-power impedance matching bandwidth of $\approx$ 101 MHz, which is about 12% in terms of fractional bandwidth. So, we could – in principle – design an antenna within this volume having impedance bandwidth that covered the entire 824–894 MHz band of operation. However, recall that this method of estimating bandwidth yields very optimistic results; this is because the bound assumes that the entire spherical volume of radius b can be used. On the other hand, were we to *begin* an antenna design project by computing this bound, we would know in advance that the monopole considered in this case would probably have a usable bandwidth much less than the 70 MHz of interest in this problem.

It is worth noting that Equation (12.36) also subtly implies one possible way to "cheat": Namely, make λ smaller, so b/λ appears larger. This is utterly practical: Simply embed the antenna in a dielectric material; then $\lambda \to \lambda/\sqrt{\epsilon_r}$, where ϵ_r is the relative permittivity of the material. The fact that the embedding material will have some loss is not necessarily a problem and may even be helpful, as pointed out in the previous section. The usual problem with this approach is that the dramatic size reductions normally of interest require material with large ϵ_r, which may not be realizable and/or may exhibit unacceptably large loss. For additional reading on this topic, [80] is suggested.

Of course, confidence that an antenna is theoretically able to achieve a specified impedance bandwidth provides no assurance that one is able to devise a matching circuit which realizes this bandwidth. To sort this out, we invoke the *Fano bound* [18]. The theory is quite complex, but a relevant special case is accessible and easy enough to follow. The situation is shown in Figure 12.20. Here, $Z_A(\omega)$ is modeled as $R_A - j/\omega C_A$, where $C_A \equiv -1/\omega_c X_A$ and ω_c is the center frequency of interest. In other words, the antenna impedance is modeled as that of a series RC circuit having precisely the same impedance at frequency ω_c. The actual antenna's impedance will be slightly different from that predicted by the model for frequencies away from ω_c, but this will not matter much since the real problem is typically the narrowbanding associated with the impedance matching circuit, which is the present concern.

$$\int_0^{\infty} \omega^{-2} \ln \left| \frac{1}{\Gamma(\omega)} \right| d\omega = \pi R_A C_A \tag{12.37}$$

The above equation constrains the extent to which $\Gamma(\omega)$ can be low (i.e., well-matched) over a specified frequency range. To see this, let us make the simplifying assumption, illustrated in Figure 12.21, that $|\Gamma(\omega)|$ is to have a constant value Γ_{min} over some range $\omega_1 \leq \omega \leq \omega_2$, and $\Gamma(\omega) = 1$ outside this range. Then evaluation of Equation (12.37) yields

$$\Gamma_{min} = e^{-\pi \omega_c R_A C_A / B} = e^{-\pi / Q(\omega_c) B} \tag{12.38}$$

where $Q(\omega_c)$ is the quality factor of the series RC circuit representing the antenna assuming a center frequency $\omega_c = \sqrt{\omega_1 \omega_2}$ corresponding to the frequency of resonance:

$$Q(\omega_c) = \frac{1}{\omega_c R_A C_A} \tag{12.39}$$

as is well known from elementary circuit theory – and B is the fractional bandwidth in this case using the geometrical mean definition of f_n:

$$B \equiv \frac{\omega_2 - \omega_1}{\sqrt{\omega_1 \omega_2}} \tag{12.40}$$

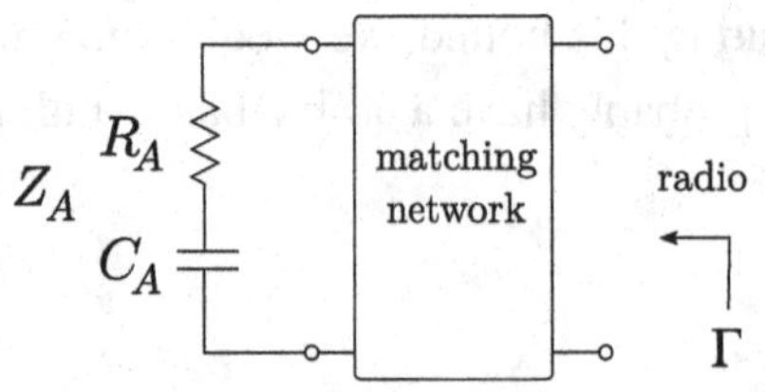

Figure 12.20. Model for an ESA interfaced to a receiver through an impedance matching network.

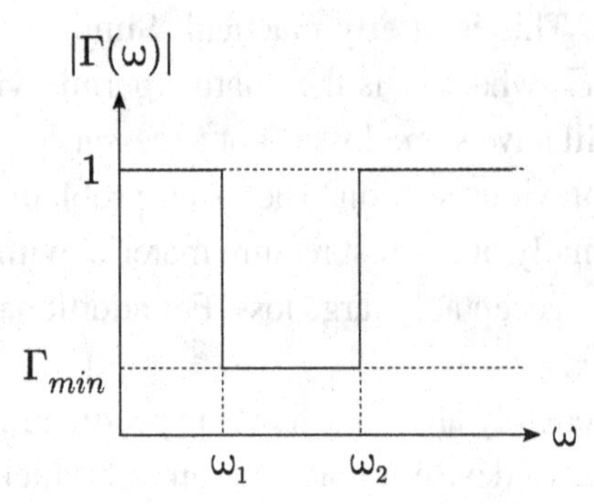

Figure 12.21. Assumed frequency response $\Gamma(\omega)$ for evaluation of the Fano bound.

Equation (12.38) reveals the fundamental tradeoff: Increasing B increases Γ_{min}; i.e. the best match achievable over the specified bandwidth is made worse. Furthermore, we see that the extent to which Γ_{min} is made worse increases with increasing Q.

Even if the Fano bound is met, one's troubles are not over. A matching circuit meeting the bound may be of any order. In particular, a matching circuit consisting of discrete reactances may require an impractically large number of components, and practical matching circuits tend to fall far short of the bound.

EXAMPLE 12.14

Use Equation (12.38) to evaluate the cellular handset antenna assumed in Example 12.10. Assume uniform VSWR = 2 is desired over some frequency range centered at 836.5 MHz.

Solution: VSWR = 2 corresponds to $\Gamma_{min} = 1/3$. The radiation resistance R_A of the antenna at $f_c = 835.6$ MHz is 1.26 Ω. The reactance X_A of the antenna at this frequency is −477.5 Ω, which corresponds to $C_A \cong 0.400$ pF. Solving Equation (12.38) we find

$$B = \frac{\pi}{\ln 3}\omega_c R_A C_A \cong 0.75\% \cong 6.3 \text{ MHz}$$

Thus it does not appear to be reasonable to expect VSWR ≤ 2 over 824–849 MHz for this antenna, or any device having comparable impedance characteristics – at least not without using loss.

Examples of practical handset antennas were shown in Figure 2.18 and 12.19. Although each example operates in multiple widely-separated frequency bands, this is accomplished using PIFA variants that exhibit multiple narrow resonances with additional parasitic elements and loss to broaden the usable bandwidth around each resonance. None comes close to achieving either of the bounds identified in this section.

If you're interested in *really* "tight" integration of antennas with systems – including antennas integrated directly into ICs – a suggested starting point is [10].

12.6 ANTENNA TUNERS

It was pointed out in Section 12.4.3 that an antenna tuner is typically required to make a non-resonant broadband antenna suitable for transmit applications. A brief introduction to the topic is provided below; for additional information a suggested starting point is [70, Ch. 15].

An antenna tuner is simply an impedance-matching circuit in which the match is implemented adaptively, as opposed to being fixed. The basic concept is illustrated in Figure 12.22. Here, the impedance match is implemented using components with variable reactances – more on those in a moment. The quality of the match is evaluated by measuring power propagating from the transmitter to the antenna, as well as the power reflected from the antenna. From these measurements VSWR can be estimated. The variable reactances are subsequently adjusted to minimize the VSWR at a specified frequency, or alternatively to increase the VSWR within acceptable limits so as to achieve larger bandwidth.

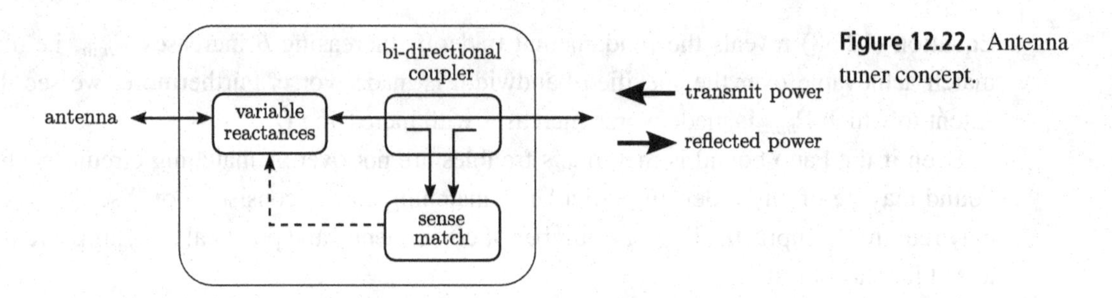

Figure 12.22. Antenna tuner concept.

Antenna tuners are commonly used with (or in) HF-band and VHF-band transceivers, in particular in military, amateur radio, and field-expedient applications in which a non-resonant and possibly electrically-short antenna must be employed. Increasingly though this approach is finding its way into other applications, as mentioned below.

One limitation of the antenna tuner architecture of Figure 12.22 is immediately obvious: Tuning is possible only when the transmitter is operating. Therefore a different approach is required for receive-only systems, or for systems which receive on frequencies significantly different from those on which they transmit. A related issue is the time required to search for an acceptable matching solution. This is, in fact, less of an issue as modern antenna tuners are typically able to identify reasonable solutions within fractions of a second.

The primary contraindication for the use of an antenna tuner is that instantaneous bandwidth is inevitably reduced. This is a consequence of the Fano bound, discussed in Section 12.5: Improving matching at one frequency or over a contiguous range of frequencies degrades impedance matching at other frequencies. Nevertheless, this may be necessary in transmit applications.

Variable reactances may be implemented in a variety of ways, depending on the frequency band, required tuning range, and power level that must be supported. Solutions include:

- Motor-driven geometry changers; i.e., devices with capacitance or inductance that is made variable by changing the shape or size of the constituent structures.
- Varactors; i.e., diodes or transistors which exhibit capacitance that varies monotonically with applied bias voltage.
- Banks of fixed reactances from which combinations are selected using switches. Switches may be electromechanical, electronic (e.g., diodes), or using *RF microelectromechanical systems (RF-MEMS)* technology.

In recent years, dynamic antenna tuning has experienced something of a renaissance, attributable largely to the emerging ability to make these systems very small using tightly-integrated combinations of RF-MEMS switching, electronically-variable reactance, RF power sampling, and control logic. As the size and power consumption of antenna tuners is reduced, it becomes increasing practical for these devices to be used to improve the performance of ESAs and in particular ESAs in handheld devices.

12.7 BALUNS

Recall from Chapter 2 that the signal at the terminals of an antenna may be either single-ended (unbalanced) or differential (balanced). This dichotomy was pointed out explicitly in

Section 2.8 for ideal monopoles, which are single-ended; whereas ideal dipoles are differential. From the antenna designer's perspective, the ideal interface between the antenna and the radio is single-ended if the antenna is single-ended, and differential if the antenna is differential. In practice the flexibility to choose the nature of the interface typically does not exist: From previous discussions (Sections 8.7 and 11.3) it is known that differential circuits offer compelling benefits; these benefits may outweigh the convenience of having the simplest possible interface with a single-ended antenna. Similarly the interface may be constrained to be single-ended for reasons pertaining to cost or convenience in the design of the radio, which may outweigh the convenience of having the simplest possible interface with a differential antenna. Both situations occur quite often, and warrant the use of a balun. Thus we now continue from where we left off at the end of Section 2.6.2.

12.7.1 Consequences of Not Using a Balun

Before addressing balun design strategies, it is worth considering the effect if a balun is not used. Figure 12.23(a) shows an ideal dipole interfaced to a balanced transmission line – the ideal situation. Figure 12.23(b) shows the transmission line replaced by a single-ended line; here a coaxial cable. Now the currents on the two arms of the dipole "see" very different transmission lines: In particular, the current on the right arm of the dipole, which is connected to the shield, sees an "inside out" coaxial cable. Thus the effective characteristic impedance associated with the current path on the shield is very different from the characteristic impedance for the proper operation of the transmission line. As a result, some current is transferred to the outward-facing surface of the shield, and some is reflected. The current on the outside of the shield can be interpreted as an extension of the antenna onto the transmission line, with an associated change in pattern and polarization characteristics. For antennas used to transmit at high power, the shield current may be a safety risk. For receive antennas, the change in current distribution may result in greater sensitivity to incident interference, which is co-polarized with the shield current, and more likely to conduct common-mode interference through proximity of the shield to interference-generating electronics.

The result for a monopole interfaced to a balanced transmission line is analogous, with the same deleterious results.

12.7.2 Balun Contraindications

The discussion of the previous section should not be interpreted as meaning that direct connection of disparate signal types, as in Figure 12.23(b), is never appropriate: Typically the undesired current mode has relatively high impedance, so the associated inefficiency may be quite small. Here are two common scenarios where a balun may be contraindicated:

- Radio systems that are receive-only or very low transmit power (for example, less than 1 W) and in which there is no particular benefit in purity of pattern or polarization. A good example is field-expedient low-power HF-band operation in which a dipole might be connected to the radio by coaxial cable, without a balun.

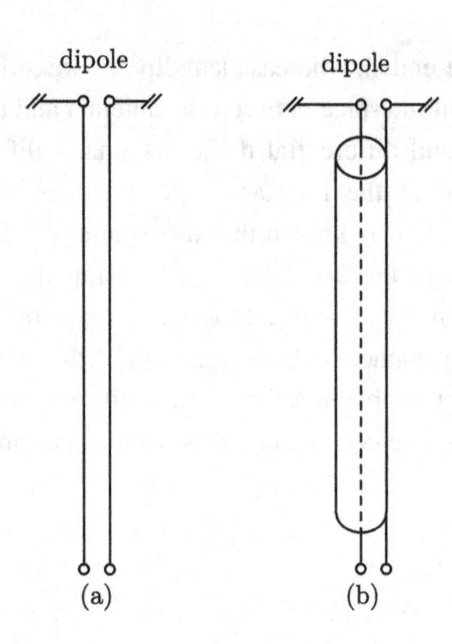

Figure 12.23. A dipoles fed by (a) parallel conductor (differential) and (b) coaxial (single-ended) transmission lines.

- Handheld transceivers that use a monopole-type antenna. In this case the monopole is far from ideal, since there is very little material to approximate a ground plane. In fact, the monopole combined with the ground plane of a printed circuit board may more closely approximate a dipole. This ambiguity in antenna topology makes a classification of the antenna as "balanced" or "unbalanced" pointless, and the signal delivered to the antenna terminals will be "mixed mode" regardless of whether a balun is employed or not.

12.7.3 Compact Baluns

If it is determined that it makes sense to use a balun, then one normally wishes for the smallest possible balun. Two possibilities in this vein are the transformer balun and the LC balun.

The use of transformers as baluns was previously discussed in Section 8.7.2. These devices typically have insertion loss of about 0.5 dB and return loss on the order of 10–20 dB at VHF and below, degrading significantly at UHF and above. This is suitable in receive applications in which the associated degradation of sensitivity is acceptable, and in transmit applications in which the associated degradation in VSWR and transmit efficiency is acceptable.

The *LC balun* was also previously discussed in Section 8.7.2. The advantage of this approach is that the balun may be both very small and have relatively low loss. The disadvantages of this approach are (1) it is difficult to achieve large bandwidth (the Fano bound on matching circuits applies here too) and (2) discrete Ls and Cs of sufficient quality may be difficult to find for frequencies in the UHF band and above.

12.7.4 Coaxial Choke Baluns

An alternative strategy which can be effective for interfacing balanced antennas to coaxial cable is to suppress ("choke") the component of the current which travels on the exterior of the shield. There are at least four common techniques that use this strategy, as follows:

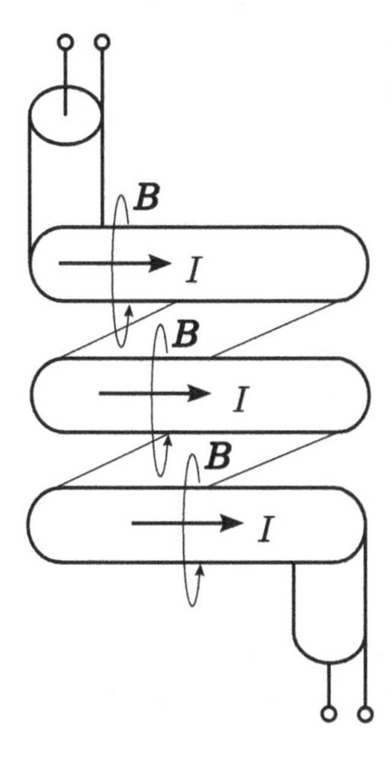

Figure 12.24. A choke balun formed by a coil of coaxial cable. Here I indicates current and B indicates the associated magnetic flux density.

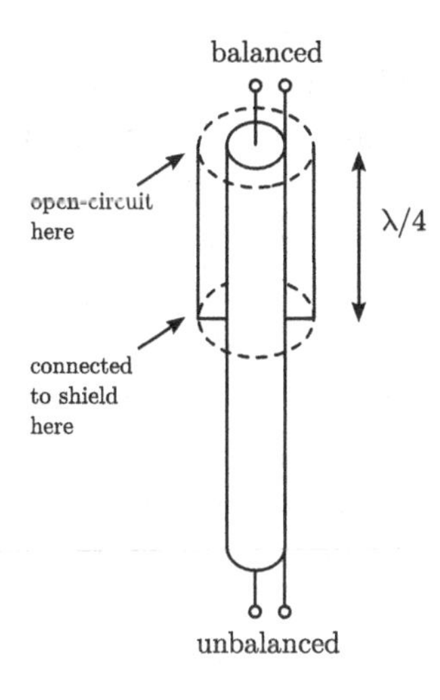

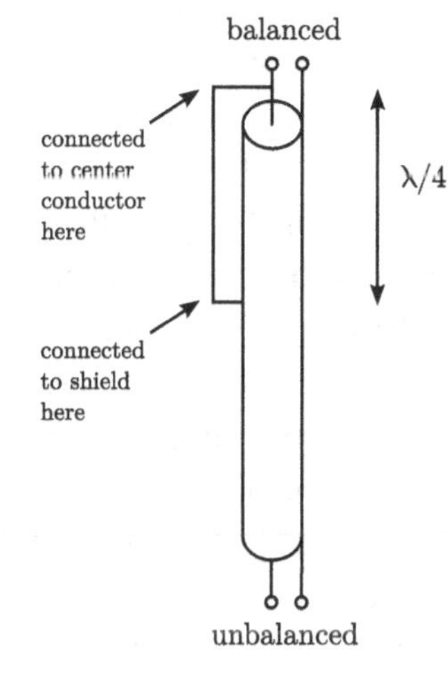

Figure 12.25. Coaxial baluns using $\lambda/4$ feedpoint structures.

A simple type of choke balun is formed by arranging a section of the coaxial cable close to the antenna into a coil, as shown in Figure 12.24. In this configuration, the magnetic field associated with the shield current on one turn of the coil opposes the magnetic field due to the shield current flowing on the adjacent turns. (To see this, just point your right thumb in the direction of current flow and realize that your fingers point in the direction of the associated magnetic field.) This effect increases the impedance perceived by the current on the shield.

Another way to suppress shield current is to use a *sleeve balun*, shown in Figure 12.25(a). In this scheme the shield is surrounded by a concentric layer of conducting material, which is shorted to the shield at a distance of $\lambda/4$ from the balanced terminals. If the distance between the exterior surface of the shield and the interior surface of the surrounding concentric layer is chosen correctly, then the shield current perceives a quarter-wave stub with short-circuit

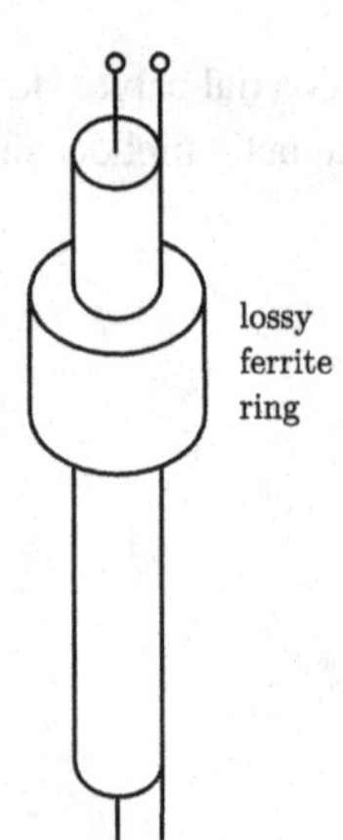

Figure 12.26. Ferrite choke balun.

termination. Therefore this structure presents infinite input impedance to the shield current at the balanced terminals (reminder: Section 9.8), and no current is able to transfer to the shield from the antenna terminals.

A similar topology is the *folded balun*, shown in Figure 12.25(b). In this case the center conductor at the balanced end is also attached to the shield at an electrical distance of $\lambda/4$ from the balanced terminals. This scheme exploits the fact that the differential currents are nominally π radians out of phase at the balanced terminals, so shorting the conductors at *any* distance from the balanced terminals causes the differential-mode currents to cancel. Since the short is made to the exterior of the shield, it is only the undesired component of the differential current mode – that is, the one traveling on the exterior of the shield – that is suppressed. By placing the short $\lambda/4$ from the balanced terminals, the proper component of the different current mode perceives an open circuit, and therefore is not affected.

A disadvantage of sleeve and folded choke baluns is that they are quarter-wavelength structures, which may be unacceptably large at UHF and below. Similarly the coil choke balun may be impractical to implement for mechanical reasons. The remaining alternative is the *ferrite choke balun*, shown in Figure 12.26. This device consists of rings of lossy ferrite material that surround the shield and interact with the magnetic field associated with shield current. This has the effect of increasing the impedance associated with the shield current path, as well as dissipating power associated with the shield current into the ferrite.

12.7.5 Other Commonly-Used Balun Types

It is worthwhile to identify two additional balun schemes, as they appear relatively often.

The $\lambda/2$ *transmission line balun* is shown in Figure 12.27. In this scheme, the dipole arms are connected by a $\lambda/2$ transmission line (shown here as coaxial cable, but any single-ended transmission line will do). The transmission line imparts a phase shift of π radians, so the currents on the dipole arms nominally add in-phase at the center conductor of the transmission line. Furthermore, the shield never makes contact with the dipole, so the impedance perceived by the shield current is very high. This balun has the additional occasionally useful property that it is also a 1:4 impedance transformer. This can be confirmed as follows: First, note that

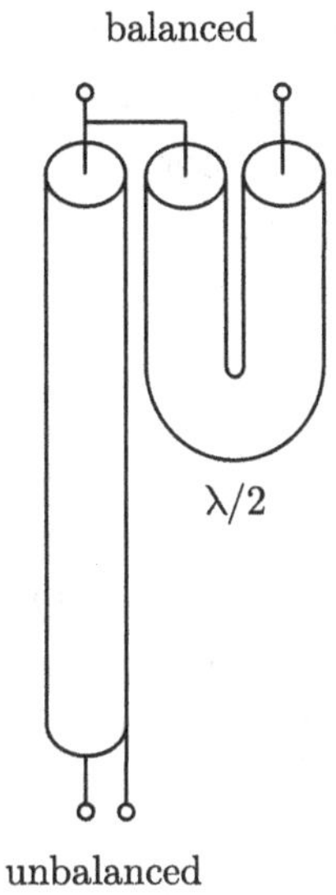

balanced

λ/2

unbalanced

Figure 12.27. The $\lambda/2$ transmission line balun. This balun also provides a 4:1 impedance conversion.

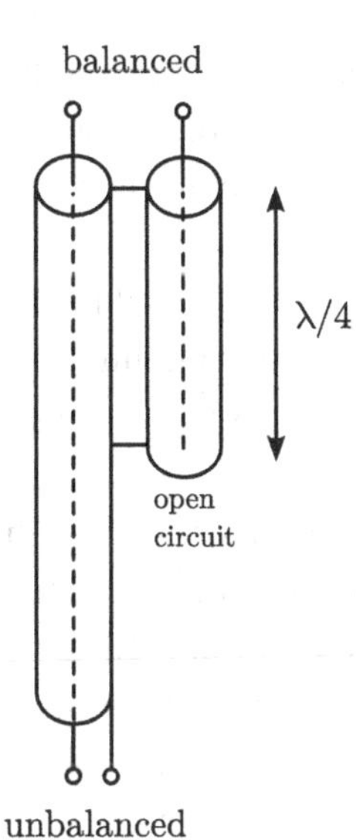

Figure 12.28. Marchand balun. Note that this structure is often depicted "unfolded"; i.e., with the transmission line sections arranged collinearly, with the unbalanced port on one end, the balanced port in the center, and the shield grounded at both ends.

the single mode current at the input to the transmission line, I_F, is twice the differential current at the dipole terminals, I_D. Second, note that the power through the conversion is nominally preserved as long as the output of the balun is impedance-matched to the characteristic impedance Z_c of the transmission line. Therefore $|I_D|^2 Z_A = |I_F|^2 Z_c = |2I_D|^2 Z_c$. Solving for Z_A, one finds $Z_A = 4Z_c$.

The other commonly-used scheme is known as the *Marchand balun*. "Marchand" has come to refer to a general method of operation that may be implemented in many possible ways. Figure 12.28 shows a relatively straightforward transmission line implementation. The essential idea here is that the unbalanced mode arriving at the balanced terminals also continues to travel into the quarter-wave stub. The stub imparts a phase shift of π radians ($\lambda/4$ out, reflection without phase shift (being an open circuit termination), $\lambda/4$ back), yielding the desired phase relationship between the balanced terminals. Furthermore, the returning shield current cancels the arriving shield current where the transmission lines come together at the balanced terminals.

Essentially the same idea may be implemented using discrete transformers and coupled microstrip structures.

Problems

12.1 Repeat Example 12.1 for a dipole with length $2h = 31.4$ cm and radius 1 mm, evaluating the result at 478 MHz only. Note that you should not need to do numerical integration to determine the effective length and you should be able to obtain the impedance from a simple modification to the existing equivalent circuit model.

12.2 Repeat Example 12.2 in the same manner as Problem 12.1.

12.3 (a) Repeat Example 12.4, but now with the goal of increasing the SNR achieved at the input to the receiver's detector/digitizer to 20 dB. Now what noise figure is required? (b) Same problem, but now let the criterion be that noise from the antenna should dominate over noise generated internally by the receiver by a factor of 10, as measured at the input to the receiver's detector/digitizer. What would be the point in making such a specification?

12.4 Revisiting Example 12.5: In some systems it is practical to select a value of transmitter output impedance. Propose a constant value that (a) optimizes minimum VSWR, (b) maximizes the bandwidth over which VSWR ≤ 2. In each case, plot the result in the same manner as in Figure 12.12.

12.5 Revisiting Example 12.6: A typical coaxial cable or attenuator itself is not exactly perfectly matched to the intended impedance. Repeat the example assuming that the VSWR looking into each port of the loss two-port is 1.1 as opposed to 1. You may assume the phase of all relevant s-parameters is zero. Explain your findings.

12.6 Calculate transmit efficiency in the following cases: (a) Problem 12.4 part (a), (b) Problem 12.4 part (b), (c) Problem 12.5. In each case, interpret your findings.

12.7 Following up Example 12.10: (a) Experiment with all four possible lossless two-reactance "L"-type matches. What are the advantages and disadvantages for each of these options? (b) How much improvement is possible using a cascade of two lossless two-reactance "L"-type matches (reminder: Section 9.5.1).

12.8 Following up Example 12.11: How much loss is required to achieve VSWR < 2 in 824–894 MHz?

12.9 Following up Example 12.13: What if we manage to embed the antenna in FR4 circuit board material? How much smaller could we expect the antenna assumed in Examples 12.10 and 12.11 to become? How does this compare to the theoretical bound on the antenna's Q? For the purposes of this problem, assume $\epsilon_r = 4.5$ and assume loss is negligible.

12.10 Estimate the theoretical minimum uniform VSWR for the cellular handset antenna from Example 12.10 if one requires that the VSWR be uniform over 824–849 MHz.

13 Analog Filters and Multiplexers

13.1 INTRODUCTION

A filter is a device that modifies the frequency response of the system that precedes it. The most common use of filters in radios is to suppress signals and noise from specified ranges of frequencies, while passing signals from other ranges with minimum attenuation and distortion. Filters are principal and essential components of any radio, providing selectivity, mitigation of spurious products, and anti-aliasing in analog-to-digital and digital-to-analog conversion.

Recall we encountered some rudimentary filters in Chapter 9 ("Impedance Matching"). We observed that impedance matching – in particular, two-reactance "L"-matching – exhibited frequency response with bandwidth that was inversely related to the ratio of the port impedances. We further observed that we could increase this bandwidth using multiple stages of impedance matching, and could decrease this bandwidth using "π"- and "T"-type matching sections. However, none of these strategies is applicable if we seek selectivity without a change in impedance, since in that case Q is zero and we start off with infinite bandwidth. What we consider in this chapter is not simply methods to achieve selectivity, but rather methods to achieve selectivity independently of the values of the port impedances.

The organization of this chapter is as follows. First, Section 13.2 introduces some basic parameters and terminology used to describe filters. Then Sections 13.3 and 13.4 describe the simplest possible filters: Namely those consisting of a single reactance and those consisting of a single resonance, respectively. Section 13.5 presents methodology for designing filters in the form of ladder-type circuits consisting of capacitors and inductors, and constrained to have a specified response. In this context we consider the topics of filter phase response and group delay. Section 13.6 introduces reflectionless multichannel filters; in particular, the diplexer. At this point (Section 13.7) we change tack and consider filters comprising transmission line structures – more broadly known as "distributed filter structures" – as opposed to discrete capacitors and inductors. Finally in Section 13.8 we review some other important filter device technologies, including crystal, SAW, and dielectric resonator filters.

13.2 CHARACTERIZATION OF FILTER RESPONSE

Before discussing particular filter types and design techniques, it is useful to establish some common terminology. Generally, filters may be classified as having one of the following five types of frequency response: *lowpass*, *highpass*, *bandpass*, *bandstop* (including *notch* filters), and *allpass*. The magnitude response of the first four types are shown in Figure 13.1.

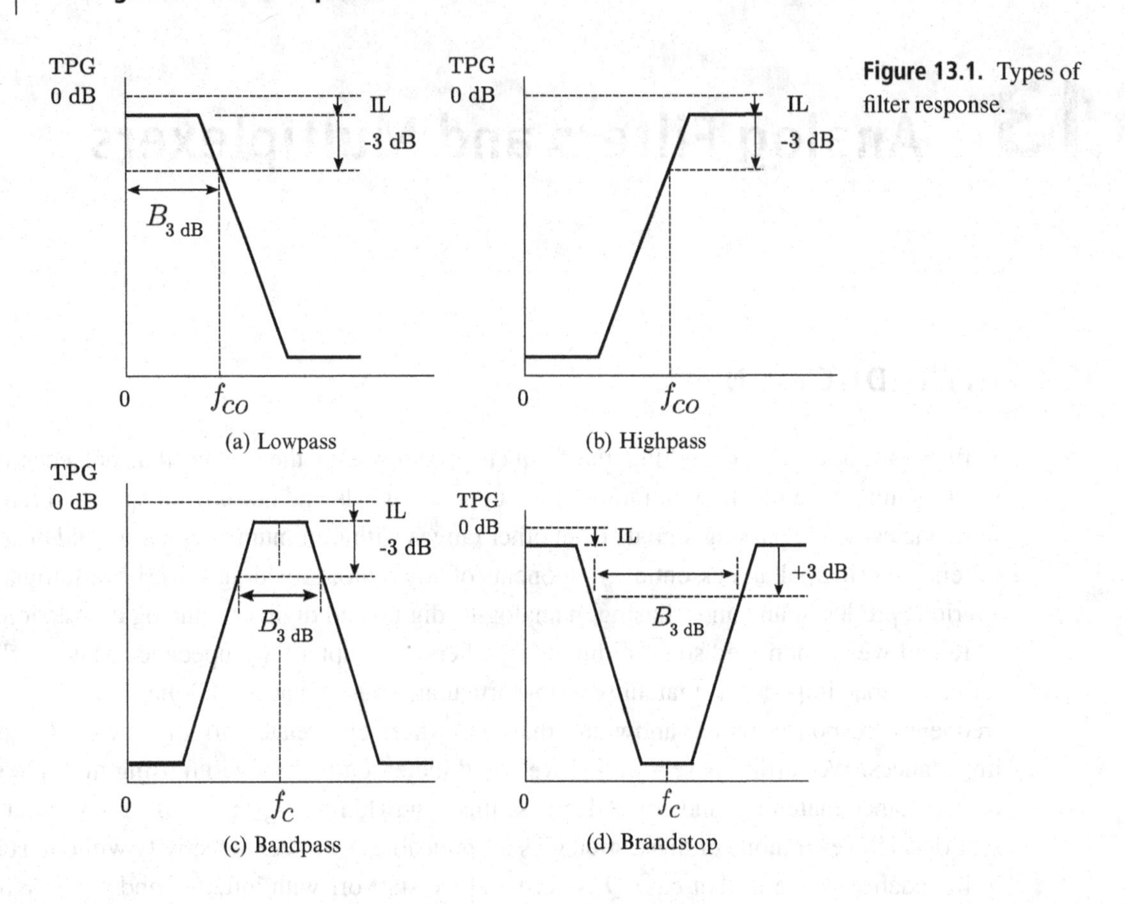

Figure 13.1. Types of filter response.

Allpass filters (*not* shown in Figure 13.1) operate primarily on phase as opposed to magnitude, and are used mainly to manipulate dispersion and delay as opposed to implementing selectivity.

For each filter type, it is common to specify a *passband*, one or two *stopbands*, and one or two *transition bands*. The passband is the contiguous range of frequencies which the filter is expected to pass without excessive attenuation or distortion. Typically, this range is defined by one or two *cutoff frequencies*, as indicated in Figure 13.1. A cutoff frequency is often defined to be a frequency at which the magnitude response has dropped to 3 dB below the nominal passband value. For example, the 3 dB bandwidth of a bandpass filter with 1 dB insertion loss (IL) is determined by the separation between the points in the frequency response at which the TPG is −4 dB.

Stopbands are contiguous ranges of frequencies that the filter is expected to block with some minimum specified level of attenuation. This leaves the transition bands, which are simply the gaps between passbands and stopbands where neither the passband nor stopband criteria are met. The principal tradeoff in filter design is between constraining the width of transition bands and *filter order*. Filter order is a measure of the complexity of a filter and is related to the number of components or discernible structures required to implement the filter.

Bandpass filters are often specified in terms of fractional bandwidth, which is the ratio of bandwidth to center frequency. (For a refresher, see Section 12.4.1.) All else being equal, smaller fractional bandwidth typically requires a higher-order filter than larger fractional bandwidth. Often the combination of fractional bandwidth and center frequency determine whether a bandpass filter can be implemented in discrete LC technologies (as described in Sections 13.3–13.5) or whether some other technology is required (such as those described in Sections 13.7 and 13.8).

A related concept is *shape factor*, which is the ratio of bandwidths measured with respect to different specified magnitude responses. For example, one might specify the shape factor of a bandpass filter as the ratio of the 30 dB bandwidth B_{30dB} to the 3 dB bandwidth B_{3dB}, in which case the shape factor is B_{30dB}/B_{3dB}. Here is an example using that definition: If $B_{3dB} =$ 1.25 MHz and $B_{30dB} = 2.5$ MHz, then the shape factor is said to be "2:1 (30 dB/3 dB)." A lower shape factor indicates narrower transition bands and therefore greater selectivity, and typically requires a higher-order filter.

In addition to magnitude response, filter performance is often described in terms of phase response, and in particular *group delay variation*. This topic is addressed in Section 13.5.6.

Filters may be either *reflective* or *absorptive*. The filters considered throughout most of this chapter are reflective, meaning suppression of frequencies in the output is achieved by reflecting power from the input. A filter that is absorptive achieves suppression by dissipating power within the filter, as opposed to reflecting the undesired power. Absorptive filters are typically implemented as *diplexers*, described in Section 13.6.

13.3 SINGLE-REACTANCE LOWPASS AND HIGHPASS FILTERS

With just one reactance, you can make a lowpass or highpass filter that is suitable for many applications. The possibilities all stem from basic circuit theory, and are presented in Figure 13.2. For example, the shunt capacitor C in Figure 13.2(a) lies in parallel with the input/output impedance Z_{in}, forming a parallel "RC" type circuit having lowpass response with 3 dB frequency $\omega_{co} = 2/Z_{in}C$ in terms of TPG.[1]

EXAMPLE 13.1

Design a single-capacitor lowpass filter having a 3 dB bandwidth of 30 MHz. The input and output impedances of the filter are to be 50 Ω.

Solution: For the shunt capacitor filter, $\omega_{co} = 2/Z_{in}C$ (see above), so we have $C = 2/\left(50\ \Omega\right)\left(2\pi \cdot 30\ \text{MHz}\right) = 212$ pF. The resulting response (TPG) is shown in Figure 13.3.

The design of the series inductor lowpass filter and single-reactance highpass filters proceeds similarly.

[1] If you feel this should be $\omega_{co} = 1/Z_{in}C$, then you are probably making the very common error of confusing TPG with voltage gain or current gain. Problem 13.1 provides you with the opportunity to sort this out for yourself.

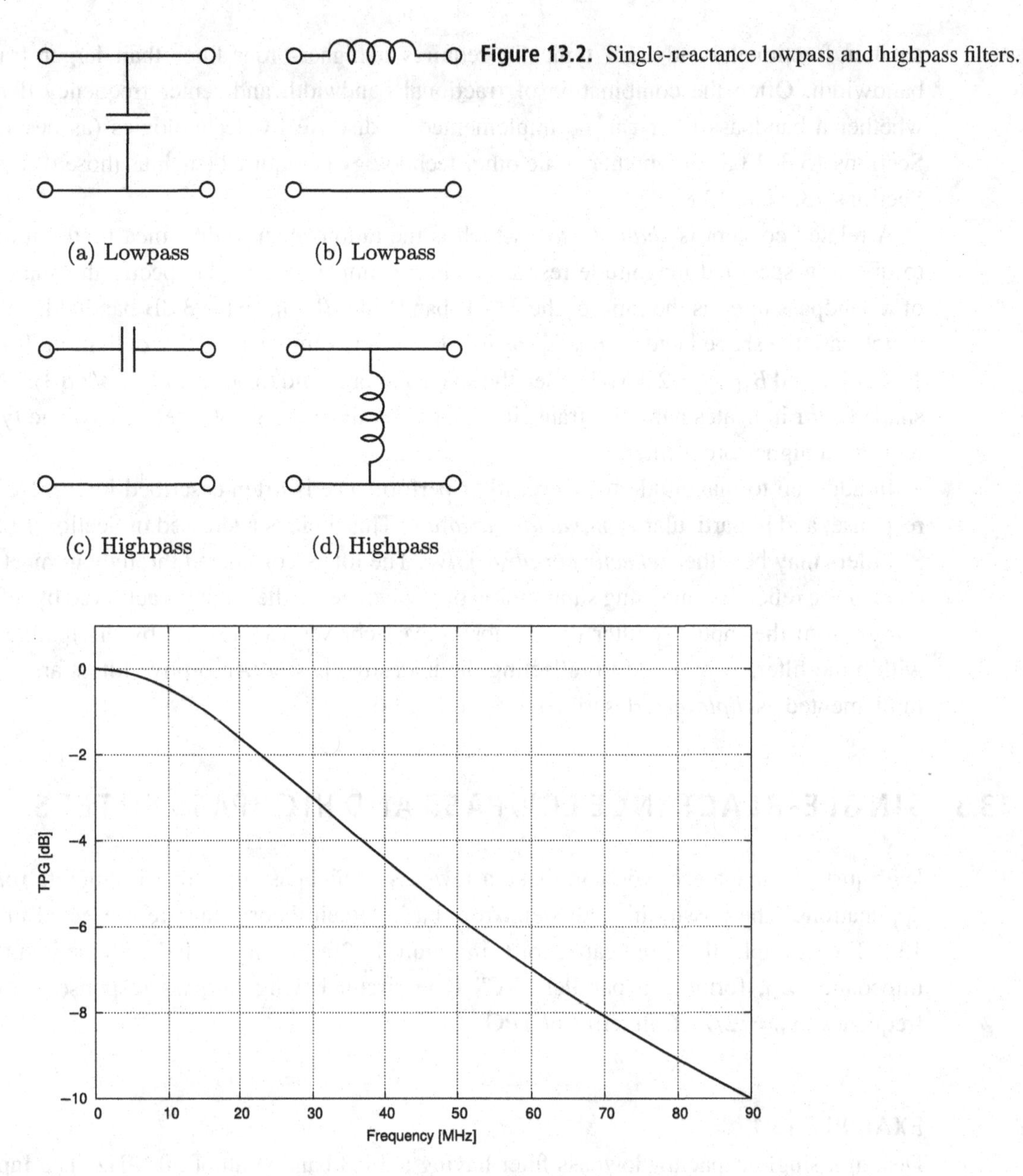

Figure 13.2. Single-reactance lowpass and highpass filters.

Figure 13.3. TPG of the shunt capacitor lowpass filter from Example 13.1.

13.4 SINGLE-RESONATOR BANDPASS AND NOTCH FILTERS

If all you need is to pass a single frequency ω_c and you are not very fussy about the overall frequency response, then one of the circuits shown in Figure 13.4 might suit you. In Figure 13.4(a), the parallel LC section is an open circuit at resonance ($\omega_0 = 1/\sqrt{LC}$), and has decreasing impedance above and below resonance, so this is a bandpass filter with $\omega_c = \omega_0$. In Figure 13.4(b), the series LC section is a short circuit at resonance and has increasing impedance above and below resonance; so this is also a bandpass filter with $\omega_c = \omega_0$. The 3 dB bandwidth of these circuits is approximately $B_Q/2$ for $Q \gg 2\pi$, where $B_Q \equiv \omega_0/Q$ (see

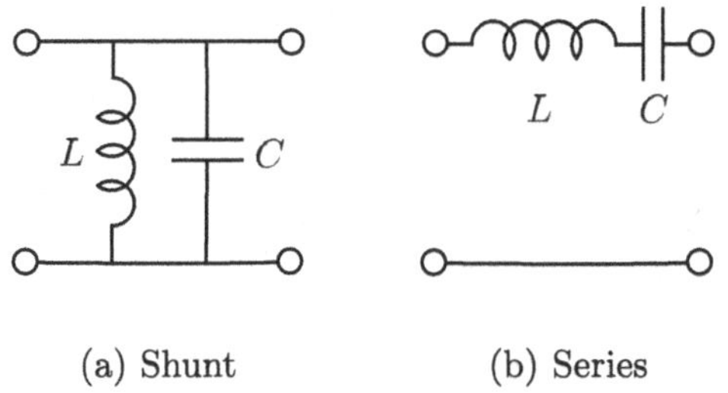

Figure 13.4. Single-resonator bandpass filters.

(a) Shunt (b) Series

Equation (9.16) and related text); and the Q of these circuits is known from basic circuit theory to be

$$Q = Z_{in}\sqrt{\frac{C}{L}} \quad \text{(parallel LC)} \tag{13.1}$$

$$Q = \frac{1}{Z_{in}}\sqrt{\frac{L}{C}} \quad \text{(series LC)} \tag{13.2}$$

assuming that Z_{in} is completely real-valued.

EXAMPLE 13.2

Design a single-resonator discrete-LC bandpass filter of the type shown in Figure 13.4(a) for 30 MHz having a 3 dB bandwidth of about 500 kHz. The input and output impedances of the filter are to be 50 Ω.

Solution: A 3 dB bandwidth of $B_{HP} = 500$ kHz corresponds to $B_Q = 2B_{HP} = 1$ MHz if the associated Q is sufficiently large. Here $Q = 2\pi f_c/B_Q = 188.5$. From the resonance relationship $\omega_c = 1/\sqrt{LC}$, we have $LC = 2.81 \times 10^{-17}$ F·H. For the Figure 13.4(a) topology, $Q = Z_{in}\sqrt{C/L}$, so we have $C/L = 14.2$ F/H; therefore $C = \sqrt{LC \cdot C/L} = 20.0$ nF and $L = LC/C = 1.41$ nH. The resulting response (TPG) is shown in Figure 13.5.

We find that the 3 dB bandwidth turns out to be about 320 kHz; a bit on the low side due to the approximate relationship between Q and 3 dB bandwidth. If a value of exactly 500 kHz were required, we would simply slightly scale down the bandwidth-derived value of Q in the above procedure.

If you need to *stop* a narrow band of frequencies, then either of the circuits shown in Figure 13.6 may be what you need. In Figure 13.6(a), the series LC section is a short circuit to ground at resonance, and has increasing impedance above and below resonance. In Figure 13.6(b), the parallel LC section is an open circuit at resonance and has decreasing impedance above and below resonance. In both cases you end up with a "notch" response with a null at ω_0, which is more or less sharp depending on the Q of the circuit.

The principal limitation of single-reactance lowpass and highpass filters described in the previous section are (1) lack of control over passband response, and (2) wide transition bands. The problem is that once you specify cutoff frequency, the design is completely determined.

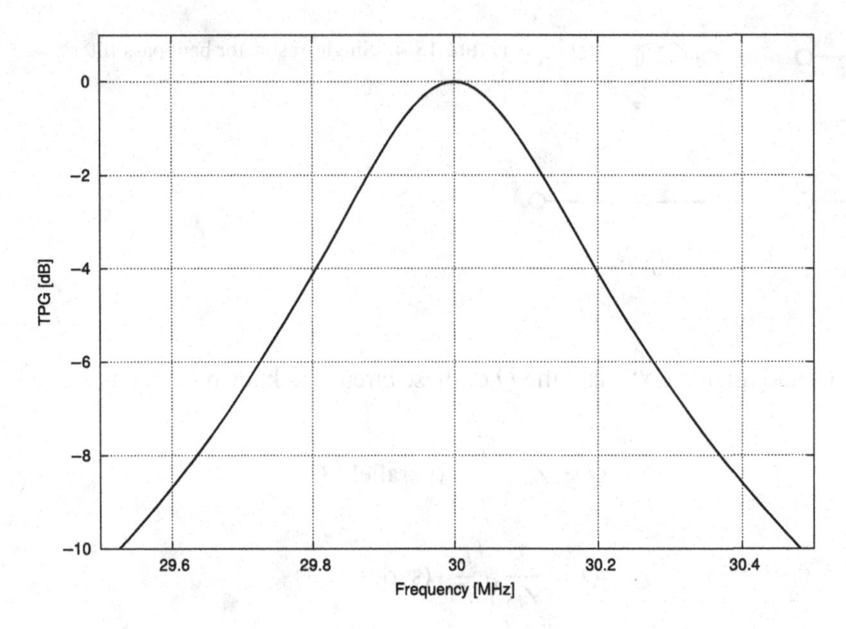

Figure 13.5. Response of the single-resonator bandpass filter from Example 13.2.

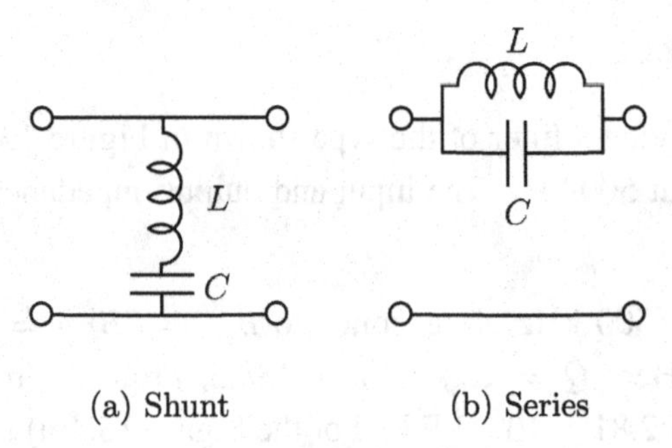

Figure 13.6. Single-resonator notch filters.

Single-resonator bandpass and notch filters suffer from the same limitations: Once you specify center frequency and bandwidth (by whatever definition), the design is completely determined and there are no "degrees of freedom" remaining to control the frequency response. To do better, a higher-order filter – that is, one with additional reactances – is required. This brings us to the topic of specified response filters.

13.5 DISCRETE (LC) FILTERS – SPECIFIED RESPONSE

In this section we present a class of methods for the design of filters having a specified response, and in which the input and output impedances are equal. These filters consist of N two-port stages, in which each stage consists of discrete capacitors and inductors in one of the topologies shown in Figure 13.7:

- Figure 13.7(a) shows a two-port stage consisting of a shunt capacitor and series inductor. This two-port will have lowpass response.

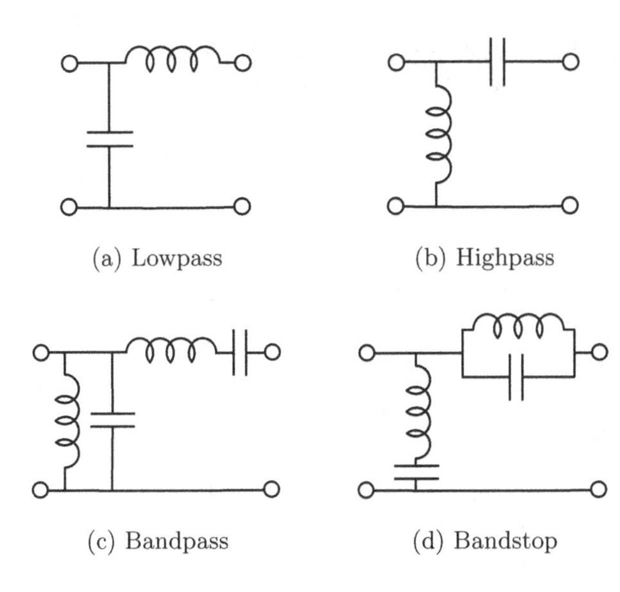

Figure 13.7. Two-port stages used to develop specified-response filters.

- Figure 13.7(b) shows a two-port stage consisting of a shunt inductor and series capacitor. This section will have highpass response.
- Figure 13.7(c) shows a two-port stage consisting of a shunt inductor and capacitor in parallel, and capacitor and inductor in series. The shunt LC section will behave as an open circuit at its frequency of resonance, whereas the series LC segment will behave as a short circuit at its frequency of resonance. If the resonances occur at the same frequency, then this stage will behave as a unity-gain two-port at resonance with overall bandpass response.
- Figure 13.7(d) shows a two-port stage consisting of a shunt combination of an inductor and capacitor in series, and series combination of a capacitor and inductor in parallel. The shunt LC section will behave as a short circuit at its frequency of resonance, whereas the series LC section will behave as an open circuit at its frequency of resonance. If the resonances occur at the same frequency, then this stage will have behave as a open-circuited two-port at resonance with overall bandstop response.

Note also that these stages have the same response when reversed left-to-right, although for simplicity we shall consider only the orientation shown in Figure 13.7.

Stages of the same type may be cascaded, with the effect that additional free parameters (i.e., more inductors and capacitors) become available for use in setting the specific characteristics of the frequency response. For example, an $N = 2$ stage version of the lowpass two-port of Figure 13.7 is shown in Figure 13.8. Note that a final shunt section is added to make the resulting two-port symmetrical, resulting in a total of $K = 2N + 1 = 5$ segments. Also note the labeling of the segments using the notation p_k, with $k = 1, 2, \ldots K$. In this notation, p_k may be either a capacitance or an inductance.

A K-segment lowpass filter design problem consists of selecting the K component values that achieve the specified response for a specified input/output impedance. Common frequency response types are known as *Butterworth*, *Chebyshev*, *Bessel*, *Gaussian*, and *elliptic*. These schemes result in frequency responses with useful distinct characteristics.

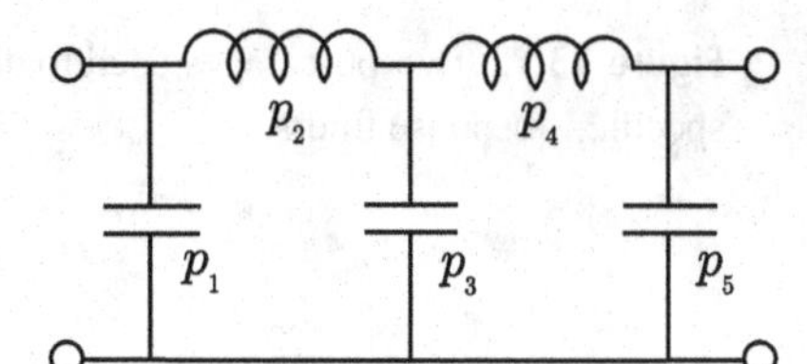

Figure 13.8. A five-stage ($N = 2$) lowpass specified-response filter, having $K = 2N + 1 = 5$ segments.

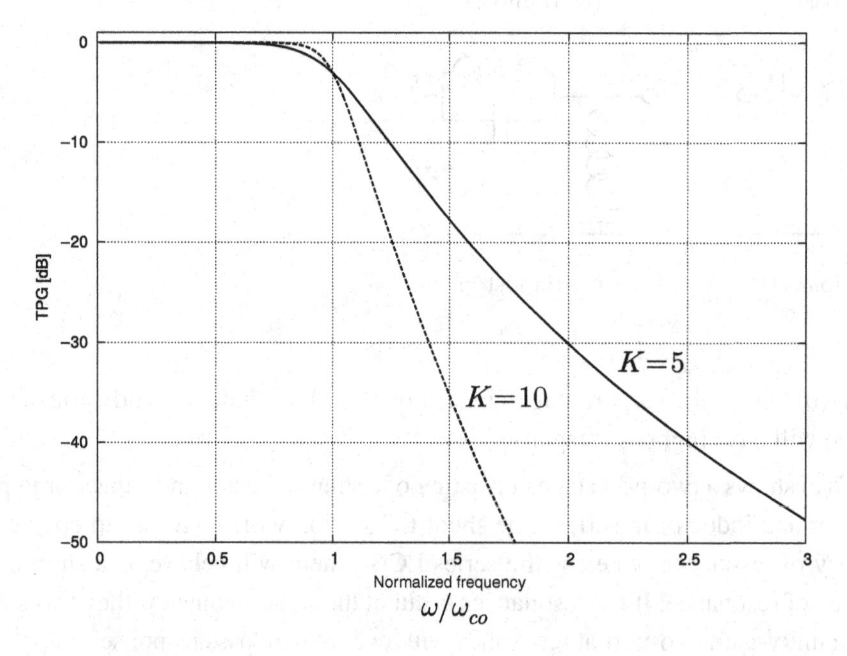

Figure 13.9. Butterworth Lowpass Response.

13.5.1 Butterworth Lowpass Filter Design

The Butterworth lowpass frequency response is defined in terms of its gain (TPG) as

$$G_T(\omega) = \left[1 + \left(\frac{\omega}{\omega_{co}}\right)^{2K}\right]^{-1} \tag{13.3}$$

This response is shown in Figure 13.9. This response is often described as being *maximally flat*, resulting in a passband which is smoothly-varying and free from inflection points. Note that for any given cutoff angular frequency $\omega_{co} = 2\pi f_{co}$, the transition from passband to stopband can be made more or less sharp by increasing or decreasing K, respectively. In other words, the selectivity of the filter is determined by number of stages.

Derivation of a procedure to determine component values for Butterworth lowpass response is beyond the scope of this text; for details, [85] and [87] are recommended. Here, we will simply show the procedure and then demonstrate that it yields the desired results. The first step is to calculate component values for a filter with $\omega_{co} = 1$ rad/s and input/output impedance $Z_{in} = 1\ \Omega$; these are

$$p_k = 2\sin\left[\frac{(2k-1)\,\pi}{2K}\right] \tag{13.4}$$

where p_k is either capacitance in units of farad (F) or inductance in units of henry (H). Next, the component values are "renormalized" to achieve a filter having the desired values of ω_{co} and Z_{in}. Cutoff frequency is renormalized simply by dividing p_k by ω_{co}. Z_{in} is renormalized by dividing capacitances by Z_{in} and multiplying inductances by Z_{in}. The following example demonstrates the procedure:

EXAMPLE 13.3

Design a maximally-flat lowpass filter with 3 dB cutoff at 30 MHz and which has TPG which is at least 40 dB down at 80 MHz. Input and output impedances are to be 50 Ω.

Solution: First, we determine the required number of stages. From Equation (13.3) we find

$$G_T(2\pi \cdot 80 \text{ MHz}) = \left[1 + \left(\frac{2\pi \cdot 80 \text{ MHz}}{2\pi \cdot 30 \text{ MHz}}\right)^{2K}\right]^{-1} = -42.6 \text{ dB}$$

for $K = 5$, and that any smaller value of K yields G_T greater than −40 dB at 80 MHz. Therefore, $K = 5$ is required.

From Equation (13.4), we obtain:

$$p_1 = 0.618 \text{ F},\ p_2 = 1.618 \text{ H},\ p_3 = 2.000 \text{ F},\ p_4 = 1.618 \text{ H},\ \text{and } p_5 = 0.618 \text{ F}.$$

Renormalizing to $f_c = 30$ MHz and $Z_{in} = 50$ Ω, we obtain:

$$p_1 = 65.6 \text{ pF},\ p_2 = 429.2 \text{ nH},\ p_3 = 212.2 \text{ pF},\ p_4 = 429.2 \text{ nH},\ \text{and } p_5 = 65.6 \text{ pF}.$$

The result is shown in Figure 13.10.

13.5.2 Butterworth Highpass Filter Design

The Butterworth highpass frequency response is defined in terms of its gain (TPG) as

$$G_T(\omega) = \left[1 + \left(\frac{\omega_{co}}{\omega}\right)^{2K}\right]^{-1} \tag{13.5}$$

Again, this is the solution if you specify the passband to be maximally flat. The design procedure is essentially the same as for Butterworth lowpass filters, with just two small differences: The odd values of k now refer to inductances whereas the even values refer to capacitances, and the component values for the $\omega_{co} = 1$ rad/s, $Z_{in} = 1$ Ω prototype highpass filter are equal to the reciprocal of the values used for the prototype lowpass filter. That is, we now have

$$p_k = \left\{2 \sin\left[\frac{(2k-1)\,\pi}{2K}\right]\right\}^{-1} \tag{13.6}$$

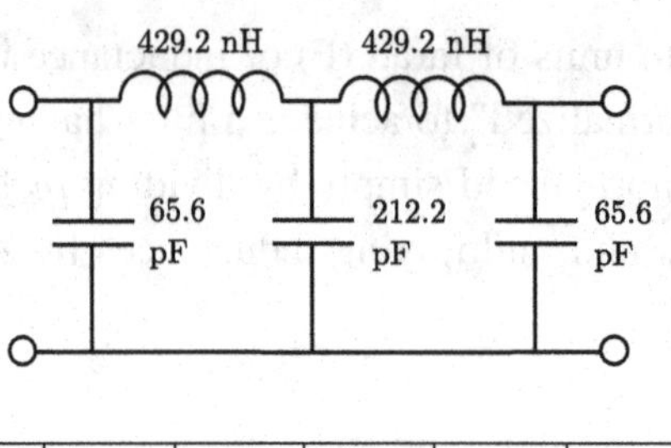

Figure 13.10. Butterworth lowpass filter from Example 13.3.

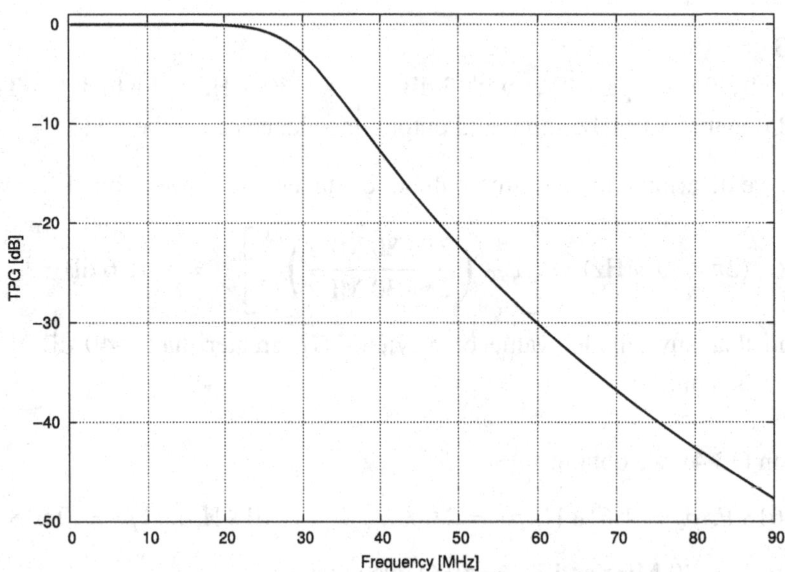

EXAMPLE 13.4

Design a maximally-flat highpass filter with 3 dB cutoff at 30 MHz and which has TPG which is at least 40 dB down at 10 MHz. Input and output impedances are to be 50 Ω.

Solution: First, we determine the required number of stages. From Equation (13.5) we find

$$G_T(2\pi \cdot 10\text{ MHz}) = \left[1 + \left(\frac{2\pi \cdot 30\text{ MHz}}{2\pi \cdot 10\text{ MHz}}\right)^{2K}\right]^{-1} = -47.7\text{ dB}$$

for $K = 5$, and that any smaller value of K yields G_T greater than −40 dB at 10 MHz. Therefore, $K = 5$ is required.

Because K and f_c are unchanged from the previous example, this design will turn out to be the highpass version of the filter from the previous example. Using Equation (13.6), we obtain:

$$p_1 = 1.618\text{ H},\ p_2 = 0.618\text{ F},\ p_3 = 0.500\text{ H},\ p_4 = 0.618\text{ F, and } p_5 = 1.618\text{ H}.$$

Renormalizing to $f_c = 30$ MHz and $Z_{in} = 50\ \Omega$, we obtain:

$$p_1 = 429.2\text{ nH},\ p_2 = 65.6\text{ pF},\ p_3 = 132.6\text{ nH},\ p_4 = 65.6\text{ pF, and } p_5 = 429.2\text{ nH}.$$

The result is shown in Figure 13.11.

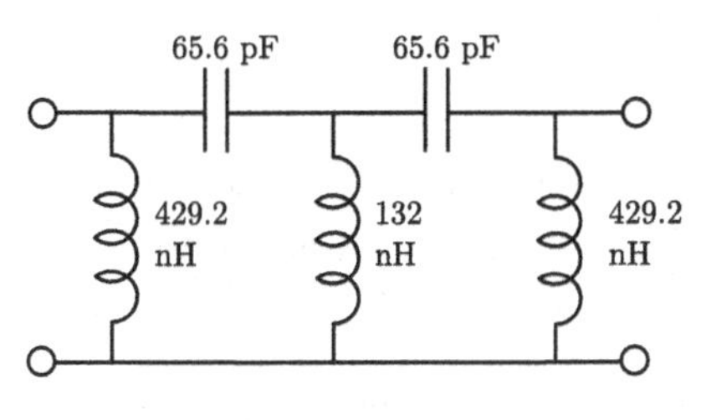

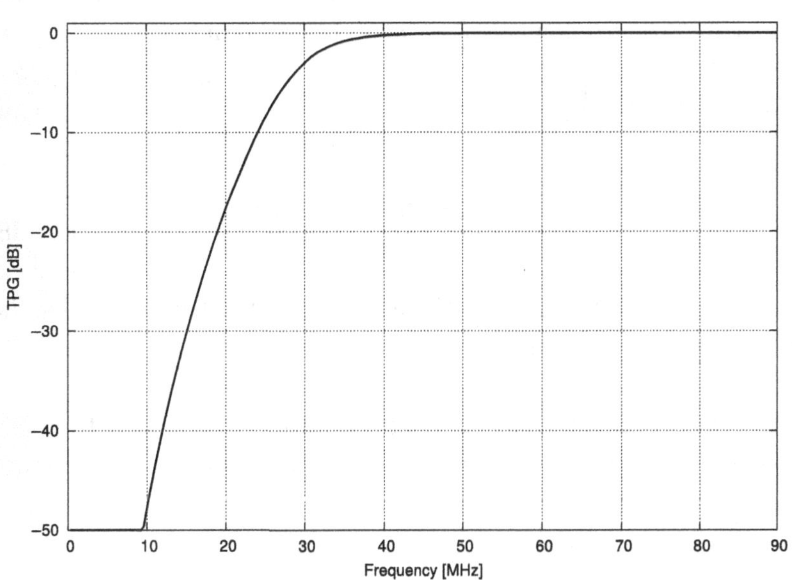

Figure 13.11. Butterworth highpass filter from Example 13.4.

13.5.3 Butterworth Bandpass Filter Design

The Butterworth bandpass frequency response is defined in terms of its gain (TPG) as

$$G_T(\omega) = \left[1 + \left(\frac{2\,|\omega - \omega_c|}{2\pi \cdot B_{HP}}\right)^{2K}\right]^{-1} \tag{13.7}$$

where B_{HP} is 3 dB bandwidth (i.e., the separation between half-power points in the bandpass response) and ω_c is the center frequency of the bandpass. Yet again, this is the solution you get when you require the passband to be maximally-flat.

The design proceeds as follows: One first designs a lowpass filter as a prototype, with $f_{co} = B_{HP}$. The bandpass filter is then obtained by adding a parallel reactance to each shunt section so that the section resonates – becoming an open-circuit – at ω_c. Similarly, a series reactance is added to each series section so that the section resonates – becoming a short-circuit – at ω_c. Each stage in the resulting filter then resembles Figure 13.7(c). The values for the additional components are determined by using the resonance relationship $\omega_c = 1/\sqrt{LC}$, in just the same way as was explained in the design procedure for single-resonance bandpass filters.

EXAMPLE 13.5

Design a maximally-flat bandpass filter with 3 dB bandwidth of 10 MHz, centered at 70 MHz, and which has TPG that is at least 40 dB down at 55 MHz and 85 MHz. Input and output impedances are to be 50 Ω.

Solution: First, we determine the required number of stages. 55 MHz and 85 MHz are both 15 MHz from f_c, so the specification is $B_{40\text{ dB}} = 30$ MHz. From Equation (13.7) we find

$$G_T(2\pi \cdot 55\text{ MHz}) = \left[1 + \left(\frac{2 \cdot 2\pi \cdot |55 - 70\text{ MHz}|}{2\pi \cdot 10\text{ MHz}}\right)^{2K}\right]^{-1} = -47.7\text{ dB}$$

$$G_T(2\pi \cdot 85\text{ MHz}) = \left[1 + \left(\frac{2 \cdot 2\pi \cdot |85 - 70\text{ MHz}|}{2\pi \cdot 10\text{ MHz}}\right)^{2K}\right]^{-1} = -47.7\text{ dB}$$

for $K = 5$, and that any smaller value of K yields G_T greater than −40 dB at the specified frequencies. Therefore, $K = 5$ is required.

Because K and the specified response (Butterworth) is unchanged from the previous two examples, this design begins with the same prototype lowpass filter (generated from Equation (13.4)):

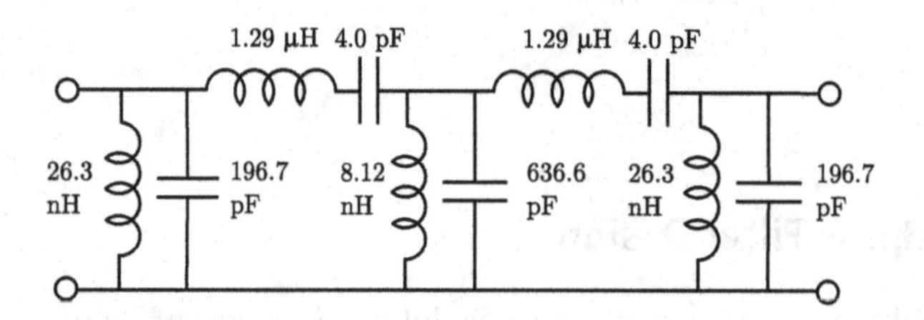

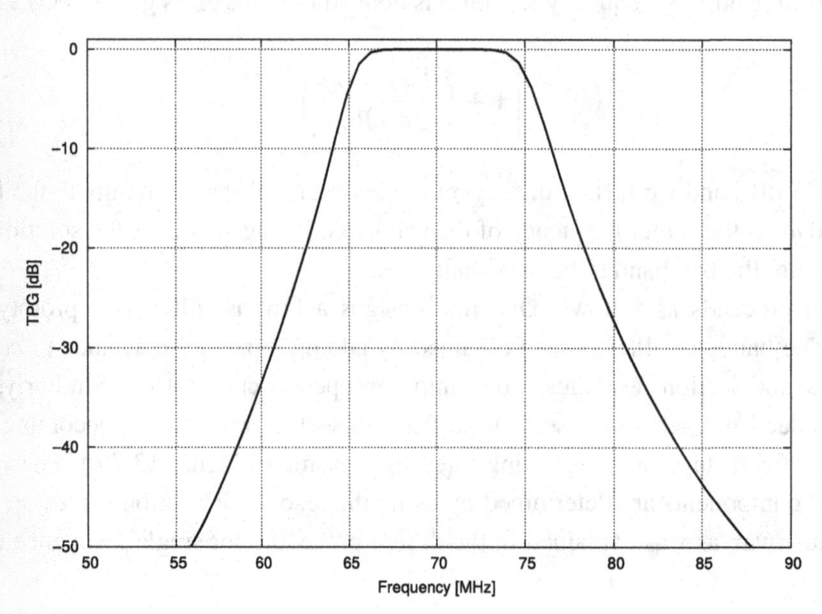

Figure 13.12. Butterworth bandpass filter from Example 13.5.

$p_1 = 0.618$ F, $p_2 = 1.618$ H, $p_3 = 2.000$ F, $p_4 = 1.618$ H, and $p_5 = 0.618$ F.

Renormalizing to a cutoff frequency of 10 MHz and $Z_{in} = 50\ \Omega$, we obtain:

$p_1 = 196.7$ pF, $p_2 = 1.29$ μH, $p_3 = 636.6$ pF, $p_4 = 1.29$ μH, and $p_5 = 196.7$ pF.

Now add components to resonate each segment at the desired center frequency of $f_c = 70$ MHz. Since p_1 is a shunt capacitor, we add a parallel inductor

$$\frac{1}{(2\pi f_c)^2\, p_1} = 26.3 \text{ nH}$$

Since p_2 is a series inductor, we add a series capacitor

$$\frac{1}{(2\pi f_c)^2\, p_2} = 4.0 \text{ pF}$$

and so on. The result is shown in Figure 13.12.

13.5.4 Butterworth Bandstop Filter Design

The Butterworth bandstop frequency response is defined in terms of its gain (TPG) as

$$G_T(\omega) = \left[1 + \left(\frac{2\pi \cdot B_{HP}}{2\,|\omega - \omega_c|}\right)^{2K}\right]^{-1} \tag{13.8}$$

where B_{HP} is now the separation between half-power points straddling the stopband region. The design procedure is essentially the same as the bandpass design procedure, except we begin with a *highpass* prototype and resonate the segments following the topology indicated in Figure 13.7(d). Here are the specific steps: One first designs a highpass filter as a prototype, with $f_{co} = B_{HP}$. The bandstop filter is then obtained by adding a series reactance to each shunt section so that the section resonates – becoming an short-circuit – at ω_c. Similarly, a parallel reactance is added to each series section so that the section resonates – becoming an open-circuit – at ω_c.

EXAMPLE 13.6

Design a $K = 5$ bandstop filter with 3 dB bandwidth of 10 MHz straddling a bandstop region centered at 70 MHz, and which is maximally-flat in the bandpass regions. Input and output impedances are to be 50 Ω.

Solution: Because K and the specified response (Butterworth) are unchanged from Example 13.4, this design begins with the same $\omega_{co} = 1$ rad/s, $Z_{in} = 1\ \Omega$ prototype highpass filter:

$p_1 = 1.618$ H, $p_2 = 0.618$ F, $p_3 = 0.500$ H, $p_4 = 0.618$ F, and $p_5 = 1.618$ H.

Renormalizing to a cutoff frequency of 10 MHz and $Z_{in} = 50\ \Omega$, we obtain:

$p_1 = 1.29$ μH, $p_2 = 197$ pF, $p_3 = 398$ nH, $p_4 = 197$ pF and $p_5 = 1.29$ μH.

Now add components to resonate each segment at the desired center frequency of $f_c = 70$ MHz. Since p_1 is a shunt inductor, we add a capacitor in series having value

$$\frac{1}{(2\pi f_c)^2 \, p_1} = 4.0 \text{ pF}$$

Since p_2 is a series capacitor, we add an inductor in parallel

$$\frac{1}{(2\pi f_c)^2 \, p_2} = 26.3 \text{ nH}$$

and so on. The result is shown in Figure 13.13.

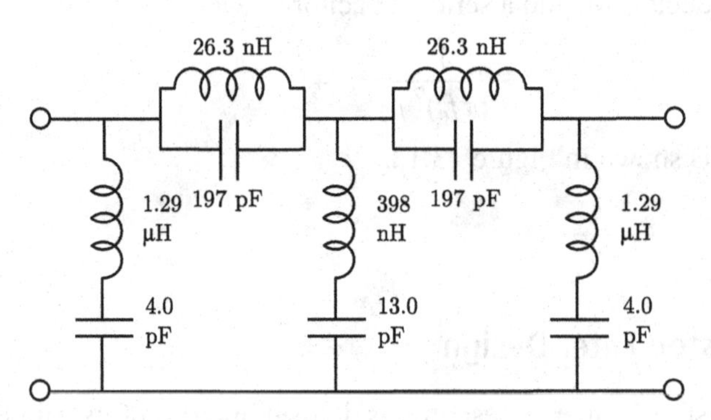

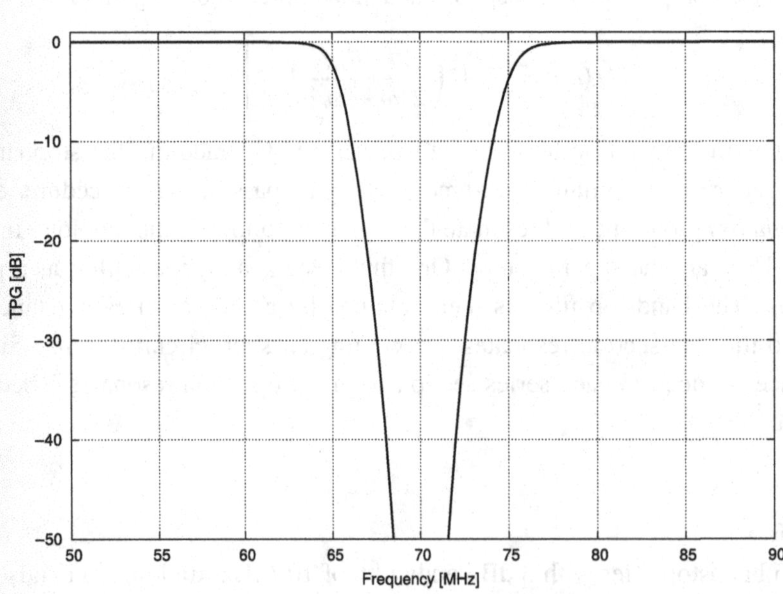

Figure 13.13. Butterworth bandstop filter from Example 13.6.

13.5.5 Chebyshev Filter Design

The primary advantage – in fact, the defining feature – of the Butterworth response is passband flatness. This comes at the expense of selectivity. A faster transition from passband to stopband for a given order N is possible if one is willing to sacrifice passband flatness. One way to do this is to require the so-called *equiripple* response; i.e., require that the maximum deviation from

Table 13.1. **Component values (p_ks) for prototype ($\omega_{co} = 1$ rad/s, $Z_{in} = 1\ \Omega$) Chebyshev lowpass filters having ripple magnitude $\leq r$ and number of segments K.**

r	K	p_1	p_2	p_3	p_4	p_5	p_6	p_7
0.1 dB	3	1.4328	1.5937	1.4328				
	5	1.3013	1.5559	2.2411	1.5559	1.3013		
	7	1.2615	1.5196	2.2392	1.6804	1.2392	1.5196	1.2615
0.5 dB	3	1.8636	1.2804	1.8636				
	5	1.8068	1.3025	2.6914	1.3025	1.8068		
	7	1.7896	1.2961	2.7177	1.3848	2.7177	1.2961	1.7896

the nominal TPG due to ripple within the passband be less than some constant. The greater the ripple allowed, the faster the transition from passband to stopband can be. This leads to the *Chebyshev* response.[2] Formal mathematical descriptions of the response (analogous to Equation (13.3)) and generating equations (analogous to Equation (13.4)) exist, but are relatively complicated (again, see [85] and [87] for details). Instead, we present in Table 13.1 the values (p_ks) for prototype Chebyshev lowpass filters having various degrees of passband ripple r, where r is the ratio of maxima to minima in the passband. The values in Table 13.1 should be sufficient for most radio engineering applications; see [87, Sec. 6.7] (among others) for additional designs.

Given a prototype lowpass filter from Table 13.1, filters for any other f_{co} and Z_{in} may be generated using the technique explained in Section 13.5.1. Similarly, Chebyshev highpass, bandpass, and bandstop filters may be generated by using the techniques explained in Sections 13.5.2, 13.5.3, and 13.5.4, respectively.

EXAMPLE 13.7

Design five-segment Chebyshev lowpass filters with 3 dB cutoff at 30 MHz, input and output impedances of 50 Ω, and $r = 0.1$ dB and 0.5 dB. Compare these to the Butterworth $K = 5$ design from Example 13.3, evaluating the response at 80 MHz.

Solution: From Table 13.1, here are component values for the prototype Chebyshev $K = 5$, $r = 0.1$ dB lowpass filter:

$$p_1 = 1.3013\text{ F},\ p_2 = 1.5559\text{ H},\ p_3 = 2.2411\text{ F},\ p_4 = 1.5559\text{ H},\ \text{and}\ p_5 = 1.3013\text{ F}.$$

Renormalizing to $f_c = 30$ MHz and $Z_{in} = 50\ \Omega$, we obtain:

$$p_1 = 138.1\text{ pF},\ p_2 = 412.7\text{ nH},\ p_3 = 237.8\text{ pF},\ p_4 = 412.7\text{ nH},\ \text{and}\ p_5 = 138.1\text{ pF}.$$

[2] There is no consensus as to how to properly translate this Russian name into English; the indicated spelling is chosen here primarily because it is both relatively common and relatively simple. Alternative spellings commonly appearing in the technical literature include *Chebychev*, *Tchebychev*, *Chebysheff*, *Tchebycheff*, and so on.

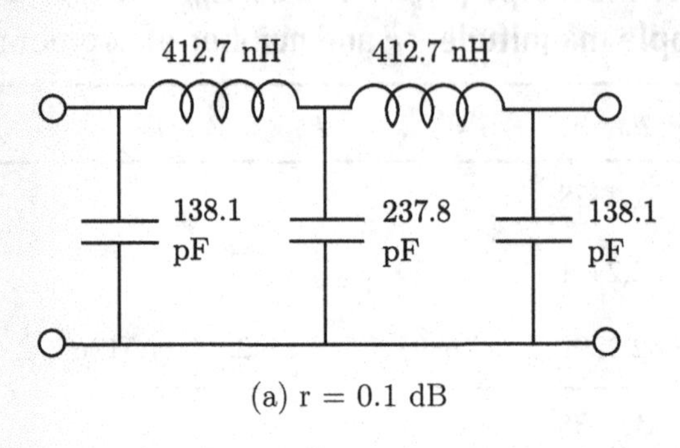

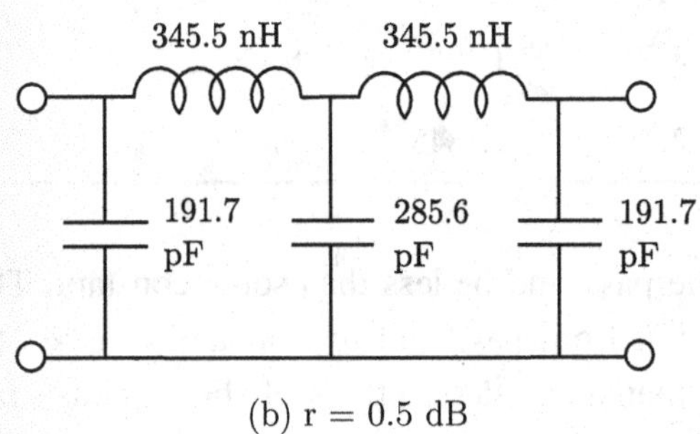

Figure 13.14. Chebyshev lowpass filters from Example 13.7.

Similarly, we obtain the following component values for the $r = 0.5$ dB version of the filter: Renormalizing to $f_c = 30$ MHz and $Z_{in} = 50\ \Omega$, we obtain:

$$p_1 = 191.7 \text{ pF},\ p_2 = 345.5 \text{ nH},\ p_3 = 285.6 \text{ pF},\ p_4 = 345.5 \text{ nH},\ \text{and } p_5 = 191.7 \text{ pF}.$$

The results are shown in Figures 13.14 and 13.15. The response of the Butterworth, Chebyshev $r = 0.1$ dB, and Chebyshev $r = 0.5$ dB filters is −42.6 dB, −54.6 dB, and −58.6 dB, respectively. This demonstrates the increased selectivity of the Chebyshev filter, as well as the tradeoff between r and selectivity for the Chebyshev response.

13.5.6 Phase and Delay Response; Group Delay Variation

An analog filter modifies the phase as well as the magnitude of the signals upon which it operates. So let us now consider the voltage frequency response $H(\omega)$, defined as the ratio of the output voltage to the input voltage; and specifically the phase of $H(\omega)$, $\angle H(\omega)$.

An important contributor to $\angle H(\omega)$ is simply delay: As causal devices, there is an irreducible delay due to the finite time required for a signal to propagate from input to output. A simple delay manifests as a phase response, which is a linear function of frequency; specifically, phase decreases as frequency increases. A typical filter exhibits this delay, plus some additional phase distortion.

To demonstrate, Figure 13.16 shows the phase response for the Butterworth and Chebyshev lowpass filters from Example 13.7. Note that this response is not exactly linear. The concern about phase variation beyond the linear variation associated with simple delay is the potential

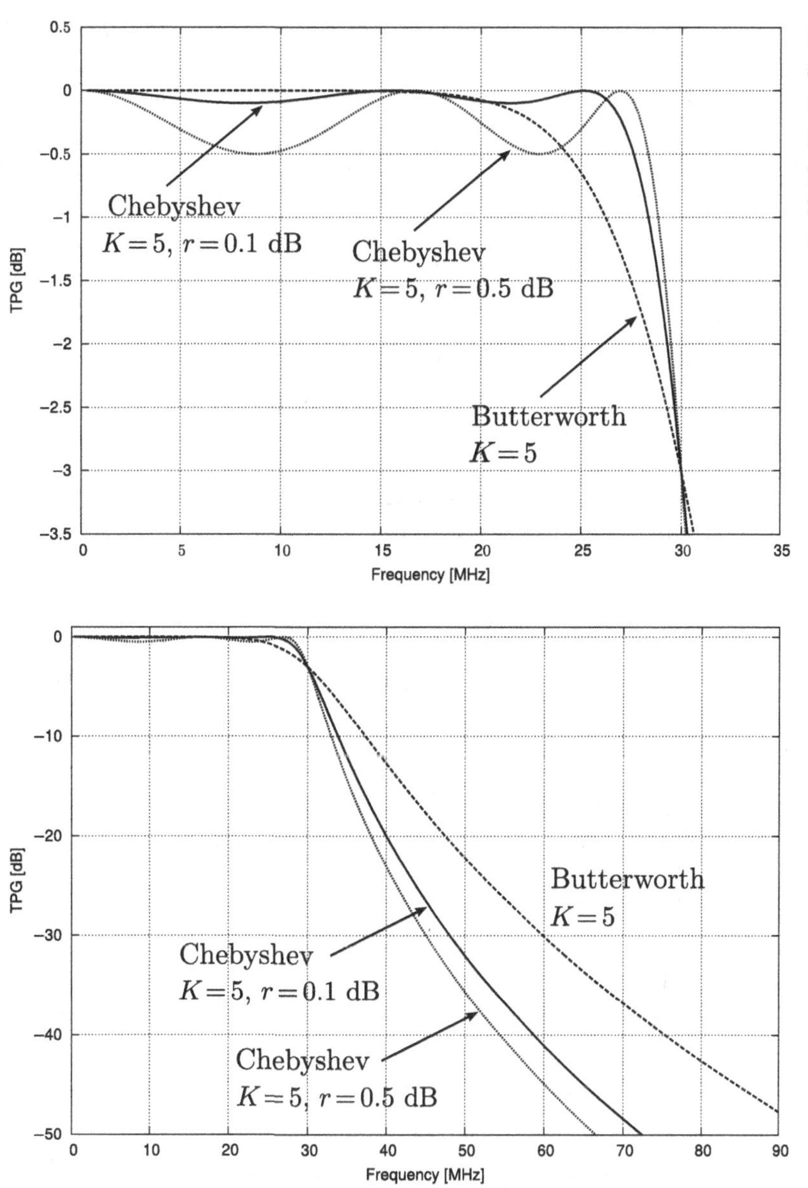

Figure 13.15. Comparison of Chebyshev and Butterworth lowpass filters from Example 13.7. Top panel shows the bandpass region from lower panel (note difference in axes).

for distortion to signals. In particular, modern high-order digital modulations are particularly sensitive to this kind of distortion. A useful characterization of the filter distortion separate from the harmless "simple delay" component is *group delay*. The group delay τ_g is the delay determined from a single point in the phase vs. frequency curve; i.e.,

$$\tau_g(\omega) = -\frac{d}{d\omega}\angle H(\omega) \tag{13.9}$$

Note that group delay is itself a function of frequency; specifically, it is the propagation delay through the filter as a function of frequency. The nominal variation in group delay over the passband is zero; that is, it is preferred that all frequencies require the same amount of time to propagate through the filter. When group delay varies significantly across the passband, the filter is said to be *dispersive*. In this case, components at different frequencies arrive at the filter

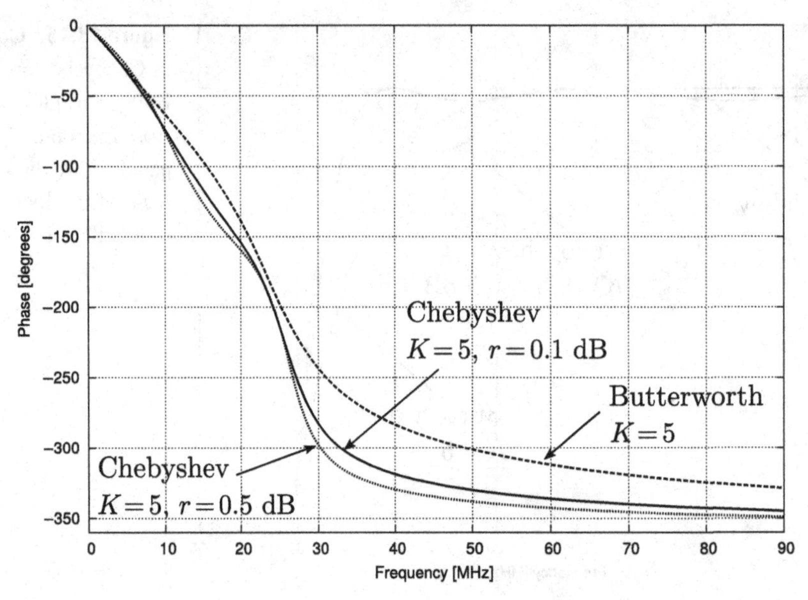

Figure 13.16. Comparison of the phase responses of the Butterworth and Chebyshev lowpass filters from Example 13.7.

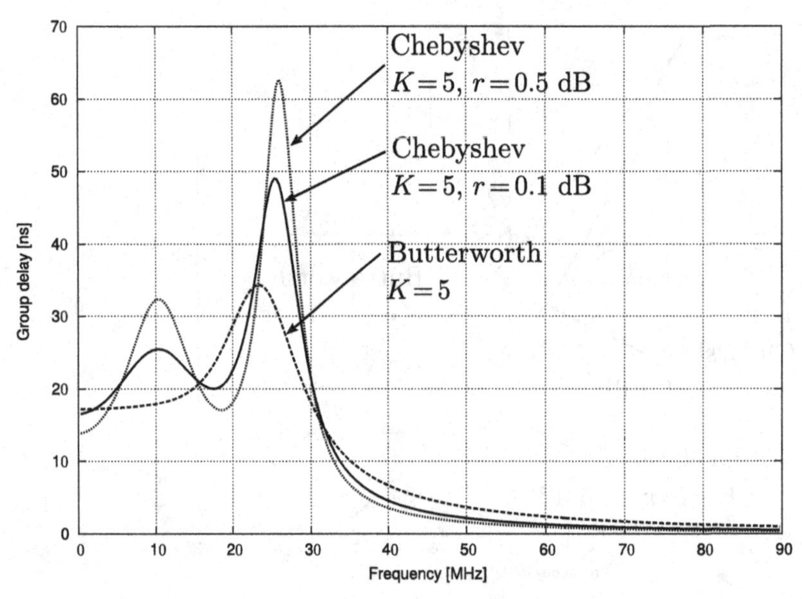

Figure 13.17. Comparison of the group delay for the filters considered in Figure 13.16.

output at different times. Figure 13.17 shows the group delay response for the Butterworth and Chebyshev lowpass filters from Example 13.7. Note that the distortion associated with the Chebyshev filter is significantly greater than that of the Butterworth filter. This is the price paid for greater selectivity.

13.5.7 Other Specified-Response Designs and Topological Variants

Butterworth (maximally-flat) and Chebyshev (equiripple) are hardly the extent of useful frequency responses that may be implemented in discrete LC filters. Here are a few other notable response types.

- The *Bessel* response, which has “maximally linear phase response”; that is, the phase component of the frequency response is free of inflection points in the passband. This also infers minimum group delay variation. The tradeoff is that the magnitude response is far less selective than the Butterworth or Chebyshev responses.
- The *Gaussian* response, which has the *minimum* possible group delay in the passband, resulting in a filter having Gaussian magnitude response.
- The *elliptic* response (also variously known as *Cauer* response or *Zolotarev* response), which is equiripple both in the passband and in the stopbands. This leads to the fastest possible transition between passband and stopband, and also leads to filters which are less sensitive to errors in component values.

It should further be noted that specified-response filters are not limited to circuit topologies comprised of stages as depicted in Figure 13.7. In fact, other topologies may offer significant advantages, including more convenient component values. For additional details, [85] and [87] are recommended. Also [23, Ch. 13] provides some useful tips.

13.6 DIPLEXERS AND MULTIPLEXERS

A common feature of all of the filters considered previously in this chapter is that they are *reflective*; that is, power is suppressed in the output by reflecting it at the input. This is undesirable in a number of applications. For example, spurious content emerging from a frequency mixer (see Section 14.3) should preferably be absorbed by the filter, as opposed to being reflected back into the mixer.

A device that may be used to implement absorptive filtering is known as a *diplexer*, and is shown in Figure 13.18. A diplexer is actually two complementary filters of the reflective type, with both ports on one side in common, and the ports on the opposite side remaining separate. Typically, but not necessarily, the individual filters are lowpass and highpass, respectively. The filter with the desired response provides the output of the diplexer, whereas the filter with the complementary/undesired response is terminated into a matched load, such as a resistor.

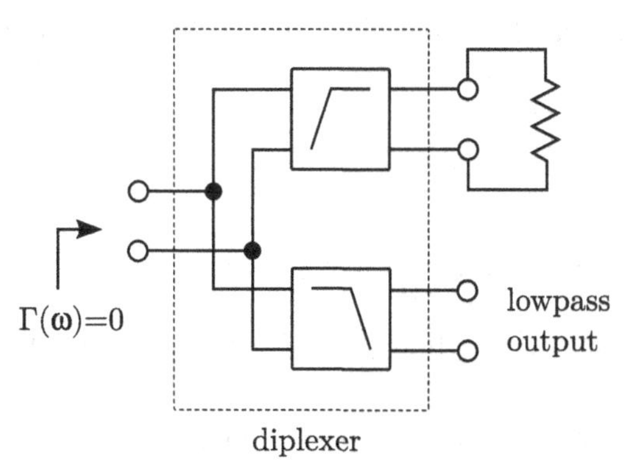

Figure 13.18. A lowpass absorptive filter implemented using a diplexer.

When properly designed, the input reflection coefficient remains very small over a very broad frequency range.

EXAMPLE 13.8

Design the simplest possible post-mixer diplexer that passes 10.7 MHz and absorbs LO feedthrough at 139.3 MHz. (To see the particular application that gives rise to these specifications, see Example 14.1.) The nominal input/output impedance is $Z_{in} = 50\ \Omega$.

Solution: The proposed topology is shown in Figure 13.19. The lowpass filter is the series inductance L, and the highpass filter is the series capacitance C. The resistance R absorbs the output of the highpass filter.

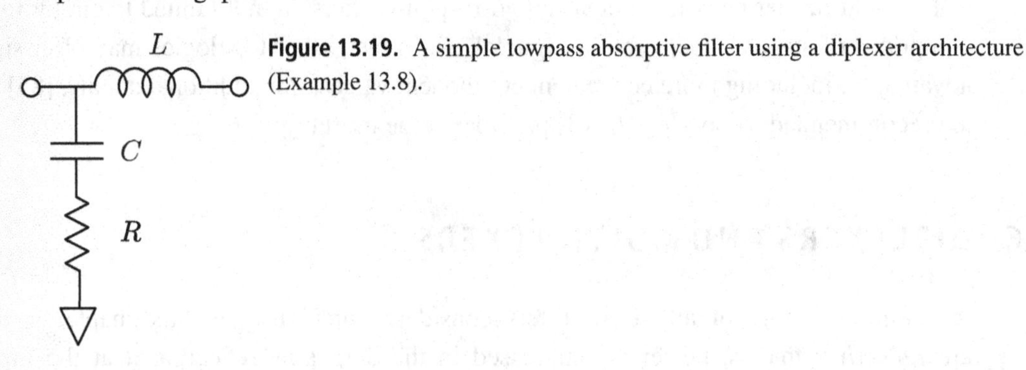

Figure 13.19. A simple lowpass absorptive filter using a diplexer architecture (Example 13.8).

The component values are determined by requiring the magnitude of the impedance looking into the lowpass section, $|j\omega L + Z_{in}|$, to be equal to the magnitude of the impedance looking into the highpass section, $|1/j\omega C + R|$, at a frequency ω_c roughly midway between the frequencies that must be passed and those that should be absorbed. Let us try $f_c = \sqrt{10.7 \cdot 139.3} = 38.6$ MHz; i.e., the geometric mean of the passband center frequency and the stopband frequency.

Finding a suitable value of L and C is relatively easy if we require $R = Z_{in}$, since then we have $\omega_c L = 1/\omega_c C$. Now we find:

$$LC = \frac{1}{\omega_c^2} = 1.70 \times 10^{-17}\ \text{H·F}$$

Values of L and C independently are determined by minimizing the embedded input reflection coefficient

$$\Gamma(\omega) = \frac{Z_i(\omega) - Z_{in}}{Z_i(\omega) + Z_{in}}$$

where

$$Z_i(\omega) = (j\omega L + Z_{in}) \parallel \left(\frac{1}{j\omega C} + Z_{in}\right)$$

Iterating over a few standard values of C, we quickly find $C = 82$ pF is nearly optimum. Thus, $L = LC/C = 207$ nH.

The resulting response is shown in Figure 13.20. The lowpass TPG and $|\Gamma(\omega)|$ are indicated in the traditional way as $|s_{21}|$ and $|s_{11}|$, respectively; both assuming reference impedance $Z_0 = Z_{in} = 50\ \Omega$. The return loss is greater than 50 dB throughout the indicated range, confirming that the filter is absorptive. The insertion loss is −0.4 dB and −11.6 dB at 10.7 MHz and 139.3 MHz, respectively. Thus, this diplexer reduces the LO feedthrough signal by 11.2 dB relative to the desired signal, without significant reflection from the input.

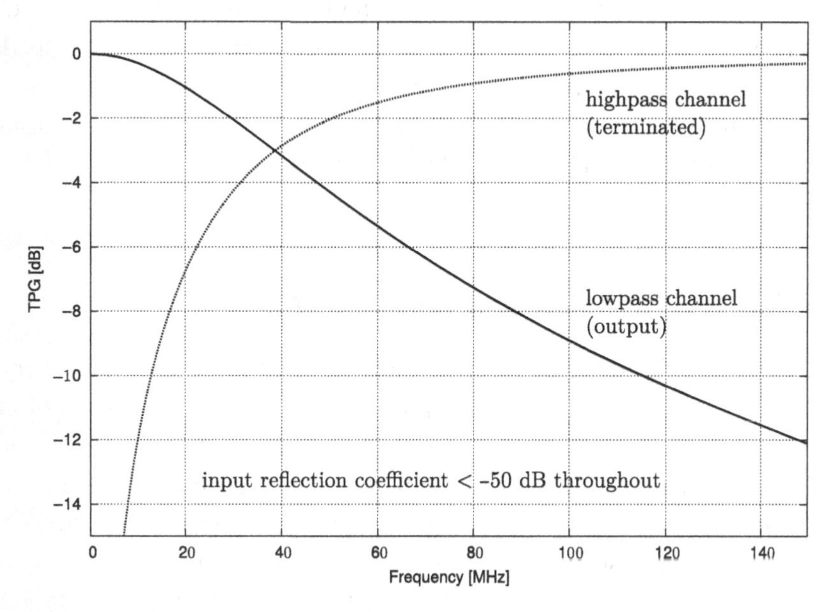

Figure 13.20. Performance of the diplexer shown in Figure 13.19 with $L = 207$ nH, $C = 82$ pF, and $R = 50\ \Omega$ (Example 13.8).

If additional selectivity is required from a reflectionless filter, then one might consider a cascade of identical diplexers, or a single diplexer of higher order. The principles of higher-order diplexer design are the same, but the specific methods are beyond the scope of this book. An easy-to-follow "recipe" for designing higher-order diplexers can be found in [69]. For recent work on this topic, see also [52].

A somewhat different use for a diplexer is to separate a segment of spectrum into two separate mutually-exclusive frequency ranges. This is achieved simply by using both outputs of the diplexer, as opposed to terminating one output into a matched load. This scheme has applications in preselector design (see Section 15.4), when it is necessary to separate the tuning range of a receiver into smaller segments that are easier to work with. This is typically superior to the alternative approach of using a two-way power splitter followed by completely independent filters, as the diplexer approach eliminates the minimum 3 dB insertion loss associated with the splitter. This is particularly important in receiver preselectors, in which the additional insertion loss may incur an unacceptable penalty on noise figure.

There is no particular reason that the number of channels processed in this manner must be limited to two. In fact, diplexers are the simplest example of a broad class of channelized filter structures known as *multiplexers*. The number of channels in a multiplexer may be arbitrarily large. (For an example of an application for a three-channel multiplexer, see Example 15.9.) For additional reading on diplexers and multiplexers, [50] is recommended.

13.7 DISTRIBUTED FILTER STRUCTURES

In Chapter 9 ("Impedance Matching") we found that discrete LC circuits were common at VHF and below, whereas distributed matching structures – in particular, microstrip lines – were common at SHF and above. At UHF, both technologies have applications, sometimes in hybrid combinations. These statements are also true for filters. Furthermore, the design of broad classes of distributed filter structures may be derived directly from the discrete LC filter designs that we have already considered. We will demonstrate this first, and then consider some additional tricks that become available in distributed filter structures.

13.7.1 Transmission Line Stubs as Single-Reactance Two-Ports

In Section 13.3, we considered single-reactance lowpass and highpass filters. Very similar filters can be achieved using open- and short-circuited stubs that are less than $\lambda/4$ in length, as shown in Figure 13.21. An open-circuited stub has a purely imaginary reactance given by Equation (9.28) and therefore behaves similarly to a shunt capacitor, or a series inductor; whereas a short-circuited stub has a purely imaginary reactance given by Equation (9.29) and therefore behaves similarly to a shunt inductor, or a series capacitor. This similarity comes with two important caveats. First, the reactance of an inductor or capacitor scales with ω and ω^{-1} respectively; whereas the impedance of open- and short-circuited stubs scales with $\cot(\omega l/v_p)$ and $\tan(\omega l/v_p)$ (where v_p is the phase velocity), respectively. Secondly, the fact that impedance of stubs depends on ω through a trigonometric function means that the response of a stub-type filter will be periodic, and only appears to be lowpass or highpass as long as $f < v_p/4l$. So, when using stubs to replace reactances we must account for the possibility of undesirable

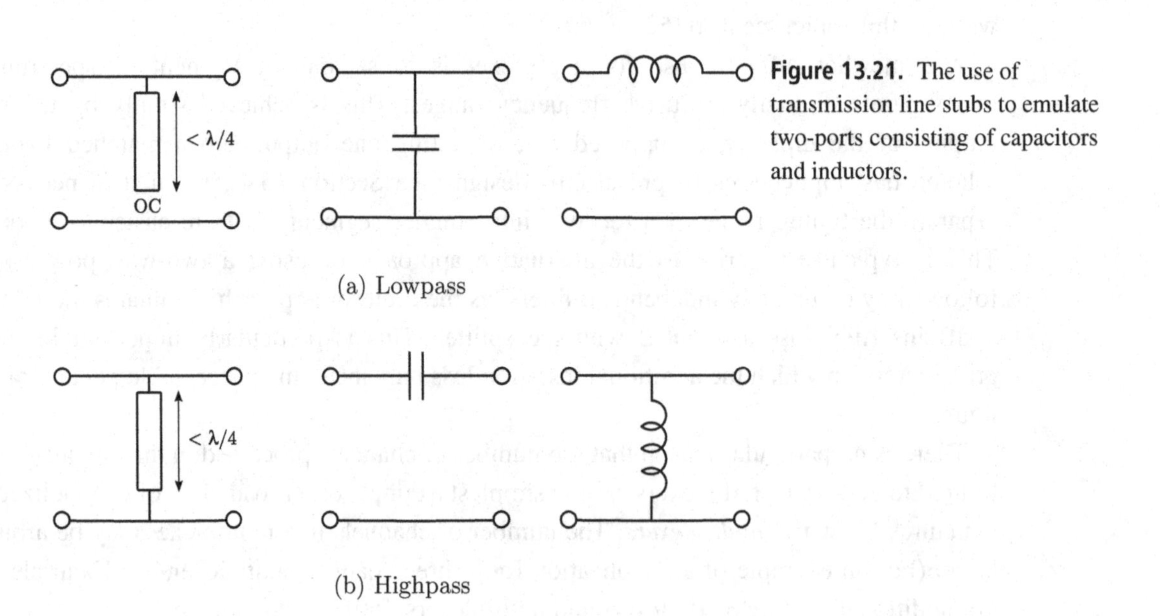

Figure 13.21. The use of transmission line stubs to emulate two-ports consisting of capacitors and inductors.

behavior at frequencies much greater than the intended cutoff frequency. Fortunately, it is often the case in radio design that some other part of the signal path is able to provide additional selectivity that makes this periodic behavior irrelevant.

EXAMPLE 13.9

Determine the TPG of a stub-type lowpass filter having TPG$= -3$ dB at a specified frequency $f_{co} = 3$ GHz. The input and output impedances of the filter are to be 50 Ω. The characteristic impedance of the transmission line comprising the stub is to be 50 Ω.

Solution: From the problem statement, $Z_{in} = 50\ \Omega$ and $Z_c = 50\ \Omega$. Lowpass response is provided by an open-circuited stub, which has an input impedance given by (Equation (9.28)):

$$Z_{stub}(l/\lambda) = -jZ_c \cot(2\pi l/\lambda)$$

The TPG is that of a shunt impedance; from Equation (8.25):

$$G_T = |s_{21}|^2 = 4\left|\frac{Z_{stub}}{2Z_{stub} + Z_0}\right|^2$$

for which TPG$= 1/2$ occurs when $|Z_{stub}| = Z_0/2$ since Z_{stub} is completely imaginary-valued. Therefore the desired stub length is

$$l = \frac{\lambda_{co}}{2\pi} \cot^{-1}\left(\frac{Z_0}{2Z_c}\right) = 0.176\lambda_{co}$$

where λ_{co} is the wavelength in the transmission line comprising the stub, at frequency f_{co}. The resulting response (TPG) is shown in Figure 13.22. Note that the selectivity is not great, since the 3 dB point is reached when the stub is $l = 0.176\lambda$, whereas the lowpass response only applies as long as $l < 0.25\lambda$.

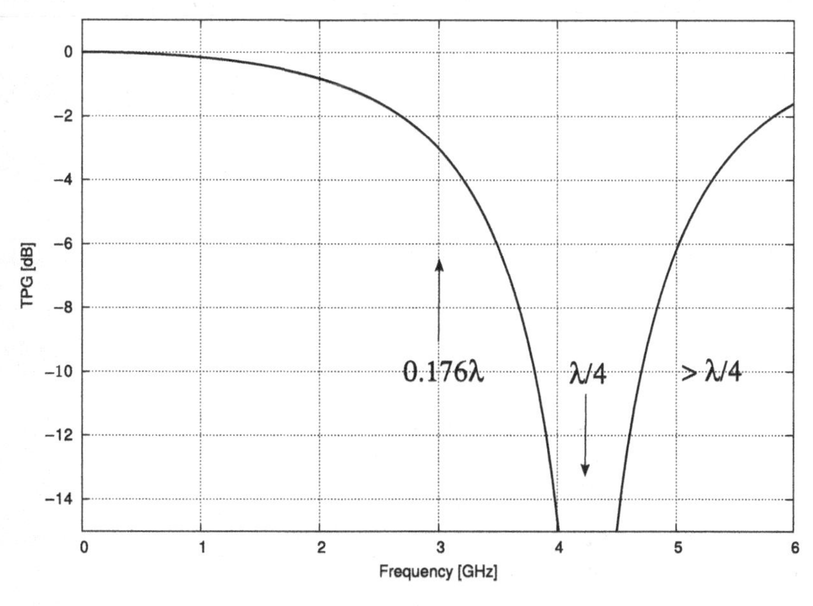

Figure 13.22. TPG of the stub-type lowpass filter from Example 13.9.

13.7.2 Quarter-Wave Stubs as Single-Resonance Two-Ports

Recall from Sections 9.7.4 and 9.8 that quarter-wave ($l = \lambda/4$) stubs have the property of impedance inversion. This is also apparent in Example 13.9, where we see the open-circuited stub behaves like a short-circuit – driving the TPG to zero – when the length reaches one-quarter wavelength. This behavior can be exploited to create transmission line filters that behave very much like the single-resonator bandpass and notch filters of Section 13.4. The concept is illustrated in Figure 13.23. A short-circuited quarter-wave stub presents an open-circuit at the design frequency, and presents an impedance less than that of an open-circuit away from the design frequency, and so behaves as a bandpass filter. Similarly, an open-circuited quarter-wave stub presents a short-circuit at the design frequency, and presents an impedance greater than that of a short-circuit away from the design frequency, and so behaves as a notch filter. As in the previous section, these structures have periodic frequency response; e.g., a quarter-wave stub used as a bandpass filter becomes a notch filter at twice the design frequency, and vice versa for quarter-wave stubs used as notch filters.

13.7.3 Filters Composed of Quarter-Wave Sections

There is an essentially endless number of schemes to combine transmission lines into useful filters. Here we will consider just one that emerges quite naturally from the designs presented in previous sections. We will have a bit more to say about the other possible ways in the next section.

The idea considered here is to construct the bandpass you desire by cascading quarter-wave stubs, with each stub representing one frequency to be either preserved or notched. Considering the findings of Sections 13.7.1 and 13.7.2, we do not need filters with broader bandwidth, but rather filters with either narrower bandwidth or greater selectivity. There are two ways to

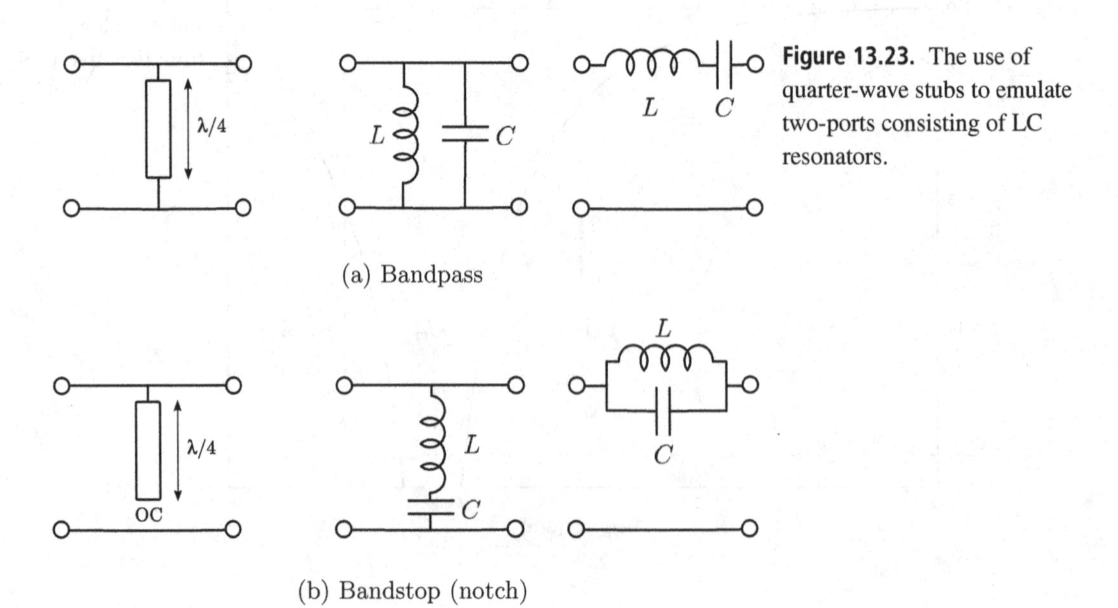

Figure 13.23. The use of quarter-wave stubs to emulate two-ports consisting of LC resonators.

achieve this. One way is simply to add quarter-wave stubs in parallel, as demonstrated in the following example:

EXAMPLE 13.10

Determine the TPG of a bandpass filter having center frequency 1.1 GHz consisting of (a) one quarter-wave stub, (b) two quarter-wave stubs in parallel. The input and output impedances of the filter are to be 50 Ω, and the characteristic impedance of the transmission line comprising the stub is to be 50 Ω.

Solution: The first filter will be a stub which is short-circuited and a quarter-wavelength long at 1.1 GHz. This stub has an input impedance given by (Equation (9.29)):

$$Z_{stub}(l/\lambda) = +jZ_c \tan(2\pi l/\lambda)$$

We worked out the TPG in terms of Z_{stub} in the previous example. The filter of part (b) is different only in that Z_{stub} is replaced by $Z_{stub} \parallel Z_{stub}$. The results are shown in Figure 13.24. Note that the parallel combination of stubs results in significantly reduced bandwidth, as expected.

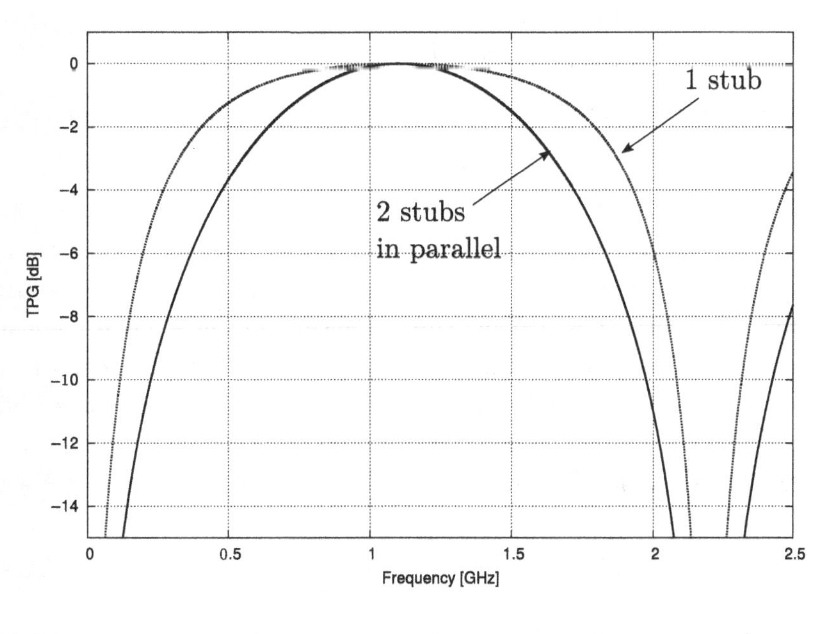

Figure 13.24. TPG of 1.1 GHz bandpass filters consisting of one quarter-wave stub and two quarter-wave stubs in parallel (Example 13.10).

The problem with putting stubs in parallel is that you quickly run out of space; i.e., there is no obvious location for additional stubs beyond two, should additional bandwidth reduction be desired. In particular, putting stubs on the same side of the transmission line and attached at the same point risks undesired interaction through electromagnetic coupling.

An alternative approach that permits any number of stubs is illustrated in Figure 13.25. Here, short-circuited quarter-wave stubs are separated by quarter-wave impedance inverters. At the design frequency, the input impedance into all stubs is infinite and the filter is essentially an uninterrupted transmission line. Away from the design frequency, the use of impedance

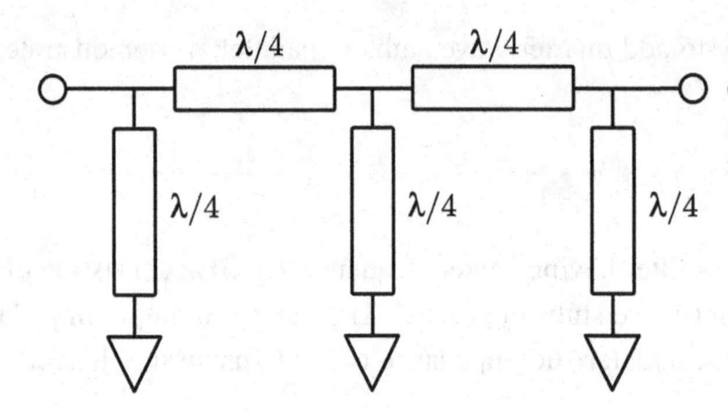

Figure 13.25. A bandpass filter composed of a cascade of three short-circuited quarter-wave stubs separated by quarter-wave impedance inverters.

inverters tends to flatten the passband and sharpen the transition bands. To see why this is, consider the impedance looking into the parallel combination of the last stub Z_{stub} and the nominal output impedance Z_{out}:

$$\frac{Z_{stub}(\omega)Z_{out}}{Z_{stub}(\omega)+Z_{out}} \tag{13.10}$$

In the neighborhood of the design frequency ω_c, the preceding quarter-wave section behaves as an impedance inverter, transforming this into the impedance

$$\approx Z_c^2\left(\frac{Z_{stub}(\omega)Z_{out}}{Z_{stub}(\omega)+Z_{out}}\right)^{-1} = \frac{Z_c^2}{Z_{out}}+\frac{Z_c^2}{Z_{stub}(\omega)} \tag{13.11}$$

If $Z_c = Z_{out}$, then this impedance is

$$Z_{out}+\frac{Z_{out}^2}{Z_{stub}(\omega)} \approx Z_{out} \tag{13.12}$$

since Z_{stub} is very large in the neighborhood of ω_c. In other words, the impedance inversion effect makes the last stub approximately "invisible" to the previous stub over the range of frequencies for which the stubs and the series transmission lines connecting them are sufficiently close to one-quarter wavelength. As a result, the TPG as a function of frequency varies quite slowly over this range. Once outside this range, impedance inversion is no longer effective and we are back to a situation in which the stubs are essentially in parallel, as in the previous example. This leads to a rapid transition to the stopbands.

EXAMPLE 13.11

Determine the TPG of a bandpass filter having center frequency 1.1 GHz using the three-stub topology of Figure 13.25. The input and output impedances of the filter are to be 50 Ω, and the characteristic impedance of the transmission line comprising the stub is to be 50 Ω.

Solution: The TPG in this case is most easily computed by determining the impedance looking into the input of the structure, calculating the associated reflection coefficient $\Gamma_{in}(\omega)$ with respect to 50 Ω, and then computing TPG as $1-|\Gamma(\omega)|^2$. The results are shown in Figure 13.26. Note that both selectivity and bandpass flatness is dramatically improved relative to either of the filters considered in Example 13.10.

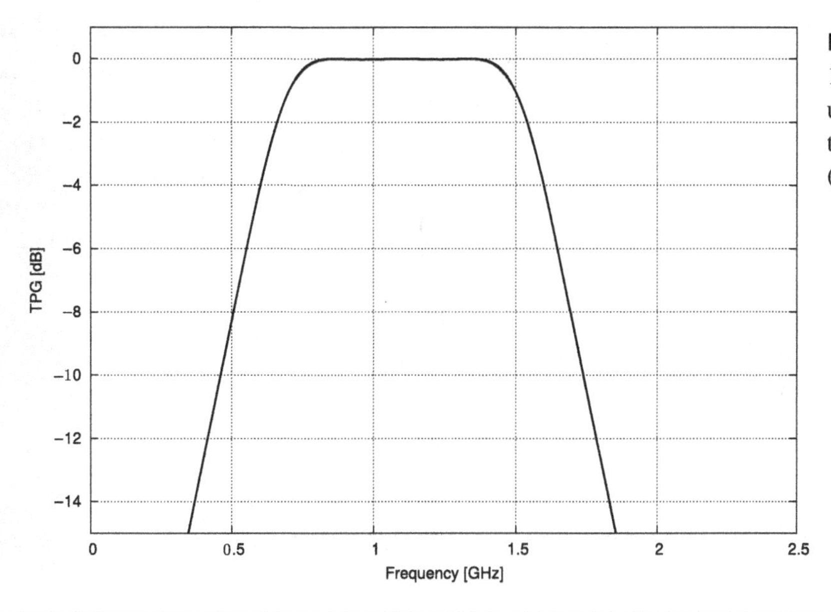

Figure 13.26. TPG of a 1.1 GHz bandpass filter using the three-stub topology of Figure 13.25 (Example 13.11).

The same technique can be used to incorporate open-circuit stubs intended to create nulls in the frequency response. Here is an example:

EXAMPLE 13.12

Modify the filter from Example 13.11 to incorporate a null at 450 MHz.

Solution: A null at 450 MHz can be implemented by using an open-circuited stub that is a quarter-wavelength long at 450 MHz. We simply concatenate this stub to the beginning of the structure, separated by another 1.1 GHz quarter-wave impedance inverter. The resulting design and TPG are shown in Figures 13.27 and 13.28, respectively. As expected, there is a null at 450 MHz. However, we also see a new null at 3 × 450 MHz = 1350 MHz – this is also as expected since we the new stub's input impedance alternates between short-circuit and open-circuit extrema with increasing frequency.

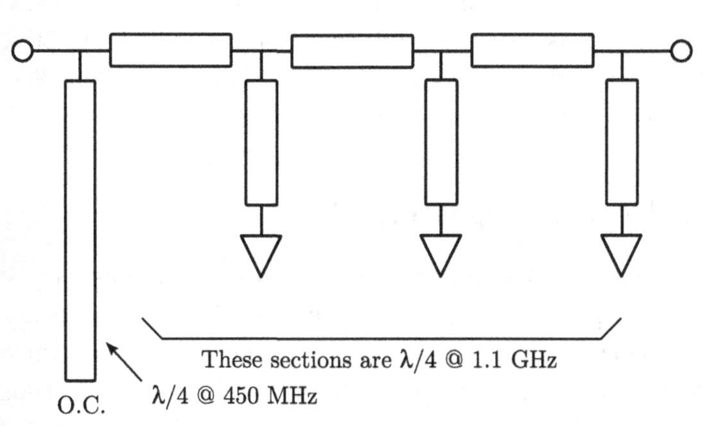

Figure 13.27. A 1.1 GHz bandpass filter modified to incorporate a 450 MHz null (Example 13.12).

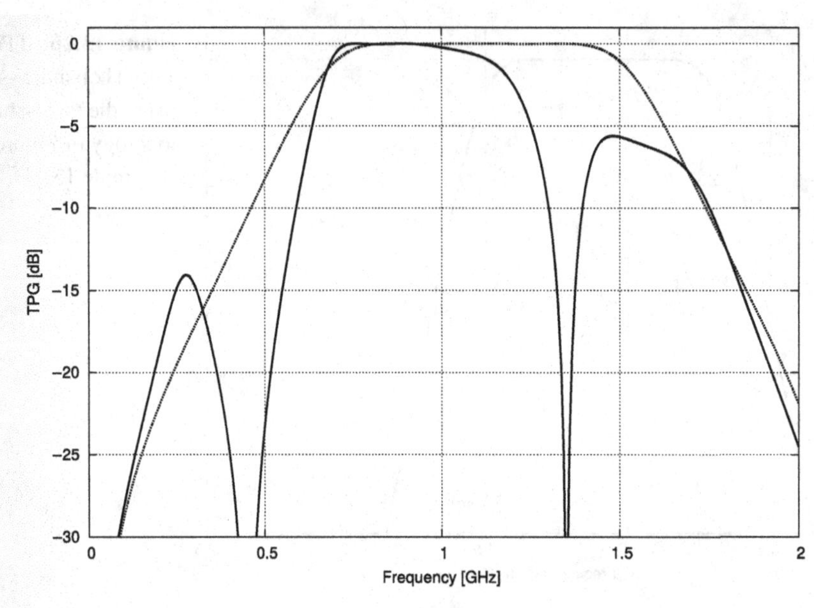

Figure 13.28. TPG of the filter structure shown in Figure 13.27 (Example 13.12). Shown for comparison is the TPG of the unmodified structure from Figure 13.26. Note change of scale.

13.7.4 Specified-Response Filters Using Transmission Line Stubs

A natural question to ask at this point is: Is there a way to realize a specified response – e.g., Butterworth, Chebyshev, etc. – as a distributed filter structure? The answer is a resounding yes, but this is a topic that requires significantly more background preparation than we can accommodate within the scope of this book. For a complete discussion including design techniques, recommended references include [50] (ancient, but seminal) and [58, Ch. 8]. Such techniques are not limited to the "stub and line" strategy described above; see, for example, Section 13.8.1 for two other ways to construct filters from microstrip lines.

13.8 OTHER FILTER DEVICE TECHNOLOGIES

So far in this chapter we have restricted ourselves to filters comprising inductors, capacitors, and transmission line stubs. This is hardly the full scope of possibilities. In principle, any structures that exhibit frequency-variable impedance can be used to create filters. What follows are a few alternative technologies that are quite common in modern radio systems.

13.8.1 Coupled Resonator and Stepped Impedance Filters

Figure 13.29 is an example of an *coupled resonator* filter. In this scheme there is no conducted flow of power from input to output; instead, power is transferred by electromagnetic coupling between the elements within the structure. This coupling is strongly frequency-dependent, and thus can be exploited to filter the signal. An important class of coupled resonator filters is known as *interdigital filters*, which are composed of multiple stub-like "fingers," which are arranged in parallel, as shown in Figure 13.29.

Figure 13.29. An L-band bandpass filter implemented using microstrip coupled resonators. The components on the left and right are surface mount capacitors, which are used to prevent DC shorting to ground.

Figure 13.30. A two-cavity 435 MHz helical bandpass filter mounted on a printed circuit board. The screws visible at the top of each cavity are used to tune the helical resonators.

Another way to implement filters in microstrip technology is the *stepped impedance* strategy. In this approach, discontinuities in the characteristic impedance of the transmission line emulate discrete inductors and capacitors.

A recommended reference for both the coupled resonator strategy and the stepped impedance strategy is [58, Ch. 8].

13.8.2 Helical Filters

Figure 13.30 is a *helical* filter, consisting of a cascade of cavities containing resonant wire helix structures. These filters appear most often in the upper VHF and lower UHF ranges; this is a regime in which discrete LC devices are becoming difficult to use, but in which microstrip technology is typically not yet sufficiently compact. Helical filters yield very high selectivity but are relatively expensive and not suitable for compact implementation.

13.8.3 Coaxial Filters

Figure 13.31 is an example of a *coaxial filter*, also known as *tubular* filter. Coaxial filters are composed of various combinations of inductive and capacitive structures that are analogous to the Ls and Cs in a discrete LC circuit. The common feature in these filters are external coaxial connectors, which makes these devices convenient to use in laboratory test fixtures and in

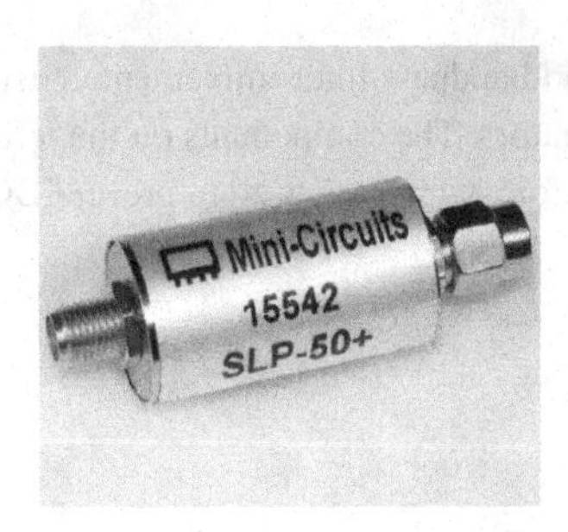

Figure 13.31. A commercially-available coaxial filter.

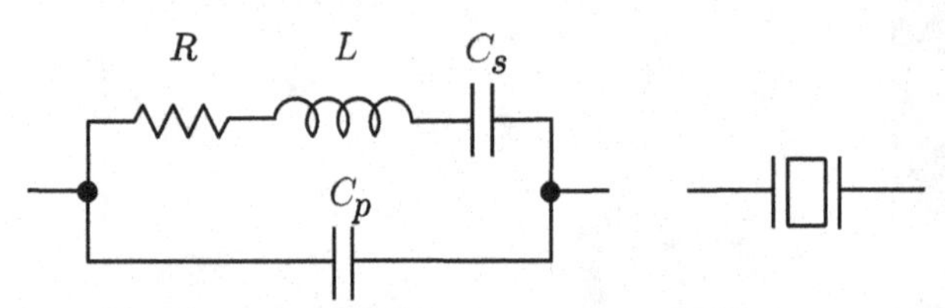

Figure 13.32. Equivalent circuit and schematic symbol for a crystal resonator. Typical values for a 10 MHz fundamental-mode crystal are $L_S \cong 9.8$ mH, $C_S \cong 0.026$ pF, $R \cong 7\ \Omega$, $C_P \cong 6.3$ pF [3, Sec. 9.5.3]. See [21, Table 4.1] for values for other frequencies and modes.

certain experimental and one-of-a-kind systems. This technology is effective from HF through SHF, but is expensive and not well-suited to compact design.

13.8.4 Crystal Filters

A *crystal filter* employs one or more *crystals*, which are resonators typically fashioned from quartz crystal material and having two leads, not unlike a capacitor or inductor. The crystal material exhibits a subtle mechanical vibration in response to electrical stimulus at frequencies determined by the molecular structure and orientation of the crystal. Each mechanical resonance is analogous to the electromagnetic resonance of the LC resonator. In fact a crystal resonator is well-modeled by the equivalent circuit shown in Figure 13.32. Note the component values mentioned in the caption of Figure 13.32: These are values that are quite difficult to achieve by using *actual* inductors and capacitors at RF frequencies, indicating the benefit of this "piezoelectric" approach to emulating electrical resonance. Commonly-available crystal resonators exhibit Q of 10 000 or greater with frequency accuracy on the order of 0.001% (10 ppm) or better and stability with respect to temperature on the order of 1 ppm/°C or better.

There is much that can be said about the design and performance of crystal resonators that is outside the scope of this book.[3] However, there are some basic principles and terminology that are useful to know when using crystals in filters. Crystal oscillators may be designed to use the *fundamental* resonances, or *overtones* (harmonics) of these resonances. Crystals designed to operate at the fundamental resonance are commonly available for frequencies from a few hundred kHz to tens of MHz. Crystals designed for higher frequencies typically use overtones, and are available for frequencies up to about 250 MHz. Owing to these frequency limitations, crystal filters are most commonly used to achieve the very high selectivity required in the low-frequency latter IF stages of superheterodyne radios (see Sections 15.6.4 and 15.6.6), and only when the signals of interest have very low fractional bandwidth.

[3] A more detailed overview of crystal filter technology can be found in [16, Secs. 4.32–4.36].

13.8.5 Surface Acoustic Wave Devices and Dielectric Resonators

At UHF and above, suitable alternatives to quartz crystal include *surface acoustic wave* (SAW) devices and *dielectric resonator* devices. SAW filters in particular provide very low shape factor (thus, high selectivity), albeit with relatively high loss. For this reason, SAW filters often appear in the early IF stages of certain receivers. A useful overview SAW technology appears in [16, Sec. 4.38].

13.8.6 Mechanical and Ceramic Filters

While discrete inductors and capacitors are easy to use at frequencies below HF, high-order filters require many such components, which creates different problems. However, it is possible to construct filters having very low shape factor and very good temperature stability in this frequency regime using mechanical resonance. These are known as *mechanical filters*. Mechanical resonance is similar to the materials-based resonance phenomena associated with the piezoelectric effect, but is *literally* mechanical – i.e., employing mechanical structures. A more compact and lower-cost alternative to mechanical filters is the *ceramic filter*, which is conceptually similar to the quartz crystal filter, but exhibits inferior selectivity and temperature stability. Mechanical and ceramic filters commonly appear as bandpass filters in the 455 kHz IF stages of certain HF receivers. For more information, useful starting points are [16, Secs. 4.31 and 4.37], respectively.

13.8.7 Electronically-Tunable Filters

In some receivers it is useful to be able to vary the center frequency of a bandpass filter. Such filters are sometimes referred to as *tracking filters*, implying that the center frequency of the filter follows the frequency to which the receiver is tuned (see Example 15.9 for a typical application). In an electronic filter composed of discrete inductors and capacitors, this requires varying the reactance of at least one of the components. A common scheme is to use varactors, which are diodes that exhibit capacitance that varies as the DC bias is varied. For more information on the design of electronically-tunable filters employing varactors, [43, Ch. 19] and [65, Sec. 3.15] are recommended.

Problems

13.1 Consider a filter two-port consisting of a single capacitor C in parallel with the input and output. The embedded input and output impedances Z_{in} are equal. (a) Derive the TPG cutoff frequency f_{co} in terms of C and Z_{in}. (b) Determine C when $f_{co} = 1$ MHz and $Z_{in} = 50\ \Omega$. (c) Check your work by plotting TPG vs. frequency for part (b).

13.2 Repeat Problem 13.1 for the following single-reactance filters: (i) series capacitor C, (ii) shunt inductor L, (iii) series inductor L.

13.3 Repeat Example 13.2 for a bandwidth of 50 kHz at 25 MHz.

13.4 Repeat Example 13.2 for the topology of Figure 13.4(b). Comment on any issues that might lead to selecting one topology over the other.

13.5 Design a notch filter for 27 MHz having (a) series-LC topology, (b) parallel-LC topology. In each case, plot TPG vs. frequency.

13.6 Design maximally-flat filters having 3 dB bandwidth of 6 MHz, $Z_{in} = 50\Omega$, $K = 7$, and the following response: (a) lowpass, (b) highpass, (c) bandpass with center frequency 21.4 MHz, and (d) bandstop with center frequency 63 MHz. Also, plot TPG vs. frequency to confirm the design.

13.7 Repeat Problem 13.6, except now for filters having equiripple response in the passband (or stopband, in the case of the bandstop filter). Let $r = 0.5$ dB.

13.8 Design the simplest possible diplexer that separates frequencies above 30 MHz from frequencies below 1.6 MHz. Use a diplexer topology with components having standard values. The nominal input/output impedance is $Z_{in} = 50\ \Omega$. Plot insertion loss for each output as well as return loss as a function of frequency.

13.9 (a) Design a stub-type filter that presents a notch at 3 GHz and can be implemented on a printed circuit board. The input and output impedances are to be 50 Ω. The characteristic impedance of the transmission line should be 50 Ω and the effective relative permittivity of the circuit board material is 2.5. (b) Plot TPG vs. frequency, indicating 3 dB and 30 dB points.

13.10 Design a stub-type bandpass filter with center frequency 2 GHz, to be implemented on FR4 circuit board material (see Section 9.7.1). The input and output impedances are to be 50 Ω. Consider designs consisting of (a) one stub, (b) two stubs in parallel, (c) two stubs separated by quarter-wave impedance inverters, and (d) three stubs separated by impedance inverters. In each case, provide a detailed diagram of the design, plot TPG vs. frequency, and determine 3 dB bandwidth.

14 Frequency and Quadrature Conversion in the Analog Domain

14.1 INTRODUCTION

It is often necessary for radios to process radio signals at *intermediate frequencies* (IFs) that are lower than the frequencies at which they are transmitted or received. In fact, it may be necessary to shift frequencies several times (i.e., employ multiple IFs), and it may be convenient or necessary for the lowest IF to be zero, resulting in a complex-valued "baseband" signal. This chapter addresses the problem of frequency conversion (conversion between RF and non-zero IFs) and quadrature conversion (conversion between complex baseband and RF or a non-zero IF). As the title indicates, this chapter is concerned specifically with *analog* methods of frequency and quadrature conversion. If we have access to a digital version of the signal, then additional considerations – most of them good news – come into play, as we shall see in Chapter 18.

This chapter is organized as follows. Section 14.2 provides a brief description of the theory of frequency conversion for physical real-valued signals. The essential device for accomplishing this is the mixer, which is described in Section 14.3. In Section 14.4 we describe the theory and implementation of quadrature conversion. Section 14.5 describes image rejection mixers, a special class of mixers that are designed to overcome the image response problem identified in Section 14.2.

14.2 FREQUENCY CONVERSION

Ideal frequency conversion entails multiplication of the input signal by a tone, commonly known as the *local oscillator* (LO), followed by filtering of undesired products. This is shown in Figure 14.1. The principle is easily demonstrated for an input signal that is itself a tone, which is simple to analyze and is a reasonable surrogate for any signal of small fractional bandwidth. Let the input signal be $x(t) = A\cos\omega_i t$, and let the LO be $B\cos\omega_L t$. Multiplication of $x(t)$ with the LO yields

$$y(t) = AB\cos\omega_i t\cos\omega_L t \tag{14.1}$$

and we may use the appropriate trigonometric identity to obtain

$$y(t) = \frac{AB}{2}\cos\left(\omega_i + \omega_L\right)t + \frac{AB}{2}\cos\left(\omega_i - \omega_L\right)t \tag{14.2}$$

Note that $y(t)$ consists of two "copies" of $x(t)$: One is shifted to the frequency $\omega_i + \omega_L$ and scaled by $B/2$, and the other is shifted to the frequency $\omega_i - \omega_L$ and scaled by $B/2$. Therefore

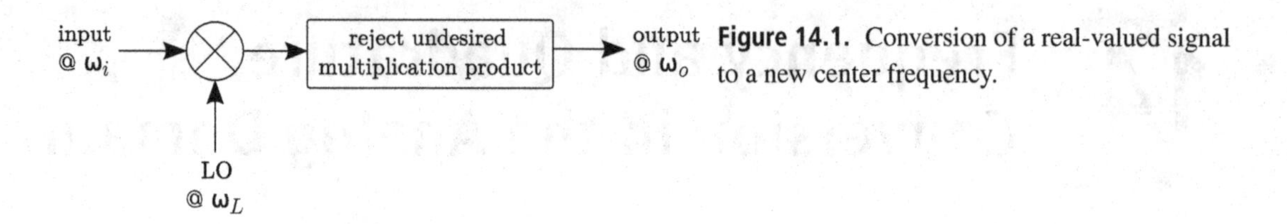

Figure 14.1. Conversion of a real-valued signal to a new center frequency.

frequency conversion is achieved by selecting ω_L such that one of these products is at the desired frequency ω_o, and then $y(t)$ is filtered so as to exclude the other product.

14.2.1 Downconversion; Low- and High-Side Injection

If the intent is for $\omega_o < \omega_i$, then this process is commonly known as *downconversion*. There are two choices for ω_L that will achieve the desired downconversion. The first possibility is simply $\omega_L = \omega_i - \omega_o$. In this case the center frequency of the "$\omega_i - \omega_L$" term in Equation (14.2) is ω_o; whereas the center frequency of the "$\omega_i + \omega_L$" term in Equation (14.2) is $2\omega_i - \omega_o$. Thus the desired output is obtained by lowpass filtering the output of the multiplier with a cutoff frequency $\omega_c < 2\omega_i - \omega_o$. The LO frequency is less than the input frequency in this scheme (i.e., $\omega_L < \omega_i$), thus it is common to refer to this choice of LO frequency as *low-side injection*.

The alternative approach is known as *high-side injection*: We choose $\omega_L = \omega_i + \omega_o$; i.e., $\omega_L > \omega_i$. In this case the center frequency of the "$\omega_i - \omega_L$" term in Equation (14.2) is $-\omega_o$, whereas the center frequency of the "$\omega_i + \omega_L$" term in Equation (14.2) is $2\omega_i + \omega_o$. What does it mean for the signal to be shifted to a frequency of $-\omega_o$ as opposed to $+\omega_o$? If the input signal is a tone, then there is no difference at all, since $\cos(-\omega_o t) = \cos \omega_o t$. If the input signal is not a tone, then the bandwidth B of the signal is greater than zero. Similarly the low-frequency edge of the signal spectrum, at $\omega_i - B/2$, is shifted to $-\omega_o - B/2$, which appears at $+(\omega_o + B/2)$ in the positive-frequency half of the spectrum since real-valued signals have conjugate-symmetric spectrum. The high-frequency edge of the signal spectrum, at $\omega_i + B/2$, is shifted to $-\omega_o + B/2$, which appears at $+(\omega_o - B/2)$ in the positive-frequency half of the spectrum. In other words, the positive-frequency spectrum of the output signal is centered $+\omega_o$ as desired, but is "flipped" with respect to the spectrum of the input signal. If the spectrum of the input signal is symmetric (e.g., an AM signal), then there is no concern. However, most modern communications signals do not have symmetric spectrum, and therefore this spectral flip must often be taken into account in subsequent processing.

Given that high-side injection flips the spectrum, why would anyone ever consider this over low-side injection? Part of the reason is that the undesired product of the multiplication appears at $2\omega_i + \omega_o$, as opposed to $2\omega_i - \omega_o$ for low-side injection. Therefore, high-side injection increases the spectral separation between the desired and undesired products, which makes it possible to achieve a specified level of suppression with a lower-performance filter. Also, we will see in Section 14.3.1 that practical implementations result in the LO leaking into the output. High-side injection puts this leakage at a higher frequency than low-side injection, again making it possible to achieve a higher-level of suppression for a given lowpass filter.

EXAMPLE 14.1

It is desired to downconvert a signal centered at 150 MHz to 10.7 MHz. The 99% bandwidth of the signal is 30 kHz. Specify the LO frequencies and filter requirements for (a) low-side injection and (b) high-side injection.

Solution: Here, $f_i \equiv \omega_i/2\pi = 150$ MHz, $f_o \equiv \omega_o/2\pi = 10.7$ MHz, and $B = 30$ kHz.
(a) For low-side injection, $f_L = f_i - f_o = 139.3$ MHz. This will result in multiplication products at 10.7 MHz (desired) and $2f_i - f_0 = 289.3$ MHz, so the filter must be lowpass with sufficiently large attenuation above $289.3 - B/2 = 289.285$ MHz. Recognizing that LO leakage is likely to be a problem in practical implementations, a cutoff frequency much greater than 10.7 MHz and much less than 139.3 MHz would be a wise choice.
(b) For high-side injection, $f_L = f_o + f_i = 160.7$ MHz. This will result in multiplication products at 10.7 MHz (desired, but spectrally flipped) and $2f_i + f_0 = 310.7$ MHz, so the filter must be lowpass with sufficiently large attenuation above $310.7 - B/2 = 310.685$ MHz. Again considering the likelihood of LO leakage, a cutoff frequency much greater than 10.7 MHz and much less than 160.7 MHz would be a wise choice.

How might we go about implementing this filtering? To answer this question, see Example 13.8, which addresses this very issue.

The phrase "sufficiently large attenuation" in the solution of course begs the question "How much attenuation is sufficient?" The answer is that this is highly dependent on the application and the details of the subsequent processing, but typically one seeks suppression of the undesired mixing product by 30 dB or greater, and suppression of LO leakage by an amount sufficient to make it comparable or negligible in magnitude compared to other spurious signals appearing in the output – more on that in Section 14.3.

14.2.2 Upconversion

If the intent is for $\omega_o > \omega_i$, then we refer to this process as *upconversion*. Once again the possibilities are low-side and high-side injection. For low-side injection, $\omega_L = \omega_o - \omega_i$. In this case the center frequency of the "$\omega_i + \omega_L$" term in Equation (14.2) is ω_o; whereas the center frequency of the "$\omega_i - \omega_L$" term in Equation (14.2) is $2\omega_i - \omega_o$. Thus the desired output is achieved by highpass filtering the output of the multiplier with a cutoff frequency ω_c that is much greater than $|2\omega_i - \omega_o|$ and much less than ω_o, perhaps also accounting for the possibility of significant LO leakage.

For high-side injection, $\omega_L = \omega_o + \omega_i$. In this case the center frequency of the "$\omega_i - \omega_L$" term in Equation (14.2) is $-\omega_o$ (thus, appearing as flipped spectrum at $+\omega_o$); whereas the center frequency of the "$\omega_i + \omega_L$" term in Equation (14.2) is $2\omega_i + \omega_o$. Thus the desired output is achieved by lowpass filtering the output of the multiplier with ω_c much greater than ω_o and much less than $2\omega_i + \omega_o$, perhaps also accounting for significant LO leakage.

EXAMPLE 14.2

It is desired to upconvert a signal centered at 10.7 MHz to 150 MHz. The 99% bandwidth of the signal is 30 kHz. Specify the LO frequencies and filter requirements for (a) low-side injection and (b) high-side injection.

Solution: Here, $f_i \equiv \omega_i/2\pi = 10.7$ MHz, $f_o \equiv \omega_o/2\pi = 150$ MHz, and $B = 30$ kHz.
(a) For low-side injection, $f_L = f_o - f_i = 139.3$ MHz. This will result in multiplication products at 150 MHz (desired) and $|2f_i - f_O| = 128.6$ MHz, so the filter should be highpass with a cutoff frequency between $128.6 + B/2 = 128.615$ MHz and $150 - B/2 = 149.985$ MHz. Recognizing that LO leakage may be a problem in practical implementations, a cutoff frequency greater than 139.3 MHz and less than 149.985 MHz would be a wise choice.
(b) For high-side injection, $f_L = f_o + f_i = 160.7$ MHz. This will result in multiplication products at 150 MHz (desired, but spectrally flipped) and $2f_i + f_O = 310.7$ MHz, so the filter should be lowpass with a cutoff frequency greater than 150.015 MHz and less than 310.685 MHz. If we also need to suppress LO leakage, then a cutoff frequency less than 160.7 MHz would be a wise choice.

14.2.3 Image Frequency

We have seen above that the basic strategy for downconversion is to select ω_L such that $|\omega_i - \omega_L| = \omega_o$, and that generally we have two possible choices of ω_L that accomplish this, corresponding to low-side and high-side injection. However, for any given choice of LO frequency, we also find that one other input frequency – different from ω_i – is also converted to ω_o. This additional frequency is known as the *image frequency*. The image frequency for downconversion is a particular concern, because signals at the image frequency will interfere with signals downconverted from ω_i at the output. This is easiest to see by example:

EXAMPLE 14.3

Returning to Example 14.1: We found that downconversion from 150 MHz to 10.7 MHz by low-side injection required an LO frequency of 139.3 MHz. What image frequency should we be concerned about?

Solution: The image frequency is 128.6 MHz: 128.6 MHz – the LO frequency (139.3 MHz) = −10.7 MHz. In other words, a signal centered at 128.6 MHz appears at the output at 10.7 MHz and is spectrally-inverted.

Generalizing the above example, we find that for any downconversion the image frequency ω_i' is given by

$$\omega_i' = \omega_i \pm 2\omega_L \tag{14.3}$$

where the "+" and "−" signs correspond to high-side and low-side injection, respectively.

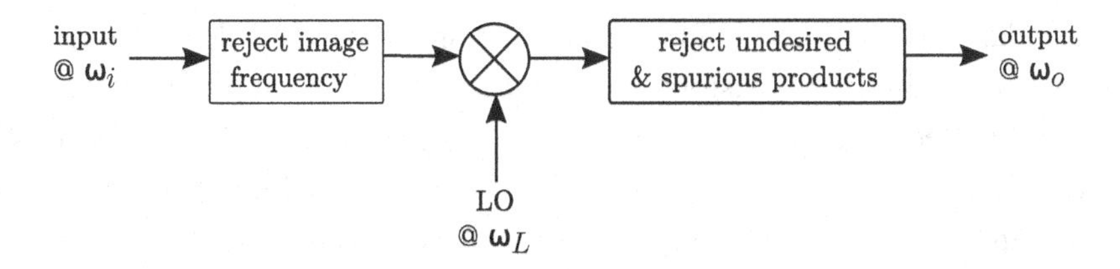

Figure 14.2. Frequency downconversion, accounting for the possibility of signals at the image frequency and spurious products from the multiplier.

Clearly, it is important that signals at the image frequency be suppressed before reaching the mixer. This is typically done using a highpass (for LSI) or lowpass (for HSI) before the mixer. Thus, the "recipe" for frequency downconversion, shown in Figure 14.2, becomes: (1) image rejection, (2) multiplication, and then (3) rejection of the undesired mixing product and spurious products including LO bleedthrough, harmonics, and intermodulation.

Image frequency response is a particular concern for downconversion, because receivers typically capture undesired signals as a well the signal of interest at the input. Image frequency response may also be a concern for upconversion, but is usually a lesser concern because transmitters, which typically upconvert, do not normally need to contend with undesired signals originating from outside the system. An important case where image frequency response *is* an important consideration in upconversion is that of "up-down" superheterodyne receivers, which upconvert (as opposed to downconverting) as the first step in processing the signal. (This class of receivers is discussed in Section 15.6.7.)

14.3 MIXERS

We have seen in the previous section that the essential element in frequency conversion is multiplication. An RF device that performs multiplication is commonly referred to as a *mixer*. Explicit, literal multiplication of analog RF signals is difficult; typically we instead use methods that generate approximations to the desired sum-and-difference multiplication products as a "side effect", and unfortunately produce additional undesired products as a consequence. The two methods commonly used to implement mixers are square-law processing and phase switching, discussed in Sections 14.3.1 and 14.3.2, respectively.

Before we get started, be advised that mixer analysis and design is a very broad and complex topic, and thus one which is impossible to fully address within the scope of this chapter. For additional reading, [19, Ch. 7], [48], and [63, Ch. 6] are suggested.

14.3.1 Square-Law Processing

One way to obtain the product of two RF signals is to sum them and apply the sum to a square-law device. A square-law device is a two-port in which the output $y(t)$ includes a term equal to $a_2x^2(t)$, where $x(t)$ is the input and a_2 is a constant. Recall from Chapter 11 that all but perfectly-linear devices have this property, and in particular diodes and transistors have this property. In particular, forward-biased diodes and FETs are square-law devices to a very good approximation, whereas bipolar transistors can be biased to behave approximately as square-law devices over a limited range of magnitudes. Summing an input RF signal with an LO is not particularly difficult, yielding $x(t) = A\cos\omega_i t + B\cos\omega_L t$. Applying this as the input to a square-law device and neglecting all but the "$x^2(t)$" term, we obtain:

$$\begin{aligned} y(t) &= a_2\left(A\cos\omega_i t + B\cos\omega_L t\right)^2 \\ &= a_2 A^2\cos^2\omega_i t + a_2 B^2\cos^2\omega_L t + 2a_2 AB\cos\omega_i t\cos\omega_L t \end{aligned} \quad (14.4)$$

This expression can be converted into the sum of tones by applying the appropriate trigonometric identities to each term. One obtains:

$$\begin{aligned} y(t) = &\frac{a_2\left(A^2+B^2\right)}{2} + \frac{a_2A^2}{2}\cos 2\omega_i t + \frac{a_2B^2}{2}\cos 2\omega_L t \\ &+ a_2AB\cos\left(\omega_i+\omega_L\right)t + a_2AB\cos\left(\omega_i-\omega_L\right)t \end{aligned} \quad (14.5)$$

Now note that the last two terms constitute a magnitude-scaled version of Equation (14.2) – therefore, we have achieved multiplication of input signal with the LO. The first term in the above expression is simply a DC term, which can be either ignored or filtered out. The second and third terms represent tones at $2\omega_i$ and $2\omega_L$ – i.e., second harmonics of the input and LO frequencies – and are potentially problematic. If sufficient space exists between these harmonics and the spectrum of the desired output signal, then it is possible to suppress them by filtering, just as the undesired product of the ideal multiplication can be suppressed.

In practice, the situation is a little more complicated. Recall from Chapter 11 that the voltage (or current, or transimpedance) characteristic of most devices can be sufficiently well-modeled using the memoryless polynomial representation of Equation (11.3):

$$y(t) = a_0 + a_1x(t) + a_2x^2(t) + a_3x^3(t) + \cdots \quad (14.6)$$

with all coefficients (a_0, a_1, a_2, a_3, and so on) being potentially non-zero. A practical square-law device must have relatively large a_2, but is likely to have non-negligible a_0, a_1, and a_3 as well. Non-negligible a_0 is of no particular concern, since it simply contributes to the DC component of the output and is easily suppressed. Non-negligible a_1 results in scaled versions of the input and LO signals appearing in the output, at the original frequencies, leading to the leakage problem noted earlier in this chapter. Non-negligible a_3 results in IM3 products at frequencies equal to $|2\omega_i - \omega_L|$ and $|2\omega_L - \omega_i|$; i.e., additional spurious products to worry about.

Figure 14.3 shows a scheme for implementing a square-law mixer in hardware. The input (RF or IF) and LO are first summed using a diplexer (see Section 13.6), and then processed through a square-law device. From the perspective of the input signal, the diplexer is a high-pass filter; from the perspective of the LO signal, the diplexer is a low-pass filter; and the signals are simply summed at the output. This scheme limits the LO power propagating backwards into the input RF path, and limits the input RF power propagating backwards into the LO path. The square-law device is typically an appropriately-biased diode or transistor. The output must be filtered to reject the undesired multiplication product and any additional spurious. A circuit of

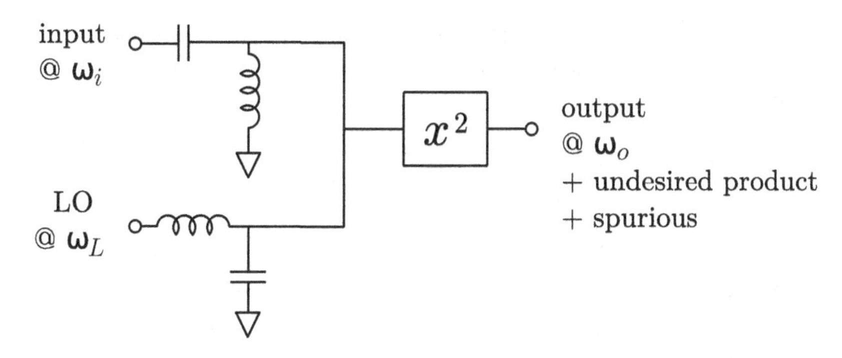

Figure 14.3. One possible implementation of a square-law mixer; in this case, configured for low-side injection.

this type which requires external power (e.g., for biasing square-law devices or to increase LO power) is often referred to as an *active mixer*.

14.3.2 Phase-Switching

The square-law approach to multiplication suffers from copious spurious output. An alternative strategy is phase switching: Simply inverting the phase of the input signal at a rate equal to twice the desired LO frequency. To see how this works, let us again consider the input tone $x(t) = A \cos \omega_i t$. Phase switching is implemented as follows: For a time interval of length $T/2 = \pi/\omega_L$, the output $y(t)$ is taken to be $x(t)$, without modification. For the next time interval of the same length, the $y(t)$ is taken to be $-x(t)$. In the next interval, $y(t)$ is back to $+x(t)$. Thus, $y(t)$ is simply $x(t)$, with a sign change every $T/2$. To see how this is related to mixing, let us first recognize that the output of the phase switching device can be written in the form

$$y(t) = x(t)w(t) \tag{14.7}$$

where $w(t)$ is a square wave, defined to have a value of $+1$ for an interval of $T/2$, and a value of -1 for an interval of $T/2$, repeating to make a square wave with period T. This is useful because a square wave can be conveniently represented as a Fourier series:

$$w(t) = \frac{4}{\pi} \sum_{n=1,3,5,...}^{\infty} \frac{1}{n} \sin\left(\frac{2\pi n}{T} t\right) = \frac{4}{\pi} \sum_{n=1,3,5,...}^{\infty} \frac{1}{n} \sin(n\omega_L t) \tag{14.8}$$

Substituting this into Equation (14.7) and separating out the first term, we obtain

$$y(t) = \frac{4A}{\pi} \cos(\omega_i t) \sin(\omega_L t) + \frac{4A}{\pi} \sum_{n=3,5,7,...}^{\infty} \frac{1}{n} \cos(\omega_i t) \sin(n\omega_L t) \tag{14.9}$$

The first term above is the desired multiplication – the fact that the LO is represented as sine as opposed to cosine is of no consequence, since they are different by a constant phase shift. The remaining terms represent spurious products at frequencies of $3\omega_L \pm \omega_i$, $5\omega_L \pm \omega_i$, and so on. These spurious products are relatively easily dealt with since they are widely separated from the desired product ($\omega_i - \omega_L$ or $\omega_i + \omega_L$) and have magnitudes proportional to $1/n$; i.e., are weaker with larger n.

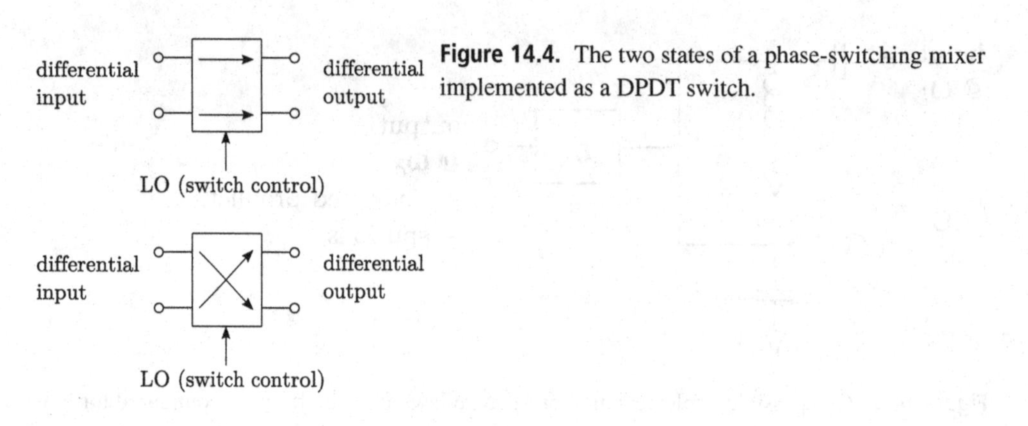

Figure 14.4. The two states of a phase-switching mixer implemented as a DPDT switch.

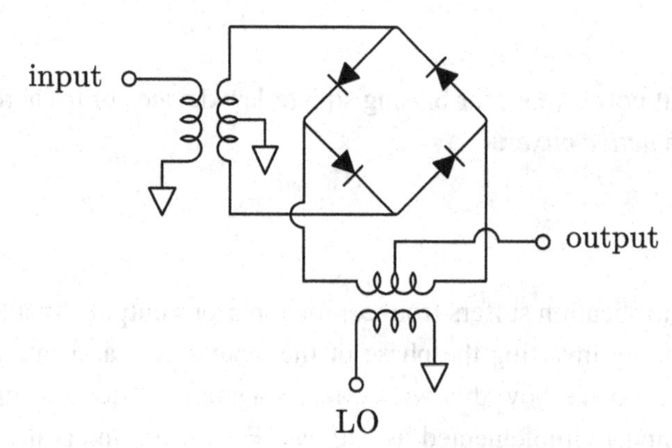

Figure 14.5. A double-balanced diode ring mixer.

Phase switching is straightforward to implement when the input and output signals are in differential form: Essentially all one needs is a sufficiently-fast double-pole, double-throw (DPDT) switch, as illustrated in Figure 14.4. The DPDT switch is used to periodically swap the connections between input and output, which inverts the phase of the signal. Note that this does not work for signals in single-ended form, and so we have yet another reason to be enthusiastic about employing differential signals.

14.3.3 Double-Balanced Diode Ring Mixers

The DPDT switch necessary for phase switching can be constructed from semiconductor devices such as diodes and transistors, and is quite suitable for integrated circuit implementation, as we shall see in Section 14.3.4. First, however, let us consider the nearly ubiquitous passive implementation of the phase-switching multiplier: the *double-balanced diode ring mixer*. In this scheme, shown in Figure 14.5, the diode ring is the DPDT switch, and balun transformers are used to convert the input (RF/IF) and LO signals from single-ended to differential form. The basic idea is that the LO provides the switching action by alternately forward- and reverse-biasing pairs of diodes, leading the input to be delivered to the output either as-is or phase-reversed.

In order for the LO to effectively control the diode bias, the LO magnitude must be quite large. Typically this scheme requires at least 4 dBm (with respect to a specified input impedance of 50 Ω) of LO power to function,[1] and typical mixers are specified for values between +7 dBm and +23 dBm.[2] Generally, spurious performance (essentially, a_2 relative to a_1 and a_3) improves with increasing LO level, resulting in lower-power spurious signals – in particular, LO leakage and IM3 – at the output.

It is useful to be able to characterize the gain, noise figure, and linearity of double-balanced diode-ring mixers, so that they may be considered in multistage analyses of the type described in Section 11.5. Here we go.

Gain. The first consideration in sorting this out is to realize that frequency conversion is non-linear, and specifically we need to be careful about how we interpret power associated with undesired mixing products. To make clear what we mean, we define the *conversion gain* to be the power in the desired output product relative to the power applied to the input.[3] Also, we define the conversion gain to be with respect to perfectly matched port impedances, which are usually (but not always) specified to be 50 Ω, single-ended. Therefore, the conversion gain of *any* passive mixer can never be greater than −3 dB, since we lose at least half the power by disposing of the undesired multiplication product. For double-balanced diode ring mixers, the maximum conversion gain is approximately −4 dB, and the conversion gain of practical devices is typically in the range −5 dB to −7 dB due to losses in internal baluns and other non-ideal characteristics.

Noise Figure. The noise figure of a double-balanced diode ring mixer is usually about 6 dB. One might guess this from the known property that the noise figure of passive devices is equal to their insertion loss when their physical temperature is equal to the noise figure reference temperature T_0. However, that is not quite the reason in this case. The main reason is that conversion loss is due to *redirection* of power, which is ideally noiseless; whereas insertion loss is due to *dissipation* of power, which generates noise. In fact, the noise arriving at the input of a double-balanced diode ring mixer is distributed across two multiplication products just as the input signal is; on this basis alone one might conclude that the noise figure should actually be closer to 0 dB, since it is not obvious that the signal-to-noise ratio should change from input to output. The actual situation is that the noise figure for this particular scheme is determined by the noise properties of the diodes and the circuit topology, resulting in a noise figure that turns out to be slightly (typically about 0.5 dB) higher than the conversion loss.

Linearity. The intercept point for double-balanced diode ring mixers is determined primarily by the LO power. The IIP_3 is typically found to be about 9 dB greater than the LO power. This provides a strong incentive to increase LO power; on the other hand, higher LO power means greater power consumption in generating the LO.

Note that 1 dB compression points vary considerably, but a typical value is 0 dBm referred to input.

Other considerations for mixers, and for double-balanced diode ring mixers specifically, include LO isolation and VSWR. LO isolation refers to the amount of LO power that "bleeds

[1] Problem 14.5 asks you to figure out what makes 4 dBm (or 7 dBm) the threshold.

[2] A diode-ring mixer that is specified for an LO level of P dBm is sometimes referred to as a "Level P" mixer.

[3] At the risk of confusing the issue, this is sometimes referred to the "single-sideband conversion gain," and the gain defined with respect to the power appearing in both multiplication products is sometimes referred to as the "double-sideband conversion gain."

through" into the RF or IF ports. VSWR is measured with respect to the nominal (specified) port impedance; it is not uncommon for hardware mixers to have VSWR greater than 2:1.

EXAMPLE 14.4

The Mini-Circuits ADE-1 surface-mount double-balanced diode ring mixer is to be used to for high-side downconversion of a signal from 150 MHz to 30 MHz. At the nominal LO power of +7 dBm, the performance of the ADE-1 is approximately as follows: Conversion loss: 5 dB, IIP_3: 15 dBm, IP_1: 1 dBm, LO-to-RF isolation: 55 dB, LO-to-IF isolation: 40 dB, and RF-to-IF isolation: 24 dB. Assume the input signal is a pair of tones located 150 MHz ± 30 kHz, both having power 10 dB below IP_1. Describe the output of the mixer, including spurious signals.

Solution: The RF input signal will be applied at $1 - 10 = -9$ dBm. The desired multiplication product at 30 MHz and undesired multiplication product at $150 + 180 = 330$ MHz will appear at about $-9 - 5 = -14$ dBm at the output. LO leakage will appear in the output as a tone at 180 MHz having power $+7 - 40 = -33$ dBm. RF leakage will appear in the output as a tone at 150 MHz having power $-9 - 24 = -33$ dBm. The limited IP_3 of the mixer will result in IM3 products at 30 MHz ± 90 kHz. The RF input level of -9 dBm is 24 dB below the IIP_3 of 15 dBm, so the IM3 level will be about $15 - 3 \cdot 24 = -57$ dBm. Additional spurious signals are, of course, not precluded, although the worst of it should be as indicated above.

14.3.4 IC Implementation

As we have noted previously, challenges in IC implementation of analog RF functions include difficulty in realizing inductive reactance, distributed impedances, and conversions between single-ended and differential representations of signals. There is no particular additional difficulty in implementing mixers and devices that include mixers, although the constraints associated with IC implementation have given rise to a few specific schemes which account for most of the designs in common use today.

In this section we will limit ourselves to consideration of two schemes that are particularly well-suited to IC implementation. The first scheme is shown in Figure 14.6, which is an active bipolar mixer. In this scheme, the input is single-ended and controls the collector current of transistor Q_1. Transistors Q_2 and Q_3 are used as switches: The LO is differential and serves to turn Q_2 on while turning Q_3 off, and vice versa. The collector resistors of Q_2 and Q_3 serve as "pull up" resistors, which cause the voltage between the output pins to alternate signs as the LO turns the transistors on and off. The output is therefore differential, proportional to current applied to the base of Q_1, and phase-switched under the control of the LO. It should be noted that there is no particular difficulty in implementing the same scheme using FETs; a CMOS IC implementation would typically utilize MOSFET transistors as opposed to bipolar transistors.

An ingenious modification to the scheme shown in Figure 14.6, and arguably the most popular scheme for IC implementation of mixers, is the *Gilbert cell mixer*, shown in Figure 14.7. This scheme is essentially two differential active mixers operated in quadrature, and

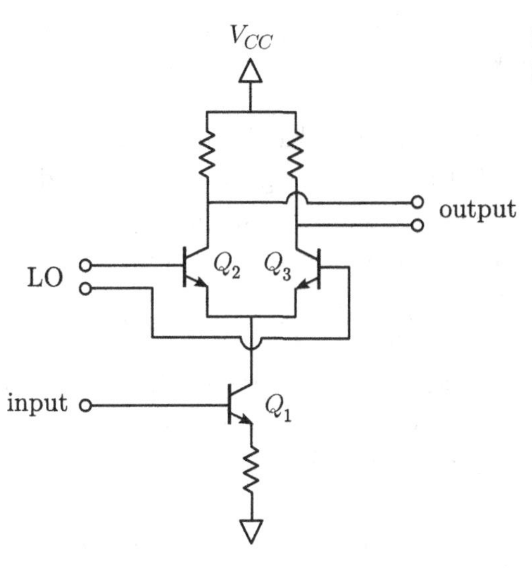

Figure 14.6. Active mixer using bipolar transistors. The FET variant is essentially the same.

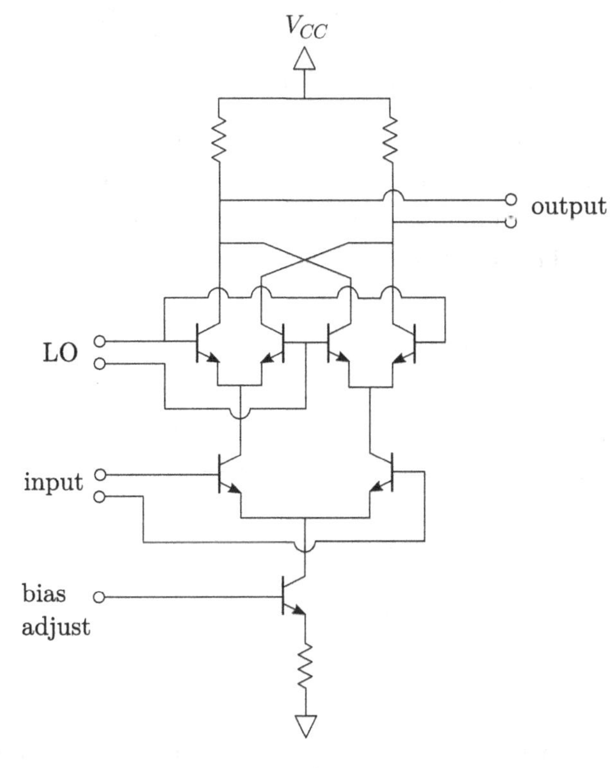

Figure 14.7. Gilbert cell mixer. Again, shown here in bipolar implementation; equally-well implemented using FETs.

is thus sometimes referred to as a *four-quadrant multiplier*. Both inputs and the output of the Gilbert cell mixer are fully differential, and this scheme has the additional advantages of relatively low power consumption and relatively low-level spurious output. A detailed analysis of the Gilbert cell is beyond the scope of this text, but it is worth becoming acquainted with Figure 14.7 so the device can be recognized when encountered.

A popular implementation of the Gilbert cell mixer has been the Phillips (now NXP) SA602A IC, which has appeared in innumerable VHF- and UHF-band receivers for decades, and is still being produced. With the continuing trend toward increased levels of integration, the functionality of this IC – including its Gilbert cell mixer – has been incorporated into other ICs.

14.4 QUADRATURE CONVERSION

In Chapters 5 and 6 we noted the advantages of "complex baseband" representation of signals for certain forms of modulation. The general idea was that a signal $s_{RF}(t)$ centered at frequency ω_c could be equivalently represented as a complex-valued signal $s(t) = s_i(t) + js_q(t)$ centered at a frequency of zero, and the relationship between these two forms of the signal is

$$s_{RF}(t) = \text{Re}\left\{s(t)e^{+j\omega_c t}\right\} = s_i(t)\cos\omega_c t - s_q(t)\sin\omega_c t \tag{14.10}$$

We are now ready to consider how we might actually go about generating $s_{RF}(t)$ from $s(t)$ – clearly, a form of upconversion – and $s(t)$ from $s_{RF}(t)$ – a form of downconversion. Because we are converting between real-valued and complex baseband form, we refer to this particular type of frequency conversion as *quadrature conversion*.

Equation (14.10) gives a practical recipe for upconversion, but things are simplified a bit with the following modification: Let us rewrite $-\sin\omega_c t$ as $\cos\left(\omega_c t + \frac{\pi}{2}\right)$. Then we have:

$$s_{RF}(t) = s_i(t)\cos\omega_c t + s_q(t)\cos\left(\omega_c t + \frac{\pi}{2}\right) \tag{14.11}$$

The typical implementation is as shown in Figure 14.8, which is an essentially literal interpretation of the above equation. This scheme requires two versions of the LO, each at frequency of ω_c but with a 90° phase difference. Figure 14.9 shows a simple circuit that accomplishes this for a fixed value of ω_c. Circuits and microwave devices that can accomplish

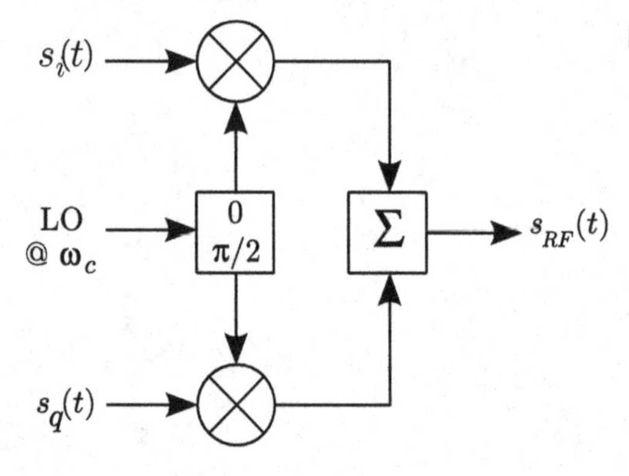

Figure 14.8. Analog quadrature upconversion.

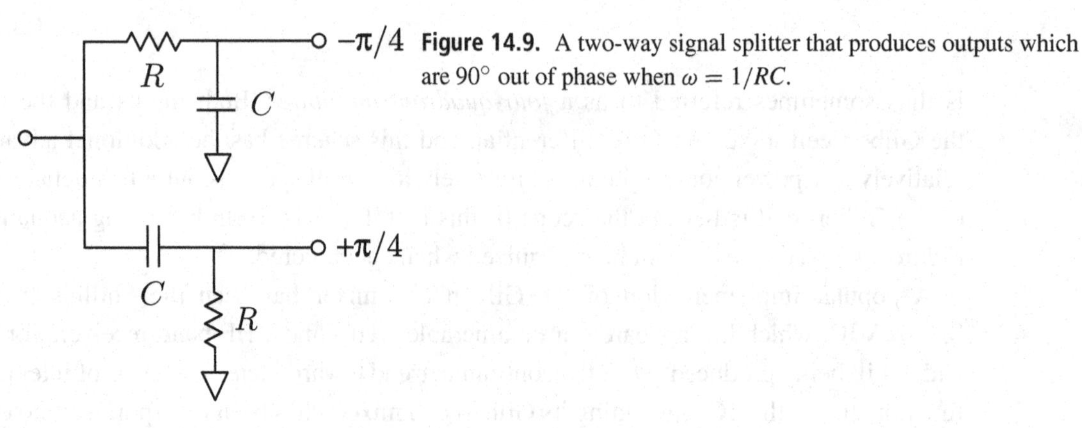

Figure 14.9. A two-way signal splitter that produces outputs which are 90° out of phase when $\omega = 1/RC$.

this over broad frequency ranges, so as to allow tuning, are quite feasible. Such circuits are part of a broader class of devices known variously as *hybrids*, *quadrature hybrids*, *quadrature splitters*, or *quadrature combiners* (since they usually work in either direction).

In practice, there are two things to look out for when considering quadratude upconverters of the type shown in Figure 14.8. First, we know that practical mixers will inevitably allow some of the baseband signal and LO to leak through to the output. The baseband signal – and most other spurious for that matter – are easily suppressed by filtering the output. The LO leakage, on the other hand, is a major concern, as the leaking LO appears dead-center in the spectrum of the upconverted signal. Thus, the mixers employed in quadrature upconverters must have very high LO-to-RF isolation. The second headache with this approach is that the LO splitter is unlikely to produce outputs that are exactly 90° separated in phase, and the conversion loss of the mixers is unlikely to be exactly equal. Thus, there will inevitably be distortion known as *quadrature imbalance*, also known as *I-Q imbalance*. In principle, quadrature imbalance can be corrected by "predistorting" $s_i(t)$ and $s_q(t)$ such that the $s_{RF}(t)$ is perfect; in practice this requires additional hardware for *in situ* calibration and so is relatively expensive and complicated to implement. More often, one seeks to make the quadrature imbalance sufficiently small that no such correction is required.

Quadrature downconversion can be interpreted mathematically as follows:[4]

$$s(t) = s_{RF}(t)e^{-j\omega_c t} = s_{RF}(t)\left[\cos\omega_c t - j\sin\omega_c t\right] \tag{14.12}$$

Again we would like replace sine with cosine and eliminate minus signs. A simple way to do that here is to first rotate the phase of the LO by $\pi/2$ radians and then rewrite the sine term, yielding:

$$s(t) = s_{RF}(t)\left[\cos\left(\omega_c t + \frac{\pi}{2}\right) + j\cos\omega_c t\right] \tag{14.13}$$

Direct implementation of this equation yields the scheme shown in Figure 14.10, minus the final stage of lowpass filters. Why are the lowpass filters required? Because the output of each mixer is the desired "difference frequency" product at a frequency of zero, in addition to the "sum frequency" product at $2\omega_L$. The latter is usually important to remove. Furthermore, we expect to see the usual spurious signals: leakage of the LO and RF signals, plus harmonics and

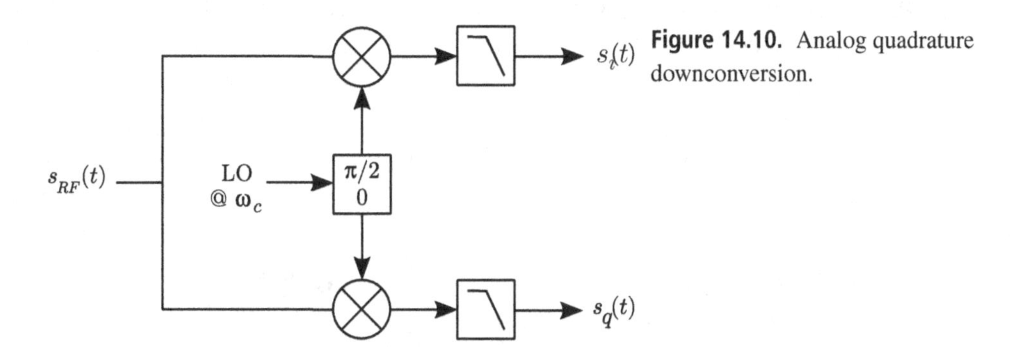

Figure 14.10. Analog quadrature downconversion.

[4] If this is not obvious, it may help to think of this in terms of the Fourier transform: Multiplication by $e^{-j\omega_c t}$ in the time domain translates to a spectral shift by $-\omega_c$ in the frequency domain.

intermodulation, all of which can be suppressed using the same filter. Also, as in the case of the quadrature upconverter, this approach is vulnerable to quadrature imbalance.

14.5 IMAGE REJECTION MIXERS

Recall from Section 14.2.3 that a major concern in frequency downconversion to a non-zero center frequency was image rejection. Also note from the previous section that in quadrature downconversion, there is no such concern. The reason there is no image rejection problem in quadrature conversion is because the LO signal is a complex exponential (i.e., $e^{-j\omega_L t}$), so multiplying the input signal by this LO shifts the *entire* spectrum (both positive and negative frequencies) to the left. This is fine as long as we are content with a complex-valued output. However, this begs the question: Could we use quadrature downconversion to achieve frequency downconversion to a non-zero center frequency with perfect image rejection, and then recover a real-valued signal from the resulting complex-valued signal? The answer is yes, and such a device is known as an image rejection mixer. There are two basic strategies, represented by the *Hartley* and *Weaver* architectures.

14.5.1 Hartley Architecture

The Hartley architecture – previously encountered in Section 5.6.1 – is depicted in Figure 14.11. In this approach, we begin with a quadrature downconversion of the real-valued signal to a complex-valued signal at the desired output frequency:

$$s(t) = s_{RF}(t)e^{-j\omega_L t} = s_{RF}(t)\left[\cos\omega_L t - j\sin\omega_L t\right] \tag{14.14}$$

with $\omega_L = \omega_i - \omega_o$. One way to recover a real-valued signal would be simply to truncate the imaginary part; however, note that this leaves us with the output $s_{RF}(t)\cos\omega_L t$, which is the same as traditional frequency downconversion, complete with the original image rejection problem. Clearly, we must use both the real and imaginary outputs of the quadrature downconversion in order to obtain image-free output.

To demonstrate the correct way to do this, let us once again (and without loss of generality) assume an input signal $A\cos\omega_i t$. For low-side downconversion, $\omega_i - \omega_L$ is positive and equal

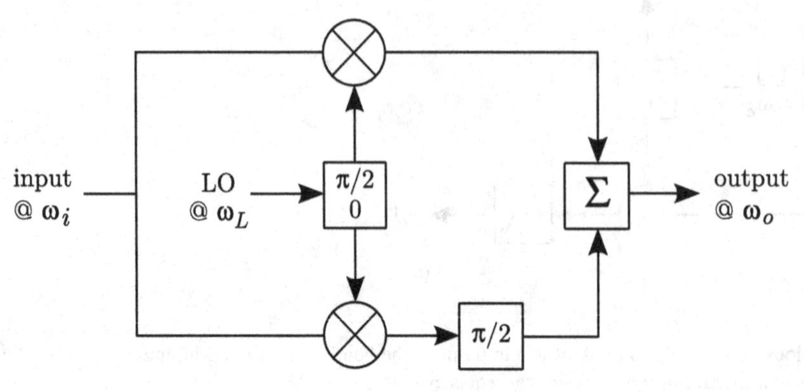

Figure 14.11. Image rejection downconversion mixer: Hartley architecture.

to ω_o. Now consider an input signal of the form

$$s_{RF}(t) = A\cos(\omega_L + \omega_o)t + B\cos(\omega_L - \omega_o)t \tag{14.15}$$

where the second term is, in this case, the undesired image frequency component: If the processing is done correctly, we should not see any factors of B in the output. Now the output of the quadrature downconversion is:

$$\begin{aligned} s(t) = {} & A\cos(\omega_L + \omega_o)t\cos\omega_L t + B\cos(\omega_L - \omega_o)t\cos\omega_L t \\ & - jA\cos(\omega_L + \omega_o)t\sin\omega_L t - jB\cos(\omega_L - \omega_o)t\sin\omega_L t \end{aligned} \tag{14.16}$$

Using basic trigonometric identities, the real and imaginary parts of $s(t) = s_i(t) + js_q(t)$ can be written as

$$s_i(t) = \frac{A}{2}\cos(2\omega_L + \omega_o)t + \frac{A}{2}\cos\omega_o t + \frac{B}{2}\cos(2\omega_L - \omega_o)t + \frac{B}{2}\cos\omega_o t \tag{14.17}$$

$$s_q(t) = -\frac{A}{2}\sin(2\omega_L + \omega_o)t + \frac{A}{2}\sin\omega_o t - \frac{B}{2}\sin(2\omega_L - \omega_o)t - \frac{B}{2}\sin\omega_o t \tag{14.18}$$

Note that we have used the fact that $\sin(-\omega_o t) = -\sin\omega_o t$ to flip the sign of the last term in the above equation.

The essential feature of the Hartley architecture is to rotate the phase of $s_q(t)$ by $+\pi/2$ radians (turning sines into cosines) and then to add the result to $s_i(t)$. Rotating the phase of $s_q(t)$ yields

$$-\frac{A}{2}\cos(2\omega_L + \omega_o)t + \frac{A}{2}\cos\omega_o t - \frac{B}{2}\cos(2\omega_L - \omega_o)t - \frac{B}{2}\cos\omega_o t$$

and adding this result to $s_i(t)$ yields:

$$A\cos\omega_o t$$

which is the ideal (real-valued, image-free) outcome. Note that we have eliminated not only the image contribution to the output, but also the undesired multiplication product is suppressed, and all of this is achieved without filtering.

For the scheme in Figure 14.11 to work, the $\pi/2$ phase shift must work at the output frequency. However, for the $2\omega_L \pm \omega_o$ terms to cancel, the phase shift must also be $\pi/2$ at these frequencies. Often such a phase shifter is not practical to implement, and so these terms must be filtered (or otherwise taken into account) at the output.

There are many variants and alternative applications of the Hartley architecture. A simple change makes this work for high-side injection (see Problem 14.7). We have described this as a downconversion operation, but essentially the same architecture works also for upconversion from a non-zero center frequency (see Problem 14.8). Finally, there are alternative implementations of the Hartley architecture which are mathematically equivalent; in particular, it is possible to apply the 90° phase shift to the input signal as opposed to the LO, which may be more convenient when the input frequency range is small but the LO needs to vary over a wide frequency range, as is typically the case in upconverters appearing in transmitters.

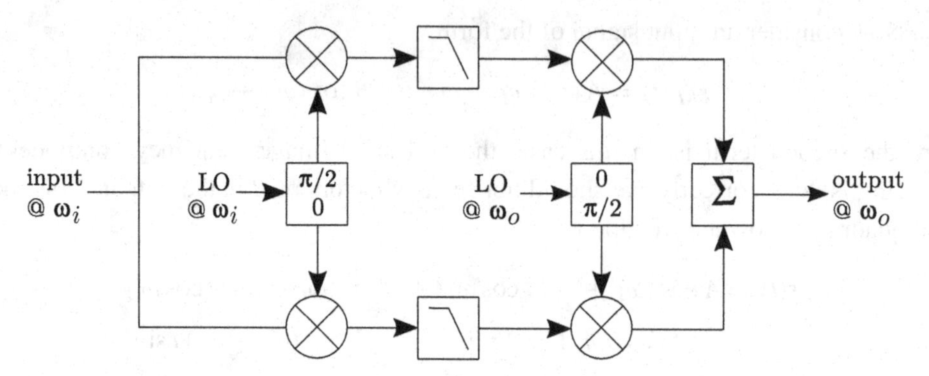

Figure 14.12. Image rejection mixer: Weaver architecture.

14.5.2 Weaver Architecture

The Weaver architecture – also previously encountered in Section 5.6.1 – is shown in Figure 14.12, and is also capable of image-free frequency conversion. Although this approach appears somewhat more complicated, the concept is quite simple: First do quadrature downconversion to zero frequency, and then do quadrature upconversion to the desired output frequency. Thus, the Weaver architecture is simply a quadrature downconverter back-to-back with a quadrature upconverter, and the result has nominally zero image response. Downconversion and upconversion are achieved using precisely the same architecture.

The principle advantage of the Weaver architecture over the Hartley architecture is that there is no need to phase-shift the information-bearing signal. This avoids the problem of achieving an accurate $\pi/2$ phase shift over the bandwidth of the signal, which can be difficult for signals with large fractional bandwidth. However, the disadvantages of the Weaver architecture are formidable: four mixers (vs. two for Hartley) and two LOs (vs. one for Hartley). Thus a Weaver image rejection mixer is considerably more expensive and consumes more power.

One additional consideration with the Weaver architecture is its utilization of an IF at zero (DC). Processing a signal that is centered at zero frequency poses several practical problems; for example, magnitude imbalance in the downconversion and upconversion manifests as a spurious signal centered in the spectrum of the output. A simple solution to this problem is to select LOs so that the center frequency of interest is slightly offset from zero in the complex baseband stage of the Weaver architecture, so that the spurious signal associated with amplitude imbalance is spectrally separated from the signal of interest, allowing it to be suppressed by filtering. This idea is also central to the concept of *near-zero IF* direct conversion – more on that in Section 15.6.9.

Problems

14.1 A cellular base station receiver downconverts the entire block of spectrum from 824 MHz to 849 MHz (i.e., $B = 25$ MHz) to a real-valued signal at an IF centered at

140 MHz. Specify the LO frequencies and filter requirements for (a) low-side injection and (b) high-side injection. For both (a) and (b), give equations indicating the IF frequency f_o at which a specified RF frequency f_i will appear, and indicate whether the spectrum will be flipped.

14.2 A cellular base station transmitter synthesizes 25 MHz of spectrum at an IF centered at 140 MHz and upconverts it to an RF band spanning 824 MHz to 849 MHz. Specify the LO frequencies and filter requirements for (a) low-side injection and (b) high-side injection. For both (a) and (b), give equations indicating the RF frequency f_o at which a specified IF frequency f_i will appear, and indicate whether the spectrum will be flipped.

14.3 Returning to Problem 14.1, specify the band of image frequencies that must be suppressed at the input for (a) low-side injection, and (b) high-side injection. For both (a) and (b), indicate whether unsuppressed signals originating from the image frequency band will be spectrally flipped.

14.4 Design a diplexer of the type shown in Figure 13.9 for the square-law mixer of Figure 14.3 assuming 50 Ω impedance terminating each port. Assume a crossover frequency of 100 MHz.

14.5 Diode ring mixers are typically designed to operate with an LO which is specified to be at least 7 dBm with respect to 50 Ω, although a mixer designed for a 7 dBm LO might work sufficiently well for an LO level as low as 4 dBm. What makes 7 dBm and 4 dBm the minimum thresholds?

14.6 Design a quadrature splitter of the type shown in Figure 14.9 for a quadrature upconverter having an output frequency of 30 MHz. Assume 50 Ω input.

14.7 Derive the Hartley architecture for downconversion using high-side injection. Show your mathematical justification (as in Section 14.5.1) and the final result (as in Figure 14.11).

14.8 Derive the Hartley architecture for upconversion using low-side injection. Show your mathematical justification (as in Section 14.5.1) and the final result (as in Figure 14.11).

15 Receivers

15.1 INTRODUCTION

We have now arrived at a point where we can reasonably consider the analysis and design of radio receivers. In Chapter 1, we considered two broad classes of radio receivers: those employing analog detection (Figure 1.10), used primarily with AM/ASK and FM/FSK modulations; and receivers employing analog-to-digital conversion (Figure 1.11), where the ADC output is subsequently processed digitally. In either case, the central problem in receiver design is to accept the signal provided by the antenna – a problem we addressed in Chapter 12 – and to deliver it to the analog detector or ADC with sufficient gain, sensitivity, selectivity, and linearity. Receivers by this definition are also sometimes referred to as *downconverters*, implying that frequency downconversion occurs; or *tuners* implying that the center frequency of the received bandpass is variable.

We have already addressed the topic of analog detectors in Chapter 5, so this chapter begins with a discussion of the requirements and characteristics of ADCs (Section 15.2). In Section 15.3 we consider the central question of gain – how much is nominal, how little is adequate, and the relationship to sensitivity. Sections 15.4 and 15.5 address techniques and requirements for selectivity. Section 15.6 provides a summary of the principal receiver architectures. Section 15.7 elaborates on the problem of frequency planning in superheterodyne receivers. In Section 15.8 we revisit the question of gain, now in the context of the various receiver architectures, and summarize the principles of automatic gain control. This chapter concludes with suggestions for receiver design case studies (Section 15.9), intended to further demonstrate the practical aspects of modern receiver design.

15.2 ANALOG-TO-DIGITAL CONVERSION

An ADC is a device that converts a continuous-time analog signal into a sequence of discrete-time digital values that approximate the input signal. After a brief introduction to ADC method of operation, the issues of sample rate versus bandwidth (Section 15.2.2) and quantization noise (Section 15.2.3) are discussed. Finally, we consider the characteristics and limitations of practical ADCs (Section 15.2.4).

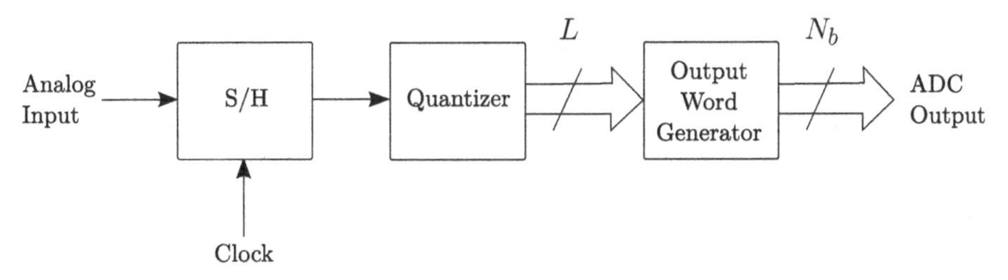

Figure 15.1. A simple ADC model.

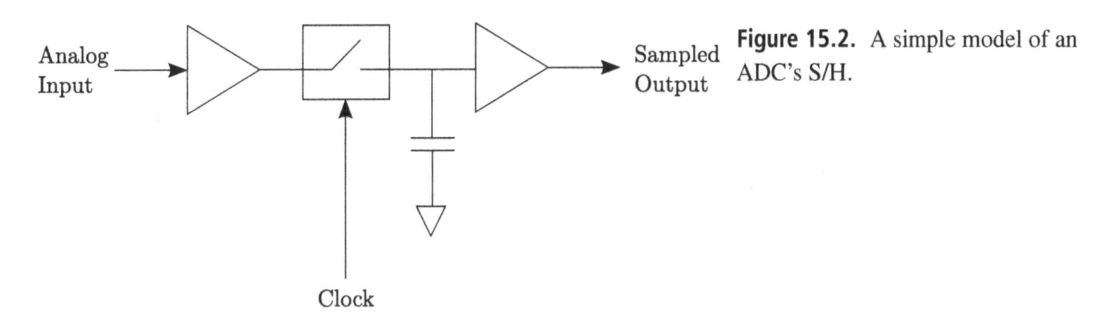

Figure 15.2. A simple model of an ADC's S/H.

15.2.1 Method of Operation

A simple model for analog-to-digital conversion is shown in Figure 15.1. This model consists of three parts: an analog *sample and hold* device (S/H), a *quantizer*, and a *output word generator*.

The S/H is a buffer which outputs the value presented to the input at a time determined by the active edge of the sample clock. A basic S/H implementation shown in Figure 15.2. In "sample" mode, an input amplifier is directly connected to an output amplifier. In "hold" mode, the switch is opened and the charged capacitor provides a nominally steady input voltage equal to the input amplifier output voltage at the time the switch was thrown. This scheme depends on the capacitor's value being large enough to effectively "freeze" the level throughout the hold interval, but also small enough to pass the bandwidth of the input signal during the "sample" interval. This tradeoff largely accounts for the bandwidth limitation of practical ADCs.

The quantizer converts the analog sample appearing at the output of the S/H to a set of two-state voltages representing that value. The simplest method in common use is known as *flash* (or *parallel*) conversion, depicted in Figure 15.3. In a flash quantizer, a reference voltage is applied to a series string of resistors. The voltages between the resistors are then compared to the input voltage using threshold-type comparators. The output of each comparator is one of two voltages; one indicating that the input voltage is higher than the voltage-divided reference signal, and the other indicating that the input voltage is lower than the voltage-divided reference signal. Thus the quantizer provides an estimate of the magnitude of the input, rounded to the closest threshold voltage which is lower than the input voltage.

The output word generator interfaces the output of the quantizer to digital logic. Generally, the L threshold decisions from the quantizer may be exactly represented as digital word consisting of $N_b = \log_2 L$ (rounding to the next largest integer number) bits. Popular digital word formats include *offset binary* – essentially, what you get by simply counting in binary from the lowest level to the highest level; and *two's complement*, which is a simple modification

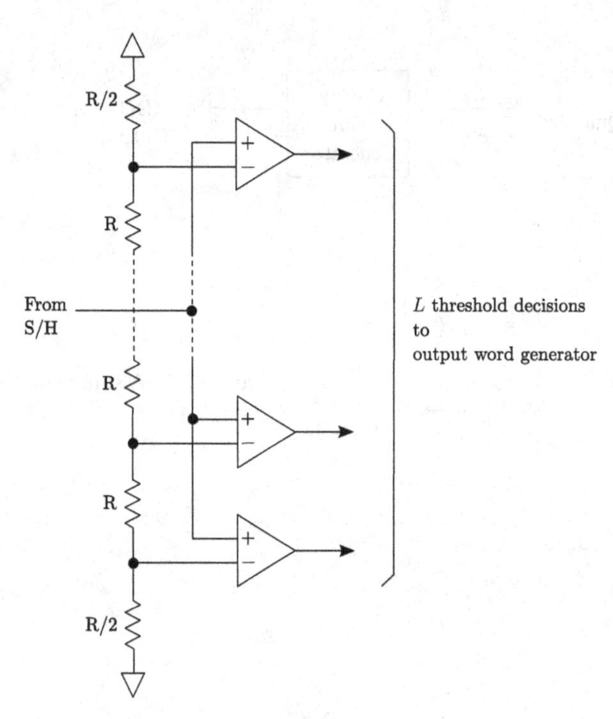

Figure 15.3. A flash quantizer.

that simplifies subsequent binary arithmetic. The output word generator may also implement error control logic and generate additional outputs to indicate various error conditions, such an "out-of-range" flag bit that is set when the signal magnitude is larger than the maximum level representable by an output word. This condition is also commonly known as *clipping*.

To be clear, the architecture implied by Figure 15.1 with flash quantization is not the only way to go. The primary advantage of flash architecture is simplicity and speed; modern flash ADCs are capable of sample rates in excess of 10 GSPS. The primary disadvantage of flash conversion is difficulty achieving acceptable linearity in the analog-to-digital transfer function, and acceptably-low generation of spurious spectral features for devices with $N_b > 8$ or so. Other architectures produce considerably better performance for larger N_b, but tend to be limited to a lower maximum sample rate. Prominent examples of such architectures in common use in radio applications are *successive approximation* and *subranging* (also known as *pipeline*) architectures.

Beyond the scope of this book but worth knowing about are Σ-Δ (pronounced "sigma-delta") ADCs. In this class of ADCs, the input is initially oversampled (digitized at a very high F_S) with very low L, and subsequently transformed into a signal with the desired (lower) F_S and desired (higher) L. This has distinct advantages with respect to anti-alias filtering and QSNR, but has historically been very rare in radio applications owing to the difficulty in oversampling sufficiently at RF/IF frequencies. Nevertheless these are starting to appear; see for example [4].

15.2.2 Sample Rate and Bandwidth

Regardless of architecture, ADCs are characterized primarily by (1) the sample rate F_S and (2) the number of bits N_b comprising the output word. The parameters F_S and N_b have profound implications for receiver system design. Let us begin with sample rate.

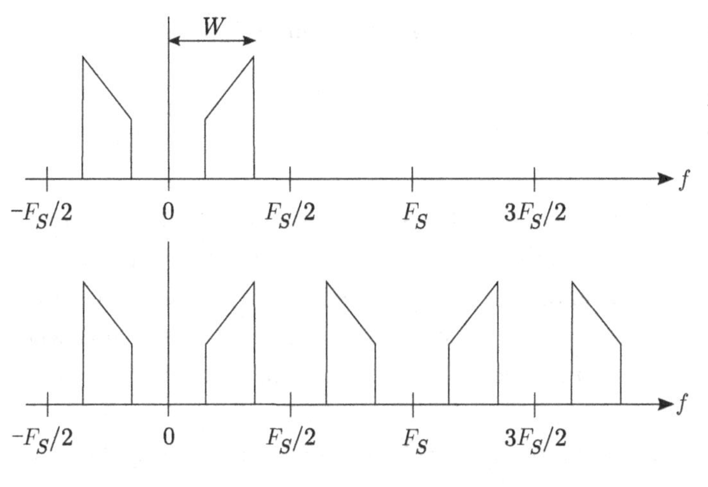

Figure 15.4. Spectrum before and after sampling. *Top*: Analog input. *Bottom*: Discrete-time (sampled) output.

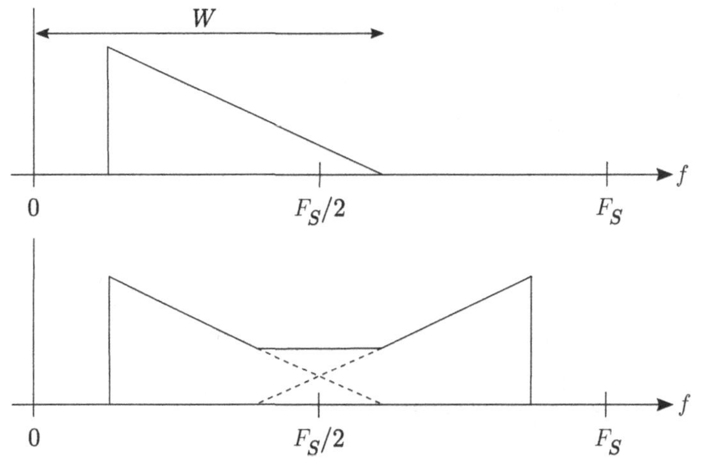

Figure 15.5. Aliasing resulting from sampling with $F_S < 2W$.

The *Nyquist–Shannon sampling theorem* states that a continuous-time signal having lowpass bandwidth W may be perfectly represented by a discrete-time signal having $F_S \geq 2W$.[1] We refer to this as the *Nyquist criterion*. The concept is illustrated in Figure 15.4. Figure 15.4 shows the spectrum of a real-valued signal prior to sampling and the spectrum of the same signal after sampling at a rate of F_S. Note that one consequence of sampling is that the spectrum in the band from 0 ("DC") to $F_S/2$ is repeated in intervals of width $F_S/2$, and that the spectrum is reversed in every other interval. We refer to these intervals as *Nyquist zones*, and the spectral content within each zone is referred to as an *alias*. The consequence of violating the Nyquist criterion is that aliases extend into adjacent Nyquist zones, as shown in Figure 15.5. This condition is known as *aliasing*, and is usually undesirable since signals of interest are distorted as a result.

Consider the implications for receiver design: If one wishes to digitize a signal of bandwidth B centered at a frequency of f_c using the scheme implied by Figure 15.4, then $W = f_c + B/2$ and the theoretical minimum sampling rate is $2W = 2f_c + B$. However, this puts the signal spectrum "flush" against the Nyquist zone boundary, and thus would require an unrealizable "brick wall" spectral response to prevent out-of-band signals from aliasing into the band of interest.

[1] A derivation of the sampling theorem is quite simple, but outside the scope of this text. However, if you would like a simple exercise to convince yourself of the truth of the theorem, try Problems (15.1).

In practice, the minimum sample rate must be greater than $2f_C + B$ in order to accommodate suitable anti-alias filtering. Here is an example of how that plays out:

EXAMPLE 15.1

The spectrum from 118 MHz to 137 MHz is allocated to AM voice communications for aviation applications. In this use the spectrum is divided into many narrowband channels. Sometimes it is useful to have access to the entire 19 MHz-wide band as opposed to monitoring a single channel at a time. Let us say we are able to specify an analog receiver that downconverts the entire band to an IF of our choosing, and outputs a real-valued signal. What is the minimum IF center frequency and minimum safe sample rate required to digitize the entire band? For the anti-aliasing filter, assume the width of the transition from passband to stopband is 10% of the passband (plausible, as we observed in Chapter 13).

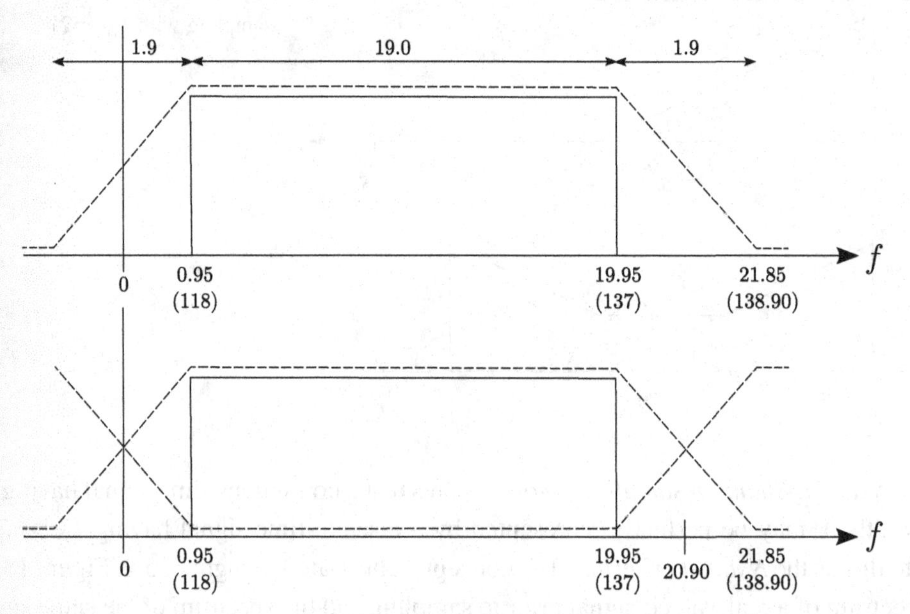

Figure 15.6. Determining minimum sample rate to digitize the 118–137 MHz aviation band, after frequency downconversion, in Example 15.1. *Top:* Downconverted spectrum (solid line) and specified anti-aliasing filter response (dashed line). *Bottom:* Same after sampling at 41.8 MSPS, demonstrating that only stopband frequencies alias into desired spectrum. All frequencies are in MHz. Numbers in parentheses are the corresponding frequencies at the input of the receiver; i.e., before downconversion.

Solution: In principle, $B = 137 - 118 = 19$ MHz of spectrum can be perfectly represented as a discrete-time signal with $F_S = 2B = 38$ MSPS as long as the band is exactly centered in a Nyquist zone by the downconverter. In practice, some margin is required to accommodate a filter to prevent out-of-band signals from aliasing into the band of interest. Referring to Figure 15.6, we see that the Nyquist zone must be at least 20.9 MHz wide in order to guarantee that out-of-band signals alias into the band of interest only from the stopbands of the anti-aliasing filter, and not from the transition bands. Therefore, F_S must be $\geq 2 \cdot 20.9$ MHz $= 41.8$ MSPS, and this works only if the center frequency after analog frequency conversion is $20.9/2 = 10.45$ MHz.

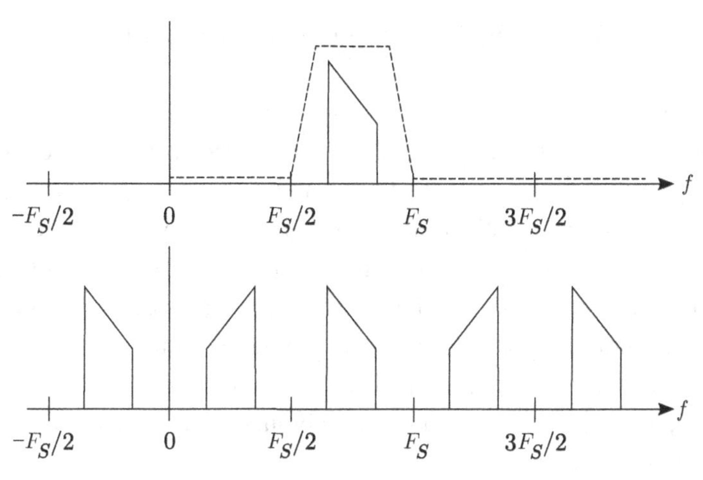

Figure 15.7. Spectrum before and after sampling; here for an input signal restricted to the second Nyquist zone. *Top*: Analog input (solid) and anti-aliasing filter response (dashed). *Bottom*: Discrete-time (sampled) output.

Whereas aliasing is avoided by confining the analog signal of interest to a single Nyquist zone, there is no fundamental reason why the input must be confined to the *first* Nyquist zone; i.e., from 0 to $F_S/2$. Figure 15.7 demonstrates this concept. Here the input signal is chosen to be identical to the alias of input signal from Figure 15.4 as it appears in the second Nyquist zone ($F_S/2$–F_S); including the spectral reversal. Note that the ADC output is identical to the result from Figure 15.4. Repeating this experiment using the third Nyquist zone (non-reversed) alias as the input signal yields the same result. Thus, unaliased sampling is possible as long as the input lies completely within any one Nyquist zone, and the resulting ADC output (interpreted as the spectrum from $-F_S/2$ to $+F_S/2$) is either identical or spectrally-reversed depending on whether an odd- or even-numbered Nyquist zone was used for input.

The use of the second or higher Nyquist zones is known as *undersampling*. Undersampling has great utility in receiver design, as we shall see in Section 15.6.2. To distinguish between undersampling and "traditional" (lowpass) sampling, we refer to digitization of signals from the first Nyquist zone as *lowpass sampling*.

15.2.3 Quantization Noise

Let us now assume that the Nyquist criterion is sufficiently well satisfied, such that perfect reconstruction of the output of an ideal S/H is theoretically possible. Now, the only thing standing in the way of perfect reconstruction of the original ADC input signal is quantization. An N_b-bit ADC is able to represent $L = 2^{N_b}$ distinct values, so the larger N_b is, the more accurately the ADC output represents the analog input.

The error associated with quantization is known as *quantization noise*, which can be interpreted as a degradation of the signal-to-noise ratio (SNR) of the analog signal. Let $x(t)$ be the ADC input, and let $x_q(t)$ be the output of the quantizer. The latter can take on only one of L levels, and the quantizer holds that level until the beginning of the next sample period. The relationship between $x(t)$ and $x_q(t)$ is

$$x(t) = x_q(t) + \epsilon(t) \tag{15.1}$$

where $\epsilon(t)$ is the quantization error; note how quantization error can be interpreted as an additive noise mechanism.

The power associated with quantization error is simply the mean square value of $\epsilon(t)$. To quantify this power, one must specify the particular scheme used to map the continuum of values of $x(t)$ into the L possible values of $x_q(t)$. Let us assume *uniform quantization.* In this scheme, a particular output from the quantizer indicates that $x(t)$ falls within one of L possible contiguous *subranges*, each $2x_p/L$ wide, where x_p is the maximum encodable input magnitude. Further, let us assume the values of $x(t)$ are distributed uniformly over each subrange; i.e., with no value within a subrange more or less common than any other value within the subrange. Then the mean square value of $\epsilon(t)$, $\langle\epsilon^2(t)\rangle$, is equal for each subrange; and $\langle\epsilon^2(t)\rangle$ taking into account all subranges is equal to $\langle\epsilon^2(t)\rangle$ within any one subrange. Using a subrange centered at zero, we may calculate this quantity as

$$\left\langle\epsilon^2(t)\right\rangle = \frac{1}{2x_p/L}\int_{-x_p/L}^{+x_p/L}\epsilon^2\, d\epsilon \tag{15.2}$$

Evaluating this integral, we find

$$\left\langle\epsilon^2(t)\right\rangle = \frac{x_p^2}{3L^2} \tag{15.3}$$

As expected, the quantization noise power increases with subrange width; i.e. is related to $2x_p/L$. The power of the original signal is $\langle x^2(t)\rangle$, so we may define a *quantization SNR* (QSNR) as follows:

$$\text{QSNR} \equiv \frac{\langle x^2(t)\rangle}{\langle\epsilon^2(t)\rangle} = 3L^2\frac{\langle x^2(t)\rangle}{x_p^2} \tag{15.4}$$

We can express this in terms of N_b by substituting $L = 2^{N_b}$, yielding

$$\text{QSNR} = 3\cdot 2^{2N_b}\cdot\frac{\langle x^2(t)\rangle}{x_p^2} \tag{15.5}$$

In decibels, this expression becomes

$$\text{QSNR} \cong 4.77 + 6.02N_b + 10\log_{10}\left[\frac{\langle x^2(t)\rangle}{x_p^2}\right]\ \text{dB} \tag{15.6}$$

Thus we see that QSNR increases by approximately 6 dB for each additional bit available to represent the output.

To go any further, we must make some assumption about the waveform $x(t)$. A highly-relevant special case is that of a sinusoid $x(t) = A\cos(\omega t + \theta)$, for which $\langle x^2(t)\rangle = A^2/2$. If we then select $A = x_p$ such that the extrema of the sinusoid fall exactly at the maximum encodable limits of the quantizer, then

$$10\log_{10}\left[\frac{\langle x^2(t)\rangle}{x_p^2}\right] = 10\log_{10}\left[\frac{A^2/2}{A^2}\right] \cong -3.01\ \text{dB} \tag{15.7}$$

so in this special case we find

$$\text{QSNR} \cong 1.76 + 6.02N_b\ \text{dB} \tag{15.8}$$

In practice, ADCs tend to have SNR that is 1–2 dB worse than the theoretical QSNR due to various imperfections associated with functions other than quantization. Thus, it is common practice to estimate the QSNR for a practical ADC encoding full-scale sinusoidal input as being about $6N_b$ dB.

It is also common practice to interpret QSNR as a form of blocking dynamic range (refer to Section 11.7 for a refresher on that concept), and thus QSNR is sometimes referred to as the *quantization dynamic range* of an ADC. This is quite reasonable since we clearly have a situation in which the presence of a strong signal may degrade the sensitivity of a receiver to weaker signals; in this case, by inundating them with quantization noise.

Now, an important note. QSNR is merely a characterization of the total power attributable to quantization error relative to the total power of the original signal. We have not yet determined the spectral characteristics of the quantization noise. A common error is to assume that the quantization noise associated with sinusoidal input has "white" spectrum: This is easily demonstrated to be false, as shown in the following example.

EXAMPLE 15.2

An ADC employs uniform four-bit quantization and is designed to accommodate input up 2 V_{pp} for a sinusoid. The sample rate is 10 MSPS. Simulate the ADC output and plot the associated spectrum when the input is $\cos 2\pi f_c t$ V with $f_c = 500$ kHz.

Solution: First note that four bits corresponds to 16 levels, and the first task is to decide how to arrange those levels and how to define the associated intervals. In this case a reasonable choice would be levels at -0.9375 V, -0.8125 V, and so on every 0.125 V up to $+0.9375$ V, and we will choose to define the intervals by rounding up from the input value to the next defined quantization level. Why not start and end at exactly -1 V and $+1$ V respectively? Because at least one of these two levels would go unused for a signal with extrema exactly at those limits. If we wish the maximum possible dynamic range, then every level should be selected at least once in every period of the sinusoid; otherwise we are losing dynamic range.

Figure 15.8 shows the first 5 μs in the time domain, beginning at $t = 0$. Note that the apparent quantization error seems pretty small even with just 16 levels. Figure 15.9 shows the spectrum of the input and quantized signals. Observe that the quantization error manifests as additional tones at multiples of the input frequency, and is certainly not noise-like as might be assumed. These tones represent the fact that the quantization error is piece-wise sinusoidal and periodic; therefore the error waveform can be interpreted as a Fourier series, which is essentially a sum of harmonics. The DC component in the ADC output is due to the fact that the quantized value is always higher than or equal to the input value; therefore the average value of the quantized waveform is positive even if the mean value of the input is zero and even though the mean value of the available quantization levels is zero.

In this case, Equation (15.8) predicts that the QSNR is $1.76 + 6.02 \cdot 4 \cong 25.8$ dB. However, we find the QSNR in the simulation is only 22.3 dB; i.e., about 3.5 dB less than predicted. The reason for the difference is that Equation (15.8) is derived assuming that the waveform samples are uniformly distributed over the quantization intervals, which is not exactly true.

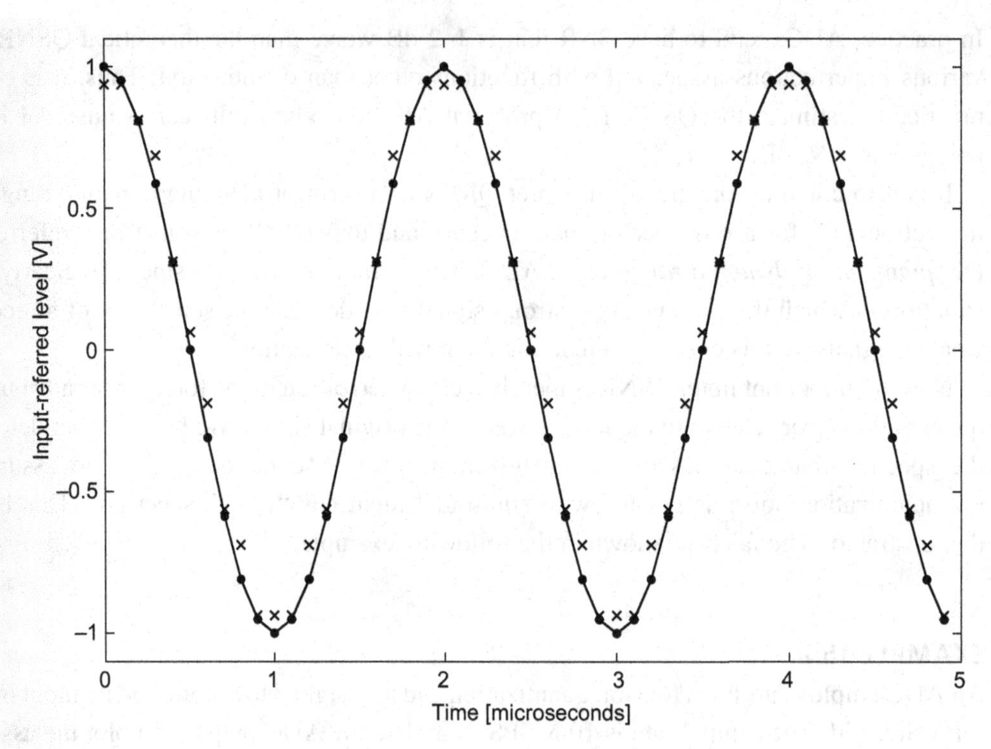

Figure 15.8. Example 15.2: *Solid line:* First 5 μs of the single-tone input to a four-bit ADC. Dot markers indicate input values at the moments of sampling. Output (quantized) values are indicated by "×" markers.

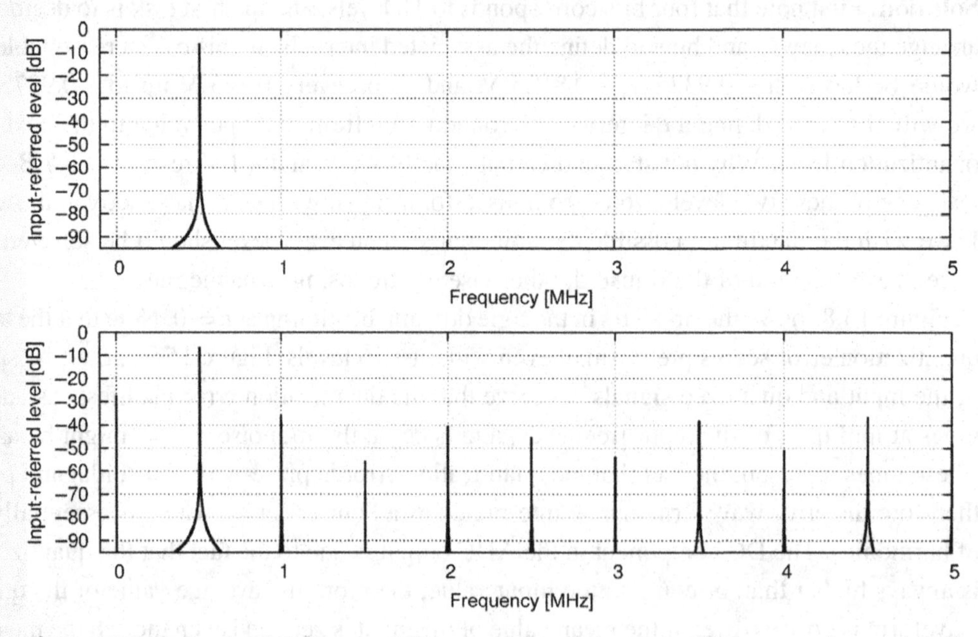

Figure 15.9. Example 15.2: *Top panel:* Input spectrum. *Bottom panel:* Spectrum after quantization.

For example, the waveform samples fall within a quantization interval close to DC about twice as often per period as they fall within the maximum or minimum quantization intervals. This issue is mitigated as the number of quantization levels is increased, since this makes the extreme quantization levels less important relative to the other levels. For example, for $N_b = 8$ one finds the difference shrinks from about 3.5 dB to about 2.6 dB. For $N_b = 12$ the difference is less than 1 dB and is typically negligible compared to other issues which would degrade the performance of a practical ADC, as will be discussed in Section 15.2.4.[2]

Here's a modification to the above example that demonstrates conditions under which the quantization error *does* have noise-like spectrum:

EXAMPLE 15.3

Repeat Example 15.2, but now let the input be

$$\cos\left[2\pi\left(f_c + 50 f_c t\right) t\right] \text{ V}$$

Note that this is essentially a "chirp" signal (also known as a "linear FM") signal which begins at a frequency of 500 kHz at $t = 0$ and increases to a frequency of 1 MHz after 1 ms. Let the length of the simulation be 1 ms.

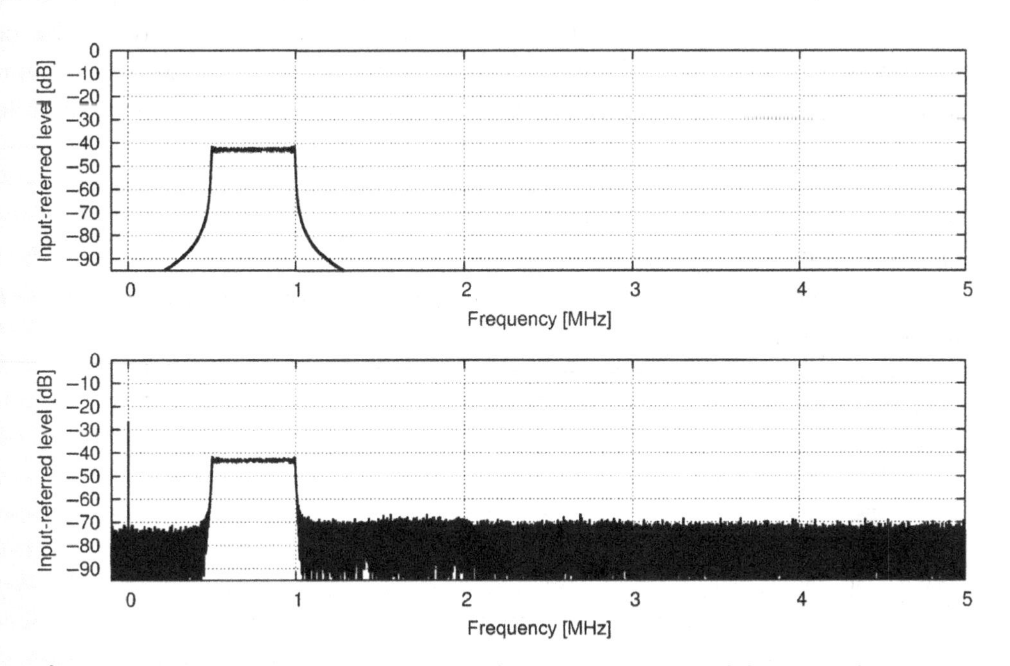

Figure 15.10. Example 15.3: *Top panel:* Input spectrum. *Bottom panel:* Spectrum after quantization.

Solution: There is not much point in showing the time-domain result in this case, since it is nearly identical to the result shown in Figure 15.8 for the first 5 μs, and the only difference at the end of the simulation is that the frequency has increased by a factor of 2. Figure 15.10 shows the

[2] It is easy to imagine a number of ways to beat this "small N_b" penalty, including non-uniform quantization levels and post-sampling correction algorithms. However, because most ADCs in radio applications have at least eight bits, a fix is not necessarily worth the effort.

spectrum of the input and quantized signals. Note that despite the similarity to the waveform in the previous example, the quantization error spectrum is utterly different; it is now very much like traditional noise. This is because the quantization error waveform is no longer periodic, because the signal itself is no longer periodic. The QSNR, however, is essentially the same as was determined in the previous example. In effect, the quantization noise power that was previously concentrated in harmonics in the previous example is now evenly distributed in frequency.

Summarizing the above example, we see that the spectrum of the quantization noise depends on the spectrum of the input signal, and in particular that the quantization noise cannot be expected to be noise-like if the input signal is periodic. The outcome also depends on the bandwidth of the signal relative to the Nyquist bandwidth. Most sinusoidal carrier modulations represent an intermediate case – certainly not periodic, but sufficiently narrowband that the resulting quantization noise spectrum may be “lumpy”.

In some applications it is important that the spectrum of the quantization noise be as white as possible, so that quantization noise artifacts do not interfere with (or mistaken for) legitimate signals. This can be achieved by adding noise at the input. This technique is known as *dithering*. The spectral distribution of the added noise is not critical. Also dithering noise need not spectrally overlap the signal of interest, making it practical to use dithering to improve the quantization noise spectrum without necessarily degrading the SNR in bandwidth occupied by the signal.

EXAMPLE 15.4

Let us repeat Example 15.2, but now we will add some lowpass filtered noise at the input to dither the signal of interest.

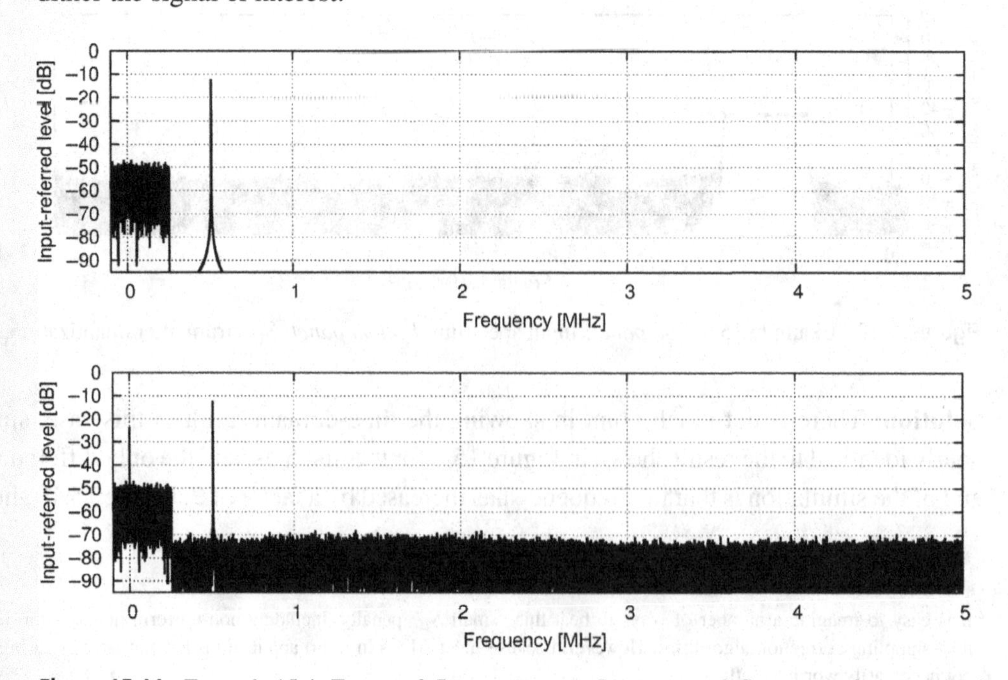

Figure 15.11. Example 15.4: *Top panel:* Input spectrum. *Bottom panel:* Spectrum after quantization.

Solution: Figure 15.11 shows the spectrum of the input and quantized signals. Note that the additional noise has resulted in the elimination of the quantization noise harmonics evident in Figure 15.9 with only a slight degradation in the signal-to-noise ratio in the spectrum close to the signal of interest.

15.2.4 Characteristics of Practical ADCs

Most modern ADCs which are suitable for receiver applications encode full-scale for signals on the order of 1 V_{pp} and have differential input impedances on the order of 100 Ω. Full-scale power P_{FS} for these ADCs is therefore typically on the order of +10 dBm.

Differential input is nearly ubiquitous, as it provides a means to deliver the highest possible second-order linearity (see Section 11.3). This is important since there can be no anti-alias filtering of intermodulation products generated within the ADC; and this is particularly important for lowpass sampling because second-order intermodulation from the lower half of the first Nyquist zone lands in the upper-half of the first Nyquist zone. These considerations strongly motivate undersampling when the passband occupies a large fraction of the Nyquist zone. For example, second harmonics of signals present in the second Nyquist zone land in the fourth Nyquist zone, where they may be mitigated by using analog anti-alias filtering preceding the ADC.

Presently, ADCs suitable for receiver applications exist for sample rates up to a few billion samples per second (gigasamples per second, or "GSPS") and up to roughly $N_b \sim 16$ bits. Sample rate is limited primarily by S/H bandwidth (as explained in Section 15.2.1), whereas N_b is limited primarily by quantizer linearity. In particular, one finds that above a certain N_b, the spurious spectral content associated with quantizer linearity becomes a bigger problem than the quantization noise.

The total impairment accounting for quantization noise, internal device noise contributions, and quantizer non-linearity is often expressed in terms of *effective number of bits* (ENOB). ENOB can be interpreted as the N_b of an ideal ADC (i.e., one limited only by quantization noise) that produces the same total impairment noise as the subject ADC. It is typical for ENOB to be 0.5–1.0 bits less than N_b for conservatively-designed ADCs, and several bits less than N_b for more aggressive designs employing larger N_b. (As we shall see in Chapter 18, it is often possible to increase ENOB more effectively by sampling with low N_b at high F_S, and then decimating (reducing sample rate) to increase ENOB.) A large difference between N_b and ENOB is simply an indication that spurious output due to analog imperfections of the ADC is dominating over the ideal (irreducible) quantization noise.

Often the primary concern is not ENOB but rather the possibility that ADC-generated spurious products can obscure *bona fide* signals, or be mistaken for *bona fide* signals. For this reason it is common to specify the spurious-free dynamic range (SFDR) of an ADC, just as one might specify the SFDR of the analog section of a receiver (recall Section 11.7). The principal difference is that the spurious output from an analog receiver section in response to input tones is a set of output tones at predictable frequencies, whereas the spurious response from an ADC in the same conditions is a combination of predictable spurious output (e.g., intermodulation products) plus non-white quantization noise and spurious associated with quantizer non-linearity and possibly other ADC imperfections.

Yet another concern with practical ADCs is *aperture uncertainty* σ_τ, which is the standard deviation (equivalently, the rms) of the timing error in the sampling of the input by the ADC. Aperture uncertainty is also known as *timing jitter*. For modern ADCs used in receiver applications, the aperture uncertainty is typically on the order of 0.5 ps. Aperture uncertainty contributes error in the discrete-time representation of the input signal which, like quantization error, can be interpreted as an additional noise contribution, and which in turn can be characterized in terms of a signal-to-noise ratio. For sinusoidal input and aperture uncertainty σ_τ, the associated SNR is

$$\text{SNR} = \frac{1}{(2\pi F_S \sigma_\tau)^2} \tag{15.9}$$

One striking feature of this relationship is the sensitive dependence on F_S: For example, we see a doubling of F_S reduces the SNR associated with aperture uncertainty by a factor of 4; i.e., about 6 dB. The same penalty is incurred regardless if the source of the error is internal to the ADC, or external to the ADC; i.e., a limitation of the sample clock. Either way, the resulting degradation in performance may be sufficient to contraindicate the use of an ADC to digitize entire tuning ranges (as was considered in Example 15.1) and may instead force the time-honored approach of using the analog section of the receiver to isolate and downconvert a single signal/channel for digitization. The following example demonstrates the point:

EXAMPLE 15.5

The Analog Devices (Inc.) AD9432 is a 12-bit ADC which, according to the datasheet, achieves ENOB = 11.0 and 0.25 ps-RMS aperture jitter when sampling a 10 MHz tone at 105 MSPS. (a) Characterize the expected SNR, (b) compare to the QSNR of an ideal 12-bit ADC, and (c) compare to the SNR associated with aperture uncertainty alone.

Solution: (a) The ENOB of 11.0 implies an SNR of $1.76 + 6.02 \cdot 11.0 \cong 68.0$ dB, which includes quantization noise as well as thermal noise and all non-ideal products.
(b) The QSNR of an ideal 12-bit ADC is 6 dB greater; i.e., $\cong 74.0$ dB.
(c) The SNR associated with aperture uncertainty is given by Equation (15.9) with F_S = 105 MSPS and $\sigma_\tau = 0.25$ ps, giving 75.6 dB. Therefore as long as the sample clock exhibits jitter less than the rated aperture uncertainty, the effective SNR of this ADC will not be limited by aperture uncertainty.

Aside from limitations in ENOB, spurious performance, and aperture uncertainty, ADCs are subject to power requirements that may be high relative to other receiver subsystems. The power required for digitization is approximately proportional to F_S, which provides a second strong motivator (the other being SNR degradation by aperture uncertainty) to minimize ADC sample rate. The tradeoff between F_S, N_b, performance metrics (ENOB, SFDR, and σ_τ), and power consumption is continuously evolving; see [82] (1999) and [38] (2005) for discussions of the associated trends.

15.3 REQUIREMENTS ON GAIN AND SENSITIVITY

We are now ready to consider two important specifications for a radio receiver: the total gain G; and sensitivity. As explained in Chapters 7 and 12, sensitivity depends fundamentally on external noise, antenna characteristics, and the internal noise of the receiver. The purpose of receiver gain is to boost the received signal to a level that is suitable for digitization or analog detection. For a receiver using analog detection, G should be sufficient that the signal of interest appears at the detector input at a power level that ensures acceptable output SNR; for example, sufficient to ensure the voltage magnitude needed for proper operation of a diode-based envelope detector.

For an ADC-based receiver, G should nominally be large enough that the total noise, including external and internal contributions, dominates over the quantization noise and any significant spurious products associated with the ADC. This latter condition is somewhat more nuanced, so let us explore it in detail.

The consequence of inadequate gain in an ADC-based receiver is that the sensitivity of the receiver will be limited by the dynamic range of the ADC, as opposed to the system temperature T_{sys} associated with the analog section. For an ideal ADC, this condition may be expressed mathematically as follows:

$$kT_{sys}G > \frac{P_{FS}}{\text{QSNR}} \frac{1}{F_S/2} \tag{15.10}$$

Here, T_{sys} is input-referred equivalent noise temperature; i.e., the sum of external noise plus receiver noise, accounting for antenna impedance mismatch and interface losses (per Equation (7.8) and associated text). The factor P_{FS}/QSNR is the total quantization noise power assuming that some other component of the input is large enough to make the ADC input full-scale, and that this signal satisfies the assumptions intrinsic to our expression for QSNR. Since the left side of the expression is in terms of power spectral density, we divide P_{FS}/QSNR by $F_S/2$, under the assumption that the quantization noise is uniformly distributed over a Nyquist zone – which is *at best* only approximately true, as demonstrated in Example 15.2. Thus, Equation (15.10) involves some bold assumptions, and should be used with great care. On the other hand, it is the best estimate of the minimum necessary gain that we have without taking into account specific characteristics of the signals of interest, and yields a convenient starting point for receiver design.

Equation (15.10) assumes that the power present at the input of the receiver is dominated by some signal in the passband, as opposed to the noise represented by T_{sys}. What if the opposite is true? This is very often the case; for example, the signal of interest might just happen to be temporarily weak; or might be weak by design, as in the case of a system like GPS. For receivers intended to process groups of narrowband signals in relatively large bandwidths (think surveillance receivers, spectrum monitoring systems, and Example 15.1) this is essentially a quiescent condition: The signal presented to the ADC is always mostly noise because a large swath of spectrum is being considered, and the sum power of any intermittent signals of interest is weak in comparison to the noise power spectral density integrated over the bandwidth. In any of these situations, the only way the power at the input of the ADC can approach P_{FS}

is if $kT_{sys}BG$ approaches P_{FS}, where B here is the noise equivalent bandwidth delivered to the ADC. Clearly G should not be much larger than the value implied by this relationship, otherwise we would expect to run into trouble with excessive clipping. Thus we have the following approximate upper bound on receiver gain:

$$kT_{sys}BG < P_{FS} \tag{15.11}$$

Now: How does one specify the necessary receiver noise figure (or, equivalently, the receiver noise temperature)? None of the considerations above changes the conclusions reached in Chapter 7. What is new is that we now realize that it is possible to spoil the sensitivity achieved by the analog section of a receiver by inappropriately setting G. Taken together, Equations (15.10) and (15.11) define the range of values of G that preserve the sensitivity achieved by the analog section of a receiver.

EXAMPLE 15.6

A particular receiver consists of an analog section having noise-equivalent bandwidth 1 MHz delivered to an ADC sampling at 10 MSPS. The antenna temperature is 300 K and the receiver's noise figure is 6 dB. The ADC outputs 12 bits and its maximum encodable input level is +10 dBm into 100 Ω (differential). Suggest a range of reasonable gain for the analog section.

Solution: First, recognize that $P_{FS} = +10$ dBm and we will need to assume that the receiver output is properly matched to the ADC input. Similarly we will need to assume that the antenna is properly matched to the receiver input. Under these conditions, the system temperature T_{sys} is the sum of the antenna temperature T_A and input-referred equivalent noise temperature of the receiver T_R:

$$T_{sys} = T_A + T_R = T_A + T_0\,(F-1) = 300\text{ K} + (290\text{ K})\,(3.98-1) \cong 1164\text{ K}$$

assuming that the noise figure $F = 6$ dB $= 3.98$ has been specified for $T_0 = 290$ K. From the problem statement, we infer $B = 1$ MHz, $F_S = 10$ MSPS, $N_b = 12$, and assume that the output bandpass of the receiver lies entirely within a Nyquist zone of the ADC. Approximating QSNR as 6 dB per bit, we have QSNR ≈ 72 dB and therefore $P_{FS}/\text{QSNR} \approx -62$ dBm. Minimum gain may now be estimated from Equation (15.10); we find $G > 38.9$ dB to ensure that noise associated with T_{sys} is greater than ADC quantization noise. Maximum gain may now be estimated from Equation (15.11); we find $G < 117.9$ dB ensures that noise associated with T_{sys} does not result in excessive clipping.

In the above example, we find that reasonable values for receiver gain span a range of $\cong 79$ dB. Considering that individual amplifier stages typically have gain no greater than about 25 dB (as discussed in Chapter 10), this amounts to an design uncertainty of at least three amplifier stages. However, it is possible to narrow this down quite a bit. First, the minimum gain we calculated merely places the power associated with T_{sys} equal to the assumed quantization noise. To ensure that T_{sys} *dominates* over quantization noise, the minimum gain should be significantly higher. Thus, we might add about 10 dB, making the minimum gain 50 dB as opposed to 38.9 dB.

Also, we might want to back off on the maximum gain to create some "headroom" to accommodate uncertainty in the antenna temperature, and – perhaps more importantly – to accommodate to possibility of signals that persistently or intermittently dominate the total power at the input of the receiver. An appropriate margin depends on the conditions in which the radio is operated. The appropriate margin also depends quite a bit on the bandwidth B since a signal of given power more easily dominates the total power if B is small, and makes less of a difference if B is large. This, in a nutshell, provides the primary motivation for larger N_b in modern ADCs; especially for those with F_S that is on the same order-of-magnitude as the bandwidth of the signal of interest. But simply making N_b as large as possible may not be enough, as illustrated in the following example:

EXAMPLE 15.7

Returning to the previous example, let us assume the receiver is for a UHF-band mobile radio, and the signal of interest is transmitted from a base station at constant power. The separation between the mobile radio and the base station varies over a range of 10:1 and is always greater than the breakpoint distance. The propagation is characterized as having a path loss exponent (n, as identified in Section 7.2) equal to 4, with Rayleigh fading. The base station transmit power has been set such that the minimum instantaneous received power (i.e., power available at maximum range and in a deep fade) is greater than the MDS of the analog (pre ADC) section of the receiver 99% of the time. Now what is the range of reasonable gain for the receiver?

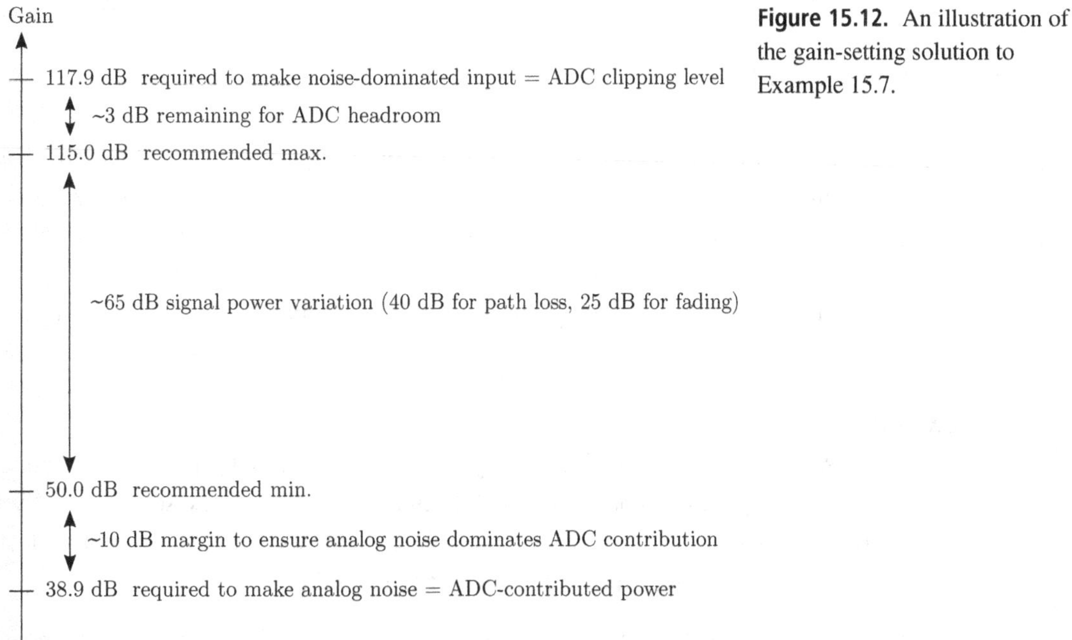

Figure 15.12. An illustration of the gain-setting solution to Example 15.7.

Solution: Recall from Chapter 3 that the power in a Rayleigh fading environment is more than 20 dB below the mean 99% of the time, and can be up to about 5 dB above the mean. Thus, the variation in signal power that the receiver must accommodate is about 25 dB from Rayleigh fading alone. On top of this, the 10:1 variation in range will result in another 40 dB of variation in signal power, assuming R^4 path loss. Thus, the total variation that must be accommodated is at least $25 + 40 = 65$ dB. We found in the previous example that the variation afforded by the dynamic range of ADC was 79 dB, so we have a margin of $79 - 65 = 14$ dB. If we set the minimum receiver gain to 50 dB as suggested above, a mere 2.9 dB of headroom remains.

This situation is shown graphically in Figure 15.12. There is not much remaining "wiggle room" in the gain since going lower reduces the 10 dB margin established for sensitivity, whereas going higher incurs into the 3 dB margin remaining and which is appropriate in order to prevent excessive clipping. Keep in mind also that the analysis so far has involved some pretty cavalier approximations, so this would already represent a very aggressive design.

The above example demonstrates a fundamental challenge in receiver design for mobile radio receivers: Merely by introducing the possibility of signal-of-interest power variation associated with path loss and multipath fading, the range of reasonable receiver gains has diminished to just a few dB. If we account for the possibility of stronger-than-expected signals (e.g., operation close to the base station), the possibility of additional (e.g., log-normal "shadow") fading, the likelihood that the ENOB is significantly less than N_b or SFDR further limits ADC dynamic range, or any number of other reasonable uncertainties, then there may be no fixed gain that simultaneously meets all of our requirements.

So, as a practical matter, gain will need to be varied dynamically in response to signal conditions. This is known as *automatic gain control* (AGC), and is discussed further in Section 15.8.

15.4 PRESELECTION

A *preselector* is a filter appearing at or near the input of a receiver, and which is intended to exclude signals or entire bands of frequencies that are not of interest. Preselectors are nearly ubiquitous in receiver designs, for one simple reason: The radio spectrum is cluttered with signals. Removing signals that are not of interest greatly reduces the linearity required from the receiver, and (as we shall soon see) greatly simplifies frequency downconversion.

EXAMPLE 15.8

In the USA, private VHF-band LMR systems have access to particular channels in the 151–155 MHz range known as "color dot" frequencies.[3] Frequency selection is by tuning an LO, so the first stage of the receiver must pass all frequencies that the user might select. Identify the characteristics of an appropriate preselector in this case.

Solution: Frequencies immediately above and below the VHF color dot frequencies are also primarily for LMR, and will typically arrive at the radio at comparable signal levels. FM

[3] This term refers to the practice of identifying available frequencies using color codes, and labeling equipment with the color codes for easy identification.

broadcast signals in the 88–108 MHz range and TV broadcast signals at 174 MHz and above may be very strong relative to LMR signals. And, there is always a chance of that nearby transmitters at much higher or lower frequencies threaten linearity. Thus, the radio being considered would typically employ a preselector with a response such as that shown in Figure 15.13. The preselector uses a fixed passband of 151–154 MHz. The FM and TV broadcast frequencies are positioned in the stopbands. Frequencies in between are subject to as much attenuation as possible, limited by the filter response. Because this filter would normally appear between the antenna and the first stage of amplification, the insertion loss in the passband contributes to the receiver noise figure. Therefore the passband insertion loss should be small enough not to significantly degrade the receiver's overall noise figure.

As we shall soon see, we may also choose to incorporate some or all of the image rejection filtering required for downconversion when specifying this response.

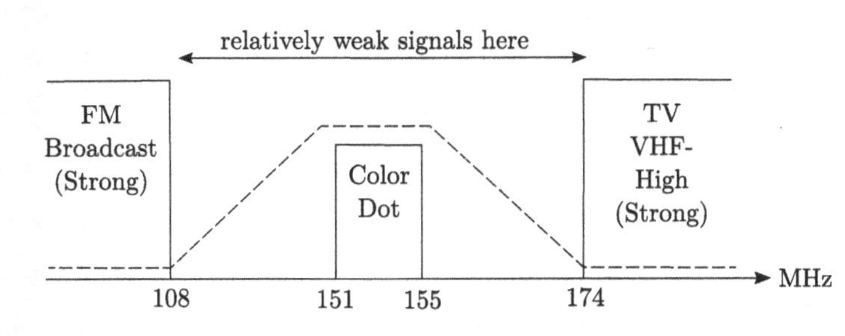

Figure 15.13. The response of a preselector (dashed line) for an LMR radio using US VHF-band color dot frequencies.

The scheme indicated in the previous example is *fixed preselection*: the preselector is simply a bandpass filter. This is suitable in many applications, in particular if the required fractional bandwidth for tuning is small. However, for certain radios fixed preselection is not possible, especially if the radio is required to access a large tuning range. For example:

EXAMPLE 15.9

In North America, broadcast television channels are arranged in contiguous 6-MHz-wide channels from 54–88 MHz (VHF-Low, five channels with a gap from 72–78 MHz), 174–210 MHz (VHF-High, seven channels), and 470–698 MHz (UHF, 38 channels). (Similar schemes are used in other regions of the world.) Identify the characteristics of an appropriate preselector in this case.

Solution: The first thing to consider in this problem is that TV broadcast signals are continuously present, and tend to be very strong. Thus, the linearity of a TV receiver is challenged primarily by TV transmitters, and in some cases also by other comparably-strong persistent signals such as FM broadcast signals. Fixed preselection with a 54–698 MHz bandpass does not provide much protection, since the bandpass would always include signals that are both not-of-interest and likely to cause problems.

Instead, high-performance TV receivers may use *tuning preselectors*, also known as *tracking preselectors*.[4] The scheme shown in Figure 15.14 uses three such preselectors. Each preselector

[4] For more about tracking filters, see Section 13.8.7.

consists of a bandpass filter with variable center frequency – a *tracking filter* – which is used to select a single 6-MHz-wide channel. Multiple tracking filters are required because it is not quite practical to design a single filter that maintains approximately constant bandwidth while tuning over a large frequency range; in particular, over the 13:1 tuning range required for TV. Therefore the receiver input is split into separate paths by band, each with a separate tracking filter. In this scheme, each filter needs only tune over a frequency range of less than 2:1, for which it is practical to achieve the desired 6 MHz bandwidth, at least approximately.

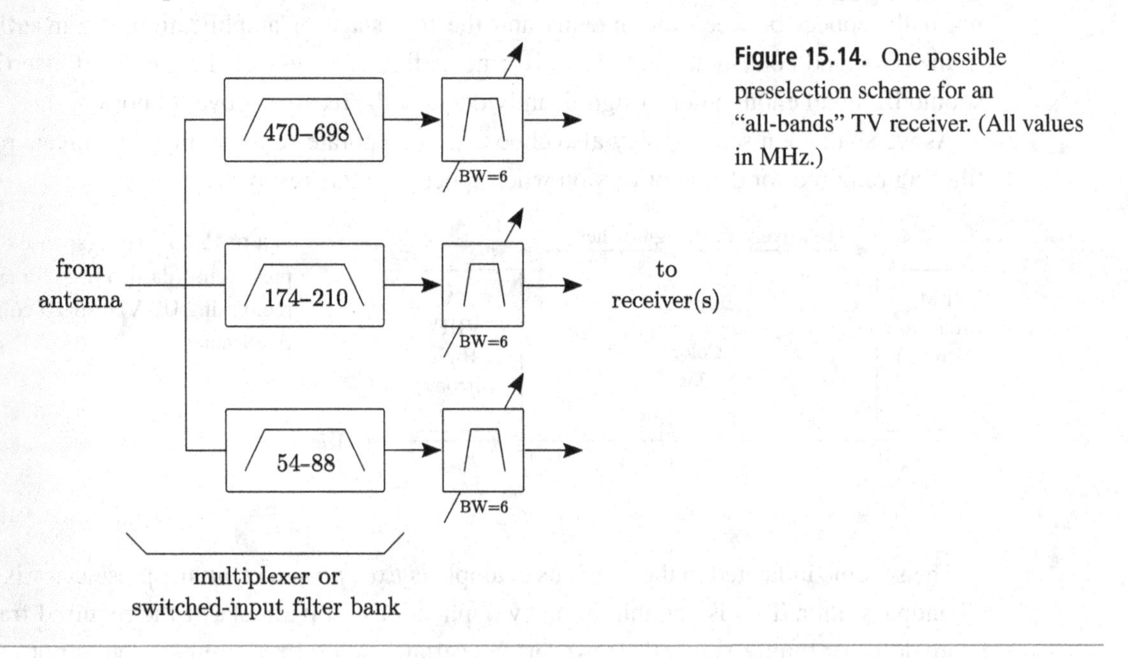

Figure 15.14. One possible preselection scheme for an "all-bands" TV receiver. (All values in MHz.)

Note that in the previous example the receiver uses not only tuning preselection, but also does a preliminary separation of the tuning range into VHF-Low, VHF-High, and UHF subbands. In principle we could implement this by using a power splitter, sending one-third of the total power (minus some insertion loss) and the entire tuning range to each of the subband tuning preselectors. However, this scheme has the drawback that the insertion loss associated with the power splitter contributes directly to the noise figure of the receiver and we lose some gain. Another solution without this limitation is simply to implement the subband splitter as a switch, such that all of the power is directed to the subband that is currently being used. This is advantageous because switches may have smaller insertion loss, near-unity gain, and may be also be smaller and less expensive than a power divider. Even better performance can be achieved by implementing the band splitting using a multiplexer (reminder: Section 13.6).[5] In this way each preselector subband receives nearly all of the power available in the subband, all the time; thus switching is not required. Furthermore, the multiplexer contributes to the selectivity of the design by suppressing all but the band of interest in the process of splitting the input.

[5] Of course this is actually a *demultiplexer*; however, RF multiplexers are typically reciprocal, so the same devices can typically be used as either multiplexers or demultiplexers.

In this chapter, our treatment of preselection is limited in scope to types of preselectors and requirements imposed on preselectors. Chapter 13 addresses the design of the filters and multiplexers that comprise preselectors.

15.5 SELECTIVITY

A preselector ordinarily does not determine the ultimate selectivity of the receiver. For example, in the VHF-High "color dot" example above, the preselector defines the 4-MHz-wide tuning range, and it is up to some combination of downstream filters to "select" the particular narrowband (25-kHz-wide) signal of interest. In the TV receiver example, there would typically be a more-selective filter in an IF stage to provide additional suppression of adjacent TV channels.

Most receivers require considerable additional filtering beyond what a preselector is able to provide. Reasons for this include:

- suppression of spurious products associated with frequency conversion and intermodulation;
- in ADC-based receivers, anti-alias filtering;
- isolation of one particular signal from the tuning range; and
- final limiting of the bandwidth to the maximum possible extent so as to maximize signal-to-noise ratio presented to the detector (be it analog or an ADC).

These tasks are typically accomplished by multiple filters distributed throughout the signal path, as we shall soon see. In receivers using an ADC, one or more of the associated filters may be implemented in the digital domain; see Chapter 18 for more on that.

15.6 RECEIVER ARCHITECTURES

We are now ready to consider the end-to-end design of receivers. As we have already established, the minimum essential ingredients for a receiver are (1) digitization or analog detection, (2) an analog section having gain sufficient to deliver a suitable signal power to the ADC or analog detector (per Section 15.3), and (3) the analog section must have appropriate level of linearity, sensitivity, and selectivity. Let us now consider a few receiver architectures that could be considered canonical.

15.6.1 Lowpass Direct Sampling

Perhaps the most obvious architecture for a receiver that must provide digital output is *direct sampling*, illustrated in Figure 15.15. In this approach, there is no frequency conversion: There is only filtering, gain, and digitization. All subsequent processing occurs in the digital domain, including isolation of individual signals from the bandpass.

Direct sampling receivers may be either *lowpass sampling* or *undersampling*. In a lowpass sampling direct conversion receiver, F_S is selected to place the entire bandpass in the first

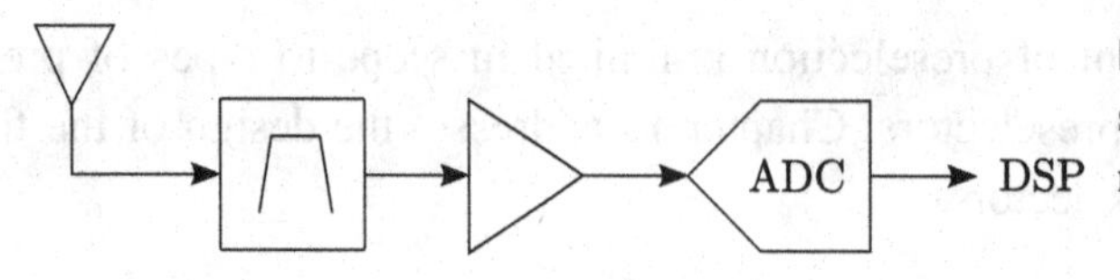

Figure 15.15. Direct sampling. Note this is a simplified block diagram of a receiver that might have multiple stages of gain and filtering.

Nyquist zone; i.e., between 0 and $F_S/2$. Lowpass direct sampling often considered, but rarely chosen.[6] To understand why, consider two commonly-proposed ideas.

- Lowpass direct sampling of the entire HF band (3–30 MHz). This requires F_S greater than 60 MSPS. This eliminates a relatively complex, expensive, and power-hungry multiple-conversion superheterodyne architecture (see Section 15.6.6) that is otherwise required.
- Lowpass direct sampling of the GPS L1 signal (1575.42 $\pm$ 20 MHz). This requires F_S greater than 3.191 GSPS. The GPS L1 signal can be received with relatively simple frequency-conversion architectures, but eliminating frequency conversion mitigates any possibility of associated signal degradation, and the resulting degradation of navigation and timing information.

The principal problem in both cases is that suitable ADCs – in particular ADCs with sufficient F_S for the required N_b (by which we really mean ENOB and SFDR) – are not available, prohibitively expensive, or have prohibitively-high power consumption. In the GPS application, the primary problem is cost and power: ADCs sampling greater than 3.191 GSPS are certainly available, but are extremely expensive and have very high power consumption. In the HF application there is no problem with F_S *per se*, but at this rate ADCs are typically not capable of achieving ENOB and SFDR sufficient to provide performance comparable to that of a multiple-conversion superheterodyne design. In the HF application there is a further complication that the second-order linearity of the receiver must be extraordinarily high. For example: a strong signal at 10 MHz generates a second-order intermodulation product at 20 MHz, which falls within the bandpass of the receiver and therefore cannot be excluded by filtering.[7]

Additional contraindications for lowpass direct sampling in many applications include (1) degradation of SNR through aperture uncertainty associated with the very high sample rate required (see Section 15.2.4); and (2) high cost and power consumption of DSP required to process the digitized signal, because the sample rate typically turns out to be orders of magnitude greater than the bandwidth of the signal of interest.

15.6.2 Undersampling

A potential solution to the sampling rate and second-order linearity issues associated with lowpass direct sampling lie in undersampling. As explained in Section 15.2.2, undersampling refers to digitization of a passband lying entirely within some Nyquist zone *other* than the first Nyquist zone, resulting in an aliased – but otherwise perfect – copy of the signal in the ADC output. This amounts to a "free" downconversion, as demonstrated in the following example.

[6] One of the rare few is the ADAT ADT-200A general-coverage HF receiver; for a technical summary see [16, Sec. 24.22].
[7] Although one might consider this approach with a tuning preselector; see Section 15.4.

EXAMPLE 15.10

Following up the discussion from Section 15.6.1, specify F_S for undersampling of the GPS L1 signal from the second Nyquist zone.

Solution: The answer is $F_S = 2.1006$ GSPS, which places the GPS L1 frequency half way between $F_S/2$ to F_S. This is shown in the middle panel of Figure 15.16. The ADC output will contain the GPS L1 signal centered at $F_S/4 = 525.15$ MHz. A minor inconvenience is the fact that the signal is spectrally-reversed, since frequencies f_2 in the second Nyquist zone map to frequencies $f_1 = F_S - f_2$ in the first Nyquist zone. However, this spectral reversal is easily accommodated in subsequent DSP.[8]

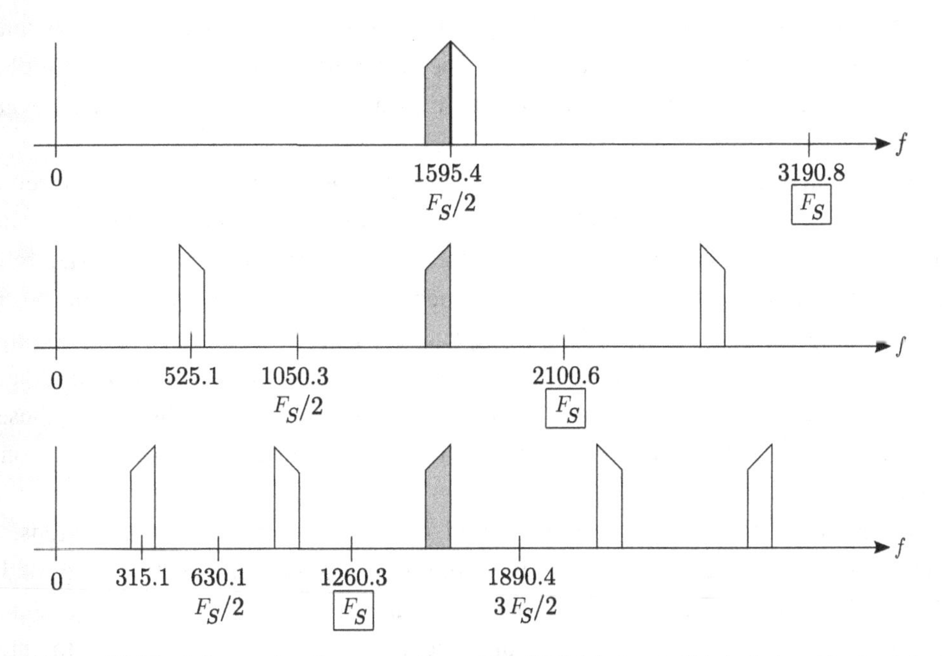

Figure 15.16. Schemes for direct sampling of the GPS L1 signal. Original signal is shaded, aliases are unshaded. *Top:* Minimum-rate lowpass direct sampling. *Middle:* Undersampling from the center of the second Nyquist zone. *Bottom:* Undersampling from the center of the third Nyquist zone.

Aliasing from the second Nyquist zone is not the only option. For example, we might choose the digitize the GPS L1 signal from the center of the third Nyquist zone (F_S to $3F_S/2$), as shown in the bottom panel of Figure 15.16. In this case $F_S = 1.2603$ GSPS, and the signal appears centered at 315.08 MHz and in this case is *not* spectrally-reversed.

The point of the preceding example is that we may undersample using any sample rate that results in a Nyquist zone that completely contains the signal of interest, as long as we remember to account for the spectral reversal associated with the even-numbered zones. (See Problem 15.8 to follow this up.)

[8] Actually, this presents an opportunity in subsequent processing: see Section 18.4.1.

Although undersampling architecture has some advantages over lowpass direct sampling, it is nevertheless not commonly used. The principle limitation is that ADCs have a lowpass analog response (this is due to their intrinsic S/H behavior), and are typically not able to receive signals at frequencies much greater than 3–5 times their maximum rated sample rate. Nevertheless it can be done: An example of a GPS L1 undersampling receiver is presented in [2].

Where undersampling really shines as a technique is not necessarily as a standalone receiver architecture, but rather as a technique for *IF sampling* in a superheterodyne receiver, as we shall see in the following sections.

15.6.3 Tuned RF

For receivers using analog detectors as opposed to ADCs, the architecture that is analogous to direct sampling is known as *tuned RF*: The difference is that the ADC is replaced by an analog detector. Like direct sampling, there is no frequency conversion; only gain and filtering.

Tuned RF architecture has three principal disadvantages. First, filters are required to have very high selectivity and very small fractional bandwidth simultaneously; this is challenging, as we saw in Chapter 13. Second, the detector must perform well at the carrier frequency. The issue here is that analog detectors are much easier to implement, and can be made to be dramatically more sensitive, at lower frequencies. Third, it is often difficult to implement the necessary gain without taking extreme measures to isolate the individual amplifier stages to prevent interstage feedback, in particular through power supply paths. All of these issues provide strong motivation for the superheterodyne architecture, described in Section 15.6.4 and beyond.

Although tuned RF is not commonly used in telecommunications applications, it does commonly appear in other applications; two prominent examples include RFID tags and certain keyless entry systems. The features of these applications that facilitate tuned RF architecture are very modest data rate and BER requirements, high intrinsic immunity to interfering signals (leading to weak selectivity requirements), and the requirement to achieve very low cost.

15.6.4 Single-Conversion Superheterodyne Architecture

Superheterodyne architecture solves many of the problems associated with direct sampling and tuned RF by first performing frequency conversion from the carrier frequency to a lower frequency that is easier to process. The simplest realization of a superheterodyne receiver is the *single-conversion* superheterodyne receiver, shown in Figure 15.17. The essential elements in this architecture are an RF section consisting of preselection and gain, downconversion by mixing, an IF section consisting of gain and filtering, and then either analog detection or digitization. The choice to digitize the IF for further processing, as opposed to using analog detection, was previously identified as "IF sampling." The IF may be either lowpass-sampled or undersampled, and all the same considerations identified in Sections 15.6.1 and 15.6.2 apply.

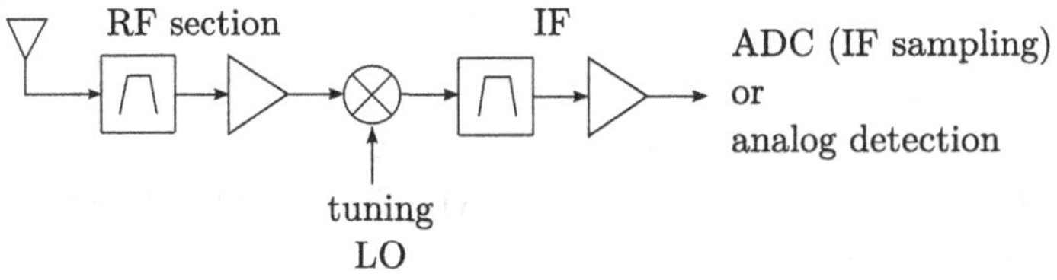

Figure 15.17. Single-conversion superheterodyne architecture. The number and order of gain and filtering stages within the RF and IF sections depend on the application.

The splitting of the receiver into "RF" and "IF" sections has two very attractive benefits. First, it becomes much easier to achieve high gain: This is because the necessary amplifier stages may be distributed between sections, and greater isolation is possible between sections because they operate in different frequency ranges. Second, processing at an IF tends to facilitate greater selectivity. This is because the fractional bandwidth required to achieve a given selectivity is increased with decreasing center frequency, and increased fractional bandwidth is typically what we need to make the desired filter realizable for a given order, as we saw in Chapter 13. Furthermore, additional filter technologies typically become available at the lower frequency used for the IF.

Thus, it is common for superheterodyne receivers to use preselectors with relatively broad bandpasses, and then to set the ultimate selectivity of the receiver using one or more filters implemented in an IF stage.

EXAMPLE 15.11

In most of the world, an FM broadcast receiver must select a single station having bandwidth of about 200 kHz from center frequencies between 87.5 MHz and 107.9 MHz. Assume a single-conversion superheterodyne architecture is used with IF center frequency 10.7 MHz (we will address why this IF makes sense in the next example.) Determine the fractional bandwidth of the filter required to achieved the necessary selectivity if implemented in (a) the RF section, and (b) the IF section.

Solution: If implemented at RF, the fractional bandwidth of a filter having bandwidth of 200 kHz ranges from 0.23% (at 87.5 MHz) to 0.18% (at 107.9 MHz). This is very small. Furthermore, the filter itself must have a tunable center frequency to be usable.

If implemented at the 10.7 MHz IF, the fractional bandwidth is 1.87% – which is much easier to achieve – and has a fixed center frequency. Thus, it makes sense to use fixed preselection to isolate the 20 MHz-wide tuning range, and then to use a fixed filter in the IF stage to isolate the particular station of interest.

The superheterodyne architecture also introduces some new problems. An LO synthesizer is required, which certainly entails significant additional cost and complexity issues. Since frequency conversion is accomplished by mixing with an LO signal, image rejection will be required before the mixer, and additional filtering following the mixer may be required to suppress undesired and spurious products – see Figure 14.2 and associated text for a reminder. The requirement for image rejection typically results in a severe limitation in tuning range or choice of IF frequencies, as we demonstrate in the following example.

EXAMPLE 15.12

Continuing the previous example: An FM broadcast receiver will use a single-conversion superheterodyne architecture to deliver a single station having bandwidth 200 kHz to an analog detector or digitizer (for the purposes of this problem, it could be either). Assuming low-side injection, find the minimum possible IF center frequency, f_{IF}. For this choice of f_{IF}, specify the LO tuning range and the requirements for a fixed image rejection filter.

Solution: A fairly common scheme is to use low-side injection with an IF center frequency of 10.7 MHz, which requires an LO which tunes from $87.5 - 10.7 = 76.8$ MHz to $107.9 - 10.7 = 97.2$ MHz. The associated image band is $76.8 - 10.7 = 66.1$ MHz to $97.2 - 10.7 = 86.5$ MHz; therefore a lowpass filter with cutoff between 86.5 MHz (high end of the image band) and 87.5 MHz (low end of the tuning range) is required. It is now evident why 10.7 MHz is a popular choice: Any significantly lower IF center frequency would require an image rejection filter that excludes part of the bottom end of the required tuning range.

EXAMPLE 15.13

Repeat the previous example, but now assume high-side injection is used.

Solution: Now we require an LO which tunes from $87.5+f_{IF}$ (all values in MHz) to $107.9+f_{IF}$. The associated image band will be $87.5 + 2f_{IF}$ to $107.9 + 2f_{IF}$; therefore a highpass filter with cutoff between 107.9 MHz (the top end of the required tuning range) and $87.5+2f_{IF}$ is required. In other words, we require $87.5+2f_{IF} > 107.9$ by enough to accommodate the transition region of the filter. Taking 1 MHz margin to be adequate (as in the previous example), we again obtain $f_{IF} \geq 10.7$ MHz. Substituting this value back into previous expressions we find the LO tuning range must be 98.2–118.6 MHz and the cutoff frequency of the highpass image rejection filter should be somewhere between 107.9 MHz and 108.9 MHz.

The two previous examples use low-side and high-side injection, respectively, and end up with about the same minimum IF frequency. So, which is preferred? Low-side injection has the advantage that the IF spectrum is not reversed, which is important for analog detection, but not so important if the IF will be digitized and processed as a digital baseband signal. The image band for low-side injection (66.1–86.5 MHz) is used for broadcast TV, whereas the image band for high-side injection (108.9–129.3 MHz) is used for aeronautical communications; the latter will generally be weaker, thus easier to suppress, and so high-side injection has an advantage from this perspective.

15.6.5 The Half-IF Problem

The tuning range of a superheterodyne receiver may be further limited – or compromised – by spurious signals. In particular, superheterodyne receivers are vulnerable to the *half-IF problem*, which is as follows: An undesired signal appearing at a frequency midway between the desired tuning frequency and the LO frequency is not sufficiently suppressed, and so is converted to a frequency equal to one-half the IF center frequency. Then the limited linearity of the associated mixer or subsequent IF amplifier results in the second harmonic of this signal, which lands at the IF center frequency. This particular harmonic is often strong enough to interfere with the desired signal.

EXAMPLE 15.14

Continuing from the previous example: Assume an FM broadcast receiver with single-conversion low-side injection to a 10.7 MHz IF. Characterize the vulnerability of this receiver to the half-IF problem.

Solution: At the low end of the tuning range, the desired frequency is 87.5 MHz, the LO frequency is 76.8 MHz, and so the half-IF frequency is (76.8+87.5)/2 = 82.15 MHz. At the high end of the tuning range, the desired frequency is 107.9 MHz, the LO frequency is 97.2 MHz, and so the half-IF frequency is (97.2+107.9)/2 = 102.55 MHz. We see that the upper portion of the tuning range is vulnerable to the half-IF problem, since when tuning these frequencies, the half-IF frequency appears in the lower portion of the tuning range and therefore cannot be eliminated by preselection.

When the half-IF problem cannot be avoided (as in the above example), the only recourse is to increase the second-order linearity of the mixer and/or IF stage. This is often not practical, and as a result the receiver may suffer from a *birdie*: a signal in the output which originates from a spurious linearity or mixing product. The half-IF problem is just one means by which birdies may appear in superheterodyne receivers.

15.6.6 Multiple-Conversion Superheterodyne Architecture

We observed in the previous section that single-conversion superheterodyne architecture is limited in how low the IF can be. This is also effectively a limitation on tuning range for a given IF. If a lower IF is required, or if a greater tuning range is required, then superheterodyne architecture with multiple frequency conversions is a possible solution. A dual-conversion superheterodyne architecture is shown in Figure 15.18.

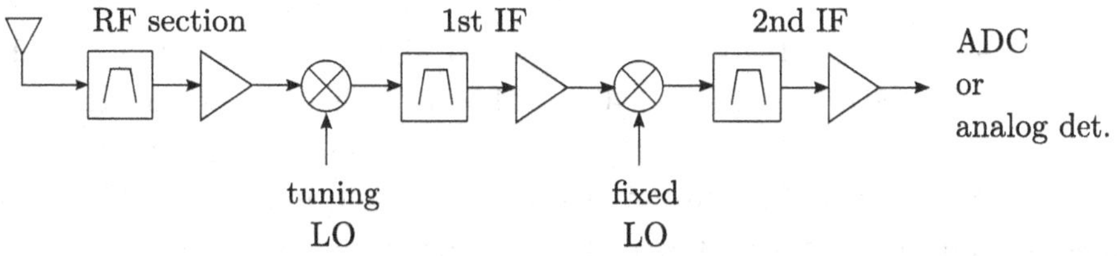

Figure 15.18. Dual-conversion superheterodyne architecture. The number and order of gain and filtering stages within the RF and IF sections depend on the application.

EXAMPLE 15.15

In Example 15.12 we ended up with a single-conversion low-side injection superheterodyne receiver having an IF of 10.7 MHz. This IF is still relatively high compared with the bandwidth of the signal of interest ($\approx$ 200 kHz), so it is reasonable to consider a second lower IF to facilitate either higher-performance analog detection or a lower-rate ADC. A common choice for this second IF is 455 kHz. Complete the design.

Solution: The second conversion could use either high-side or low-side injection. If spectral reversal is a concern, then low-side injection should be selected. In this case, the frequency of the second LO will be $10.7 - 0.455 = 10.245$ MHz and the image band for the second conversion will be centered at $10.7 - 2 \cdot 0.455 = 9.79$ MHz. Therefore a well-designed radio will include a filter before the second mixer that provides adequate suppression of a bandwidth greater than 200 kHz around 9.79, as well as accounting for any other mixing or spurious products emanating from the first mixer. The fractional bandwidth of the final IF filter is $0.200/0.455 \cong 44\%$, a dramatic improvement in terms of implementation effort over the 1.87% fractional bandwidth if all the selectivity must be achieved at the 10.7 MHz IF.

Here is an example of how a multiple-conversion superheterodyne architecture becomes necessary to support a specified tuning range:

EXAMPLE 15.16

In North America, 869–894 MHz is used for GSM cellular uplinks (i.e., mobile to base station communications). Assume that it has been decided that a final IF of 10.7 MHz will be used in a new receiver for this application. Can this be achieved using single-conversion? Double-conversion? Propose a frequency plan.

Solution: For low-side single conversion, the image band for the first conversion is $869 - 2 \cdot 10.7 = 847.6$ MHz to $894 - 2 \cdot 10.7 = 872.6$ MHz. Note that this cannot be achieved using a fixed image rejection filter, since the low end of the tuning range would have to be abandoned in order to ensure adequate (actually, any) image rejection. For high-side single conversion, the image band for the first conversion is $869 + 2 \cdot 10.7 = 890.4$ MHz to $894 + 2 \cdot 10.7 = 915.4$ MHz: Same problem, except now the high end of the tuning range is affected. Therefore if image rejection filtering must be fixed, then single-conversion is not reasonable.

A typical scheme for double conversion from this frequency band is via low-side injection to an IF of 240 MHz, as shown in Figure 15.19. In this case, the tuning range of the first LO is $869 - 240 = 629$ MHz to $894 - 240 = 654$ MHz, and the associated image band is $869 - 2 \cdot 240 = 389$ MHz to $894 - 2 \cdot 240 = 414$ MHz. Either low-side or high-side injection would be suitable, since there is a large gap between the tuning range and the image band in either case; here low-side is selected simply because it results in lower LO frequencies, which are somewhat easier to synthesize. The frequency of the second LO is $240 - 10.7 = 229.3$ MHz, again using low-side injection. This is not the only reasonable solution, but it is typical.

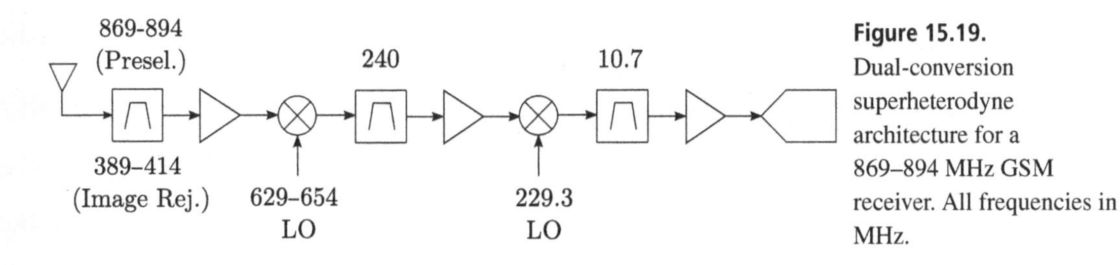

Figure 15.19. Dual-conversion superheterodyne architecture for a 869–894 MHz GSM receiver. All frequencies in MHz.

Summarizing: Advantages of double conversion relative to single conversion are: (1) potential for significant improvements in the tuning range that can be achieved with fixed preselection, (2) improved selectivity, (3) more convenient/flexible distribution of gain (since now there are three sections operating at different frequencies), and (4) a greater range of options for image band placement vs. spectral reversal. Concerning the last point: Note that in a dual-conversion receiver, it is possible for both conversions to be high-side, so in the FM radio example we could have the desirable high-side image band for the first conversion *and* a relatively low IF without spectral reversal.

Disadvantages of dual-conversion relative to single-conversion architecture are significantly increased power consumption due to the need for two LO synthesizers, increased cost and size owing to the additional stage of conversion, and greatly increased potential for generation of troublesome spurious signals. Mitigating birdies in multiple-conversion superheterodyne receivers requires careful planning of IF frequencies and bandwidths, and often requires limitations in tuning range, substitution of components to increase linearity, and additional filtering: More on this in Section 15.7.

The question might now arise: Why stop at two conversions? There is not much advantage in a third conversion for an FM broadcast receiver, since the proposed 455 kHz IF resulting from two conversions in Example 15.15 already has a fractional bandwidth of about 44%. However, the GSM base station receiver considered in Example 15.16 could very well benefit from a third conversion to (for example) 455 kHz. Generalizing, systems that commonly employ triple-conversion superheterodyne architecture are those designed for relatively-small bandwidth signals at relatively high carrier frequencies, and for which relatively high selectivity is required: Besides 800 MHz GSM, other prominent examples include UHF-band LMR and certain satellite telecommunications systems, such as Iridium.

15.6.7 Other Superheterodyne Architectures

In previous sections we noted that tuning range may be constrained by the choice of IF: In particular, we found a higher minimum IF is required to accommodate a specified tuning range with fixed image rejection filtering. However, some receivers require tuning over large contiguous frequency ranges where a sufficiently low first IF is not possible. Prominent examples are HF-band general-coverage receivers, for which a typical tuning range would be 3–30 MHz; and spectrum analyzers, which commonly tune from just a few kHz to frequencies in the SHF range.

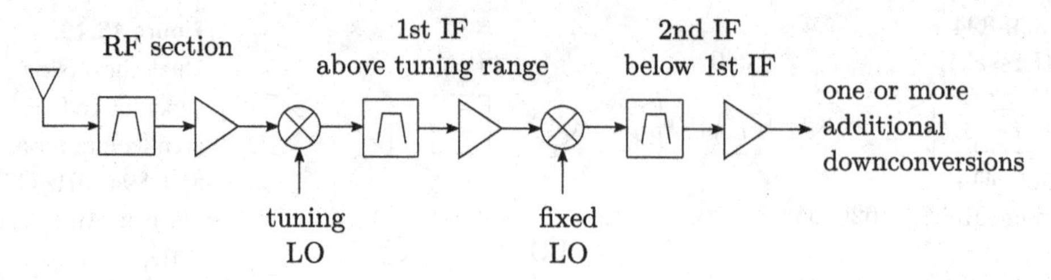

Figure 15.20. "Up-down" superheterodyne architecture.

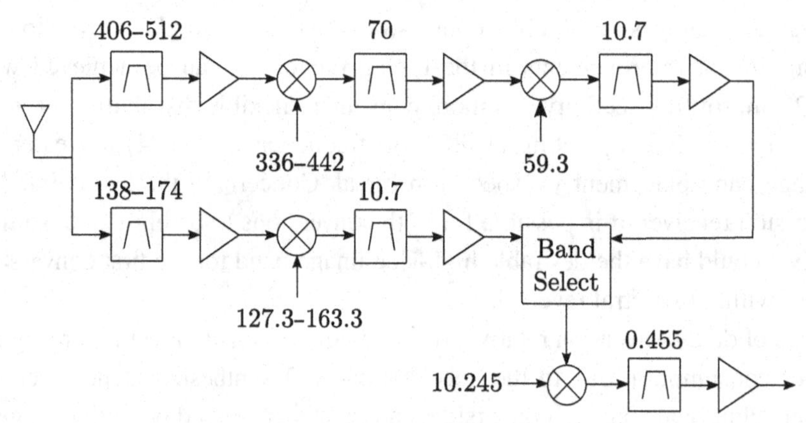

Figure 15.21. An example of a multiband receiver that is able to operate at both 138–174 MHz and 406–512 MHz using shared IFs at 10.7 MHz and 455 kHz. All frequencies in MHz.

Up-Down Architecture. Assuming direct sampling is not acceptable, the solution is an "up-down" superheterodyne architecture. As shown in Figure 15.20, the first IF of an up-down superheterodyne receiver lies *above* the tuning range, and the remainder of the receiver is essentially a conventional single- or multiple-conversion superheterodyne receiver. The first LO is chosen such that the image band lies at higher frequencies with arbitrarily large separation from the desired tuning range, facilitating image rejection filtering which is fixed, yet highly effective.

"Divide and Conquer" Architecture. In principle there is no tuning range too large to be accommodated by up-down superheterodyne architecture. However, the required first IF frequency may turn out to be unacceptable, either because the required LO frequency is too high, or because the desired selectivity cannot be achieved because the required fractional bandwidth is too small at the first IF. Prime examples include TV receivers (recall Example 15.9) and certain multiband radios designed for military, LMR, and amateur radio applications. Like TV receivers, radios in the latter category require the ability to tune over multiple bands; e.g. a police radio that may be required to access LMR systems at both 138–174 MHz and 406–512 MHz. In this case it may be necessary to implement each band as a separate receiver, as shown in Figure 15.21. In this example, the two bands are handled using double-conversion and triple-conversion architectures, respectively; and then merged to use a single common final IF at 455 kHz to facilitate the use of a single detector or ADC.

Image Rejecting Architectures. We have noted repeatedly above that image rejection looms large as a challenge in superheterodyne architecture. A potential work-around is to implement the first conversion using the Weaver or Hartley techniques described in Section 14.5. Taking into account practical limitations – in particular, gain and phase imbalance – such techniques can be expected to achieve on the order of 30 dB of image rejection, potentially eliminating the need for fixed image rejection filtering, and thereby removing the associated constraints on tuning range. Unfortunately this level of image rejection is not adequate for most applications. Furthermore, these techniques entail a considerable increase in the complexity of the design. For these reasons, this tactic is only occasionally employed.

15.6.8 Direct Conversion

We noted in Section 14.4 that it is possible to convert a bandpass signal directly to an IF of zero, resulting in a complex baseband representation of the signal (see Figure 14.10 and associated text). We referred to this as "quadrature downconversion." When used as a receiver architecture, this scheme is commonly known as *direct conversion*. A direct conversion receiver architecture is shown in Figure 15.22. Big advantages of direct conversion over any superheterodyne scheme are arbitrarily large tuning range, simplicity (and subsequent happy implications for cost, size, and weight), and the utilization of the lowest possible IF frequency – i.e., zero – which facilitates the lowest possible ADC sampling rate – i.e., twice the low-pass bandwidth, which can be approximately equal to the bandwidth of the signal of interest. The fact that the signal is delivered in complex baseband form may also be a significant advantage over IF sampling by eliminating the need for quadrature conversion in the digital domain.

Unfortunately direct conversion also entails many disadvantages, most of which are serious enough to make its use inadvisable in applications where this architecture might otherwise be attractive. Here they are:

(1) The requirement for the LO frequency to be in the center of the RF passband creates two additional problems. First, LO power may leak out the input, and is impossible to suppress by filtering. This creates interference for other devices. Second, the LO leaked into the RF path is downconverted with the signal of interest and appears as a "tone jammer" at DC.
(2) As previously explained (Section 14.4) the performance of the receiver becomes critically dependent on the balance of gain and phase between baseband output paths. In practice, this balance will always be imperfect, owing to imbalance in the LO quadrature splitter and small differences in the gain and phase of the baseband amplifiers. The consequences of imbalance include generation of a tone at DC – i.e., exactly in the center of the passband – and distortion of the signal constellation, which is particularly problematic for higher-order phase-amplitude modulations. Mitigation of this problem typically requires *in situ* calibration, which dramatically increases the complexity of the receiver.

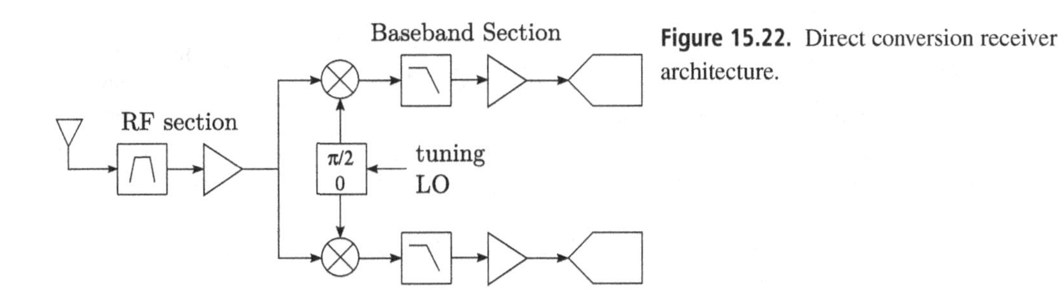

Figure 15.22. Direct conversion receiver architecture.

(3) The placement of the final IF at zero exposes the signal of interest to a multitude of internally-generated interference signals which are strongest at frequencies near DC; prominent among these being power supply noise and digital switching noise.
(4) All gain must be either at RF, or at baseband. Baseband gain is a concern because the second-order intermodulation generated by baseband amplifiers falls in-band, and therefore these amplifiers must have very high second-order linearity.

Despite this intimidating list of problems, two technology trends are driving toward increased adoption of direct conversion architecture. The first is increased use of differential circuits and signals, which goes a long way toward increasing the realizable second-order linearity that is at the root of problem (4), and leads to improved common-mode rejection and thereby addresses problem (3). (See Sections 8.7 and 11.2.5 for a reminder as to why this is.) The second trend is the ever-improving ability to implement receiver functionality in RFIC form; and in particular in CMOS. The integration of RF and digital functionality on the same IC make it feasible to consider the kind of calibration required to tackle problem (2).

For additional reading on the analysis and design of direct conversion receivers, [47] and [62] are recommended.

15.6.9 Near-Zero IF

A slight tweak on the direct conversion concept is *near-zero IF* architecture, previously mentioned at the very end of Chapter 14. A near-zero IF receiver is essentially a direct conversion receiver which is tuned such that the center frequency is just above or below the signal of interest. Thus, the LO does not overlap the signal of interest at RF or baseband. This provides some relief from problems (1), (3), and (4) at the expense of increasing the required output low-pass bandwidth and incurring some additional digital processing to account for the off-center signal of interest.

15.6.10 Superheterodyne Architecture with Quadrature-Conversion Final Stage

A very common approach in modern receivers is shown in Figure 15.23; this consists of a superheterodyne receiver followed by quadrature conversion. The single- or multiple-conversion superheterodyne section converts the signal to a fixed IF frequency, providing most of the selectivity and gain in the process. Quadrature conversion is then used in lieu of IF sampling to reduce the effective IF to zero, facilitating the lowest possible ADC sampling rate

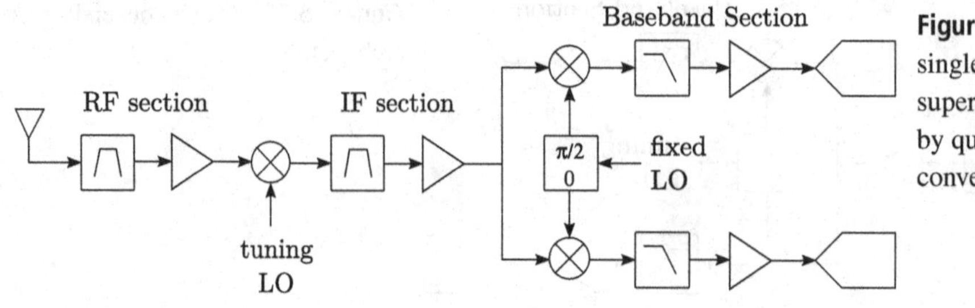

Figure 15.23. A single-conversion superheterodyne followed by quadrature down-conversion.

and eliminating the need for quadrature conversion in DSP. Thus, this architecture combines the advantages of superheterodyne and direct conversion architectures, and goes a long way toward mitigating the disadvantages of each architecture as used alone. The benefits are so compelling that this architecture is nearly ubiquitous in some applications, including in particular most mobile telecommunications applications.

15.7 FREQUENCY PLANNING

Frequency planning refers to the process of selecting IF center frequencies and injection modes (i.e., low-side or high-side) in a superheterodyne receiver. As we noted in Section 15.6, these selections have implications for preselection, image rejection filtering, and post-mixer spurious rejection filtering. A particular challenge in frequency planning is anticipating all relevant spurious signals as they propagate through the receiver and generate new spurious products through mixing and intermodulation. Once identified, these spurious signals determine specifications for filter passband and stopband characteristics, and linearity requirements for amplifier stages.

For receivers with tuning ranges in crowded sections of the radio spectrum, the associated frequency planning and assignment of filter and amplifier specifications can become extraordinarily difficult. This is especially true for multiple-conversion superheterodyne receivers. In such cases it is common to employ graphical techniques to track and identify spurious products, or to use software to similar effect. Recommended reading on this topic is [65, Sec. 3.5].

In Section 15.6, certain IF center frequencies appeared repeatedly; in particular, 10.7 MHz and 455 kHz. In the FM broadcast receiver example, we noted that 10.7 MHz emerges as a reasonable choice because it is the minimum frequency which facilitates a single-conversion design with fixed image rejection filtering. Reasons for choosing particular IF center frequencies in other problems were not as specifically well-motivated. In practice, RF designers tend to choose IF center frequencies from a small set of commonly-used values, as opposed to seeking values which minimize IF center frequency. The principal reason for this is that the design of IF filters can be relatively challenging, in particular if these filters are used to set the selectivity of the receiver. It is often possible to purchase a ready-made bandpass filter at a standard IF frequency with performance that is superior to a custom design, and perhaps also at less cost and in a more desirable form factor. Often this is because a particular technology is required to realize the filter, and which is implementable using specialized design and manufacturing techniques, but not practical to construct from discrete components or typical distributed filter techniques.

It is not cost-effective for manufacturers to provide such filters for a continuum of possible values. Therefore, the industry has converged on a discrete set of center frequencies that accommodate most needs. The lowest such value is 455 kHz, which is commonly used as the final IF in single-channel-output superheterodyne receivers for modulations that are on the order of 250 kHz or less. In addition to 10.7 MHz, other common IF center frequencies include 45 MHz, 70 MHz, and 140 MHz. Beyond these values there are others that are popular for particular applications.

15.8 GAIN CONTROL

Now that we have identified the canonical receiver architectures, let us return to the issue of required gain (Section 15.3) and specifically to the issue raised by Example 15.7 and in particular by Figure 15.12. In that example, we identified the possibility that 50 dB gain would be adequate, but only after making quite a few assumptions. It follows that there are some good reasons to allow the gain to be variable. Of course if we do this, then the gain must be *automatically* variable, which raises the question of how the receiver will determine the correct amount of gain and then how it will be varied. This concept is known as *automatic gain control* (AGC).

Appropriate AGC strategies depend on the architecture of the receiver, and receivers that employ AGC typically require different strategies for different stages of the receiver. This is another application for stage-cascade analysis as presented in Section 11.5, since the various AGC strategies incur different tradeoffs between sensitivity and linearity.

15.8.1 AGC Strategy for a Single-Channel-Output Receivers

Let us first consider a superheterodyne receiver in which the bandwidth of the final IF stage is approximately equal to the bandwidth of a channel of interest, and in which the output of this IF stage is then delivered to either an analog detector or an IF sampling ADC. Here "channel" refers to the smallest assignable division of spectrum in the band, so this is a single signal in some systems, but may be multiple signals in a system using CDMA. It makes sense to allow the total gain of the receiver to vary in such a way that the output power is nearly constant and equal to the power that optimizes the performance of the analog detector, or which makes the best use of the dynamic range of the ADC. There may also need to be some mechanism that prevents the gain from lingering at the maximum setting in the absence of a signal, as this could result in a signal overload condition when the signal first appears.

This leaves the question of how to distribute the variation of total gain. For example, one possibility is to simply to allow the gain of the final IF stage to vary as necessary. However, this may not be practical, as the final IF stage may not have maximum gain sufficient to accommodate the range of variation. Also, this requires earlier RF/IF stages to accommodate the full range of variation of the signal, as opposed to being able to benefit from a reduced range of signal power variation that a distributed AGC strategy would provide.

A more common strategy is something like Figure 15.24. In this double conversion superheterodyne receiver, all three stages participate in the gain control, as follows:

- The second IF implements a "fast" gain control mechanism, which implements variations at a rate comparable to the coherence time of the propagation channel. (Reminder: Section 3.5.3.) When the channel bandwidth is "narrowband" in the propagation sense, then this mechanism is essentially tracking propagation channel magnitude. In Example 15.7, this would be tracking Rayleigh fading.
- The first IF implements a "slow" gain control mechanism, which tracks slower variations, such as those associated with path loss and shadowing. It also makes sense to implement a slower form of gain control in this stage when the wider bandwidth of the first IF stage is likely to include multiple signals.

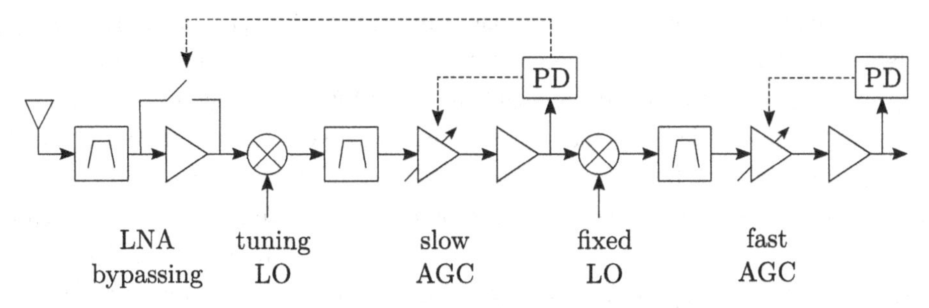

Figure 15.24. Typical AGC strategy for a double-conversion, single-channel-output superheterodyne receiver. "PD" stands for "power detector". Dashed lines indicate gain setting decisions.

- The RF stage implements an even slower gain control mechanism: A two-state control, which allows the LNA to be bypassed when the total power is very high. The advantage of LNA bypassing is that it is simple to implement, and facilitates a tradeoff between high linearity (most important when total power is high) and high sensitivity (most important when total power is relatively low).

The considerations for single-channel-output receivers using direct sampling, tuned RF, and direct conversion architectures are essentially the same. However, such receivers tend to have fewer gain stages, which tends to limit options for implementation of AGC.

Finally it is important to note that AGC considerations do not necessarily stop at the final analog IF or baseband output of the receiver: We have identified numerous instances of AGC mechanisms in the demodulators described in Chapters 5 and 6. These may also be taken into account as part of the radio's overall AGC strategy.

15.8.2 AGC Strategy for Multiple-Channel-Output Receivers

Some receivers are designed to digitize multiple channels or an entire band of signals at once, with all subsequent processing – including subsequent stages to separate the channels – to occur in DSP. For example, some base station receivers in modern cellular telecommunications systems operate in this manner. Example 15.1 is another instance. In such receivers fast AGC is likely to backfire; for example, fast AGC response to a strong signal that is currently dominating the total power might degrade performance for weaker signals of interest that are independently varying in strength. Thus, superheterodyne receivers that deliver multiple channels in a single output tend to have only slow gain control with a relatively small range of gain variation, or, in some cases, no AGC at all. Because such receivers must accommodate the strongest and weakest possible signals simultaneously, and because only slow AGC is possible, the receiver is required to have relatively good linearity and an ADC with relatively high ENOB and SFDR.

15.8.3 AGC Strategy for Cellular CDMA Receivers

In cellular CDMA systems a similar problem arises. In these systems multiple transmitters use the same channel simultaneously. At the receiver, some of these will be relatively strong and some will be relatively weak. If AGC responds only to power in a particular frequency channel, then it is in effect optimizing the reception of the strongest signals to the detriment of the weaker signals. Furthermore there is little point in dynamically-varying gain in the IF stages, since the individual signals fade *independently*, because they experience different channel impulse responses; and fade *less*, because CDMA signals are relatively broadband compared to the channel coherence bandwidth. For these reasons AGC in CDMA receivers is typically limited to controlling the gain in the first RF stage of the receiver, often just controlling the gain of the LNA. Because this may not be entirely adequate, control of the transmit power of mobile stations is a much greater concern in CDMA cellular systems.

15.8.4 Power Measurement for AGC

In order to implement AGC, there must be some mechanism for measuring power at the appropriate places within the receiver. The power at the output of the receiver is straightforwardly determined in receivers using ADCs and in single-channel-output receivers using analog amplitude modulation (including CW and SSB) detection. However, most receivers have significantly greater bandwidth in earlier stages than in later stages. When this is the case, it is possible to improve AGC performance by measuring power at multiple stages as indicated in Figure 15.24. This can be achieved by using a coupler to tap a small amount of power at the point of interest, from which an analog measurement of power may be made.

The principal challenges in analog power measurement for AGC are sensitivity and dynamic range. In this case, "sensitivity" refers to the lowest power that can be determined accurately, and "dynamic range" refers to the range of input power that can be determined accurately. These requirements often preclude the use of envelope detection (e.g., Figure 5.9), which typically has terrible sensitivity and dynamic range. Modern "RF detectors" typically use amplifier cascades to improve sensitivity, followed by logarithmic amplifiers ("log amps") for high dynamic range power detection.

A fundamental tradeoff in the design of an AGC power measurement circuit is integration time (i.e., averaging) vs. level-tracking accuracy. When the power being measured is constant, the measurement device reports a value that is nominally also constant, but which also includes measurement noise. The variance of this noise can be reduced by integration (i.e., averaging). However, increasing averaging reduces the effective "time constant" of the measurement device; e.g., a discrete jump in input power is reported as a gradual change in input power, so the AGC is slower to react. The effective time constant of the power measurement device must be appropriate for the desired rate of AGC variation, which is application-specific.

15.8.5 Schemes for Varying Gain

Implementation of AGC requires a means to vary gain. There are three common techniques: bypassing, digital step attenuators, and variable gain amplifiers.

Bypassing is a binary form of AGC in which the two states are (1) signal routed through an amplifier, and (2) signal routed around the amplifier. As shown in Figure 15.24, this is most often used at the input of a receiver, and in particular to bypass the LNA. LNA

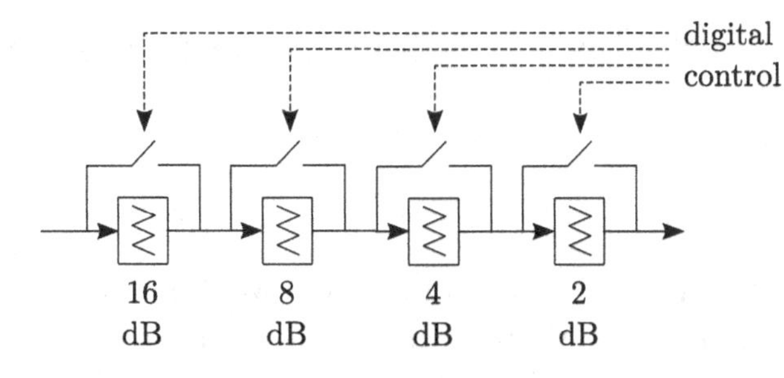

Figure 15.25. Block diagram of a typical digital step attenuator.

bypassing facilitates a straightforward tradeoff between a high-sensitivity, low-linearity state (LNA in-line) and a lower-sensitivity, higher-linearity state (LNA bypassed). LNA bypassing is often seen in direct conversion receivers, where there may be few other options; and in HF-band general coverage receivers, where an LNA is not strictly necessary and the linearity improvement resulting from bypassing may be more useful in some signal environments. In some HF-band general-coverage receivers, this choice is not implemented automatically, but rather is made available to the user. In both applications, bypassing may be considered to be a form of preselection dynamic range optimization, as opposed to AGC *per se*.

Digital step attenuators are cascades of one or more fixed attenuators, any of which may be bypassed under digital control. A typical scheme is shown in Figure 15.25. Here, selectable values range from 0 dB to 30 dB in 2 dB steps, plus a small amount of irreducible insertion loss associated with the bypass switches. Individual fixed attenuators are typically implemented as impedance-matched "π" resistor networks, which achieve attenuation through power dissipation. To avoid spoiling the noise figure of the receiver, these devices are typically employed in an IF stage as opposed to the RF stage.

Variable gain amplifiers are amplifiers in which gain is varied by directly varying the gain of one or more transistor amplifier stages. This can be done by making some aspect of the bias variable; e.g., in bipolar transistor amplifiers this can be accomplished by varying the DC base voltage, varying the effective resistance used for emitter degeneration, or modifying collector current.

15.9 CASE STUDIES

Although we have now covered the fundamentals, receiver design is a broad topic that is difficult to summarize in a comprehensive way. The considerations that lead to particular architectures, frequency plans, and gain control schemes are complex and application-dependent.

One way to gain an appreciation of the relevant issues in practical design is by studying examples of successful designs. In a few cases this is possible by studying the research or educational literature in a certain application area. In other cases this is more readily accomplished by studying datasheets, application notes, and associated design examples. Well-documented receiver design examples are invaluable for both students and experts as examples of "known good" design strategies and techniques, as well as a source of ideas generally.

What follows are some suggestions on where to look next for accessible design examples, organized by application area.

15.9.1 AM/FM Broadcast Receivers

Modern audio broadcast receivers typically consist of one or two ICs that handle both AM (about 500–1800 kHz) and FM (about 88–108 MHz) signals.[9] Despite the disparate frequency bands (MF vs. VHF), large tuning ranges (137% and 20% in fractional bandwidth, respectively), and different modulations, both AM and FM can be accommodated on the same ICs with only a few remaining functions – in particular, antenna interfacing, preselection, IF filtering, and a frequency reference crystal[10] – implemented off-chip. To be clear, these functions can also be implemented on-chip, but the limitations of RFIC technology are such that the performance is relatively poor, and so this is normally not done unless there is no alternative.

FM broadcast receivers typically use dual-conversion superheterodyne architecture with IFs at 10.7 MHz and 455 kHz, as explained via Examples 15.11–15.15 and in Section 15.7. Since users normally do not listen to AM and FM simultaneously, it is possible to use the same signal path for AM. In this case the receiver may be operated as an up-down superheterodyne system (Section 15.6.7), which conveniently accommodates the very large fractional bandwidth of the AM broadcast band.

AM and FM are relatively simple to demodulate (refresher: Chapter 5). As a result all-analog (i.e., no digitization) designs are relatively common as this approach is somewhat less expensive and has potentially lower impact in terms of size, power consumption, and thermal management. An example of a modern all-analog AM/FM broadcast receiver IC is NXP TEF6901A [54]. This IC is designed for car radios and is typical in that LO synthesis and AGC are implemented on-chip. LO synthesis is by on-chip phase-locked loop synthesizer using an off-chip 20.5 MHz crystal for the reference oscillator. (See Section 16.8 for more about this scheme.)

An example of a modern AM/FM broadcast receiver that uses IF sampling architecture (Section 15.6.4) is the STMicroelectronics TDA7528 IC [75]. This IC is a single-conversion superheterodyne receiver that outputs a 10.7 MHz analog IF intended to be processed by a companion IC containing the ADC and digital signal processing. An incentive for DSP processing is to accommodate the various information-bearing subcarriers that can be embedded in AM and FM broadcast signals.

15.9.2 Television Tuners

For historical reasons, the analog section of a television receiver, terminating in an analog IF, is known as a *tuner*. Like AM/FM broadcast receivers, television tuners are commonly implemented in IC form with on-chip LO synthesis and AGC, and off-chip antenna interfacing,

[9] These are the approximate North American frequency ranges; the ranges in other regions differ.
[10] More on this in Section 16.5.

preselection, IF filtering, and a frequency reference crystal. A representative example of a modern television tuner IC is the Maxim MAX3544 [49], which tunes 47–862 MHz using a single-conversion superheterodyne architecture with a tracking filter as the primary means of providing channel selectivity. The chip outputs an analog IF centered at 36 MHz. Since modern television signals use relatively complex digital modulations (see e.g. Section 6.17) a "baseband processing" IC is required to digitize and demodulate the signal. In principle subsequent processing could begin with either IF sampling or analog quadrature downconversion; however, IF sampling is currently favored since the 6 MHz bandwidth at an IF of 36 MHz (44 MHz is also common) is well within the capabilities of modern ADCs. Other television tuner ICs include ADCs and digital demodulation on the same chip as the tuner in order to achieve a higher degree of integration. Other architectural approaches are applicable with various tradeoffs; see e.g. [86] for a representative example of the considerations associated with tuner design.

15.9.3 HF Receivers

HF receivers present a unique set of design considerations and challenges. Generally HF receivers can be categorized as either "general coverage" or "special purpose." A *general coverage* HF receiver is one which is designed to cover most or all of the 3–30 MHz HF band and is designed to demodulate most or all of the disparate forms of modulation found there. An accessible discussion of HF general coverage receiver design can be found in [70, Sec. 4.7]. A technical summary of the design of the popular high-performance Kenwood TS-870 triple-conversion IF sampling general coverage transceiver can be found in [65, Sec. 10.2.2].[11]

For a completely different architectural approach, the ADAT ADT-200A provides a useful case study in direct sampling architecture. A technical summary of this commercially-available receiver – originally designed by amateur radio experimenters – can be found in [16, Sec. 24.22].[12]

A *special purpose* HF receiver might be defined as one which is not "general coverage"; these appear in myriad applications including long-range government and military communications (see e.g. Section 7.9), long-range telemetry, and amateur radio. Rutledge (1999) [68], an undergraduate-level textbook on radio electronics, describes the design of the Norcal 40A 40-m band ($\approx$ 7 MHz) amateur radio as a detailed tutorial.

15.9.4 Cellular, WLAN, and Global Navigation Satellite Systems (GNSS) Receivers

The common features in these applications are (1) UHF–SHF band operation, (2) extremely large market, and subsequently (3) extreme levels of hardware integration. As a result, the design of receivers for these applications has become an increasingly highly-specialized topic, and it is uncommon to encounter design details in the open literature. However, the

[11] See also the detailed description of the Rhode & Schwartz EB200 HF through UHF general coverage receiver in the same book [65, Sec. 1.5.4].

[12] This book [16] is an interesting read for those interested in experimental receiver design.

principles described in this chapter are certainly relevant and applicable. Some examples that are readily accessible (although potentially dated, given the pace of technical development in these applications) are as follows: Descriptions of 5 GHz WLAN transceivers can be found in Lee (2004) [39, Sec. 19.6.2–4] and Razavi (2012) [63, Ch. 13]. Lee (2004) also includes a GPS receiver example [39, Section 19.6.1]. Besser & Gilmore (2003) [6, Sec. 3.3] include an example of receiver for a CDMA mobile phone.

15.9.5 Quadrature Conversion RF/IF Receivers

As noted in Section 15.6.10, analog quadrature conversion from an IF to complex baseband has some compelling features and is therefore very common. Applications include cellular and mobile satellite receivers; DBS set-top boxes, which typically receive an L-band IF created by an LNB at the dish feed (see Section 7.10); and cellular base station receivers and other receivers for signals with bandwidths larger than can conveniently be accommodated by IF sampling. We also noted that direct conversion architecture – i.e., quadrature conversion directly from the carrier frequency – is increasingly common (Section 15.6.8). A representative example of a modern single-IC quadrature-downconversion receiver that might be employed in either case is the Linear Technology LTM9004 [45], which includes internal single-ended-to-differential conversion, quadrature downconversion, reconfigurable lowpass filters, and 14-bit ADCs that operate at rates up to 125 MSPS.

Problems

15.1 Consider the conversion of the sinusoid $x(t) = A\cos\omega t$ to a discrete-time digital representation. (a) According to the Nyquist–Shannon sampling theorem, what is the requirement on sample rate F_S for subsequent perfect reconstruction of of $x(t)$? (b) Given values for the minimum number of samples in one period of $x(t)$, provide a specific mathematical algorithm to recreate the original continuous-time sinusoid. Your algorithm should be specific to sinusoids; do not use the general expression for analog signal reconstruction, which involves an infinite sum of "sinc" functions. (c) Explain specifically what goes wrong if you attempt part (b) when the sample rate does not satisfy the Nyquist criterion. Use a time-domain argument; i.e., your answer should not simply be "aliasing." (d) Repeat part (c) using a frequency-domain argument; i.e., explain using diagrams similar to those shown in Figures 15.4 and 15.5.

15.2 Repeat Example 15.1, this time for (a) the 88–108 MHz FM broadcast band, (b) a single station from the FM broadcast band, assuming $B < 200$ kHz bandwidth.

15.3 Show that the integral in Equation (15.2) leads to Equation (15.3).

15.4 In Example 15.2, the magnitude of the sinusoidal input is 1 V (2 V_{pp}). Repeat this example using the same quantization levels, but now using a sinusoid having magnitude (a) 0.9 V (1.8 V_{pp}), (b) 0.5 V (1.0 V_{pp}), (c) 1.5 V (3 V_{pp}). In each case, determine the QSNR.

15.5 In Example 15.5, determine the level of aperture uncertainty that results in the associated SNR being equal to the SNR implied by the specified ENOB. At this level of aperture uncertainty, what is the expected overall SNR of the ADC?

15.6 In Example 15.6, what is the result if N_b is (a) decreased to eight bits, (b) increased to 16 bits.

15.7 A typical bandwidth for a single-channel-output UHF-band LMR receiver would be 25 kHz. Repeat Examples 15.6 and 15.7 assuming this bandwidth (as opposed to 1 MHz), $F_S = 100$ kSPS, and with N_b equal to (a) 12 bits (as before), (b) eight bits, (c) 16 bits. Include a diagram similar to Figure 15.12.

15.8 Following up Example 15.10, what is the possible minimum sample rate for direct sampling of the GPS L1 signal? Assume 20 MHz bandwidth and assume that an additional 10% is required on either side of the spectrum to accommodate the roll-off of an anti-aliasing filter. Indicate the Nyquist zone used and whether the ADC output will be spectrally-reversed.

15.9 Determine the minimum IF center frequency, LO tuning range, and image band for a single-conversion superheterodyne receiver implementing the following tuning ranges: (a) the 118–137 MHz aviation band, (b) 137–174 MHz, (c) 174–210 MHz. In each case, work out the results for both low-side and high-side injection, and express a preference, if any. Do not assume that there is an advantage in using a standard IF frequency.

15.10 Repeat Example 15.14, now assuming high-side injection.

16 Frequency Synthesis

16.1 INTRODUCTION

Frequency synthesis is the generation of a periodic waveform having a specified frequency. Here we concern ourselves primarily with the generation of sinusoids, which are commonly required as LO signals in frequency conversion, and as carriers in analog modulations such as AM and FM. In this chapter we consider the synthesis of sinusoids using feedback oscillators (Sections 16.2–16.6), phase-locked loop synthesizers (Section 16.8), and direct digital synthesis (Section 16.9). The problem of phase noise and its effect on radio system performance is introduced in Section 16.4. Also, Section 16.10 summarizes additional considerations that arise in IC implementation of frequency synthesis devices.

16.2 LC FEEDBACK OSCILLATORS

In Section 8.5.1, it was noted that two-ports that are active and include internal feedback are prone to instability, and that one particular consequence of instability could be oscillation. In that context, oscillation was undesirable; now, however, let us consider how we might exploit this type of mechanism for frequency synthesis.

16.2.1 The LC Resonator

Let us begin with a well-known passive circuit that exhibits oscillatory behavior: The *LC resonator*, which is an inductor (the "L") and a capacitor (the "C") in parallel as shown in Figure 16.1.[1] One first establishes a steady-state condition (Figure 16.1(a)) by applying a DC voltage source set to some non-zero value $v_0 > 0$, which charges the capacitor to a voltage of v_0. At this point no current flows through the capacitor; subsequently, the voltage across the inductor is zero.

Oscillation begins when the applied voltage is removed, as indicated in Figure 16.1(b). No longer constrained by the source voltage, the capacitor begins to equalize the charge between its plates by sending current through the inductor. The resulting current through the inductor creates a magnetic field which grows stronger the longer the current flows. The voltage v across the capacitor decreases because the net difference in charge across the plates is diminishing.

[1] One might argue that this is actually a *series* LC circuit; in that case just think of the term "parallel" as referring to the fact that the output voltage v is the voltage across the parallel arrangement of the inductor and capacitor.

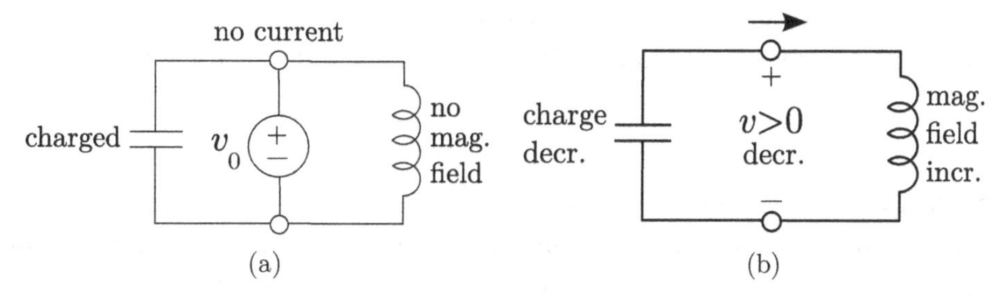

Figure 16.1. LC resonator: (a) Initial steady-state condition, (b) condition immediately after the applied voltage is removed.

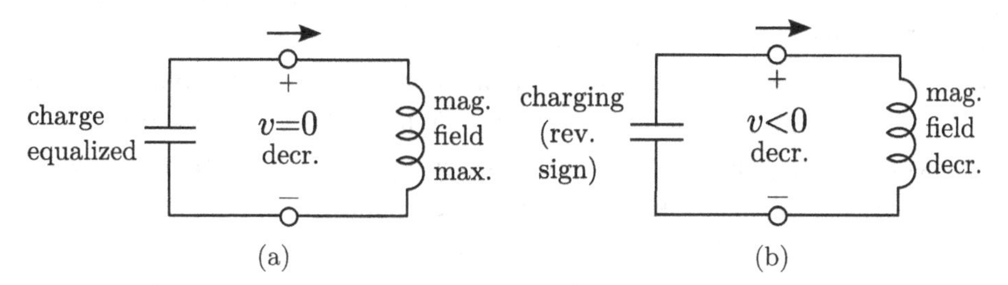

Figure 16.2. LC resonator (continued): (a) Voltage v across the capacitor reaches zero, (b) v becomes negative and continues to decrease, now "powered" by the inductor.

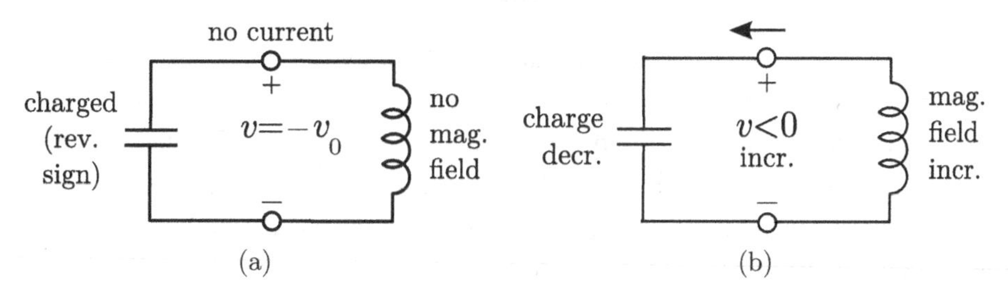

Figure 16.3. LC resonator (continued): (a) Voltage v across the capacitor reaches $-v_0$, (b) The process continues in reverse.

Once the charge is evenly distributed between the plates of the capacitor (i.e., the capacitor is fully discharged), $v = 0$ (Figure 16.2(a)). However, the magnetic field that has built up in the inductor keeps the current flowing, and so the redistribution of charge across the capacitor plates continues. A short time later we have the situation shown in Figure 16.2(b): The voltage across the capacitor is negative and decreasing, the flow of current is decreasing, and the magnetic field of the inductor is decreasing.

Current continues to flow until the charge distribution on the capacitor is completely reversed, as shown in Figure 16.3(a). At this point the capacitor is once again fully charged (albeit with the opposite sign), so $v = -v_0$, $i = 0$, and the magnetic field of the inductor has completely dissipated. This is exactly the situation we began with, except for the change in sign. Thus, the process now continues in reverse (Figure 16.3(b)): The current flows in the opposite direction and v begins increasing from $-v_0$, eventually returning the circuit to its original state with $v = +v_0$. With no external stimulus or internal resistance, this cycle continues indefinitely.

Once the LC resonator is started, the voltage across the capacitor (and inductor) will be sinusoidal. From elementary circuit theory, the frequency of this sinusoid is

$$\omega_0 = \frac{1}{\sqrt{LC}} \tag{16.1}$$

where L and C are the inductance and capacitance, respectively. The quantity ω_0 is said to be the frequency of *resonance*. Any discontinuity in voltage or current (such as turning off the voltage source in Figure 16.1(a)) results in a sinusoid having frequency ω_0.

This is not quite what we need for a practical oscillator, however. In a practical LC resonator, internal resistance is inevitable and will result in "damped" sinusoidal behavior; that is, the magnitude of the sinusoid will decay as the resistance dissipates the energy that was initially provided. Therefore, a practical LC oscillator must have some mechanism for overcoming the internal resistance. One such mechanism could be to repurpose the voltage source shown in Figure 16.1(a) to apply a periodic signal that "nudges" the circuit as needed to cancel the deleterious effect of the internal resistance.[2] The principal problem with this approach is that it is difficult to periodically activate the voltage source in a way that does not significantly distort the output with respect to the desired sinusoidal waveform. However, we can achieve a very similar effect with relatively low distortion by using feedback, which we shall now consider.

16.2.2 Sustaining Resonance Using Feedback

We now consider how we might sustain stable oscillation using feedback. First, let us be clear: Here we mean "stable" in a very different sense than we used the term in Chapter 8: Here "stable" means simply that oscillation is sustained with constant frequency, magnitude, and phase; at least approximately. To underscore the distinction, note that stable oscillators are routinely developed using transistors that are unstable at their operating point.

The concept underlying the feedback oscillator is shown in Figure 16.4. We will describe the action here in terms of voltages, but the quantities could just as easily be currents, or a mix of voltages and currents. The basic idea is that the output $y(t)$ is the amplified sum of an input $x(t)$ with feedback taken from the output and processed through a two-port device having frequency response $H(\omega)$. Taking the gain of the amplifier to be A, we can describe the output as follows:

$$Y(\omega) = A\left[X(\omega) + H(\omega)Y(\omega)\right] \tag{16.2}$$

where $X(\omega)$ and $Y(\omega)$ are the Fourier transforms (i.e., frequency domain representations) of $x(t)$ and $y(t)$, respectively. Thus we may write the system frequency response as follows:

$$\frac{Y(\omega)}{X(\omega)} = \frac{A}{1 - AH(\omega)} \tag{16.3}$$

Now consider the following question: Under what conditions may the output be non-zero even when the input is zero? This requires that the system frequency response be infinite, which requires

[2] This is essentially the definition of a *relaxation oscillator*. Such oscillators are relatively rarely used in radio applications; see Section 16.10 for a little more about this.

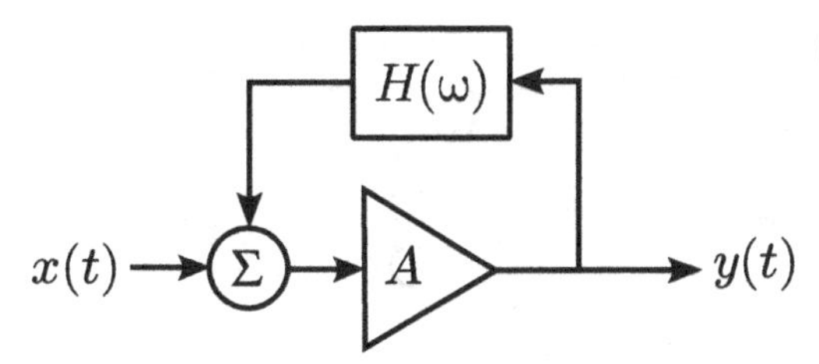

Figure 16.4. Feedback oscillator block diagram.

$$1 - AH(\omega) = 0 \tag{16.4}$$

This in turn implies the following requirements:

$$|AH(\omega)| = 1 \quad \text{and} \tag{16.5}$$

$$\angle AH(\omega) = 2\pi n \,, \quad n = 0, 1, 2, ... \tag{16.6}$$

These two requirements are known as the *Barkhausen criteria*. The left side of Equation (16.5) is the nominal gain along any path beginning and ending at the same place along the loop formed by the amplifier and the feedback path; this should nominally be 1 for stable output. Similarly, the left side of Equation (16.6) is the nominal change in phase along any such path; this should be zero (modulo 2π) in order for the feedback not to interfere with the output.

Now, a very important point: The Barkhausen criteria are not sufficient conditions for stable oscillation. The criteria address only the issue of how to sustain output in the absence of input. We have not yet determined how we start the oscillation: This is known as the "starting problem." Further, we have not yet considered what specifically is going to oscillate, and how we are going to get sinusoidal output, since the structure shown in Figure 16.4 does not have anything to say about oscillation itself, and does not appear to indicate any particular preference for waveform. In fact, we shall see shortly that practical feedback oscillators typically do *not* satisfy the unity-gain criterion (Equation (16.5)), but instead manage stable oscillation by allowing the amplifier gain A to be input-dependent.

Let us now consider the problem of what is going to oscillate. Any solution must lie in the so-far unspecified feedback processing block $H(\omega)$. A large and important class of possible solutions involve the use of an LC resonator in this block. Note that the LC resonator addresses both aspects of the problem: First, it tends to oscillate; and second, it resonates specifically at one frequency (the resonant frequency, ω_0), and therefore produces sinusoidal output, at least in principle.

Now, how to incorporate the LC resonator into a feedback oscillator? Figure 16.5 shows two schemes that will *not* work. In Figure 16.5(a), the feedback path has infinite impedance at resonance, so there can be no feedback at resonance. In Figure 16.5(b), the feedback structure will simply "freeze" the resonator in its starting condition. What is needed instead is a way to integrate the LC resonator that allows the voltage or current across the parallel LC structure to vary independently of the feedback structure, and limits the role of the feedback to that of overcoming the intrinsic resistance.

There are in fact two very simple solutions, shown in Figure 16.6. In the *Colpitts* strategy (Figure 16.6(a)), we split the capacitor into two series capacitors, and connect the feedback path

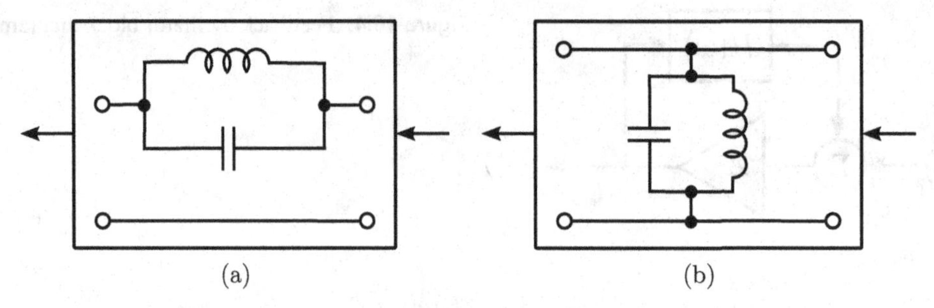

Figure 16.5. Two choices for $H(\omega)$ that will *not* result in stable oscillation.

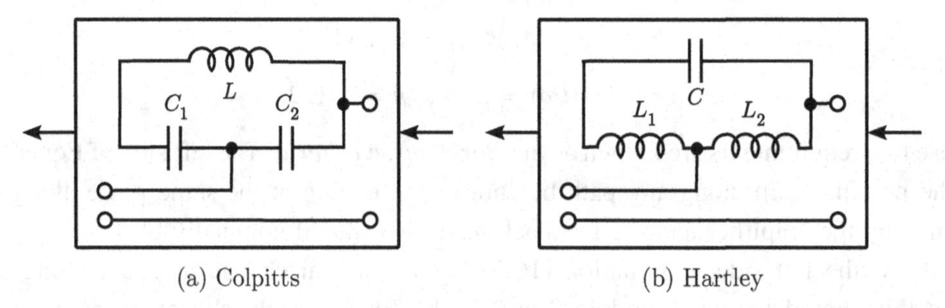

Figure 16.6. Two schemes for $H(\omega)$ that can be used to develop an LC feedback oscillator.

through the new junction. This allows the voltage (or current) across the parallel LC structure to be different from the voltage (or current) associated with the feedback path. The *Hartley* strategy (Figure 16.6(b)), is identical in all respects, except that we split the inductor as opposed to splitting the capacitor. Nearly all practical feedback oscillators employing LC resonators can be classified as either Colpitts or Hartley, including a large number of variants that go by different names. The most prominent of these will be discussed in Section 16.3.3. In practice, the choice between Colpitts and Hartley typically depends on issues associated with obtaining or controlling the necessary values of capacitance and inductance. However, it is worth noting that Hartley oscillators have a significant disadvantage in most applications because it may be difficult to avoid (or account for) mutual inductance due to linking of magnetic fields between the two inductors in close proximity.

Before moving on, a note on terminology: When used in this way, the LC resonator is commonly referred to as a *tank* or *tank circuit*. For the purposes of this chapter, the terms "tank" and "resonator" are interchangeable.

16.3 DESIGN OF LC FEEDBACK OSCILLATORS

We are now ready to consider specific techniques for the analysis and design of oscillators using LC resonators. We begin in Sections 16.3.1 and 16.3.2 with an implementation using the popular Colpitts resonator in an oscillator using a common base bipolar transistor amplifier. This is a special case, which is relatively easy to analyze, and yet is representative of the much broader class of LC feedback oscillators addressed in Section 16.3.3.

16.3.1 Colpitts Topology

Continuing from Figure 16.6(a), a simple Colpitts oscillator may be implemented using a bipolar transistor in common base configuration as a voltage amplifier, as shown in Figure 16.7. Although hardly the only possible choice (more on this in Section 16.3.3), the common base configuration is convenient because the output is in-phase with the input, and the gain is close to unity, all consistent with the Barkhausen criteria. The capacitor C_B is used to reference the base to ground at RF. For this reason, this particular configuration is commonly referred to a *grounded base* Colpitts oscillator.

Figure 16.8 shows a practical implementation of the scheme conceived in Figure 16.7. Here, the transistor is biased using the "base voltage divider with emitter degeneration" scheme (Section 10.3; specifically, Figure 10.7). In the present circuit the collector bias resistor (R_C)

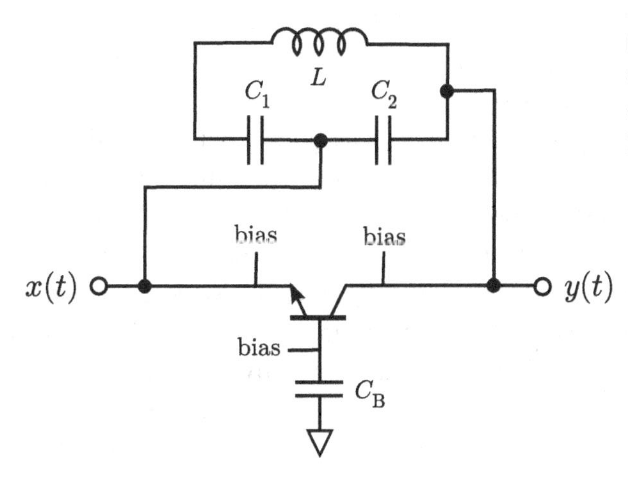

Figure 16.7. Scheme for implementing a Colpitts oscillator using a bipolar transistor in common base configuration as the gain block. Drawn here to resemble Figure 16.4.

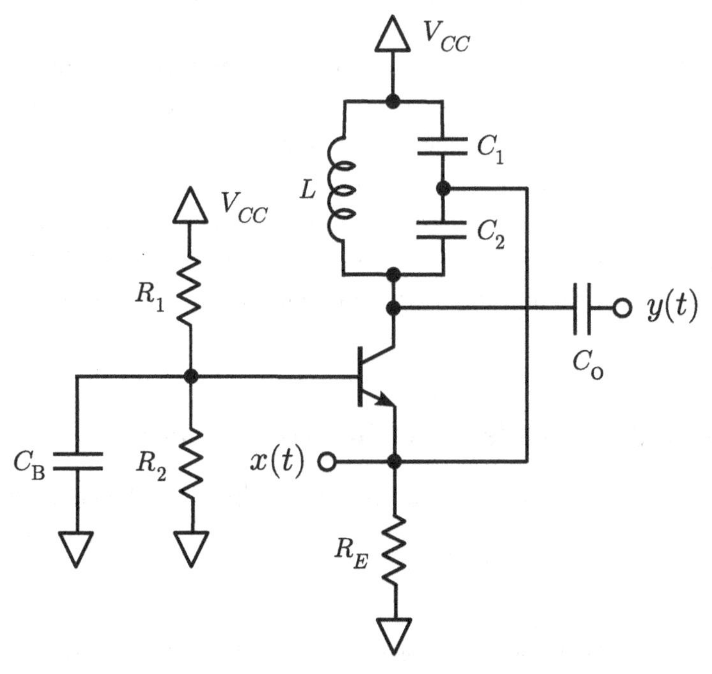

Figure 16.8. Fleshing out the schematic in Figure 16.7: A practical Colpitts oscillator using a common base bipolar transistor.

is omitted (i.e., chosen to be zero) so as to avoid loading the LC resonator or reducing the loop gain. Also, the emitter bypass capacitor indicated in Figure 10.7 is eliminated, since the emitter is the *input* in the common base configuration; and in fact it is now the *base* that is being grounded at RF, using C_B. Capacitor C_o is used to decouple the collector bias from the output.

One issue remains: How do we start the oscillation? Apparently a starting signal $x(t)$ could be applied to the emitter as indicated in Figure 16.8, since this corresponds to the input in the original feedback structure. In practice, there is normally no need for an explicit starting signal. Instead, any signal present in the circuit, even if quite weak, is potentially able to start the oscillation. One such signal is the ramping up of V_{CC} from zero when the circuit is first turned on; another is the intrinsic noise of the resistors and transistor. All that is needed is for the spectrum of the starting signal to have non-zero power at the desired frequency of oscillation. Once this spectral content finds its way into the LC resonator, the feedback structure should be able to sustain the nascent oscillation.

This brings us back to the Barkhausen criteria, which we will now find convenient to violate. In a well-designed oscillator, the "start-up" loop gain should be significantly larger than 1 so as to guarantee that a weak starting oscillation is able to quickly build up to the desired level. At that point, the oscillator must have some mechanism to cause the "effective" loop gain to drop back to 1, so as to maintain stable oscillation. The mechanism for this was suggested in Section 16.2.2: We make the gain A of the feedback amplifier – in this case, the common base transistor – dependent on the magnitude of the input. Specifically, we need A to decrease as the amplifier input increases. Happily, we get this for free, since – as we noted in Section 11.2.3 – transistor amplifiers experience gain compression at sufficiently high signal levels. Now, the "starting problem" is the challenge of making the start-up loop gain large enough to reliably start the oscillator, while making the effective steady-state loop gain close enough to 1 so as to satisfy the Barkhausen criteria and thereby sustain stable oscillation.[3]

Although Figure 16.8 is an implementable circuit, let us make a few minor modifications so as to display the schematic in the form in which it is usually presented. The modified circuit is shown in Figure 16.9. Here, one terminal of C_1 has been moved from V_{CC} to ground. This in principle has no effect on the behavior of the circuit, since C_1 plays no role in the bias and because V_{CC} and ground should be at the same potential in the small-signal AC analysis. To ensure the latter is the case, we explicitly bypass V_{CC} to ground using an additional capacitor C_C, which should have sufficiently high capacitance to function as a short circuit at the desired frequency of oscillation. Finally, we now explicitly represent the load impedance as a resistor R_L attached to the output.

16.3.2 Analysis and Design of the Grounded Base Colpitts Oscillator

The problem we now face in designing a grounded base Colpitts oscillator is determining appropriate values of L, C_1, and C_2. Assuming all other capacitors have sufficiently large values

[3] For a more general overview of the "starting problem" and solutions, a recommended starting point is [72, Sec. 13.1].

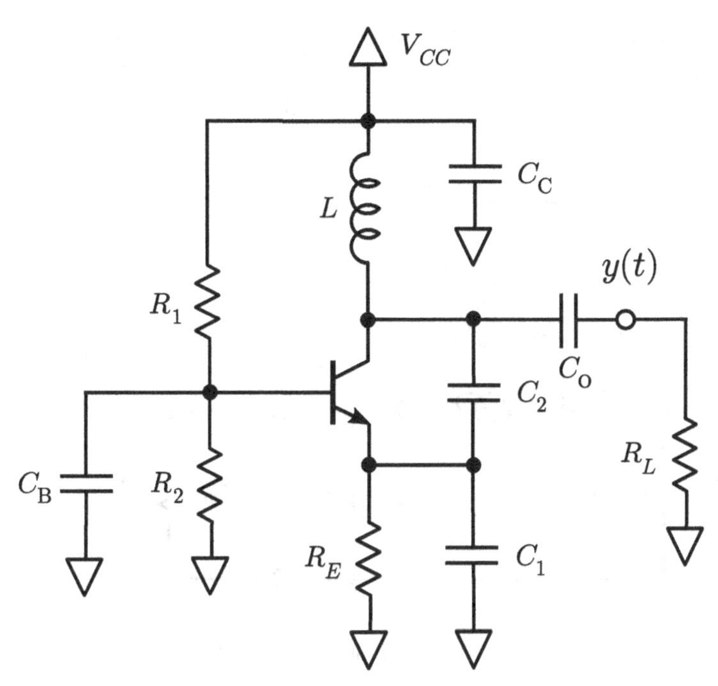

Figure 16.9. A practical Colpitts grounded base oscillator using a common base bipolar transistor. This is essentially the same as the circuit shown in Figure 16.8, except now in a form in which it is more commonly implemented. R_L represents the load impedance.

to be treated as short circuits at the frequency of resonance $\omega_0 = 2\pi f_0$, and neglecting any loading by the transistor, bias resistors, and R_L, we have

$$\omega_0 = \frac{1}{\sqrt{LC_T}} \tag{16.7}$$

where the tank capacitance is defined as the total (series) capacitance

$$C_T \equiv \frac{C_1 C_2}{C_1 + C_2} \tag{16.8}$$

The tank inductance L is constrained only by Equation (16.7). C_1 and C_2 are constrained by both Equations (16.7) and (16.8), as well as by the need to sustain stable oscillation and to overcome the starting problem.

Let us deal with the problem of how to select C_1 and C_2 to achieve stable oscillation.[4] For this problem we will use the well-known hybrid-π model, shown in Figure 16.10, as a simple AC model for a bipolar transistor. In this model, the transistor is modeled as a voltage-dependent current source $g_m v_{be}$, where v_{be} is the base–emitter voltage and g_m is the *transconductance* defined as

$$g_m \equiv \frac{I_C}{V_T} \tag{16.9}$$

where I_C is the collector bias current and V_T is the thermal voltage, which is usually about 25 mV (see Equation (10.2)). The impedance looking into the base–emitter junction is r_π, given by

$$r_\pi = \frac{\beta}{g_m} \tag{16.10}$$

[4] It should be noted that there are many approaches to this problem. The particular strategy shown here is adapted from [23, Sec. 11.5]. Another approach (and recommended additional reading) is [34, Sec. 5-5] and [21, Sec. 3.3.4].

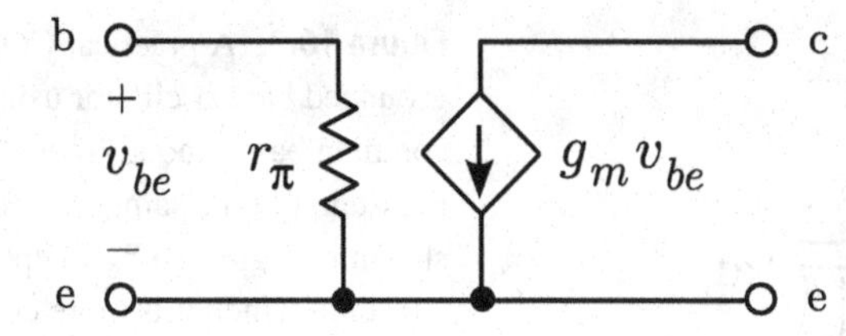

Figure 16.10. Hybrid-π model for a bipolar transistor. Nodes "c," "b," and "e" refer to collector, base, and emitter, respectively.

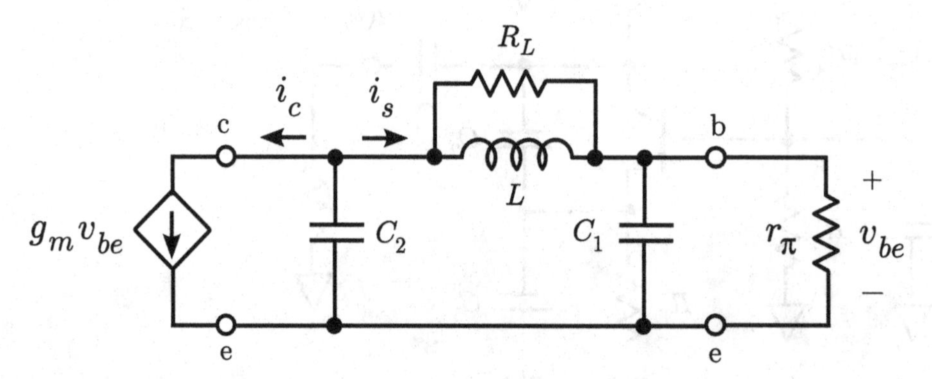

Figure 16.11. AC model for the Colpitts grounded base oscillator of Figure 16.9.

where β is the DC current gain (i.e., I_C/I_B). To apply the model, we begin with the original circuit (Figure 16.9) and short-circuit C_B, C_C, and C_0 under the assumption that these all have sufficiently large capacitance at ω_0, and open-circuit R_1, R_2, and R_E under the assumption that they have sufficiently large resistance. We then replace the transistor with the hybrid-π model, resulting in the circuit of Figure 16.11.

Analysis proceeds as follows: Referring to the figure, we note

$$i_c = -i_s - v_{ce}\,(j\omega C_2) \tag{16.11}$$

Further, we note

$$i_c = g_m v_{be} \tag{16.12}$$

$$i_s = v_{be}\,(j\omega C_1 + 1/r_\pi) \tag{16.13}$$

$$v_{ce} = v_{be} + i_s\,(j\omega L \parallel R_L) \tag{16.14}$$

Now the algebra: We substitute Equations (16.12)–(16.14) into Equation (16.11), divide through by v_{be}, and collect real and imaginary terms on the right. The result is

$$g_m = \left[-1/r_\pi + \omega C_2 X_S/r_\pi + \omega^2 C_1 C_2 R_S\right] - j\omega\left[C_2 + C_1\,(1 - \omega C_2 X_S) + C_2 R_S/r_\pi\right] \tag{16.15}$$

where R_S and X_S are defined to be the real and imaginary parts of $j\omega L \parallel R_L$, respectively.

Before we continue, it should be noted that it is not formally necessary for the terms in square brackets to be real-valued as we have assumed. In particular, frequency ω may be complex-valued. For example: Should ω be defined as $\omega' + j\omega''$ where ω' and ω'' are real-valued, the time dependence of the associated waveforms would have the form

$$e^{j\omega t} = e^{j\omega' t} e^{-\omega'' t} \tag{16.16}$$

that is, the imaginary part of ω'' indicates waveforms which are exponentially increasing or decreasing with time. Thus for stable steady-state oscillation, $\omega'' = 0$ is nominal. We shall proceed under this assumption, but keep this in mind as we will need to return to this concept shortly.

Transconductance is a real-valued quantity, so if ω is also real-valued it must be true that

$$C_2 + C_1 \left(1 - \omega C_2 X_S\right) + C_2 R_S / r_\pi = 0 \tag{16.17}$$

and then solving for C_2:

$$C_2 = \frac{C_1}{\omega C_1 X_S - R_S/r_\pi - 1} \tag{16.18}$$

Now returning to Equation (16.15) and solving for C_1 we obtain:

$$C_1 = \frac{1}{2}\frac{X_S}{\omega R_S} g_m \left[1 \pm \sqrt{1 - \frac{4R_S}{X_S^2 g_m^2}\left(g_m + \frac{1}{r_\pi}\right)\left(1 + \frac{R_S}{r_\pi}\right)}\right] \tag{16.19}$$

Which completes the derivation.

To summarize: Given R_L, L, and ω, the values for C_1 and C_2 obtained from Equations (16.18) and (16.19) will nominally result in an oscillator that supports oscillation which neither diminishes nor increases in magnitude.

Finally, an important note: Nowhere in this derivation have we explicitly required that $\omega = 1/\sqrt{LC_T}$; we have required only that ω be real-valued. The frequency is used only to determine the impedance of components, and there is no specific constraint on the waveform, so it should not be surprising that the solution turns out to be resonant at a different frequency. Fortunately, the frequency implied by the solution cannot be very different from the desired frequency, since the loop gain must be much less than 1 when the tank is not resonating. However, we must expect that some adjustment will be required to achieve *both* the desired frequency of resonance *and* stable oscillation. Further, whatever adjustment we make should address the starting problem, which the above derivation does not consider. These issues are most easily understood by considering a practical example, which follows.

EXAMPLE 16.1

Design a grounded base Colpitts oscillator that delivers a 1 MHz tone to a 50 Ω load. Use the 2N3904 NPN BJT ($100 \leq \beta \leq 300$) biased at $I_C = 10$ mA using a 9 VDC supply.[5]

Solution: The circuit schematic is shown in Figure 16.9; we need merely to determine the component values.

Let us begin by considering the tank circuit. We choose $L = 1$ μH since this is a standard value which is easily realizable at 1 MHz. Using Equation (16.7) we find the required tank capacitance is then $C_T = 25.3$ nF, which is also easily realizable at 1 MHz and thus further

[5] The popular 2N3904 transistor and relatively low frequency of 1 MHz is selected so that the student may experiment with this example in hardware using inexpensive, simple-to-use components and equipment. If you are able, please give it a try!

validates our choice for L. When designing the bias circuitry, it will be useful to know that the reactance of L and C_T at 1 MHz is +6.3 Ω and −6.3 Ω, respectively. We could now proceed to work out values for C_1 and C_2, but instead let us save this for last, for reasons that will become apparent later.

The bias design proceeds as follows: Since there is no collector bias resistor, the collector voltage will vary around $V_{CC} = 9$ V. A reasonable strategy is to begin by dividing the supply voltage approximately equally between V_{CE} and V_E. Since I_C is specified, this will result in a relatively large value of R_E, which is useful since this will reduce the extent to which R_E loads the tank. Just to have round numbers, let us make $V_E = 4$ V and $V_{CE} = 5$ V. $R_E = V_E/I_E$; so assuming β is sufficiently large, $R_E \cong V_E/I_C = 400\ \Omega$, which is much larger than the tank reactances and so should not significantly load the resonator. In this example we prefer to use standard values so as to facilitate subsequent laboratory testing of the design. The nearest standard value is 390 Ω, so V_E is revised to 3.9 V. Subsequently $V_B = V_E + V_{BE} \cong 4.6$ V assuming $V_{BE} \cong 0.7$ V.

For a "stiff" base bias, I_2, the current through R_2, should be much greater than I_B. On the other hand, I_2 should not be made arbitrarily large (and thus R_2 should not be made too low) since this will unnecessarily increase power consumption. For $I_2 \geq 10 I_B$ and again using $I_E \cong I_C$, we find

$$\frac{V_B}{R_2} \geq 10\frac{I_C}{\beta}$$

and solving for R_2:

$$R_2 \leq \frac{\beta}{10}\frac{V_B}{I_C}$$

Using the minimum β of 100 so as to obtain a conservative value, we obtain $R_2 = 4.6$ kΩ; we round up to the nearest standard value of 4.7 kΩ. A value for R_1 may be determined by neglecting the base current; then $V_B \cong V_{CC}R_2/(R_1 + R_2)$, which yields $R_1 = 4.3$ kΩ. These values are much greater than the tank reactances and therefore should not significantly load the resonator.

The values of the bypassing and DC-blocking capacitors are not critical. Following typical practice, we select these as follows: Capacitance C_o provides DC-blocking; it must have impedance at 1 MHz which is much less than R_L; we choose 0.1 μF ($-j1.59$ Ω at 1 MHz). The purpose of capacitance C_B is to short the base to ground at 1 MHz; we choose 1 μF ($-j0.16$ Ω at 1 MHz). The purpose of capacitance C_C is to bypass the supply voltage (i.e., shunt anything from the power supply that is not at DC to ground) and also to ensure that C_1 and L are connected with a virtual short-circuit at resonance; we choose 10 μF ($-j0.02$ Ω at 1 MHz).

Now we return to the tank circuit. The transistor has $g_m = I_C/V_T \cong 0.385$ mho and $r_\pi = \beta/g_m$, giving values between 260 Ω and 780 Ω for $\beta = 100$ and 300, respectively. $R_L = 50\ \Omega$ from the problem statement, so we have everything we need to calculate C_1 and C_2 using Equations (16.18) and (16.19). For the "+" option in Equation (16.19), we obtain $C_1 = 460.2$ nF, $C_2 = 27.3$ nF; and for the "−" option we obtain $C_1 = 27.4$ nF, $C_2 = 422.0$ nF. In either case, $C_T \cong 25.7$ nF, which is somewhat higher than the nominal value of 25.3 nF, corresponding to a frequency of operation which is about 1% lower than desired.

So, as promised earlier, the solution obtained from Equations (16.18) and (16.19) is close, but not quite what we need.

Now, we could simply tune one or both capacitors to achieve the desired value of C_T; however, recall that we also have the starting problem to deal with. A convenient way to address both problems at once is to arrange for C_1 to be the larger capacitor (i.e., use the "+" solution), and then to reduce its value to simultaneously reduce C_T and increase loop gain. By allowing loop gain greater than 1 we will be violating the Barkhausen criteria; however, we can rely on the gain compression present in the actual transistor to keep the oscillation peak magnitude from growing. So, employing standard values, we begin with $C_1 = 390$ nF and $C_2 = 27$ nF ($C_T = 25.3$ nF), and the completed schematic is shown in Figure 16.12.

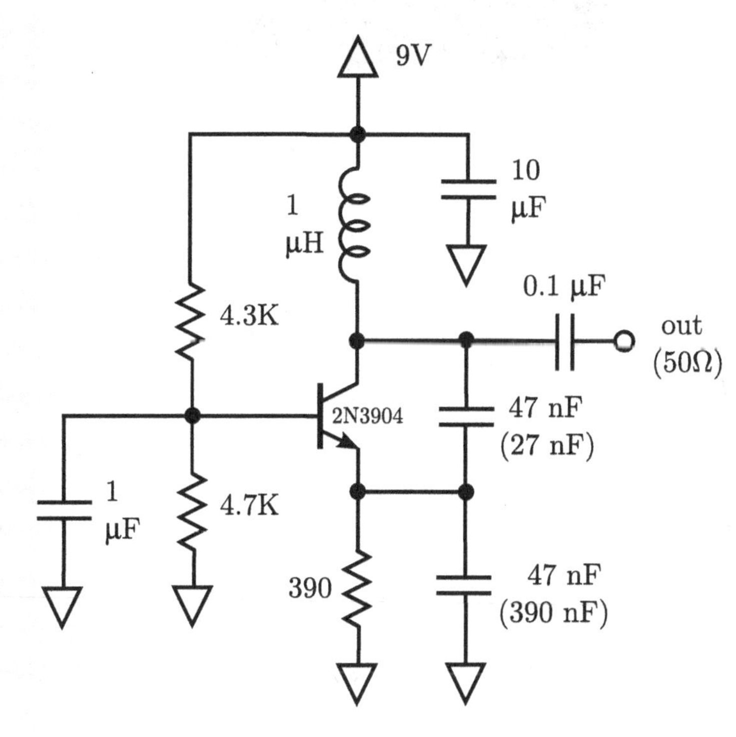

Figure 16.12. Completed grounded base Colpitts oscillator from Example 16.1. Values in parentheses are those determined in the first pass design.

This solution rests on quite a few assumptions: In particular, we assumed a very simple transistor model and have neglected the effect of bias resistors and blocking and bypass capacitors in loading the tank. Before building this circuit, it would be wise to test the design in a simulation that accounts for these issues and allows us to confirm that the oscillator has reasonable starting properties. For the present problem, SPICE is particularly well-suited to this task.[6] Figure 16.13 shows the result from the author's simulation.[7] In this simulation, noticeable

[6] Most students have some familiarity with SPICE; also, very good and full implementations of SPICE are freely available. Dedicated RF CAD software packages that are capable of transient (time domain) analysis are also suitable, and probably better-suited for higher-frequency design.

[7] If you are using SPICE and you find your simulated oscillator doesn't start at all, make sure that you use the "`startup`" argument when you run the simulation so that there is a start-up transient, and make sure that the `timestep` parameter is sufficiently small; i.e. small fraction of the anticipated period. I used 10 ns.

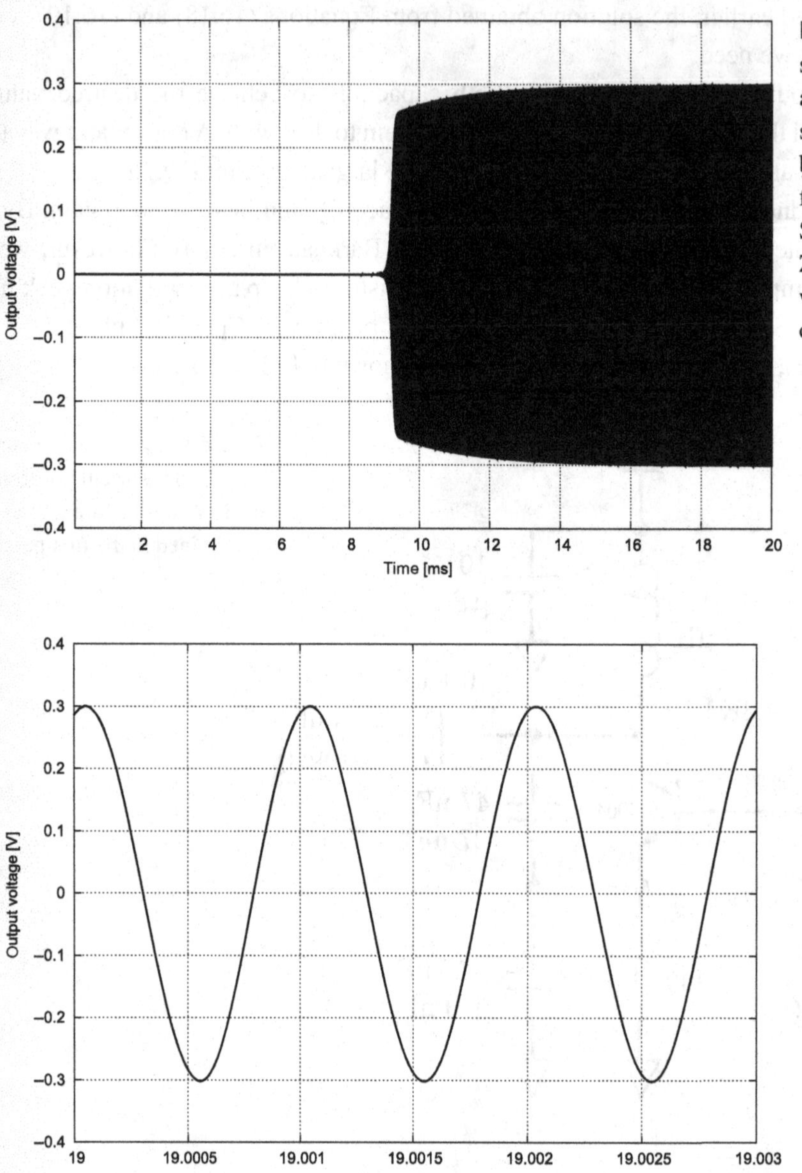

Figure 16.13. SPICE simulation of the voltage across R_L for the first-pass solution for the grounded base Colpitts oscillator from Example 16.1. *Top*:. Starting transient. *Bottom*: Zooming in to see the waveform for stable oscillation.

oscillation begins after about 8 ms and settles into stable 1 MHz oscillation of about 0.6 V_{pp} within about 5 ms of starting.

The SPICE results confirm that our analysis and design methodology are reasonable; however, if you build and test this circuit in the lab, you will most likely find that it will not start. The reason is that we have still not accounted for all non-ideal component and interconnect characteristics that exist in practice, and the available starting transient may not be sufficient to start the oscillator at its nominal loop gain. The solution, as we have indicated before, is to increase the loop gain by decreasing C_2. C_2 may have to be decreased to as little as 47 nF (again, restricting ourselves to standard values) before the oscillator starts reliably in lab conditions. Unfortunately, this results in the frequency shifting to about 1.22 MHz. The remedy now is to increase C_1. Both SPICE and laboratory experiments indicate a stable 1.03 MHz oscillation of about 1 V_{pp} is obtained for $C_1 = C_2 = 47$ nF, with the oscillator starting within 1 ms. (Problem 16.4 gives you the opportunity to verify this.) This result is indicated as the final design in Figure 16.12. Note that as a result of our post-design modifications the output magnitude has significantly increased – possibly a plus.

The final comment in the above example raises the following question: How does one design such an oscillator for a particular output level? In the above example, we specified only the load resistor R_L and then simply took whatever output level resulted from the process of adjusting the loop gain and frequency. In practice, a better approach would be to design the oscillator for a relatively high value of R_L, so that the output impedance has reduced effect on the LC resonator, and then use a second stage of amplification to set both the output impedance and the output level. An amplifier used in this way – i.e., to present a desired constant input impedance in lieu of a undesired or variable impedance – is known as a *buffer amplifier*.

With or without a buffer amplifier, the performance of feedback oscillators using LC tank circuits is generally quite poor by modern RF standards. The problems fall fundamentally into three categories: frequency accuracy, frequency stability, and spectral purity. The frequency of the oscillator is in principle determined by the values of the inductor and capacitors used in the tank; however, the accuracy of these values is limited by manufacturing tolerances, which are typically on the order of 1% at best. Furthermore, practical inductors have significant intrinsic capacitance, practical capacitors have significant intrinsic resistance, practical resistors have significant intrinsic inductance, and the completed circuit will have hard-to-predict amounts of all of these owing to layout and packaging effects. These values exhibit additional variation due to time-varying temperature and humidity, which leads to the frequency stability issue. The spectral purity issue arises from the limited Q of the LC resonator,[8] which allows spectral content at frequencies close to the nominal frequency to being sustained by the feedback loop, as opposed to be suppressed by the response of $H(\omega)$. The solution to all three of these issues is to replace or augment the tank circuit with devices that have higher Q, better stability with respect to environmental conditions, and which can be manufactured with improved tolerances – more on this in Section 16.5.

16.3.3 Alternative Implementations and Enhancements

The grounded base Colpitts oscillator developed in previous sections is a good example of the broad class of feedback oscillators, but it is hardly the only approach. The common base BJT amplifier, which serves as the active device in this design, may be replaced with common emitter or common collector amplifier configurations, each which may have various advantages and disadvantages in a particular application. Similarly, the BJT may be replaced by another type of bipolar transistor, or by a FET in common gate, common source, or common drain configuration; or for that matter even an op-amp might be employed. As mentioned previously,

[8] Review Section 9.4 for a refresher on quality factor.

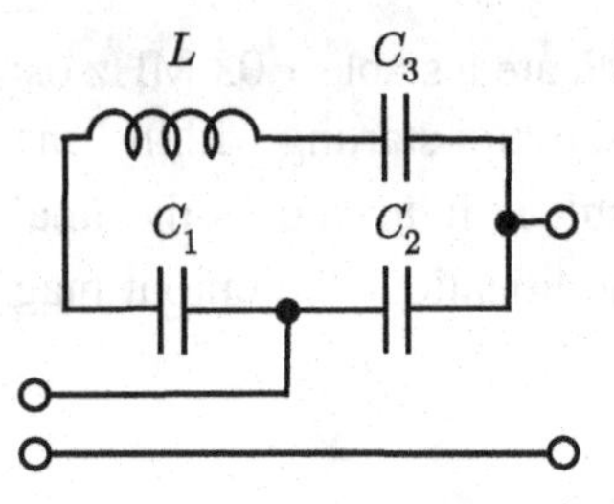

Figure 16.14. The Clapp resonator.

it is common to include an output buffer to isolate the output of the oscillator from the load impedance, making it possible to design the oscillator for a nominal constant output impedance that is different from the required load impedance.

Any of these approaches applies as well to the design of oscillators using the Hartley (as opposed to Colpitts) scheme for extracting feedback from the LC resonator. Furthermore, Colpitts and Hartley are not the only possibilities. One frequently-encountered alternative is the *Clapp resonator*, shown in Figure 16.14, which is simply the Colpitts resonator with an additional capacitor in series with the inductor. The frequency of resonance is still given by Equation (16.7), but with the tank capacitance now the capacitance of the series combination of all three tank capacitors:

$$\frac{1}{C_T} = \frac{1}{C_1} + \frac{1}{C_2} + \frac{1}{C_3} \tag{16.20}$$

The additional capacitor C_3 contributes another degree of freedom in fine-tuning the design, and making the C_3 *variable* leads to a very common design for a variable-frequency oscillator (see Section 16.6).

In principle, any circuit that exhibits resonance and accommodates a feedback tap can be used as the resonator in a feedback oscillator – which leads naturally to the idea of oscillators using crystals and other devices, as will be discussed in Section 16.5.

Additional considerations that apply in choosing from these possibilities include constraints imposed by specific combinations of performance requirements, the nature of the implementation (e.g., discrete components vs. IC; more on this in Section 16.10), the availability of suitable design and modeling tools, and of course the experience and skill of the designer. Recommended additional reading addressing the full panoply of possible LC feedback oscillator circuits is [21].

16.4 PHASE NOISE, SPURIOUS, AND RECIPROCAL MIXING

We noted in the previous section that it was difficult to constrain a feedback oscillator to output precisely one frequency so as to produce purely sinusoidal output. All practical frequency synthesis devices – not just feedback oscillators – produce output over a range of frequencies. This undesired spectral output can be characterized as being a combination of waveform distortion (e.g., that associated with the non-linearity of gain devices), spurious signals, and *phase noise*. In this section we consider phase noise and spurious output in a little more detail, including some of the consequences.

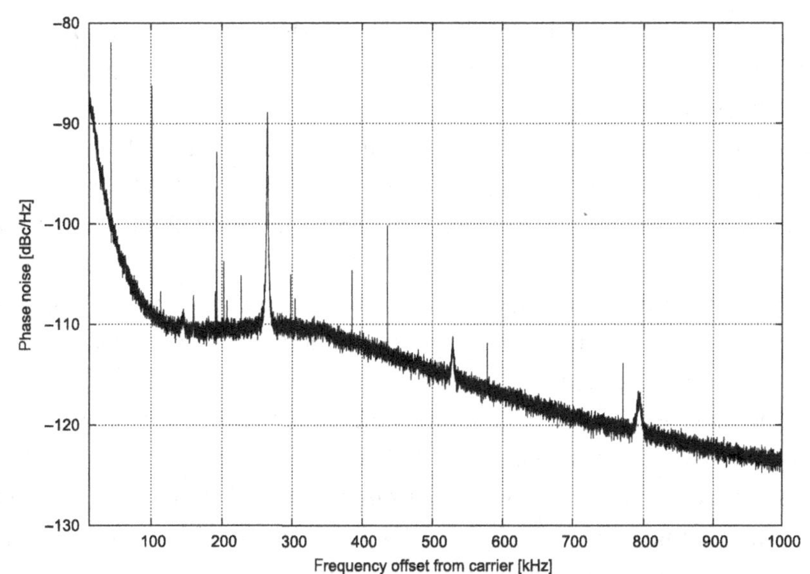

Figure 16.15. Measured steady-state spectrum from a typical PLL synthesizer, here generating a -13 dBm tone at 3 GHz. The plot begins 10 kHz to the right of 3 GHz, so the intended tone is slightly beyond the left edge of the plot.

Figure 16.15 shows the typical spectrum of a frequency synthesis device. This spectrum represents the steady-state condition; i.e., any start-up transients are assumed to have settled out and any remaining imperfections are associated with the steady-state operation of the device. Note that power is present at all frequencies but in diminishing amounts with increasing distance from the peak output (nominally, the desired frequency of operation). This is the phase noise. The spectral shape of the phase noise can be attributed to particular non-ideal mechanisms in frequency synthesizers; for more on this topic, a useful starting point is [40].

Phase noise is typically characterized as power spectral density (PSD) at a specified offset from the nominal frequency, relative to the total power. In Figure 16.15, the phase noise at a particular frequency separation from the intended tone is quantified as the power in a 1 Hz bandwidth relative to the total power of the synthesizer output. For example, the phase noise in Figure 16.15 is -108 dBc/Hz at 100 kHz offset, with the "c" in "dBc" standing for "carrier," and "dBc/Hz" being shorthand for "the ratio of power in 1 Hz relative to total power, in dB." This is referred to as a "single sideband" (SSB) measurement, because we have considered the phase noise on only one side of the nominal frequency, whereas phase noise exists on both sides.

In the time domain, phase noise can also be interpreted as *timing jitter*, and vice versa. (Recall the concept of timing jitter was previously introduced in Section 15.2.4.) The characterization of phase noise as timing jitter is more common when describing the quality of clock signals, digitally-modulated signals, and in digital systems generally.

We have alluded to the origins of phase noise in previous sections. For a feedback oscillator, phase noise is partially due to the non-zero bandwidth of the feedback path; that is, the bandwidth of $H(\omega)$. To the extent that the bandwidth of the feedback resonator is greater than zero, the loop gain of the oscillator is non-zero for frequencies different from the frequency of resonance, and the output waveform is thereby able to "wobble" within that bandwidth. This is sometimes referred to as *random walk FM*. Random walk FM dominates the close-in phase noise and provides strong motivation for oscillators using high-Q resonators, as will be discussed in Section 16.5. A second contribution to the phase noise is *flicker FM*, which is associated with the semiconductor noise originating from the gain device. Many other factors may contribute to the phase noise spectrum.

Obviously phase noise is not ideal behavior, but how in particular is the performance of a radio affected? Let us assume that the purpose of the desired sinusoid is to serve as the LO in a frequency conversion. The phase noise spectrum can be interpreted as a continuum of LO signals that coexist with the desired LO signal. Thus, the output signal will have the form of a continuum of copies of the input signal, and in this sense frequency-converted signals inherit the phase noise of the LO. The impact depends on the modulation, with analog modulations being relatively less affected, and digital modulations – in particular high-order phase-amplitude modulations – being potentially severely affected. Figure 16.16 shows an example.

A separate and particularly insidious effect is known as *reciprocal mixing*, illustrated in Figure 16.17. Reciprocal mixing occurs when a strong interfering signal is present in the input close to, but separated from, the desired signal. Both signals inherit the phase noise of the LO, and the relatively stronger phase noise of the interferer overlaps and thereby degrades the SNR of the weaker desired signal in the output. Reciprocal mixing poses a particularly difficult problem for receivers designed to accommodate multiple signals simultaneously, and especially for broadband software radio applications. For example, reciprocal mixing is a serious problem for cellular base station receivers that attempt to downconvert all frequency channels in the receive band simultaneously, as opposed to one channel at a time. The "all-channels" VHF airband receiver of Example 15.1 could be expected to be similarly affected.

So how good is "good enough" for phase noise? The answer depends on the application, and (as noted above) the architecture of the radio. In wireless communications systems in which spectrum is divided into contiguous channels, minimum requirements depend on the signal spectrum with respect to the spacing of the channel center frequencies. Table 16.1 shows some typical values.

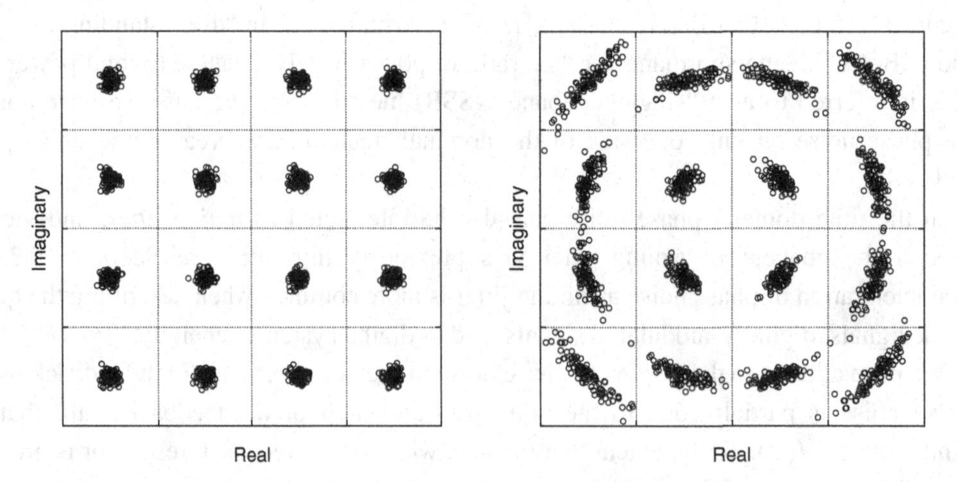

Figure 16.16. Degradation of the constellation of a 16-QAM signal resulting from phase noise. *Left:* Received constellation with no phase noise. *Right:* Same received signal with phase noise. Note symbols are smeared into adjacent decision regions, increasing the probability of symbol error independently of the original SNR of the received signal.

Table 16.1. **Phase noise specifications for some common wireless applications.**

Application	Specification (SSB)	
WCDMA	−90 dBc/Hz at 10 kHz,	−113 dBc/Hz at 100 kHz
GSM	−111 dBc/Hz at 100 kHz,	−143 dBc/Hz at 3 MHz
IS-54	−115 dBc/Hz at 60 kHz	
DECT	−85 dBc/Hz at 100 kHz	
Bluetooth	−119 dBc/Hz at 3 MHz	
WiFi	−116 dBc/Hz at 3 MHz	

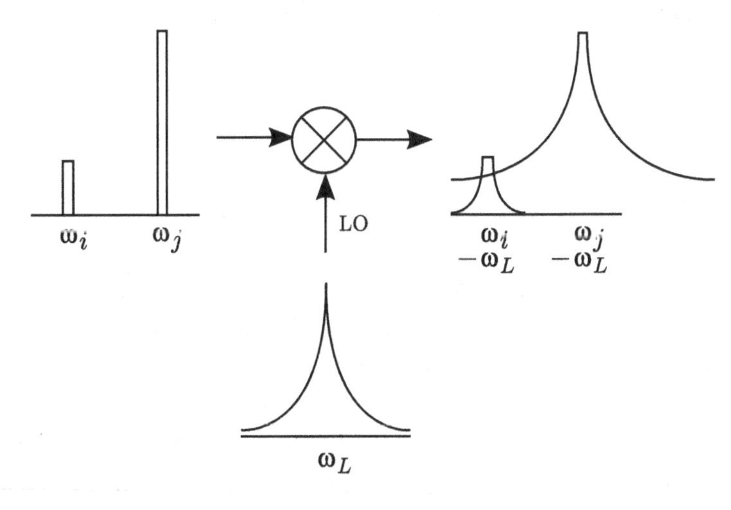

Figure 16.17. Reciprocal mixing. Here, the desired signal is at ω_i and the stronger undesired signal is at ω_j. Assuming downconversion with low-side LO injection.

Finally, we note that practical frequency synthesis devices typically output discrete spurious signals in addition to the continuum of spectral content, as depicted in Figure 16.15. These spurious signals are typically due to power supply and clock signals and their harmonics, and subsequent mixing with the output tone. Spurious signals in the output of PLL synthesizers (coming up in Section 16.8) may also be due to non-ideal frequency division, especially that associated with dual-modulus prescalers used in fractional-N techniques.

16.5 OSCILLATORS USING CRYSTALS AND OTHER HIGH-Q RESONATORS

At this point it is clear that LC feedback oscillators have three primary limitations. First, the low Q of the LC resonator (equivalently, the relatively wide bandwidth of the feedback path $H(\omega)$) allows the oscillator to output spectral content over a range of frequencies around the nominal frequency. The resulting phase noise is unacceptable in many applications, and to do better requires a resonator with higher Q.

Second, we found that it is difficult to precisely set frequency, as discussed in Section 16.3.2 and demonstrated in Example 16.1. Generally it is quite difficult to achieve better than 5% accuracy in ω_0 without either very elaborate models (essentially precluding a simple design procedure) or manual tuning of the component values after the circuit is constructed. This amount of error is unacceptable in many applications, and in particular in transmit applications: For example, 5% error in an FM broadcast transmitter intended to operate at 100 MHz is 5 MHz error, which is gigantic compared to the 200 kHz grid of center frequency assignments in the FM band. To do better requires a resonator utilizing components with more predictable and repeatable values. In the FM broadcast example, an acceptable level of precision might be 1% of the channel spacing (200 kHz), which is 2 kHz or about 0.002% of the carrier frequency. Such values are typically expressed in units of *parts per million*; e.g., 0.002% is 20 ppm.

The third problem with LC oscillators of the type considered in previous sections is *frequency stability*.[9] The resonant frequency in a feedback oscillator will vary as the constituent inductors and capacitors change values in response to temperature and aging. This is often unacceptable for the same reasons that poor predictability, as discussed in the previous paragraph, is unacceptable. Thus, we seek a resonator that is much less vulnerable to the effects of temperature and aging.

16.5.1 Crystal Oscillators

An effective and popular solution to all three problems identified above is to replace or augment the LC resonator with a crystal. As explained in Section 13.8.4, crystal resonators offer frequency accuracy, stability, and Q that are far superior to LC resonators. Figure 16.18 shows a crystal oscillator implemented as a modification of the Colpitts oscillator of Figure 16.9. Here the crystal is inserted in the feedback path between the LC resonator and the emitter, and the series resonance is used. The series resonance of the crystal makes the frequency response of feedback path $H(\omega)$ dramatically narrower around the desired frequency of operation, yielding the benefits identified above.

How does one go about designing such an oscillator? Obviously, the formal procedure developed in Section 16.3.2 must be modified to account for the crystal. However, we noted in Example 16.1 that this particular procedure is actually useful only as a starting point, and that some experimentation is required to obtain the desired result. Following on our experience in Example 16.1, the following procedure is suggested:[10] First, select L (the tank inductance) to have a reactance of a few hundred ohms at ω_0. Then, choose C_2 and $C_1 \gg C_2$ such that the tank resonant frequency is ω_0. The crystal should of course be series-resonant at ω_0, and the resonance of the LC tank should be wide enough that the resonance of the crystal falls well within this useful range even if there is significant error in tank resonance center frequency. Finally, R_E and possibly C_1 are adjusted to achieve prompt starting with acceptable distortion.

[9] Once again the term here has a completely different meaning than in previous definitions where we referred to "amplifier stability" and "stable oscillation."

[10] See [3, Sec. 9.5.4] for useful elaboration on this approach.

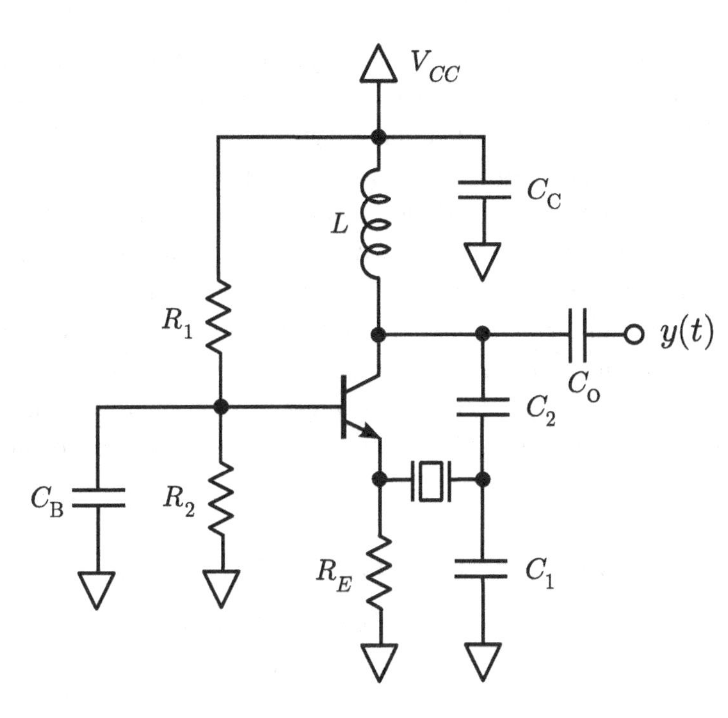

Figure 16.18. A series-resonant crystal oscillator implemented as a modification of the Colpitts oscillator of Figure 16.9.

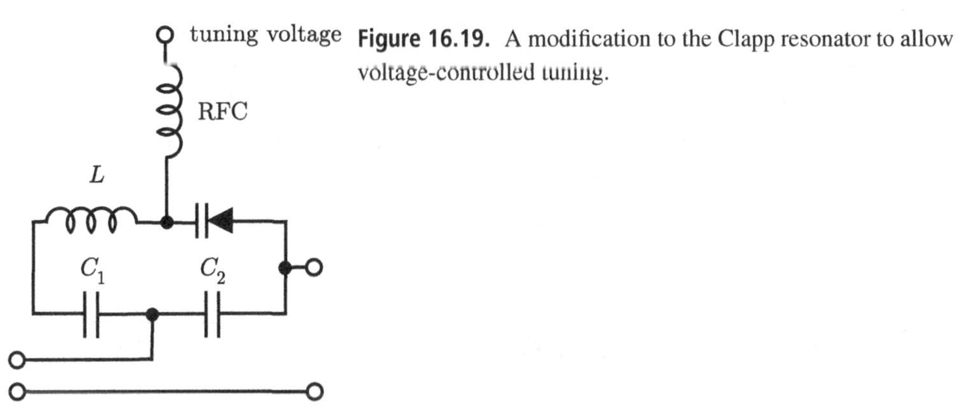

Figure 16.19. A modification to the Clapp resonator to allow voltage-controlled tuning.

Many other popular designs can be developed as modifications to the oscillator of Figure 16.18. A version of this circuit using the crystal in parallel-resonant fashion is obtained by installing the crystal in parallel with L (or the series combination of C_1 and C_2) and replacing the short connection from the tank to the emitter. A Clapp version of this circuit can be obtained by using the crystal in its series-resonant mode as a replacement for the tank inductance. A *Butler oscillator* is essentially the oscillator of Figure 16.18 with an emitter-follower amplification stage inserted between the collector and the low input impedance presented by the series-resonant mode of the crystal [3, Sec. 9.5.4].

16.5.2 Temperature-Stabilized Crystal Oscillators

It is useful to be aware of some common enhancements to fixed-frequency crystal oscillators, and the associated nomenclature. An *oven-controlled crystal oscillator* (OCXO) is exactly what the term implies: a crystal oscillator in an oven – albeit a very small one. The oven is set to a temperature much greater than ambient (typically about 75 °C), such that the temperature of the crystal is essentially constant and independent of the rest of the system. Since temperature is the primary source of variation in the electrical properties in the crystal, this leads to greatly improved frequency accuracy and stability; 0.1 ppm is routinely obtained; stability as fine as 0.001 ppm is possible. However, OCXOs have the disadvantages of being mechanically-awkward and requiring additional power to operate the oven.

A compromise is the *temperature-compensated crystal oscillator* (TCXO). A TCXO uses no oven; instead the circuit accounts for the effect of temperature on the crystal by adjusting the values of other components in the feedback path. Stability is much improved relative to an uncompensated crystal oscillator, and the best TCXOs approach the performance of an OCXO.

16.5.3 Resonator Technologies for Higher Frequencies

Crystals typically are not useable above a few hundred MHz. At higher frequencies, alternatives are PLL synthesizers (Section 16.8) and different resonator technologies. At UHF and above, alternative resonator technologies include *surface acoustic wave (SAW) resonators*, and *piezoelectric ceramic resonators* fashioned from lead zirconate titanate ("PZT"). These devices behave much like crystal resonators, but have Q which is not nearly as good. They are nevertheless typically superior to LC resonators in terms of Q, accuracy, and stability at the frequencies at which they are used. For additional reading on the use of these devices, [21, Secs. 4.8 and 4.9] is recommended. For higher-Q at higher frequencies, other possibilities include dielectric resonators and yttrium iron garnet (YIG) sphere resonators.

16.6 VARIABLE-FREQUENCY OSCILLATORS AND VCOS

Sometimes you need the ability to vary the frequency of an oscillator; e.g., for use as a tuning LO. For a LC feedback oscillator, the obvious solution is to devise a way to vary the inductances and/or capacitances of the resonator. In most applications it turns out to be more convenient to vary capacitance than inductance. In early radios, oscillator tuning was accomplished using a variable capacitor whose value was set by manually rotating a knob. In some modern applications, the manually-variable capacitor is replaced by a *digital step capacitor* or some other device that generates discrete values of capacitance in response to a digital control signal.

However, the far more common solution in modern systems is to implement some part of the tank capacitance in the form of a *varactor*. A varactor typically consists of a reverse-biased diode that exhibits a capacitance which is proportional the bias (DC) voltage across the terminals. Varactors are also known as *tuning diodes*, *voltage-variable capacitance diodes*, or *varicaps*. Varying the bias voltage across the diode varies the capacitance, which varies the resonant frequency of the tank, which changes the frequency of the oscillator. Such an oscillator is known as a *voltage-controlled oscillator* (VCO).

There are several possible schemes for integrating a varactor into an LC resonator. Let us consider how we might go about this with a Colpitts-type oscillator. It is usually wise to avoid implementing either C_1 or C_2 (referring to Figure 16.6(a)) as the variable capacitance, since these components have a big impact on the amplitude and starting characteristics of the oscillator. Figure 16.19 shows a more common approach: We begin with the Clapp resonator of Figure 16.14, and replace C_3 with a varactor. The additional inductor (labeled "RFC" for "RF choke") is intended to isolate the tuning voltage circuit from the oscillator; blocking capacitors may also be required depending on how this circuit is integrated with the bias circuitry. An alternative is to add the varactor in parallel with the Colpitts resonator.

The three principle challenges in varactor-based VCO design are (1) achieving the desired tuning range (i.e., sufficient range of capacitance values) while keeping the varactor reverse-biased, (2) achieving a linear (or at least acceptable) frequency vs. voltage characteristic,[11] and (3) integration of the varactor bias circuitry with the oscillator.

16.7 NEGATIVE RESISTANCE OSCILLATORS

So far in this chapter we have restricted our attention to LC oscillators, consisting of a resonant circuit plus some active feedback circuitry to sustain the oscillation. This paradigm is most useful for the design and analysis of oscillators when the active devices – usually transistors – can be described in terms of equivalent circuits. This paradigm becomes awkward at high frequencies, especially at SHF and above, where it is more convenient to describe gain devices in terms of their two-port parameters. Thus it becomes more convenient to view an oscillator as a two-port device whose terminations ensure instability, subsequently resulting in oscillation. For historical reasons, the term *negative resistance oscillator* is applied in this case. However, the term is apt, since a one-port consisting of an unstable gain device exhibits negative resistance; that is, it creates current in the direction opposite the direction expected for a passive one-port. Thus the one-port is able to overcome resistance in the resonator, which is a requirement for sustained oscillation. The analysis and design of negative resistance oscillators is outside the scope of this book; [58, Sec. 13.2] and [21, Ch. 5] are recommended starting points.[12]

16.8 PHASE-LOCKED LOOP (PLL) SYNTHESIZERS

One often requires a tone that can be tuned in frequency, is very accurate and stable, and which has low phase noise. On the basis of techniques presented so far, these requirements lead us in somewhat different directions: A VCO is tunable, but typically has poor accuracy and stability;

[11] VCO circuits will often include two varactors arranged "nose-to-nose" – i.e., with cathodes attached – as a means to address this issue.

[12] Section 5.11 of [21] demonstrates the application of the negative resistance concept to the traditional LC oscillator, which underscores the point that the "negative resistance oscillator" is not necessarily a "microwave" concept, but just a more convenient way to design oscillators at microwave frequencies.

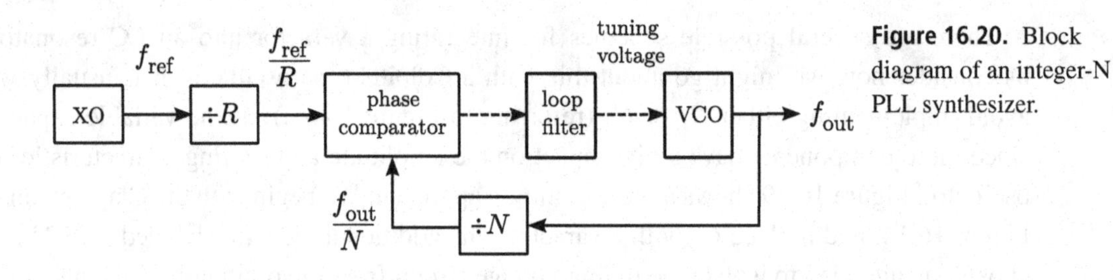

Figure 16.20. Block diagram of an integer-N PLL synthesizer.

whereas a crystal oscillator (for example) may be accurate and stable but is difficult to make tunable over a wide range and is typically limited to a maximum frequency of a few hundred MHz. A phased-locked loop (PLL) synthesizer is a device that combines the desirable features from both approaches.

16.8.1 Integer-N Synthesizers

An *integer-N PLL synthesizer* is shown in Figure 16.20. The synthesizer consists of a VCO, which creates the desired output frequency f_{out} and is phase-locked to a crystal oscillator (indicated as "XO" in the diagram), typically operating at a different frequency f_{ref} (for "reference"). In the scheme shown in Figure 16.20, the VCO output (which is also the synthesizer's output) is processed through a *frequency divider*, which outputs a periodic signal at frequency f_{out}/N. At the same time, the XO output is processed through a different frequency divider, which outputs a periodic signal at frequency f_{ref}/R. For the purposes of explanation, let us initially assume $R = N = 1$, so there is no frequency division. The outputs of the VCO and XO are then input to a *phase comparator*, which produces a signal which is proportional to the difference in phase between the inputs. Again for the purposes of explanation, let us temporarily ignore the loop filter shown in the diagram. The output of the phase comparator is used as the VCO control voltage. In this way, the VCO output is phase-locked to the XO: If a phase difference develops, it is automatically detected by the phase comparator and the VCO control voltage is adjusted accordingly. Since frequency is simply the derivative of phase, locking phase also locks frequency; in this way the VCO inherits the stability and accuracy of the crystal oscillator.

The "$R = N = 1$" scenario described above is not very useful: After all, if we wanted $f_{out} = f_{ref}$ we could have just used the XO as the frequency source. However, f_{out} can be varied by allowing R and N to take on other values. This is demonstrated by the example below:

EXAMPLE 16.2

An integer-N PLL synthesizer is to be designed such that it is able to tune 140 MHz, 150 MHz, and 160 MHz (i.e., three frequencies). The frequency of the crystal oscillator is 40 MHz. Determine R and N.

Solution: We choose $R = 4$, so the frequency-divided output of the crystal oscillator is 10 MHz, which is equal to the desired spacing between frequencies. When $f_{out} = 140$ MHz, the frequency-divided VCO output is 10 MHz when $N = 14$, so this is the value of N that will lock the VCO at 140 MHz. Similarly, $N = 15$ and 16 will cause the VCO output to lock at 150 MHz and 160 MHz, respectively.

Note that for an integer-N PLL synthesizer, the step size for frequency tuning is f_{ref}/R, and the output frequency is then set by N, and is $f_{out} = Nf_{ref}/R$. The big advantage of this synthesizer over the VCO alone is frequency accuracy and stability; the primary disadvantage is that we may only tune discrete frequencies. However, signals in radio communications systems are commonly arranged into discrete channels anyway, so discrete frequency tuning is not necessarily a disadvantage.

It should also be noted that the *absolute* accuracy of the PLL synthesizer's output is not normally the same as the absolute accuracy of the crystal oscillator. This is because the frequencies are normally different. For example, let us say the frequency of the crystal oscillator in Example 16.2 is accurate to ± 1 ppm. The frequency of the synthesizer will then be nominally accurate to ± 1 ppm; however, the *absolute* frequency error is about 150 Hz at 150 MHz (the center tuning), whereas it is only 40 Hz at 40 MHz (the reference frequency).

16.8.2 Fractional-N Synthesizers

A common problem is that the frequencies required from the synthesizer do not correspond to integer multiples of f_{ref}/R for convenient values of f_{ref}. One way to sidestep this issue is to make R sufficiently large to reduce the error relative to the desired frequencies to an acceptable amount. Alternatively, a *fractional-N* synthesizer might be considered. A fractional-N synthesizer is essentially an integer-N synthesizer in which the "divide-by-N" operation is time-variable; e.g., "divide by N" for a given number of clock ticks, and then switching to "divide by $N + 1$" for a given number of clock ticks. Such a scheme is referred to as having *dual modulus*. The *average* division factor for a dual modulus divider will be somewhere between N and $N + 1$, and the fact that the divider is actually switching between two states will not be apparent as long as the switching period is relatively small. The advantage of this approach is that R need only be large enough to set the necessary step size (as in the integer-N scheme). Disadvantages of this approach are significant increase in the complexity of the output frequency divider, and some additional spurious. The latter is attributable to switching between the N and $N + 1$ states, which generates a periodic transient that must be negotiated by the PLL. This is one of the mechanisms that may give rise to the discrete spurious tones depicted in Figure 16.15.

16.8.3 Dividers, Phase Comparators, Loop Filters, and Prescalers

PLL synthesizers are relatively complex systems, consisting of several disparate subsystems. Of these subsystems, we have already discussed VCOs (Section 16.6). Let us now briefly consider the remaining subsystems.

Frequency Divider. Frequency division presents no particular challenge; for example, we may divide by 2 using a digital "flip-flop" device, and we may divide by powers of 2 using a digital counter circuit. The fact that the output of these devices is digital (i.e., a square wave as opposed to sinusoidal) is of no consequence, since we are subsequently interested in only the frequency and phase of the waveform, and not the waveform itself. In fact, many popular phase comparison schemes work best when – or simply assume – that the input is a square wave.

Loop Filter. The loop filter is a passive low-pass device that serves two purposes. First, it limits the rate at which the VCO control voltage may be changed; i.e., a relatively narrow loop filter bandwidth allows only slowly-varying change, whereas a relatively wide loop filter bandwidth allows relatively faster change. The latter condition allows the PLL to respond quickly to changing conditions; in particular, the transients associated with changing N, changing between N and $N + 1$, and initial start-up. However, such a loop filter allows more noise to reach the VCO frequency control, resulting in relatively high phase noise for essentially the same reason a low-Q resonator results in high phase noise in a feedback oscillator. Reducing the loop bandwidth improves phase noise, but may cause the synthesizer to "loose lock" in response to transient conditions, and may prevent start-up entirely. Thus, the loop filter bandwidth is a tradeoff between the spectral purity of the synthesizer output, and the lock range for start-up and tuning.

The second purpose of a loop filter is to smooth out the potentially ragged output from the phase comparator. This makes it possible to use relatively crude phase detection and comparison techniques operating on square-wave representations of the frequency inputs. This in turn simplifies implementation of PLL synthesizers as integrated circuits.

Phase Comparator. Common schemes for implementing phase comparators include analog phase detectors and digital phase-frequency detection. *Analog phase detectors* are essentially multipliers (perhaps implemented as mixers) in which the inputs are the signals to be compared, and the phase difference can be inferred from the product. The principles of this approach have already been covered is Section 5.5.4 (pertaining to coherent demodulation) and so do not need to be repeated here. Recall from Chapter 14 that signal multiplication entails considerable complexity and relatively high power consumption; thus modern PLL synthesizers typically use this approach only when alternative methods are not feasible.

The predominant alternative to analog phase detection is shown in Figure 16.21, which consists of a *phase-frequency detector* (PFD) followed by a *charge pump*. This scheme assumes square-wave signals, so if either input is sinusoidal it will normally be "limited" (clipped) to approximate a square wave, and possibly also buffered through a logic gate to produce a clean square wave. The first stage of the PFD uses type-D flip-flops to output digital signals Q_1 and Q_2, which are asserted high on the leading edge of the associated input signals D_1 and D_2, and then reset when both Q_1 and Q_2 are high. If D_1 leads D_2, then the resulting duty cycle (i.e., the fraction of time the signal is asserted) of Q_1 will be greater than zero and the duty cycle of Q_2 will be very close to zero. If instead D_2 leads D_1, then the resulting duty cycle of Q_2 will be greater than zero and the duty cycle of Q_1 will be very close to zero.

Q_1 and Q_2 are then delivered to the charge pump, where they are used to control switches which connect and disconnect the current sources I_1 and I_2 to the output. Also connected to the output is a capacitance C_o, which may be part of the loop filter, but is included here for clarity. When Q_1 is asserted, I_1 deposits charge into C_o and the output voltage increases; when Q_2 is asserted I_2 extracts charge from C_o and the output voltage decreases. Thus, the output of the

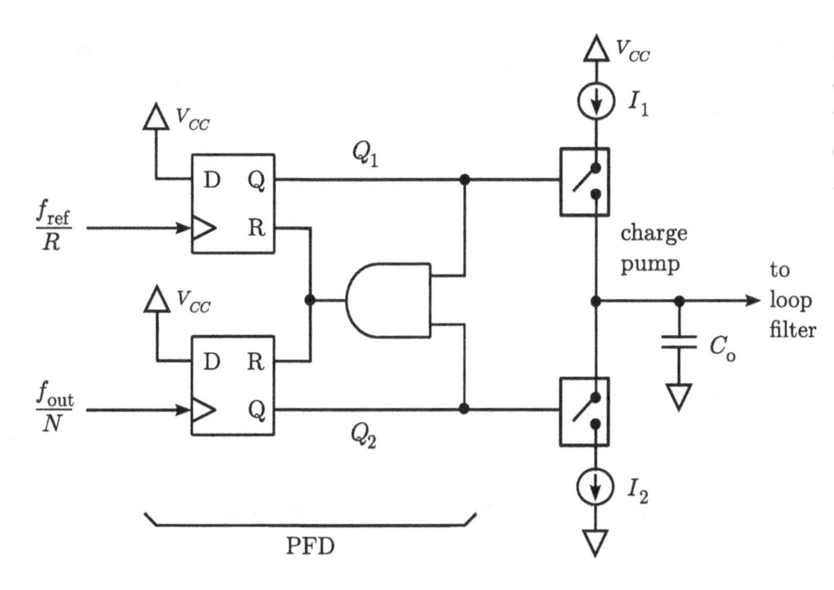

Figure 16.21. A phase comparator using square wave phase-frequency detector (PFD) followed by a charge pump.

charge pump is a quantized version of the desired VCO control signal. As mentioned above, the loop filter is used to smooth this signal so that the quantization does not "jerk" the VCO control frequency and thereby create additional spurious output. It should be noted that charge pumps may be implemented in a number of ways, some of which are somewhat different from the scheme depicted here; but the general strategy is essentially the same.

Prescaler. Finally, we note that it may be that the output frequency divider (i.e., the divide-by-N) must operate at RF frequencies, which may be beyond the maximum clock rate of the digital technology used to implement the divider. In this case, the divider is commonly proceeded by a *prescaler*, which is essentially a digital frequency divider implemented in analog RF technology. Fractional-N synthesizers are commonly implemented using dual-modulus prescalers, combining prescaling and fractional-N division into a single functional block.

16.8.4 PLL Design Considerations

This section has provided a brief qualitative summary of PLL synthesizers. One thing we have not discussed is the topic of loop design; that is, how to select the "gain" of the phase detector – i.e., how many volts are output in response to a given phase difference – and the loop filter bandwidth to achieve the desired tradeoff between lock range and phase noise (and spurious). This problem shares some features with the design of feedback oscillators, and has much in common with the mathematical theory of control systems. Recommended references on these topics, as well as recommended additional reading on PLL synthesizers generally, include [3, Sec. 9.7] and [27, Secs. 9.27–29].

16.9 DIRECT DIGITAL SYNTHESIS

A recurring theme in this chapter has been that traditional methods of frequency synthesis – feedback oscillators, VCOs, and PLL synthesizers – entail considerable complexity and

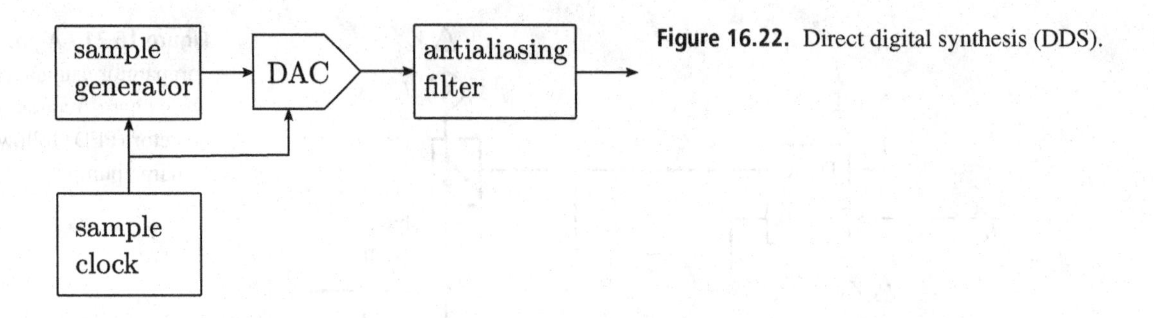

Figure 16.22. Direct digital synthesis (DDS).

uncertainty in both design and analysis. Given the availability of digital-to-analog converters (DACs) operating at clock rates in the RF regime, one wonders: Why not just generate sinusoids using DACs? The answer is you can, and this approach is known as *direct digital synthesis* (DDS). In contemporary radio design, DDS turns out to be quite useful in some respects, and not so great in others. In this section we consider the technique and some of its pros and cons.

The typical structure of a DDS frequency synthesizer is shown in Figure 16.22. Generally, one needs only to generate a digital representation of the waveform (more on this in a moment), convert it to analog by using the DAC, and then filter to remove undesired aliases and as much quantization noise and as many spurious signals as possible.[13] The big advantage of this approach is, of course, that the waveform can be generated with frequency accuracy, stability, and phase noise, which is limited only by the sample clock, and so can be *very* good; for example, if a crystal oscillator is used. Unlike a PLL synthesizer, a DDS device can be retuned instantaneously and continuously as desired.

In principle, it is possible to generate a sinusoid at any frequency between near zero and just below $F_S/2$, where F_S is the frequency of the sample clock. Modern DACs designed for this purpose are available with sample rates extending into the GHz range. The DAC produces another copy of the output as an alias between $F_S/2$ and F_S, yet another copy between F_S and $3F_S/2$, and so on; therefore in principle it is possible to obtain a sinusoid tunable over a range of width $F_S/2$ centered at frequencies many times greater than F_S.

There are at least two principal strategies for implementation of the sample generator in Figure 16.22. The brute force approach is to use a *waveform table*. In this approach, the waveform samples are generated in advance and are simply "played out" into the DAC. Care must be taken to ensure the first sample of the waveform sample set is consistent with the last sample, so as to avoid a glitch that repeats at the waveform period. This in turn provides strong motivation to make F_S an integer multiple of the frequency of the generated waveform. Thus, the waveform table approach can be awkward if continuous or rapid tuning is desired.

An alternative to the waveform table playback strategy is to generate the waveform samples "on the fly", using a mathematical algorithm. The principal advantages of this strategy are that (1) the required memory is reduced to nearly zero, and (2) frequency and other parameters of the sinusoid may be varied dynamically – even sample by sample – since there

[13] The fundamentals of digital-to-analog conversion are covered in Section 17.3, but are not really necessary to grasp the material in this section.

is no need to compute the waveform in advance. The primary difficulty in this strategy is calculating the function $\sin \omega t$ and/or $\cos \omega t$ in real time. The problem is not that accurate algorithms are not available; after all, the simplest computers – even handheld calculators – can do this. The problem is instead that one typically wishes to implement the calculation in programmable logic, such as an FPGA, as opposed to a CPU, since synthesizers are more likely to be implemented in the former technology for reasons pertaining to cost, power, size, and issues related to system integration – see Section 18.6 for more about this. To make the computation of a sinusoid waveform suitable for efficient implementation in programmable logic, the algorithm should preferably not require multiplication, looping or recursion, or external memory. The best-known of these algorithms is known as *CORDIC* ("COordinate Rotation DIgital Computer") [81].

DDS has some significant limitations. The first thing to note is that a suitable DAC might not be available for frequencies greater than a few hundred MHz, so modern DDS devices often require analog frequency conversion to reach frequencies in the UHF range and above. Second, the sample generator and DAC require a sample clock, and that sample clock may itself be the output of an oscillator or synthesizer whose performance may be limiting. To the extent that the sample clock suffers from issues of frequency accuracy, stability, and phase noise, so too will the output from the DAC suffer. DACs themselves contribute additional spurious content as a result of non-linearity in the digital-to-analog conversion process and DAC imperfections, analogous to those of ADCs. The ability to exploit aliases to obtain output at frequencies greater than $F_S/2$ is limited by the usable bandwidth of the analog portion of the DAC, which may be as little as $3F_S$ at RF frequencies. Finally, we note the most significant barriers to widespread adoption of DDS over feedback oscillators and synthesizers: Cost and power consumption, both of which are typically considerably greater for DDS at radio frequencies.

16.10 IC IMPLEMENTATION OF OSCILLATORS AND SYNTHESIZERS

So far in this chapter we have considered feedback oscillators, PLL synthesizers, and DDS. Despite the formidable disadvantages of DDS identified in the last paragraph of the previous section, it has two big things going for it: (1) it is very well-suited to integrated circuit (IC) implementation, since the programmable logic and DAC are usually already in this form; and (2) design is relatively simple, especially considering the complexity (and, frankly, guesswork) that is typically associated with the design of feedback oscillators and PLL synthesizers. On the other hand, the advantages of feedback oscillators and PLL synthesizers – in particular, cost, power consumption, and frequency range – are formidable. By implementing these technologies in IC form, one might achieve the best of both worlds; and, in particular, make the design of systems using frequency synthesis devices considerably easier.

The principal challenge in IC implementation of non-DDS frequency synthesizers is integration of a resonator onto the IC. For LC feedback oscillators, the primary problem is the inductance: On-chip inductors must typically be in implemented as flat spirals, which require overlapping traces and are therefore difficult to design and integrate, and which can manifest only relatively small values of inductance compared to available discrete inductors. As a result, fully-integrated feedback oscillators using LC resonators are typically limited to frequencies of a few hundred MHz or higher. At lower frequencies, inductors and high-Q components of resonators are typically left off-chip, intended to be provided as discrete components connected through dedicated pins on the IC. This does not entirely solve the problem, however, as the necessary interconnects themselves contribute both inductance and phase, and introduce uncertainty in the phase shift through the resonator. Higher-Q resonators employing crystal, SAW, or ceramic resonators cannot be integrated into conventional RFIC technologies such as CMOS. On the bright side, varactors required for VCOs are relatively easy to implement in semiconductor IC technologies; for example, CMOS VCOs typically utilize varactors fashioned from MOSFETs.

The remaining circuit elements required for a PLL synthesizer – i.e., PFD, charge pump, loop filter, frequency dividers, and prescalers – are straightforward to implement in modern IC technologies; thus fully-integrated single-chip PLL synthesizers are now routinely available for frequencies from a few hundred MHz into the SHF range. These chips typically require only an external reference oscillator (typically a crystal oscillator), bypass capacitors, and a just a few additional discrete passive components for control and interfacing purposes. The fact that PLL synthesizers can be implemented in a single chip with only these few remaining external components accounts in large measure for the ability to implement complete radios – e.g., cell phones – in compact hand-held packages.

Some other types of oscillators – not previously identified in this chapter – are easier than feedback oscillators to implement in IC technologies such as CMOS. An important class of these are *phase-shift oscillators*, in which a feedback path is used to introduce a particular phase shift (or, what is essentially the same, a delay) between the output and input of the gain device (typically, an op-amp) that results in oscillation.[14] An important category of phase-shift oscillators are *ring oscillators*, which consist of cascades of logic buffers or op-amps producing oscillation at a frequency determined by the delay or phase shift through the cascade, and is ideally suited to differential implementation.[15] Other classes include *relaxation oscillators*[16] and *Wein bridge oscillators*.[17] A common feature of all three of these classes is that inductance is not required, which largely accounts for their popularity in IC design. However, ring oscillators and relaxation oscillators have particularly poor phase noise and waveform distortion performance, and Wein bridge oscillators are difficult to implement at frequencies much above the MF band. Thus other types of frequency synthesis tend to prevail in contemporary radio designs.

For additional reading including detailed treatments of oscillators from the perspective of modern (in particular, CMOS) RFIC design, [63, Ch. 8] and [39, Ch. 17] are suggested reading.

[14] Suggested references: [72, Sec. 13.2.2] (basic) and [21, Secs. 1.7 and 3.5] (advanced).
[15] Suggested reference: [21, Sec. 6.2.3].
[16] Suggested reference: [21, Sec. 6.2].
[17] Suggested references: [72, Sec. 13.2.1] (basic) and [21, Sec. 1.6] (advanced).

Problems

16.1 Why do oscillators require gain devices as part of their design?

16.2 In Figure 16.9, $C_1 = C_2 = 100$ pF and $L = 130$ nH. Estimate the anticipated frequency of oscillation. Specifically identify all assumptions required for this estimate to be valid.

16.3 In Figure 16.9, $C_1 = 100$ pF, $C_2 = 10$ pF, and the desired frequency of oscillation is 100 MHz. Estimate an appropriate value for L, and briefly summarize the assumptions required to make this estimate.

16.4 Using SPICE, confirm the difference in the "first pass" ($C_1 = 390$ nF, $C_2 = 27$ nF) and "final" ($C_1 = C_2 = 47$ nF) designs obtained in Example 16.1. Provide your answer in the form of plots as shown in Figure 16.13 for both cases, and comment on the differences.

16.5 Repeat Example 16.1 with the following changes: (a) desired frequency is 3 MHz, (b) output impedance is 1000 Ω, (c) $I_C = 20$ mA. In each case, use standard values for all components.

16.6 A receiver has a nominal noise figure of 7 dB using LOs having negligible phase noise. At the input to this receiver, an interfering signal has a total power equal to −13 dBm with bandwidth of about 100 kHz. The LO for this mixer is replaced with one having −145 dBc/Hz (SSB) phase noise at an offset of 3 MHz. Estimate the effective noise figure as function of frequency offset from the interfering signal.

16.7 A particular LMR system utilizes a channel spacing of 30 kHz around 450 MHz. What frequency accuracy is required to ensure tuning to within 5% of the channel separation? Give your answer in percent of the carrier frequency and in ppm.

16.8 Varactor-tuned VCOs using crystals – sometimes referred to as "VCXOs" – are possible, and sometimes appear in specific applications. What are the pros and cons of a VCXO relative to a fixed-frequency crystal oscillator and a VCO without a crystal resonator?

16.9 Determine R and N for an integer-N PLL synthesizer (Figure 16.20) that tunes from 77.4 MHz to 97.2 MHz in 200 kHz steps. The frequency of the crystal oscillator is 1 MHz.

16.10 In the previous problem, assume the crystal has a frequency stability of 0.5 ppm. What is the expected range of frequency error in the synthesizer output?

17 Transmitters

17.1 INTRODUCTION

Radio transmitters typically consist of a modulator and a power amplifier, and usually also involve digital-to-analog conversion and frequency upconversion. Of these elements, we have previously covered modulation (Chapters 5 and 6) and frequency conversion (Chapter 14) in detail. In this chapter we address the remaining elements and the associated architectural considerations. Architectural considerations are presented in Section 17.2. Digital-to-analog conversion is presented in Section 17.3. Sections 17.4–17.6 address power amplifiers (PAs), which are a largely distinct enterprise from the small-signal amplifiers considered in Chapter 10. This chapter concludes with a discussion of methods for combining amplifiers in parallel (Section 17.7), which has applications for the design of transmitter PA stages.

17.2 ARCHITECTURES

Section 1.4 presented two broad architectural strategies for the design of radio transmitters. In this section we elaborate on this with some extra detail accumulated from the intervening chapters. In fact, there are at least six common transmitter architectures, with many more variants that are less commonly used. The common strategies are depicted in Figure 17.1. Essentially the only common element in all six architectures is the final stage, which is a PA.

In the first two schemes (Figure 17.1(a) and (b)), sinusoidal carrier modulation is generated using an oscillator operating at the carrier frequency. In Figure 17.1(a), AM (including ASK) is generated by varying the amplitude of an oscillator, or FM (including FSK) is generated by varying the frequency of the oscillator. In Figure 17.1(b), AM is generated by controlling the output power of the PA; this is a technique we shall have more to say about in Section 17.4.6. The principal advantage of these schemes is simplicity. These architectures appear primarily in low-cost/low-power applications such as wireless keyless entry; and in very high-power applications, such as AM and FM broadcasting.

In Figure 17.1(c), the modulator generates output at an intermediate frequency (IF), which is subsequently upconverted to the carrier frequency using a superheterodyne frequency converter. Once ubiquitous, this architecture is steadily declining in popularity due to the many advantages of digital baseband processing, which is employed in schemes (d)–(f). Nevertheless, such radios certainly exist and in fact are prevalent in some applications requiring simple, low-cost communications using AM, SSB, or FM. Examples of these applications include Citizen's

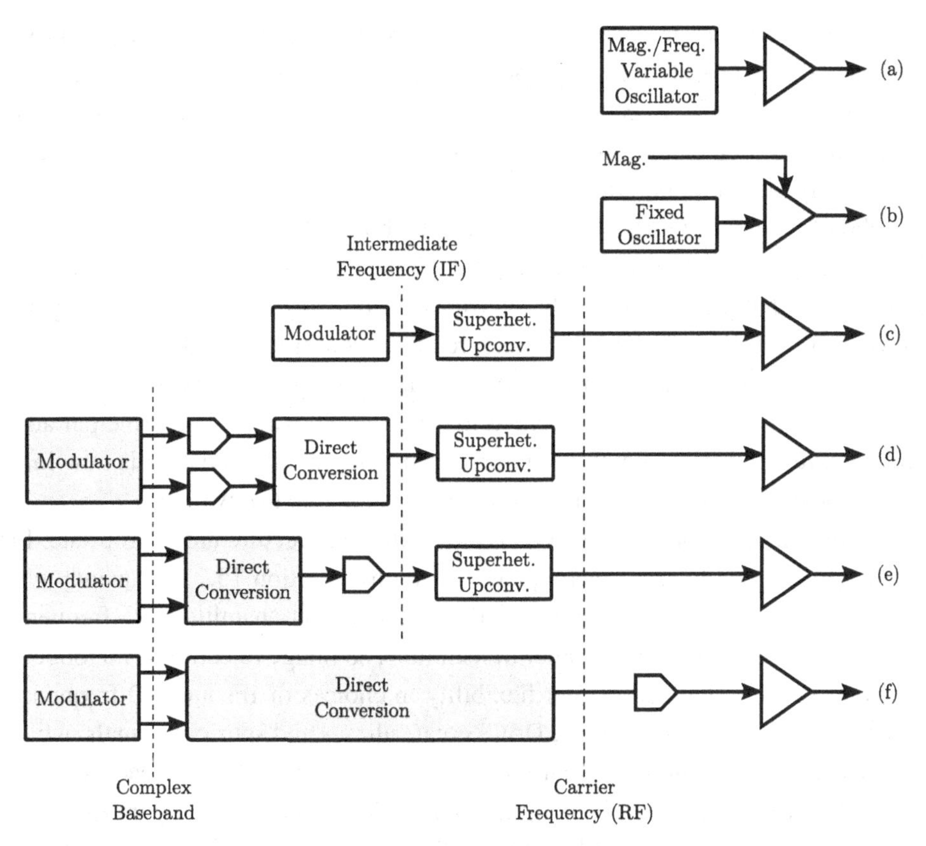

Figure 17.1. Some common architectures for transmitters. Shaded blocks indicate digital-domain implementation.

Band (DSB-AM at HF), air-to-ground ("airband," DSB-AM at VHF) and certain narrowband FM-based UHF-band mobile radio applications.

Figure 17.1(d) is probably the most common architecture employed in modern radio transmitters, and is nearly ubiquitous in cellular mobile (user) phones. In this scheme, modulation is implemented in the digital domain and the resulting complex baseband waveform is converted to analog. The analog complex baseband signal is then upconverted by quadrature conversion, resulting in a real-valued signal (see Section 14.4 for a reminder of this method). The output of the direct conversion stage can be at the carrier frequency, but more often the output is at an IF, which is subsequently upconverted to the carrier frequency using a superheterodyne frequency converter. The latter scheme is typically preferred because most problems associated with direct conversion are easier to mitigate at lower frequencies. (There is another reason associated with architectural symmetry between receiver and transmitter, which we shall address a bit later.) As employed here, this strategy is known as *direct conversion*, and is the transmit version of the similarly-named receiver architecture discussed in Sections 15.6.8 and 15.6.10.

The scheme shown in Figure 17.1(e) is essentially the same as scheme (d), except now direct conversion is implemented in the digital domain and the DAC outputs a signal at the IF frequency. This scheme yields all the advantages of direct conversion, with essentially none of the disadvantages since the digital implementation is not vulnerable to analog-domain limitations such as gain and phase imbalance and LO bleed-through. The disadvantage is that the DAC must now operate at rates comparable to the IF frequency, whereas in scheme (d) the baseband DACs operated at rates comparable to the signal bandwidth, which is typically orders of magnitude less. Because the IF DAC scheme results in increased cost and power consumption, this architecture appears primarily in non-mobile or base station transmitters, and is relatively rare in mobile radios.

The scheme shown in Figure 17.1(f) pushes the border between digital and analog from IF to RF: Now all processing occurs in the digital domain, and the DAC outputs a signal at the RF frequency. Note that this is essentially the transmit equivalent of direct sampling architecture (Sections 15.6.1 and 15.6.2). This architecture is not commonly used as it is quite difficult to obtain acceptable performance from a DAC operating at RF frequencies, and cost and power consumption are greatly increased relative to other architectures. The principal advantage of this architecture is complete flexibility in processing, since all processing is now implemented entirely in some combination of software and programmable firmware.

At this point, a few more comments on transmitter architecture are appropriate. First, some good news: In contrast to receiver frequency planning (Section 15.7), frequency planning for transmitters is typically quite straightforward. The input to a transmitter's RF frequency conversion system is free of signals from other transmitters, so image rejection is no longer a primary concern, and so there is much greater flexibility in choices of IFs and LO frequencies. Some issues remain, however: For example, DACs create aliases and spurious signals, which must be controlled, as must undesired and spurious multiplication products from each stage of mixing.

Second: In the case of transceivers (i.e., radios with both receivers and transmitters), there is no particular imperative for the architecture of the transmitter to be the same as the architecture of the receiver. However, this typically happens anyway, so that LOs and sampling clocks may be shared. This sharing reduces the cost and complexity of the transceiver. Even half-duplex transceivers using different frequency plans for transmit and receiver will frequently share LOs, as it is often more attractive from a systems perspective to have a single LO synthesizer that is able to switch between two different frequencies than two synthesizers where each is required only a fraction of the time.

The majority of effort unique to transmitter design is in addressing two problems: digital-to-analog conversion (in those systems that use DACs), and PAs. We now tackle these issues specifically.

17.3 DIGITAL-TO-ANALOG CONVERSION

A DAC is a device that converts a digital (discrete-time and quantized) representation of a signal into a continuous-time analog representation of the same signal. DACs have much in common with ADCs, so a review of Section 15.2 may be useful before taking on this section.

17.3.1 Method of Operation

A simple model for digital-to-analog conversion is shown in Figure 17.2. This model consists of *decoding* and *waveform assembly*. The decoder plays a role analogous to that of the output

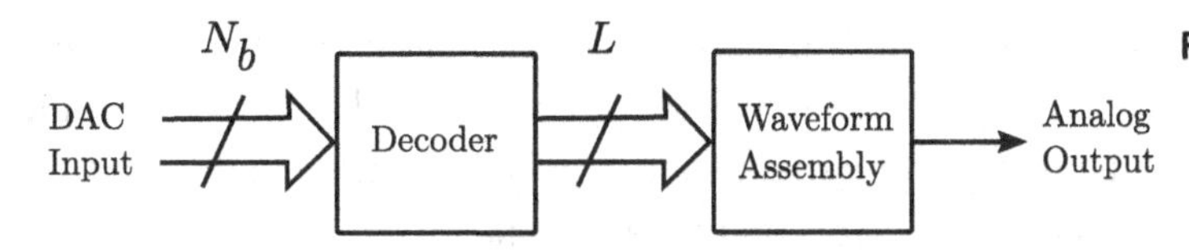

Figure 17.2. A simple DAC model.

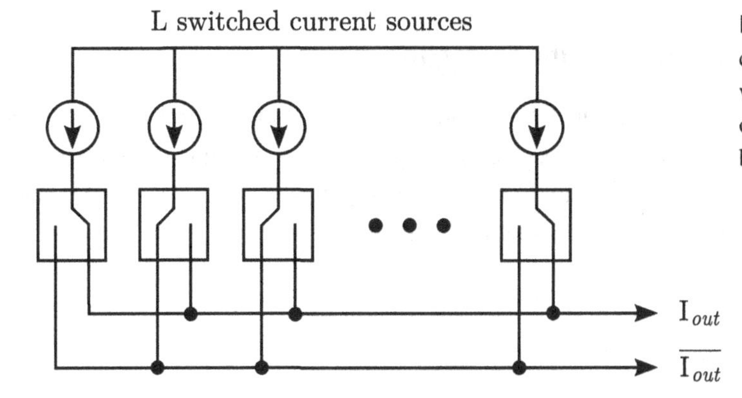

Figure 17.3. A simple model of a current source array used for waveform assembly in a DAC. Each of the L switches is controlled by one bit from the decoder.

word generator (Figure 15.1 and associated text) in an ADC, but in reverse: The N_b-bit digital word delivered to the input of the decoder is translated to a set of $L = 2^{N_b}$ bits, which represent threshold decisions. As in an ADC, the input representation takes various forms (e.g., offset binary, two's complement, and so on). However, the L output bits are typically simple threshold decisions; i.e., each bit indicates whether the desired value is larger or smaller than the value represented by the bit.

Waveform assembly is analogous to the quantizer (e.g., Figure 15.3 and associated text) in an ADC: L bits representing threshold decisions are accepted, and an analog signal having one of L magnitudes is output. Note that this is not quite "dequantization," since the output waveform remains quantized, taking on only one of L values at a time. All that has happened is that L bits have been converted to one analog signal representing the same information.

There are many ways to do waveform assembly in DACs, but a common method for radio applications is using a structure known as a *current source array*. A rudimentary current source array is shown in Figure 17.3. The structure consists of L identical current sources followed by single-pole dual-throw switches. Each of the L input bits controls the state of one of these switches, so the output is simply the current provided by a single source times the number of input bits that are asserted. The output of this structure is typically designed to be differential (as shown in Figure 17.3), so as to take advantage of the many benefits of differential signals and devices as identified throughout this book.[1]

The current source array is certainly not the only way to do waveform assembly, and the structure shown in Figure 17.3 is certainly not the only way to implement a current source array. One of the reasons current source arrays turn out to be popular is because they are relatively easy to implement in available integrated circuit technologies; in particular, in CMOS. Each source in Figure 17.3 can be implemented as a single MOS transistor, and each switch can be

[1] If you find the differential topology of Figure 17.3 confusing, simply imagine $\overline{I_{out}}$ being tied to ground; then you have a single-ended output (and a lot of wasted current!).

implemented as a pair of identical MOS transistors, so the entire array can be implemented using $3L$ identical MOS transistors. MOS transistors are very small and easy to implement, so it is common to increase the number of current source/switch elements in the array from L to a larger number as a means to mitigate errors and improve linearity.

17.3.2 Sample Rate, Bandwidth, and sinc Distortion

The output of a DAC is analog, but it is not quite continuous-time. Rather, the output of an ideal DAC appears as shown in Figure 17.4(e): a contiguous sequence of pulses, with each pulse having width equal to T_S where $1/T_S$ is the sample rate. This has some consequences which are important to understand: aliasing and sinc distortion.

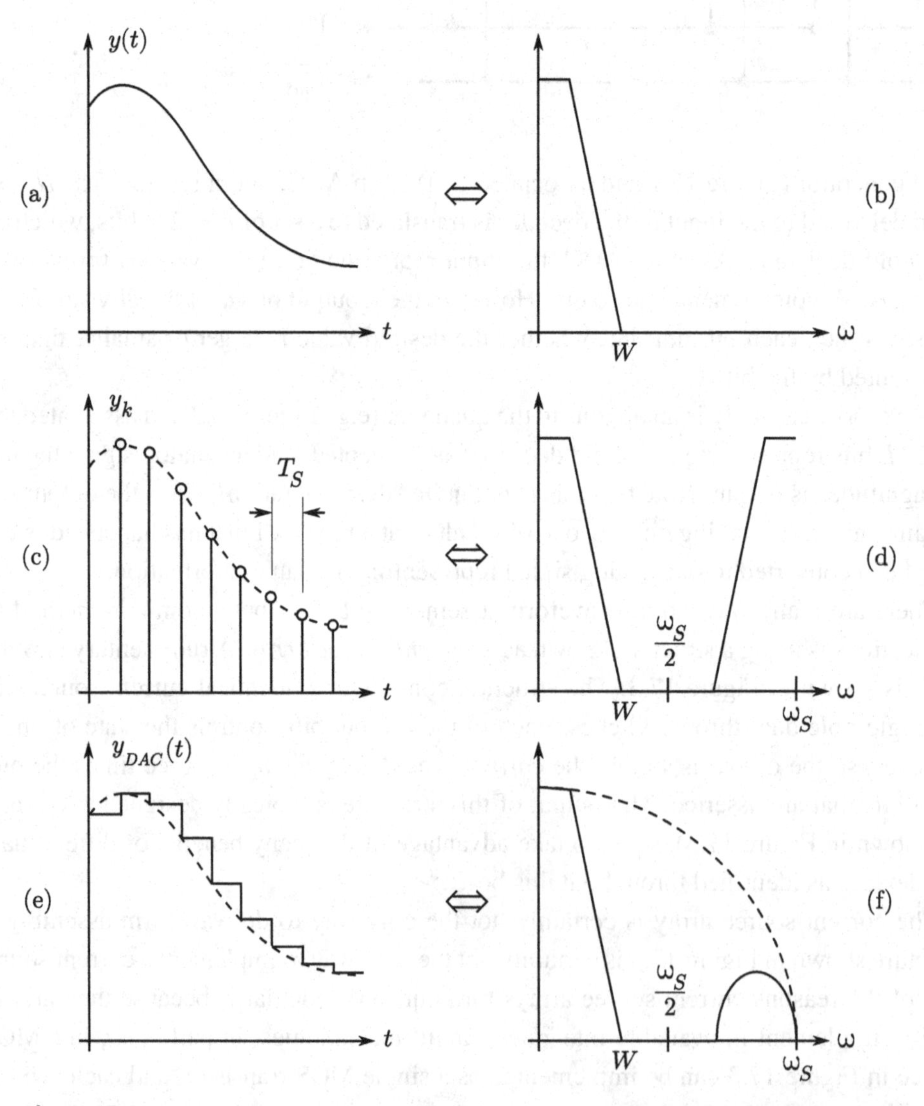

Figure 17.4. Time and frequency domain waveforms associated with digital-to-analog conversion: (a) Intended signal $y(t)$, (c) discrete-time representation of $y(t)$, (e) DAC output; (b), (d), and (f) are the associated spectra.

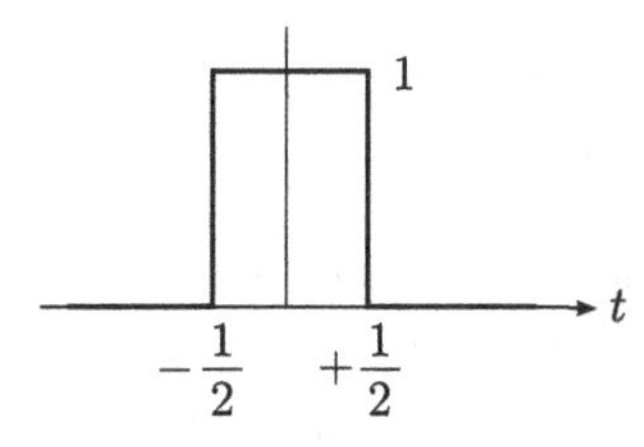

Figure 17.5. The rectangular pulse function $\Pi(t)$.

To sort this out, first consider the DAC input y_k (Figure 17.4(c)) which consists of samples from the desired continuous-time analog signal $y(t)$ (Figure 17.4(a)). Because each DAC input sample represents a fixed value realized over the sampling period, we define

$$y_k \equiv \frac{y(kT_S)}{T_S} \tag{17.1}$$

Let us temporarily ignore quantization error, since it will not affect the findings of the present analysis. The output of the DAC (Figure 17.4(e)) may then be written in the form:

$$y_{DAC}(t) = \sum_{k=-\infty}^{+\infty} y_k \, \Pi\left(\frac{t - kT_S - T_S/2}{T_S}\right) \tag{17.2}$$

where $\Pi(t)$ is the rectangular pulse function shown in Figure 17.5. Now we note the following relationships between time-domain functions and their Fourier transforms:

$$\Pi(t) \quad \longleftrightarrow \quad \mathrm{sinc}(f) \equiv \frac{\sin \pi f}{\pi f} \tag{17.3}$$

$$\Pi\left(\frac{t}{T_S}\right) \quad \longleftrightarrow \quad T_S\,\mathrm{sinc}(fT_S) \tag{17.4}$$

(from the scaling property of the Fourier transform)

$$\Pi\left(\frac{t - kT_S - T_S/2}{T_S}\right) \quad \longleftrightarrow \quad T_S\,\mathrm{sinc}(fT_S)\, e^{-j2\pi f T_S(k+1/2)} \tag{17.5}$$

(from the delay property of the Fourier transform). Recall that $\omega \equiv 2\pi f$ and $\omega_S \equiv 2\pi/T_S$. Making these substitutions and expanding the exponential into separate factors, we obtain

$$\Pi\left(\frac{t - kT_S - T_S/2}{T_S}\right) \quad \longleftrightarrow \quad T_S\,\mathrm{sinc}\left(\frac{\omega}{\omega_S}\right) e^{-j\omega k T_S}\, e^{-j\omega T_S/2} \tag{17.6}$$

Now returning to Equation (17.2) we find the Fourier transform of the DAC output is

$$Y_{DAC}(\omega) = \sum_{k=-\infty}^{+\infty} y_k T_S\,\mathrm{sinc}\left(\frac{\omega}{\omega_S}\right) e^{-j\omega k T_S}\, e^{-j\omega T_S/2} \tag{17.7}$$

Rearranging factors and using Equation (17.1) we obtain

$$Y_{DAC}(\omega) = \left[\sum_{k=-\infty}^{+\infty} y(kT_S) e^{-j\omega k T_S}\right] \mathrm{sinc}\left(\frac{\omega}{\omega_S}\right) e^{-j\omega T_S/2} \tag{17.8}$$

The factor in square brackets is the discrete-time Fourier transform (DTFT) of $y(t)$. The DTFT of a continuous function (such as $y(t)$) is equal to the Fourier transform of the continuous function (here, $Y(\omega)$), repeated in the frequency domain with periodicity ω_S. Therefore we may write

$$Y_{DAC}(\omega) = \left[\sum_{k=-\infty}^{+\infty} Y(\omega - k\omega_S) \right] \operatorname{sinc}\left(\frac{\omega}{\omega_S}\right) e^{-j\omega T_S/2} \tag{17.9}$$

Figure 17.4(f) demonstrates $|Y_{DAC}(\omega)|$. Note the first factor in the above equation is simply the spectrum of $y(t)$ within $[-\omega_S/2, +\omega_S/2]$, plus all of the aliases of $y(t)$ repeating outside of this interval. The second factor in the above equation is *sinc distortion*, which is an artifact of the "sample and hold" property of the output waveform. The last factor in the above equation is merely the $T_S/2$ delay required to make the pulse edges align with sampling instants, and so has no particular significance.

We may now address the consequences of digital-to-analog conversion from a signal processing perspective. First, it is clear that DACs, like ADCs, are vulnerable to aliasing. The Nyquist sampling theorem should be satisfied, i.e., $F_S \geq 2W$, where W is the lowpass bandwidth of the signal of interest. The consequence of violating this criterion is that aliased copies of the desired output will overlap in the frequency domain, with associated distortion in the time domain. Therefore, DACs are normally followed by anti-alias filters to remove the undesired aliases.

Note further that DACs – again, like ADCs – may be undersampled. That is, one is not obligated to select the low-frequency (lowpass) alias; one might alternatively select a higher-frequency alias. Just as ADC undersampling yields a "free" downconversion in receivers, DAC undersampling yields a "free" upconversion in transmitters.

Sinc distortion, on the other hand, is unique to DACs. This distortion imposes an additional lowpass response onto the first and second Nyquist zones, going to zero at $n\omega_S$ with $n = \pm1, \pm2, \pm3, \ldots$. If the intended output is the first Nyquist zone alias, then this is not altogether bad, since the sinc response partially suppresses all higher- and lower-frequency aliases and reduces the suppression required from an anti-aliasing filter. However, the first Nyquist zone alias is also distorted by the sinc response. If $W \ll 1/T_S$, then this distortion is small and is typically neglected. If W is not small relative to $1/T_S$, then "sinc compensation" may be required. This is simply a filter that has frequency response nominally equal to $1/\operatorname{sinc}(\omega/\omega_S)$. Sinc compensation filtering, when required, can be combined with anti-aliasing filtering. Alternatively, the compensation may be applied prior to the DAC, facilitating a convenient digital implementation and thereby simplifying the analog hardware. See Section 18.2.3 for further discussion about this idea.

EXAMPLE 17.1

The direct conversion architecture shown in Figure 17.1(d) is used in a CDMA mobile phone capable of bandwidth up 1.25 MHz. Specify the minimum reasonable DAC sample rate and calculate the sinc distortion at this sample rate. Assume 10% excess bandwidth to accommodate anti-alias filtering.

Solution: A signal having 1.25 MHz bandwidth has a lowpass bandwidth of 1.25/2 = 0.625 MHz in complex baseband form. Adding 10% to accommodate anti-alias filtering gives 0.6875 MHz lowpass bandwidth. To satisfy the Nyquist criterion, the minimum sample rate is $F_S = 1/T_S = 2 \times 0.6875$ MHz $= 1.375$ MSPS. Sinc distortion is maximum at the edge of the signal bandwidth, where $\omega/\omega_S = f/F_S = 0.625$ MHz/1.375 MSPS $= 0.455$. Since $\text{sinc}(0.455) = 0.693$, sinc distortion results in about 3.2 dB of attenuation at the edges of the signal spectrum relative to the center.

17.3.3 Quantization Noise and Dynamic Range

DACs, like ADCs, generate quantization noise. Quantization noise is attributable to the fact that the output may take on only certain values, and the resulting non-uniform distribution of sample magnitudes manifests as noise power distinct from the intended output signal. The signal-to-noise ratio and associated dynamic range considerations are as presented in Section 15.2.3: The bottom line is that increasing N_b (i.e., more bits, so greater L) reduces quantization noise relative to the desired signal power, and thereby increases the associated dynamic range. Also as is the case for ADCs, the limitation in decreasing quantization noise by increasing N_b is that non-linear effects eventually become apparent, typically manifesting as spurious signals.

17.4 POWER AMPLIFIERS

A PA is the component of a transmitter that delivers the desired RF waveform at the desired power, and is the last active component in the signal chain of a transmitter. Note that this definition makes no mention of gain, which is a clue that there can be some significant differences between PAs and amplifiers as presented in Chapter 10. To emphasize the point, note that PAs may have specified output power as little as 1 mW, whereas "gain" amplifiers may in some cases exhibit output power up to a few watts. That is considerable overlap.

In Chapter 7 (Figure 7.6 and associated text) we identified radio communications applications requiring transmit power in the range 100 mW to 1 kW. Not explicitly considered were broadcasting applications; these sometimes require power on the order of 1 MW. PAs up to about 1 kW may be implemented using transistors in the types of circuits described in this chapter. This accounts for PAs used in most modern radio communications applications. At significantly higher output power, the power-handling capability of available transistors is exceeded, and alternative devices – beyond the scope of this book – are required.

17.4.1 Efficiency vs. Linearity

The primary specifications for a PA are *output power* P_L delivered to a specified load resistance R_L, and *efficiency*. Various definitions of efficiency are in common use, each having some application. Presently, the efficiency of concern is the ratio of P_L to power provided by the DC power supply, P_{DC}. This efficiency is known variously as either the *collector efficiency* or *drain efficiency*, depending on whether the transistors delivering the power are of the bipolar or FET type, respectively. In this chapter the collector/drain efficiency will be indicated using the variable ϵ_P; thus:

$$\epsilon_P \equiv \frac{P_L}{P_{DC}} \tag{17.10}$$

Note that for non-PA amplifiers – including essentially all of the amplifiers considered in Chapter 10 – ϵ_P is very low, and we did not care. Here is an example that explains why the issue suddenly leaps to prominence when dealing with PAs:

EXAMPLE 17.2

The final stages of a particular transmitter consists of a non-PA amplifier followed by a PA. The first amplifier outputs 1 mW while consuming 10 mW from the power supply, and so has $\epsilon_P = 10\%$. Let us say the PA subsequently outputs 1 W at the same efficiency; thus the PA consumes 10 W from the power supply. Therefore the PA consumes 100 times as much power from the DC power supply as the preceding amplifier. Even if the transmitter consists of 10 non-PA amplifiers, the overall efficiency of the transmitter is approximately that of the PA, and so the efficiency of all preceding amplifiers is essentially irrelevant.

For the reason identified in the example, one typically focusses on optimizing ϵ_P for the PA, and the efficiencies for all preceding amplifiers get at most a cursory check. Further, this realization motivates PA design strategies that achieve the highest possible ϵ_P. Values of ϵ_P in practical

Table 17.1. **Summary of common PA classes of operation. "Practical" ϵ_P values are adopted from [68, Fig. 10.1]. As explained in Section 17.4.8, non-linear PAs are sometimes employed as subsystems in more complex PAs suitable for linear modulations.**

Class	Linearity	Conduction Angle	ϵ_P Ideal	Practical	Typical Uses
A	Linear	360°	$\ll 50\%$		Rare as PA
A	Quasi-linear	360°	50%	$\approx 35\%$	SSB, High-order QAM
AB	Quasi-linear	180°–360°	50%–78%	35%–60%	AM, Low-order digital
B	Quasi-linear	180°	78%	$\approx 60\%$	SSB
C	Non-linear	$\leq 180°$	100%	$\approx 75\%$	AM,[a] CM[b]
D	Switched[c]	180°	100%	$\approx 75\%$	High-power CM[b]
E	Switched[c]	$\approx 180°$	100%	$\approx 90\%$	Low-power CM[b]
F	Non-linear	$\approx 180°$	$\approx 100\%$	$\approx 80\%$	UHF and higher CM[b]

[a] Amplitude modulation by varying bias voltage.
[b] Constant modulus; e.g., FM, MSK.
[c] To be clear, definitely non-linear.

systems range from about 35% to about 90% (see Table 17.1). In the above example, increasing the efficiency of the power amplifier from 10% to just 40% reduces the power consumed from 10 W to 2.5 W – a dramatic improvement in terms of the DC power system requirements, facilitating lower voltage levels and (in the case of mobile radios) longer battery life.

Note that at least one other metric of efficiency is in common use: *power-added efficiency* ϵ_{PA}, defined as

$$\epsilon_{PA} \equiv \frac{P_L - P_{in}}{P_{DC}} \tag{17.11}$$

where P_{in} is the input power; i.e., power delivered to the PA by the preceding stage. This metric makes a bit more sense for PAs that have very low gain, since it accounts for the RF power "invested" at the input.

EXAMPLE 17.3

A PA outputs 10 W into 50 Ω at a drain efficiency of 70%. To do this, the PA must be driven by 10 V_{pp} input signal. The input impedance of the PA is 50 Ω. What is the power-added efficiency of the PA?

Solution: Here, $P_L = 10$ W and $\epsilon_P = 0.7$, so $P_{DC} = P_L/\epsilon_P = 14.3$ W. 10 V_{pp} = 5 V_p = 3.54 V_{rms}, so $P_{in} = (3.54\ V_{rms})^2/(50\Omega) = 0.25$ W. Therefore the power-added efficiency is

$$\epsilon_{PA} = \frac{10\text{ W} - 0.25\text{ W}}{14.3\text{ W}} = 0.682 = 68.2\%$$

As we shall see shortly, the primary route toward improvement of efficiency requires compromising on linearity. We have alluded to this issue frequently in previous sections of this book: in Chapter 5 the need for PA linearity and associated poor efficiency was cited as an disadvantage of AM and SSB relative to FM; in Chapter 6 essentially the same criticism was made about all modulations in which the RF waveform magnitude varied significantly from a constant value. Boiled down to its fundamentals, the issue is this: If the information in a modulation is entirely in phase or frequency, then *linear* operation is not required, and if a PA can be non-linear, then dramatic improvements in efficiency are possible. This tradeoff will be laid out a bit more explicitly as we introduce the various classes of PA operation in the following sections. Table 17.1 is provided as an aid in discriminating between these disparate schemes.

A final note before moving on to a discussion of PA classes of operation: To be concise the discussion will presume implementation in bipolar transistor technology. This should not be interpreted as meaning that implementation using FET-type devices is contraindicated, not possible, or is significantly different! As was the case in Chapter 10, the general principles are quite independent of the transistor technology, and only when the particulars of an application are considered does it become more clear whether a bipolar or a FET implementation is preferred. Both technologies continue to find frequent application in modern radio communications systems.

17.4.2 Class A; Linear vs. Quasi-Linear Operation

A simple and generic model for understanding Class A operation is shown in Figure 17.6. Here we assume a bipolar transistor in common emitter configuration with the simplest possible bias scheme: Collector tied to supply voltage V_{CC} through an RF choke (RFC) inductor, and emitter tied to ground. The output is taken from the collector through a DC-blocking capacitor C_{block}. The input V_{in} is applied directly to the base, and so serves as both base bias as well as RF input signal. The harmonic filter will be explained a bit later. For the moment, just assume it is in resonance and so appears as an open circuit to both the transistor and the load.

A suitable approximation of the I_C vs. V_{BE} (=V_{in}) characteristic of the transistor is available via the Ebers–Moll model as explained in Chapter 10 (Equation (10.1)). Note here that we consider I_C, like V_{BE}, to be an instantaneous quantity; i.e., not merely the DC (steady-state) component of the instantaneous quantity.

The defining characteristic of a Class A PA is that the transistor operates in active mode over the entire range of the waveform; essentially operating in accordance with assumptions made in Chapter 10. This is illustrated in Figure 17.7. Note that V_{in} remains well above 0.7 V

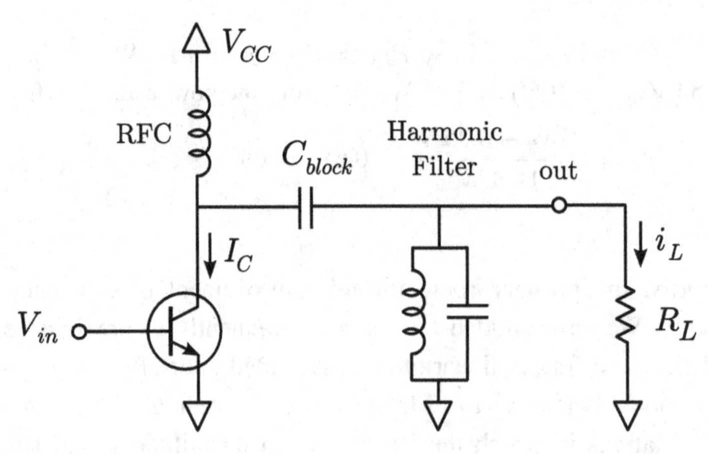

Figure 17.6. A generic Class A amplifier model (also applies to Classes B, C, and D).

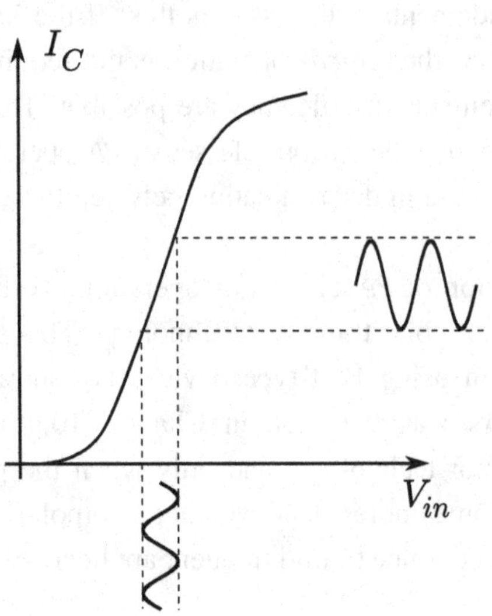

Figure 17.7. Linear Class A operation.

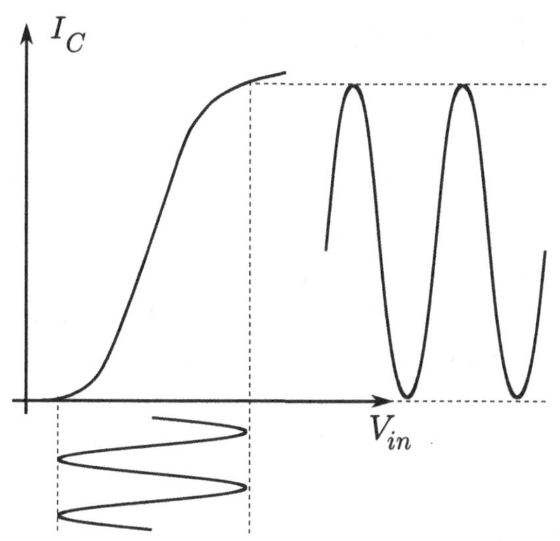

Figure 17.8. Quasi-linear Class A operation.

(assuming bipolar technology), and the variation in V_{in} is small enough so that the associated segment of the I_C vs. V_{BE} curve is linear to a good approximation.

It is immediately clear that ϵ_P must be very poor under these conditions. This is because there is a continuous flow of current through the transistor. In particular, the mean value of I_C is large compared with the time-varying component that represents the RF signal of interest. In fact, the DC component of I_C remains "on" even if the desired RF signal goes to zero.

Efficiency can be improved by increasing the voltage swing associated with the desired RF signal, so that the RF signal represents a greater portion of I_C compared to the DC component. This is shown in Figure 17.8. Here, one is limited to $V_{in} \geq 0.7$ V in order to sustain active region operation, and also V_{in} must remain sufficiently small so as not to damage the transistor. By extending the voltage swing to such limits, the amplifier can no longer be considered to be approximately linear, but instead is only "quasi-linear." Note there is no rigorous distinction between "linear" and "quasi-linear" operation, and in neither case is the amplifier *perfectly* linear. The purpose in making the distinction of "quasi-linear" operation is mainly to remind ourselves that we are allowing the linearity to be significantly degraded so as to accommodate other design goals; presently, improving efficiency.

The maximum theoretical collector efficiency of the Class A amplifier is about 50%, as we shall now show. Let us assume the RFC is completely effective in blocking RF, and that the blocking capacitor is completely effective in blocking DC. Then we may write I_C in terms of its DC and RF components:

$$I_C = I_{DC} - i_L \cos(\omega_c t) \tag{17.12}$$

where I_{DC} is the current through the RFC, and i_L is the magnitude of the sinusoidal carrier current through the load.[2] The power delivered to R_L may be written as

$$P_L = \frac{1}{2} i_L^2 R_L \tag{17.13}$$

[2] Formally we should account for the modulation as well; however, the action of interest here all occurs within one period of the carrier, so there is essentially no loss of generality by neglecting modulation as we have in Equation (17.12).

The power consumed from the power supply to accomplish this is simply

$$P_{DC} = V_{CC} I_{DC} \tag{17.14}$$

which yields

$$\epsilon_P \equiv \frac{P_L}{P_{DC}} = \frac{1}{2} \frac{i_L^2 R_L}{V_{CC} I_{DC}} \tag{17.15}$$

This equation applies generally, but what we really want to know is the *maximum* efficiency; therefore we should evaluate the above expression for the largest possible voltage swing that can be accommodated in a quasi-linear mode of operation. In order to remain in active region operation,

$$i_L R_L < V_{CC} - 0.7 \text{ V} \approx V_{CC} \tag{17.16}$$

where we have made the simplifying approximation that V_{CC} is large compared to the $V_{BE} = 0.7$ V threshold for active region operation. At this threshold, $i_L \approx I_{DC}$. Substituting $i_L R_L \approx V_{CC}$ and $i_L \approx I_{DC}$ into Equation (17.15) we find

$$\epsilon_P \approx 1/2 = 50\% \text{ (approx. maximum)} \tag{17.17}$$

In practice, quasi-linear Class A amplifiers typically achieve ϵ_P in the 30%–40% range; i.e., significantly less than 50% due to various losses not accounted for in the above simple analysis and also because of the need to "back off" from the maximum voltage swing so as to meet distortion requirements.

EXAMPLE 17.4

Design a Class A PA to deliver 5 W to a 50 Ω load at 20 MHz. Use the lowest possible standard power supply voltage and calculate the resulting power efficiency and consumption metrics. (Neglect the harmonic filter for now.)

Solution: Referring to Figure 17.6, the design process consists of setting V_{CC} and choosing the values of the RFC and blocking capacitor. Note that in this topology, V_{CC} is the DC (mean) value of the collector voltage V_C, and V_C varies symmetrically above and below V_{CC} with peak deflection $|v_L| = |i_L| R_L$. For quasi-linear operation, $V_{CC} - v_L \approx 0.7$ V puts the negative peak at the threshold of cutoff, so $V_{CC} > v_L + 0.7$ V is required. Since $P_L = v_L^2 / (2R_L)$ we have

$$v_L = \sqrt{2 P_L R_L} \cong 22.4 \text{ V}$$

so V_{CC} must be greater than ≈ 23.1 V. The next largest commonly-available "standard" power supply voltage is 24 V, so let us set $V_{CC} = 24$ V.

The inductance of the RFC is determined by the requirement that the associated reactance X_{RFC} be large compared to R_L. A reasonable value to propose is $X_{RFC} = 10 R_L$, which gives ≅ 4 μH at 20 MHz. Note also this inductor must be rated for the appropriate amount of current (determined below). The blocking capacitance is determined by the requirement that the associated reactance X_{block} be small compared to R_L. A reasonable value to propose is $X_{block} = R_L/10$, which gives ≅ 1.6 nF at 20 MHz.

Now let us calculate the collector efficiency:

$$\epsilon_P \equiv \frac{P_L}{P_{DC}} = \frac{5\text{ W}}{V_{CC}I_{DC}}$$

I_{DC} is equal to the average value of the collector current, which varies sinusoidally between $(V_{CC} + |v_L|)/R_L$ and $(V_{CC} - |v_L|)/R_L$, and is therefore $V_{CC}/R_L = 0.48$ A. Returning to the above equation, $\epsilon_P = 0.434 = 43.4\%$; somewhat less than the nominal Class A efficiency of 50% as expected. The power supply consumption for this design is 11.52 W, which can be computed by using either $V_{CC}I_{DC}$ or P_L/ϵ_P. Therefore the transistor is dissipating $P_{DC} - P_L = 6.52$ W.

The above example alerts us to another manner in which PA design is different from conventional amplifier design: The problem of how to handle large amounts of power that are consumed, but not delivered to the load. In the example, the transistor dissipates 6.52 W of the power provided by the power supply, which means that it will become very hot. Suitable mechanical arrangements will be required to transfer the heat safely away from the transistor and thereby prevent damage. Thermal management is an important consideration in PA design and provides additional incentive to increase ϵ_P.

17.4.3 Harmonic Filtering

Let us now address the harmonic filter appearing in Figure 17.6. This filter is depicted as a bandpass filter consisting of a parallel LC circuit arranged in parallel with the load. Component values are chosen such that the frequency of resonance $\omega_0 = \omega_c = 1/\sqrt{LC}$, so the LC circuit is in resonance at the desired frequency of operation. The purpose of the harmonic filter is to suppress harmonics of desired signal; that is, components at frequencies equal to $2\omega_c$, $3\omega_c$, and so on. Why the special interest in harmonics? Because amplifiers which are not strictly linear produce large harmonics. (This was explored in detail in Section 11.2.) Therefore by bandpass filtering at ω_c with appropriate bandwidth, the higher order terms are suppressed and the distortion in the waveform is reduced. For Class A, harmonic filtering can be interpreted as simply an enhancement. We shall see that in some other classes of operation, harmonic filtering is integral to the fundamental operation of the amplifier and not optional.

Finally, note that there are many filter designs and topologies that are able to suppress harmonics while preserving the fundamental. Here, the choice of a parallel LC in parallel with the load is intentional: We wish the current associated with the potentially large harmonic terms to be sent safely to ground, and not reflected into the transistor.[3] It is not necessary to always use a single parallel LC shunted to ground, but whatever is used must be able to withstand large voltages and currents, and should handle power associated with the harmonic components in a safe way.

[3] Actually there is a special case where we would like to do exactly this: Class F operation (Section 17.4.7).

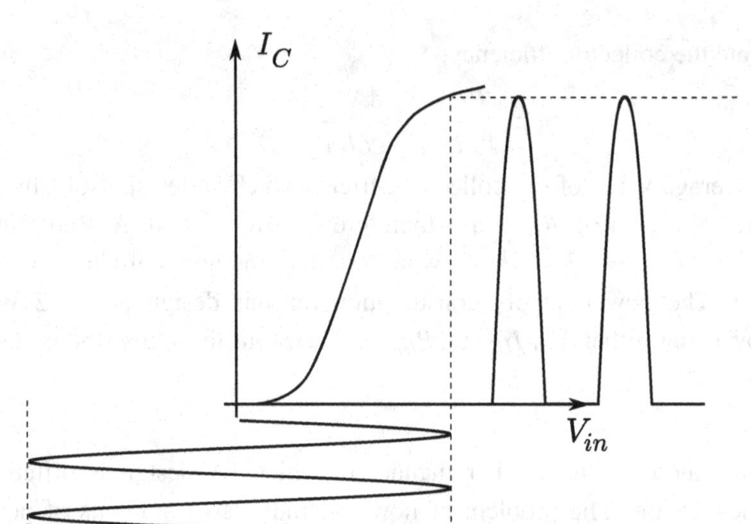

Figure 17.9. Class B operation.

17.4.4 Class B

The reason for the severely-limited efficiency of Class A operation is that the supply current is always flowing. This can be improved dramatically by centering the mean value of V_{in} at the threshold of active region operation, as shown in Figure 17.9. This is the defining characteristic of Class B operation. One's first reaction may be that this result is unusable, since there is only output when the transistor is active-biased, and so half of the waveform is missing. We will return to that issue in a moment. First, however, let us confirm that the efficiency is actually improved.

To determine the theoretical maximum collector efficiency in this case, first note that P_L is reduced by a factor of 2 relative to Equation (17.13) simply because half the waveform is essentially set to zero. Thus Equation (17.13) is now:

$$P_L = \frac{1}{4} i_L^2 R_L \tag{17.18}$$

The power consumed from the power supply is slightly more difficult to calculate, because I_{DC} is not immediately obvious. But it is not difficult to calculate. Without loss of generality we may write

$$I_{DC} = \frac{1}{T_c} \int_{-T_c/4}^{+T_c/4} i_L \cos \omega_c t \, dt \tag{17.19}$$

where $T_c = 2\pi/\omega_c$ is one period of the carrier. In other words, we recognize that the mean value is simply the integral over one-half of a period, divided by the period. Evaluating this expression, we obtain simply

$$I_{DC} = i_L/\pi \tag{17.20}$$

and subsequently

$$\epsilon_P \equiv \frac{P_L}{P_{DC}} = \frac{\pi}{4} \frac{i_L^2 R_L}{i_L V_{CC}} \tag{17.21}$$

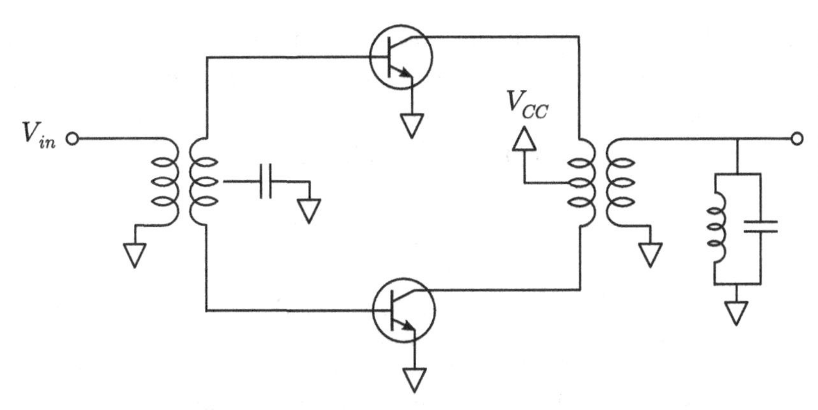

Figure 17.10. A generic Class B amplifier in "push-pull" configuration. (Also may be employed in Class D operation.)

To determine the maximum efficiency, we evaluate the above expression for the largest possible voltage swing that can be accommodated without clipping. Assuming V_{CC} is large relative to 0.7 V, we have

$$i_L R_L < V_{CC} \tag{17.22}$$

Substituting $i_L R_L = V_{CC}$ into Equation (17.21) we find

$$\epsilon_P \approx \pi/4 = 78\% \text{ (approx. maximum)} \tag{17.23}$$

So the efficiency of Class B operation is about 28% better relative to that of Class A. In practice, high-performance Class B amplifiers typically achieve ϵ_P of roughly 60%; Like Class A, significantly less than the maximum due to various losses not accounted for in the above simple analysis and also because of the need to "back off" from the maximum voltage swing so as to meet distortion requirements.

Now, back to the problem of missing one-half of the waveform. There are at least two approaches to fix this: harmonic filtering and push-pull topology. Let us consider harmonic filtering first. It is important to realize that the harmonic filter does not merely dispose of undesired harmonics. In fact, a harmonic filter nominally contains no loss (e.g., just reactances), so it should not dissipate power; it may only store and release energy. The harmonic filter stores energy when the transistor is conducting, and then releases it when the transistor turns off. The transient response of an LC circuit is sinusoidal with frequency equal to the resonant frequency ω_0, so by making $\omega_0 = \omega_c$, the transient excited by the transistor's turn-off transient is a sinusoid that conveniently fills in the missing half-cycle. For this reason, the harmonic filter is sometimes alternatively referred to as a *tank* or a *tank circuit*.

The classic Class B *push-pull* architecture is shown in Figure 17.10.[4] In this scheme, an input balun[5] is used to generate two versions of the input: A zero-mean single-ended version delivered to the upper transistor, and a zero-mean *and inverted* single-ended version delivered to the lower transistor. Each transistor is operated as described above, so the lower transistor produces its non-zero half-cycle when the upper transistor is off, and vice versa. These two outputs are combined using a balun, which in effect adds the two output waveforms. Power

[4] Just to be clear, the balun input approach shown here is not the only way to go. Push-pull architecture can also be implemented by using, for example, one NPN and one PNP transistor in the proper arrangement; see [72, Sec. 14.3] for a primer.

[5] Reminder: Figure 8.19 and associated text.

consumption is doubled because two transistors are operating simultaneously; however, power delivered to the load is also doubled, so the maximum theoretical collector efficiency remains at 78%. The primary benefit of the push-pull architecture over the single-transistor architecture is that the same distortion specification may be met at higher power.

EXAMPLE 17.5

A Class-B push-pull PA is to deliver 25 W to a 50 Ω load. Determine a reasonable minimum power supply voltage and calculate the resulting efficiency.

Solution: Note that in this topology, V_{CC} is the maximum possible value of the collector voltage $V_{C,max}$, and so for best efficiency $V_{C,max} = V_{CC}$ such that the voltage across the transistor side of the output balun varies sinusoidally within the range $-V_{CC}$ to $+V_{CC}$. The output power is

$$P_L = \frac{V_{C,max}^2}{2R_L'}$$

where R_L' is the load resistance as transformed by the output balun. Let $\alpha \equiv R_L/R_L'$ be the impedance ratio of the output balun. Then

$$V_{C,max} = \sqrt{\frac{2R_LP_L}{\alpha}} = \frac{50 \text{ V}}{\sqrt{\alpha}}$$

Now 50 V is a pretty big supply voltage – in fact, the closest commonly-available "standard" supply voltage is 48 V. However, we can use the transformer action of the output balun to reduce this: Choosing $\alpha = 2$ gives $V_{C,max} = 35.3$ V, $\alpha = 4$ gives $V_{C,max} = 25$ V, and so on. In principle, we could reduce V_{CC} to a value just a few times the cutoff voltage using this approach. However, moderation is required because (1) reducing the voltage swing increases the relative impact of the distortion associated with the region around cutoff, and (2) decreasing V_{CC} while holding P_L constant means I_{DC} is increasing. The latter is a concern because the power supply must be able to deliver increased current, and then the components must be rated to withstand this current. A reasonable choice here is $\alpha = 4$, giving $V_{C,max} = 25$ V. This maximum collector voltage is conveniently close to, and less than, a commonly-available standard power supply voltage: 28 V. Thus we choose $V_{CC} = 28$ V and $\alpha = 4$, which is conveniently implemented using a transformer with turns ratio 1:2 (input : output). Now let us calculate the collector efficiency:

$$\epsilon_P \equiv \frac{P_L}{P_{DC}} = \frac{25 \text{ W}}{V_{CC}I_{DC}}$$

Adapting Equation (17.20),

$$I_{DC} = 2\frac{I_L'}{\pi}$$

where I_L' is the maximum current flowing through the input side of the output balun and the factor of 2 accounts for the fact that two transistors are operating simultaneously. I_L' is simply $V_{C,max}/R_L' = 2$ A; therefore $I_{DC} = 1.27$ A and $\epsilon_P = 0.70 = 70\%$. Note again that the failure to reach the nominal Class B efficiency of ≈78% in this case is due in part to choosing V_{CC} to be a standard value of 28 V as opposed to the nominal value of 25 V.

To avoid any possibility of confusion, we here note that a Class B push-pull PA is *not* a differential amplifier. The transistor pair may be operating on a differential signal; however, the output is assembled from time-division multiplexed versions of the nominal output waveform. The output balun is not performing differential-to-single ended conversion; rather it is being used to add the time-division multiplexed outputs. It is true that a very slight modification to the input and output baluns would facilitate differential input and output, respectively. However, this amplifier would still not enjoy the same fundamentally good second-order linearity associated with differential amplifiers. It *is* possible to arrange *pairs* of Class A, single-transistor Class-B, and Class-B push-pull amplifiers into differential amplifiers.

A disadvantage of Class B operation with respect to Class A operation is increased distortion, since the output waveform is centered on the least linear portion of the input-output characteristic; i.e., in the region around the threshold between active operation and cutoff. Above we assumed for simplicity that transistors were either conducting (active region operation) or off; in fact what happens when V_{BE} is between 0 and 0.7 V is considerably messier, resulting in significant distortion over this part of the waveform. This issue is somewhat mitigated by making the output swing as large as possible, so that a correspondingly smaller fraction of the overall swing is associated with this less-linear region. The harmonic filter also helps. Note the contrast with respect to Class A operation: In Class B operation, linearity gets better as the output swing gets *larger*, at least until the non-linearity associated with saturation becomes significant.

Class B PAs are employed in applications where Class A quasi-linear operation would normally be the first choice, but in which the improvement in efficiency outweighs the degradation of linearity. Class B push-pull amplifiers are relatively common in high-power HF and VHF applications using SSB.

17.4.5 Class AB and Conduction Angle

The usual definition of Class AB operation is similar to Class B operation, but with the input voltage bias modified such that the transistor conducts through more than 50% of the carrier period. If zero RF input corresponds to a value of V_{BE} significantly greater than the threshold for the active region operation, and yet small enough so that no portion of the waveform is clipped due to saturation, then the PA is operating in Class AB. The advantage of this approach over Class B is that the distortion associated with the "dead zone" between $V_{BE} = 0$ V and ≈ 0.7 V can be largely eliminated. The disadvantage is that efficiency is degraded, since now each transistor is "on" more than 50% of the time.

Class AB operation includes a continuum of possibilities between Class A operation and Class B operation. This range of possibilities is commonly quantified by *conduction angle*, which is simply the portion of the carrier period during which the constituent transistors conduct, expressed in degrees. Thus, the conduction angles for Class A and Class B operation are 360° and 180° respectively, and Class AB falls in between. Conduction angle parameterizes a tradeoff between maximum linearity (Class A) and maximum efficiency (Class B); thus one chooses Class AB operation when quasi-linear operation is desired and system-level constraints dictate a solution intermediate to Classes A and B. The concept of conduction angle may be applied to any PA, as indicated in Table 17.1.

Presently Class AB amplifiers are a popular choice for most cellular applications not employing constant modulus modulation, and are particularly prevalent in CDMA systems.

17.4.6 Class C

Class C is motivated by further improving efficiency. A precise definition is elusive, because there are a number of disparate ways in which Class C operation is implemented. However, these implementations can be broadly characterized as combinations of two particular strategies, which we will now discuss. Both strategies begin with the idea of "Class B single transistor" operation, and make further modifications to further increase efficiency.

In the first strategy, the conduction angle is reduced significantly below 180°, such that the transistor conducts over a smaller fraction of the carrier period. This can be achieved simply by reducing the mean value of V_{in} to increasingly negative values. Efficiency improves as conduction angle decreases, so in principle this method of operation can achieve 100% efficiency; however, it accomplishes this by being switched on 0% of the time: obviously not a very useful condition. A typical conduction angle is 150°, yielding theoretical $\epsilon_P \approx 85\%$. The carrier is severely mangled in this process, but is subsequently restored by the harmonic filter, which suppresses all Fourier components other than the fundamental.

In the second strategy, the conduction angle is maintained between 150° and 180°, but the transistor is deliberately driven into saturation at each positive peak of the collector voltage. The resulting waveform now looks more like a square wave than a sinusoid; however, that can be cleaned up by harmonic filtering, as in the previous strategy. The payoff is the improvement in efficiency. To see this, consider the extreme case in which the transistor output waveform is well-modeled as a square wave. In this case,

$$P_L = \frac{1}{2} i_L^2 R_L$$

since the load current is constant, but only on 50% of the time. The power drawn from the power supply is:

$$P_{DC} = V_{CC} I_{DC} = \frac{1}{2} V_{CC} i_L$$

And finally we note $i_L R_L = V_{CC}$. Therefore under the square wave approximation, $\epsilon_P \equiv P_L / P_{DC} = 1 = 100\%$.

In practical Class C designs, efficiencies of 70%–80% are typically achieved by using some combination of these two strategies. In other words, practical Class C PAs achieve efficiencies comparable to theoretical Class B efficiency. This largely accounts for why RF engineers are willing to put up with brazenly non-linear performance. However, the fact that envelope magnitude variation is degraded means that Class C is not suitable for all modulations. Here is a quick summary of how this plays out:

- Class C is an excellent choice for FM and M-FSK, and in particular for MSK and especially for GMSK. In all of these modulations, there is no information conveyed by carrier magnitude, so the carrier envelope is nominally constant. All relevant information is conveyed in frequency and phase, which is preserved in Class C operation. It is probably the case that the very high efficiency offered by Class C operation, combined with the fact that Class C is suitable only for near-constant-modulus modulations, has made FM and MSK singularly popular in modern radio communications systems. It is only in recent years that communications systems previously employing these modulations have begun to migrate to other modulations having better spectral efficiency.

- Class C is unusable for SSB, M-ASK, and M-QAM, since these modulations convey information in the carrier magnitude variation. (For AM, read on ....)
- Class C can be used for QPSK if measures are taken to reduce carrier magnitude variation. "Raw" QPSK exhibits momentary near-zero carrier magnitude between certain symbol carrier-phase states, which in the PA manifest as undesirable transient currents that would not otherwise be present. This problem is largely mitigated in OQPSK and $\pi/4$-QPSK, and so these modulations can be supported by Class C PA albeit with a small amount of distortion due to magnitude clipping.
- Class C is poorly suited to BPSK and M-PSKs with $M > 4$, since there is no avoiding zero or near-zero crossings in these modulations.
- Class C can be used for AM, with a slight architectural modification. The modification is that an unmodulated carrier is applied to the input, and the message signal is used to vary $V_{C,max}$. (This is illustrated in Figure 17.1(b).) When used in this manner, the Class C PA is both PA and modulator. This is quite advantageous in many applications, as this yields quasi-linear output (in the sense that the output carrier magnitude is proportional to the message signal) with much higher efficiency than could be obtained by a quasi-linear PA.

In Classes A, B, and AB, design procedures yield specified combinations of power supply voltage, efficiency, and dissipation. Analogous procedures for Class C PA design are not easily articulated, especially when the second (saturation) strategy is employed. In practice, Class C amplifiers can be designed by beginning with a Class B single transistor design, and then reducing mean V_{in} to reduce conduction angle (i.e., first strategy) and increasing $V_{C,max}$ to achieve saturation (second strategy). Class C design must also take into account the limits of harmonic filtering. To a greater extent than in Class B, harmonic filtering is required to restore a proper sinusoidal carrier.

17.4.7 The Rest of the Alphabet: High-Efficiency Non-linear PAs

The historical popularity of the Class C PA demonstrates that PA efficiency is frequently more important than elegant modulation design, at least from a system perspective. This has resulted in additional classes of PAs also having theoretical optimum efficiency, but which are able to achieve higher efficiencies in practice as well. None is linear – or even quasi-linear – so if a modulation is not suitable for Class C, then neither will it be suitable for any of the following classes of PAs.

Class D: Class D is essentially Class B in which the transistors are operated as switches – similar to Class C "saturation mode" operation, but here switching as "hard" as possible so as to minimize transition time and thereby maximize realized efficiency. The realized efficiency in practice falls short of 100% due to dissipation of power within the transistor associated with switching transients; that is, the extra current that is required to transition the transistor between the conduction and non-conducting states. Class D, like Class B, often appears in a push-pull configuration (Figure 17.10).

Class E: Class E improves on the efficiency of Class D by mitigating the power dissipation in the transistor associated with switching transients. This involves reactive output loading so as to zero the power required to switch the transistor. The resulting waveform is far from square; nevertheless when properly designed, the overall efficiency is significantly improved over Class D, and a harmonic filter is once again relied upon to restore a sinusoidal waveform. Practical Class E PAs achieve the highest efficiencies of all; $\epsilon_P \approx 90\%$ being routinely achieved. For this reason, Class E is steadily gaining in popularity over the previously ubiquitous Class C PA for modern non-linear PA applications. The primary contraindication for Class E over Class C is a requirement for the transistor to survive very high peak collector (or drain, for FETs) voltages and currents. Whereas the peak collector voltage in Class C is $V_{C,max}$ (normally set just below V_{CC}), the peak voltage for Class E is $3.6V_{C,max}$. This has traditionally limited Class E implementation in RF CMOS, thereby limiting use in RFICs. In contrast to Class C, straightforward design procedures exist for Class E PAs, although the details are beyond the scope of this book. The reader is referred to [73], which is also the seminal paper on the Class E PA concept.

Class F: Class F, like Class E, is all about reducing power dissipation associated with switching, so that the transistor more closely resembles an ideal switch. However, the tactic in this case is to repurpose the undesired harmonics of the nominal (maximally efficient) square wave. A generic representation of the scheme is shown in Figure 17.11. Here an impedance inverter is inserted between the blocking cap and the harmonic filter. Since the harmonic filter presents a relatively small impedance away from the design frequency of the impedance inverter, and appears in parallel with the load, the impedance seen by the collector is relatively high at frequencies away from the nominal center frequency. The carrier (fundamental) frequency is able to propagate to the load, whereas higher-order harmonics see a much larger impedance and are thereby halted at the collector. When properly designed, the effect of the trapped harmonic current is to "square up" the waveform, which subsequently reduces the power dissipated by the transistor during transitions. In practice efficiency of about 80% can be achieved – not as good as Class E, but typically better than Classes C and D. The primary contraindication for Class F is difficulty in implementing the impedance inverter at VHF and below. At UHF and above, this is relatively straightforward as the impedance inverter may simply be a quarter-wavelength section of transmission line (again see Section 9.8). At lower frequencies, the necessary transmission line structure becomes too large and must be replaced with discrete reactances. This brings in various non-ideal effects that are difficult to mitigate at high power levels.

Other Classes: Other PA classes of operation that are occasionally cited are G, H, and S. These have relatively limited application and are beyond the scope of this book. For additional information, an excellent source of information is [34, Secs. 14.5–14.6] and references therein.

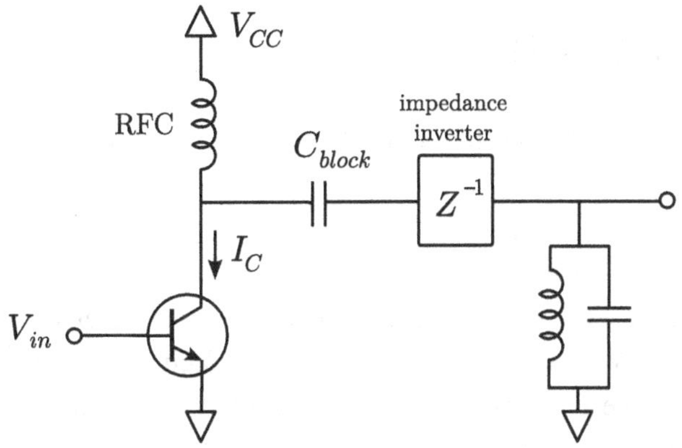

Figure 17.11. A generic Class F amplifier.

17.4.8 Repurposing Non-Linear PAs as Quasi-Linear PAs

A final note on the dichotomy between (quasi-)linear PAs and non-linear PAs: Because the latter provide such spectacularly improved efficiency, engineers are strongly motivated to find ways to make non-linear amplifiers work with linear modulations. This leads to three concepts, briefly discussed below, that strive to accomplish this goal. All three have their origins in early high-power HF applications, and were essentially abandoned as design trends in the mid-to-late twentieth century favored constant-modulus modulations at higher frequencies. Thus practical implementations of these methods in modern radio communications systems are relatively rare, but seem to be experiencing a significant revival of interest. For further reading, a recommended starting point is [60] and references therein.

- *Kahn EER*: In Section 17.4.6 we noted the possibility of using a Class C PA as an AM modulator. A generalization of this idea is known as *envelope elimination and restoration* (EER), also sometimes known as *Kahn modulation* (after the inventor). In the original analog implementation of EER, the input signal is decomposed into a phase/frequency-modulated sinusoidal carrier having constant modulus, and a separate signal representing the desired carrier magnitude. The carrier is then amplified using an efficient non-linear PA, with the carrier magnitude signal being used as power control. In DSP implementations, these two signals are digitally synthesized (with the carrier being synthesized at complex baseband or an IF), so as to avoid the distortion associated with analog deconstruction of the signal. Kahn EER facilitates the use of Class C amplifiers to transmit SSB as well as AM.
- *Chireix*: In this method (also known variously as *outphasing*, *ampliphase*, and *linear amplification using non-linear components* (LINC)), the modulated signal is decomposed into two constant-modulus signals that are amplified separately and then combined. Since the component signals are constant-modulus, non-linear PAs may be used. DSP implementation facilitates convenient digital synthesis of the constituent constant-modulus signals, avoiding issues with imperfect analog decomposition of a modulated carrier.
- *Doherty*: In this method, a quasi-linear amplifier is augmented with a non-linear "peaking" amplifier, which contributes power in when the desired output power exceeds a specified threshold.

17.5 CONSIDERATIONS IN PA DESIGN

In this section we address some considerations that tend to be unique to PA design.

17.5.1 Supply Voltage

In PA design, supply voltage becomes a primary design parameter. Consider that to deliver 1 W to a 50 Ω load requires $\sqrt{1\ \text{W} \cdot 50\ \Omega} = 7.07$ V (RMS), which corresponds to 10 V (peak) and 20 V_{pp}. Regardless of the PA class, V_{CC} (or V_{DD}) $>$ 10 V would be required to support this. This poses no particular problem for base station applications, but for mobile and portable applications this lands us in quite a bit of trouble because in those applications we are limited to what we are able to summon from batteries.[6] Mobile radios in some applications routinely go to 5 W, implying $V_{CC} > 22.4$ V. In base station and broadcasting applications, the voltages required to deliver 10 W or more to a 50 Ω load become inconveniently large. Thus, there is a strong incentive to design PAs for relatively low load impedances – on the order of a few ohms – and then to use impedance matching techniques to transfer power to the higher-impedance load with low VSWR. Devices that are designed to drive low-impedance loads can fairly be characterized as current sources; thus PAs are essentially current sources.

17.5.2 Load Impedance Matching

The reader may have noticed that we managed to get through all of Section 17.4, including several design examples, without once mentioning the output impedance of a PA. In fact, what we did was to implicitly assume the condition we stumbled upon in Section 17.5.1: That for the PA to have a reasonable supply voltage, it is convenient to implement it as current source. We must then consider the performance of the transistor as a current source. Although we may ignore the limitations of the transistor in order to get a "first pass" design, each transistor has an output impedance at which this performance will be somehow optimal; i.e., an output impedance that maximizes some aspect of linearity and/or power handling capability while delivering the desired current. This then becomes an important consideration in choosing a load impedance and associated matching network.

This brings us to impedance matching. It is a common misconception that the purpose of a load impedance matching network at the output of a PA is to *maximize* power transfer into the load. This is incorrect for two reasons. First, as we decided in Section 17.5.1, it is either convenient or necessary to reduce the required supply voltage from that required to drive 50 Ω at particular power level, so in this sense the goal of impedance matching is to reduce voltage, not increase power transfer. The second issue is that maximizing power transfer at the output of a PA typically degrades the performance of the PA. Maximum power transfer implies conjugate

[6] Of course modern DC–DC conversion technology facilitates stepping up battery voltages, but then we contend with a new layer of efficiency and PA-generated spurious signal issues associated with this conversion. As it turns out, DC–DC conversion is frequently required just to bridge the gap between what is available from batteries to *minimum* usable PA supply voltages.

matching, and conjugate matching implies dissipation of power equally between source (in this case, the PA) and load. Thus conjugate matching results in an immediate 50% penalty in ϵ_P with the wasted power serving only to increase the heat generated by the PA. A better strategy is usually to implement impedance matching so as to minimize power dissipated in the PA. Often this means minimizing VSWR, but in general each transistor will have some output impedance which results in an useful balance between competing design goals.

Although the general principles of impedance matching remain those presented in Chapter 9, in PA design we have two additional concerns: load pull and power handling. *Load pull* refers to variation in PA output impedance as a function of load current. This is an inevitable consequence of quasi-linear operation, since the ratio of voltage to current varies over the range spanned by the collector (or drain) current. As a result, there may be no single nominal output impedance, but rather a set of output impedances associated with each possible power level. On the Smith chart (i.e., when Γ_{out} is plotted in a plane using its real and imaginary components as Cartesian coordinates), these appear as concentric contour lines, with higher power represented by the outermost curves. Since the output matching circuit is normally fixed, one must identify an impedance that serves as a best fit and then verify that the resulting excursions do not unacceptably degrade performance. Because load pull is associated with the non-linear behavior of the PA, it is quite difficult to predict in advance and often must be measured instead. Exacerbating the problem is that the load itself, or interfacing circuitry, may become non-linear at high power levels such that the load impedance seen by the PA also varies with output power.

The second additional concern in PA output impedance matching is power handling. It is common for voltages and currents present at the PA output to be greater than what most discrete components (including capacitors and inductors) are able to withstand. At VHF and below, transformers – appropriately selected – are frequently better suited; at UHF and above, distributed matching structures are a common choice. Beyond the ability to handle voltage and current independently, there is also the issue of heat generation through loss dissipation. A tiny amount of resistive loss in a component used within the output matching network can lead to unacceptable heating and catastrophic failure; for example, 1 Ω of resistive loss in series with a 50 Ω load means the loss dissipates about 2% of the power; for $P_L = 100$ W this is 2 W: more than enough to destroy discrete reactances of the type you might consider using in an impedance matching network. Transmission line transformers offer a relatively good combination of voltage/current handling and power handling, and so frequently appear as components of baluns and impedance matching networks in modern PA designs.

17.5.3 Source Impedance Matching, Buffers, and Drivers

As is the case at output, certain transistors will exhibit input impedances that yield relatively good performance. The combination of the desired collector (or drain) voltage swing and input impedance suggests a nominal input voltage swing, often referred to as a *drive level*. To ensure that the intended drive level is delivered to the PA input, the PA is often preceded by a *buffer amplifier* followed by a *driver amplifier*. The buffer amplifier is normally a low-gain device designed to isolate the impedance at its input from the impedance at its output. This calls for high input impedance and low output impedance, and so is normally accomplished by an emitter follower (in bipolar technology; which is a source follower in FET technology); see Section 10.3.3. This normally involves little or no gain: The buffer amplifier serves merely as a current source for the driver amplifier. The purpose of the driver amplifier is to generate a waveform which is optimum – or at least well-suited – to the input requirements of the PA, and at an appropriate impedance. This includes setting the average voltage level needed for biasing in various classes of operation, setting a nominal input voltage swing, delivering the signal at the nominal impedance, and perhaps also performing single-ended to differential conversion. In linear and quasi-linear PAs, buffer and driver amplifiers may also be used to provide some of the total gain, so that the PAE required from the final (PA) stage can be reduced. "Buffering" and "driving" functions might also be combined into a single amplifier, perhaps even into a single transistor.

17.5.4 PAPR and Back Off

PAs are often not operated at maximum rated power, but rather are set to operate with varying degrees of "back off". This is particularly important for quasi-linear PAs used with modulations having *peak-to-average power ratio* (PAPR) significantly greater than 1. As the name suggests, PAPR is the peak power of the RF waveform relative to the average power of the waveform. PAPR is a fundamental property of a modulation and independent of the power level at which it is transmitted. For modulations which are strictly constant modulus such as FM and MSK, PAPR $= 1$ and there is no issue. Low-order PSKs and phase-amplitude modulations exhibit PAPR in the range 3–6 dB, requiring the PA to be backed off by at least this amount so that intervals of peak power do not result in excessive distortion. PAPR is higher for multicarrier signals such as the downlink signal transmit from a CDMA base station, where the back off required to accommodate PAPR is commonly in the range 3–9 dB. PAPR is highest for multicarrier modulations such as OFDM; for example, PAs for IEEE 802.11a/g (WLAN) require back off up to 17 dB.

PAPR is sometimes alternatively expressed in terms of *crest factor*. Crest factor is simply the ratio of peak magnitude relative to the average magnitude of the modulation, and so is simply the square root of PAPR.

It should be noted that back off normally works against efficiency, since efficiency is normally maximized by maximizing output waveform magnitude.

17.5.5 Power Control

Mobile radios in cellular telecommunications systems typically require stringent power control. This requirement is most extreme in CDMA-based networks, in which variation in output power up to 71 dB (from −50 dBm to +21 dBm) is required to achieve an approximation of the nominal condition of equal signal-to-interference ratio for users sharing a channel. This variation is awkward to implement at baseband, for the simple reason that this requires a DAC with large N_b. It is more convenient to implement most or all of the required power variation at the PA. In non-linear PAs this is easily accomplished simply by varying the power supply voltage, since the output power is proportional to supply voltage squared. A more generally-applicable approach is *subranging*, in which multiple PAs are employed, with each PA designed for a different power level and which is switched in or out of the signal path as needed. For other modulations using quasi-linear PAs, a combination of techniques might be required.

17.6 PA LINEARIZATION

It is sometimes the case that the PAs available for a particular application do not meet linearity specifications. Here we address two common shortcomings and the methods that are sometimes employed to overcome them.

17.6.1 Consequences of PA Non-Linearity

A particularly onerous consequence of limited linearity is *spectral regrowth*, in which distortion gives rise to signal power outside the intended bandwidth. An example is shown in Figure 17.12. This particular issue has risen to prominence with the technical trend away from constant modulus modulations (FM, MSK) toward linear modulations (e.g., QAM) for greater spectral efficiency, and the simultaneous tightening of channel bandwidth and adjacent channel interference specifications motivated by the desire to increase the number of simultaneous uses of a given segment of spectrum. Spectral regrowth is commonly quantified as *adjacent channel power ratio* (ACPR), which is the ratio of power delivered into an adjacent spectral channel relative to the power delivered in the desired spectral limits.

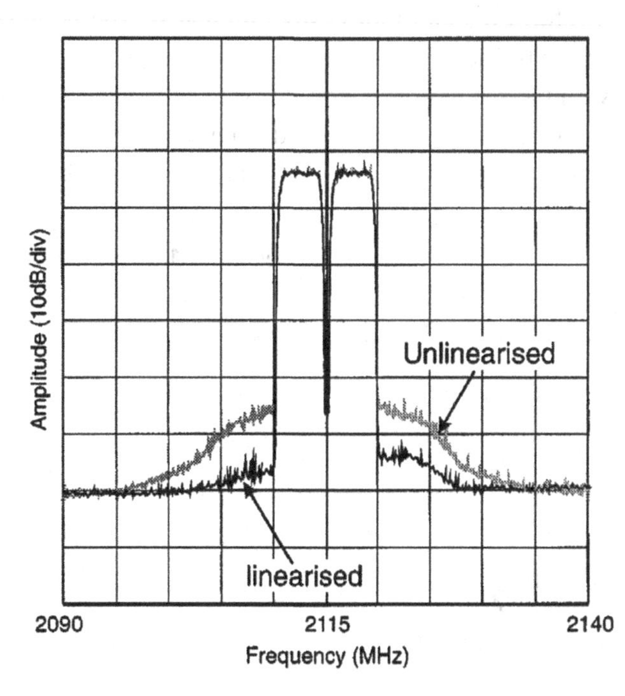

Figure 17.12. Two adjacent CDMA signals as they appear at the output of a PA without linearization; spectral regrowth is evident as the "shoulders" appearing about 40 dB below the peak spectrum. Also shown is the output for a PA using a digital predistortion linearization technique (see Section 17.6.2). © 2002 IEEE. Reprinted, with permission, from [60].

Another consequence of distortion from limited linearity is characterized by *error vector magnitude* (EVM), which quantifies the degree to which the carrier magnitude and phase is in error with respect to the nominal symbol magnitude/phase at the nominal sampling time.

Before leaping into a discussion of mitigation techniques, let us first note that we have already encountered a number of schemes which might be applied to improve linearity. The first is simply back off (Section 17.5.4), which is effective with quasi-linear amplifiers because the interior region of the input–output curve (e.g., I_C vs. V_{BE} for bipolar transistors) is more linear than the more extreme regions of the curve. The primary contraindication for back off as a linearization technique is reduction in efficiency. Second, the "repurposing" techniques identified in Section 17.4.8 might also be viewed as linearization techniques.

These methods aside, deliberate approaches to linearization of PAs currently fall into three categories: predistortion, feedforward correction, and feedback correction. A brief introduction to each follows.

17.6.2 Predistortion

Predistortion involves anticipating the effect that PA non-linearity will have on the signal, and modifying the input signal to account for this effect such that the output is closer to the intended linear output. This is illustrated in Figure 17.13. Predistortion turns out to be quite practical because the nature of the non-linear distortion of many PAs is sufficiently well-characterized by the third-order memoryless polynomial model presented in Section 11.2.2. The model is simply this:

$$y(t) \approx a_0 + a_1 x(t) + a_2 x^2(t) + a_3 x^3(t) \tag{17.24}$$

where $y(t)$ and $x(t)$ are the output and input respectively (in this case, either could be interpreted as either voltage or current). As before, we are unconcerned with a_0 because it represents DC, which is either intrinsic to the waveform definition or easily eliminated. The nominal linear output occurs when $a_2 = a_3 = 0$, since this leaves only the desired term $a_1 x(t)$. For the present discussion, we are unconcerned with the second-order non-linearity associated with a_2, since this normally produces output only at frequencies far removed from the carrier frequency (see e.g. Figure 11.1), and in any event is usually much less of a problem than the third-order

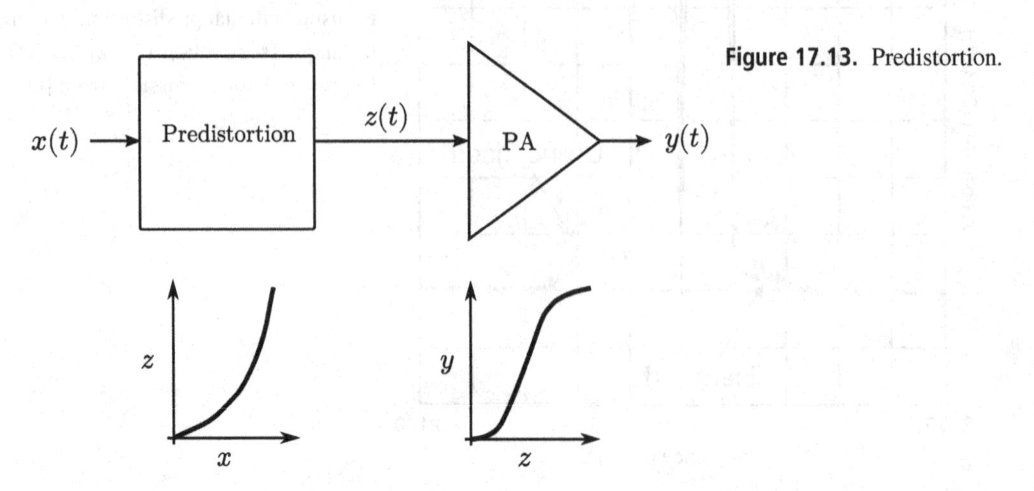

Figure 17.13. Predistortion.

non-linearity associated with a_3. (See Sections 11.2.4 and 11.2.5 for a refresher.) Ideal predistortion in this case consists of identifying the alternative input $z(t)$ for which

$$a_1 z(t) + a_3 z^3(t) = a_1 x(t) \tag{17.25}$$

This is a cubic equation that can be solved for z with x as an independent variable, given a_1 and a_3 as measurable parameters of the PA. Thus, the predistortion subsystem calculates $z(t) = f(x(t))$ and delivers $z(t)$ as the input to the PA.

At first glance the predistorter might appear to require a very sophisticated implementation. However, note that Equation (17.25) may be written in the form of the following cubic equation:

$$z^3 + \frac{a_1}{a_3} z - \frac{a_1}{a_3} x = 0 \tag{17.26}$$

which depends on just one parameter, a_1/a_3. For typical PAs, the resulting function $z = f(x)$ ends up looking something like that shown in Figure 17.13. This characteristic may be implemented with pretty good accuracy simply as the ratio of current to voltage across a square-law device, such as a diode or a FET transistor (see Equation (10.3) and associated text). This requires careful biasing, but linearization using this simple predistortion approach has been known to deliver ACPR improvements on the order 4 dB or so for CDMA implementations. Further improvements typically require digital predistortion, which poses no particular problem in radios in which the baseband is digitally synthesized anyway. A demonstration of the efficacy of digital predistortion can be seen in Figure 17.12.

There are many issues which conspire to limit the effectiveness of predistortion. First, the memoryless third-order polynomial response model of Equation (17.24) is only a model. Although the method can be adapted with various degrees of difficulty to account for a_2 as well as non-zero a_4, a_5, and so on, the real problems turn out to be that these parameters also vary in response to temperature and bias (which itself might be variable as part of power control, for example); and, worse, the PA might not be as memoryless as we would have hoped. This leads to an "arms race" of sorts in which higher and higher levels of linearity require increasingly sophisticated modeling of the PA's input–output characteristics. Nevertheless, simple PA linearity models sometimes prove adequate to reach system level goals for ACPR and EVM.

17.6.3 Feedforward Linearization

Predistortion is "open loop"; one must determine the required predistortion characteristic in advance, and there is no accommodation if the actual PA input–output characteristic is not as expected. In principle one can do better by somehow determining the actual distortion in real time, and then mitigating that. This is the central concept in feedforward linearization.

A generic description of this scheme is shown in Figure 17.14. Here we tap the input to the PA as a reference for what the signal is supposed to look like. Then we tap the output from the PA, and attenuate this signal such that the linear gain applied by the PA is removed. Subtraction of this signal from the previous signal yields an estimate of the non-linear distortion contributed by the PA. Linearization is achieved by applying the correct amount of gain to this error signal (nominally not requiring a PA since the error signal should be small compared with the PA

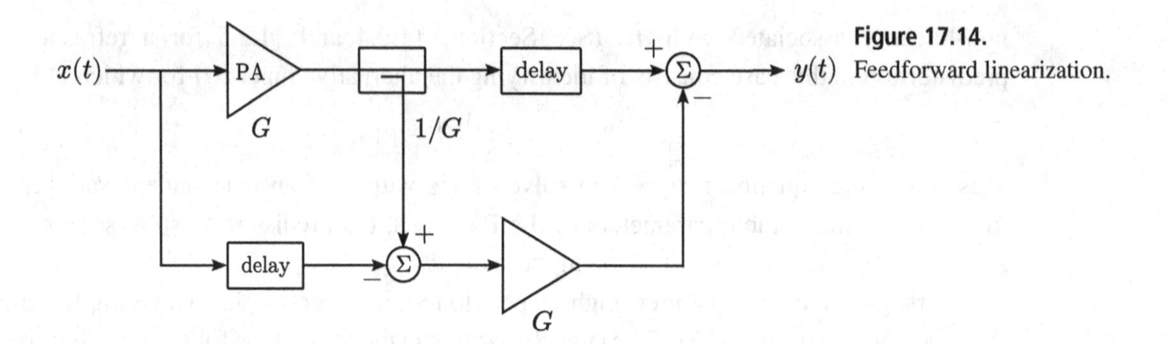

Figure 17.14. Feedforward linearization.

output) and then subtracting this from the PA output. The delay blocks are intended to prevent the delay associated with the propagation time through the PA and error amplifier from being interpreted as part of the distortion.

The obvious advantage of feedforward linearization is that no *a priori* knowledge or modeling of PA behavior is required; the system automatically does whatever is required to mitigate the distortion that is present. The disadvantage of feedforward linearization is that a host of implementation issues must be well-enough mitigated to achieve the desired benefit. For starters, the delays must be quite precise over a sufficiently large bandwidth. This of course requires the actual delay – possibly including significant group delay variation (see Section 13.5.6) be known and remain constant. The error signal amplifier must have excellent linearity, as any distortion in this amplifier directly degrades the cancellation of distortion achieved at the output. Despite these challenges, feedforward linearization is often effective and is routinely used in modern PAs.

17.6.4 Feedback Linearization

A natural alternative to feedforward linearization is feedback linearization. The central concept remains to compare the distorted PA output to a reference, but now the correction is made to the input, as opposed to the output, of the PA. Here two architectures emerge, known as *polar feedback* and *Cartesian feedback*.

Polar feedback is shown generically in Figure 17.15. This can be interpreted as a modification of the Kahn EER technique (Section 17.4.8) to accommodate feedback. In this scheme the PA operates on a constant modulus (i.e., frequency/phase only) signal obtained by limiting the input signal, and the output carrier magnitude is controlled by bias modulation. The output of the PA is sampled, attenuated, and decomposed into an estimate of the envelope (typically by some form of envelope detection) and a constant modulus signal representing frequency and phase (typically by limiting). The output envelope is compared to the input envelope, resulting in an error signal, which subsequently controls the PA bias. Similarly, the magnitude-limited PA output is multiplied with the magnitude-limited input signal, resulting in a phase error control signal, which is subsequently applied to a phase shifter at the PA input. The latter should be recognizable as a phase-locked loop (PLL; for a refresher, see Section 16.8). The magnitude control system also forms a loop, analogous to a PLL. The phase and magnitude control loops allow the PA to adapt to its own non-linearity. Like feedforward linearization,

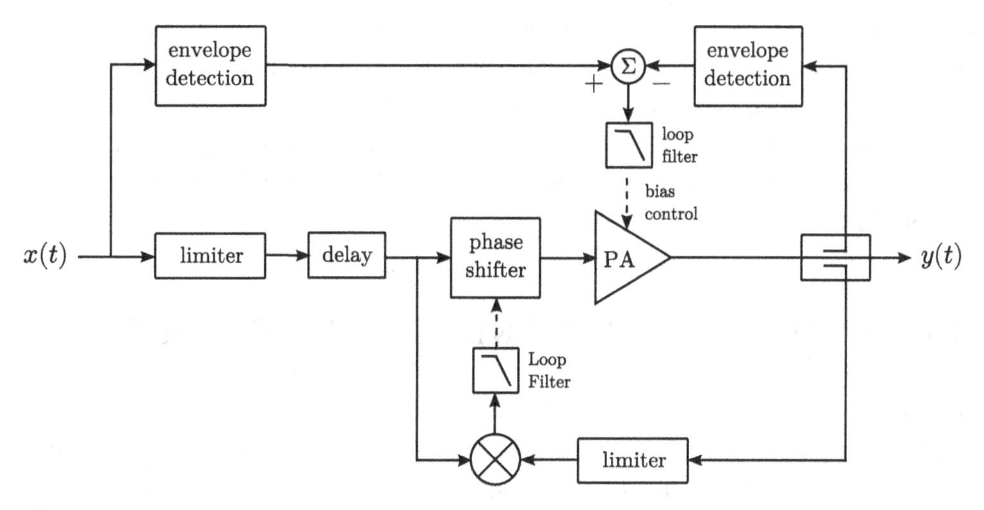

Figure 17.15. Polar implementation of feedback linearization.

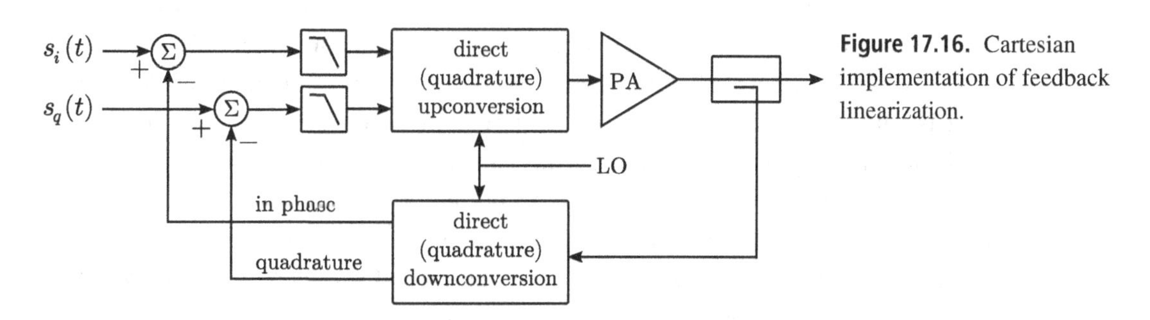

Figure 17.16. Cartesian implementation of feedback linearization.

the principal contraindication for polar feedback linearization is implementation difficulty: It is complex and entails many of the headaches associated with PLLs; in particular, trading off between tracking fidelity and loop bandwidth. Precise synchronization of the magnitude and phase corrections is also challenging. Nevertheless, this scheme appears in practical implementations.

Cartesian feedback is conceptually similar to polar feedback linearization, although the schemes appear at first glance to be very different. Cartesian feedback is shown in Figure 17.16. In this scheme, the PA output is attenuated and then decomposed into baseband in-phase and quadrature (hence, "Cartesian") components, using quadrature downconversion. As in polar feedback linearization, the components are compared, forming a negative feedback loop. Note the required quadrature downconversion to complex baseband is tantamount to implementing a receiver. The quadrature downconversion task can be shared with the actual receiver if Cartesian feedback linearization is being implemented in a half-duplex or time division duplex transceiver (see Section 1.3 for a refresher); otherwise the transceiver requires a second, dedicated receiver to do this. If the radio does waveform synthesis at baseband – as in Figures 17.1(d)–(f) – it may be well-worth the trouble to implement a second receiver, since then the required comparison operation is conveniently implemented in the "native" complex baseband format, perhaps even in DSP, and the RF implementation is simplified relative to the polar feedback scheme. A distinct advantage of Cartesian feedback over polar feedback is that the two feedback loops in the former are architecturally identical, which simplifies the problem of coordinating them. Cartesian feedback linearization seems to be steadily increasing in popularity, which can probably be attributed to the prevalence of transmitters using direct conversion architecture combined with the relative ease with which the necessary quadrature downconverson functionality can now be compactly and inexpensively implemented using modern RFIC technology.

17.7 QUADRATURE-COUPLED AND PARALLELIZED AMPLIFIERS

Sometimes it is simply not possible to achieve the required output power from a single PA while meeting other specifications. Then it makes sense to consider ways in which PAs might be combined in order to function as a single PA having greater output power. An obvious approach is simply to split the input into two identical channels, apply identical PAs to each channel, and then sum the output. In principle this approach is quite reasonable, and in fact has some compelling advantages over a single-PA architecture: Not only does each PA need to deliver less power, but also each PA now dissipates a smaller amount of power, so the associated heat is naturally spread over a larger area, making it much easier to manage. The principal difficulty in this approach has to do with output matching. Recall the VSWR at the PA output (and perhaps also the PA input) needs to be pretty good if the PA is to handle high power levels. However, it can become quite challenging to design matching circuits between splitters and PAs, and between PAs and combiners, that exhibit sufficiently low VSWR over the entire bandwidth, especially when each PA's output impedance may be dynamically varying (recall Section 17.5.2). It would be far better if the splitting/combining scheme was insensitive to the PA's port impedance. Precisely this capability is provided by a *quadrature hybrid*, discussed next.

17.7.1 Quadrature Hybrids

A *quadrature hybrid* is a passive four-port device depicted in Figure 17.17. All four ports are designed to function at the same impedance Z_0. When all four ports are properly matched, an input applied to port 1 appears at port 2 reduced by $1/\sqrt{2}$ and rotated $-90°$ in phase, and appears at port 3 reduced by $1/\sqrt{2}$ and rotated $180°$ in phase. Thus, the quadrature hybrid may be used as a two-way power divider. The arrows in Figure 17.17 indicate the results for inputs applied to each of the other three ports, in the same manner.

Figure 17.18 shows the use of two quadrature hybrids to parallelize two identical PAs. First, let us assume that everything is perfectly matched; i.e., $\Gamma_i = \Gamma_o = \Gamma_L = 0$. In this case a signal applied to port 1 of the first hybrid appears at the inputs of the top and bottom PAs multiplied by $-j/\sqrt{2}$ and $-1/\sqrt{2}$, respectively. These components subsequently appear at port 4 of the second hybrid multiplied by $+j/2$ and $+j/2$ respectively, and so add in phase as delivered to the load. Thus when there are no impedance mismatches, this scheme is essentially equivalent to a non-quadrature power splitting/combining scheme except that here the total phase shift through the system is $+90°$.

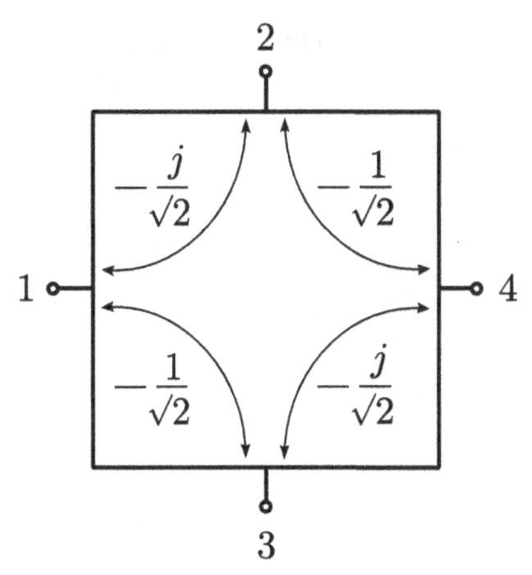

Figure 17.17. Quadrature hybrid.

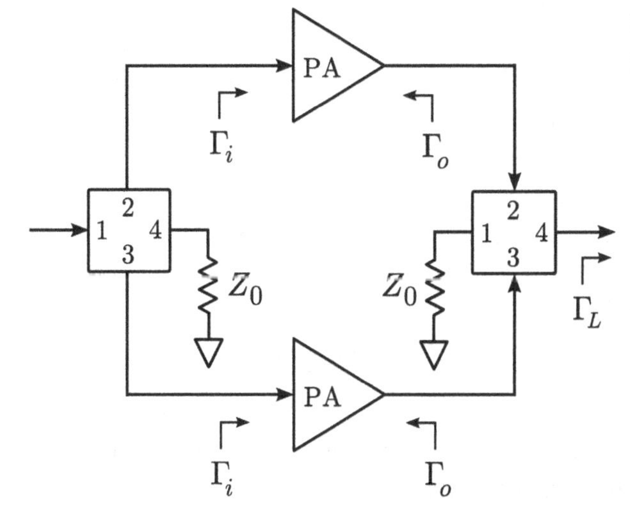

Figure 17.18. Two identical PAs operating in parallel using quadrature hybrids.

However, a remarkable thing happens when Γ_i is no longer zero. Waves reflected from the top and bottom PAs return to port 1 of the first hybrid multiplied by $-\Gamma_i/2$ and $+\Gamma_i/2$ respectively, and therefore cancel. Waves reflected from the top and bottom PAs also come together at port 4 of the first hybrid multiplied by $j\Gamma_i/2$ and $j\Gamma_i/2$ respectively, and so do not cancel, but are harmlessly dissipated in the load attached to port 4. Therefore there is no reflection from port 1, regardless of the degree to which the PA input impedances might be mismatched. Since there is no reflection from port 1, the impedance looking into port 1 remains equal to Z_0 – just as if the PA inputs were perfectly matched.

A similar phenomenon takes place in the second (combining) hybrid. Let us assume that both Γ_o and Γ_L are non-zero. In this case power reflected from the load re-enters the second hybrid, is split, and travels toward the outputs of the top and bottom hybrids. Waves reflected from the top and bottom PAs return to port 4 of the second hybrid multiplied by $\Gamma_o/2$ and $-\Gamma_o/2$ respectively, and therefore cancel. Waves reflected from the top and bottom PAs also come together at port 1 of the second hybrid multiplied by $j\Gamma_o/2$ and $j\Gamma_o/2$ respectively, and so do not cancel, but are harmlessly dissipated in the load attached to port 1. Therefore there is no reflection from port 4, regardless of the degree to which the PA input impedances might be

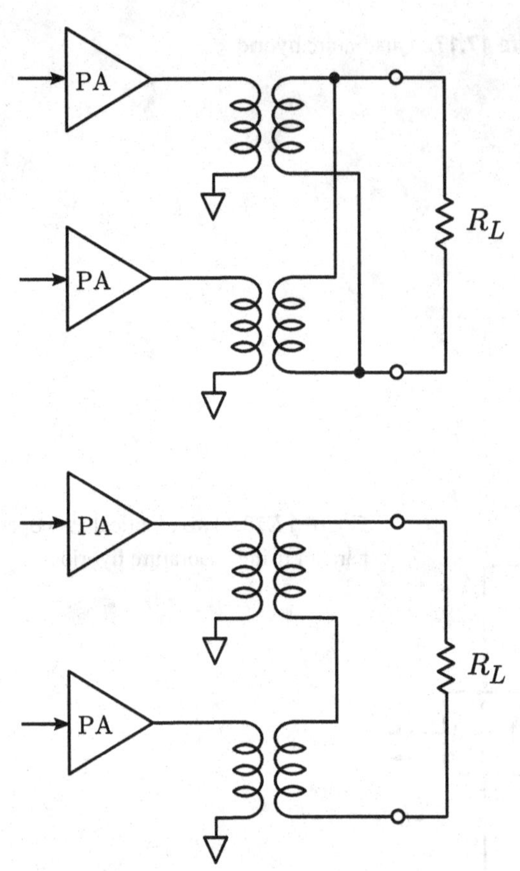

Figure 17.19. PA combining using transformers. *Top:* Current combining. *Bottom:* Voltage combining.

mismatched. Since there is no reflection from port 4, the impedance looking into port 4 remains equal to Z_0 – just as if the PAs' outputs were perfectly matched.

This approach turns out to be surprisingly effective in practice, yielding return losses typically in excess of 15 dB with over large bandwidths even when the PAs' inputs and outputs are poorly matched. Best of all, the quadrature hybrids themselves can be implemented as quarter-wavelength-scale transmission line structures with low loss and intrinsically high power-handling capability. Two examples of suitable structures are the *branch line coupler* and the *Lange coupler*.[7] Finally, quadrature hybrids are cascadeable, so it is quite practical to construct power splitter/combiner networks that accommodate larger numbers of PAs in parallel.

17.7.2 Combining Using Transformers

Quadrature combining is certainly not the only practical way that PAs may be parallelized. Another approach is transformer coupling, illustrated in Figure 17.19. The output of each PA is connected to the primary of a transformer. In Figure 17.19(top), the secondaries are wired in parallel, resulting in a summing of current. Subsequently the voltage across the load increases,

[7] A recommended starting point for details on these devices is [58].

so power is increased. In Figure 17.19(bottom), the secondaries are wired in series, resulting in a summing of voltage. Subsequently the current through the load increases, so power is increased. This scheme may also be used to implement power control, as it is relatively easy to vary output power simply by varying the number of PAs that are turned on, by using switches to properly terminate the primary sides of the transformers associated with those PAs that are turned off.

Problems

17.1 Following up Example 17.1, determine the minimum sample rate that reduces sinc distortion to 0.5 dB at the edges of the signal spectrum.

17.2 Consider a mobile radio whose PA outputs 1 W at a collector efficiency of 50%. The radio's power supply consists of a 9 V battery that is regulated to 6 V, which is subsequently applied to the PA. (a) How much current is required from the DC power supply to operate the PA? (b) What is the power-added efficiency if the PA must be driven with 100 mW input?

17.3 Design a Class-A PA using a bipolar transistor to deliver maximum power at 50 MHz using a 3.3 VDC power supply. Then calculate output power, collector efficiency, DC power consumption, and power dissipated in the transistor. Perform the design for the following cases: (a) Load impedance equal to 50 Ω. (b) Largest load impedance that facilitates 1 W output.

17.4 Following up Example 17.5, determine: (a) Power consumed from the power supply for this design, (b) power dissipated by each transistor, (c) peak collector current, (d) peak current delivered to the load.

17.5 A PA is being designed for a "software defined" GSM cellular base station. In this application, the PA must accommodate N MSK carriers simultaneously. Determine peak-to-average power ratio, crest factor, and suggest a back off margin for the following cases: (a) $N = 1$, (b) $N = 2$, (c) $N = 10$.

18 Digital Implementation of Radio Functions

18.1 INTRODUCTION

Although this book has hardly been reticent on the important role of digital signal processing (DSP) in modern radio systems, we have bypassed many of the details so far. In this chapter we circle back to the topic of digital implementation of some principal radio functions. There are essentially three broad categories of functions we must address: filtering, frequency conversion, and modulation.

Section 18.2 ("Single-Rate Filters") provides a primer on digital implementation of lowpass filtering, concentrating on finite impulse response (FIR) filters. Such filters are used in a number of different ways. They are prevalent especially in direct conversion architecture and in architectures which employ direct conversion as the interface between complex baseband and a real-valued IF. Changes in sample rate are typically required as part of this operation, and are specifically addressed in Section 18.3 ("Multirate Filters"). Then Section 18.4 ("Quadrature Upconversion and Downconversion") addresses techniques for combining filtering, rate changing, and frequency conversion in a computationally-efficient manner.

We are then in a position to tie up some loose ends from Chapter 6 ("Digital Modulation"): Specifically, how to go about implementing pulse shaping in transmitters, and symbol timing recovery in receivers. These topics are addressed in Section 18.5.

Finally, Section 18.6 addresses the hardware technologies available for DSP, including general-purpose computing, programmable logic devices, and various other possibilities. A major theme here is the need for parallelism and the resulting tradeoff between sample rate, reconfigurability, and size/power/cost.

18.2 SINGLE-RATE FILTERS

Digital filters can be classified according to two broad categories: *Finite impulse response* (FIR) and *infinite impulse response* (IIR).[1] The distinction is exactly as implied by the nomenclature, but implementation considerations introduce some subtleties which are potentially confusing to the uninitiated.

To begin, let us assume we wish frequency response which is as close to "brick wall" performance as possible. Exactly this idea was considered in Section 6.4.2 (Figure 6.7 and

[1] Reminder: Impulse response is the response of a system to an impulse; that is, an input of vanishingly small duration and finite (unit) energy. The frequency response of the system is equal to the Fourier transform of the impulse response.

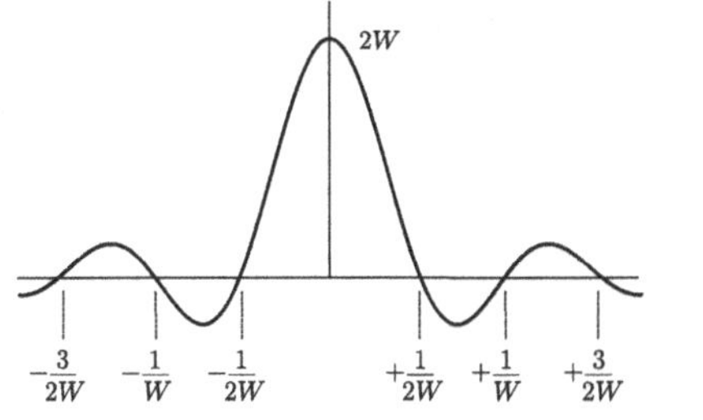

⇔

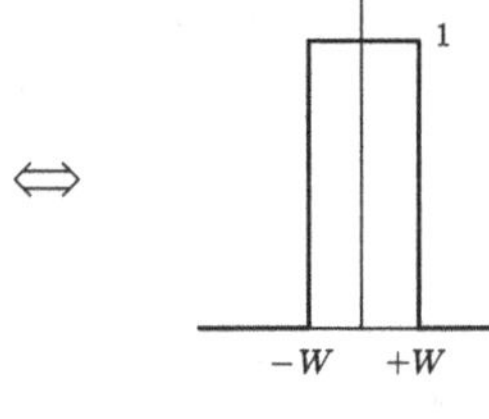

Figure 18.1. Ideal "brick wall" lowpass filter (*right*) and associated sinc impulse response (*left*).

associated text). We now re-introduce the concept as shown in Figure 18.1, with only a slight revision in that the lowpass bandwidth W is not necessarily related to any properties of the modulation within this bandwidth. The associated impulse response is

$$h(t) = \text{sinc}\,(2Wt) = \frac{\sin\,(\pi 2Wt)}{\pi 2Wt} \tag{18.1}$$

An exact implementation of this filter formally is of the IIR type, since sinc is non-zero outside any finite time span considered. However, this impulse response may be truncated, yielding a useable approximation that is formally of the FIR type: More on that in the next section.

In common practice, the term "IIR" usually implies an *implementation* that cannot be described in terms of a finite set of samples from a particular impulse response. Typically this is because IIR filter implementations involve some form of feedback; that is, the present output depends on previous values of the output. In contrast the output of a FIR filter depends only on the input, and only on the input over a bounded period of time.

18.2.1 FIR Filter Fundamentals

FIR filters account for a very large majority of digital filters in radio applications, so let us proceed assuming FIR implementation. In a continuous time system, the time-domain output $y(t)$ is related to the time-domain input $x(t)$ by convolution; i.e.,

$$y(t) = \int_{-\infty}^{+\infty} h(\tau)x(t-\tau)d\tau \tag{18.2}$$

where $h(t)$ is the impulse response. A discrete-time implementation of this operation is obtained by sampling the above equation with $t = kT_S$, where k enumerates samples and $T_S = 1/F_S$ is the sample period. This yields:

$$y_k = \sum_{n=-\infty}^{+\infty} h_n x_{k-n} \tag{18.3}$$

where $y_k \equiv y(kT_S)$, $h_n \equiv h(nT_S)$, and $x_k \equiv x(kT_S)$. If the impulse response is finite, then we need only consider the N contiguous values of h_n that are possibly non-zero. Without loss of generality, we may arrange for these N values to be h_0 through h_{N-1}; thus:

$$y_k = \sum_{n=0}^{N-1} h_n x_{k-n} \tag{18.4}$$

This operation is typically implemented with the aid of a shift register, as shown in Figure 18.2.

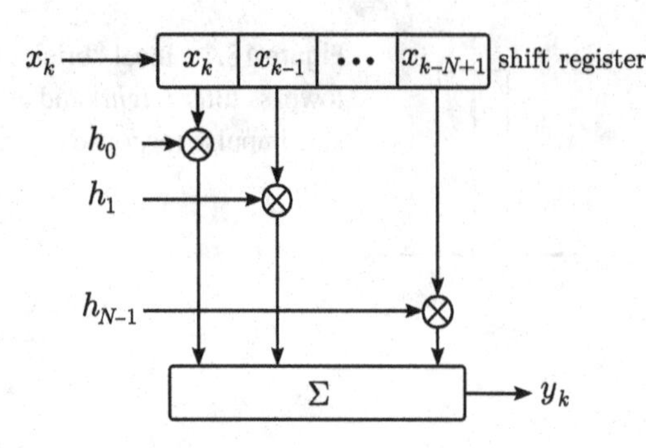

Figure 18.2. Shift register implementation of a digital FIR filter.

Now, a bit of nomenclature: An FIR filter employing an N-sample impulse response is said to be of order $N - 1$. Here we will use the symbol M to indicate order, and so $M = N - 1$.

The frequency response of an FIR filter is given by

$$H(\omega) = \sum_{n=0}^{M} h_n e^{-j\omega n T_S} \tag{18.5}$$

which is periodic in the frequency domain, repeating every F_S corresponding to Nyquist zones extending from 0 to F_S, F_S to $2F_S$, and so on for both negative and positive frequencies.

One way to design an FIR filter is to identify the desired impulse response $h(t)$, sample this response to obtain h_n, and discard all but N samples. For "brick wall" response, we might try sampling Equation (18.1) as follows:

$$h_n = \text{sinc}\left(2W\left[n - \frac{M}{2}\right]T_S\right) \tag{18.6}$$

with M even (N odd) so as to obtain samples which are symmetrical around $t = 0$, including $t = 0$.[2] To satisfy the Nyquist sampling criterion, we select $T_S < 1/2W$.

This leaves the problem of how to select M. Ideally M should be as large as possible, since the ideal impulse response is infinite. However, an order-M FIR requires as many as $M + 1$ multiplications per output sample;[3] therefore increasing M implies increasing computational burden. In practice, one chooses the smallest value of M which yields acceptable performance. The following example demonstrates the tradeoff.

EXAMPLE 18.1

Modern LMR communications systems use a variety of modulations and operate with bandwidth as small as 6.25 kHz. Consider an LMR receiver with a direct conversion stage yielding output in complex baseband form. The in-phase and quadrature components are sampled at 100 kSPS; quite large relative to the lowpass bandwidth so as to reduce the selectivity required from the analog anti-aliasing filters. Design a digital FIR filter that limits the complex baseband signal to 6.25 kHz bandwidth.

[2] This particular symmetry is not strictly necessary, but is convenient here.

[3] We shall soon consider a few ways to reduce the number of multiplications without affecting the order, hence the words "as many as."

Solution: Here identical filters are used for the in-phase and quadrature components. Each filter has lowpass bandwidth $W = 6.25/2 = 3.125$ kHz. The sample rate $F_S = 100$ kSPS.

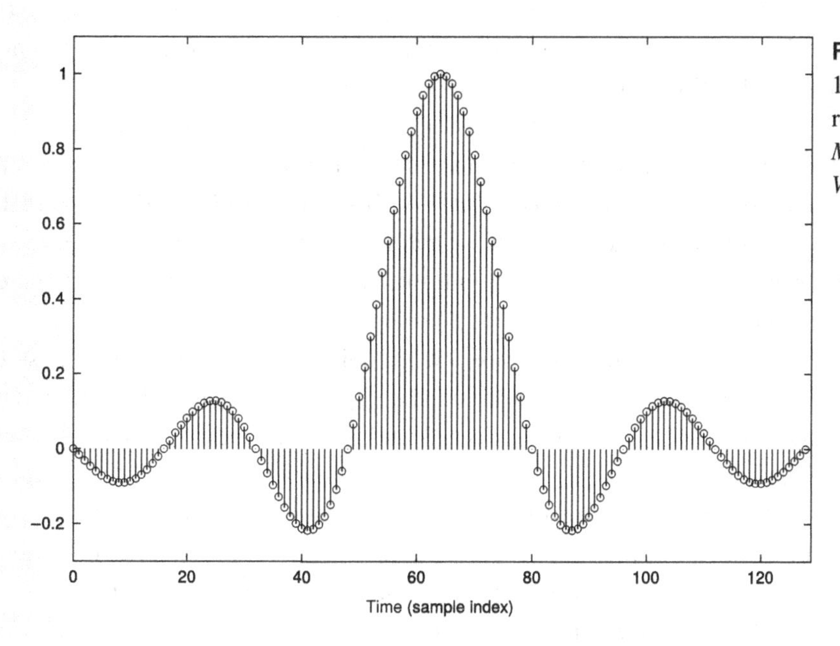

Figure 18.3. Example 18.1, first pass: Impulse response for sinc with $M = 128$ and $W = 3.125$ kHz.

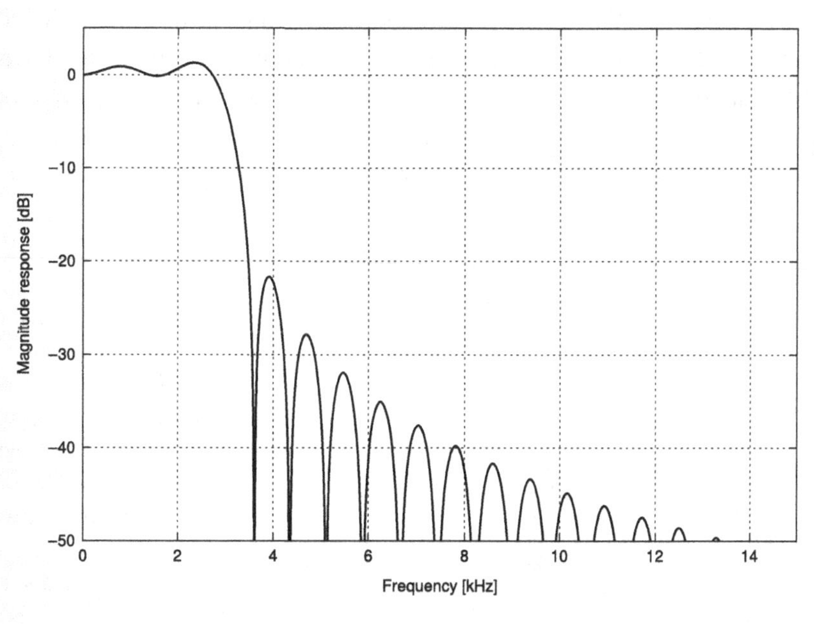

Figure 18.4. Example 18.1, first pass: Frequency response for sinc with $M = 128$ and $W = 3.125$ kHz.

A first pass design might consist of adopting the sinc impulse response of Equation (18.6) with an educated guess for M, which determines how much of the nominal impulse response is retained after truncation. Figure 18.3 shows the sampled impulse response for $M = 128$, which seems reasonable since values are relatively small at the edges and several oscillations of the sinc waveform are captured. Figure 18.4 shows the corresponding frequency response determined using Equation (18.5). Four issues are apparent. (1) The effect of truncating the ideal impulse response is unbounded frequency response: In particular, sidelobes extending to the edge of the Nyquist zone. (2) Ripple in the passband (the so-called Gibbs phenomenon), also due to truncation. (3) The transition from passband to stopband is not "brick wall," but rather a smooth transition into a null at a frequency significantly greater than W. (4) The response at $f = W$ is about −6 dB relative to the nominal passband response, which is probably too much in most applications and is certainly too much in the LMR application.

Items (1)–(3) may be mitigated by increasing M while holding F_S constant, since this is equivalent to increasing the span of the sinc function that is included before truncation. However, the response at $f = W$ remains about −6 dB regardless of M. (Problem 18.1 offers you the chance to confirm this for yourself.) Therefore simply increasing the filter order is not a universal solution, and is especially undesirable owing to the associated increase in computational burden.

With these observations in mind, we make two modifications in the second pass design. First, we increase W by 20% to 3.75 kHz (an educated guess) so as to reduce the attenuation seen at 3.125 kHz. Second, we keep $M = 128$ and instead modify the impulse response to reduce the effect of truncation. This strategy is essentially the same as was employed in Section 6.4.3 – and for precisely the same reason. In this case, however, we are (for now) unconcerned with intersymbol interference, and so there is no need to restrict our attention to the raised cosine response. Instead, we adopt the somewhat more flexible approach known as *windowing*: Multiplying the sinc impulse response with a "window" which *gradually* turns off the impulse response. For this example, we adopt the simplest imaginable window: the triangular window, as shown in Figure 18.5. Figure 18.6 shows the corresponding frequency response. Now the attenuation at 3.125 kHz is about −1.3 dB, which would be considered to be acceptable in most communications applications. Furthermore, the frequency domain artifacts of truncation – passband spectral ripple and stopband sidelobes – are much improved relative to the first pass design.

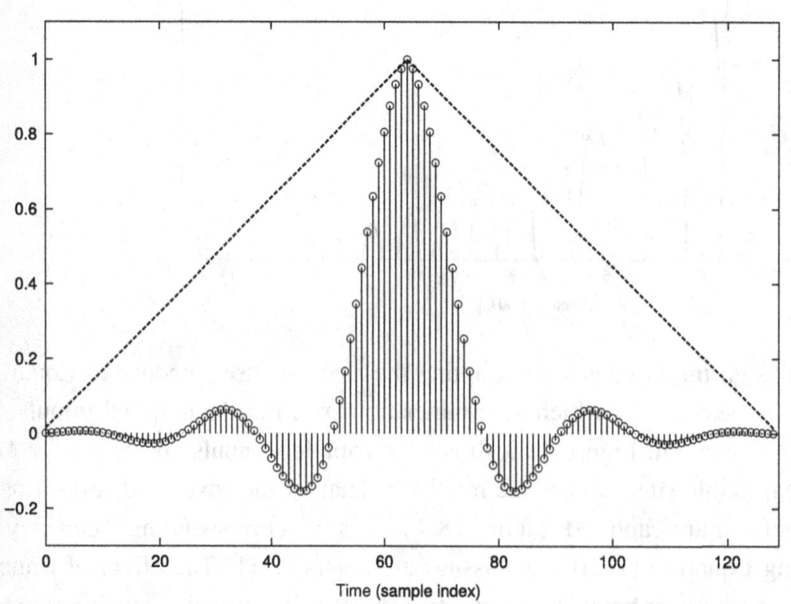

Figure 18.5. Example 18.1, second pass. *Stem plot:* Impulse response for triangular-windowed sinc with $M = 128$ and $W = 3.75$ kHz. *Dashed line:* Triangular window function that was applied to the sinc function to obtain this impulse response.

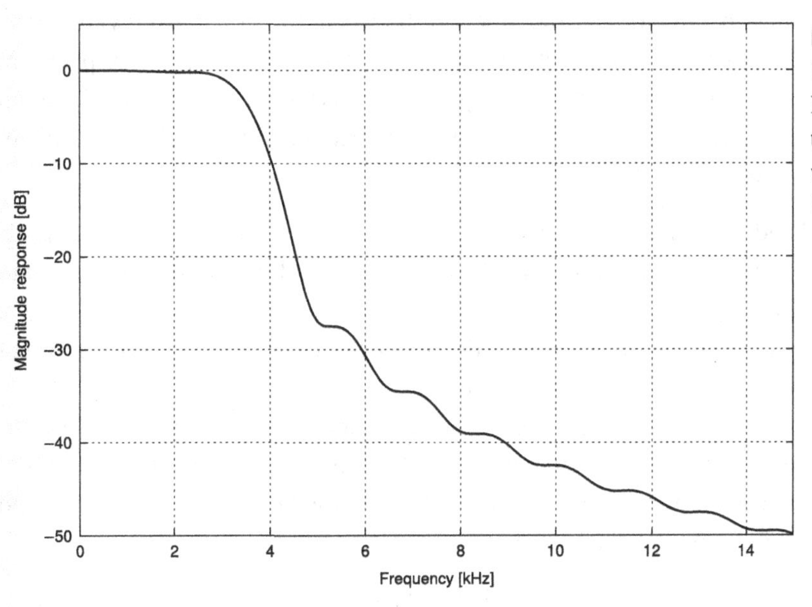

Figure 18.6. Example 18.1, second pass: Frequency response for triangular-windowed sinc with $M = 128$ and $W = 3.75$ kHz.

Subsequent iterations of design might take into account the possibility of reducing M and/or using alternative window functions so as to tailor the frequency response to meet specific requirements on passband ripple and stopband response. (See Problem 18.2.)

18.2.2 FIR Filter Design Using Windows; The Kaiser Method

The example above uses one of a broad class of FIR filter design techniques based on window functions. In the example we simply applied a triangular window to the sinc impulse response so as to mitigate truncation effects. Alternatively, we could have used one of the many other window functions, each of which somehow tapers from a maximum at the center to zero at the ends: A few common windows aside from the triangular window include the *Hann window*, the *Hamming window*, and the *Blackman window*. It should also be noted that not using a window is equivalent to applying a *rectangular window*, and that the triangular window is sometimes referred to as the *Bartlett window*. The various window functions represent different tradeoffs between fractional bandwidth and sidelobe level; in general, the former cannot be decreased without increasing the latter. For additional details, [56, Sec. 7.5] is recommended.

Although the above strategy is quite reasonable, it is also somewhat awkward from a system design perspective since quite a bit of the design process consists of guessing: i.e., selection of a window, selection of M, and selection of W so as to constrain the response to fit the desired bandwidth. This involves quite a bit of trial and error over three degrees of freedom, and so one is frequently left wondering whether M for any given design that meets the frequency response requirements might be larger than necessary. From this perspective, a better method would allow one to specify passband flatness and/or stopband response, and deliver a filter with the minimum M required to achieve these goals. Precisely this capability is offered by the Kaiser window method, which we shall describe next.

The principal difference between the *Kaiser window* and the windows cited previously is that the Kaiser window is actually a family of window functions characterized by the variable β, which parameterizes the tradeoff between fractional bandwidth and stopband response. Furthermore, the relationship between β, passband ripple, and stopband response is expressible in relatively simple empirical equations, as is the relationship between minimum M and these filter specifications. Thus, given the filter specifications, one may calculate β and M. Here is the procedure:[4]

(1) The passband specification is given by the parameters δ_p and W_p. W_p is the highest frequency at which the passband ripple is $\leq \delta_p$. δ_p is in linear units normalized to $H(0)$, so $|H(\omega)/H(0) - 1| \leq \delta_p$.

(2) The stopband specification is given by the parameters δ_s and W_s. W_s is the lowest frequency for which $|H(\omega)/H(0)|$ is required to be $\leq \delta_s$.

(3) This procedure does not allow the passband response to be specified independently of the stopband response, so a parameter δ is defined to be the smaller of δ_p and δ_s. (Note, however, that some tuning of the passband and stopband responses will be possible by adjusting W in Step (7).) The parameter A is defined as

$$A \equiv -20 \log_{10} \delta \tag{18.7}$$

(4) The parameter β is determined from A using one of the following empirically-derived equations:

$$\beta = 0.1102\,(A - 8.7)\ ,\ \ A > 50 \tag{18.8}$$

$$\beta = 0.5842\,(A - 21)^{0.4} + 0.07886\,(A - 21)\ ,\ \ 21 \leq A \leq 50 \tag{18.9}$$

$$\beta = 0\ ,\ \ A < 21 \tag{18.10}$$

(5) M is determined from the following empirically-derived equation:

$$M = \frac{A - 8}{2.285\ \Delta\omega}\ , \ \ \text{where}\ \ \Delta\omega \equiv \pi \frac{F_s - F_p}{F_S/2} \tag{18.11}$$

Here, note that $\Delta\omega$ quantifies the widt h of thc transition region between F_p and F_s relative to the width of a Nyquist zone, which is π rad/s in normalized frequency units. Also note that M increases with increasing A and decreasing $\Delta\omega$: Therefore sharper transition region, reduced passband ripple, or lower stopband response all require greater M. Finally, keep in mind that this is an empirical relationship; in fact the necessary value of M might be more or less by as much as 2.

(6) The $N = M + 1$ samples of the discrete-time sampled Kaiser window are

$$w_n = I_0\left(\beta\sqrt{1 - \left(\frac{n - M/2}{M/2}\right)^2}\right)\ , 0 \leq n \leq M \tag{18.12}$$

where $I_0(\cdot)$ is the zero-order modified Bessel function of the first kind.

(7) The central N samples of the prototype (sinc) impulse response of Equation (18.6) are calculated. A suggested starting value for W is $(F_s - F_p)/2$; i.e., the center of the specified transition region.

[4] The central idea is articulated in [33]; for a somewhat more accessible discussion see [56, Sec. 7.5.3].

(8) The Kaiser window is applied to the prototype (sinc) impulse response to obtain the impulse response of the desired FIR filter.

As a follow-up step, adjustments in the response may be made by varying W in Step (7).

EXAMPLE 18.2

Repeat Example 18.1, now using the Kaiser method.

Solution: Following the procedure outlined above:

(1) The intended passband is 3.125 kHz lowpass, so we set $W_p = 3.125$ kHz. In the previous example we decided we could live with 1.3 dB attenuation at this frequency, which a linear factor of 0.861, so let us set $\delta_p = 1 - 0.861 = 0.139$.

(2) In the previous example no stopband specification was provided, nor was there an obvious opportunity to impose one other than through trial and error. In a mobile radio application, we might define the stop band to begin at the far edge of the adjacent channel, thereby obtaining a constraint on spectral leakage on all channels not immediately adjacent to the channel of interest. Taking that approach, we set $W_s = 3W_p = 9.375$ kHz. The response of the filter obtained in Example 18.1 was down about 42 dB at 9.375 kHz, so let us adopt this as a specification. Attenuation of -42 dB is a linear factor of 0.00794, so we set $\delta_s = 0.00794$.

(3) Here $\delta_s < \delta_p$, so we set $\delta = \delta_s = 0.00794$ and $A = -20 \log_{10} \delta = 42$.

(4) Since $A = 42$, we use Equation (18.9) to calculate β:

$$\beta = 0.5842\,(A - 21)^{0.4} + 0.07886\,(A - 21) \cong 3.63$$

(5) Since $F_S = 100$ kSPS we have

$$\Delta\omega = \pi \frac{F_s - F_p}{F_S/2} = 0.125\pi$$

and the filter order is:

$$M = \frac{A - 8}{2.285\ \Delta\omega} = 37.9$$

Here we shall round up to $M = 38$, so the length of the filter will be $N = M + 1 = 39$ coefficients.

(6) The N samples of the discrete-time sampled Kaiser window w_n are obtained from Equation (18.12), and shown in Figure 18.7.

(7) As suggested in the procedure description, we begin with $W \leftarrow (F_s - F_p)/2 = 6.2495$ kHz. Using this value of W, the central N samples of the prototype (sinc) impulse response are calculated from Equation (18.6) and shown in Figure 18.7.

(8) This window is applied to the prototype (sinc) impulse response of Equation (18.6) to obtain the desired impulse response, shown in Figure 18.7.

The resulting frequency response is shown in Figure 18.8. Note the order of this filter ($M = 38$) is dramatically less than in the previous example ($M = 128$), although both meet requirements in some sense. In fairness, one should consider experimenting with M and different types of windows in the previous example before declaring the current method superior. However, it is quite convenient to have a procedure which *begins* with a good estimate of the minimum number of taps required to meet specifications.

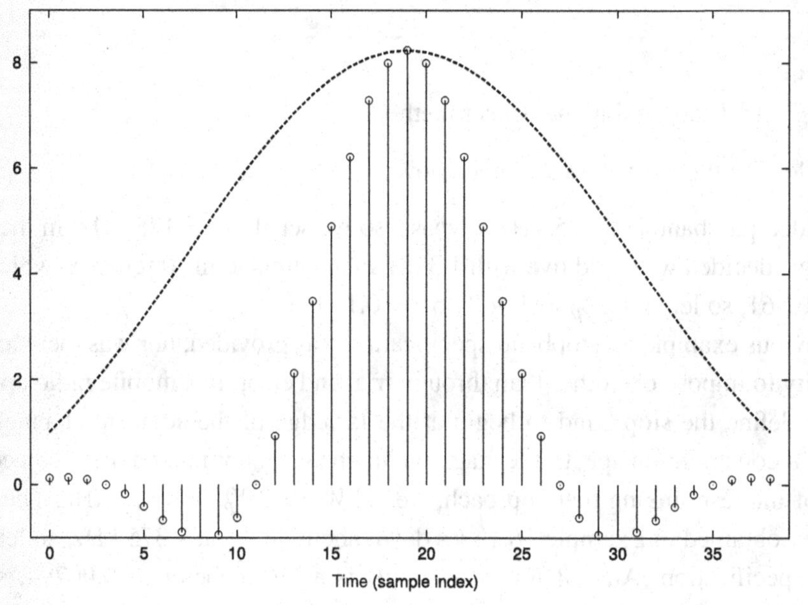

Figure 18.7. Example 18.2: *Solid:* Impulse response for windowed sinc with $M = 38$ and $W = 6.2495$ kHz. *Dashed:* Kaiser ($\beta = 3.63$) window function that was applied to the sinc function to obtain this impulse response.

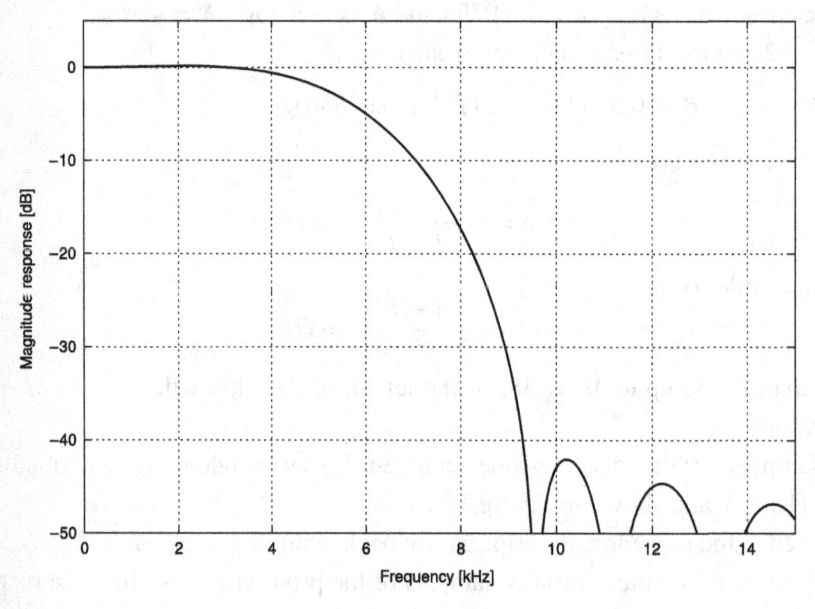

Figure 18.8. Example 18.2: Frequency response using the Kaiser method.

18.2.3 Other Methods for FIR Filter Design and Applications

Another approach to FIR filter design is the *FFT method.* This approach can be useful if a particular frequency response is desired, but is not well-represented by sinc with one of the well-known window functions. In the FFT method one samples the desired *frequency* response (as opposed to the impulse response) and applies the inverse discrete Fourier transform (IDFT) to obtain the associated impulse response. This version of the impulse response is then truncated (possibly after windowing) to obtain the desired filter coefficients.[5] Although this method is quite simple, the results are unlikely to be optimum in any particular sense. In particular, filters designed using this approach may be unnecessarily long, or may have impulse or frequency responses that vary from the desired responses in unacceptable ways.

Better alternatives are certainly available. There exists a broad class of methods in which one specifies *constraints* on the desired frequency response – as opposed to forcing a specific response – and then computes an FIR filter that meets the constraints and perhaps optimizes some other aspects of the response. For example, one can formulate the FIR filter design problem explicitly as a *constrained optimization* problem in which the goal is to find the discrete-time impulse response that optimizes passband flatness or peak stopband response under a constraint on passband ripple or transition bandwidth. Prominent among these methods is the *Parks–McClellan algorithm*, which is relatively complicated but is commonly employed in FIR filter design software. For additional details on this and related methods, [56, Sec. 7.7] is recommended.

The above descriptions assume that the filter's frequency response is of primary importance. However, in some applications one is more concerned with the form of the output as opposed to the response of the filter. An example is filtering intended for passband equalization, as opposed to stopband suppression. Particular examples in this vein are pulse-shaping filters (coming up in Section 18.5.1), allpass delay filters used for symbol timing recovery (coming up in Section 18.5.2), and filters used to predistort the input to a DAC in order to compensate for DAC-induced sinc distortion (reminder: Section 17.3.2). Although any of these can be handled by using the FFT method, an explicit constrained optimization procedure is likely to yield a shorter filter with better overall performance.

18.2.4 Digital Filters with Butterworth, Chebyshev, and Elliptic Responses

In Section 13.5 (in the context of analog filters) we noted that it is sometimes desirable for a filter to have one of the canonical frequency responses: e.g., Butterworth (maximally flat passband), Chebyshev (equiripple passband or stopband), elliptic, and so on. These particular responses may be implemented exactly as IIR filters; see e.g., [56, Sec. 7.3]. They may also be implemented approximately as FIR filters using the FFT method or other methods in which the frequency response may be appropriately constrained.

18.2.5 Reducing Computational Burden

Before moving on to multirate filters, it is useful to identify two simple "tricks" that may be employed to reduce the number of mathematical operations required to compute the output of a digital FIR filter.

[5] The name "FFT" is attached to this method because typically the IDFT is done using an inverse FFT, which is mathematically equivalent.

First, recall the basic convolution equation that describes the operation of an FIR filter:

$$y_k = \sum_{n=0}^{M} h_n x_{k-n} \tag{18.13}$$

We have seen that FIR filters typically have an impulse response that is symmetric in the time domain; that is, $h_n = h_{M-n}$. This suggests a modification to the implementation implied by the equation above (and shown in Figure 18.2) to reduce the number of multiplications. In the modified scheme, we first add input samples that are to be multiplied by the same coefficient, and then multiply the *sum* by the coefficient. For N odd (M even), this is:

$$y_k = h_{M/2}\, x_{k-M/2} + \sum_{n=0}^{M/2-1} h_n \left(x_{k-n} + x_{k-M+n}\right) \tag{18.14}$$

which reduces the number of multiplications required from N to $(N-1)/2+1$. For N even (M odd), we have

$$y_k = \sum_{n=0}^{N/2-1} h_n \left(x_{k-n} + x_{k-M+n}\right) \tag{18.15}$$

which reduces the number of multiplications required from N to $N/2$. Although it is true that this scheme introduces a number of additions equal to the number of multiplications that are eliminated, multiplication requires far greater computational effort, and so this scheme results in a considerable reduction in computational burden.

The second idea concerns the presence of zeros in the impulse response. Examination of the sampled prototype (sinc) response in Figure 18.3 reveals zeros every sixteenth sample, which is due to the fact that $F_S/2W = 16$ in this example. The result of multiplication with zero is zero, so there is no point in even doing the multiplication for these samples, which means that every sixteenth multiplication (and the associated addition after the multiplication) is unnecessary. There is not much saved in this example, but the savings become a big deal when $F_S/2W$ is smaller. For example, we might consider a sample rate $F_S = 12.5$ kSPS with $W = 3.125$ kHz, which still satisfies Nyquist by a large margin but results in $F_S/2W = 2$, and so results in a prototype impulse response in which every other sample is a zero. In this case the number of multiplications could be reduced by 50% simply by eliminating the multiplications that will always output zero. Such a filter is known as a *halfband filter*. The dramatic improvement in computational effort associated with halfband filters provides strong motivation for arranging F_S to be equal to $4W$ when possible.

18.3 MULTIRATE FILTERS

A multirate filter is a discrete-time filter for which the output sample rate is different from the input sample rate; i.e., the sample rate changes within the filter. Two common applications of such filters are decimation and interpolation, so let us begin with an introduction to these concepts.

18.3.1 Integer-Rate Decimating FIR Filters

In Examples 18.1 and 18.2 we considered a filter with $F_S/W = 32$, which represents a very large degree of overkill with respect to the Nyquist sample rate criterion $F_S/W \geq 2$. This is often justifiable as it allows one to make do with low-performance (less selective) analog anti-aliasing filters. However, there is no value in maintaining a sample rate so much greater than $F_S/W = 2$ once the signal has been digitized. Therefore the discrete-time filter in this application would typically be followed by *decimation*: A reduction in sample rate accomplished by discarding samples. In this case decimation by a factor of 8 would be quite appropriate: This would entail keeping only every eighth sample that emerges from the lowpass filter, and discarding the intervening seven samples, resulting in a new sample rate $F'_S = F_S/8 = 12.5$ kSPS. This results in $F'_S/W = 4$, which still satisfies the Nyquist criterion. In this operation, the factor 8 is known as the *decimation factor R*.

Upon reflection, we see that this is a colossal waste of computational effort, since $(R-1)/R = 87.5\%$ of the multiplications required to generate the 100 kSPS output are simply discarded to obtain the 12.5 kSPS output. A far better approach is to anticipate which output samples are not required, and to avoid computing those samples.

In principle this can be done using the same shift register structure shown in Figure 18.2, clocking the shift register at the input rate F_S but computing the output at output rate F'_S. This eliminates the unnecessary computation, but still requires a clock that operates at rate F_S.

A better way is to reorganize the filter into the mathematically equivalent form shown in Figure 18.9, which employs R FIR filter "segments" each of length N/R (assuming for the moment that N/R is an integer). In this scheme, the entire structure, including the filter segments, is clocked at the output rate. The input device that distributes the input samples to the filter segments is known as a *commutator*. Although the commutator appears to be operating at the input rate, in fact the usual implementation is in the form of a "fan out" by R followed by latches preceding each filter segment. Thus, the commutator requires no additional effort to implement beyond what is already required for clocking of the input samples.

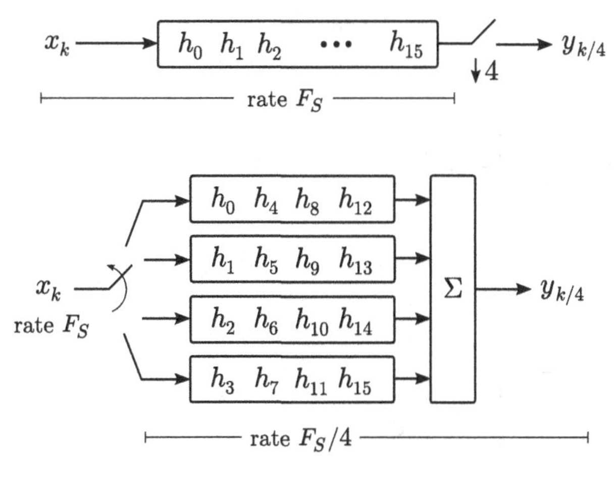

Figure 18.9. Rearranging an $N = 16$ single-rate FIR filter with $R = 4$ (*top*) into a multirate structure that operates at the output rate (*bottom*). These values of N and R are selected to make it possible to write out all coefficients in a compact diagram; for the example addressed in the text, $R = 8$ and N is the number of taps in the original single rate FIR filter.

In Figure 18.9, the FIR segments are of length $N/R = 4$; an integer. What if N/R is not an integer? This is no problem: One simply "zero pads" the impulse response to obtain the smallest N for which N/R *is* an integer. This results in a few multiplications by zero in the resulting multirate structure, which can be anticipated and not computed. Thus there is no penalty in terms of computational burden.

18.3.2 Integer-Rate Interpolating FIR Filters

Interpolation means increasing the sample rate while filtering. There are at least four cases in which one may want to increase the sample rate while filtering: (1) In digital-to-analog conversion, for example to reduce the selectivity required from an anti-aliasing filter following a DAC; (2) in certain transmit architectures in which digital-to-analog conversion occurs at an IF (or at the carrier frequency) as opposed to complex baseband; (3) non-integer rate conversion (see Section 18.3.3); (4) sometimes in symbol timing acquisition in digital modulations (see Section 18.5).

Non-multirate implementations of an interpolation filter are shown in Figure 18.10. In the *zero fill* strategy, one increases the sample rate by a factor of R by inserting $R - 1$ zeros following each input sample, and then applies the desired filtering operation at the output rate. The *zero-order hold* (ZOH) strategy is identical except the $R - 1$ padding samples are equal to the input sample.

At first glance, it might appear that the zero fill strategy is unusably crude, and that ZOH must be a superior approach. However, this is not the case. Let us begin with ZOH: This should remind you of sample-and-hold in DACs (reminder: Figure 17.4 and associated text). The consequence of holding the value of the previous sample between samples is a sinc response in the output. The sinc response is a double-edged sword: It has the advantage of reducing aliases after interpolation, but also imposes sinc distortion on the passband. The zero fill approach does not exhibit the sinc response problem, and in fact produces essentially no distortion. This is because bandlimiting filters (including lowpass filters and bandpass filters) are essentially averaging operations, so the zero samples are properly filled in – their values are literally

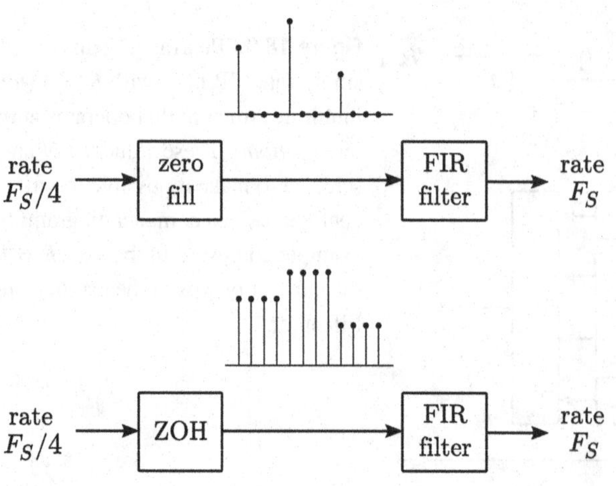

Figure 18.10. Non-multirate implementations of interpolation. *Top*: Zero fill. *Bottom*: ZOH.

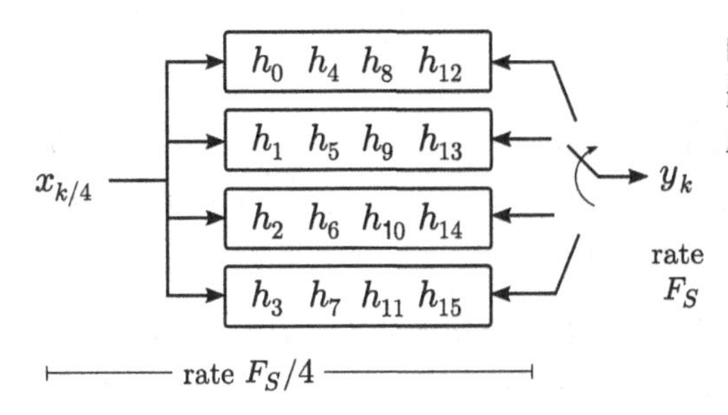

Figure 18.11. A multirate implementation of a zero fill interpolate-by-R FIR filter, in this case for $N = 16$ and $R = 4$.

interpolated between the non-zero samples. Another way to look at this is that the frequency content associated with the sudden transition from non-zero samples to zero samples lies mostly outside the bandwidth of any subsequent filtering, and so the zero values are instead properly filled in after filtering.

We should be clear that the passband distortion associated with ZOH is not really much of a problem either, since subsequent filtering could be modified to correct for this distortion. Where the zero fill approach really comes in handy, and ZOH less so, is in reducing the computational burden of the filter. The zero fill approach provides $R - 1$ zeros for every R samples delivered to the input of the filter, which means that only $N \cdot R/(R - 1)$ multiplications are required to compute each output sample output from an order M FIR filter. Figure 18.11 shows an implementation of such a filter in which the structure operates at the input rate, even though the output appears at R times the input rate.

18.3.3 Non-Integer and Large-*R* Techniques

The preceding discussion has assumed that the decimation or interpolation factor R has been an integer. Although this is typically the case, this is not a fundamental restriction. Non-integer-rate decimation and interpolation can be effected using various combinations of integer-rate interpolation followed by integer-rate decimation, or vice versa.

In some modern applications R becomes very large. For example, some receivers digitize swaths of spectrum which are tens of MHz wide, with the intent of processing individual signals that might be only tens of kHz wide, resulting in R on the order of 1000. This leads to FIR filters of very high order and which may be intractable even taking into account the efficiency afforded using an appropriate structure, running at the lower output rate.

An alternative in this case is the *cascaded integrator comb (CIC) filter* [26]. The CIC filter is a multirate FIR filter consisting only of delays and additions, and no multiplications, resulting in a spectacular reduction in computational burden relative to a conventional FIR filter structure. However, the selectivity of the CIC structure is poor for low R, and the response cannot be customized. Furthermore the CIC filter imposes a sinc-type passband response that warrants correction in most applications. Therefore practical systems typically distribute large-R rate conversions between an initial CIC filter and a conventional multirate FIR, with the latter providing rate conversions in the range 4–32 as well as compensation for the CIC's sinc passband response. The application to interpolation is exactly analogous.

EXAMPLE 18.3

In a particular receiver a single complex baseband signal having bandwidth 200 kHz and centered at 0 Hz must be extracted from a complex baseband signal at a sample rate of 64 MSPS. Suggest an architecture for accomplishing this.

Solution: Since the signal of interest has 200 kHz bandwidth and is centered at 0 Hz, the desired lowpass bandwidth is $W = 100$ kHz. To satisfy Nyquist, the output sample rate $F'_S \geq 2W =$ 200 kSPS; although a somewhat higher rate is probably required in order to accommodate the non-brick-wall nature of the filter responses. A decimation factor of 256 reduces $F_S =$ 64 MSPS to $F'_S = 250$ kSPS, which is 25% greater than Nyquist and so provides a comfortable margin for anti-aliasing. A reasonable design would be to use a CIC multirate filter to first decimate by $R_1 = 64$, and then a traditional multirate FIR filter to decimate by $R_2 = 4$ and to perform the sinc response compensation. Other schemes in which $R_1 R_2 \sim 256$ with $R_1 \geq 10$ or so would also be appropriate.

18.4 QUADRATURE UPCONVERSION AND DOWNCONVERSION

Chapter 14 summarized the various methods for analog frequency conversion in radios. This included quadrature conversion, which is the defining element of direct conversion architecture. One may implement the associated multiplications and filters in the digital domain as discrete-time implementations of the same sequence of operations. However, this is nearly always a spectacularly inefficient use of digital-domain resources. In this section, we outline some methods for implementing these functions that are quite unlike the analog implementations, and are possible only when the signals are expressed in discrete-time form.

We have already encountered at least one of these methods: Undersampling (Section 15.6.2), in which we effect a "free" frequency downconversion simply by allowing the input of an ADC to arrive from a Nyquist zone other than the first. We have also seen in previous sections of this chapter that exploiting characteristics of the discrete-time impulse response (e.g., using symmetry to reduce multiplication count), strategic selection of sample rate (e.g., selecting $F_S = 4W$ in order to obtain a halfband filter), and reorganization of processing when sample rate changes are required, all result in dramatic improvements in computational efficiency. We now extend those ideas to quadrature frequency conversion.

18.4.1 $F_S/4$ Quadrature Downconversion

When the desired frequency shift is equal to $F_S/4$ – that is, equal to one-quarter of the sample rate – it is possible to implement quadrature conversion with *no* multiplications or additions. This is now shown by example for quadrature conversion from a center frequency of $f_c = F_S/4$ to complex baseband: First we write the complex baseband output y_k in terms of the real-valued input x_k with $t = kT_S$:

$$y_k = x_k \, e^{-j2\pi f_c k T_S} \tag{18.16}$$

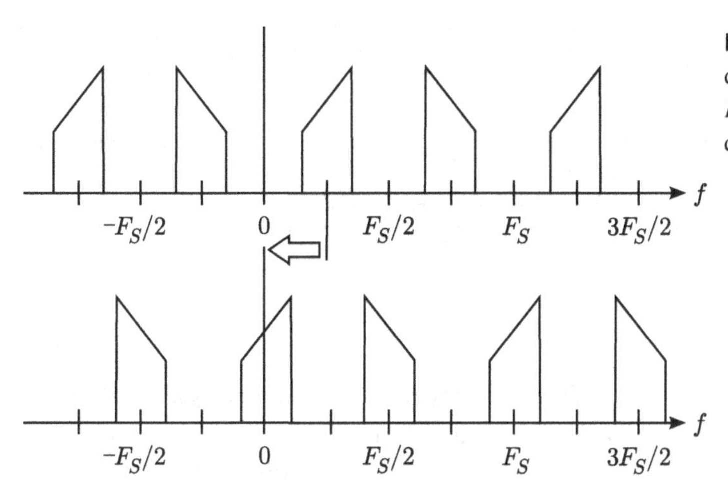

Figure 18.12. Quadrature downconversion implemented as a $F_S/4$ shift-left operation; frequency domain view.

For $f_c = F_S/4 = 1/4T_S$, we have

$$y_k = x_k \, e^{-jk\pi/2} \tag{18.17}$$

However, $e^{-jk\pi/2} = +1, -j, -1, +j, +1, \ldots$ in a repeating sequence. Therefore we have

$$y_0 = x_0 + j0$$
$$y_1 = 0 - jx_1$$
$$y_2 = -x_2 + j0$$
$$y_3 = 0 + jx_3$$
$$y_4 = x_4 + j0$$

and so on, repeating with period 4 samples. Note that each value of y_k is obtained simply by assigning x_k to be either the real or imaginary part of y_k and possibly changing sign. Not only does this not require any multiplication or addition,[6] but also every other sample arriving at a subsequent lowpass filter operating on the in-phase and quadrature components will be zero, so one-half of the multiplications in these filters may be eliminated.

Figure 18.12 shows what has happened in the frequency domain: The entire spectrum is shifted left by $F_S/4$. Examination of this figure reveals that this simple technique also facilitates undersampling: For example, a $F_S/4$ shift-*right* operation is accomplished simply by changing sign and would move the spectral component located at $3F_S/4$ to complex baseband, since this component would also appear at $-F_S/4$ in the input spectrum.

To take full advantage of the $F_S/4$ spectral-shifting method, the in-phase and quadrature lowpass filters should be implemented as multirate structures; otherwise we would still have multiplications occurring at the input rate. The lowpass filters are required to have $W \approx F_S/4$ in order to eliminate the portion of the adjacent alias that now appears at the high-frequency edge of the first Nyquist zone. If we select $W = F_S/4$ exactly, then every other coefficient of

[6] Note sign changes are essentially "free" in terms of computational effort, since they may be effected by flipping a bit. That is, a sign change is not necessarily implemented as multiplication by -1.

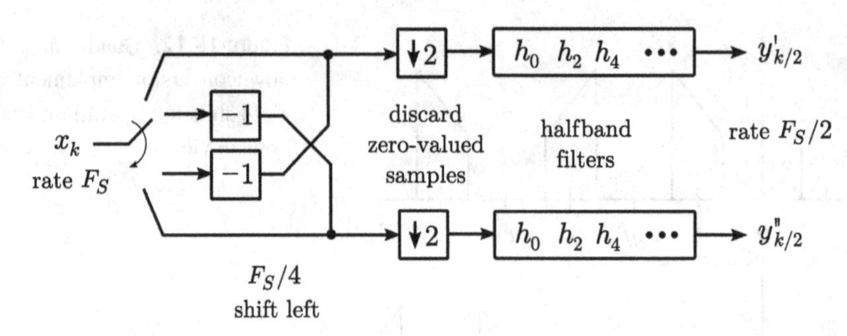

Figure 18.13. $F_S/4$ shift-left quadrature downconverter with $R = 2$ decimation and halfband filtering.

the impulse response will be zero. Therefore any implementation of such a filter (single rate or multirate) will require only one-half of the multiplications that would be required of a filter for which F_S/W was not exactly 4. Filtering with $F_S/W = 4$ facilitates $R = 2$, as noted in Section 18.2.5. When we reorganize the $R = 2$ halfband filter as a multirate structure of the type shown in Figure 18.9, we obtain two parallel FIR filter segments in which all the coefficients in one of the segments are zero; this segment can therefore be eliminated and the commutator may be replaced by a simple decimator in which the zero-valued input samples are discarded. The result is shown in Figure 18.13.

EXAMPLE 18.4

A particular receiver uses a 1 GSPS ADC to undersample a real-valued analog IF centered at 750 MHz with bandwidth 400 MHz. From the ADC output it is desired to extract a complex baseband signal for subsequent processing. Suggest a DSP architecture that accomplishes this.

Solution: Figure 18.14(a) and (b) shows the spectrum at the input and output, respectively, of the ADC. Note that we are undersampling from the center of the second Nyquist zone, so the desired signal appears at 250 MHz in the ADC output and is spectrally reversed. Alternatively – preferably in this case – we may view the correctly-oriented alias at −250 MHz as the desired signal, and apply a $F_S/4$ shift-right to recover it. The resulting signal must be subsequently lowpass-filtered to exclude the portions of the aliases that will intrude into the first Nyquist zone after decimation by 2. The result after filtering and decimation is shown in Figure 18.14(d).

Figure 18.15 shows the "brute force" approach: a direct conversion receiver which is essentially a digital implementation of the analog direct conversion receiver. In this architecture, a digital complex-valued LO is generated with frequency $f_0 = -250$ MHz (negative so as to shift the spectrum to the right). In this architecture the LO synthesis and the two multiplications required to implement the mixer are clocked at 1 GSPS, which is difficult to achieve in modern DSP hardware.[7] The fact that the FIR filters may be implemented as multirate halfband filters is useful, but provides no relief from the problem that the LO synthesizer and input mixer multiplications must be clocked at the input rate.

[7] As of the year 2015, anyway! Although the threshold for "difficulty" here may evolve in future years, it is quite likely that practical digitization rates will continue to out-pace practical rates for multiplication for some time.

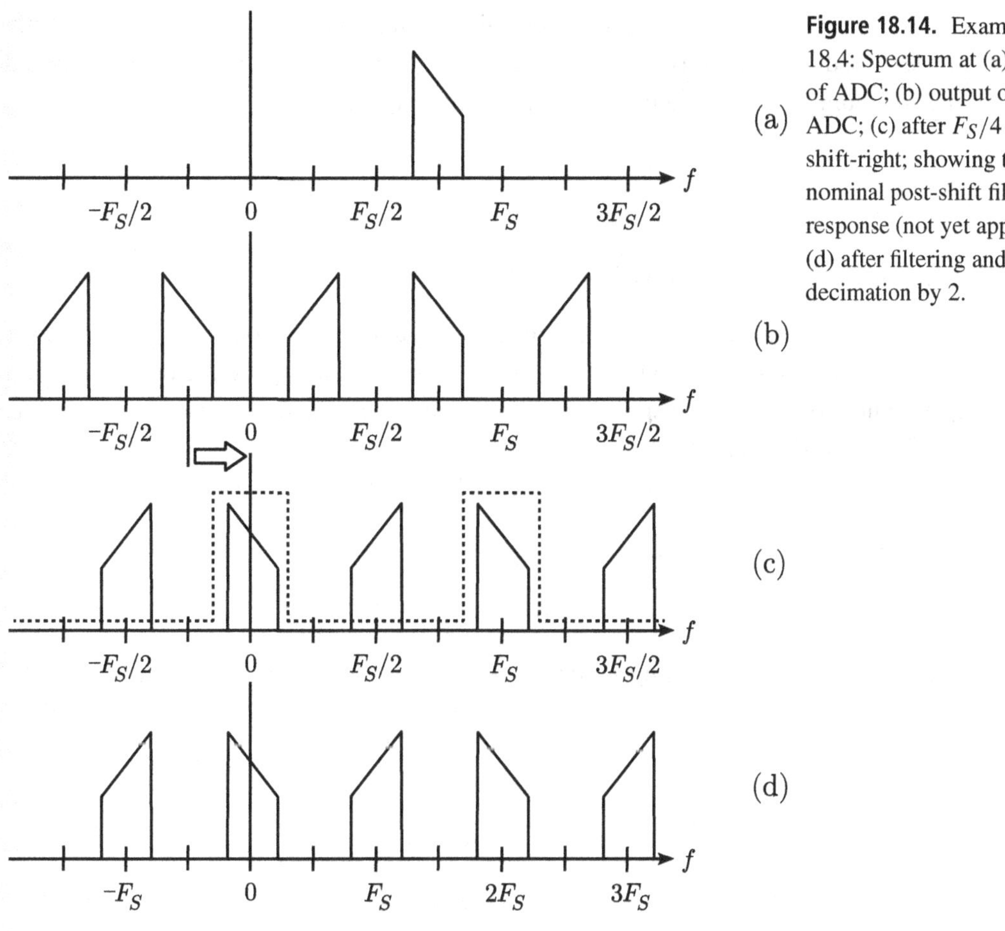

Figure 18.14. Example 18.4: Spectrum at (a) input of ADC; (b) output of ADC; (c) after $F_S/4$ shift-right; showing the nominal post-shift filter response (not yet applied!); (d) after filtering and decimation by 2.

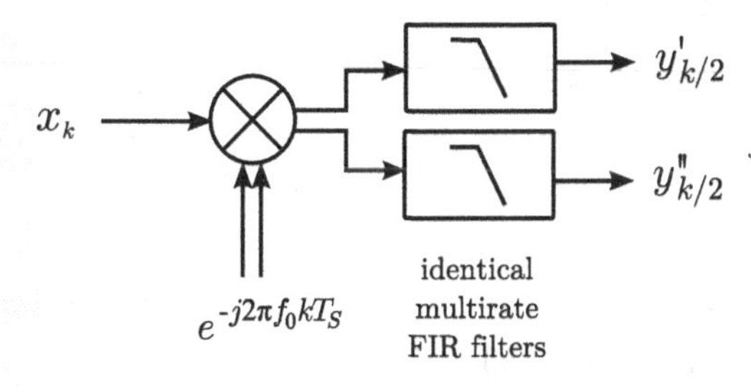

Figure 18.15. Example 18.4: "Brute force" direct conversion approach (not recommended). $f_0 = -250$ MHz.

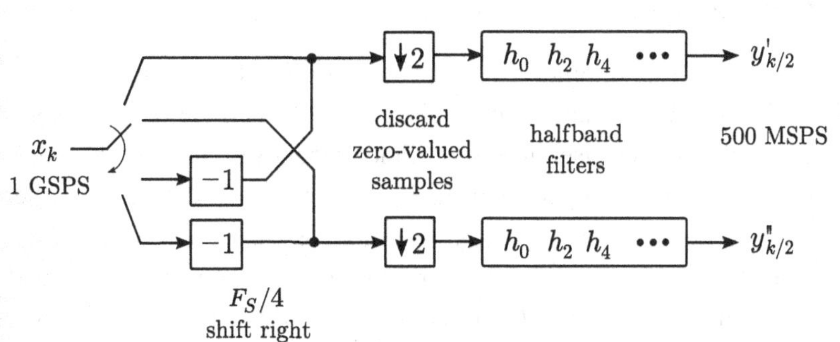

Figure 18.16. Example 18.4: $F_S/4$ shift-right quadrature downconverter with $R = 2$ decimation and halfband filtering.

So, we replace the LO synthesis and mixer with a multiplier-free $F_S/4$ shift-right structure as described above and as shown in Figure 18.16. Now the only operations required to obtain the spectrum shown in Figure 18.14(c) are sign changes.

18.4.2 $F_S/4$ Quadrature Upconversion

Essentially the same trick can be applied to quadrature upconversion. Consider a complex baseband discrete-time signal $x_k = x'_k + jx''_k$, where x'_k and x''_k are the real-valued in-phase and quadrature components, respectively. Quadrature upconversion to a real-valued IF centered at f_c is accomplished as follows:

$$y_k = \text{Re}\left\{x_k\, e^{+j2\pi f_c k T_S}\right\} \tag{18.18}$$

For $f_c = +F_S/4 = +1/4T_S$, we have

$$y_k = \text{Re}\left\{x_k\, e^{+jk\pi/2}\right\} \tag{18.19}$$

where $e^{+jk\pi/2}$ is the repeating sequence $+1, +j, -1, -j, +1, \ldots$. Therefore we have

$$y_0 = +x'_0$$

$$y_1 = -x''_1$$

$$y_2 = -x'_2$$

$$y_3 = +x''_3$$

$$y_4 = +x'_4$$

and so on, repeating with period 4 samples. Thus each value of y_k is either the in-phase or quadrature component of x_k with possibly a sign change, and no multiplication or addition is required. The associated structure consists of interpolating lowpass multirate FIR filters whose outputs are combined by a multiplier-free $F_S/4$ shift-right operation. As in downconversion, the lowpass FIR filters will be halfband structures if $R = 2$, since every other sample from either filter (e.g., x''_0, x''_2, and so on for the in-phase filter) is discarded.

18.4.3 Multirate Quadrature Downconversion From Other IFs

The stunning reduction in computational burden afforded by $F_S/4$ quadrature conversion raises the question of whether similar reductions might be possible when the IF is not equal to $F_S/4$. In one sense the answer is no, because the $F_S/4$ "trick" depends on the fact that multiplication by an LO reduces to a simple manipulation of the input samples. But there is another way that one can avoid the higher-rate frequency mixing operation, as we shall now describe.

To begin, consider the architecture shown in Figure 18.17. Here the real-valued discrete-time IF signal x_k is centered at frequency f_c and sampled at rate F_S. To obtain the corresponding complex baseband representation, one multiplies by the discrete-time LO signal $l_k = e^{-j2\pi f_c k T_S}$,

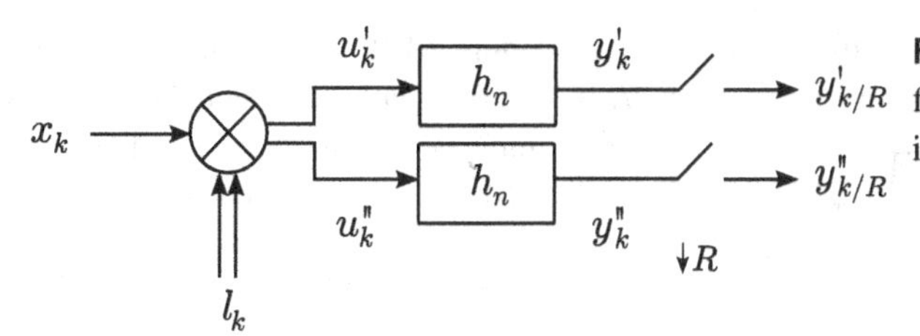

Figure 18.17. Quadrature downconversion, filtering, and decimation: First pass implementation.

yielding a signal $u_k = u'_k + ju''_k$ at the same sample rate. One subsequently applies lowpass filters to obtain $y_k = y'_k + jy''_k$ – again at the input rate F_S – and then finally decimates by R to obtain a sample rate commensurate with the new bandwidth, yielding $y_{k/R}$.

Although the architecture shown in Figure 18.17 will certainly yield the desired result, it is computationally inefficient because $R - 1$ of R samples computed are subsequently discarded. As was the case in Section 18.3.1, the better approach is to rearrange the processing so that the unneeded samples are not computed, resulting in a structure that operates at the output sample rate F_S/R. The key to making the mixer operate at the output rate is to reorganize the architecture of Figure 18.17 such that mixing follows filtering. This is most easily explained mathematically. Note:

$$u'_k = \text{Re}\{l_k x_k\} = x_k \cos(2\pi f_c k T_S) \quad (18.20)$$

$$u''_k = \text{Im}\{l_k x_k\} = -x_k \sin(2\pi f_c k T_S) \quad (18.21)$$

Subsequently:

$$y'_k = \sum_{n=0}^{M} h_n u'_{k-n} = \sum_{k=0}^{N} h_n \cos(2\pi f_c [k-n] T_S) x_{k-n} \quad (18.22)$$

$$y''_k = \sum_{n=0}^{M} h_n u''_{k-n} = -\sum_{k=0}^{N} h_n \sin(2\pi f_c [k-n] T_S) x_{k-n} \quad (18.23)$$

Therefore $y_k = y'_k + jy''_k$ may be expressed as follows:

$$y_k = \sum_{n=0}^{M} h_n e^{-j2\pi f_c [k-n] T_S} x_{k-n} = \left[\sum_{n=0}^{M} h_n e^{+j2\pi f_c n T_S} x_{k-n} \right] e^{-j2\pi f_c k T_S} \quad (18.24)$$

Recognizing the final factor as l_k, we have

$$y_k = \left[\sum_{n=0}^{M} h_n e^{+j2\pi f_c n T_S} x_{k-n} \right] l_k \quad (18.25)$$

and so now the mixer appears at the output of the filtering operation, as opposed to the input. A consequence of doing this is that the filter coefficients are no longer h_n, but rather $h_n e^{+j2\pi f_c n T_S}$. This is a bit awkward as we would prefer the coefficients to be real-valued. To achieve this we define

$$p'_n \equiv h_n \cos(2\pi f_c n T_S) \quad (18.26)$$

$$p''_n \equiv h_n \sin(2\pi f_c n T_S) \quad (18.27)$$

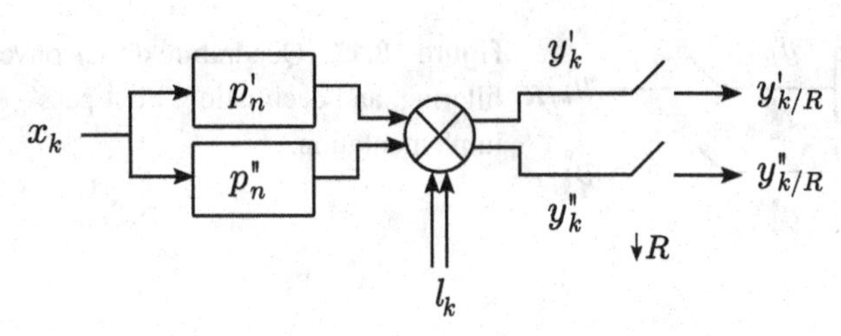

Figure 18.18. Quadrature downconversion, filtering, and decimation: Second pass implementation. The order of filtering and mixing have been exchanged.

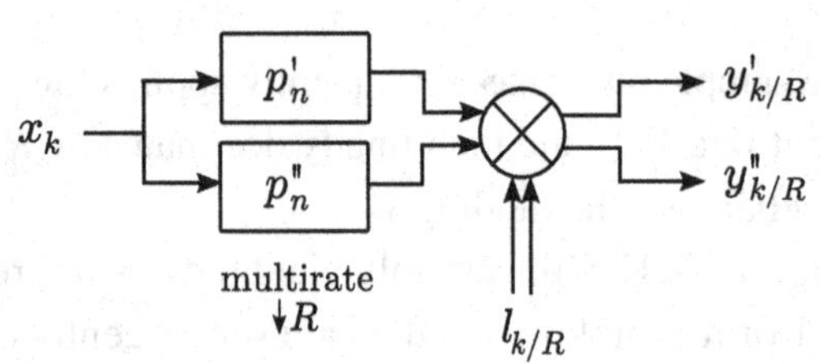

Figure 18.19. Quadrature downconversion, filtering, and decimation: Final implementation. The order of mixing and decimation have been exchanged, and the FIR filters have been replaced with multirate equivalents. Now the entire structure is operating at the output rate.

So that

$$y_k = \left[\sum_{n=0}^{M} p'_n x_{k-n} + j \sum_{n=0}^{M} p''_n x_{k-n} \right] l_k \tag{18.28}$$

This implementation is shown in Figure 18.18.

At this point, the mixer (in fact, the entire structure) is still operating at the input rate. However, we now exchange the order of mixing and decimation, and then subsequently implement each FIR filter as a multirate structure using precisely the same technique described in Section 18.3.1. The final implementation is shown in Figure 18.19. Now the entire structure operates at the output rate. It should be noted, however, that the overall number of multiplications required to implement the FIRs has been doubled, so $R \geq 4$ is required for the overall computational burden to be reduced relative to a structure in which the mixer precedes multirate FIRs.

EXAMPLE 18.5

We consider a modification to Example 18.4: As before, the receiver uses a 1 GSPS ADC to undersample a real-valued analog IF centered at 750 MHz with bandwidth 400 MHz. However, now it is desired to extract a complex-baseband signal having 200 MHz bandwidth for subsequent processing, and this 200 MHz-wide swath of spectrum may come from any part of the 400 MHz-wide IF signal. Suggest a DSP architecture that accomplishes this.

Solution: First, note that the $F_S/4$ version considered in Example 18.4 could be modified to work in this case: We could just multiply the output by a second complex LO to move the desired frequency to zero, and then lowpass filter again with W slightly greater than 100 MHz. Choosing the output sample rate to be 250 MHz results in a second-stage filter that could be implemented as a multirate halfband structure with $R = 2$. However, this approach is considerably more complicated than necessary, and leaves the first-stage filter operating at 500 MHz as opposed to the output rate of 250 MHz.

So, we use the architecture of Figure 18.19 so that the entire structure may operate at the output rate. To begin, we determine the coefficients h_n for the non-multirate lowpass FIR filters in Figure 18.17. The edge of the desired 200 MHz passband corresponds to a lowpass cutoff frequency of 100 MHz, whereas the Nyquist zone boundary at 250 MSPS is 125 MHz. Therefore the primary requirement on the lowpass filter is that the stopband begin at 150 MHz with sufficient attenuation to make aliasing from frequencies above 150 MHz acceptable. In the absence of other guidance, we require that the attenuation be at least 30 dB at 150 MHz.[8] Any of the filter design methods described in Section 18.2 might be employed to get the h_n; here we employ the Kaiser window method of Section 18.2.2. In this case, we set $W_p = 100$ MHz, $W_s = 150$ MHz, and $\delta = \delta_s = -30$ dB=0.0316. Thus, $A = 30$, $\beta = 2.12$, $\Delta\omega = 0.1\pi$ rad/s, and M must be at least 30. Anticipating that the multirate structure will consist of $R = 4$ parallel FIR filters each consisting of N/R coefficients, we make $N = 32$ so each FIR filter will consist of an equal integer number (8) coefficients, and the associated $M = N - 1 = 31$. The resulting response is as shown in Figure 18.20; note that this is the response in either the single-rate implementation or the multirate implementation.

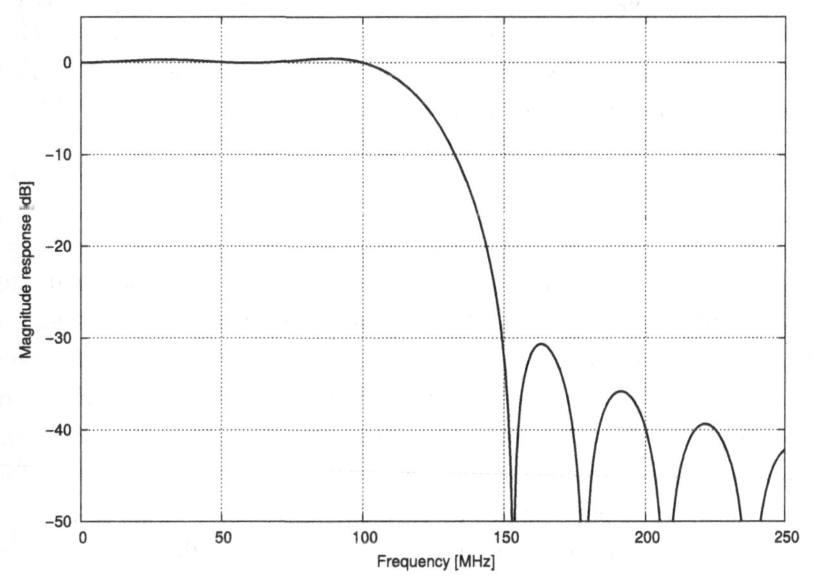

Figure 18.20. Example 18.5: Output frequency response. Note that this is the result in both the prototype single-rate design and the final multirate design.

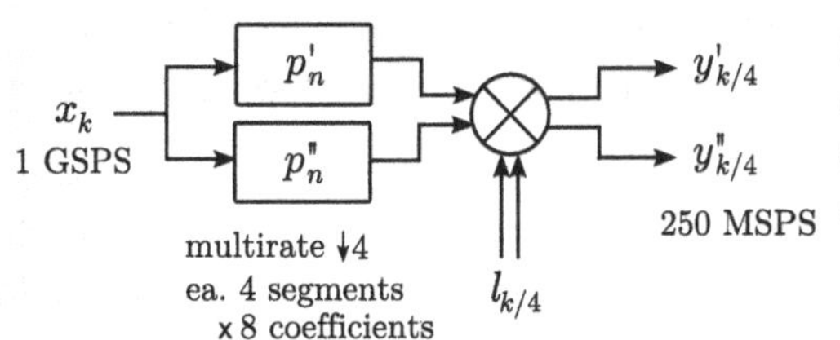

Figure 18.21. Example 18.5: Completed design. The entire structure operates at the output rate of 250 MSPS. Center frequency is selected by adjusting both the filter coefficients (p'_n, p''_n) and the decimated-rate LO ($l_{k/4}$).

The completed structure is shown in Figure 18.21. The in-phase and quadrature lowpass filters are both multirate with $N = 32$ and $R = 4$, consisting of coefficients which have been "heterodyned" by the desired tuning frequency as specified by Equations (18.26) and (18.27). As in Example 18.4 the desired alias from the ADC output is centered in the first

[8] A better way to set this specification would be to take into account the selectivity of the analog receiver, which might also be contributing to the attenuation at 150 MHz, and of course the expected range of signal levels in the passband.

negative-frequency Nyquist zone. For example a center frequency of 760 MHz appears at 240 MHz (spectrally-reversed) in the first positive-frequency Nyquist zone, and at −240 MHz (not spectrally-reversed) in the first negative-frequency Nyquist zone. Therefore $l_k = e^{-j2\pi f_c k T_S}$ with $f_c = -240$ MHz. The complex mixer multiplies the complex-valued output of the multirate filters by $l_{k/R}$; i.e., by every fourth sample of the "original" input-rate complex-valued LO signal. Thus tuning (that is, changing the selected center frequency in the IF input) is implemented by changing the filter coefficients and the frequency of the complex sinusoid multiplying the output.

18.5 APPLICATIONS IN DIGITAL MODULATION

As pointed out in previous chapters of this book, modulation and demodulation are performed in the digital domain in nearly all modern radio systems using digital modulation. Given some of the insights from previous sections in this chapter, we now circle back to the topic of DSP implementation of digital modulation and point out some useful ideas.

18.5.1 Pulse Shaping

As pointed out in Section 6.4.1 (Figure 6.5 and associated text), the usual recipe for pulse-shaped digital modulation is to first obtain the symbols (carrier magnitude-phase states) and then to apply the desired pulse shape using a pulse-shaping filter. Given a symbol rate of $1/T$, the output rate F_S of such a filter must be significantly greater than $1/T$; hence such filters are inevitably interpolation filters. Implementation as integer-rate zero-fill interpolation facilitates the use of a computationally-efficient multirate structure as described in Section 18.3.2 (Figure 18.11), in which the input samples are symbols at rate $1/T$ and the output samples represent the desired modulated waveform at rate R/T. Recall also that the impulse response of the zero-ISI pulse shaping filter has zeros at intervals of $1/T$; thus one out of every R coefficients of the digital FIR filter implementing this response is zero. This can be exploited to further reduce the computational burden associated with pulse shaping and the associated interpolation.

18.5.2 Symbol Timing Recovery

In a receiver we have the associated issue of symbol timing recovery (Section 6.16.3): Given a signal which is sampled at a rate much greater than the symbol rate $1/T$, how does one recover a signal in which the samples represent symbols sampled at the maximum effect time (MET, as we referred to this). Here we have two solutions that are worth mentioning in light of the findings of this chapter. The first approach is to use something like the structure shown in Figure 6.37 in which the signal is oversampled by a factor R (i.e., at R times the symbol rate), and then one selects from among the R possible downsampled "phases" of the input signal.

However, Figure 6.37 is essentially a decimating multirate structure, and there is considerable advantage in combining the "q-metric calc." operation with matched filtering itself. This reduces the computational burden by eliminating the initial (single-rate) computation of the matched filter.

Further reduction in computational burden can be achieved by computing only the "prompt" (currently best) timing solution as well as the "early" and "late" versions; i.e., just three of the R paths. When the computed metrics indicate that either the early or late paths are better than the prompt path, then that path is selected as prompt, new early and late paths are identified, and the process continues. This is an example of a broad class of schemes known as *early-late synchronizers*; here facilitated by the fact that multirate decimation naturally provides a set of symbol timing phases to choose from.

In some applications it is not convenient or acceptable to have to choose from just R discrete symbol timing phases. Instead, one might prefer a delay-locked loop (DLL) architecture in which the symbol timing can be continuously varied with resolution finer than the output sample rate. Such delays can be implemented using an *allpass delay filter*. The concept is developed as follows: A continuous-time signal $x(t)$ that is strictly bandlimited to lowpass bandwidth W can be reconstructed from discrete-time samples $x(nT_S)$ as follows:

$$x(t) = \sum_{n=-\infty}^{+\infty} x(nT_S)\,\mathrm{sinc}\left(\frac{t - nT_S}{T_S}\right) \tag{18.29}$$

assuming, of course, that the Nyquist criterion $1/T \geq 2W$ is satisfied. Let $y(t)$ be a version of $x(t)$ that is delayed by τ; i.e., $y(t) = x(t - \tau)$. Then we have

$$y(t) = \sum_{n=-\infty}^{+\infty} x(nT_S)\,\mathrm{sinc}\left(\frac{t - \tau - nT_S}{T_S}\right) \tag{18.30}$$

We now sample $y(t)$ at rate $1/T_S$, yielding

$$y(kT_S) = \sum_{n=-\infty}^{+\infty} x(nT_S)\,\mathrm{sinc}\left(\frac{kT_S - \tau - nT_S}{T_S}\right) \tag{18.31}$$

switching to discrete-time notation ($y(kT_S) \rightarrow y_k$ and so on), we have

$$y_k = \sum_{n=-\infty}^{+\infty} x_n\,\mathrm{sinc}\left(k - n - \frac{\tau}{T_S}\right) = \sum_{n=-\infty}^{+\infty} x_n h_{k-n} \tag{18.32}$$

where the h_n are samples of the sinc function as indicated above. This is a filtering operation expressed in the form of digital convolution (compare with Equations (18.2) and (18.3) to see this). Since commutation is commutative, we may rewrite this as

$$y_k = \sum_{n=-\infty}^{+\infty} h_n x_{k-n} \tag{18.33}$$

where the h_n now function as filter coefficients. Since the h_n are samples from a sinc function, the infinite sum is well-approximated by the same sum truncated to a finite length around the maximum of the sinc function. This results in a FIR filter exactly of the form of Equation (18.4).

Thus, such a filter can be used as a continuously-variable delay operation, with the delay determined by the coefficients h_n, which in turn are samples of the delayed sinc function.

Of course the truncation of Equation (18.33) to obtain a FIR filter may lead to unacceptable changes to the response. We identified this problem and the possibility of an alternative constrained optimization approach in Section 18.2.3. For more information on allpass delay filters including additional design methods, a recommended starting point is [79].

18.5.3 Adaptive Equalization

Adaptive equalizers used in mitigation of time-resolved multipath in frequency-selective fading channels (Section 6.15) – in particular in MMSE and MLSE receiver architectures – are fundamentally FIR filters. In practical receivers, these filters typically do double- or triple-duty as timing synchronizers and decimators. (The Viterbi algorithm typically implemented in MLSE receivers is of course also implemented in DSP.)

18.5.4 Carrier Frequency Tracking

Carrier frequency tracking (Section 6.16) typically involves the use of digital filters. IIR filters are commonly used as part of the scheme for sensing the presence of a signal at a certain frequency; here the IIR architecture makes sense since there are typically no special passband requirements – only that the filter be narrow – and an IIR filter that is narrow but otherwise unconstrained can be implemented with much less computation than an FIR filter having comparable bandwidth. Digital filtering also appears in more sophisticated schemes such as band-edge tracking (Figure 6.35), which is a natural fit for a multirate filter implementation.

18.6 DSP HARDWARE TECHNOLOGIES

Finally it is appropriate to discuss hardware implementations of the DSP described throughout this chapter. Because the capabilities of DSP hardware are always rapidly evolving, comments will be limited to a few general observations that are most likely to be applicable for the foreseeable future.

18.6.1 CPUs, Their Limitations, and Alternatives

Bandwidths up to a few hundred kHz in complex baseband form can usually be accommodated using "general purpose" computing, which includes personal computers and lower-performance processors typically found in mobile devices. Such devices follow a *central processing unit* (CPU) paradigm, in which processing is defined in software and implemented as a sequence of mathematical operations performed one at a time. A class of these devices that commonly appears in radios employs dedicated DSP processors or coprocessors, which can be used to speed up certain types of operations, such as FIR filters and FFTs. The advantages of

CPU-based DSP in radios is software reconfigurability and ease of integration with peripheral functions such as source coding/decoding, networking, and physical interfaces.

The principal limitation of CPU-based DSP is the essentially serial execution of operations. Since modern digital modulations require thousands to hundreds of thousands of mathematical operations per output sample, the serial instruction-based processing paradigm of CPU architecture becomes an increasingly intractable bottleneck for bandwidths greater than a few MHz, even when the CPU's clock rate is on the order of a few GHz. Even at lower bandwidths where CPUs are easily able to keep up, the associated size, weight, power, and cost are typically unattractive compared to other alternatives (identified below) *unless* one of the advantages identified above is pretty important.

To do better, it becomes necessary to abandon sequential-instruction architecture in favor of schemes which allow many operations to be performed in parallel. For example: Rather than computing an order-M FIR filter as $M + 1$ multiplications one at a time, one might compute all $M + 1$ multiplications simultaneously. This requires M hardware multipliers; on the other hand, integer multipliers are very cheap, compact, and power-efficient compared to CPUs. This parallelization allows the clock rate to be on the order of the sample rate, as opposed to the CPU approach, which requires a clock rate on the order of the sample rate times the number of multiplications per sample, times the number of clock cycles required to perform a multiplication. Devices falling into this category include special-function integrated circuits (ICs), *field programmable gate arrays* (FPGAs), and *application specific integrated circuits* (ASICs).[9] Each of these is considered briefly below.

18.6.2 Special-Function ICs

A special function IC is precisely what the name implies: Digital processing implemented on an IC which has been designed to implement a specified function. Currently most wireless communications standards are available for implementation in this form. The advantages of using such devices are good size/weight/power characteristics, relatively low cost, and convenience because design effort is reduced to the problem of interfacing to the IC. The primary disadvantage of this approach is lack of flexibility: For example, a IEEE 802.11b special function IC typically cannot be reconfigured to use a newer variant of 802.11, or reconfigured to implement ZigBee (which shares the same frequency band), and so on.

18.6.3 FPGAs

An FPGA is essentially a reconfigurable special-function IC. FPGA ICs consist of large numbers of units (variously referred to as "cells" or "blocks") performing basic operations (addition, multiplication, memory, and so on) that may be combined in various ways to implement FIR filters, FFTs, and so on. The connections between these units is determined by

[9] And then there are *graphics processing units* (GPUs), which are essentially arrays of rudimentary CPUs. Although there is currently much experimentation with GPUs as custom high-performance DSP engines for radio applications, GPUs do not now appear in "mainstream" radio applications. But in the future? Maybe.

firmware downloaded to the FPGA after power-up; hence the term "field programmable." Thus the principal advantage of FPGAs is flexibility: Both in terms of implementing entirely new DSP, as well as in being able to redefine/reconfigure DSP without modifying the hardware, and perhaps even without cycling power. The disadvantage of FPGAs is that they cannot be optimized for size, weight, power, and cost to the degree that a special-function IC can, and a large fraction of the innate processing capability is inevitably left unused in any given application. Nevertheless the use of FPGAs in modern radios is steadily increasing, driven in part by demand for reconfigurability, and in part due to technical innovations that allow FPGAs to emulate and thereby replace CPUs (i.e., *softcore processing*). Present-day FPGAs routinely operate at clock rates of 100s of MHz and beyond, and – with certain caveats – are typically capable of achieving output sample rates equal to the clock rate.

18.6.4 ASICs

ASICs are essentially optimized hardware implementations of FPGA firmware. It is a fairly common occurrence that the processing implemented in FPGA form does not require the reconfigurability of an FPGA IC; but rather only the high degree of parallelism offered by FPGA architecture. In this case it is possible to synthesize a special-function IC that implements the design defined by the FPGA firmware, using hardware that is optimized for that particular design. This results in a device that is typically much less expensive than the original FPGA (presuming large-quantity production) and with superior size, weight, and power characteristics. Thus, ASICs combine the advantages of FPGAs and special-function ICs.

Problems

18.1 Following up Example 18.1: Recreate Figures 18.3 and 18.4 (sinc impulse response, rectangular window, $W = 3.125$ kHz) except now for (a) $M = 64$, (b) $M = 32$, (c) $M = 256$. In each case comment on the effect of increasing/decreasing filter order while holding all else constant.

18.2 Following up Example 18.1: Recreate Figures 18.5 and 18.6 (sinc impulse response, triangular window, $W = 3.75$ kHz) except now for (a) $M = 64$, (b) $M = 32$. In each case comment on the effect of increasing/decreasing filter order while holding all else constant. Specifically, indicate the response at the nominal band edge $f = 3.125$ kHz.

18.3 According to Equation (18.10), the Kaiser window for $A < 21$ has $\beta = 0$. Characterize this filter. What is special about $A \approx 21$, and why is that the threshold for $\beta = 0$?

18.4 A certain system has a sample rate of 1 MSPS. It is desired to implement a lowpass FIR filter with passband cutoff frequency 100 kHz and stopband start frequency 200 kHz. Assuming the Kaiser window method, make a plot of the required filter order as a function of the desired stopband attenuation in dB. Values of interest for the latter are up to 40 dB.

18.5 (a) Use the Kaiser window method to design the filter specified in Problem 18.4. Let the required passband ripple be less than 1 dB and the required stopband attenuation be 30 dB. Plot the impulse response. (b) Plot the associated frequency response in dB.

18.6 Write out expressions for the output of an FIR in the form of Equations (18.14) and (18.15) in the case of (a) $N = 7$, (b) $N = 6$.

18.7 In the first paragraph of Section 18.3.1 we propose reducing sample rate from 100 kSPS to 12.5 kSPS ($R = 8$), yielding $F'_S/W = 4$. (a) What specifically is the argument against increasing R to 16? (b) Why not increase R to some intermediate value between 8 and 16, such as 12?

18.8 It is desired to use the multiplier-free $F_S/4$ quadrature downconversion method with $R = 2$ to convert the largest possible bandwidth to complex baseband form using a sample rate of 1.6 GSPS. (a) Using the shift-right approach, what would be the center frequency of the analog passband and maximum bandwidth assuming oversampling by 20% (relative to the Nyquist rate) at the output? (b) Repeat (a) for the shift-left approach.

APPENDIX A

Empirical Modeling of Mean Path Loss

Models for mean path loss $\overline{L_p}$, including those which are empirically derived, can usually be expressed in a common functional form. Such models are especially important in the analysis and design of mobile radio systems in the VHF and UHF bands, where propagation effects lead to very complex dependence of $\overline{L_p}$ on distance, frequency, and antenna heights. In this appendix we derive this fundamental form, demonstrate that this form is consistent with the simple forms of propagation discussed in Chapter 3, and then present the popular Hata and COST231-Hata empirical path loss models, which are expressed in this form.

A.1 LOG-LINEAR MODEL FOR MEAN PATH LOSS

Let us begin with a brief review of two basic modes of propagation from Chapter 3. Free space path loss was expressed as

$$\overline{L_p} = L_{p0} \equiv \left(\frac{\lambda}{4\pi R}\right)^{-2} \tag{A.1}$$

where R is distance and λ is wavelength. (Since we are restricting our attention to frequencies below SHF, atmospheric losses and rain fade are assumed negligible.) Wavelength is equal to c/f, where c is the speed of light and f is frequency, so the above equation can be rewritten as

$$\overline{L_p} = \left(\frac{4\pi}{c}\right)^2 R^2 f^2 \quad \text{(free space)} \tag{A.2}$$

A comparable expression can be obtained for two-ray (flat Earth) path loss. We begin with Equation (3.37) for the power density:

$$S_{ave} \cong \frac{|V_0|^2}{\eta} \frac{16\pi^2}{\lambda^2} \frac{h_1^2 h_2^2}{R^4} \quad \text{for } R \gg h_1, h_2 \tag{A.3}$$

where h_1 and h_2 are the heights of the transmit and receive antennas above ground, V_0 is in RMS units and an isotropic antenna was assumed. The power captured by an antenna with receive gain G_R assuming an antenna with transmit gain G_T is

$$P_R \cong \frac{|V_0|^2}{\eta} \frac{16\pi^2}{\lambda^2} \frac{h_1^2 h_2^2}{R^4} \cdot G_T \cdot \left(\frac{\lambda^2}{4\pi} G_R\right) \tag{A.4}$$

where the factor in parentheses is the effective aperture of an antenna having gain G_T. Separately, the transmitted power P_T can be determined by integrating the power density in Equation (3.4) (adjusted for V_0 in rms units) over a sphere of radius r:

$$P_T = \int_{\theta=0}^{\pi}\int_{\phi=0}^{2\pi} \frac{|V_0|^2}{\eta r^2} \cdot r^2 \sin\theta \, d\theta \, d\phi = \frac{|V_0|^2}{\eta} 4\pi \tag{A.5}$$

Therefore

$$P_R \cong P_T \cdot \frac{4\pi}{\lambda^2} \frac{h_1^2 h_2^2}{R^4} \cdot G_T \cdot \left(\frac{\lambda^2}{4\pi} G_R\right) \tag{A.6}$$

Putting this in the form of Equation (3.5):

$$P_R \cong P_T G_T \left(\frac{h_1^2 h_2^2}{R^4}\right) G_R \tag{A.7}$$

Therefore the two-ray path loss may be written as

$$\overline{L_p} \cong \frac{R^4}{h_1^2 h_2^2} \quad \text{(two-ray)} \tag{A.8}$$

Note that in both cases, $\overline{L_p}$ is related to R through a "power law" $\overline{L_p} \propto R^n$, where $n = 2$ for free space propagation and $n = 4$ for two-ray propagation. Similarly, $\overline{L_p}$ is related to f through the power law $\overline{L_p} \propto f^m$, where $m = 2$ for free space propagation and $m = 0$ for two-ray propagation. Furthermore, we note that the dependence on antenna heights can be described as multiplicative factors through some function which may depend on the specific mode of propagation. This function is simply equal to 1 for free space propagation (i.e., no dependence on antenna height) and equal to $h_1^{-2} h_2^{-2}$ for two-ray propagation. Taken together, these considerations suggest that mean path loss might be generally described using the following functional form

$$\overline{L_p} = L_0 R^n f^m F_1(h_1) F_2(h_2) \tag{A.9}$$

where $F_1(h_1)$ and $F_2(h_2)$ are functions describing the dependence on h_1 and h_2, and L_0 is a constant of proportionality, absorbing any remaining constants and having whatever units are necessary to make the expression dimensionally correct (i.e., unitless).

Because mean path loss can span over many orders of magnitude, it is traditional to express $\overline{L_p}$ using expressions in logarithmic form; specifically, in decibels. Doing so results in a "log-linear" form which better facilitates empirical analysis and is often easier to interpret. Taking 10 times the base-10 logarithm of each side of Equation (A.9), we obtain

$$\overline{L_p} = L_0 + 10n \log_{10} R + 10m \log_{10} f + L_1(h_1) + L_2(h_2) \ \text{[dB]} \tag{A.10}$$

where we have made the definitions $L_1(h_1) \equiv 10 \log_{10} F_1(h_1)$, and $L_2(h_2) \equiv 10 \log_{10} F_2(h_2)$. In principle we may choose any units we like for R, f, h_1, and h_2, since the conversion factors necessary to obtain the correct (unitless) value for $\overline{L_p}$ can be absorbed into L_0. Traditionally, however, one chooses units of km for R, MHz for f, and m for h_1 and h_2. These choices are represented by adding subscripts to the symbols for R and f, as follows:

$$L_p = L_0 + 10n \log_{10} R_{\text{km}} + 10m \log_{10} f_{\text{MHz}} + L_1(h_1) + L_2(h_2) \ \text{[dB]} \tag{A.11}$$

where h_1 and h_2 are understood to be in meters.

EXAMPLE A.1

Express free space propagation in the form of Equation (A.11).

Solution: Let us begin by expressing R and f explicitly in km and MHz, respectively, in Equation (A.2):

$$\overline{L_p} = \left(\frac{4\pi}{3\times 10^8 \text{ m/s}}\right)^2 \left(\frac{R_{km}}{10^{-3}}\right)^2 \left(\frac{f_{MHz}}{10^{-6}}\right)^2 \cong 1750\, R_{km}^2 f_{MHz}^2 \tag{A.12}$$

Taking $10\log_{10}$ and expanding factors into terms, we obtain

$$\overline{L_p} = 32.4 + 20\log_{10} R_{km} + 20\log_{10} f_{MHz} \text{ [dB]} \tag{A.13}$$

Comparing to Equation (A.11), we find $L_0 = 32.4$, $n = 2$, $m = 2$, and $L(h_1) = L(h_2) = 0$.

EXAMPLE A.2

Express two-ray propagation in the form of Equation (A.11).

Solution: Beginning with Equation (A.8):

$$\overline{L_p} = \left(\frac{R_{km}}{10^{-3}}\right)^4 h_1^{-2} h_2^{-2} = 10^{12}\, R_{km}^4 h_1^{-2} h_2^{-2} \tag{A.14}$$

Taking $10\log_{10}$ and expanding factors into terms, we obtain

$$\overline{L_p} = 120 + 40\log_{10} R_{km} - 20\log_{10} h_1 - 20\log_{10} h_2 \text{ [dB]} \tag{A.15}$$

(h_1 and h_2 assumed to be in meters). Comparing to Equation (A.11), we find $L_0 = 120$, $n = 4$, $m = 0$, $L(h_1) = -20\log_{10} h_1$ and $L(h_2) = -20\log_{10} h_2$.

A.2 HATA MODEL

The form of Equation (A.11) lends itself readily to empirical modeling of mean path loss. The fundamental strategy in empirical modeling is to collect a large number of measurements of mean path loss as a function of R, f, h_1, and h_2, and then to determine values of n and m, and functions $L(h_1)$ and $L(h_2)$, that result in the best fit of Equation (A.11) to these data. There is no expectation that the resulting model will yield accurate estimates of $\overline{L_p}$ on a case-by-case basis; however, the model will typically be good enough on a statistical basis to permit informed design of specific radio links and to facilitate engineering analyses of radio links generally.

The Hata model provides parameters for Equation (A.11) based on an empirical analysis of measurements of VHF and UHF mobile radio path loss in Japan in the 1960s [22]. The Hata model is widely used in the design of mobile radio systems, and also serves an important role as a "standard" or "reference" model facilitating comparative analyses of radio technologies. The model is summarized in Table A.1.

Table A.1. **Summary of the Hata and COST231-Hata models for mean path loss in VHF/UHF mobile radio systems. The top section of the table indicates applicability of each model, whereas the bottom section gives the parameters to be used in Equation (A.11).**

Model	Hata	COST231-Hata
f	100–1500 MHz	1500–2000 MHz
R	1–20 km	Same as Hata
h_1	30–200 m	Same as Hata
h_2	1–10 m	Same as Hata
L_0	(Equations (A.16)–(A.18))	49.3 if "metropolitan"; 46.3 otherwise
n	4.49	Same as Hata
m	2.616	3.39
$L_1(h_1)$	(Equation (A.19))	Same as Hata
$L_2(h_2)$	(Equations (A.20)–(A.22))	Same as Hata Equation (A.20)

In the Hata model, the parameter L_0 depends on terrain classification. Categories of terrain are specified to be "urban," "suburban," and "open rural." The values of L_0 are:

$$L_0 = 69.55 \text{ , urban} \tag{A.16}$$

$$L_0 = 64.15 - 2\left[\log_{10}\left(\frac{f_{\text{MHz}}}{28}\right)\right]^2 \text{ , suburban} \tag{A.17}$$

$$L_0 = 28.61 - 4.78\left[\log_{10} f_{\text{MHz}}\right]^2 + 18.33\log_{10} f_{\text{MHz}} \text{ , open rural} \tag{A.18}$$

There is no particular definition that precisely identifies the difference between "urban," "suburban," and "open rural" environments; the user decides which form best suits the intended application.

The primary application of the Hata model is to mobile radio systems consisting of mobile stations that communicate with base stations. The parameter h_1 is taken to be the height of the base station antenna, which is assumed to be at least 30 m. The corresponding antenna height function for Equation (A.11) is:

$$L_1(h_1) = -\left(13.82 + 6.55\log_{10} R_{\text{km}}\right)\left(\log_{10} h_1\right) \tag{A.19}$$

In the original Hata model, the mobile antenna is assumed to be 1–10 m above ground, and the corresponding antenna height function is specified only for the "urban" case and for $100 < f_{\text{MHz}} < 200$ and $400 < f_{\text{MHz}} < 1500$. Within these limitations, Hata further distinguishes between various types of urban environments. The specified antenna height function for "small/medium cities" is:

$$L_2(h_2) = 1.56\log_{10} f_{\text{MHz}} - 0.8 + \left(0.7 - 1.1\log_{10} f_{\text{MHz}}\right) h_2 \tag{A.20}$$

The specified antenna height function for "large cities" and $f_{MHz} \geq 400$ is:

$$L_2(h_2) = 4.97 - 3.2\left[\log_{10}(11.75h_2)\right]^2 \tag{A.21}$$

The specified antenna height function for "large cities" and $f_{MHz} \leq 200$ is:

$$L_2(h_2) = 1.1 - 8.29\left[\log_{10}(1.54h_2)\right]^2 \tag{A.22}$$

Again, there is no definition that specifically identifies the difference between "small/medium cities" and "large cities"; the user decides which form best suits the intended application. A common (but nevertheless arbitrary) criterion used to make this distinction is mean building height: If greater than 15 m, the environment is considered to be "large city"; otherwise "small/medium city."

The original Hata specification for $L_2(h_2)$, as described above, leaves a few gaps. First, $L_2(h_2)$ is not specified for environments that are not considered "urban." Traditional work-arounds for this issue are to use the "small/medium city" height function, or simply to set $L_2(h_2)$ to zero. There is usually not much harm done in either case, since $L_2(h_2)$ is typically small compared to the other terms in Equation (A.11) for non-urban environments. Second, the original Hata model provides no guidance on how to choose $L_2(h_2)$ for $200 < f_{MHz} < 400$ in a "large city" environment. The traditional workaround is simply to use Equation (A.21) for $f_{MHz} < 300$ and Equation (A.22) for $f_{MHz} > 300$. Although these choices are somewhat arbitrary, the impact on practical problems is generally small, and in any case one must keep in mind that the Hata model is never expected to yield accurate "site specific" predictions anyway.

A.3 COST231-HATA MODEL

The COST231-Hata model emerged in the 1990s as an extension of the Hata model to support the development of mobile radio systems operating at frequencies up to 2 GHz [36]. The modifications relative to the original Hata model are minor, and are summarized in Table A.1. First, the dependence on frequency changes from $f^{2.616}$ to $f^{3.39}$; i.e., somewhat greater loss with increasing frequency. Second, L_0 is specified to be 46.3, increasing to 49.3 in environments deemed to be "metropolitan." Finally, the Hata "small/medium city" form for $L_2(h_2)$ (Equation (A.20)) is adopted.

A.4 OTHER MODELS

While the Hata and COST231-Hata models are quite popular in VHF/UHF mobile radio applications, various other models are also in common use. Many can be shown to be minor variants of the Hata model, or can otherwise be shown to be log-linear in the form of Equation (A.11). Models in this form also exist for *indoor* propagation in the UHF and SHF bands.

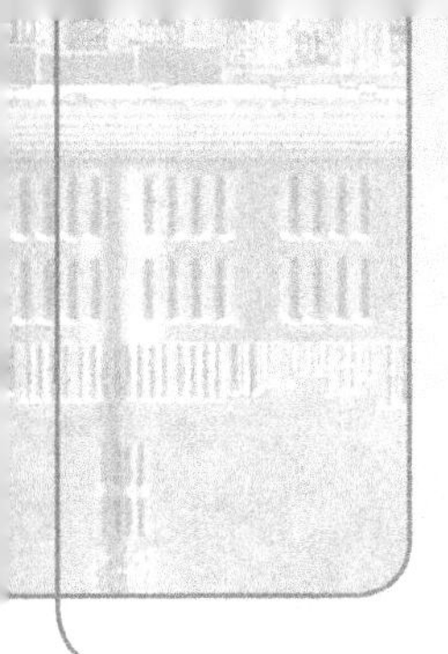

APPENDIX B
Characteristics of Some Common Radio Systems

This appendix summarizes the technical characteristics of some commonly encountered radio systems. This summary is by no means comprehensive – the number of systems that could be addressed is simply too large. Also the information provided here is necessarily at a high level, with only the principal or defining details provided. Nevertheless this information should be useful as quick/initial reference, and as a guide in relating concepts described in the numbered chapters of this book to the technologies employed in practical radio systems.

Here are a few things to keep in mind before beginning. First: Frequency bands, where given, are approximate. This is because there are typically variations between countries and/or between regions of the world. Second: In some cases additional information on the systems identified in this appendix can be found in the numbered chapters of the book. All such instances may not necessarily be pointed out, but the index of this book should be effective for locating the additional information. Of course for definitive technical information on these systems, the reader should consult the relevant standards documents.

B.1 BROADCASTING

Broadcasting is the delivery of audio, video, and/or data from a single transmitter to multiple receivers, primarily for purposes of entertainment and public information. Broadcasting has enormous historical significance as the original application of radio technology, and the legacy of technology developed for early broadcast radio is evident in modern radio systems of all types.

A few of the most common terrestrial broadcast systems now in common operation are identified in Table B.1. Some additional details follow.

It should be noted that modern broadcast FM is not simply FM-modulated audio. Modulated along with the audio are a number of *subcarriers* that permit *stereo* audio and a rudimentary data service. The subcarriers are essentially transparent to a minimal (monaural audio) receiver, but are easily recovered with a bit of additional signal processing.

Modern broadcast AM and FM transmissions often include a *digital overlay* signal. The overlay consists of OFDM-modulated signals that lie adjacent to existing channel boundaries; see Figure B.1 for an example. In the USA, the *In Band On Channel* (IBOC) system (commonly known by the brand name "HD Radio") is common. The digitally-modulated signal offers higher audio quality and additional content and data services for suitably-equipped receivers. The power in the digital overlay is limited to be between 1% and 10% of the power in

Table B.1. **Some common terrestrial broadcast systems.**

	Audio AM	Audio Shortwave	Audio FM	Television ATSC	Television DVB-T
Freq. Band	MF	HF	VHF	VHF–UHF	VHF–UHF
Channel Width	10 kHz	5 kHz	200 kHz	6 MHz	5–8 MHz
Modulation	DSB-LC	OFDM	WBFM	8VSB	OFDM
Typ. Max. ERP	50 kW	500 kW	100 kW	1 MW	varies
Comments	530–1600 kHz, digital overlay	DRM	88–108 MHz, digital overlay, subcarriers	North/ Central America & South Korea	Europe, Asia, Australia

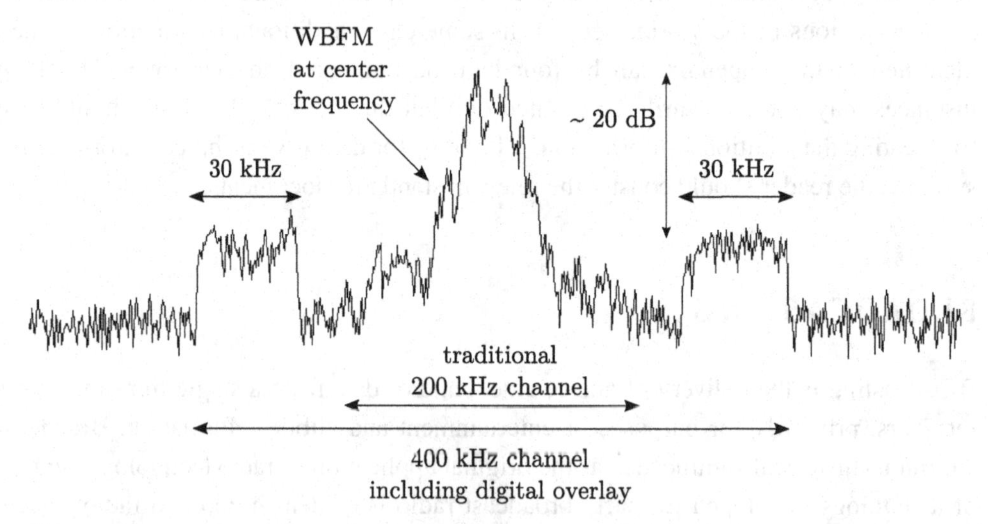

Figure B.1. Spectrum of an analog FM broadcast signal (WVTF 89.1 MHz in the USA) that transmits with an IBOC digital overlay. Actual measured spectrum, vertical scale in dB.

the primary WBFM signal. This, combined the fact that adjacent channels are typically not assigned in the same geographical region, mitigates the risk that the overlay interferes with an a WBFM transmitter in an adjacent 200 kHz channel.

Table B.1 identifies the digital television standards ATSC,[1] which is used in North America, Central America, and South Korea; and DVB-T, which is used in much of the rest of the world. Another important digital television system is ISDB-T, which is a distinct OFDM-based digital television standard used primarily in South America and Japan. These standards replace the

[1] See Section 6.17 for a lot more on ATSC.

now largely-deprecated analog standards NTSC, PAL, and SECAM – although many stations using these standards remain in operation in various places around the world.

A similar transition is underway in the "shortwave" (HF) broadcast bands, where Digital Radio Mondiale (DRM) is now prevalent, but many DSB-LC AM broadcast stations and a few SSB broadcast stations remain operational.

An important class of broadcast systems not included in Table B.1 is direct broadcast satellite (DBS), which is essentially broadcast television transmit from a satellite in geostationary orbit. DBS is typically implemented at frequencies in the SHF (predominantly Ku) band.[2]

B.2 LAND MOBILE RADIO

Land mobile radio (LMR) is a very broad class of radio systems intended for voice and data communications between mobile terrestrial users. Applications include public safety (police, fire, etc.), business operations (taxi dispatch, security, etc.), personal/recreational use, and amateur radio. LMR is distinct from cellular telecommunications (Section B.3) in that there is no "telecommunication" – that is, users communicate with other users either directly or through a repeater, and no network switching or routing is required. However, that distinction is likely to become ambiguous in the future as some LMR systems migrate to technologies originally intended for telecommunications – again, see Section B.3 for more on this.

A few of the most common LMR systems now in common operation appear in Table B.2. As in broadcasting, a transition to digital systems is underway, but is happening slowly. Analog FM systems remain very common, and often the technical emphasis is on reducing bandwidth (e.g., US efforts to transition to 6.25 kHz channels) as opposed to implementing new digital systems.

P25 and TETRA are the digital standards most commonly employed in the USA and in Europe, respectively. P25 uses a version of $M = 4$ FSK (4FSK) known as *continuous 4-level FM* (C4FM), where the word "continuous" refers to the pulse shaping used to optimize the spectral efficiency. TETRA uses $\frac{\pi}{4}$-DQPSK, which yields slightly better spectral efficiency but requires better (more linear) power amplifiers. TETRA channels are wider, but TETRA employs an explicit TDMA scheme with four users per channel; thus the spectral efficiency is similar to P25.

Many LMR systems employ *trunking*, which is a multiple access scheme in which frequency channels are used in groups that are dynamically assigned, as opposed to being used individually. This improves overall capacity by distributing traffic more uniformly over the available frequencies.

Typical transmit power for Analog FM, P25, and TETRA is on the order of 1–10 W for handheld units and 10–100 W for fixed/vehicular units and repeaters. Typical maximum range is on the order of kilometers, although systems are often interconnected using repeaters with high antennas to provide coverage over a much larger area.

[2] See Section 7.10 for more about DBS.

Table B.2. **Some common LMR systems.**

	Analog FM	P25	TETRA	CB	FRS
Freq. Band	VHF–UHF	VHF–UHF	UHF	HF (~ 27 MHz)	UHF (~ 460 MHz)
Channel Width	5–25 kHz	6.25 or 12.5 kHz	25 kHz	10 kHz	12.5 kHz
Modulation	NBFM	4FSK (C4FM)	$\frac{\pi}{4}$- DQPSK	DSB-LC AM	NBFM
Comments		USA	Europe, TDMA	USA	USA

The US Citizens Band (CB) and Family Radio Service (FRS) are included in Table B.2 as examples of unlicensed LMR systems used primarily for personal and recreational communications. Maximum transmit power for these systems is 4 W and 500 mW, respectively. This yields range on the order of kilometers using low-gain antennas close to the ground over level terrain, although CB signals intermittently reach much greater distances when ionospheric conditions permit. Systems that are similar, although perhaps with slight differences in technical parameters, exist throughout the world.

B.3 MOBILE TELECOMMUNICATIONS

In mobile telecommunications systems, users are connected to the network by radio, as opposed to wires. Mobile telecommunications combined with wireless data networks (the topic of the next section) account for the majority of contemporary economic activity – and, subsequently, new technology development – in the field of radio.

B.3.1 General Characteristics

Whereas the emphasis in LMR is "one-to-many" communication, the emphasis in mobile telecommunications is "one-to-one" communications between users, facilitated by network switching with user-specific addressing; i.e., phone numbers and IP addresses.[3] Thus, mobile telecommunications systems are nominally seamless extensions of the worldwide

[3] This is not always so clear-cut: Modern mobile telecommunications systems may also provide a "push-to-talk" (PTT) capability that emulates LMR, and some modern LMR systems use technologies that would normally be considered telecommunications systems, such as LTE.

telecommunications system. Nearly all mobile telecommunications systems employ cellular architecture (reminder: Figure 1.6 and associated text) in some form. This category is not limited to terrestrial systems, since we can also count satellite-based systems such as Iridium and Inmarsat as mobile telecommunications systems.

Existing terrestrial mobile telecommunications exist primarily in the UHF band, with most systems implemented between 700 MHz and 1 GHz, and within a few hundred MHz of 2 GHz. Certain segments of the latter are called the *Personal Communications Services* (PCS) and *Advanced Wireless Services* (AWS) bands.[4] The reasons why mobile cellular telecommunications ended up in these bands are primarily historical. Although these terrestrial mobile telecommunications systems could potentially perform much better in the VHF and lower UHF regime (where path loss is much lower), these frequencies had been in use for television broadcasting for decades before the first modern cellular systems were were implemented in the 1980s, and remaining spectrum, primarily in the VHF band, was generally too narrow. Similarly, legacy applications precluded implementation of cellular systems between 1 GHz and about 1.8 GHz. With a few exceptions, satellite-based mobile telecommunications systems operate primarily in narrow bands between 1.4 GHz and 1.7 GHz, which is shared with a variety of other satellite systems that are considered to be compatible from an interference perspective.

What follows is a brief summary of terrestrial cellular systems, organized historically. It is quite difficult – and far beyond the scope of this book – to summarize all potentially-relevant aspects of modern cellular systems. Such a summary would probably have limited utility anyway, given the rapid evolution of these systems in terms of both technology development and deployment. Additional current information on these systems is probably best obtained from journals – *IEEE Communications Magazine* is recommended in particular – and from the official documents that define the standards.

B.3.2 First-, Second-, and Third-Generation Cellular Systems

It is common to classify terrestrial cellular telecommunications by "generation." First-generation systems emerged in the 1980s, employed primarily analog modulations (NBFM in particular), and are now essentially extinct. Second-generation ("2G") technologies emerged in the mid 1990s and employed primarily low-order digital sinusoidal carrier modulations. Notable among the 2G technologies are GSM and IS-95. GSM uses GMSK with eight users per carrier in a TDMA scheme to manage access to 200 kHz channels. IS-95 uses QPSK with DSSS, allowing a variable number of users to share a 1.25 MHz channel via CDMA.

Although operational 2G systems are becoming rare, the associated technologies gave rise to a panoply of enhanced and hybridized versions that are known collectively as third-generation ("3G") systems, in which the primary emphasis is increased rates for data applications. The example in Section 7.8 is representative of the downlink (base station to mobile user link) in a PCS-band 3G cellular system, demonstrating how downlink rates between a few hundred kb/s to a few Mb/s are achieved.

[4] The names do not infer any particular features aside from identifying specific frequency ranges.

B.3.3 Fourth-Generation Cellular Systems ("4G") and LTE

Presently, 3G systems are nearly ubiquitous but are rapidly being overtaken by fourth-generation ("4G") systems. The hallmarks of 4G technologies are OFDM, MIMO, carrier aggregation,[5] and a transition from circuit-switched networking (a legacy of the early emphasis on voice communications) to IP-based networking (to facilitate easier integration with data).

Prominent among 4G technologies is *Long Term Evolution* (LTE).[6] LTE itself is not really a single standard, but rather a broad family of standards that continues to evolve and is being adapted to a variety of applications that have traditionally been outside the domain of cellular telecommunications – this includes LMR (especially in the public safety arena) and wireless data networking (the topic of the next section).

The emerging primacy of LTE makes it appropriate to identify a few of the defining technical characteristics. The LTE downlink uses OFDM with QPSK, 16QAM, or 64QAM. The atomic unit of time-frequency utilization in LTE is the *resource block*, which consists of seven contiguous symbols $\times$ 12 contiguous subcarriers. Data are assigned to resource blocks on a dynamic basis, which facilitates relatively efficient utilization of available spectrum. In this sense LTE uses TDMA and FDMA simultaneously. This offers considerable advantages over earlier technologies which are primarily FDMA (albeit perhaps with TDMA within frequency channels), since in traditional FDMA systems it is possible for capacity to be limited because some frequency channels can remain idle while others are running at maximum capacity.

The symbol period T for the OFDM modulation in LTE is about 66.7 μs; therefore the subcarrier spacing is $1/T \cong 15$ kHz and subsequently a resource block is about 0.5 ms $\times$ 180 kHz. In the time domain, consecutive pairs of resource blocks form 1 ms *subframes*, and 10 subframes are organized into 10 ms *frames*. In the frequency domain, contiguous groups of resource blocks are combined to form channels having widths between 1.25 MHz and 20 MHz. In principle, downlink data rates up to about 100 Mb/s can be achieved in ideal conditions.

For the uplink (mobile-to-base station link), LTE uses single-carrier QPSK or 16QAM with FDMA, commonly referred to as "SC-FDMA", as opposed to OFDM. This avoids the problem of implementing efficient power amplifiers with sufficient linearity to accommodate the high PAPR of OFDM. The maximum possible uplink data rate is about half the maximum downlink rate, which is appropriate since data rate requirements in telecommunications systems tend to be highly asymmetrical in this way.

B.3.4 Fifth-Generation Cellular Systems ("5G")

What is next? The answer is, of course, fifth-generation ("5G") technology: 5G standards are being developed as this book is being written, but features will certainly emphasize increased data rates and continued integration of voice and data applications of all types across a variety of physical-layer standards. Significant increases in data rates require additional spectrum

[5] Reminder: See the last paragraphs of Section 7.8.

[6] Technically, early versions of LTE are not considered 4G because they do not meet some of the performance specifications that are considered to be required for 4G. Later versions *are* considered 4G.

which is simply unavailable in existing cellular telecommunications bands; therefore a new emphasis in 5G will be operation at frequencies in the SHF (3–30 GHz) and EHF (primarily 30–80 GHz) bands, where wider bandwidths can be accommodated. It is also likely that 5G will accommodate emerging *internet of things* (IoT) and *machine-to-machine* (M2M; alternatively known as *device-to-device* (D2D)) communications applications.

B.4 WIRELESS DATA NETWORKS

Wireless data networks connect computers to a data network by radio, as opposed to wires or cables. Such networks are referred to as wireless local area networks (WLANs), which are indoor or short-range outdoor networks; or wireless metropolitan area networks (WMANs), which implies longer links, perhaps spanning a few kilometers. As noted in the previous section, wireless data networks along with mobile telecommunications account for the majority of contemporary economic activity and new technology development in the field of radio.

B.4.1 IEEE 802.11 and 802.11b

In WLAN applications, the IEEE 802.11 family of standards currently dominate.[7] The earliest realization of this standard, known simply as "802.11" (alternatively as "802.11-1997" or "802.11 Legacy"), achieved data rates up to about 2 Mb/s using single-carrier DSSS or FHSS approaches in the 2.4 GHz unlicensed band. This standard became popular in the late 1990s but was quickly superseded by 802.11b.

IEEE 802.11b, despite itself being superseded by multiple generations of standards (summarized below), remains ubiquitous. It is so popular that its marketing name – "Wi-Fi" – has become a synonym for WLAN in general. IEEE 802.11b uses a single-carrier modulation scheme known as *complementary code keying* (CCK) to transmit data at rates of 5.5 Mb/s or 11 Mb/s in channels up to 22 MHz wide in the 2.4 GHz unlicensed band.[8] In CCK, groups of 4 bits (for 5.5 Mb/s) or 8 bits (for 11 Mb/s) are mapped to a sequence of 8 QPSK symbols; this mapping is essentially a form of error control coding. Transmit power is up to 100 mW, which yields a usable range of about 20 m indoors and about 100 m outdoors assuming low-gain antennas.

Data rates perceived by the user are roughly half the theoretical maximum data rates, attributable in part to the multiple access scheme used in 802.11b. This scheme is known as *carrier sense multiple access with collision avoidance* (CSMA/CA). In this scheme, a radio may transmit if it is unable to detect a transmission already underway in the same channel. If the transmitting radio does not receive an acknowledgement from the receiving radio – perhaps due to a "collision" with another radio that began transmitting at the same time – then the

[7] A pretty good summary of the IEEE 802.11 family of standards through 2010 (so, including 802.11b, -a, -g, and -n), containing many more details than are conveyed here, is [25].

[8] The exact frequencies vary from country to country, but generally tens of MHz somewhere just above 2.4 GHz.

transmitter retransmits the affected data.[9] Although there is considerable inefficiency associated with this unscheduled form of channel access, this scheme is simpler to implement than a rigid TDMA system and permits higher throughput when the number of users per channel is low.

B.4.2 IEEE 802.11a, -g, and -n

By the early 2000s, advancements in technology permitted greatly improved data rates, leading to the widespread adoption of the standards 802.11a and 802.11g. IEEE 802.11a uses OFDM to send rates up to 54 Mb/s in channels of about 20 MHz in the 5 GHz unlicensed band.[10] The primary benefit of 5 GHz is increased availability of spectrum, allowing more RF channels and thus more simultaneous users. IEEE 802.11g is essentially 802.11a implemented in the 2.4 GHz band, providing a higher-rate alternative to 802.11b. Like 802.11b, transmit power for 802.11a and 802.11g is about 100 mW. IEEE 802.11a/g yields roughly twice the range of 802.11b indoors, and improvement of tens of percent outdoors. Since the mid 2000s it has been practical and common for Wi-Fi radios to support all three standards: i.e., 802.11b, -a, and -g.

The principal barrier to further increases in data rates was lack of available spectrum in the unlicensed 2.4 GHz and 5.8 GHz bands. This led to the IEEE 802.11n standard in 2009, which incorporates MIMO as well as a number of other technical refinements that facilitate rates up to 600 Mb/s in 40 MHz channels in the 2.4 GHz and 5 GHz bands. Like 802.11a/b/g, 802.11n uses about 100 mW transmit power, but achieves about double the range; nominally up to about 70 m indoors and 250 m outdoors.

B.4.3 IEEE 802.11ac and -ad

The two major emerging WLAN standards as this book is being written are IEEE 802.11ac and IEEE 802.11ad, which continue the OFDM+MIMO strategy but with significantly increased rates owing to combinations of wider channels, higher-order modulations, beamforming, and increased number of MIMO streams. IEEE 802.11ac – also currently referred to *Very High Throughput* (VHT) – is designed for 5 GHz (only) and uses subcarrier modulations up to 256-QAM and channel bandwidths up to 160 MHz to achieve data rates up to a few Gb/s.

IEEE 802.11ad – also currently referred to as "WiGig" – is designed for the 60 GHz unlicensed band,[11] and is expected to consistently achieve rates above 2 Gb/s using 2.16 GHz-wide channels. In contrast to all previous standards, 802.11ad maximum transmit power will be about 10 mW (as opposed to about 100 mW), owing to the greater difficulty of efficient power amplifier design in this frequency regime. In addition, there is a factor of about 22 dB additional path loss associated with the increase in frequency relative to 5 GHz. The outcome is that range is limited to about 10 m in line-of-site conditions, and poor penetration into adjacent rooms in indoor deployments. On the other hand, any attempt to increase range significantly beyond tens of meters would be complicated by oxygen absorption in this band (reminder: Section 3.7.3),

[9] The "CA" in "CSMA/CA" refers to an feature in the protocol that allows a transmitter to reserve time for transmission; however, this feature is rarely implemented. Thus 802.11 networks are typically just "CSMA."

[10] Varies from country to country, but generally hundreds of MHz between 5.1 GHz and 5.8 GHz.

[11] The exact frequencies vary from country to country, but generally several GHz in 57–66 GHz.

which becomes dominant over free-space path loss for ranges greater than about 50 m. In fact, this last issue is considered a possible strength of IEEE 802.11ad from the perspective that interference between nearby systems (e.g., two systems located at opposite ends of a building, or in adjacent buildings) is reduced by a much greater extent than would be possible by relying on free-space path loss alone.

B.4.4 Longer-Range Systems: IEEE 802.16 (WiMAX) and 802.11af (TVWS)

Although there are many more wireless data network technologies that are currently in use or are currently emerging, we now limit ourselves to just two more: IEEE 802.16 and 802.11af. The IEEE 802.16 family of standards – also known by the marketing title "WiMAX" – is properly classified as a WMAN technology. The principal distinction between WiMAX and the standards mentioned above is the intended use for "metropolitan area" outdoor data networks, delivering data rates up to tens of Mb/s over ranges up to several kilometers in various frequency bands in the SHF regime.[12] As this book is being written, the future of WiMAX is unclear; LTE seems to be emerging as a more popular alternative.

IEEE 802.11af facilitates 802.11-type data networking in unused broadcast television channels below 1 GHz. This is a response to two issues: Limited spectrum in the 2.4 GHz and 5 GHz unlicensed bands, and desire for longer range facilitated by the reduced path loss of lower frequencies. Data rates of tens to hundreds of Mb/s at ranges up to 1 km or so are anticipated using single carrier and OFDM modulations in unused 6 MHz-wide television channels. IEEE 802.11af is also known by a number of other names, including *Television White Spaces* (TVWS),[13] *White-Fi*, and *Super Wi-Fi*. The future of IEEE 802.11af is presently a bit murky, as it depends on co-existence with broadcast television – i.e., deciding what constitutes an "unused" television channel – and the nominal data rates are relatively low.

B.4.5 Future Trends

Finally, one should note the potential for 4G and 5G mobile telecommunications systems to partially subsume WLAN and WMAN systems in the future. In a large number of applications, emerging 4G systems are already suitable to serve as wireless data networks. However, since the spectrum available to mobile telecommunications systems will remain relatively small, it currently seems likely that distinct WLAN/WMAN systems will continue to be required to accommodate applications requiring the highest data rates.

B.5 SHORT-RANGE DATA COMMUNICATIONS

A distinct class of wireless data communications systems comprises systems that are optimized for relatively low data rates over short range, and for which WLAN technology is

[12] For a reminder about why it is so hard to deliver greater data rates at these distances, see the example in Section 7.8.

[13] It should be noted that "TVWS" really refers to *all* schemes (not just 802.11af) that use fallow television channels.

typically overkill. Examples described below are Bluetooth, ZigBee, and the simplex systems used for remote keyless entry and tire pressure monitoring in automobiles. The purpose of these systems is to move relatively small amounts of data in scenarios in which a wired connection would be more expensive or inconvenient, and at the lowest possible power and device cost.

B.5.1 Bluetooth

Bluetooth is a scheme for point-to-point networking in support of devices such as wireless computer peripherals, wireless microphones and headphones; and for connections between smart phones, computers, and entertainment devices. Such systems are sometimes referred to as personal area networks (PANs). Bluetooth operates in the 2.4 GHz unlicensed band, using FHSS multiple access at 1600 frequency hops per second. Data rates up to hundreds of kb/s over ranges up to a few meters are achieved using channels of 1 MHz or 2 MHz with transmit power in the range 1–100 mW. Bluetooth currently employs a variety of modulations including BFSK with Gaussian pulse shaping (commonly referred to as GFSK), $\frac{\pi}{4}$-DQPSK, and 8-DPSK. The latest versions of the standard support data rates up to 3 Mb/s. Bluetooth was originally conceived as IEEE 802.15.1, but the standard is now managed by an industry consortium.

B.5.2 ZigBee

ZigBee – also known as IEEE 802.15.4 – serves a role similar to Bluetooth but is more commonly found in industrial applications, such as networks interconnecting large numbers of sensors. ZigBee is specified to operate in a number of unlicensed UHF bands including 2.4 GHz. Data rates up to 250 kb/s over ranges up to tens of meters are achieved using DSSS OQPSK modulation in 5 MHz channels with transmit power in the range 1–100 mW. A notable feature of ZigBee is *ad hoc* mesh networking, which allows ZigBee radios that would normally be out of range to communicate by relaying data through other ZigBee radios.

B.5.3 Automotive Applications: RKE and TPMS

Remote keyless entry (RKE) is the system that allows one to unlock an automobile from a distance, usually by pressing a button on a "fob" attached to one's key ring. There is no single standard for RKE; however, most systems operate in unlicensed spectrum in the lower UHF range (433 MHz is common in the USA), typically sending bursts of a few kb/s using either binary ASK or binary FSK at a level of a few milliwatts. RKE is strictly simplex.

A *tire pressure monitoring system* (TPMS) is the simplex system that conveys tire pressure data from a transmitter located in the valve stem of a tire to a receiver located elsewhere in the vehicle. In all other respects, TPMS closely resembles RKE: Both use the simplest possible binary digital modulations in unlicensed spectrum below 1 GHz to send bursts of very low rate data.

The primary considerations in both RKE and TPMS are low power (for long battery life), rugged implementation, and low cost. For RKE there is, of course, the additional matter of security, which requires that RKE data be well-encrypted.

B.6 RADIO FREQUENCY IDENTIFICATION (RFID)

RFID encompasses a broad class of technologies for determining the identity and (sometimes) the status of objects over ranges of tens to hundreds of meters. The "killer app" for RFID is inventory control; i.e., tracking the location of large numbers of items such as products in retail stores, medical instruments and pharmacy items in hospitals, and so on. The fundamental elements of an RFID system are *tags*, which are attached to the objects of interest; and *readers*, whose locations are known and are able to wirelessly communicate with the tags. The communication is typically initiated by the reader, which broadcasts a radio signal that all tags within range should be able to detect. This is referred to as *interrogation*. In response, the tags transmit a signal which includes information that allows the reader to uniquely identify the tag. Detection of a tag's signal indicates that the associated object must within range of the reader.

RFID systems can be broadly categorized as *active*, *passive*, or *semi-passive*. In an active RFID system, the tags are battery-powered and the system operates more or less as a traditional half-duplex radio communications system. Active RFID systems typically operate in the lower UHF band (e.g., unlicensed spectrum around 430 MHz), which is a tradeoff that accommodates the relatively small size of the tags, which can accommodate only small antennas. The principal drawbacks of active RFID systems pertain to the tags, which are too bulky and expensive for tracking large numbers of small objects. Therefore active RFID is used mainly for tracking relatively small numbers of large or valuable objects; e.g., shipping containers.

In a passive RFID system, the tag is powered by the received signal. When a signal is received, the power in the signal is converted to DC, used to power-up the electronics (typically a small IC), which in turn controls the response. Because the available power is extremely limited, passive tags typically do not actually transmit a response signal. Instead, the tag's antenna is connected to a varying impedance under the control of the tag's electronics. The tag's antenna re-radiates the incident power with a magnitude and phase that is determined by the connected impedance. A common scheme employs two states: One state in which the antenna is severely mismatched, resulting in most of the incident signal being re-radiated; and another state in which the antenna is well-matched, resulting in relatively low re-radiation. Thus the response signal received by the reader is the "backscatter" from the tag, with the information from the tag represented as variation in the magnitude and/or phase of this backscatter. This strategy offers the huge advantage that passive RFID tags can be made very inexpensively and with nearly-negligible bulk. For example, Figure B.2 shows a passive RFID tag designed for the operation in the 900 MHz unlicensed band.[14] The tag is small, flat, and flexible so that it is easily attached to curved surfaces.

The backscatter strategy employed by passive RFID systems imposes severe constraints on the modulation. The 900 MHz band systems now in common use employ a form of binary ASK known as *pulse interval encoding* (PIE). In this scheme, each bit period begins with an interval during which the carrier is always on, and the difference between a "0" and a "1" is represented by the total fraction of the bit period during which the carrier remains on. This scheme ensures that the tag receives enough power to remain energized while responding. The backscattered

[14] Exact frequencies vary between countries.

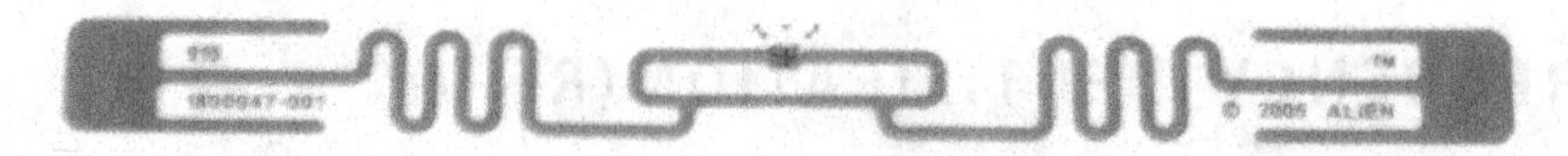

Figure B.2. A tag used in a 900 MHz-band passive RFID system. This tag is about 10 cm × 1 cm. The IC is the small rectangular device located near the center of the tag. The antenna is essentially a resonant dipole that has been compressed in length using a combination of "meandering" and end loading.

signal from tag to reader is typically modulated using one of two schemes: "FM0" or *Miller modulation*. In these schemes the bits are represented by the number of state changes during a bit period. In FM0, for example, zero changes in a bit period indicates a "1" whereas one change in a bit period indicates a "0".

Data rates in passive RFID are typically in the range of tens to hundreds of kb/s. The incentive for high data rates is to reduce the length of transmissions, so as to minimize contention for spectrum. Range for passive systems operating in the 900 MHz band is typically tens of meters with reader transmit levels of a few watts. Also common are HF-band (13.56 MHz in the USA) passive RFID systems, which have much shorter range but also have advantages in some scenarios where UHF-band propagation is problematic; this includes in particular tags affixed to containers of liquid. Active and passive RFID systems also appear in bands around 150 kHz, 2.4 GHz, and in the 3–10 GHz range; each having various advantages and disadvantages.

Semi-passive RFID tags are passive RFID tags that also include a battery, which is used to implement functions that must be maintained even when the tag is not within range of a reader, or for which the power provided by the reader's signal is inadequate. Typical applications include tags used as sensors to monitor temperature or other environmental conditions.

B.7 GLOBAL NAVIGATION SATELLITE SYSTEMS (GNSS)

A GNSS is a system of satellite-based transmitters that can be used to determine the location of a receiver with high accuracy. As this book is being written, two such systems – GPS and GLONASS – are fully operational; see below for more information on these. At least two other systems are semi-operational and still in development: These are European Union's *Galileo* system and China's *Beidou* system.

All four systems rely on essentially the same principle, known as *time of arrival* (TOA) ranging. The idea is that each satellite contains an accurate clock and encodes the current time in its transmission. Upon the receipt of this signal, the receiver is able to use this information to determine the travel time from the satellite to the receiver, and subsequently the distance between the satellite and the receiver. The receiver also knows the location of the satellite at the time of transmission, which constrains the possible locations of the receiver to the surface of a sphere surrounding the satellite. Repeating this process with at least three other satellites results in at least three such spheres whose surfaces intersect at one location; i.e., that of the receiver. In practice there are many issues that contribute uncertainty in this process; however, modern commercial GNSS receivers routinely deliver location accuracy on the order of just tens of meters or better.

The US Global Positioning System (GPS) consists of about 30 satellites, such that eight or so are always within view of a receiver on the surface of the Earth. The primary GPS frequency band is centered at 1575.42 MHz, also known as the "L1" frequency. Right-circular polarization is used. The primary signal transmitted on the GPS L1 frequency is known as the *coarse/acquisition* (C/A) signal. The C/A signal is a DSSS BPSK signal which is transmitted from each satellite using a unique spreading code; thus GPS uses CDMA to allow all satellites to share the same spectrum (see Example 6.9). The data conveyed in the C/A signal are a 50 b/s "navigation message" that contains the information required by the receiver to determine absolute time and the locations of the satellites. The 50 b/s data are spread using a 1023-chip sequence at 1.023 million chips per second; thus 20 460 chips per bit. This facilitates TOA estimation to an accuracy of no worse than one chip period, which is about 1 μs, corresponding to a distance of about 300 m. In practice the realizable accuracy is typically a small fraction of a chip period, corresponding to ranging errors of just tens of meters. GPS satellites transmit a number of additional signals, including signals with higher chip rates to facilitate higher TOA estimation accuracy. The L1 frequency in particular contains not only the C/A signal, but a number of additional CDMA signals occupying a total bandwidth of about 24 MHz.

The Russian Federation's GLONASS system uses a similar number of satellites and a similar set of frequencies. Restricting attention to the C/A signal, the principle difference is that the GLONASS C/A signal is spread at 511 thousand chips per second, resulting in a bandwidth of about half the GPS C/A signal bandwidth. Also, each satellite transmits on a unique center frequency, and the center frequencies are equally spaced by 562.5 kHz beginning at 1602 MHz. Thus GLONASS is an FDMA system: The DSSS modulation is used only for ranging, and is not needed for multiple access.

B.8 RADAR, REMOTE SENSING, AND RADIO ASTRONOMY

Radar is the sensing, ranging, and characterization of objects by reflection or scattering of radio signals. Applications of radar are well known; it suffices here to point out that radars are radio systems that have a great deal in common with communications and navigation systems. The principal distinction is that a radar is designed to detect its own transmissions, as opposed to those of other transmitters.[15]

Radio systems also find applications in geophysical remote sensing and astronomy. In these applications, radio is used to measure the intensity, polarization, spectrum, or waveform of naturally-occurring radio emission, with the goal of learning something about the source or the intervening medium. The principles of antennas and receivers described in this book apply equally-well to these applications. An important consideration is that most sources of natural radio emission are extraordinarily weak in comparison to anthropogenic radio emission, and useful frequencies span the radio spectrum. This leads to considerable difficulty in managing the use of the radio spectrum, since frequencies that are useful for remote sensing and radio astronomy overlap frequencies that are allocated to communications, navigation, and radar systems. A recommended introduction to remote sensing and radio astronomy from the perspective of spectrum management and coexistence is [53].

[15] Although not even this is quite true: In *multistatic radar*, the transmitter and receiver are in separate locations.

REFERENCES

[1] Advanced Television Systems Committee, *ATSC Digital Television Standard – Part 2: RF/Transmission System Characteristics*, Doc. A/53 Part 2:2011, 15 December 2011. http://www.atsc.org/cms/index.php/standards/standards/50-atsc-a53-standard.

[2] D.M. Akos, M. Stockmaster, J.B.Y. Tsui & J. Caschera, "Direct Bandpass Sampling of Multiple Distinct RF Signals," *IEEE Trans. Comm.*, Vol. **47**, No. 7, July 1999, pp. 983–988.

[3] Amateur Radio Relay League (ARRL), *2014: The ARRL Handbook for Radio Communications*, 91st Edn, 2013. *Note this reference is updated annually; section and figure numbering is subject to change.*

[4] Analog Devices Inc., "Wideband IF Receiver Subsystem AD6676" (datasheet), Rev. 0, 2014.

[5] H.L. Bertoni, *Radio Propagation for Modern Wireless Systems*, Prentice Hall PTR, 2000.

[6] L. Besser & R. Gilmore, *Practical RF Circuit Design for Modern Wireless Systems: Vol. 1: Passive Circuits and Systems*, Artech House, 2003.

[7] D.E. Bockelman & W.R. Eisenstadt, "Combined Differential and Common-Mode Scattering Parameters: Theory and Simulation," *IEEE Trans. Microwave Theory & Techniques*, Vol. **43**, No. 7, July 1995, pp. 1530–1539.

[8] C.M. Butler, "The Equivalent Radius of a Narrow Conducting Strip," *IEEE Trans. Ant. & Prop.*, Vol. **30**, Jul. 1982, pp. 755–758.

[9] R.A. Burberry, *VHF and UHF Antennas*, Peter Peregrinus Ltd., 1992.

[10] H.M. Cheema & A. Shamim, "The Last Barrier: On-Chip Antennas," *IEEE Microwave Magazine*, Vol. **14**, No. 1, Jan./Feb. 2013, pp. 79–91.

[11] C. Cho *et al.*, "IIP3 Estimation From the Gain Compression Curve," *IEEE Trans. Microwave Theory & Techniques*, Vol. **53**, No. 4, Apr. 2005, pp. 1197–1202.

[12] D.C. Cox & D.P. Leck, "Distributions of Multipath Delay Spread and Average Excess Delay for 910-MHz Urban Mobile Radio Paths," *IEEE Trans. Ant. & Prop.*, Vol. AP-23, No. **2**, Mar. 1975, pp. 206–213.

[13] S. Cripps, "The Intercept Point Deception," *IEEE Microwave Mag.*, Feb. 2007, pp. 44–50.

[14] K. Davies, *Ionospheric Radio*, Peter Peregrinus, 1990.

[15] D.M.J. Devasirvatham, "Radio Propagation Studies in a Small City for Universal Portable Communications," *Proc. IEEE Vehicular Tech. Conf.*, Philadelphia, June 1988, pp. 100–104.

[16] C. Drentea, *Modern Communications Receiver Design & Technology*, Artech House, 2010.

[17] M.L. Edwards & J.H. Sinsky, "A New Criterion for Linear 2-Port Stability Using a Single Geometrically Derived Parameter," *IEEE Trans. Microwave Theory & Techniques*, Vol. **40**, No. 12, December 1992, pp. 2303–2311.

[18] R.M. Fano, "Theoretical Limitations on the Broadband Matching of Arbitrary Impedances," *J. Franklin Inst.*, vol. **249**, Jan.–Feb. 1950.

[19] R. Gilmore & L. Besser, *Practical RF Circuit Design for Modern Wireless Systems, Vol. II: Active Circuits and Systems*, Artech House, 2003.

[20] G. Gonzalez, *Microwave Transistor Amplifiers: Analysis & Design*, 2nd Edn, Prentice-Hall, 1997.

[21] G. Gonzalez, *Foundations of Oscillator Circuit Design*, Artech House, 2007.

[22] M. Hata, "Empirical Formula for Propagation Loss in Land Mobile Radio Services," *IEEE Trans. Vehicular Tech.*, vol. **VT-29**, no. 3, Aug. 1980, pp. 317–325.

[23] J. Hagen, *Radio-Frequency Electronics: Circuits and Applications*, 2nd Edn, Cambridge University Press, 2009.

[24] M. Hamid & R. Hamid, "Equivalent Circuit of Dipole Antenna of Arbitrary Length," *IEEE Trans. Ant. & Prop.*, Vol. **45**, No. 11, Nov. 1997, pp. 1695–1696.

[25] G.R. Hiertz *et al.*, "The IEEE 802.11 Universe," *IEEE Communications Mag.*, Vol. **48**, No. 1, Jan. 2010, pp. 62–70.

[26] E.B. Hogenauer, "An Economical Class of Digital Filters for Decimation and Interpolation," *IEEE Trans. Acoustics, Speech and Signal Proc.*, Vol. **29**, No. 2, April 1981, pp. 155–162.

[27] P. Horowitz & W. Hill, *The Art of Electronics*, 2nd Edn, Cambridge University Press, 1989.

[28] G.A. Hufford, A.G. Longley & W.A. Kissick, "A Guide to the Use of the ITS Irregular Terrain Model in the Area Prediction Mode," NTIA Rep. 82-100, Apr. 1982;

[29] International Telecommunications Union, ITU-R R.372-10, *Radio Noise*, 2009.

[30] International Telecommunications Union, ITU-R P.838-2, *Specific Attenuation Model for Rain for Use in Prediction Methods*, 2003.

[31] W.C. Jakes (Ed.), *Microwave Mobile Communications*, Wiley, 1974.

[32] R.C. Johnson (Ed.), *Antenna Systems Handbook*, McGraw-Hill, 1993.

[33] J.F. Kaiser, "Nonrecursive Digital Filter Design Using the I_0 – sinh Window Function," *Proc. IEEE Int'l Symp. Circuits & Systems*, San Francisco CA, 1974.

[34] H.L. Krauss, C.W. Bostian & F.H. Raab, *Solid State Radio Engineering*, Wiley, 1980.

[35] J. Kucera & U. Lott, "Low Noise Amplifier Design," Ch. 3 from *Commercial Wireless Circuits and Components Handbook*, M. Golio, ed., CRC Press, 2002.

[36] T. Kürner, "Propagation Models for Macro-Cells," Sec. 4.4 of *Digital Mobile Radio Towards Future Generation Systems: COST 231 Final Report*, 1999. Available on-line: 146.193.65.101/cost231/.

[37] K. Kurokawa, "Power Waves and the Scattering Matrix," *IEEE Trans. Microwave Theory & Techniques*, Vol. **13**, No. 2, March 1965, pp. 194–202.

[38] B. Le, T.W. Rondeau, J.H. Reed & C.W. Bostian, "Analog-to-Digital Converters," *IEEE Signal Processing Magazine*, Vol. **22**, No. 6, Nov. 2005, pp. 69–77.

[39] T.H. Lee, *The Design of CMOS Radio-Frequency Integrated Circuits*, 2nd Edn, Cambridge University Press, 2004.

[40] T.H. Lee & A. Hajimiri, "Oscillator Phase Noise: A Tutorial," *IEEE J. Solid-State Circuits*, Vol. **35**, No. 3, March 2000, pp. 326–336.

[41] W.C.Y. Lee, *Mobile Communications Engineering*, McGraw-Hill, 1982.

[42] C.A. Levis, J.T. Johnson & F.L. Teixeira, *Radiowave Propagation: Physics & Applications*, Wiley, 2010.

[43] R. C.-H. Li, *RF Circuit Design*, 2nd Edn, Wiley, 2012.

[44] H.J. Liebe, "An Updated Model for Millimeter Wave Propagation in Moist Air," *Radio Sci.*, Vol. **20**, No. 5, Sep.–Oct. 1985, pp. 1069–1089.

[45] Linear Technology Corp., "LTM9004 14-Bit Direct Conversion Receiver Subsystem" (datasheet), 2011.

[46] A.G. Longley & P.L. Rice, "Prediction of Tropospheric radio transmission over irregular terrain, A Computer method," ESSA Tech. Rep. ERL 79-ITS 67, U.S. Government Printing Office, Washington, DC, July 1968.

[47] A. Loke & F. Ali, "Direct Conversion Radio for Digital Mobile Phones – Design Issues, Status, and Trends," *IEEE Trans. Microwave Theory & Techniques*, Vol. **50**, No. 11, Nov. 2002, pp. 2422–2435.

[48] S.A. Maas, *The RF and Microwave Circuit Design Cookbook*, Artech House, 1998.

[49] Maxim Integrated Products, "Multiband Digital Television Tuner MAX3544" (datasheet), Rev. 0, July 2010.

[50] G. Matthaei, L. Young & E.M.T. Jones, *Microwave Filters, Impedance-Matching Networks, and Coupling Structures*, Artech House, 1980.

[51] J.S. McLean, "A Re-Examination of the Fundamental Limits on the Radiation Q for Electrically-Small Antennas," *IEEE Trans. Ant. & Prop.*, Vol. **44**, May 1996, pp. 672–675.

[52] M.A. Morgan & T.A. Boyd, "Reflectionless Filter Structures," *IEEE Trans. Microwave Theory & Techniques*, Vol. **63**, No. 4, April 2015, pp. 1263–1271.

[53] National Research Council, *Spectrum Management for Science in the 21st Century*, National Academies Press, 2010.

[54] NXP B.V., "TEF6901A Integrated Car Radio" (datasheet), Rev. 03, March 20, 2008.

[55] Office of Engineering and Technology (OET), "Longley–Rice Methodology for Evaluating TV Coverage and Interference," Federal Communications Commission Bulletin No. 69, July 2, 1997.

[56] A.V. Oppenheim & R.W. Schafer, *Discrete-Time Signal Processing*, 3rd Edn, Prentice-Hall, 2010.

[57] Philips Semiconductors, "400 MHz Low Noise Amplifier with the BFG540W/X" (Application Note), 1999.

[58] D.M. Pozar, *Microwave Engineering*, 4th Edn, John Wiley & Sons, 2012.

[59] J.G. Proakis & M. Salehi, *Digital Communications*, 5th Edn, McGraw-Hill, 2008.

[60] F.H. Raab *et al.*, "Power Amplifiers and Transmitters for RF and Microwave," *IEEE Trans. Microwave Theory & Techniques*, Vol. **50**, No. 3, March 2002, pp. 814–826.

[61] T.S. Rappaport, *Wireless Communications: Principles and Practice*, 2nd Edn, Prentice Hall PTR, 2002.

[62] B. Razavi, "Design Considerations for Direct-Conversion Receivers," *IEEE Trans. Circuits & Systems II*, Vol. **44**, No. 6, June 1997, pp. 428–435.

[63] B. Razavi, *RF Microelectronics*, 2nd Edn, Prentice-Hall, 2012.

[64] J.H. Reed (Ed.), *An Introduction to Ultra Wideband Communication Systems*, Prentice-Hall, 2005.

[65] U. Rhode & J. Whitaker, *Communications Receivers: DSP, Software Radios, and Design*, 3rd Edn, McGraw-Hill, 2001.

[66] J.M. Rollett, "Stability and Power-Gain Invariants of Linear Twoports," *IRE Trans. Circuit Theory*, Vol. **9**, No. 1, March 1962, pp. 29–32.

[67] C. Rowell & E.Y. Lam, "Mobile-Phone Antenna Design," *IEEE Ant. & Prop. Mag.*, Vol. **54**, No. 4, Aug. 2012, pp. 14–34.

[68] R.B. Rutledge, *The Electronics of Radio*, Cambridge University Press, 1999.

[69] W.E. Sabin, "Diplexer Filters for an HF MOSFET Power Amplifier," *QEX*, July/August 1999, p. 20.

[70] W.E. Sabin & E.O. Schoenike (Eds.), *HF Radio Systems & Circuits*, Noble Publishing, 1998.

[71] D.L. Schilling, *Meteor Burst Communications: Theory and Practice*, Wiley, 1993.

[72] A.S. Sedra & K.C. Smith, *Microelectronic Circuits*, 5th Edn, Oxford University Press, 2004.

[73] N.O. Sokal & A.D. Sokal, "Class E, A New Class of High-Efficiency Tuned Single-Ended Power Amplifiers," *IEEE J. Solid State Circuits*, Vol. **10**, June 1975, pp. 168–176.

[74] A.D. Spaulding & R.T. Disney, "Man-made Radio Noise, Part 1: Estimates for Business, Residential, and Rural Areas," Office of Telecommunications Report 74-38, June 1974.

[75] STMicroelectronics, "TDA7528 FM/AM car-radio receiver front-end for IF-sampling systems with fully integrated VCO" (datasheet), Rev. 7, Sept. 2013.

[76] W.L. Stutzman & G.A. Thiele, *Antenna Theory & Design*, 3rd Edn, Wiley, 2013.

[77] T.G. Tang, Q.M. Tieng & M.W. Gunn, "Equivalent Circuit of a Dipole Antenna Using Frequency-Independent Lumped Elements," *IEEE Trans. Ant. & Prop.*, Vol. **41**, No. 1, Jan. 1993, pp. 100–103.

[78] US Department of Commerce NTIA/Institute for Telecommunication Sciences, "Irregular Terrain Model (ITM) (Longley–Rice)," [on-line] flattop.its.bldrdoc.gov/itm.html;

[79] V. Valimaki & T.I. Laakso, "Principles of Fractional Delay Filters," *IEEE Int'l Conf. Acoustics, Speech, and Sig. Proc.*, Istanbul, Turkey, June 2000, pp. 3870–3873.

[80] J.L. Volakis, C.-C. Chen & K. Fujimoto, *Small Antennas: Miniaturization Techniques & Applications*, McGraw-Hill, 2010.

[81] J.E. Volder, "The CORDIC Trigonometric Computing Technique," *IRE Trans. Electronic Computers*, Sep. 1959, pp. 330–334.

[82] R.H. Walden, "Analog-to-Digital Converter Survey and Analysis," *IEEE J. Selected Areas in Communications*, Vol. **17**, No. 4, April 1999, pp. 539–550.

[83] M.M. Weiner, *Monopole Antennas*, Marcel Dekker, 2003.

[84] H.A. Wheeler, "Transmission Line Properties of a Strip on a Dielectric Sheet on a Plane," *IEEE Trans. Microwave Theory & Techniques*, Vol. **25**, No. 8, Aug. 1977, pp. 631–647.

[85] A. Williams & F. Taylor, *Electronic Filter Design Handbook*, 4th Edn, McGraw-Hill, 2006.

[86] J. Xiao *et al.*, "Low-Power Fully Integrated CMOS DTV Tuner Front-End for ATSC Terrestrial Broadcasting," *VLSI Design*, Vol. **2007**, Article ID 71974, 13 pp.

[87] A.I. Zverev, *Handbook of Filter Synthesis*, Wiley, 1967.

INDEX

Printed in the United States
by Baker & Taylor Publisher Services